Symbol	Meaning
F	F ratio; the ratio of the between-column variance to the within-column variance of a sample of observations (9.4)
$F(\nu_1, \nu_2)$	F ratio in which ν_1 and ν_2 are the numbers of degrees of freedom for the numerator and denominator, respectively (9.4)
$F(x) = P(X \leq x)$	Cumulative probability that a random variable X is equal to or less than the value x (2.1)
f	Number of observations (frequency) in a class interval of a frequency distribution (3.4)
$f = \frac{n}{N}$	Sampling fraction (6.5)
f_0	Observed frequency in a χ^2 goodness of fit test (9.1)
f_t	Theoretical (or expected) frequency in a χ^2 goodness of fit test (9.1)
$f(x)$	Value of the function f at x (1.2)
$f(x) = P(X = x)$	Probability that a random variable X is equal to the value x (2.1)
$f(x_1, x_2, \ldots, x_k)$	A value of the hypergeometric probability distribution; the probability of x_1 occurrences of type 1, x_2 of type 2, . . . , x_k of type k where sampling is without replacement (2.6)
$f(x_1, x_2, \ldots, x_k)$	A value of the multinomial probability distribution; the probability of x_1 occurrences of type 1, x_2 of type 2, . . . , x_k of type k where sampling is with replacement (2.5)
$f(x, y)$ $= P(X = x \cap Y = y)$	Joint probability that X takes on the value x and Y takes on the value y (2.8)
$f(x\|y)$ $= P(X = x \mid Y = y)$	Conditional probability that a random variable X is equal to the value x given that the random variable Y is equal to y (2.8)
G	Geometric mean (3.14)
$g(y) = P(Y = y)$	Probability that a random variable Y is equal to the value y (2.8)
$g(y\|x)$ $= P(Y = y\|X = x)$	Conditional probability that a random variable Y is equal to the value y given that the random variable X is equal to x (2.8)
H	Value of the highest observation (3.1)
H_0	"Null" hypothesis; basic hypothesis which is being tested (7.1)
H_1	Alternative hypothesis; rejection of null hypothesis H_0 implies tentative acceptance of the alternative hypothesis H_1 (7.1)
I	Effect of the irregular factors in time series analysis (11.4)
k	Number of classes in a frequency distribution (3.1)
k	Number of columns in a contingency table or in an arrangement of data to which an analysis of variance is applied (9.2)
L	Value of the lowest observation (3.1)
L	Number of strata (8.6)
$L(a_1\|\theta_1)$	Conditional opportunity loss of act a_1 given state of nature θ_1 (17.1)
$L(\hat{\theta}; \theta)$	Loss involved in estimating $\hat{\theta}$ when the parameter value is θ (17.4)
MA	Moving average figures in seasonal variations analysis (11.5)
Md	Median (3.10)
MS_b	Between-column mean square (9.4)
MS_w	Within-column mean square (9.4)
m_k	kth moment around the mean of a frequency distribution of observations (4.3)
m'_k	kth moment around the origin of a frequency distribution of observations (4.3)
μ	Arithmetic mean of a probability distribution (2.7)
$\mu = \mu_x$	Arithmetic mean of a population (3.8)
$\mu_k = E(X - \mu)^k$	kth moment around the mean of a probability distribution (4.3)
$\mu'_k = E(X^k)$	kth moment around the origin of a probability distribution (4.3)
$\mu_{\bar{p}}$	Mean of the sampling distribution of a proportion (7.2)
$\mu_{\bar{p}_1 - \bar{p}_2}$	Mean of the sampling distribution of the difference between two sample proportions (7.3)
$\mu_{\bar{x}} = \mu_x$	Mean of the sampling distribution of the arithmetic mean; it is equal to the mean of the population (6.5)
$\mu_{\bar{x}_1 - \bar{x}_2}$	Mean of the sampling distribution of the difference between two sample means (7.3)
μ_Y	Arithmetic mean of a population of Y values (10.5)
$\mu_{Y.X}$	Computed value of Y from a population two variable regression equation; it is the population value which corresponds to the $\bar{Y}_X$ value in the regression equation computed from sample observations (10.6)
$\mu_{Y.12}$	Computed value of Y from a population regression equation involving the dependent variable Y and the independent variables X_1 and X_2 (10.11)
N	Number of observations in a population (3.8)
N_i	Number of elements in the ith stratum (8.6)
n	Number of observations in a sample (3.4)
$n(A_1)$	Number of elements in event (set) A_1 (1.3)
$n - a : a$	Odds are $n - a$ to a (1.2)
n_i	Number of elements in the sample from the ith stratum (8.6)
$\binom{n}{x}$	Number of combinations of n objects taken x at a time (1.5)
ν	Number of degrees of freedom (7.4)
$\varnothing$	Null set (empty set) (1.2)
$OL(A_i\|p)$	Conditional opportunity loss of act A_i given p (14.3)
$P(A)$	Probability of the event A (1.1)
$P(A_1 \cup A_2)$	Probability of the occurrence of either the event A_1 or A_2 (or both) (1.2)

STATISTICAL ANALYSIS FOR DECISION MAKING

The Harbrace Series in Business and Economics

Statistical Analysis for Decision Making

MORRIS HAMBURG

Wharton School of Finance and Commerce
University of Pennsylvania

HARCOURT, BRACE & WORLD, INC.

New York *Chicago* *San Francisco* *Atlanta*

ISBN: 0-15-583760-5

Library of Congress Catalog Card Number: 77-113706

Printed in the United States of America

Illustrations by BMA Associates, Inc.

Foreword

This book is an introduction to probability theory and statistical analysis. It presents a solid theoretical base for applications in business and public administration. Descriptive statistics and statistical inference are fully examined so that a sound foundation for the classical approach is developed. Significantly, the book consistently relates classical theory to situations where inferential ability is essential for reaching decisions based on realistic economic reasoning.

The decision-making character of management activities is recognized as being critical for dealing intelligently with social and industrial systems. This text shows repeatedly the connection between statistical methodology and the decisions of managers. Every chapter projects this interest and, more than that, five chapters are devoted to a direct examination of Bayesian decision methods and their implications.

The author's ability to communicate effectively is apparent. He focuses on applications and evaluations of methodology. In the development of interdependent ideas and techniques the sequence of presentation creates a flow of uninterrupted reasoning. However, good organization of materials is not enough in a field as highly developed as this one. Total competence is merely a *sine qua non*. What matters is the ability to infuse communication with vitality. In this Professor Hamburg has succeeded by his style and his examples. In terms of the qualities that differentiate one book from another, this one represents an outstanding contribution to the textbook literature of probability and statistics.

Martin K. Starr

Preface

This book is designed for a first course in statistics for students of business and public administration, the social sciences, and liberal arts. It gives a noncalculus presentation of basic statistics, with emphasis on fundamental concepts and methods. It may be used for a one- or two-semester course or for corresponding numbers of quarters, depending upon appropriate selections of chapters.

Many students take only one course in statistics. Hence, the broad coverage of topics included in this text should aid such a student in gaining an appreciation of the scope of the field of statistical analysis. Furthermore, the book provides the necessary foundation for the use of quantitative methods in other courses, particularly in business and economics, as well as for further work in statistics and other quantitative disciplines.

A central concern is the presentation of basic theory and specific quantitative techniques for application to problems of managerial decision making and analysis. The topics of probability and random variables, descriptive statistics, statistical inference, and Bayesian decision theory are covered in that sequence, which is roughly the historical order of the flow of ideas in statistics. A modern approach is taken, with inclusion of a set theory definition of probability and greater emphasis on probability, random variables, and Bayesian decision theory than is found in many basic texts. A large number of worked out problems are presented. Both faculty and students at the Wharton School who used mimeographed versions of the text indicated that these illustrative problems were extremely helpful in cementing an understanding of ideas in both theory and methodology.

The viewpoint of this text is that the problems attacked by statistical analysis — particularly in fields such as business administration, economics, and public administration — are so varied and complex that no single approach to the subject is likely to provide adequate preparation for a student in any of these fields. Therefore, a balanced viewpoint has been adopted in which, as noted earlier, elements of probability, descriptive statistics, statistical inference, and statistical decision theory have all been included. These different modes of emphasis in statistics build upon one another.

However, even a composite conception of the subject matter of statistics must, in the nature of the case, be somewhat selective. Consequently, in this book, relatively little space has been devoted to subjects such as methods of collection, tabulation, and graphic presentation of data. On the other hand, the subjects of economic time series and index numbers, which are usually not discussed in general texts on probability and statistics or mathematical statistics, are presented because they are of considerable interest and importance to students of business administration, public administration, and economics. Also a chapter on chi-square tests and the analysis of variance has been included, because these subjects represent logical follow-ups to the topic of two-sample hypothesis testing in statistical inference.

In recent time, virtually every aspect of our lives has been changing rapidly, and new problems demand fresh methods of solution. As time goes on, newer theories and methods of statistical analysis will be developed and the mix of topics in a basic statistics text will correspondingly shift. In this book I have tried to convey the feeling that statistics is an exciting, fluid field which deals with a means of inquiry, a scientific method for acquiring, processing, and using knowledge in decision making.

My grateful appreciation is expressed to the many persons and organizations that have rendered assistance to me in the writing of this book. Many helpful suggestions, comments, and criticisms were offered by the reviewers of the manuscript — Martin Starr, William J. Wrobleski, and Neil Weiss. William F. Matlack and Robert J. Atkins also gave valuable reviews and critiques of portions of the book. My sincere gratitude is expressed to Bernard Siskin and Kevin Stitt for developing end-of-chapter problems and for their work on the problem solutions. In that connection, my thanks also go to my colleagues in the Statistics and Operations Research Department of the Wharton School of

Finance and Commerce, University of Pennsylvania, who were responsible for many of the original problem ideas.

I am indebted to the Literary Executor of the late Sir Ronald A. Fisher, F.R.S., to Dr. Frank Yates, F.R.S., and to Oliver & Boyd Ltd., Edinburgh, for permission to reprint Tables III and IV from their book *Statistical Tables for Biological, Agricultural and Medical Research.* I am also indebted to the other authors and publishers whose generous permission to reprint tables or excerpts from tables is acknowledged at the appropriate places. My gratitude also goes to Mrs. Sylvia Balis for her efficient, loyal, and tireless assistance in secretarial and typing chores, and to Mr. Ernest J. Browne, Director of the Wharton Duplicating Center, and his staff for typing assistance.

Lastly, I would like to dedicate this book to my wife, June, and my children, Neil and Barbara. They helped me in many of the innumerable tasks connected with the production of the book, and endured, cheerfully and with forbearance, the many hours I stole from our family life for the writing task.

Morris Hamburg

Contents

Introduction

The Nature of Statistics

This is a book about statistics. But what is statistics? In one use of the term, statistics refers simply to numerical data. We are all familiar with collections of statistics pertaining to sports, population, the economy, the stock market, etc. However, in another use, with which we are more concerned in this book, statistics refers to a body of theory and methods of analysis. The subject matter of statistics is very broad, extending from the planning and design of experiments, surveys, and other studies which generate data to the collection, analysis, presentation, and interpretation of the data. Hence, numerical data constitute the raw material of this subject. The most widely known statistical methods are those which summarize such data in terms of averages and other measures for purposes of description. For example, if interest is centered upon the incomes of a group of 1000 families drawn at random in a particular city, important characteristics of these incomes may be described by calculating an average income and a measure of the spread or dispersion of these incomes around the average. However, the essence of modern statistics is the theory and methodology for the drawing of inferences which extend beyond the particular set of data examined and for the making of decisions based on appropriate analyses of such data. Thus, in the preceding illustration, interest is probably not centered upon the incomes of the particular 1000 families included in the sample, but rather in an *inference* about the income of *all* families in the city from which the sample was drawn. Or, the marketing department of a company may want the income data in order to *decide* on the type of advertising program to use in promoting a certain product.

For example, the inference may be in the form of a *test of a hypothesis* that the average income of all families in the city is $7500 or less against an alternative hypothesis that the average income exceeds $7500. On the other hand, the inference may be in the form of a single figure which represents an *estimate* of the average income of *all* families in the city based on the average income observed in the sample of 1000 families. Or the marketing department of the aforementioned company may wish to choose among different types of advertising programs depending upon whether it concludes that the city is a low-, medium-, or high-income area. The mathematical *theory of probability* provides the logical framework for the mental leap from the sample of data studied to the inference about all families in the city and for decisions such as the type of advertising program to be used.

A few points may be noted concerning the preceding example. An inference may have been desired about the incomes of all families in the city. However, since it would have been too expensive and too time-consuming to obtain the income data for every family in the city, only the sample of 1000 families was drawn. All families in the city or, more generally, the totality of the elements about which the inference is desired is referred to in statistics as the "universe" or "population." The 1000 families, which represent a collection of only some of the universe elements, as we have seen, are referred to as a "sample." In statistics, *sample data* are observed in order to make *inferences* or *decisions* concerning the *populations* from which samples are drawn.

Another point to note is that the sample was referred to as having been drawn "at random" from the population. A random sample is one drawn in such a way that the probability or likelihood of inclusion of every element in the population is known. However, even though these probabilities of inclusion may be known, the average income that would be observed for a random sample of 1000 families would vary from sample to sample. These sample to sample variations are known as *chance sampling fluctuations*. Although we cannot predict with certainty what the average income will be for any particular sample, the theory of probability, a branch of mathematics, enables us to compute the long-run relative frequency of occurrence of these differing sample results. It is an intriguing and remarkable fact that, even though there is *uncertainty* concerning which particular sample may have been drawn, probability theory provides a rational basis for inference and decision making about the population or larger group from which the sample was taken. Much of this text is directed toward the theory and methods by which such inferences and decisions are made.

The Role of Statistics

Statistical concepts and methods are widely applied in many areas of human activity. They are extensively used in the physical, natural, and social sciences, in business and public administration, and many other fields.

In the sciences, the applications are far-ranging, extending from the design and analysis of experiments to the testing of new and competing hypotheses. In industry, statistics makes its contributions in short- and long-range planning and decision making and in day-to-day operational decision making and control. Hence, many firms use statistical methods for analyzing patterns of change and to forecast future movements in economic activity for the firm, the industry, and the economy as a whole. Such forecasts often provide the foundation for corporate planning and control, with areas such as purchasing, production, and inventory control being dependent upon short-range forecasts, and capital investment and long-term development decisions being dependent upon longer range forecasts. In addition to their use of forecasts, areas such as production control, inventory control, and quality control often employ statistical methods on a standard basis as well. For example, many firms use statistical techniques for controlling the quality of their manufactured products. In this type of application, statistical methods are used to differentiate between variation attributable to chance causes and variation too great to be considered as resulting from chance. This latter type of variation can be analyzed and remedied. A large number of cases have been recorded in which applications of these statistical quality control methods have resulted in substantial improvements in the quality of product and in lower costs because of reduction in rework and spoilage. It is of interest that such statistical quality control methods have been given credit as being a major factor in the vast improvement in the quality of Japanese-manufactured products in the post-World War II period.

Over the past couple of decades in the fields of business and government, there has been an expanding development of a body of quantitative techniques and procedures whose purpose is to aid and improve managerial decision making. The field of statistics has provided many of the fruitful ideas and techniques in this development. Currently, applications of statistics pervade virtually every area of activity of the business firm including production, financial analysis, distribution analysis, market research, research and development, manpower planning, and accounting. There has been a rapid acceleration in the use and in the degree of sophistication of statistical methods in virtually every field in which they have been introduced. These methods constitute an integral part of the general development of more rational and quantitative approaches to the solution of business problems. One outstanding characteristic of this development has been the increased adoption of scientific decision-making approaches using mathematical models. These models are mathematical formulas or equations which state the relationship among the important factors or variables in a problem or system. Thus, an equation may be developed which represents the relationship between a company's sales and the economic and other variables that influence sales. Chapter 10 of this text discusses the methods for deriving one such type of mathematical model. The statistical methods discussed in that and other chapters bring a logical, objective, and systematic approach to decision making in business and other fields. They assist in structuring a problem and in bringing the application of judgment to it.

Vast governmental statistical activities are conducted at federal, state, and local levels. There are many applications of statistical ideas and methods in the administration of governmental affairs, and a great deal of governmental statistical activity in the collection and dissemination of statistical data. The most highly organized and extensive statistical information systems are those of the federal government. Systems such as national economic accounts are virtually indispensable as orderly frameworks of analysis for public and private planning and decision making. Such information systems, which include national income and product accounts, input-output accounts, flow of funds accounts, balance of payments accounts, and national balance sheets, depend upon massive statistical collection and distribution systems. Statistical methods are applied to the resulting data to assess past trends and current status and to project future economic activity. These methods provide measures of human and physical resources, economic growth, well-being, and potential. They are essential tools for appraising the performance and for analyzing the structure and behavior of an economy.

Massive data collection and dissemination activities are also carried out by federal and other governmental and private agencies in fields such as population, vital statistics, education, labor force, employment and earnings, business and trade, prices, housing, medical care, public health, agriculture, natural resources, welfare services and resources, crime and law enforcement, area and industrial development, construction, manufacturing, transportation, and communications.

Statistical analysis constitutes a body of theory and methods which plays an important role in this wide variety of areas of human activity. It is extremely useful for communicating information, for drawing conclusions and inferences from data, and for the guidance of rational planning and decision making.

We begin our discussion with the theory of probability to lay the groundwork for the study of statistics.

CHAPTER ONE

Elementary Probability Theory

1.1 The Meaning of Probability

Probability theory is a fascinating subject which can be studied at a variety of intellectual and mathematical levels. There is an on-going debate about the foundations of this theory and about the meaning of probability. However, there is no question about the general value of the theory and its usefulness in a variety of applications in the physical, natural, and behavioral sciences as well as in many areas of business and governmental activity. Probability lies at the foundation of statistical theory and application. Thus, we find that a knowledge of probabilistic methods has become increasingly essential in quantitative analyses of business and economic problems. In particular, probability theory is a basic component of the formal theory of decision making under uncertainty. Probability measures provide the decision maker in business and in government with the means for quantifying the uncertainties which affect his choice of appropriate actions.

When a business executive makes statements such as, "We probably will get the contract on which we bid last week; the demand for our product this year will probably exceed last year's level; or the price of our stock will probably rise above $40 per share by the end of the year," he undoubtedly feels he has made meaningful assertions. Furthermore, persons to whom such statements have been made usually feel they have understood their meaning. However, careful analysis reveals that there are considerable difficulties attached to specifying the precise meaning of each of these statements. Even if the executive states that the probability is two-thirds or the odds are two-to-one of obtaining the contract on which a bid was made last week, a host of problems still remains concerning the exact meaning of the statement. It would be a useful exercise for the student to return to the above statements after reading this chapter, and to determine what precise meaning should be attached to them, assuming a probability measure (say 2/3) were specified in each case.

The mathematical theory of probability originated during the seventeenth century when the French nobleman Antoine Gombauld, known as the Chevalier de Méré, raised certain questions about games of chance. Specifically, he was puzzled about the probability of obtaining two sixes at least once in twenty-four rolls of a pair of dice. This is a problem with which you should have little difficulty after reading this chapter. De Méré posed the question to Blaise Pascal, a young French mathematician, who solved the problem. Subsequently, Pascal discussed this and other puzzlers raised by de Méré with the famous French mathematician, Pierre de Fermat. In the course of their correspondence the mathematical theory of probability was born.

Several different methods of measuring probabilities will be presented. They represent different conceptual approaches and reveal some of the current intellectual controversy concerning the foundations of probability theory. In this section, we discuss *a priori* probability, *relative frequency of occurrence*, and *subjective* concepts of probability. In Section 1.2, we give a definition of a measure of probability from the viewpoint of set theory. It may be noted that regardless of the definition of probability used, the same mathematical rules apply in performing the calculations (i.e., measures of probability are added or multiplied under the same general circumstances, irrespective of the definition employed).

A Priori Probability

Since probability theory had its origin in gambling games, it is not surprising that the method of measuring probabilities which was first developed was particularly appropriate for gambling situations. A so-called *classical* or *a priori* concept of probability defines the probability of an event as follows: if there are a possible outcomes favorable to the occurrence of an event A, and b possible outcomes unfavorable to the occurrence of A, and all of these possible outcomes are equally likely and mutually exclusive, then the probability that A will occur, denoted $P(A)$, is

$$P(A) = \frac{a}{a+b} = \frac{\text{number of outcomes favorable to occurrence of } A}{\text{total number of possible outcomes}}$$

Thus, if a fair coin with two faces, denoted head and tail, is tossed into the air in such a way that it spins end over end a large number of times, the probability that it will fall with the head uppermost is $P(\text{Head}) = 1/(1 + 1) = 1/2$. In this case, there is one outcome favorable to the occurrence of the event "head" and one outcome unfavorable. The extremely unlikely situation that the coin will stand on end is defined out of the problem, i.e., it is not classified as an outcome for the purpose of the probability calculation.

Another example is the result obtained in rolling a true die. A die is a small cube with a number of dots on each of its six faces denoting 1, 2, 3, 4, 5, or 6, respectively. A "true" die is uniform in shape and density, and is therefore equally likely to show any of the six numbers on its uppermost face when rolled or tossed. The probability of obtaining a "one" if such a die is rolled is $P(1) = 1/(1 + 5) = 1/6$. Here, there is one outcome favorable to the event "one" and five outcomes unfavorable. Untrue dice, which are not uniform in density, are said to be "loaded"; such dice are outside the scope of the examples discussed in this text, and hopefully will remain outside your experience as well.

As a third illustration involving games of chance, if a deck of 52 cards is shuffled and a card is randomly drawn from the deck, the probability that it is a spade is $P(\text{Spade}) = 13/52 = 1/4$. In this situation, there are 13 cards (outcomes) favorable to the occurrence of the event "spade" and 39 cards unfavorable.

Some of the terms used in the classical or a priori concept of probability require further explanation. The "event" whose probability is sought consists of one or more possible outcomes of the given activity of tossing a coin, rolling a die, or drawing a card. These activities are referred to in modern terminology as "experiments," which is a term that refers to processes which result in different possible outcomes or observations. An event will be more formally defined in Section 1.2, where a more modern definition of probability is given in terms of set theory. The term "equally likely" in referring to possible outcomes is undefined and is considered to be an intuitive foundation concept. All branches of mathematics start with analogous undefined terms, "primitive ideas," or "primitive propositions." Two or more outcomes are said to be "mutually exclusive" if when one of the outcomes occurs, the others cannot. Thus the appearance of a "one" and the appearance of a "two" are mutually exclusive events, since if a "one" results, a "two" cannot. The results of an experiment are conceived of as a complete or exhaustive set of mutually exclusive outcomes.

These classical or a priori probability measures have two very interesting characteristics. First, the objects referred to as *fair* coins, *true* dice, or *fair* decks of cards are abstractions in the sense that no real world object exactly possesses the features postulated. For example, in order to be a *fair* coin, thus equally likely to fall "head" or "tail," the object would have to be a perfectly flat, homogeneous disc. Secondly, in order to determine the probabilities in the above examples, no coins had to be tossed, no dice rolled, nor cards shuffled. That is, no experimental data were required to be collected; the probability calculations were based entirely upon logical prior (thus, *a priori*) reasoning.

In the context of this definition of probability, if it is *impossible* for an event A to occur, the probability of that event is said to be zero. For example, if the event A is the appearance of a seven when a single die is rolled, then $P(A) = 0$. A probability of one is assigned to an event which is *certain* to occur. Thus, if the event A pertains to the appearance of any one of the numbers 1, 2, 3, 4, 5, or 6 on a single roll of a die, then $P(A) = 1$. In the a priori method of measurement, as well as in all other methods, the probability of an event A is a number, such that $0 \leq P(A) \leq 1$, and the sum of the probability that an event will occur and the probability that it will not occur is equal to one.

Relative Frequency of Occurrence

The classical or a priori concept of probability, while useful for solving problems involving games of chance, encounters serious difficulties when confronted with a wide range of other types of problems. For example, it is inadequate for answering questions such as: What are the probabilities that (a) a black male American, age 30 will die within the next year, (b) a consumer in a certain metropolitan area will purchase a company's product during the next month, (c) a production process used by a particular firm will produce a defective item? In none of these situations is it feasible to establish a set of complete and mutually exclusive outcomes, each of which is equally likely to occur. For example, in (a), there are only two possible occurrences, the individual will die during the ensuing year or he will live. The likelihood that he will die is, of course, much smaller than that he will live. How much smaller? This is the type of question that requires reference to empirical data. The probability that a black male American, age 30 will live through the next year is greater than the corresponding probability that a male inhabitant of India, age 30 will survive the year. However, how much greater is it and precisely what do these probabilities mean?

We know that the life insurance industry establishes mortality rates by observing, how many of a sample of, say 100,000 black American males, age 30, die within a one year period. The number of deaths divided by 100,000 is then the relative frequency of occurrence of death for the 100,000 individuals studied. It may also be viewed as an estimate of the probability of death for Americans in the given color-sex-age group. This relative frequency of occurrence concept will be further illustrated by a simple coin tossing illustration and will then be specifically defined.

Suppose you are given a coin to toss which is known to be biased, i.e., it is not a true coin. You are not told whether it is more likely that a head or a tail will be obtained if the coin is tossed. However, you are asked to determine the probability of the appearance of a head by means of a large number of tosses of the coin. Assume that 10,000 tosses of the coin result in 7000 heads and 3000 tails. Another way of stating the results is that the relative frequency of occurrence of heads is 7000/10,000 or 0.70. It certainly seems reasonable to assign a probability of 0.70 to the appearance of a head with this particular coin. On the other hand, if the coin had been tossed only three times and one head resulted, you would have little confidence in assigning a probability of 1/3 to the occurrence of a head.

This illustration highlights a number of essential characteristics of the relative frequency approach to probability. Consider an experiment in which there are independently repeated trials. The number of outcomes a of an event A in which we are interested is recorded in n trials of the experiment. Then, the relative frequency of occurrence of A, denoted $R(A)$ is given by

$$R(A) = \frac{a}{n}$$

We postulate that the probability $P(A)$ of the event A is given by the limit of $R(A)$, as n increases indefinitely (tends to infinity). In fact, no rigorous mathematical demonstration can be given that such a limit exists. However, the tendency for relative frequencies to fluctuate considerably in small numbers of trials and then settle down and approach some fixed value as the number of trials increases has been observed experimentally many times. Also, the concept of the existence of such a limit is intuitively appealing. In practice, if a large number of trials has been made, then the relative frequency $R(A)$ is often referred to as the probability of A and the symbol $P(A)$ is used instead.

Returning to the mortality illustration, if 800 of the 100,000 individuals of the given group died during the year, then the relative frequency of death is

800/100,000. This may be thought of as an estimate of the probability of death of individuals, in the color-sex-nationality-age group. At later points in time, with changes in environmental conditions, obviously there would be changes in the ratios for the same color-sex-nationality-age group. It is not clear, therefore, that any limit exists around which the relative frequency of death of black male Americans, age 30 fluctuates and which it approaches. Thus, it may be observed that the relative frequency concept, as is true for every definition of probability, is not free from philosophical difficulties. Often, probability is defined in this approach as relative frequency in the long run under what are termed uniform or stable conditions or under a "constant cause system."

The specific nature of these conditions is usually very difficult to define. In many interesting cases, relative frequency of occurrence of an event has been observed in a finite number of trials for a situation where the conceptually possible number of trials is infinite; the uniformity of the set of conditions under which these observations have been made is a matter of judgment.[1]

Also, the type of inference that is warranted from relative frequency data is a matter of subjective judgment and depends upon how the investigator structures his problem. For example, the relative frequency of mortality figure earlier referred to, 800/100,000, may be thought of as an estimate of a population value for a time period, e.g., the year in which the deaths occurred. The 100,000 individuals represent a sample drawn from this past statistical universe. The variation over time of such ratios would then be estimates of changing population parameters. From another viewpoint, mortality may be considered to be a random process; the observed relative frequency is then a sample of observations from the random process which generates the variation. Of course, a life insurance company's use of the relative frequency figure is not centered on the mortality rate in the year in which the deaths occurred. That is, the company is not primarily interested in an estimate of a past population value, but rather in an estimate of the mortality rate for future years. Thus, the company generally uses the observed relative frequency figure as an estimate of a future population or as an estimate of a future observation from the random process which produces the variation in mortality. The fact that judgment is involved in the establishment and interpretation of relative frequency of occurrence estimates of probabilities is a good point to keep in mind for the subsequent discussion of subjective probabilities. These subjective probabilities are particularly useful in the case of unique

[1]There are methods for determining whether a "constant cause system" or a "state of statistical control" exists in certain situations, e.g., through the use of control charts. Nevertheless, it remains a matter of personal judgment in most practical situations as to how much change in environmental conditions is required in order to conclude that uniform conditions no longer prevail.

future events, where neither past relative frequency data nor classical probabilities appear to be relevant.

Subjective Probability

The subjective or personalistic concept of probability is relatively recent.[2] Its application to statistical problems has occurred virtually entirely in the post-World War II period, particularly in connection with statistical decision theory. According to this concept, the probability of an event is the degree of belief or degree of confidence placed in the occurrence of an event by a particular individual based on the evidence available to him. This evidence may consist of relative frequency of occurrence data and any other quantitative or nonquantitative information. If the individual believes it is unlikely an event will occur, he assigns a probability close to zero to its occurrence; if he believes it is very likely the event will occur, he assigns it a probability close to one.

Those who accept subjective probability argue that in assigning probabilities to events, other information in addition to past relative frequencies of occurrence should be taken into account. To make this point clear, let us consider an oversimplified, somewhat artificial example. Suppose a company which purchases a product from a certain supplier has had the following experience with shipments from that firm: 1% defective items in each of ten shipments, 2% defectives in each of 85 shipments and 3% defectives in each of five shipments. Assume all shipments contained the same number of items. These data are displayed in Table 1.1.

Suppose the purchasing company wants to know the probability that the next shipment from this supplier will contain 2% defective items. In the

Table 1.1 Percentage of Defective Items in One Hundred Shipments.

% Defectives	*Number of Shipments*
1	10
2	85
3	5
	100

[2]The concept was first introduced in 1926 by Frank Ramsey who presented a formal theory of personal probability in F. P. Ramsey, *The Foundation of Mathematics and Other Logical Essays* (London: Kegan Paul; New York: Harcourt, Brace, & World, 1931). The theory was developed primarily by de Finetti, Koopman, I. J. Good, and L. J. Savage.

absence of any further information, it seems reasonable to assign a probability of 0.85 to that event. That is, since shipments with 2% defectives occurred in 85% of the past cases, the relative frequency of occurrence would seem to be a good estimate of the probability in question. However, suppose the purchasing company acquires some additional information. It learns that the engineer who has been in charge of production for the supplier, and who has been the key person responsible for the maintenance of the quality level of the product has just resigned his position with the company. Furthermore, it is known that his knowledge has not been passed on to a suitable replacement. Therefore, a deterioration in quality of the product, at least until suitable remedial measures can be instituted, seems reasonable. Should a probability of 0.85 still be assigned?

In this case it certainly seems reasonable that the assignment of probabilities should no longer depend solely on past relative frequency data. The purchaser, as a practical business man, should undoubtedly anticipate that shipments in the near future will display quality levels different from those indicated by the data in Table 1.1. For example, percentages of defectives in excess of three percent are possibilities for shipments in the near future, and somehow or other, for decision making purposes, the purchaser must reckon with the likelihood of such shipments. What the purchaser now needs is a new distribution of all the percentage defectives he feels are possible with probability assignments attached to each. It might be argued that the purchaser should wait until conditions within the supplier company are again stable and reasonable assurance is given that acceptable quality levels will be maintained. However, suppose the purchaser cannot delay his decisions for that period, and must take appropriate action now.

Subjective probabilities should be assigned now on the basis of all objective and subjective evidence currently available. These probabilities should reflect the decision maker's current degree of belief. Reasonable persons might arrive at different probability assessments because of differences in experience, attitudes, values, etc. Furthermore, in general, these probability assignments may be made for events which will occur only once, in situations where the concept of a repetitive sequence of trials under uniform conditions does not appear to be a useful model.

This approach is thus a very broad and flexible one, permitting probability assignments to events for which there may be no objective data, or for which there may be a combination of objective and subjective data. These events may occur only once and may lie entirely in the future. However, the assignments of these probabilities must be consistent. For example, if the purchaser in the illustration above assigns a probability of 0.40 to the event that a shipment will have 2% or less defective items, then a probability of 0.60 must be assigned to the event that a shipment will have more than 2% defective items.

In this book we accept the concept of subjective or personal probability as a reasonable and useful one, particularly in the context of situations in business decision making.

1.2 Elementary Set Theory and Operations

Set theory is a unifying and fundamental group of concepts which are used in virtually all of mathematics. Children in grammar school are now taught set theoretic concepts and often learn much of their elementary mathematics using the language, concepts, and notation of sets. On the other hand, some of the most advanced concepts of mathematics are also expressed in this framework. Modern approaches to probability and random variables generally employ set theory, and that apparatus will be used here for the development of some fundamental concepts and tools.

A set is any well-specified collection of distinct objects. The objects which comprise the set are usually referred to as *elements* or *members* of the set, and are said to *belong* to the set or to be *contained* in it. The set must be *well-specified* or *well-defined* in the sense that it must be perfectly clear whether a given object does or does not belong to it. The members of the set are distinct in the sense that repetition of elements is not permitted in specifying the set. The *collection*, or *aggregation*, or *totality* of elements is referred to simply as a "set." Thus the following collections are all examples of sets:

(1) the students enrolled in a certain college
(2) the books in a library
(3) the employees of a company
(4) the accounts receivable of a company
(5) the assets of a company
(6) the citizens of the United States
(7) the odd numbered positive integers
(8) the possible outcomes of the roll of a single die

For convenience, only brief verbal specifications of sets have been given in these examples. However, in order to have well-defined sets, it may be necessary to elaborate on some of the above statements. For instance, we might restrict example (1) to "the fully-matriculated undergraduates enrolled in a certain college as of September 30 of a certain year."

Ways of Specifying Sets

There are two common methods for specifying or designating sets:

(1) the *listing* or *roster* method
(2) the *defining property* method

The listing or roster method for specifying a set simply provides a listing of all the elements in the set. The elements are usually enclosed within braces. For example, the set consisting of the possible outcomes (tail = T, head = H) of a single toss of a coin may be expressed as

$$S = \{T, H\}$$

The set of possible outcomes of two tosses of a coin may be written

$$S = \{(T, T), (T, H), (H, T), (H, H)\}$$

The names of the vice-presidents of finance, manufacturing, and production of a corporation comprise a set which may be expressed as

$$S = \{\text{A. H. Jones, C. F. Smith, P. J. McGillicuddy}\}$$

The order in which the elements of a set are listed is of no importance. It is important, however, that each element be listed only once. Note in the second example above that there are four elements in the set, namely, (T, T), (T, H), (H, T), and (H, H). Thus, each element consists of an ordered pair, the first and second components in the pair denoting respectively the outcomes of the first and second toss of the coin. Since each element consists of an ordered pair, the order of the items within the pair is important to distinguish clearly between an outcome of the first or the second toss. However, note that the order of listing of the four elements of the set is of no consequence.

Each of the preceding illustrations is an example of a *finite* set; i.e., each of these sets consists of a finite number of elements. It is convenient to use the listing method for finite sets with few elements.

Now, let us consider the set composed of all odd-numbered positive integers given earlier as example (7). This is an illustration of an infinite set, and may be denoted

$$S = \{1, 3, 5, \ldots\}$$

More specifically, this is an example of a countably (or denumerably) infinite set. A countably infinite set is one in which a one-to-one relationship can be established between the members of the set and the set of all integers. In our example, this means that each of the odd-numbered positive integers may be paired off with one of the elements of the set of all integers.[3] Where a finite but large number, or an infinite number, of elements is included in a set the defining property method of specifying the set is more satisfactory.

---

[3]The one-to-one relationship may be illustrated as follows:

$$\begin{array}{cccccc} 1 & 3 & 5 & 7 & 9 & \\ \cdot & \cdot & \cdot & \cdot & \cdot & \cdots \\ 1 & 2 & 3 & 4 & 5 & \end{array}$$

The designation "defining property" is very descriptive of the method; i.e., rather than explicitly listing the elements of a set they are specified by a defining property which permits one to decide whether or not a given object belongs to the set. Under the usual notation a symbol representing a general description is given on the left-hand side of a vertical line or colon and the defining property for inclusion of members of the set on the right-hand side. Thus, in the first example above, we might have

$$S = \{x \mid x \text{ is an } H \text{ or } T \text{ denoting head or tail as the outcome of the toss of a coin}\}$$

or

$$S = \{x : x \text{ is an } H \text{ or } T \text{ denoting head or tail as the outcome of the toss of a coin}\}$$

The fourth example might read

$$S = \{x \mid x \text{ is an odd numbered positive integer}\}$$

The symbol | or : is read "such that." Therefore, the first designation would be expressed in words "S is the set of all elements x such that x is an H or T denoting head or tail on the toss of a coin." Note that the symbol x may or may not represent a number.

Sample Spaces

A *sample space* is a set whose elements represent the possible outcomes of an experiment. (The elements are often referred to as "points" in the sample space). The experiment may be real or conceptual. Thus, the sets of outcomes of tossing a coin once, twice or any number of times are all sample spaces. The notation is the same as that for specifying sets, i.e.,[4]

$$S = \{T, H\}$$

and

$$S = \{(T, T), (T, H), (H, T), (H, H)\}$$

for tossing a coin once and twice, respectively.

In these examples, a physical experiment may actually be performed, or we may easily conceive of the possibility of such an experiment. In the first case, the experiment consists of *one trial*, a single toss of the coin; in the second case, the experiment contains *two trials*, the two tosses of the coin.

In other situations, although no sequence of repetitive trials is involved, we may conceive of a set of outcomes as an experiment. These outcomes may

---

[4]The mnemonic symbol, S, is conventionally used to denote a sample space.

simply be the result of an observational process and need not bear any resemblance to a laboratory experiment. It suffices that the outcomes be well defined. Thus, we may think of each of the following two-way classifications as constituting sample spaces:

0	1
Customer was granted credit.	Customer was not granted credit.
Employee elected a stock purchase plan.	Employee did not elect a stock purchase plan.
The merger will take place.	The merger will not take place.
The company uses direct mail advertising.	The company does not use direct mail advertising.

The elements in these two-point sample spaces may be designated in a binary classification as zero and one, respectively, as indicated by the column headings. Therefore, each of the four illustrative sample spaces may be conveniently symbolized as

$$S = \{0, 1\}$$

These are all examples of sample spaces with a finite number of discrete elements. Similarly, we can conceive of observational processes which might generate countably infinite sample spaces.

There are at least two methods of graphically depicting sample spaces: (1) graphs using the conventional rectangular Cartesian coordinate system and (2) "tree diagrams." These methods are most feasible in the case of finite sample spaces with relatively small numbers of sample points. Tree diagrams are more tractable because of the obvious graphic difficulties encountered by the coordinate system method beyond three dimensions. These methods may be illustrated by the following simple examples:

Example 1-1 Depict graphically the sample space generated by the experiment of tossing a coin twice.

Solution: The sample space was earlier designated as

$$S = \{(T, T), (T, H), (H, T), (H, H)\}$$

Let $T = 0$ and $H = 1$. The sample space may now be written

$$S = \{(0, 0), (0, 1), (1, 0), (1, 1)\}$$

This may be graphed in two-dimensional coordinates as shown in Figure 1-1. A tree diagram for this sample space is shown in Figure 1-2.

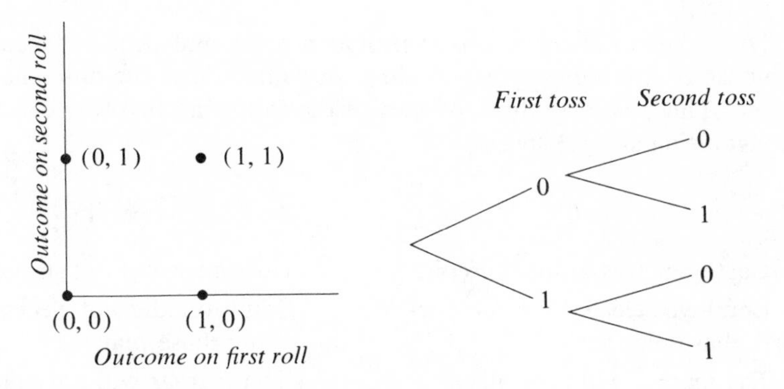

Figure 1-1 Graph for coin-tossing experiment.

Figure 1-2 Tree diagram for coin-tossing experiment.

Example 1-2 A market research firm studied the differences among consumers by income groups in a certain city in terms of their purchase of or failure to purchase a given product during a one month period. The income groups used were low, middle, and high. Consumers were classified as (a) failed to purchase, and (b) purchased the product at least once. Show this situation graphically.

Solution: Let 0 stand for "low" income, 1 for "middle" income and 2 for "high" income. For purchase behavior let 0 stand for "failed to purchase" and 1 stand for "purchased at least once." The sample space may be expressed as

$$S = \{(0, 0), (1, 0), (2, 0), (0, 1), (1, 1), (2, 1)\}$$

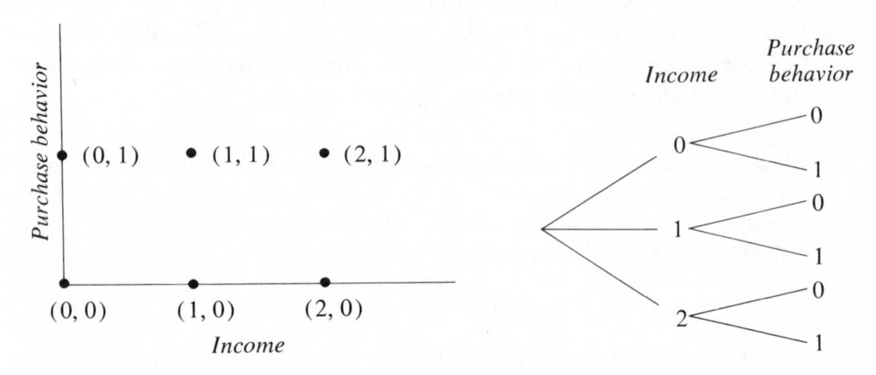

Figure 1-3 Graph for classification of consumers by income and purchase behavior.

Figure 1-4 Tree diagram for classification of consumers by income and purchase behavior.

The graph is given in Figure 1-3, and the tree diagram in 1-4. To illustrate the way the tree diagram is read in this case, let us consider the element (0, 1). This element is depicted in the tree by starting at the left-hand side and following the uppermost branch to the "0," then continuing from this fork down the branch leading to a "1." This (0, 1) element denotes a consumer classified as "low" for income and "purchased at least once" for buying behavior.

Events

The term "event" is used in ordinary conversation and there is usually no ambiguity as to its meaning. However, since the concept of an event is fundamental to probability theory, it deserves explicit definition. This definition will be given in terms of set theory. Before we can do this, however, we must first define a *subset*. Once a sample space S has been specified, a *subset* may be defined as a set each of whose elements is also an element of S. Since the set S satisfies the definition, S is a subset of itself. An *event* is simply any subset of the sample space S. An *elementary event* is a single possible outcome of an experimental trial. It is thus an event which cannot be further subdivided into a combination of other events.

Example 1-3 A single roll of a die constitutes an experiment. The sample space generated is

$$S = \{1, 2, 3, 4, 5, 6\}$$

The occurrence of any particular outcome of the experiment is an event. We may define an event E as the "appearance of a 2 or a 3." This may be expressed as

$$E = \{2, 3\}$$

If either a 2 or a 3 appears on the upper face of the die when it is rolled, the event E is said to have "occurred." Note that E is not an elementary event, since it can be subdivided into the two elementary events "2" and "3."

Example 1-4 A single card is drawn from a well-shuffled deck of 52 cards. If an "ace" is drawn, the event "ace" may be said to have "occurred." However, this is not an elementary event, since it can be decomposed into the four events, "ace of spades," "ace of clubs," "ace of hearts," and "ace of diamonds," each of which constitutes an elementary event, since no further subdivision is possible. Thus, the drawing of a single card from the deck is an experiment which generates a sample space consisting of fifty-two elementary events represented by the different cards which could possibly have been drawn.

Set Operations and Definitions

If a sample space has been defined, a number of operations may be performed on subsets of this sample space. These operations introduce other sets or events of interest. It is important to realize in studying the definitions which follow that the terms *event* and *set* are synonymous. That is, since an event is simply a subset of a sample space, it is a *set*, or a *collection of elements of a sample space.* Often terms such as *union, intersection,* and *complement,* which are introduced in the paragraphs which follow, are defined solely in terms of sets. However, since our emphasis here is on probability and we are interested in the assignments of probabilities to events, we will use the terminology of events as well as sets.

Two of the most important set operations, *union* and *intersection*, involve combinations of events. Assume that a sample space S has been defined and that A_1 and A_2 are events (subsets) in this sample space. The *union* of the two events A_1 and A_2, denoted $A_1 \cup A_2$, is the set whose elements belong to A_1 or A_2, or both. In general, for n events $A_1, A_2, \ldots, A_n$ defined on a sample space S, the union of these n events is the set of elements in S which are members of at least one of the events $A_1, A_2, \ldots, A_n$. Various notations are used for the union of two events, A_1 and A_2: $A_1 \cup A_2$, $A_1 + A_2$, A_1 or A_2. The notation is read "the union of A_1 and A_2," "A_1 cup A_2," or simply "A_1 or A_2." Despite the fact that the notation "A_1 or A_2" is used for the union of two events, it is important to note that the meaning of union is the same as the *and/or* of legal terminology rather than the *or* of ordinary speech. That is, an element is a member of the union of two events, $A_1 \cup A_2$, if it is a member of either A_1 *or* A_2 or is a member of both A_1 *and* A_2.

Set definitions and operations are often illustrated by means of Venn diagrams.[5] In these diagrams, the sample space is generally represented by the interior of a rectangle. Events in the sample space are displayed as the interiors of plane figures; circles are frequently used for this purpose. Cross hatching or tinting is often employed to emphasize events of interest. The union of the two events A_1 and A_2 is shown as the tinted region in the Venn diagram in Figure 1.5.

The *intersection* of two events A_1 and A_2 in the sample space S is the set whose members are elements of both A_1 and A_2 (Figure 1-6). In general, for n events $A_1, A_2, \ldots, A_n$ defined in a sample space S, the intersection of these n events is the set of elements in S which are members of all of the n events. As in the case of the union of events, various notations are employed for the intersection of events. The following notations are commonly used: $A_1 \cap A_2$; A_1, A_2; A_1A_2; A_1 and A_2. The notation is read "the intersection of A_1 and A_2," "A_1 cap A_2," or "A_1 and A_2." In this text, the union and intersections of A_1 and A_2 will be written $A_1 \cup A_2$ and $A_1 \cap A_2$, respectively.

The *complement* of an event A in the sample space S is the set consisting of all elements which do not belong to A. Notations in use for the complement of an event A include $\bar{A}$, $\sim A$, A', and A^*. We will use the symbol $\bar{A}$, read "A bar" or "not-A" for the complement of A. The complement of the event A is the tinted region in Figure 1-7.

---

[5]John Venn (1834–1923) was an English logician and author. Among his works are *The Logic of Chance* (1866), *Symbolic Logic* (1881), and *The Principles of Empirical Logic* (1889).

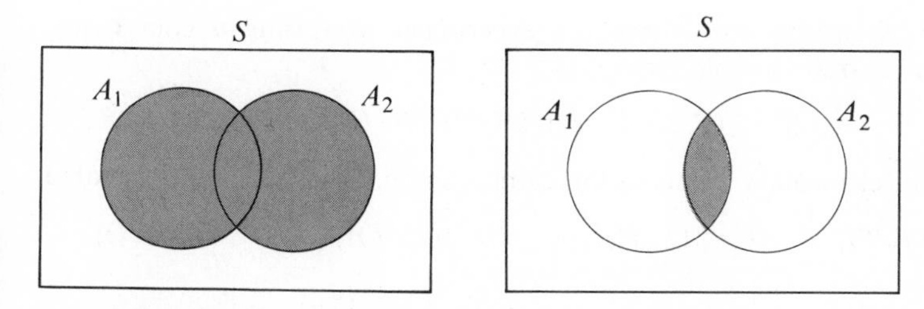

Figure 1-5 Union of A_1 and A_2, $(A_1 \cup A_2)$.

Figure 1-6 Intersection of A_1 and A_2 $(A_1 \cap A_2)$.

Two events are of particular importance, the *certain event* and the *impossible event*. When a sample space S has been defined, S itself is referred to as the *certain event* or the *sure event*, since in any single trial of the experiment which generates S, one or another of its elements must occur. The complement of S is referred to as the *impossible event*, *null set*, or *empty set*. The empty set is usually denoted by the special symbol $\varnothing$. It is the set which contains no elements or the set which is empty. By convention, the empty set is a subset of every sample space S, and by the definition of complement, it is also the complement of the certain event S, with respect to a sample space S, i.e., $\bar{S}, = \varnothing$. The union of any event A and its complement $\bar{A}$ is equal to the certain event; thus, $A \cup \bar{A} = S$. If A is the empty set $\varnothing$, we have $\varnothing \cup S = S$.

Two events A_1 and A_2 are said to be *mutually exclusive events* or *disjoint events*, if when one of these events occurs, the other cannot occur. In set theoretic language, two sets A_1 and A_2 are *mutually exclusive* or *disjoint* if no element of either set is a member of the other, i.e., the sets contain no elements in common. Therefore, the intersection of the two sets contains no elements, and is said to be equal to the empty set. This may be written $A_1 \cap A_2 = \varnothing$ A Venn diagram of two mutually exclusive events (or sets) is given in Figure 1-8.

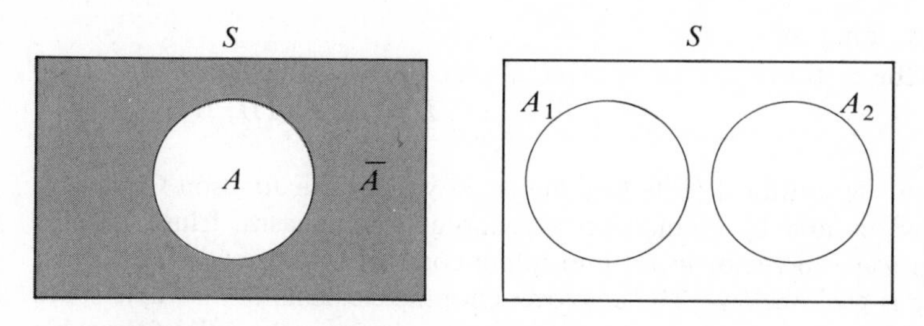

Figure 1-7 Complement of A.

Figure 1-8 Two mutually exclusive events, A_1 and A_2.

Example 1-5 Consider once more the experiment of tossing a coin twice, which gives rise to the sample space

$$S = \{(T, T), (T, H), (H, T), (H, H)\}$$

There are four elementary events in this sample space. They may be designated

$$A = \{(T, T)\}, \quad B = \{(T, H)\}, \quad C = \{(H, T)\}, \quad D = \{(H, H)\}$$

It will be important in subsequent work in probability theory and statistics to be able to convert verbal statements about events into their mathematical counterparts. The following are illustrative examples:

Verbal Statements Concerning Events	*Set Theoretic Counterpart*
No heads	$A = \{T, T\}$
One head (exactly)	$B \cup C = \{(T, H), (H, T)\}$
Two heads (exactly)	$D = \{H, H\}$
At least one head	$\bar{A} = B \cup C \cup D = \{(T, H), (H, T), (H, H)\}$
"No heads" and "at least one head" are complementary events	$A \cup \bar{A} = S$
One or no heads (less than two heads)	$A \cup B \cup C = \{(T, T), (T, H), (H, T)\}$
More than one head	$D = \overline{A \cup B \cup C} = \{H, H\}$
Intersection of "at least one head" and "exactly one head"	$\bar{A} \cap (B \cup C) = \{(T, H), (H, T)\} = B \cup C$
"No heads" and "two heads" are mutually exclusive events	$A \cap D = \varnothing$
Either zero, one, or two heads (the certain event)	$S = A \cup B \cup C \cup D = \{(T, T), (T, H), (H, T), (H, H)\}$

Example 1-6 A statistical table lists the employees of the Johnson Company, classified by sex and by opinion on a proposal to emphasize fringe benefits rather than wage increases in an impending contract discussion. The format of the table is given below. The employees are the elements of a sample space S. Instead of showing numbers of employees, the entries in the cells of the table give the appropriate set theoretic notations. The set of male employees has been

designated by A_1, the set of female employees by A_2, the set of employees in favor of the proposal by B_1, and so forth.

Verbal statements concerning selected events and their set theoretic equivalents are given below the table.

	Opinion			
Sex	*In favor*	*Neutral*	*Opposed*	*Total*
Male	$A_1 \cap B_1$	$A_1 \cap B_2$	$A_1 \cap B_3$	A_1
Female	$A_2 \cap B_1$	$A_2 \cap B_2$	$A_2 \cap B_3$	A_2
Total	B_1	B_2	B_3	S

Verbal Statements Concerning Events[a]	*Set Theoretic Counterpart*
Male employees	A_1
Male employees in favor of the proposal	$A_1 \cap B_1$
All employees other than males in favor of the proposal	$\overline{A_1 \cap B_1}$
All male employees in favor, neutral, or opposed to the proposal	$(A_1 \cap B_1) \cup (A_1 \cap B_2) \cup (A_1 \cap B_3) = A_1$
Employees not in favor of the proposal	$\bar{B}_1 = B_2 \cup B_3$
Employees in favor of the proposal	$B_1 = (A_1 \cap B_1) \cup (A_2 \cap B_1)$ or $B_1 = \overline{B_2 \cup B_3}$
All employees	$S = A_1 \cup A_2$ or $S = B_1 \cup B_2 \cup B_3$

Question: If numbers of employees were given in the cells of the table, how would you calculate the number of employees in the set $A_1 \cup B_1$? (Hint: In general, it is incorrect simply to add the number of employees in set A_1 to the number in B_1.)

[a]Strictly speaking, the words "the set of" should precede each statement in this column, e.g., "the set of male employees."

Answer: Let $n(A_1 \cup B_1)$, $n(A_1)$, $n(B_1)$, and $n(A_1 \cap B_1)$ denote the numbers of employees in sets $(A_1 \cup B_1)$, A_1, B_1, and $(A_1 \cap B_1)$, respectively. Then $n(A_1 \cup B_1) = n(A_1) + n(B_1) - n(A_1 \cap B_1)$. The number of employees in the intersection $A_1 \cap B_1$ must be subtracted because of the double-counting of $n(A_1 \cap R_1)$ that occurs when $n(A_1)$ is added to $n(B_1)$.

Probability as a Function of Events on a Sample Space

In Section 1.1 probability was defined from several different points of view. Now, probability will be defined in terms of set theory; specifically as a so-called "*non-negative additive set function*." Because the concepts of set theory provide so fundamental and convenient a framework for our thinking, it is particularly fruitful to approach probability in the context of these ideas.

Recall that a *function* in mathematics is simply a rule which assigns one and only one number to each member of a set of elements known as the *domain* of the function. The elements of the domain are numbers in the most familiar cases. The set of numbers assigned to the set of elements in the domain is referred to as the *range* of the function. A simple example will illustrate this concept. Consider the function

$$f(x) = x^2;\ x = -2, -1, 1, 2$$

In the interest of simplicity, the values of x have been restricted to the numbers -2, -1, 1, and 2. This is referred to as the *admissible region* of the domain. The set of values of $f(x)$ constitutes the range of the function. The function is portrayed schematically in Figure 1-9 and tabulated in Figure 1-10.

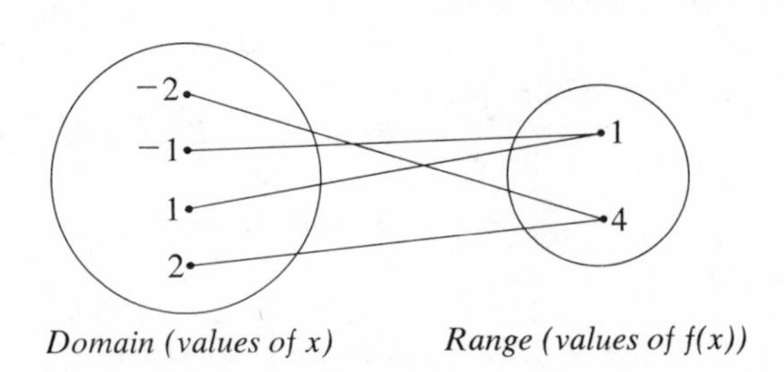

Figure 1-9 Schematic representation of the function $f(x) = x^2;\ x = -2, -1, 1, 2$.

The general principle that only one element in the range can be assigned to each element in the domain, although an element in the range may be paired with more than one element in the domain can be observed in this illustration.

x	$f(x)$
-2	4
-1	1
1	1
2	4

Figure 1-10 Tabular representation of the function $f(x) = x^2$; $x = -2, -1, 1, 2$.

Probability may be defined as a function of events on a sample space. We restrict the present discussion to finite sample spaces, although it may be extended to the infinite case as well if appropriate modifications are made in the third axiom presented below. Given a sample space S, probability may be defined as a function which assigns a particular real number to every event A in S; this number is denoted $P(A)$ and is referred to as the "probability of the event A." Thus the events (subsets) of the sample space constitute the domain of the function; the probabilities assigned to these events comprise the range. To be a "probability function," this function must obey the following three axioms:

Axiom 1 $0 \leq P(A) \leq 1$ for every event A.

Axiom 2 $P(S) = 1$.

Axiom 3 If A and B are events in S, and $A \cap B = \varnothing$, that is, if A and B are mutually exclusive events, then $P(A \cup B) = P(A) + P(B)$.

If probability is viewed purely as a formal mathematical system, all of the properties of probability theory can be derived from these three axioms. The first axiom states that probabilities are non-negative, real numbers lying between the values of zero and one, inclusive. The second axiom refers to the fact that the probability of the "certain event" is equal to one. This means that some one or another of the possible mutually exclusive outcomes of the experiment which generated the sample space must occur on a given trial. The third axiom states that if A and B are mutually exclusive (or disjoint) events in the sample space S, then the probability of the union of these two events is obtained by adding the probabilities of A and B. If we consider n mutually exclusive events $A_1, A_2, \ldots, A_n$ in S, the third axiom is easily extended to

$$P(A_1 \cup A_2 \cup \cdots \cup A_n) = P(A_1) + P(A_2) + \cdots + P(A_n)$$

Mathematically, any set of numbers may be paired with elements in a sample space, and if the resulting function obeys these axioms, it may be called a "probability function."

The third axiom indicates the reason for the term *additive set function*, since the probability assigned to the union of two or more mutually exclusive events is obtained by *adding* the probabilities assigned to the *sets* representing these events. Kyburg and Smokler state, "It has been said (facetiously) that there is no problem about probability: it is simply a non-negative, additive set function, whose maximum value is unity."[6] However, if probability theory is to be viewed not merely as an abstract mathematical system, but is to be used to help solve real-world problems, then clearly the numbers which are assigned as probabilities must bear some relationship to real-world phenomena. In the context of business decision-making problems, particularly as regards non-repetitive events, it seems inevitable that judgment must play a part in the assignment of probabilities to the outcomes which affect the decisions to be made.

The axioms of probability theory are illustrated in the following two examples.

Example 1-7 Consider the conceptual experiment of rolling a *true* die one time. The sample space is

$$S = \{1, 2, 3, 4, 5, 6\}$$

If probability assignments are made to these sample points according to the classical definition, we have the following probability function:

Events A_i	*Probability of Events* $P(A_i)$
1	1/6
2	1/6
3	1/6
4	1/6
5	1/6
6	1/6
	1

A brief comment about the notation is in order. There are six events, designated A_i, namely, $A_1, A_2, \ldots, A_6$, with the corresponding probabilities

[6] H. E. Kyburg, Jr. and H. E. Smokler, *Studies in Subjective Probability* (New York: Wiley, 1964), p.3.

$P(A_1), P(A_2), \ldots, P(A_6)$. Thus, the events may be specified as A_i, $i = 1, 2, \ldots$, 6, and the probabilities of these events as $P(A_i)$, $i = 1, 2, \ldots, 6$.

The first axiom is satisfied, since each probability lies between zero and one, inclusive; specifically, each $P(A_i) = 1/6$. The second axiom is met, since the probability of the certain event is equal to one, i.e., $P(S) = 1$. This, as indicated earlier, is interpreted to mean that on any roll of the die it is certain that one of the outcomes 1, 2, 3, 4, 5, or 6 must occur. The third axiom may be shown to be satisfied by referring to the two mutually exclusive events A_1 (the "occurrence of a 1") and A_2 (the "occurrence of a 2"). The probability that either a 1 or a 2 will occur in a single roll is $P(A_1 \cup A_2) = P(A_1) + P(A_2) = 1/6 + 1/6 = 1/3$. It should be noted that the events referred to in the axioms need not be elementary events. Thus, if the event E refers to the appearance of a 1 or 2 and event F to the appearance of a 3 or 4, then $P(E \cup F) = P(E) + P(F) = 1/3 + 1/3 = 2/3$. However, the events defining the sample space must constitute a *partition* of that space, i.e., they must be an exhaustive (complete) set of mutually exclusive events. It may be noted that the probability of occurrence of any event E or its complement $\bar{E}$ is equal to one, i.e., $P(E \cup \bar{E}) = P(S) = 1$.

Example 1-8 A corporation economist was concerned about the surging of credit demand and the effect of inflationary pressures in the economy upon his company's activities during the next six-month period. He established the following subjective probability distribution concerning the possible action of the Federal Reserve System in the area of monetary policy during the next half year.

Events E_i	*Probability of Events* $P(E_i)$
Tighten monetary policy	0.5
Leave current monetary situation unchanged	0.4
Relax monetary policy	0.1
	1.0

This economist assigned the value 0.9 to the probability that the Federal Reserve System will either tighten monetary policy or leave the current monetary situation unchanged.

$$P(E_1 \cup E_2) = P(E_1) + P(E_2) = 0.5 + 0.4 = 0.9$$

Odds Ratios

Regardless of the definition used, we sometimes prefer to express probabilities in terms of odds. Thus, in the examples given above, if the probability that a six will appear on the roll of a die is $1/6$, then the odds that a six will appear are one-to-five, written 1:5. If the economist in Example 1-8 assesses the probability that the Federal Reserve System will adopt a tightened monetary policy as $0.5/1 = 1/2$, then in terms of odds his probability assignment is 1:1 that the Federal Reserve System will follow this policy as opposed to not following it.

If the probability that an event A will occur is

$$P(A) = \frac{a}{n}$$

then the odds in favor of the occurrence of A are:

$$\text{odds in favor of } A = \frac{a}{n - a} = a : n - a$$

The odds unfavorable to A are:

$$\text{odds against } A = \frac{n - a}{a} = n - a : a$$

Another way of viewing the odds ratio is as a ratio of the probability of the occurrence of an event, A, to the probability of its complement, $\bar{A}$. Thus,

$$\text{odds in favor of } A = \frac{P(A)}{P(\bar{A})} = \frac{a/n}{(n - a)/n} = \frac{a}{n - a}$$

$$\text{odds against } A = \frac{P(\bar{A})}{P(A)} = \frac{(n - a)/n}{a/n} = \frac{n - a}{a}$$

1.3 Fundamental Probability Rules

In most applications of probability theory, we are interested in combining probabilities of events that are related in some meaningful way. In this section, we discuss two of the fundamental ways of combining probabilities. They are *addition* and *multiplication*.

Let us assume a group of ten individuals each of whom has been classified on two bases, (1) male or female and (2) under 40 years of age or 40 years and over. Let S represent the sample space consisting of the ten individuals, A_1, the set of males, and A_2, the set of persons under 40 years of age.

We may represent this sample space diagrammatically by an Euler diagram, which is (in addition to Venn diagrams) another way of depicting sets and sample spaces. In an Euler diagram,[7] the elements of a sample space are represented by points placed in the interior of a rectangle. Our present

---

[7]Named after Leonhard Euler (1707–1783), a Swiss mathematician and physicist. He was the author of many works on analytic mathematics, algebra, and other mathematical subjects, hydrodynamics, astronomy, optics, and acoustics.

discussion is restricted to finite sample spaces containing equally likely sample points. An event is represented by the number of sample points lying within a plane figure denoting the event. Thus, in Figure 1-11, the sample space S of ten individuals classified by sex and age is shown.

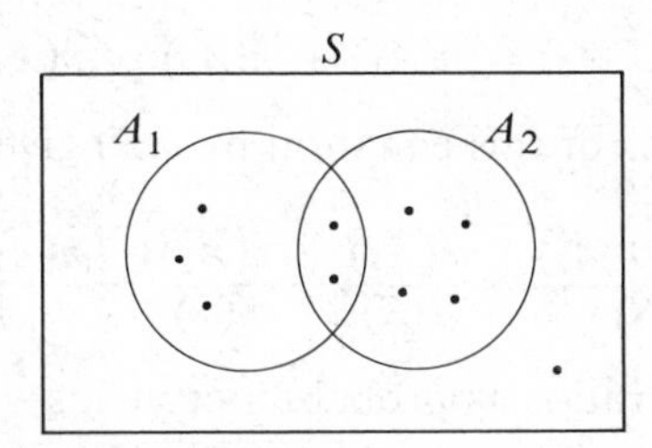

Figure 1-11 Euler diagram of sample space S; A_1, the set of males; and A_2, the set of persons under 40 years of age.

Addition Rule

Let $n(A_1)$ represent the number of elements in event A_1, $n(A_2)$ the number of elements in event A_2, etc. Then, in this example,

$n(A_1) = 5 =$ the number of males
$n(A_2) = 6 =$ the number of persons under 40 years of age
$n(A_1 \cup A_2) = 9 =$ the number of persons who are either male or under 40 years of age, or both
$n(A_1 \cap A_2) = 2 =$ the number of persons who are both male and under 40 years of age.

It can be seen that

$$n(A_1 \cup A_2) = n(A_1) + n(A_2) - n(A_1 \cap A_2)$$

or

$$9 = 5 + 6 - 2$$

The reason that the number of points in the intersection of A_1 and A_2 had to be subtracted is that otherwise double-counting would have resulted, since the points in the intersection belong to both A_1 and A_2.

In this context, the probability of an event may be denoted by the ratio of the number of sample points lying within that event to the total number of points in the sample space. Therefore, $P(A_1) = n(A_1)/n(S) = 5/10$, $P(A_2) = n(A_2)/n(S) = 6/10$, etc.

We can now prove the addition rule for any two events A_1 and A_2 in a sample space S.

Addition Rule

$$P(A_1 \cup A_2) = P(A_1) + P(A_2) - P(A_1 \cap A_2) \tag{1.1}$$

This rule is proved simply by first observing that

$$n(A_1 \cup A_2) = n(A_1) + n(A_2) - n(A_1 \cap A_2)$$

Dividing both sides of this equation by $n(S)$ gives

$$\frac{n(A_1 \cup A_2)}{n(S)} = \frac{n(A_1)}{n(S)} + \frac{n(A_2)}{n(S)} - \frac{n(A_1 \cap A_2)}{n(S)}$$

Interpreting these ratios as probabilities yields

$$P(A_1 \cup A_2) = P(A_1) + P(A_2) - P(A_1 \cap A_2)$$

It follows as a corollary of the addition rule that if A_1 and A_2 are mutually exclusive events, $P(A_1 \cap A_2) = 0$, and

$$P(A_1 \cup A_2) = P(A_1) + P(A_2) \tag{1.2}$$

The addition rule can, of course, be extended to the case of more than two events. If there are three events A_1, A_2, and A_3, it can readily be shown that

$$\begin{aligned} P(A_1 \cup A_2 \cup A_3) = {} & P(A_1) + P(A_2) + P(A_3) - P(A_1 \cap A_2) \\ & - P(A_1 \cap A_3) - P(A_2 \cap A_3) + P(A_1 \cap A_2 \cap A_3) \end{aligned} \tag{1.3}$$

(Hint: In order to prove this relation, let $B = A_2 \cup A_3$, and evaluate $P(A_1 \cup B)$.)

It may be observed that $P(A_1 \cup A_2 \cup A_3)$ is equal to the sum of the probabilities of the three events A_1, A_2, A_3 grouped into all possible distinct combinations, i.e., singles, pairs, and triples. Plus signs appear before the odd combinations (singles, triples), and minus signs before the even combination (pairs). This generalizes for the probability of n events, $P(A_1 \cup A_2 \cup \cdots \cup A_n)$, in the obvious way; all possible distinct combinations are established for singles, pairs, triples, . . . , with plus signs attached to probabilities of odd combinations and minus signs to probabilities of the even ones. If the n events $A_1, A_2, \ldots, A_n$ are mutually exclusive, then

$$P(A_1 \cup A_2 \cup \cdots \cup A_n) = P(A_1) + P(A_2) + \cdots + P(A_n) \tag{1.4}$$

The addition rule evaluates the probability of the union of two or more events. It is applicable whenever we are interested in the probability that any one of several events will occur. For example, in the case of two events A_1 and A_2, $P(A_1 \cup A_2)$ gives the probability of occurrence of either A_1 or A_2 (or both). On the other hand, in many applications, we are interested in the probability of the intersection of two or more events. This situation prevails whenever we want to compute the probability of the *joint occurrence* of two or more events. These events may occur simultaneously or successively. For example, we may be interested in computing the probability of obtaining two kings if two cards are drawn simultaneously from a well shuffled deck of cards. Or, we may be interested in the probability of the appearance of two kings if two cards are drawn, one after the other from the deck. These probabilities will be computed as an illustration after the *multiplication rule* is proved.

Just as the probability of the union of events was arrived at by adding up appropriate probabilities, the probability of the intersection of events is obtained by multiplying *appropriate* probabilities. As was the case for the addition theorem, a rule will be derived for the multiplication of probabilities for two events; then this rule will be generalized for n events.

Conditional Probability

In considering joint probabilities, it is necessary to introduce the concept of *conditional probability*. The meaning of any probability depends upon the set of elements to which the discussion is limited, that is, to some sample space S. In that sense, the probability is conditional upon the definition of S. For example, if we are interested in the probability that a manufacturing firm will have 100 employees or more, that figure will depend on whether the sample space consists of the manufacturing firms in an industry, a city, a state, etc. Thus, every probability statement may be viewed as a conditional probability. However, the term "conditional probability" is usually reserved for probability statements about a reduced sample space, within a given sample space S. To illustrate this concept, consider the sample space S of ten individuals depicted in Figure 1-11. If a person is selected at random from this group, the probability that the individual is under 40 years of age is

$$P(A_2) = \frac{n(A_2)}{n(S)} = \frac{6}{10}$$

On the other hand, suppose we are interested in the probability that an individual is under 40 years of age, given that he is a male. In this case, the relevant sample space is restricted to the number of sample points lying within the event A_1, "male," and we now want to know the proportion of such points which possess the property "under 40 years of age." This conditional probability is denoted $P(A_2 \mid A_1)$ and is read "the probability of A_2, given A_1." The required probability is

$$P(A_2 \mid A_1) = \frac{n(A_1 \cap A_2)}{n(A_1)} = \frac{2}{5}$$

The reduced sample space consists of the five points lying within A_1, representing the event "male." The two points lying in the intersection

$A_1 \cap A_2$ represent the number of those who are male, who also possess the property of being under 40 years of age. Thus, the *conditional probability*, $P(A_2 \mid A_1)$, is given by the ratio of the number of points in the subset $A_1 \cap A_2$ to the number of points in the reduced sample space A_1. Probabilities such as $P(A_1)$ and $P(A_2)$ are usually referred to as "*unconditional probabilities*" and are so designated in this text. This nomenclature will be used to differentiate clearly between probabilities such as $P(A_2)$ and $P(A_2 \mid A_1)$, although as indicated at the outset of the discussion of conditional probabilities, all probabilities are "conditional" or "depend upon" the sample space within which they are defined.

We can now prove the multiplication rule for two events A_1 and A_2.

Multiplication Rule

$$P(A_1 \cap A_2) = P(A_1)\,P(A_2 \mid A_1) \tag{1.5}$$

This rule is proved in one step by simply writing down the definition of the probability of the intersection of two events in terms of sample points and expressing its equality with two factors into which this definition can be broken. Thus,

$$\frac{n(A_1 \cap A_2)}{n(S)} = \frac{n(A_1)}{n(S)} \cdot \frac{n(A_1 \cap A_2)}{n(A_1)}$$

Identifying each of these expressions in terms of the probability it represents gives

$$P(A_1 \cap A_2) = P(A_1)\,P(A_2 \mid A_1)$$

Conditional probabilities are often computed by means of this multiplication rule. Solving for $P(A_2 \mid A_1)$, we obtain

$$P(A_2 \mid A_1) = \frac{P(A_1 \cap A_2)}{P(A_1)}, \text{ where } P(A_1) \neq 0 \tag{1.6}$$

Interchanging the designations A_1 and A_2 gives

$$P(A_1 \mid A_2) = \frac{P(A_1 \cap A_2)}{P(A_2)}, \text{ where } P(A_2) \neq 0$$

We can observe from (1.6) that a conditional probability is a ratio of probabilities, which in terms of sample points is equal to the expression given earlier for the reduced sample space. Thus,

$$P(A_2 \mid A_1) = \frac{P(A_1 \cap A_2)}{P(A_1)} = \frac{\dfrac{n(A_1 \cap A_2)}{n(S)}}{\dfrac{n(A_1)}{n(S)}} = \frac{n(A_1 \cap A_2)}{n(A_1)}$$

If $P(A_2 | A_1) = P(A_2)$, then the multiplication rule becomes

(1.7) $$P(A_1 \cap A_2) = P(A_1)\, P(A_2)$$

and events A_1 and A_2 are said to be *statistically independent*. If two events A_1 and A_2 are independent, then knowing that one of the events has occurred does not affect the probability that the other will occur. Thus, if A_1 and A_2 are independent, then $P(A_2 | A_1) = P(A_2)$ and $P(A_1 | A_2) = P(A_1)$. If the events A_1 and A_2 are not independent, they are said to be *dependent* or *conditional* events.

Example 1-9 As a simple illustration of these ideas, let us consider the following example. What is the probability of obtaining two kings in drawing two cards from a shuffled deck of cards? The first thing to note is that this is a poorly defined question. We must specify how the two cards are to be drawn from the deck. That is, are the two cards drawn simultaneously? If not, if the cards are to be drawn successively, is the first card replaced prior to drawing the second?

Solution: First, let us assume that the two cards are drawn successively without replacing the first card. Denoting the events "king on first card" and "king on second card" by A_1 and A_2, respectively, it is clear that these are dependent events. That is, the probability of obtaining a king on the second draw is dependent upon whether a king was obtained on the first. The required joint probability for the successive occurrence of A_1 and A_2 is then $P(A_1 \cap A_2) = P(A_1)\, P(A_2 | A_1)$. Considering the draw of the first card, since there are 52 cards in the deck, of which four are kings, $P(A_1) = 4/52$. Given that a king was obtained on the first draw, the probability of drawing a king on the second is $P(A_2 | A_1) = 3/51$. This follows since there are only 51 cards remaining, three of which are kings, given that a king was obtained on the first draw. Therefore the required probability is

$$P(A_1 \cap A_2) = \frac{4}{52} \cdot \frac{3}{51} = \frac{1}{221}$$

What is the probability of obtaining two kings if the two cards are drawn simultaneously from the deck? The answer to this question is exactly the same as in the preceding computation. Here, we think of one of the two cards as the "first" and the other as the "second" although they have been drawn together. A_1 is again used to denote the event "king on first card" and A_2 "king on second card."

Now let us assume the two cards were drawn successively, but that the first card was replaced into the deck, and the deck was reshuffled prior to drawing the second card. Since "king on first card" and "king on second card" are now independent events, their joint probability is given by the formula $P(A_1 \cap A_2) = P(A_1)\, P(A_2)$. Therefore,

$$P(A_1 \cap A_2) = \frac{4}{52} \cdot \frac{4}{52} = \frac{1}{169}$$

In this case, of course, the composition of the deck is the same on both drawings.

This simple example has interesting implications for sampling theory. If the deck of cards is viewed as a statistical universe, we have here a situation of drawing a sample of two items at random from this universe (1) *without replacement*, and (2) *with replacement* of the sampled elements. In the case of human populations, sampling is usually carried out without replacement. That is, when an individual is drawn into the sample, after the necessary data are obtained, he is usually not replaced (either conceptually or actually) into the universe prior to the drawing of another individual. Sometimes it is convenient, for purposes of simplicity, to calculate probabilities in a "sampling without replacement" situation as though "replacement" had taken place. These types of situations will be discussed in the next chapter.

The generalization of the multiplication rule is straightforward, although some subtleties arise. The joint probability of the occurrence of n events, $A_1, A_2, \ldots, A_n$, may be expressed as

$$\textbf{(1.8)} \quad P(A_1 \cap A_2 \cap \cdots \cap A_n) = P(A_1)\,P(A_2 \mid A_1)\,P(A_3 \mid A_2 \cap A_1) \cdots P(A_n \mid A_{n-1} \cap \cdots \cap A_1)$$

This notation means that the joint probability of the n events is given by the product of the probability that the first event A_1 has occurred; the conditional probability of the second event A_2, given that A_1 has occurred; the conditional probability of the third event A_3, given that both A_2 and A_1 have occurred; etc. Of course, the n events can be numbered arbitrarily; therefore any one of them may be the first event, any of the remaining $n - 1$ may be second, and so forth.

The joint probability of the n events, $A_1, A_2, \ldots, A_n$, in the special case where these events are independent, is[8]

$$\textbf{(1.9)} \qquad P(A_1 \cap A_2 \cap \cdots \cap A_n) = P(A_1)\,P(A_2)\,P(A_3) \cdots P(A_n)$$

For simplicity of exposition, it was assumed in the derivation of the addition and multiplication rules that the sample space S contained a countable

---

[8]If the n events are independent, this statement holds. However, a *necessary* and sufficient condition for n events to be mutually independent is that the probability of every combination that can be established of the n events and their complements must be equal to the product of the probabilities of the components of the respective combinations. For example, for three events to be mutually independent, all of the following must be true:

$$\begin{aligned} P(A_1 \cap A_2 \cap A_3) &= P(A_1)P(A_2)P(A_3) \\ P(\bar{A}_1 \cap A_2 \cap A_3) &= P(\bar{A}_1)P(A_2)PA_3) \\ P(A_1 \cap \bar{A}_2 \cap A_3) &= P(A_1)P(\bar{A}_2)P(A_3) \\ &\vdots \\ P(\bar{A}_1 \cap \bar{A}_2 \cap \bar{A}_3) &= P(\bar{A}_1)P(\bar{A}_2)P(\bar{A}_3) \end{aligned}$$

It is possible to construct situations involving three events, where $P(A_1 \cap A_2 \cap A_3) = P(A_1)P(A_2)P(A_3)$, yet some of the other probability equations indicated above would not hold.

and finite number of equiprobable sample points. If the probability axioms (suitably modified) are obeyed, the same rules can be shown to apply to events defined on sample spaces with a countably infinite number of elements and to continuous sample spaces as well. Therefore, in the illustrations that follow and in the exercises at the end of the chapter, a few problems involving countably infinite sample spaces are introduced. Continuous sample spaces will be discussed in subsequent chapters.

It is possible to solve a large variety of probability problems using only the addition and multiplication rules, either separately or together. A number of illustrative examples follow.

Example 1-10 A fair coin is tossed twice. What is the probability of obtaining exactly one head?

Solution: One way of arriving at the solution is through the combined use of the addition and multiplication theorems. Denote the appearance of a head by H and a tail by T. The event "exactly one head" in two trials may occur by obtaining a head on the first trial followed by a tail on the second or a tail followed by a head. These two events are mutually exclusive. Thus, by the addition rule,

$$P(\text{exactly one head}) = P((H \cap T) \cup (T \cap H)) = P(H \cap T) + P(T \cap H)$$

The appearance of a head on the first toss and a tail on the second are independent events, as are a tail on the first toss and a head on the second. Thus, by the multiplication rule,

$$P(H \cap T) = P(H)\,P(T) = 1/2 \times 1/2 = 1/4$$

and

$$P(T \cap H) = P(T)\,P(H) = 1/2 \times 1/2 = 1/4$$

Hence,

$$P\text{ (exactly one head)} = 1/4 + 1/4 = 1/2$$

Another way of obtaining the solution is to consider the sample space of equiprobable points, as in Exercise 1-5,

$$S = \{(T, T), (T, H), (H, T), (H, H)\}$$

Since two of the four sample points represent the occurrence of exactly one head, we have

$$P\text{ (exactly one head)} = 2/4 = 1/2$$

Example 1-11 A batch of transistors contains 10% defectives. Three transistors are drawn at random from the batch one at a time, each being replaced prior to the next draw. What is the probability of obtaining at least one defective transistor?

Solution: Let D and G represent the appearance of a defective and a good transistor, respectively, when a transistor is drawn from the batch. Then, since the sampling is done with replacement, on any draw of one transistor

$$P(D) = 0.1 \quad \text{and} \quad P(G) = 0.9$$

Let E denote the event "at least one defective in a sample of three transistors." Then $\bar{E}$, the complement of E, denotes the event "zero defectives in a sample of three transistors." Since the event $\bar{E}$ represents the successive occurrence of three good transistors, and since the appearances of good transistors on the first, second, and third drawings are independent events, we obtain by the multiplication rule

$$P(\bar{E}) = P(G \cap G \cap G) = P(G)\,P(G)\,P(G) = (0.9)(0.9)(0.9) = 0.729$$

Therefore,

$$P(E) = 1 - P(\bar{E}) = 0.271$$

Example 1-12 The probability of hitting a target on a single trial is 0.05. Assume that the trials are independent. How many trials are required for the probability of hitting the target at least once to be 0.99?

Solution: Let n stand for the unknown number of trials and E denote the event "at least one hit in n trials." Then $\bar{E}$ represents the event "zero hits in n trials." Since the probability of a miss on any trial is 0.95, the probability of obtaining zero hits in n trials (n misses) is by the multiplication rule

$$P(\bar{E}) = (.95)^n$$

The probability of at least one hit is 0.99. In symbols,

$$P(E) = 0.99$$

Therefore,

$$P(\bar{E}) = 1 - 0.99 = 0.01$$

Equating the two expressions for $P(\bar{E})$ gives

$$(0.95)^n = 0.01$$

Solving for n by taking logarithms of both sides of this equation yields

$$n(\log 0.95) = \log 0.01$$

$$n = \frac{\log 0.01}{\log 0.95} = \frac{-2}{9.9777 - 10} = \frac{-2}{-0.0223} \approx 90 \text{ trials}$$

Example 1-13 What is the probability of obtaining a six at least once in two rolls of a die?

Solution: It is instructive to solve this problem in three different ways. Let the appearance of a six and non-six on a given roll be denoted by 6 and $\bar{6}$, respectively. Then, $P(6) = 1/6$ and $P(\bar{6}) = 5/6$. The following events constitute "successes," i.e., the appearance of a six at least once: $(6 \cap \bar{6})$, $(\bar{6} \cap 6)$, and $(6 \cap 6)$. It may be noted here that the intersection of the two events 6 and $\bar{6}$ refers to the *successive occurrence* of a six and a non-six on two trials of the experiment of rolling a die. By the multiplication rule, since the events on first and second trial are independent,

$$P(6 \cap \bar{6}) = P(6)\,P(\bar{6}) = (1/6)(5/6) = 5/36$$
$$P(\bar{6} \cap 6) = P(\bar{6})\,P(6) = (5/6)(1/6) = 5/36$$
$$P(6 \cap 6) = P(6)\,P(6) = (1/6)(1/6) = 1/36$$

Since $(6 \cap \bar{6})$, $(\bar{6} \cap 6)$, and $(6 \cap 6)$ are mutually exclusive events, we have by the addition theorem

$$\begin{aligned} P(\text{six, at least once}) &= P[(6 \cap \bar{6}) \cup (\bar{6} \cap 6) \cup (6 \cap 6)] \\ &= P(6 \cap \bar{6}) + P(\bar{6} \cap 6) + P(6 \cap 6) \\ &= 5/36 + 5/36 + 1/36 = 11/36 \end{aligned}$$

Another way of solving this problem is to obtain the probability of the complement of the desired event and subtract that figure from one. The complement of the event "six at least once in two trials" is "no sixes in two trials." This latter event is denoted $(\bar{6} \cap \bar{6})$. Since

$$P(\bar{6} \cap \bar{6}) = P(\bar{6})\,P(\bar{6}) = (5/6)(5/6) = 25/36$$
$$P(\text{six, at least once}) = 1 - P(\bar{6} \cap \bar{6}) = 1 - 25/36 = 11/36$$

A third way of solving the problem is in terms of a sample space of equally likely elements; these elements are listed below in columns.

1,1	2,1	3,1	4,1	5,1	6,1
1,2	2,2	3,2	4,2	5,2	6,2
1,3	2,3	3,3	4,3	5,3	6,3
1,4	2,4	3,4	4,4	5,4	6,4
1,5	2,5	3,5	4,5	5,5	6,5
1,6	2,6	3,6	4,6	5,6	6,6

This sample space has 36 elements. The first and second numbers in each element represent the outcomes on the first and second rolls, respectively. There are eleven points in this sample space which represent the event "six at least once in two trials." Thus,

$$P(\text{six, at least once}) = 11/36$$

If the problem had been framed in terms of two dice being rolled once, with the question being "What is the probability that at least one six would appear?" exactly the same solution would apply. In this case, $6 \cap \bar{6}$ would refer to the simultaneous occurrence of a six on "die one" and a non-six on "die two," and so forth. It is sometimes helpful in this type of problem to think of the two dice as being of different colors, say, green and red, in order to highlight the fact that (say) a six on the first die and a five on the second is a different event from a five on the first die and a six on the second. Thus, e.g., a six on the green die and a five on the red is a different event from a five on the green die and a six on the red.

Example 1-14 The game of "craps" is played as follows. A player whom we shall refer to as the "roller" rolls two dice. There are one or more other players whom we can collectively refer to as the "non-roller." The roller wins if he obtains a 7 or 11 on the first roll, (i.e., a total of 7 or 11 on the two dice). The non-roller wins if a 2, 3, or 12 is obtained. If neither a 7, 11, 2, 3, nor 12 is obtained, the number rolled is referred to as the "point." When the roller obtains a point on the first roll, he continues to roll the dice. The game is then decided only by the occurrence of the point or a 7. If the roller repeats the point before obtaining a 7, he wins. Otherwise, the nonroller wins. What is the probability that the non-roller will win the game?

Solution: A useful viewpoint in this problem is that there are two mutually exclusive ways for the non-roller to win: (1) on the first roll, and (2) thereafter. The total probability that the non-roller will win is the sum of these two probabilities. For convenience, the basic computations have been summarized in the following table.

(1) *Possible Outcome*	(2) *Probability*	(3) $P(A_1)$	(4) $P(A_2 \mid A_1)$	(5) $P(A_1 \cap A_2)$
2	1/36			
3	2/36			
4	3/36	3/36	6/9	18/324 = 55/990
5	4/36	4/36	6/10	24/360 = 66/990
6	5/36	5/36	6/11	30/396 = 75/990
7	6/36			
8	5/36	5/36	6/11	30/396 = 75/990
9	4/36	4/36	6/10	24/360 = 66/990
10	3/36	3/36	6/9	18/324 = 55/990
11	2/36			
12	1/36			
	1			392/990 = 196/495

Columns (1) and (2) give the possible outcomes on a roll of two dice and their respective probabilities (See Example 1-13). Let $P(E_1)$ represent the probability that the non-roller wins on the first roll. Then, since the non-roller wins on the occurrence of a 2, 3, or 12,

$$P(E_1) = 1/36 + 2/36 + 1/36 = 4/36 = 1/9$$

Let $P(E_2)$ denote the probability that the game is not decided on the first roll, and the non-roller wins. The calculations for $P(E_2)$ are given in Columns (3), (4), and (5). The meaning of the entries in these columns is given by the following definitions:

$P(A_1)$ = probability that the specified number is the point.
$P(A_2 \mid A_1)$ = probability that non-roller wins, given that the specified number is the point.
$P(A_1 \cap A_2)$ = probability that the given number is the point and the non-roller wins.

Then, $P(E_2) = \Sigma\ P(A_1 \cap A_2)$, where the symbol Σ (Greek capital sigma) means "the sum of." Hence $P(E_2)$ is equal to the sum of the terms of form $P(A_1 \cap A_2)$.

Column (3) is self-explanatory. As an example of the computations in Column (4), consider the first entry, $P(A_2 \mid A_1) = 6/9$. If 4 is the point obtained on the first roll, the game can only be decided by the appearance of a 4 or a 7. These probabilities are $P(4) = 3/36$ and $P(7) = 6/36$. A rough intuitive argument will be given. On any given roll, the odds that a 7 rather than a 4 will appear are 6/36 to 3/36 or 6 to 3. The long run interpretation is that the odds are 6 to 3 that a 7 will appear *before* a 4. Thus, given that 4 is the point, the odds that the non-roller will win are 6:3, and the probability that the non-roller will win is 6/9. Another way of looking at it is that there is a reduced sample space consisting of nine equally likely points, six of which represent a win by the non-roller and three of which represent a win by the roller. Again, given that 4 is the point, the probability that the non-roller wins is 6/9.

The joint probabilities given in Column (5), $P(A_1 \cap A_2)$, are obtained by multiplying the respective entries in Columns (3) and (4). The total probability that the game is not decided on the first roll, and the non-roller wins, is the sum of these Column (5) probabilities, thus

$$P(E_2) = 196/495$$

Therefore, the probability that the non-roller will win is $P(\text{non-roller wins}) = P(E_1) + P(E_2) = 1/9 + 196/495 = 251/495$. Stating this in terms of odds, the odds in favor of the non-roller are 251 to 244 or 1.03 to 1.00.

Example 1-15 Every so often one reads in the newspapers that someone has played (?) Russian roulette and has killed himself. This barbaric game will be considered as an illustration of a series of experimental trials which generates an infinite sample space.

Evidently, Russian roulette proceeds as follows. An individual with a revolver (assume a six-shooter) places one bullet in the chamber, spins the revolving cylinder, points the gun to his forehead and fires. Let us consider this as an experiment which results in two possible outcomes on a given trial, (1) the individual kills himself, or (2) he does not kill himself. If the individual plays this game for an infinite number of trials, what is the probability that he will kill himself? Assume that the revolving cylinder is spun again after each trial. The intuitive feeling that one is virtually certain to kill himself in the long run is verified by the probability calculations.

Because of the structure of a six shooter, the probability assigned to the event that the individual will kill himself on any single trial is 1/6 and the probability that he will not kill himself is 5/6. Let $P(A_n)$ represent the probability that he will kill himself in n trials. Then,

$$\begin{aligned} P(A_n) &= (1/6) + (5/6)(1/6) + (5/6)^2(1/6) + \cdots + (5/6)^{n-1}(1/6) \\ &= (1/6)[1 + (5/6) + (5/6)^2 + \cdots + (5/6)^{n-1}] \end{aligned}$$

The equation for $P(A_n)$ is an application of the multiplication and addition rules. The probability that the player will kill himself on the first trial is 1/6; the probability that he will fail to kill himself on the first trial followed by killing himself on the second trial is $(5/6) \times (1/6)$. Thus, the probability that he will be killed by the end of the second trial is $1/6 + (5/6)(1/6)$, and so forth.

The expression within the brackets in the above equation is the sum of a geometric progression with a common ratio of 5/6. The sum of a geometric progression of n terms is equal to

$$S_n = a\left(\frac{1 - r^n}{1 - r}\right)$$

where a is the value of the first term, n is the number of terms, and r is the common ratio, that is, the ratio between any term and the preceding one.

In any geometric progression, if $|r|$ is less than one, as it is in this case, and if S denotes the value of S_n as n increases beyond any preassigned fixed quantity, then[9]

$$S = \frac{a}{1 - r}; \ |r| < 1$$

The symbol $|r|$ means the "absolute value" of r, that is, the positive numerical value of r without regard to sign. For example, $|+r| = r$ and $|-r| = r$, also.

The sum of the expression within the brackets in the equation for $P(A_n)$ as n increases without limit is

$$S = \frac{1}{1 - 5/6} = 6$$

Letting P represent the limit of $P(A_n)$ as n increases without limit, we have

$$P = (1/6) \times 6 = 1$$

Thus, if the player repeats the game indefinitely, it is certain he will kill himself.

Example 1-16 The following table refers to the 2500 wage employees of the Johnson Company, classified by sex and by opinion on a proposal to emphasize

---

[9]A more formal way of expressing this idea in terms of infinite series is that S denotes the limit of S_n as n increases without bound, i.e., $S = \lim_{n \to \infty} S_n$.

fringe benefits rather than wage increases in an impending contract discussion. This is the same table shown in Example 1-6 as an illustration of set theory. In the present example, frequencies of occurrence have been entered in each cell of the table.

Sex	*Opinion* *In Favor*	*Neutral*	*Opposed*	*Total*
Male	900	200	400	1500
Female	300	100	600	1000
Total	1200	300	1000	2500

(a) Calculate the probability that an employee selected from this group will be
  (1) a female opposed to the proposal.
  (2) neutral.
  (3) opposed to the proposal, given that the employee selected is a female.
  (4) either a male or opposed to the proposal.
(b) Are opinion and sex statistically independent for these employees?

Solution: We use the following representation of events

A_1: Male B_1: In favor
A_2: Female B_2: Neutral
    B_3: Opposed

In part (a), we have

(1) $P(A_2 \cap B_3) = 600/2500 = 0.24$
(2) $P(B_2) = 300/2500 = 0.12$
(3) $P(B_3 \mid A_2) = \dfrac{P(B_3 \cap A_2)}{P(A_2)} = \dfrac{600/2500}{1000/2500} = 0.60$
(4) $P(A_1 \cup B_3) = P(A_1) + P(B_3) - P(A_1 \cap B_3)$

$$= \frac{1500}{2500} + \frac{1000}{2500} - \frac{400}{2500}$$

$$= \frac{2100}{2500} = 0.84$$

In part (b), in order for opinion and sex to be statistically independent, the joint probability of the intersection of each pair of A events and B events would have to be equal to the product of the respective unconditional probabilities. That is, the following equalities would have to hold:

$$P(A_1 \cap B_1) = P(A_1)\,P(B_1) \qquad P(A_2 \cap B_1) = P(A_2)\,P(B_1)$$
$$P(A_1 \cap B_2) = P(A_1)\,P(B_2) \qquad P(A_2 \cap B_2) = P(A_2)\,P(B_2)$$
$$P(A_1 \cap B_3) = P(A_1)\,P(B_3) \qquad P(A_2 \cap B_3) = P(A_2)\,P(B_3)$$

Clearly, these equalities do not hold, as, for example,

$$P(A_1 \cap B_1) \neq P(A_1)\,P(B_1)$$
$$\frac{900}{2500} \neq \left(\frac{1500}{2500}\right)\left(\frac{1200}{2500}\right)$$

Another way of viewing the problem is that each conditional probability would have to be equal to the corresponding unconditional probability. Thus, the following equalities would have to hold:

$$\begin{array}{lll} P(A_1 \mid B_1) = P(A_1) & P(A_1 \mid B_2) = P(A_1) & P(A_1 \mid B_3) = P(A_1) \\ P(A_2 \mid B_1) = P(A_2) & P(A_2 \mid B_2) = P(A_2) & P(A_2 \mid B_3) = P(A_2) \end{array}$$

These equalities do not hold, as, for example,

$$P(A_1 \mid B_1) \neq P(A_1)$$
$$\frac{900}{1200} \neq \frac{1500}{2500}$$

The nature of the dependence (lack of independence) can be summarized briefly as follows: The proportion of males declines as we move from favorable to opposed opinions. This type of relationship is sometimes referred to in the following terms: There is a direct (inverse) relationship between the proportion of males (females) and favorableness of opinion.

1.4 Bayes' Theorem

The Reverend Thomas Bayes (1702–1761), an English Presbyterian minister and mathematician, considered the question of how one might make inductive inferences from observed sample data about the populations that gave rise to these data. Mathematicians had previously concentrated on the problem of deducing the consequences of specified hypotheses. Bayes was interested in the inverse problem of making statements about hypotheses from observations of consequences. He developed a theorem which calculated probabilities of "causes" based on the observed "effects." Since World War II, a considerable body of knowledge has developed known as *Bayesian decision theory* whose purpose is the solution of problems involving decision making under uncertainty. Bayes' theorem is the essential tool by means of which the difficult problem of uncertainty is handled. The theorem simply states a rule for computing conditional probabilities; as such, it is not debatable. However, considerable controversy surrounds the ways in which the theorem has been used. This controversy will be discussed after the proof of the theorem is given.

The structure of the following problem illustrates very clearly the nature of the probability calculation made by Bayes' theorem.

Assume there are three identical urns which contain two balls each.

The balls are also identical except as to color. One urn contains two red balls; the second contains a red and a white ball; the third contains two white balls. An urn is selected at random, and a ball is drawn from it at random. It is red. What is the probability that the urn originally selected was the one with the two red balls? We suggest that the reader attempt to answer this question intuitively at this point. The solution will be given after the statement and proof of Bayes' theorem.

The type of result required by the above question is often referred to as an "a posteriori probability" or "posterior probability." That is, the probability is computed *posterior* to the observation of sample evidence; in this case, the evidence is the red ball drawn from the urn. The probability that the urn with the two red balls was the one selected before any sampling was done is an "a priori probability" or "prior probability."

Assume a set of complete and mutually exclusive events $A_i, i = 1, 2, \ldots, n$. The appearance of one of the A_i events is a necessary condition for the occurrence of another event B, which is observed. The conditional probabilities $P(B \mid A_i)$ and $P(A_i)$ are known. The posterior probability of event A_1 given that B has occurred is given by Bayes' theorem.

Bayes' Theorem

(1.10)
$$P(A_1 \mid B) = \frac{P(A_1)\,P(B \mid A_1)}{\sum_{i=1}^{n} P(A_i)\;P(B \mid A_i)}$$

To prove this theorem, we write the joint probability $P(A_1 \cap B)$ in two different ways:

$$P(A_1 \cap B) = P(A_1)\,P(B \mid A_1)$$

and

$$P(A_1 \cap B) = P(B)\,P(A_1 \mid B)$$

Setting the right-hand sides of these two equations equal to one another gives

$$P(B)\,P(A_1 \mid B) = P(A_1)\,P(B \mid A_1)$$

Dividing both sides of this equation by $P(B)$, we have

(1.11)
$$P(A_1 \mid B) = \frac{P(A_1)\,P(B \mid A_1)}{P(B)}$$

where $P(B) \neq 0$.

Since the A_i are mutually exclusive events, and since in order for event B to occur, one of the events A_i must occur,

$$P(B) = P(A_1)\,P(B \mid A_1) + P(A_2)\,P(B \mid A_2) + \cdots + P(A_n)\,P(B \mid A_n),$$

or, in summation notation,

$$P(B) = \sum_{i=1}^{n} P(A_i)\, P(B \mid A_i) \tag{1.12}$$

Equation 1.12 states that $P(B)$ is given by an expression which adds up the probabilities of all of the mutually exclusive ways in which event B can occur.

Substituting Equation 1.12 into the denominator of the right-hand side of Equation 1.11 yields Bayes' theorem.

$$P(A_1 \mid B) = \frac{P(A_1)\, P(B \mid A_1)}{\sum_{i=1}^{n} P(A_i)\, P(B \mid A_i)}$$

Although this theorem was proved for a particular one of the A_i, namely A_1, it is perfectly general, since any of the n events A_i can be designated A_1. Another point worthy of note is that there is no necessary time sequence in the A_i and B events insofar as the above mathematical proof is concerned. However, in most of the applications of the theorem to decision problems, the A_i represent events which precede the occurrence of the observed event B. In this connection, we can think of the theorem as answering the question, "Given that event B has occurred, what is the probability that it was preceded by the event A_1?" Another way of looking at it is that $P(A_1 \mid B)$ is the revised probability assignment to event A_1 after observing event B.

We now apply Bayes' theorem to solve the problem of the balls in the urns. Let the events A_1, A_2, and A_3 represent the selection of the urns with the two red balls, one red and one white ball, and two white balls, respectively. Then $P(A_1)$, $P(A_2)$, and $P(A_3)$ denote the prior probabilities of selection of these urns. The event B, which in this problem occurs after the selection of an urn, refers to the observation of a red ball. The three expressions $P(B \mid A_1)$, $P(B \mid A_2)$, and $P(B \mid A_3)$ denote the probabilities that the red ball was drawn, given that urns one, two, and three, respectively, were the ones sampled. $P(A_1 \mid B)$ is the posterior probability that A_1 was the urn originally selected, given that a red ball has been drawn. For convenience, we present the solution by Bayes' theorem in tabular form in Table 1-2, which gives the solutions for $P(A_2 \mid B)$, $P(A_3 \mid B)$, and $P(A_1 \mid B)$.

The answer, therefore, to the question "What is the probability that the urn originally selected was the one with two red balls, given that a red ball was observed in a random draw of one ball?" is

$$P(A_1 \mid B) = \frac{P(A_1)\, P(B \mid A_1)}{\sum_{i=1}^{n} P(A_i)\, P(B \mid A_i)} = \frac{1/3}{1/2} = \frac{2}{3}$$

which is given as the first entry in Column (5) of Table 1-2. The prior prob-

Table 1-2

(1) A_i	(2) $P(A_i)$	(3) $P(B \mid A_i)$	(4) $P(A_i)P(B \mid A_i)$	(5) $P(A_i \mid B)$
$A_1:(R, R)$	1/3	1	1/3	$\frac{1/3}{1/2} = 2/3$
$A_2:(R, W)$	1/3	1/2	1/6	$\frac{1/6}{1/2} = 1/3$
$A_3:(W, W)$	1/3	0	0	$\frac{0}{1/2} = 0$
Total	1		1/2	1

ability assignment to the selection of this urn in the absence of any experimental evidence concerning drawing of balls from urns was 1/3.

Bayes' theorem weights together prior information with experimental evidence. The manner in which it does this is interestingly revealed by making modifications in the urn problem. Suppose that instead of one ball being selected from the urn, two balls are randomly drawn, *with replacement*, and both balls are red. What is now the probability assigned to the event that the urn originally selected was the one with the two red balls? In this problem the event B is the occurrence of the two red balls. Again, the solution is presented in tabular form and is shown in Table 1-3. The only column requiring comment is (3) in which the values of $P(B \mid A_i)$ are given. Taking the second entry in that column, we see that given the selection of the box with the red and white ball, the probability of obtaining two red balls in a row in

Table 1-3

(1) A_i	(2) $P(A_i)$	(3) $P(B \mid A_i)$	(4) $P(A_i)P(B \mid A_i)$	(5) $P(A_i \mid B)$
$A_1:(R, R)$	1/3	1	1/3	$\frac{1/3}{5/12} = 4/5$
$A_2:(R, W)$	1/3	1/4	1/12	$\frac{1/12}{5/12} = 1/5$
$A_3:(W, W)$	1/3	0	0	$\frac{0}{5/12} = 0$
Total	1		5/12	1

sampling with replacement is 1/4. This represents an application of the multiplication theorem for two independent events, the obtaining of red balls on the first and second draws, each of which has a probability of 1/2. The posterior probability of 4/5 for $P(A_1 \mid B)$ in this problem indicates that having drawn two red balls in a row from this urn, it is now more likely that the urn originally selected was the one with two red balls than in the previous example where only one red ball was observed.

In an experiment of sampling with replacement, which resulted in drawing ten red balls in a row, $P(A_1 \mid B)$ would be very close to one. This would mean that the experimental data indicated so strongly that the sampling was being done from the urn with the two red balls that this evidence heavily outweighed the prior probability assignments in determining the posterior probability, $P(A_1 \mid B)$. In fact, it would be found that the prior probabilities, $P(A_i)$, could be changed over a fairly wide range, and the $P(A_1 \mid B)$ value would still remain quite close to one.

On the other hand, let us observe how the posterior probability would change if the experimental evidence were the same as originally assumed, but the numbers of the different types of urns were changed. Assume there was one urn with two red balls, two urns with one red and one white ball each, and one urn with two white balls. The original experiment is repeated. That is, an urn is selected at random and a ball is randomly drawn from it. The ball is red. What is the probability that the urn from which the ball was drawn was the one with the two red balls? The solution to this problem is given in Table 1-4. Compared to the posterior probability of 2/3 in the original problem, with the same sample evidence of one red ball, the $P(A_1 \mid B)$ has now dropped to 1/2. This, of course, reflects the effect of the different prior probabilities assigned to the events A_i.

In this problem, the prior probabilities were assumed to be known. In

Table 1-4 Solution to the Four Urn Problems Where the Experimental Evidence is One Red Ball

(1) A_i	(2) $P(A_i)$	(3) $P(B \mid A_i)$	(4) $P(A_i)P(B \mid A_i)$	(5) $P(A_i \mid B)$
$A_1:(R, R)$	1/4	1	1/4	$\frac{1/4}{1/2} = 1/2$
$A_2:2(R, W)$	1/2	1/2	1/4	$\frac{1/4}{1/2} = 1/2$
$A_3:1(W, W)$	1/4	0	0	$\frac{0}{1/2} = 0$
Total	1		1/2	1

many other types of problems, the $P(A_i)$ values would not be known. Therein lies the source of much of the controversy concerning the use of Bayes' theorem. When objective prior probabilities are available, there is no dispute with the appropriateness of the application. Objective probabilities in this context are usually taken to mean simple relative frequencies of occurrence. Many of the early applications of Bayes' theorem by members of the classical school involved questionable assumptions of equal prior probabilities. Some of the situations in which this equal likelihood assumption was made were so dubious that for some time the theorem itself fell into disrepute. However, there is clearly no question concerning the *mathematical validity* of the theorem.

In modern Bayesian decision theory, subjective or personalistic probability assignments are made in many applications. It is argued that it is meaningful to assign prior probabilities concerning hypotheses based upon degree of belief. Bayes' theorem is then viewed as a means of revising these probability assignments. In business applications, this has meant that executives' intuitions, subjective judgments and present quantitative knowledge are captured in the form of prior probabilities; these figures undergo revision as relevant empirical data are collected. This procedure seems sensible and fruitful for a wide variety of applications, many of which will be discussed in later chapters. A few examples will be given here to suggest some of the many possible types of applications of this very interesting theorem.

Example 1-17 A man regularly plays darts in the recreation room of his home, observed by his eight year-old son. The father has a history of making bulls-eyes 30% of the time. The son, who habitually sits injudiciously close to the dart board, reports to his father whether or not bulls-eyes have been made. However, as is often the case with eight year-old boys, the son is an imperfect observer. He reports correctly 90% of the time. On a particular occasion, the father throws a dart at the board. The son reports that a bulls-eye was made. What is the probability that a bulls-eye was indeed scored?

Solution: Let A be the event "bulls-eye scored" and $\bar{A}$ "bulls-eye not scored." Let B represent "son reports that a bulls-eye has been scored." Then,

$$P(A \mid B) = \frac{P(A)\,P(B \mid A)}{P(A)\,P(B \mid A) + P(\bar{A})\,P(B \mid \bar{A})} = \frac{(.3)(.9)}{(.3)(.9) + (.7)(.1)} = .79$$

This is illustrative of a class of problems involving an information system which transmits uncertain knowledge. That is, the son may be thought of as an information system of 90% "reliability." A more colorful way of expressing it is that the son has "error in him" or is a "noisy" information system (where the term "noise" refers to random error).

Example 1-18 Assume that 1% of the inhabitants of a country suffer from a certain disease. A new diagnostic test is discovered which gives a positive indication 97% of the time when an individual has this disease and gives a negative indication 95% of the time when the disease is absent. An individual is selected at random, is given the test, and reacts positively. What is the probability that he has the disease?

Solution: Let D represent "has the disease" and $\bar{D}$ "does not have the disease." Let $+$ and $-$ represent positive and negative indications, respectively. Then, by Bayes' theorem,

$$P(D \mid +) = \frac{(.01)(.97)}{(.01)(.97) + (.99)(.05)} = .16$$

This is a startlingly low probability which doubtless runs counter to intuitive feelings based on the 97% and 95% figures given above. If the "cost" incurred is high when a person is informed that he has a disease (e.g., such as cancer) on the basis of such a test when in fact he does not, it would appear that action based solely on conditional probabilities such as the above 97 and 95% might be seriously misleading and costly.

The implications of this application are quite subtle. Actually, correct decisions would be made a very high proportion of the time on the basis of this test. The probability of correct decisions is

$$\begin{aligned} P\binom{\text{correct}}{\text{decisions}} &= P(D \cap +) + P(\bar{D} \cap -) \\ &= P(D)\,P(+ \mid D) + P(\bar{D})\,P(- \mid \bar{D}) \\ &= (.01)(.97) + (.99)(.95) = .9502 \end{aligned}$$

However, from the $P(D \mid +) = .16$ calculation given above it follows that if a person has a positive indication on this test and is informed he has the disease, the probability of error is .84. On the other hand, a corresponding Bayes' theorem calculation shows that the probability is only .00032 that a person would have the disease, given a negative indication. Thus, only very rarely would an error be committed if a negative indication were obtained on the test.

In summary, the Bayes' theorem computations indicate the following:

$$\begin{aligned} P(D \mid +) &= .16 \qquad & P(D \mid -) &= .00032 \\ P(\bar{D} \mid +) &= .84 \qquad & P(\bar{D} \mid -) &= .99968 \end{aligned}$$

The method by which optimal decisions are made, taking into account all of these probabilities in addition to the seriousness of the two types of error implied by $(\bar{D} \mid +)$ and $(D \mid -)$, is a basic topic of statistical decision theory.

Example 1-19 A corporation uses a "selling aptitude test" to aid it in the selection of salesmen. Past experience has shown that only 65% of all persons applying for a sales position achieved a classification of "satisfactory" in actual selling, whereas the remainder were classified "unsatisfactory." Of those classified as "satisfactory," 80% had scored a passing grade on the aptitude test. Only 30% of those classified "unsatisfactory" had passed the test. On the basis of this information, what is the probability that a candidate would be a "satisfactory" salesman given that he passed the aptitude test?

Solution: If S stands for a "satisfactory" classification as a salesman and P stands for "passes the test," then the probability that a candidate would be a "satisfactory" salesman, given that he passed the aptitude test, is

$$P(S \mid P) = \frac{(.65)(.80)}{(.65)(.80) + (.35)(.30)} = .83$$

Thus, the tests are of some value in screening candidates. Assuming no change in the type of candidates applying for the selling positions, the probability that a random applicant would be satisfactory is 65%. On the other hand, if the company only accepts an applicant if he passes the test, this probability increases to .83.

Example 1-20 A certain company is contemplating the marketing of a new product. The company's marketing vice-president is particularly concerned about the product's superiority over the closest competitive product, which is sold by another company. The marketing vice-president assessed the probability of the new product's superiority to be .7. This executive then ordered a market survey which indicated that the new product was superior to its competitor. Assume the market survey has the following reliability: If the product is really superior, the probability that the survey will indicate "superior" is .8. If the product is really worse than its competitor, the probability that the survey will indicate "superior" is .3. After the completion of the market survey, what should be this executive's revised probability assignment to the event "new product is superior to its competitor"?

Solution: Let S represent the event "new product is superior to its competitor" and s the event "market survey indicates that the new product is superior to its competitor." Then

$$P(S \mid s) = \frac{(.7)(.8)}{(.7)(.8) + (.3)(.3)} = .86$$

1.5 Counting Principles and Techniques

In the problems we have encountered so far, the pertinent sample spaces have been comparatively simple. However, in many situations the numbers of points in the appropriate sample spaces are so great that efficient methods are needed to count these points, in order to arrive at required probabilities or answer other questions of interest. In this connection, it is useful to return to the concept of sequences of trials of experiments to specify a simple but important fundamental principle.

The Multiplication Principle

If an experiment can result in n_1 distinct outcomes on the first trial, n_2 distinct outcomes on the second trial, and so forth for k sequential trials, then the total number of different sequences of outcomes in the k trials is $(n_1)(n_2) \cdots (n_k)$.

It is sometimes helpful to think in terms of the sequential performance of tasks rather than trials of an experiment. Thus, using somewhat different language than was employed in the context of experimental trials, if the first of a sequence of tasks can be performed in n_1 ways, the second in n_2 ways, and so forth for k tasks, then the sequence of k tasks can be carried out in $(n_1)(n_2) \cdots (n_k)$ ways.

Thus, if a coin is tossed and then a card is drawn at random from a standard deck of cards, there are $2 \times 52 = 104$ possible different sequences. For example, one such sequence of outcomes might be head, king of spades.

If a die is rolled three times, there are $6 \times 6 \times 6 = 216$ different sequences.

If it is possible to go from Philadelphia to Baltimore in two different ways and from Baltimore to Washington in three different ways, then there are $2 \times 3 = 6$ ways of going from Philadelphia to Washington via Baltimore.

A tree diagram is often helpful in thinking about the total possible number of sequences. For example, in the case of the trip from Philadelphia to Washington, if A_1 and A_2 denote the two ways of going to Baltimore and B_1, B_2, and B_3 the three ways of proceeding from Baltimore to Washington, then the total number of possible sequences is indicated by the total number of different paths through the tree, going from left to right (See Figure 1-12).

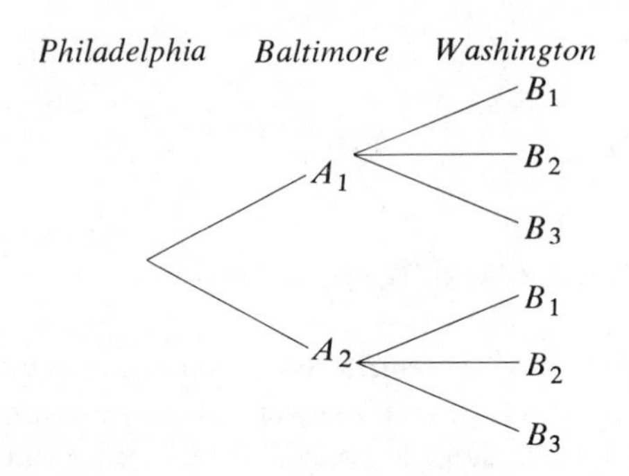

Figure 1-12 Tree diagram for trip from Philadelphia to Washington.

In the following sections, the multiplication principle is used in many different types of problems.

Permutations

In order to handle the problem of counting points in complicated sample spaces, the counting techniques of combination and permutation are used. In this connection, it is helpful to think in terms of objects which occur in groups. These groups may be characterized by type of object, the number belonging to each type, and the way in which the objects are arranged. For example, consider the letters a, b, c, d, e. There are five objects, one of each type. If we have the letters a, b, b, c, c, there are five objects, one of type a, two of b, and two of c. Returning to the first group of objects, $a\ b\ c\ d\ e$, $b\ a\ c\ d\ e$, and $c\ d\ e\ a\ b$ differ in the *order* in which the five objects are arranged, but each of these groups contains the same number belonging to each type.

Suppose we have a group of n different objects. In how many ways can these n objects be arranged in order in a line? Applying the multiplication principle, we see that any one of the n objects can occupy the first position, any of the $n - 1$ remaining objects can occupy the second position and so forth until we have only one possible object to occupy the nth position. Thus, the number of different possible arrangements of the n objects in a line consisting of n positions is

$$n! = (n)(n-1)\cdots(2)(1)$$

The symbol $n!$ is read "n factorial." We shall only be concerned with cases for which n is a non-negative integer. By definition $0! = 1$.

Some examples of factorials are

$$\begin{aligned} 1! &= 1 \\ 2! &= 2 \times 1 = 2 \\ 3! &= 3 \times 2 \times 1 = 6 \\ &\cdot \\ &\cdot \\ &\cdot \\ 10! &= 10 \times 9 \times \cdots \times 2 \times 1 = 3{,}628{,}800 \end{aligned}$$

It is useful to note that $n! = n(n-1)!$. Thus, $10! = 10 \times 9!$, etc. We see from this relation that it makes sense to define $0! = 1$, since if we let $n = 1$, we have

$$1! = 1(0)!$$

and $0! = 1$. This enables us to maintain a consistent definition of factorials for all non-negative integers.

Factorials obviously increase in size very rapidly. For example, how many different arrangements can be made of a deck of 52 cards if the cards are placed in a line? The answer, 52! is a number which contains 68 digits.[10]

Frequently, we are interested in arranging in order some subgroup of n different objects. If x of the objects ($x \leq n$) are to be selected and arranged in order, as in a line, then each such arrangement is said to be a *permutation* of the n objects taken x at a time. The number of such permutations is denoted ${}_nP_x$. For example, suppose there are 50 girls competing in a beauty contest for three rankings, first, second, and third. How many permutations of the 50 girls taken three at a time are possible, and thus, how many different rankings are possible? The answer is

$${}_{50}P_3 = 50 \times 49 \times 48 = 117{,}600$$

This follows, since any one of the 50 girls can occupy the first position, any of the remaining 49 the second position, and any of 48 could fill the third place. By the multiplication principle the number of different sequences of first, second, and third rankings is obtained by the indicated product.

We can now generalize this procedure to obtain a convenient formula for the number of permutations of n different objects taken x at a time.

$$\begin{aligned} {}_nP_x &= (n)(n-1)\cdots[n-(x-1)] \\ &= (n)(n-1)\cdots(n-x+1) \\ &= \frac{(n)(n-1)\cdots(n-x+1)(n-x)!}{(n-x)!} \end{aligned}$$

(1.13) $${}_nP_x = \frac{n!}{(n-x)!}$$

It can be seen, in general, that if there are x positions to be filled, the first position can be filled in n ways; after one object has been placed in the first position, $x - 1$ positions remain. The second can be filled in $n - 1$ ways, the third in $n - 2$ ways and so forth down to the xth or last position, which can be filled in $n - (x - 1)$ ways. In writing down the factors which must be multiplied together, zero is subtracted from n in the first position, one is subtracted from n in the second position and so forth down to $(x - 1)$ subtracted from n in the xth position. The formula $n!/(n - x)!$ follows from the definition of a factorial since $(n - x)!$ cancels out all factors after $(n - x + 1)$ in the numerator. Thus, in the beauty contest problem, where $n = 50$ and $x = 3$, we have

$${}_{50}P_3 = \frac{50!}{(50-3)!} = \frac{50!}{47!} = 50 \times 49 \times 48$$

---

[10]Warren Weaver points out in *Lady Luck* (Garden City, N.Y.: Doubleday, 1963), p. 88, about the number of possible arrangements in 52!, that, "If every human being on earth counted a million of these arrangements per second for twenty-four hours a day for lifetimes of eighty years each, they would have made only a negligible start in the job of counting all these arrangements — not a billionth of a billionth of one percent of them!"

A special case of the formula for permutations occurs when all of the n objects are considered together. In this situation, we are concerned with the number of permutations of n different objects taken n at a time, which is

$$(1.14) \qquad {}_nP_n = \frac{n!}{(n-n)!} = \frac{n!}{0!} = n!$$

For example, if a consumer were given one cup of coffee of each of five brands and were asked to rank these according to preference, the total number of possible rankings, excluding the possibility of ties, would be

$${}_5P_5 = 5! = 120$$

Combinations

In the case of *permutations* of objects, the *order* in which the objects are arranged *is of importance*. *Where order is not important*, we are concerned with *combinations* of objects rather than permutations. A simple example will illustrate the difference. Suppose the president of a company is interested in setting up a very small finance committee of two men. He plans to select these two men from a group of three executives named Brown, Jones, and Smith. How many possible committees could be formed? It is clear that order is of no importance in this situation. That is, a committee consisting of Brown and Jones is no different from a committee of Jones and Brown. Using first letters to symbolize the three names, there are three possible committees, BJ, BS, and JS.

This is an example of *combinations* of three objects taken two at a time. The terminology is similar to that used for permutations, so in general, we refer to the number of combinations that can be made of n different objects taken x at a time.

Using the same group of letters and treating them merely as symbols, if order of arrangement were important, it is clear that the number of permutations of the three objects taken two at a time would exceed the number of combinations of three objects taken two at a time. The following six permutations can be made in this case:

BJ JB
BS SB
JS SJ

To develop a formula for combinations, we need merely to consider the relationship between numbers of combinations and numbers of permutations for the same group of n objects taken x at a time. Thus, fixing attention for the moment on any particular combination, there are x objects which fill x positions. How many permutations can be made of these x objects in the x positions? Clearly, any one of the x objects may fill the first position, $x - 1$

the second, and so forth down to one object for the xth position. Thus, $x!$ distinct permutations can be formed of the x objects in x positions. Therefore, the number of permutations that can be formed of n different objects taken x at a time is $x!$ times the number of combinations of these n objects taken x at a time. The symbol for the number of combinations of n objects taken x at a time is $\binom{n}{x}$. Thus,

$$\binom{n}{x}x! = {}_nP_x$$

Solving for $\binom{n}{x}$ yields the following formula for the number of combinations that can be made of a group of n different objects, taken x at a time:

(1.15)
$$\binom{n}{x} = \frac{{}_nP_x}{x!} = \frac{n!}{(n-x)!x!}$$

Returning to the committee illustration, the number of combinations of the three men taken two at a time is

$$\binom{3}{2} = \frac{3!}{2!1!} = 3$$

which was the number previously listed. Similarly, the number of permutations of three objects taken two at a time is seen to be

$${}_3P_2 = \frac{3!}{1!} = 6$$

which was the number of ordered arrangements listed earlier.

Permutations Involving Elements That Are Alike

Up to this point, the discussion of permutations and combinations has pertained to groups of elements, all of which are different. We now turn to the problem of determining the number of distinguishable arrangements that can be formed when some of the objects are identical. Let us consider a simple illustration to aid us in arriving at a formula for the number of different permutations that can be made of n objects, n_1 of which are of type 1, n_2 of type 2, . . . , n_k of type k.

How many distinguishable arrangements can be made of four balls, two of which are white and two of which are black, if the balls are placed in a line? We find there are six possible arrangements, as follows:

WWBB   BBWW
WBWB   BWBW
WBBW   BWWB

Had the balls been of different colors, the number of possible permutations of the four balls taken four at a time that could have been formed would have been

$$_4P_4 = 4! = 24$$

What is the relationship between these two numbers of permutations, namely 6 and 24? If we consider any one of the permutations of the two white balls and two black balls, say

WWBB

we can imagine each of the balls as having a number printed on it to make it distinguishable from the other ball of the same color. We shall indicate the number by a subscript. Thus, suppose we imagine the numbers 1 and 2 as being entered on the balls; we can now imagine the arrangement

$$W_1 \; W_2 \; B_1 \; B_2$$

As compared to the previous six permutations when some of the balls were indistinguishable, we now have

$$6 \times 2! \times 2! \text{ permutations}$$

since 2! arrangements can be made of the two white balls by permuting their subscripts while keeping the black balls unchanged, and 2! orders can be made of the two black balls keeping the white balls unchanged. Thus, the relation between the 4! permutations when all balls were distinguishable and the six possible arrangements when there are two indistinguishable white and two indistinguishable black balls is

$$4! = 6 \times 2! \times 2!$$

Generalizing this rationale, if $P_m(n; n_1, n_2, \ldots, n_k)$ denotes the number of distinguishable arrangements that can be formed of n objects, taken n at a time, where n_1 are of type 1, n_2 of type 2, . . . , n_k of type k, and $n = n_1 + n_2 + \cdots + n_k$, we have the following relationship,

$$n! = P_m(n; n_1, n_2, \ldots, n_k) n_1! \, n_2! \cdots n_k!$$

Solving for $P_m \, (n; n_1, n_2, \ldots, n_k)$ yields the following formula:

(1.16) $$P_m(n; n_1, n_2, \ldots, n_k) = \frac{n!}{n_1! \, n_2! \cdots n_k!}$$

A very important special case of this formula occurs when there are just two types of objects, as in the example of the colored balls, x of one type and $n - x$ of the other. In this situation,

(1.17) $$P_m(n; x, n - x) = \frac{n!}{x!(n - x)!} = \binom{n}{x}$$

Thus, we see that the number of different permutations that can be made of n objects, x of which are of one type and $n - x$ of a second type, is equal to the number of combinations of n different objects, taken x at a time. It is also apparent that

(1.18) $$\binom{n}{x} = \frac{n!}{x!(n-x)!} = \binom{n}{n-x}$$

The expression $\binom{n}{x}$ is often referred to as a binomial coefficient because of the way it appears in the binomial expansion, discussed in Chapter 2. It would be useful if the reader returned to this section after reading the material on the binomial expansion (binomial probability distribution). Equation (1.17) will aid him in understanding the meaning of the binomial coefficients; Equation (1.18) explains the symmetry in these coefficients.

Example 1-20 A brief market research questionnaire requires the respondent to answer each of ten successive questions with either a "yes" or a "no." The sequence of ten yes-no responses is defined as the respondent's "profile." How many different possible profiles are there?

Solution: There are two possible responses for each question. Therefore, by the multiplication principle, there are $2 \times 2 \times \cdots \times 2 = 2^{10} = 1024$ different profiles.

Example 1-21 A manufacturing firm wants to locate five warehouses in the 48 continental states of the U.S. It wants only one warehouse to be located in any state. If it wanted to examine the desirability of every possible combination of locations, how many locations would the firm have to consider?

Solution:

$$\binom{48}{5} = \frac{48!}{5!\,43!} = \frac{48 \times 47 \times 46 \times 45 \times 44}{5 \times 4 \times 3 \times 2 \times 1} = 1{,}712{,}304$$

An important practical principle emerges from a consideration of this example. Obviously it is not feasible to examine explicitly every possible combination of locations. Most geographical locations which would work out advantageously for the placement of warehouses are at or near concentrations of demand. Therefore, even assuming the company is correct in wanting to locate one warehouse in each of five states, the search for a solution can be reduced to a small fraction of the total possible number of locations.

Example 1-22 There are six different operations in a manufacturing process. Let us refer to them as A, B, C, D, E, and F. A must be performed first and F must be performed last. All other operations may be performed in any order. How many different sequences of operations are possible?

Solution: Since A and F are fixed, we need only be concerned with B, C, D, and E. Any of these four may be performed first, any of the remaining three may come second, etc. Therefore, the number of possible different sequences is given by ${}_4P_4 = 24$.

Example 1-23 An underwriting syndicate is to be formed from a group of investment banking firms, each of which is classified either as type A or type B. There are six type A and eight type B firms. In how many ways can a syndicate of six firms be formed if

(a) it must consist of three firms of type A and three of type B?
(b) it must consist of at least three firms of type A and at least one of type B?

Solution: In (a), we can think of the sequential selection of three firms from the six type A firms followed by a selection of three firms from the eight type B firms, where the order of selection is unimportant. Hence, the number of possible syndicates is

$$\binom{6}{3} \times \binom{8}{3} = 1120$$

The tabular arrangement below gives the solution to part (b):

Different Methods of Forming the Syndicate	*Number of Ways of Forming the Syndicate*
$3A$, $3B$	$\binom{6}{3}\binom{8}{3} = 1120$
$4A$, $2B$	$\binom{6}{4}\binom{8}{2} = 420$
$5A$, $1B$	$\binom{6}{5}\binom{8}{1} = 48$
	Total 1588

Example 1-24 A rug manufacturer is planning a display for a furniture show. He intends to hang a set of narrow strips of a new type of carpeting in a lengthwise manner with some space between the strips. The strips differ only in color; three are red, two are grey, four are green, and one is blue. How many distinguishable arrangements can be made of the ten strips of carpet?

Solution: This is an application of formula (1.16), where $n = 10$, $n_1 = 3$, $n_2 = 2$, $n_3 = 4$, and $n_4 = 1$

$$P_m(10; 3, 2, 4, 1) = \frac{10!}{3!\,2!\,4!\,1!}$$
$$= \frac{10 \times 9 \times 8 \times 7 \times 6 \times 5}{3 \times 2 \times 1 \times 2 \times 1 \times 1} = 12{,}600$$

Example 1-25 A shipment of 1000 articles contains 10% defectives and 90% good articles. If a sample of five articles is drawn at random from the shipment, what is the probability of observing one or less defectives if (a) the sampling is carried out *with replacement* or (b) the sampling is carried out *without replacement?*

Solution: (a) *Sampling with replacement*

We first note that P(one or less defectives) = P(zero or one defectives). Let 0 denote zero defectives and 1 denote one defective. Then, by the addition rule,

$$P(0 \cup 1) = P(0) + P(1)$$

We calculate $P(1)$ first. Suppose the defective item occurred on the first item drawn, and was followed by four good items. We represent this event by the symbols $D\ G\ G\ G\ G$, where D and G represent the occurrence of a defective and good item, respectively. Then, by the multiplication rule,

$$P(D \cap G \cap G \cap G \cap G) = \left(\frac{1}{10}\right)\left(\frac{9}{10}\right)\left(\frac{9}{10}\right)\left(\frac{9}{10}\right)\left(\frac{9}{10}\right) = \left(\frac{1}{10}\right)\left(\frac{9}{10}\right)^4$$

However, the defective item could occur on the first, second, third, fourth, or fifth draws; i.e., any of the following events represents the occurrence of one defective:

$$\begin{matrix} D & G & G & G & G \\ G & D & G & G & G \\ G & G & D & G & G \\ G & G & G & D & G \\ G & G & G & G & D \end{matrix}$$

It is easy to see in this problem that the number of ways in which exactly one defective could occur is five. In more complex problems, the listing becomes tedious and time consuming. We observe that there are $P_m(5;1,4)$ distinct permutations of the five items. Each of these permutations has a probability of $\left(\frac{1}{10}\right)\left(\frac{9}{10}\right)^4$. Since $P_m(5;1,4) = \frac{5!}{1!\,4!}$, then

$$P(1) = \frac{5!}{1!\,4!}\left(\frac{1}{10}\right)\left(\frac{9}{10}\right)^4 = 5\left(\frac{1}{10}\right)\left(\frac{9}{10}\right)^4 = 0.328$$

By a similar line of argument,

$$P(0) = \frac{5!}{0!\,5!}\left(\frac{1}{10}\right)^0\left(\frac{9}{10}\right)^5 = \left(\frac{9}{10}\right)^5 = 0.590$$

Hence,

$$P(0 \cup 1) = 0.590 + 0.328 = 0.918$$

(b) *Sampling without replacement*

Since the sampling is without replacement, we must take account of the partial exhaustion of the shipment (statistical universe) because of the sampling process. We may think of the five items as being drawn sequentially, although, of course, the calculation would be the same if the articles were drawn simultaneously. Again, let us calculate $P(1)$ first.

We can obtain $P(1)$ by computing an appropriate ratio. The denominator of the ratio consists of the total number of sample points in the experiment of drawing a sample of five items without replacement from 1000 items; the numerator consists of the number of sample points which represent the occurrence of one defective and four good items. Computing the denominator first, we observe that the total number of points in the sample space is the number of samples of five items that can be selected without replacement from a population of 1000. This is the number of combinations that can be formed of 1000 items taken five at a time, i.e.,

$$\binom{1000}{5}$$

The numerator is computed by thinking in terms of a sequence of two tasks. First, we draw four good items from the 900 good items and then one defective from the 100 defective items. Using combinations and the multiplication principle gives for the numerator

$$\binom{900}{4}\binom{100}{1}$$

Therefore,

$$P(1) = \frac{\binom{900}{4}\binom{100}{1}}{\binom{1000}{5}} = \frac{\dfrac{900!}{4!\,896!} \times \dfrac{100!}{1!\,99!}}{\dfrac{1000!}{5!\,995!}}$$

$$= \frac{5!}{4!\,1!} \times \frac{900 \times 899 \times 898 \times 897 \times 100}{1000 \times 999 \times 998 \times 997 \times 996}$$

$$= 5\left(\frac{900}{1000}\right)\left(\frac{899}{999}\right)\left(\frac{898}{998}\right)\left(\frac{897}{997}\right)\left(\frac{100}{996}\right) \approx 0.328$$

The meaning of this calculation may be deduced from the last line. Skipping the factor 5 for the moment, the next factor 900/1000 is the probability of a good item on the first draw; 899/999 is the probability of a good item on the second draw, given a good item was obtained on the first draw, and so forth until 100/996, which is the probability of a defective item, given that four good items have preceded it. The factor 5 appears, as in part (a), because the defective item may have occurred on the first through fifth draws. Thus, an alternative method of obtaining this probability would have been simply to use the multiplication rule for conditional probabilities at the outset.

The two methods of solution represent different conceptual ways of formulating the problem and are mathematically equivalent.

The probabiity of zero defectives is similarly[11]

$$P(0) = \frac{\binom{900}{5}\binom{100}{0}}{\binom{1000}{5}} = \frac{\frac{900!}{5!\,895!}\;\frac{100!}{0!\,100!}}{\frac{1000!}{5!\,995!}}$$

$$= \left(\frac{900}{1000}\right)\left(\frac{899}{999}\right)\left(\frac{898}{998}\right)\left(\frac{897}{997}\right)\left(\frac{896}{996}\right) \approx 0.590$$

Therefore,

$$P(0 \cup 1) = 0.590 + 0.328 \approx 0.918$$

The results in (a) and (b) are so close because the probability of drawing a defective or good item changes very little as each of the five sampled items is withdrawn.

Example 1-26 Data concerning 45 workers in an industrial plant were punched on tabulating cards, with one card pertaining to each individual. The following table indicates the sex and skill classification of these workers:

Skill Classification	*Sex*		*Total*
	Male	*Female*	
Unskilled	14	9	23
Semi-skilled	10	11	21
Skilled	1	0	1
Total	25	20	45

(a) If a card is selected at random, what is the probability that it pertains to
 (1) an unskilled male worker?
 (2) either a female or semi-skilled worker?

(b) If ten cards are selected at random, what is the probability that there are cards for at least two semi-skilled workers among the ten?

Solution: Using the following representation of events

A_1 : male B_1 : unskilled
A_2 : female B_2 : semi-skilled
 B_3 : skilled

[11]Tables of factorials and logarithms of factorials are useful in carrying out calculations in more complex problems of this type.

we have for part (a)

(1) $P(A_1 \cap B_1) = 14/45$
(2) $P(A_2 \cup B_2) = 20/45 + 21/45 - 11/45 = 2/3$

For part (b), we will set up the required probabilities in terms of combination formulas, but will not evaluate the arithmetic.

(1) $P(\text{at least two } B_2\text{'s}) = 1 - [P(0B_2\text{'s}) + P(1B_2)]$

$$= 1 - \frac{\binom{21}{0}\binom{24}{10} + \binom{21}{1}\binom{24}{9}}{\binom{45}{10}}$$

Here, the most expedient way to arrive at the answer is to compute the probability that the specified event "at least two B_2's" will *not* occur and subtract that probability from one. We observe that the probability of at least two semi-skilled workers is equal to one minus the probability of zero or one semi-skilled worker (one or less). The probabilities of zero and one semi-skilled worker, respectively, are

$$P(0) = \frac{\binom{21}{0}\binom{24}{10}}{\binom{45}{10}} \quad \text{and} \quad P(1) = \frac{\binom{21}{1}\binom{24}{9}}{\binom{45}{10}}$$

by the same line of reasoning as in Example 1-25 for sampling without replacement.

PROBLEMS

1. If 0 represents a non-response to a mailed questionnaire, and 1 represents a response, depict the set of outcomes representing responses to three out of four questionnaires.
2. A certain manufacturing process produces parachutes. A worker tests each parachute as it is produced and continues testing until he finds one defective. Describe the sample space of possible outcomes for the testing process.
3. An investment company is planning to add two new stocks to its portfolio. Its research group recommends five stocks, Ranox, BIM, Bavo, Park Mining, and Goldflight. Describe the sample space representing the possible choices.
4. An econometric model predicts whether the GNP will increase, decrease, or remain the same the following year. Let X represent "the model's prediction, coded" and Y the "actual movement of GNP, coded." Graph the possible outcomes of X and Y.
5. A special electronic part is ordered by a firm in Youngstown, Ohio, from a

firm in London. The London firm can fly the part to either New York, Philadelphia, or Chicago. Once it reaches one of these cities, it can be sent by train or truck to Youngstown. Use a tree diagram to find all possible shipping routes.

6. Draw a tree diagram depicting the possible outcomes resulting from flipping a coin three times.
7. A farmer decides to rotate his crops every two years among corn, wheat, and hay, with only one crop being planted every two years. Use a tree diagram to depict all the possible patterns he could use.
8. Let $S = \{1, 2, 3, 4, 5\}$
   $A = \{1, 2, 3\}$
   $B = \{3, 4\}$
   $C = \{2, 3, 4, 5\}$

   Define the following sets:
   (a) $A \cap B$
   (b) $A \cup C$
   (c) $\overline{(A \cup C)} \cap B$
   (d) $A \cap \bar{C}$
9. A bin contains ten pens, seven of which write and three of which do not write. A person picks pens until he finds one that writes. Draw the tree diagram depicting the possible outcomes.
10. Which of the following statements are correct? If incorrect, indicate why or give the correct answer.
    (a) $\{1, 2, 2, 1\}$ is a set
    (b) $\{1, 2\} \cup \{2, 3\} = \{1, 2, 2, 3\}$
    (c) $\{1, 2\} \cap \{2, 3\} = \{2\}$
    (d) $\{1, 2, 3\} \cup \{1, 2, 3\} = \{1, 2, 3\}$
    (e) $\{1, 2, 3\} \cap \{3, 4, 5\} = \varnothing$
11. A firm promoted two of the following four salesmen to sales managers.
    W = Mr. Wilson
    G = Mr. Green
    R = Mr. Ravin
    B = Mr. Brown

    State whether each of the following statements is correct or incorrect and why.
    (a) The sample space concerning the promotion of two salesmen is

    $$\{B, R, G, W\}$$

    (b) The set representing the event Mr. Wilson or Mr. Brown is promoted is

    $$\{WG, WR, WB, RW, RG, RB\}$$

    (c) The set representing the event that both Mr. Brown and Mr. Green are promoted is

    $$\{BG\}$$

(d) The set representing the event that neither Mr. Wilson nor Mr. Green was promoted is

$$\{RB\}$$

(e) The set representing the event that Mr. Ravin and Mr. Green were not both promoted is

$$\{WB\}$$

(f) The set representing the event that neither Mr. Ravin, nor Mr. Wilson, nor Mr. Green was promoted is

$$\{B\}$$

12. Which of the following pairs of events are mutually exclusive?
    (a) (1) Park New common stock closes higher on a given day; (2) Park New common stock closes lower on the same day.
    (b) In a shipment of two relays, (1) exactly one is defective, (2) two are defective.
    (c) Three people A, B, C apply for two job openings. (1) A is hired, (2) B is hired.
    (d) On two rolls of a die, (1) a six occurs, (2) the sum of the two faces is five.
    (e) On two rolls of a die, (1) a four occurs, (2) the sum of the two faces is six.

13. A prospective buyer tests three polyethylene bags at a pressure of 18 pounds. One bag is from manufacturer A, one from manufacturer B, and one from manufacturer C. State whether the following events are elementary events. If an event is not elementary, list the elementary events of which it is composed.
    (a) Only manufacturer A's bag burst.
    (b) Manufacturer A's bag burst.
    (c) Two of the three bags burst.
    (d) None of the bags burst.

14. A language school has 490 instructors. It is known from the records that

    300 speak French
    200 speak German
    50 speak Russian
    20 speak French and Russian
    30 speak German and Russian
    20 speak German and French
    10 speak all three languages

    How many teachers
    (a) speak two but only two of the three languages?
    (b) speak at least one of the three languages?
    (c) speak German if they speak French?
    (d) speak French or German?

    (Note: Of the 300 who speak French, some also speak German or Russian. Some of the 300 speak all three languages.)

15. A small parcel delivery service company presently uses 2-ply rated tires and averages 15,000 miles per tire. Roll-More Tire Company sends the firm two tires to test. Let 0 represent the event "a tire lasted 15,000 miles or less" and 1 represent the event "a tire lasted more than 15,000 miles." We define four elementary events as

$$A = \{0, 0\} \qquad C = \{0, 1\}$$
$$B = \{1, 0\} \qquad D = \{1, 1\}$$

and the sample space as

$$S = \{(0, 0), (1, 0), (0, 1), (1, 1)\}$$

The first digit in each element represents the outcome for the first tire and the second digit represents the outcome for the second tire.
Express the following statements verbally:
(a) $B \cup C \cup D$
(b) $A \cup D$
(c) $B \cup C$
(d) $A \cup \bar{D}$
(e) $\bar{A} \cup \bar{D}$
(f) $A \cap D$

16. Let $P(A) = .5$; $P(B) = .4$; $P(A \cap B) = .2$
(a) Are A and B mutually exclusive events? Why?
(b) Are A and B independent events? Why?

17. In a group of 20 adults there are eight males and nine Republicans. Further, there are five male Republicans. If a random selection is made, what is the probability of selecting a female who is not a Republican?

18. A certain family has three children. Male and female children are equally probable and successive sexes are independent. Let M stand for male, F for female.
(a) List all elements in the sample space.
(b) What are the probabilities of the ordered events MMM and MFM?
(c) Let A be the event that both sexes appear. Find $P(A)$.
(d) Let B be the event that there is at most one girl. Find $P(B)$.
(e) Prove that A and B are independent.

19. Events A and B have the following probability structure:

$$P(A \cap B) = 1/6$$
$$P(A \cap \bar{B}) = 2/9$$
$$P(\bar{A} \cap B) = 1/3$$

(a) What is the probability of $\bar{A} \cap \bar{B}$?
(b) Are A and B independent events?

20. If the probability that company A will buy company B is .6, what are the odds that company A will not buy company B?

21. In roulette there are 38 slots in which a ball may land. There are numbers 0, 00, and 1 through 36. The odd numbers are red, even numbers are black, and the zeroes are green. A ball is thrown randomly into a slot.

(a) What is the probability that it is red?
(b) What is the probability it is number 27?
(c) What is the probability it is either red or the number 27?
(d) What are the odds in favor of black?
(e) What are the odds in favor of the number 27?
(f) If you play black an infinite number of times what fraction of times will you win? What fraction of times will you lose?

22. An investment firm purchases three stocks for one-week trading purposes. It assesses the probability that the stocks will increase in value over the week as .9, .7, and .6, respectively. What is the probability that all three stocks will increase, assuming that the movements of these stocks are independent? Is this a reasonable assumption?

23. The probability that a life insurance salesman following up a magazine lead will make a sale is .3. A salesman has two leads on a certain day. Assuming independence, what is the probability that

    (a) he will sell both?
    (b) he will sell exactly one policy?
    (c) he will sell at least one policy?

24. There are two major reasons for classifying a bottle of soda defective; the filler (a machine) either overfills or underfills the bottle. Two percent of the time it underfills and 1% of the time it overfills. What is the probability that a bottle will be rejected because of the filler?

25. A firm has five engineering positions to fill and is trying to recruit recent graduates for the positions. In the past, 40% of the college students who were offered similar positions have turned them down. The firm offers positions to six graduates. Is the firm justified in doing so? Explain, assuming independence.

26. A national franchising company is interviewing prospective buyers in the Tulleytown area. The probability that an interviewee will buy the franchise is .1. Assuming independence, what is the probability the firm will have to interview more than five people before making a sale?

27. Two politicians, A and B are set to debate. In order to see who should speak first, it is decided that A then B, then A then B, and so forth, will roll a die until one of them rolls the number 1. The one who rolls the number 1 first, speaks first. What is the probability that

    (a) A speaks first?
    (b) B speaks first?

28. The assembly line of a Poca Cola Bottling Company has three points of inspection on each returned bottle before it is filled. The first inspection point sorts out and rejects all bottles that are not Poca Cola's. The second sorts out all bottles which are chipped or cracked. The third, located after the bottles come out of the bottle sterilizer, rejects bottles which have any objects left in them. The probability that each of the inspection points will incorrectly accept or reject a bottle is .05.

(a) What is the probability that
  (1) a chipped Poca Cola bottle with gum in it will be filled?
  (2) a Poca Cola bottle which should be filled will be filled?
  (3) a chipped empty Poca Cola bottle will be filled?

(b) If a Poca Cola bottle which should be filled passes the first inspection, what is the probability it will *not* be filled?

29. A census of a company's 500 employees in regard to a certain proposal showed 125 of its 150 white-collar workers in favor of the proposal, and a total of 125 workers opposed to the proposal. (All the workers can be classified as either white-collar or blue-collar.)
    (a) What is the probability that an employee selected at random will be a blue-collar worker opposed to the proposal?
    (b) What is the probability that an employee picked at random will be in favor of the proposal?
    (c) What is the probability that if a blue-collar worker is chosen, he will be in favor of the proposal?
    (d) Are job type (blue- or white-collar) and opinion on the proposal independent? Prove your answer and give a short statement as to the implication of this finding.

30. The following information pertains to new-car dealers in the United States:

Type of Dealership	*Region of Dealer*				
	North	*South*	*Midwest*	*West*	*Total*
Admiral Motors	155	50	135	110	450
Bord Motors	90	65	40	90	285
Shysler Motors	50	50	30	85	215
U. S. Motors	35	35	15	65	150
Total	330	200	220	350	1100

If a name is selected randomly from the American Automobile Dealer's Association list of all United States dealers handling the four American manufacturers' brands, what is the probability that
    (a) it is a Southern Admiral Motors dealer?
    (b) it is a Southern dealer?
    (c) it is an Admiral Motors dealer?
    (d) it is a Southern dealer if he is known to be an Admiral Motors dealer?
    (e) Are type of dealership and region independent?

31. A lobbyist for the Open Pit Mining Corporation has entertained 100 of the members of a state legislature which had considered three bills of interest to Open Pit. For each legislator

    X = the number of votes cast by the legislator which were favorable to Open Pit
    Y = the amount of money spent by the lobbyist entertaining the legislator

A statistician at Open Pit analyzed the voting and obtained the following table:

			X		
Y	0	1	2	3	*Total*
\$ 0	2	2	2	2	8
10	2	2	2	6	12
20	0	24	24	2	50
50	10	6	4	10	30
	14	34	32	20	100

(a) Find $P(X = 0/Y = \$50)$, $P(X = 3/Y = \$50)$
$P(X = 0/Y = \$0)$, and $P(X = 3/Y = \$0)$.

(b) The company comptroller claims the entertainment expenditure was a waste of money, since the votes and expenditures were clearly independent. Are they independent?

32. A certain proposal was put forth by a company's management to all of its sales representatives in different sales regions. Questionnaires were sent to each salesman and the results were:

		Region		
Opinion	*East*	*Middle West*	*Pacific Coast*	*Total*
Opposed	20	15	15	50
Not opposed	80	85	285	450
Total	100	100	300	500

(a) What is the probability that a questionnaire selected at random is that of an Eastern salesman opposed to the proposal?

(b) What is the probability that a questionnaire selected at random is that of a Midwestern salesman?

(c) If a questionnaire is selected from the group which responded unfavorably to the proposal, what is the probability that the salesman comes from the Pacific Coast region?

(d) Are the salesman's regional district and his opinion on the proposal independent? If yes, prove it. If no, specify what the numbers in the cells of the table would have been had the two factors been independent.

33. The probability that an airplane accident which is due to structural failure is diagnosed correctly is 0.85, and the probability that an airplane accident which is not due to structural failure is diagnosed incorrectly as being due to structural failure is 0.35. If 30% of all airplane accidents are due to structural failures, find the probability that an airplane accident is due to

structural failure, given that it has been diagnosed as being due to structural failure.

34. A firm is contemplating changing the packaging sizes of its product, eliminating the three-ounce size and offering at a slightly higher price a four-ounce size. The marketing manager feels the probability that this change will increase profits is 70%. The change is tried in a limited test area and results in reduced profits. The probability that this result would occur even if the change would actually increase profits nationally is .4, while if it would not increase profits nationally, the probability is .8. What should be the manager's revised probability of the profitability of the change?

35. According to accident reports, 25% of all accidents which occurred while equipment was being used were caused by faulty equipment and 75% by improper use of the equipment. The probability that on a given day an accident will occur while equipment is being used is .05. What is the probability that on any given day there will be an accident

    (a) caused by faulty equipment?
    (b) caused by improper use of equipment?

36. An investor feels the probability that a certain stock will go up in value during the next month is .7. Value-Dow, an investment advisory firm predicts the stock will not go up over the period. Value-Dow over time has proven to be correct 80% of the time. What probability should the investor assign to the stock going up in light of Value-Dow's prediction?

37. Twenty percent of the items produced by a certain machine are defective. The company hires an inspector to check each item before shipment. The probability the inspector will incorrectly ship an item is .3. If an item is shipped, what is the probability that it is not defective?

38. TEC Exploration Company is involved in a mining exploration in Northern Canada. The chief engineer originally feels that there is a 50-50 chance that a significant mineral find will occur. A first test drilling is completed and the results are favorable. The probability that the test drilling would give misleading results is .3. What should be the engineer's revised probability that a significant mineral find will occur?

39. In a study of past records, it has been found that 25% of all shirts manufactured had an imperfection in them. Two persons are hired to inspect the shirts before they are shipped from the factory. The probability that either inspector will misclassify a shirt is 10%, and their decisions are independent.

    (a) If it is decided to class as imperfect any shirt which either inspector rejects, what is the probability that if a shirt is classified as good, it is actually good; actually imperfect?
    (b) If it is decided to class as imperfect only the shirts that both inspectors reject, what is the probability that a shirt classified as good is actually good; actually imperfect? Also, if it is classified as imperfect, what is the probability it is actually good; actually imperfect?

40. A motivational researcher shows a woman 15 projected colors for new fall clothes and asks her to pick her five favorites.
    (a) Give a specific outcome of the experiment.
    (b) How many such outcomes are there?
    (c) How many outcomes will contain the color russet?
    (d) What is the probability she will choose russet, as one of her five favorites?
41. In a determination of preference of package design, a panel of housewives was given four different packaging designs and asked to rate them. How many different possible rankings could the panel have given (excluding ties)?
42. A college professor anticipates teaching the same course for the next few years. So as not to become bored with his own jokes, he decides to tell a set of exactly three jokes each year. He may repeat one or two jokes from year to year, but he vows never to repeat the same set of three jokes. How many years can he last with a repertoire of seven jokes?
43. A company has three supervisors and seven regular employees. Each week, a skeleton force of one supervisor and three regular employees is chosen at random to work on Saturday.
    (a) If one particular supervisor and three particular regulars always play bridge during their lunch hour and never at any other time, what is the probability they will play bridge on a given Saturday?
    (b) How many possible sets of Saturday work forces will contain a particular supervisor?
    (c) How many possible sets of Saturday work forces will contain a particular regular employee?
44. A committee consists of five union and four nonunion men. In how many ways can a subcommittee of five be formed consisting of three union and two nonunion men?
45. A committee consists of ten people. It is decided to appoint a chairman, a vice-chairman, and a secretary-treasurer. How many different ways can this be done?
46. Admiral Motors orders eight different upholstering colors for its cars, and ten different body paint colors.
    (a) How many different color combinations of body and upholstering are available to the customer?
    (b) If Admiral Motors allows the customer to order a roof color different from the basic body color, how many different color coordinations of body, roof, and upholstering are available to the customer?
47. Rittleman Furs, Inc. has just purchased an electronic computer to handle its accounts receivable. Each data card contains 80 columns in which a number from 0 to 9 or a letter may be punched to represent information about an account. It is decided that each account will be assigned four identification symbols and these symbols will be punched in the first four columns of the data card to identify each card with a particular account.

(a) If numbers only are to be used, how many accounts can be handled by this method?
(b) If the first column is to be a letter and the next three are to be numbers, how many accounts can be handled?
(c) If either a letter or a number may be punched in each column, how many accounts can be handled?

48. A student faces a ten-question true-false examination fully aware that he is unprepared. His strategy is to mark five questions true and five questions false. In how many ways can he mark his examination, using five T's and five F's?

49. A sales manager calls a meeting of his twelve salesmen. The conference room consists of twelve chairs laid out in a line all facing a large blackboard.
(a) How many seating arrangements are possible?
(b) The manager does not want Mr. Wilson and Mr. Jones to sit together. How many possible arrangements are there if these two people do not sit together?

50. A firm desires to build five new factories, one in the 13 Southern states, two in the six Middle Atlantic states, one in the four Far Western states, and one in either Michigan, Indiana, Illinois, or Wisconsin. If it wants to study the desirability of each possible combination of locations, how many combinations would the firm have to consider?

51. A lunch counter has ten seats. Five patrons sit at the lunch counter.
(a) If they sit randomly, how many different orderings of empty and full seats are possible?
(b) If they sit randomly, what is the probability that all five will sit together?

52. A certain organization consists of five men and five women. A committee of five people is to be formed.
(a) What is the probability that there will be exactly one woman on the committee?
(b) Given that Mr. Wilson is to be chairman of the committee, what is the probability that of the four remaining to be selected, exactly one will be a woman?

53. A lathe operator needs a certain size fitting to complete a job, but he does not know its stock number. There are fourteen fittings in the stock room of which six are the correct size. The lathe operator orders five fittings to be picked at random from the stock room. What is the probability that at least one of the five will be the correct size?

54. An economist claims that he can tell from reading a company's annual report whether its gross sales will increase the following year. The economist is given the reports of ten companies from last year and is told that five had higher gross sales this year and five did not. He is asked to separate the ten reports into two groups, those he believes will increase, and those he believes will not increase gross sales. Three of the five that he places in the increase pile actually did increase gross sales. If one were to make the selections

randomly, what is the probability that one would place at least three correctly in the increase group? Based on the results, do you believe the economist can actually predict increases in gross sales of companies?

55. A mechanic needs a certain size bolt and nut to finish a job. He has in a bin ten bolts and eight nuts, of which eight bolts and six nuts are of the correct size. He reaches in and selects a bolt and a nut at random. What is the probability that he will have to select again to finish the job?

56. The probability of hitting a target on a single trial is .05. How many trials are required to make the probability of hitting the target at least once .99?

C H A P T E R   T W O

Discrete Random Variables and Probability Distributions

2.1 Random Variables

Decisions must often be made in an environment which is continually changing. In order for intelligent actions to be taken, the nature of this variation in the environment must be understood and estimated in some way. For example, differences in political, social, and economic conditions affect the operation of a business enterprise, and executives must be able to estimate the outcomes of decisions made under different combinations of the background conditions. We have already seen how variations in outcomes may be described in terms of the sample space of an experiment. In this chapter we consider two essential concepts: (1) the *random variable*—a *function* which assigns numerical values to the different outcomes defined by a sample space, and (2) the *probability function*—a *function* which assigns probabilities to these numerical values. These concepts are central to all of statistics, and although we introduce them here in the context of the business decision problem, they are used in every field in which statistical methods are applied.

We will introduce the formal definitions of *random variable* and *probability function* by the following previously discussed simple example. Other examples will follow.

Example 2-1 Consider again the experiment of tossing a fair coin two times. We shall concentrate on the number of heads obtained on the two tosses and their respective probabilities of occurrence. The sample space for this experiment is

$$S = \{(T, T), (T, H), (H, T), (H, H)\}$$

The elements of the sample space and the number of heads corresponding to each are listed in Table 2-1.

Table 2-1 Two Tosses of a Coin Experiment. Random Variable—Number of Heads.

Elements of the Sample Space	*Number of Heads*
T, T	0
T, H	1
H, T	1
H, H	2

This listing defines a function. The domain of the function consists of the elements of the sample space; the range consists of the numbers assigned to each of the sample points. This *function* is referred to as a *random variable.* Formally, a *random variable* is a *real-valued function defined on a sample space.* The term "real-valued" simply means that the elements of the range of the function are real numbers. Ordinarily, a shortcut method of referring to the concept of a random variable is used. In the above situation, we simply refer to the random variable "number of heads obtained in two tosses of a fair coin." Thus, we say that this random variable takes on three values, 0, 1, and 2. Note, however, that strictly speaking, the random variable is the indicated function, and not merely the numbers in the range of the function.

We usually are interested in the probabilities associated with the various values of a random variable. A probability of 1/4 is assigned to each point in the sample space of the two coin experiment as shown in Table 2-2.

Table 2-2 Sample Space, Number of Heads, and Probabilities in the Two Tosses of a Coin Experiment.

Elements of the Sample Space	*Number of Heads*	*Probability*
T, T	0	1/4
T, H	1	1/4
H, T	1	1/4
H, H	2	1/4

The last two columns of Table 2-2 can be more compactly shown by adding up the probabilities associated with each distinct number of heads. The resulting distribution is known as a "*probability function*" or "*probability distribution*" of the random variable of interest. In keeping with general practice, we will usually use the term "probability distribution" throughout the remainder of this text. The probability distribution for the random variable "number of heads" for the two coin experiment is shown in Table 2-3.

Table 2-3 Probability Distribution for Number of Heads in the Two Coin Experiment.

Number of Heads x	*Probability* $f(x)$
0	1/4
1	1/2
2	1/4

A *probability distribution* may be defined as *a function in which the domain consists of the possible values that a random variable can take on, and the range is composed of the probabilities associated with those values.* If we let the symbol X stand for the random variable (number of heads, in this example), then we can represent the values that the random variable can assume by x. The probability that the random variable X will take on the value x is symbolized as $P(X = x)$ or simply $f(x)$. In Table 2-3, the values of the random variable are listed under the column headed x and the probabilities at these values are shown under $f(x)$. Thus, in summary,

$$P(X = 0) = f(0) = 1/4$$
$$P(X = 1) = f(1) = 1/2$$
$$P(X = 2) = f(2) = 1/4$$

Note that these probabilities sum to one.

When a probability distribution is graphed, it is conventional to display the values of the random variable on the horizontal axis and their probabilities on the vertical scale. The graph of the probability distribution for the two tosses of a coin experiment is shown in Figure 2-1.

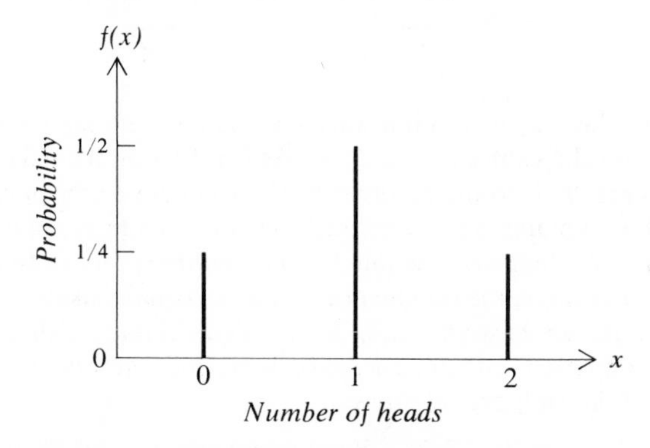

Figure 2-1 Graph of probability distribution of number of heads obtained in two tosses of a coin.

Example 2-2 A department store has classified its list of charge customers into two mutually exclusive categories: (1) high volume purchasers, and (2) non-high volume purchasers. There are twenty percent high volume purchasers. Assume that a sample of four customers is drawn at random from the list. What is the probability distribution of the random variable "number of high volume purchasers"? It may be assumed that the list of charge customers is so large that even though the sample is drawn without replacement, it is sufficiently accurate for computations to be performed as though the sampling were carried out with replacement. That is, the partial exhaustion of the list because of the drawing of the items in the sample is so small that, for practical purposes, the probabilities of obtaining the two types of purchasers remain unchanged.

Let A represent the occurrence of a high volume purchaser and $\bar{A}$ the occurrence of a non-high volume purchaser. The elements of the sample space for the experiment involved in the drawing of the sample of four customers are listed in Table 2-4.

Table 2-4 Elements of the Sample Space for the Experiment of Drawing a Random Sample of Four Customers.

$\bar{A}\ \bar{A}\ \bar{A}\ \bar{A}$	$A\ A\ \bar{A}\ \bar{A}$	$A\ A\ A\ \bar{A}$
	$A\ \bar{A}\ A\ \bar{A}$	$A\ A\ \bar{A}\ A$
$A\ \bar{A}\ \bar{A}\ \bar{A}$	$A\ \bar{A}\ \bar{A}\ A$	$A\ \bar{A}\ A\ A$
$\bar{A}\ A\ \bar{A}\ \bar{A}$	$\bar{A}\ A\ A\ \bar{A}$	$\bar{A}\ A\ A\ A$
$\bar{A}\ \bar{A}\ A\ \bar{A}$	$\bar{A}\ A\ \bar{A}\ A$	
$\bar{A}\ \bar{A}\ \bar{A}\ A$	$\bar{A}\ \bar{A}\ A\ A$	$A\ A\ A\ A$

We denote by X the random variable "number of high volume purchasers." X can take on the values 0, 1, 2, 3, 4. As can be seen from Table 2-4, one sample point corresponds to the occurrence of zero high volume purchasers; four points to one high volume purchaser; six points to two high volume purchasers; four points to three high volume purchasers; and one point to four high volume purchasers. However, the sample points are not equally likely. The probability of a high volume purchaser is 0.2; of a non-high volume purchaser, 0.8. Taking one sample point each for zero, one, two, three, and four high volume purchasers, we have the following probabilities:

$$P(\bar{A}\ \bar{A}\ \bar{A}\ \bar{A}) = (0.8)(0.8)(0.8)(0.8) = (0.8)^4$$
$$P(A\ \bar{A}\ \bar{A}\ \bar{A}) = (0.2)(0.8)(0.8)(0.8) = (0.2)(0.8)^3$$
$$P(A\ A\ \bar{A}\ \bar{A}) = (0.2)(0.2)(0.8)(0.8) = (0.2)^2\,(0.8)^2$$
$$P(A\ A\ A\ \bar{A}) = (0.2)(0.2)(0.2)(0.8) = (0.2)^3\,(0.8)$$
$$P(A\ A\ A\ A) = (0.2)(0.2)(0.2)(0.2) = (0.2)^4$$

Considering any given value of the random variable "number of high volume purchasers," each elementary event has the same probability. For example, in the case of one high volume purchaser, each of the indicated four elementary events (sample points) has a probability of $(0.2)(0.8)^3$. Multiplying the specified probabilities for elementary events by the number of such points in the composite events "zero high volume purchasers," "one high volume purchaser," and so forth, we have

$$P(X = 0) = f(0) = 1(0.8)^4 = 0.4096$$
$$P(X = 1) = f(1) = 4(0.8)^3(0.2) = 0.4096$$
$$P(X = 2) = f(2) = 6(0.8)^2(0.2)^2 = 0.1536$$
$$P(X = 3) = f(3) = 4(0.8)(0.2)^3 = 0.0256$$
$$P(X = 4) = f(4) = 1(0.2)^4 = 0.0016$$

These values are summarized in Table 2-5 in the form of a probability distribution. It may be noted that the probabilities add up to one.

Table 2-5 Probability Distribution of Number of High Volume Purchasers.

x	$f(x)$
0	0.4096
1	0.4096
2	0.1536
3	0.0256
4	0.0016

Of course, it was not necessary to list all the points in the sample space in order to derive this probability distribution. We could have determined the probabilities for one sample point each, for zero, one, through four high volume purchasers as above, and then determined the number of elementary events in each of these by means of permutations. Thus, in the case of zero high volume purchasers, we must have zero such purchasers occurring with four non-high volume purchasers. The number of distinct arrangements that can be made in this case is the number of permutations of four objects, zero of type one, and four of type two. Thus,

$$P_m(4; 0, 4) = \frac{4!}{0!\,4!} = 1$$

Therefore, there is only one elementary event in "zero high volume purchasers," that is, $\bar{A}\,\bar{A}\,\bar{A}\,\bar{A}$.

Similar considerations in the case of one, two, three, and four high volume purchasers yield the following:

$$P_m(4; 1, 3) = \frac{4!}{1!\,3!} = 4$$

$$P_m(4; 2, 2) = \frac{4!}{2!\,2!} = 6$$

$$P_m(4; 3, 1) = \frac{4!}{3!\,1!} = 4$$

$$P_m(4; 4, 0) = \frac{4!}{4!\,0!} = 1$$

These are the same numbers as were obtained earlier by counting points in the sample space.

A graph of the probability distribution for this problem is given in Figure 2-2.

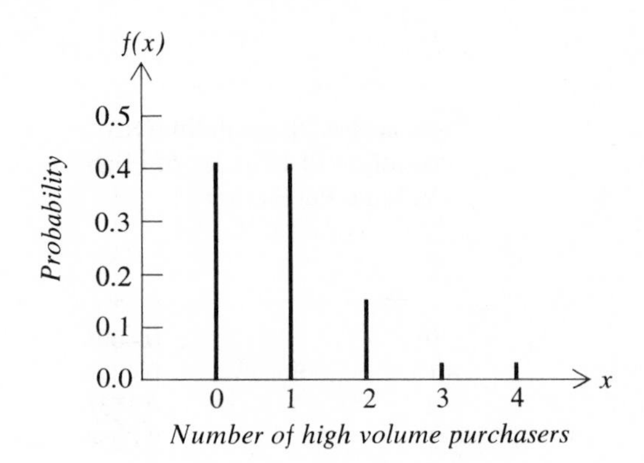

Figure 2-2 Graph of probability distribution of number of high volume purchasers.

Example 2-3 A corporation economist develops a subjective probability distribution for the change that will take place in the following year in gross national product (GNP). He establishes the following five categories of change and defines a random variable by letting the indicated numbers correspond to each category.

Change in GNP	*Number Assigned*
Down more than 5%	−2
Down 5% or less	−1
Unchanged	0
Up 5% or less	+1
Up more than 5%	+2

On the basis of all information available to him, the economist assigns probabilities to each of these possible events as indicated in Table 2-6.

Types of Random Variables

Random variables are classified as either *discrete* or *continuous*. A *discrete random variable* is one which can take on only a finite or countably infinite number of distinct values. The three preceding examples illustrated probability distributions of discrete random variables.

A random variable is said to be *continuous* in a given range if it can assume any value in that range. The term "continuous random variable" implies that variation takes place along a *continuum*. These types of variables

Table 2-6 Subjective Probability Distribution of Change in Gross National Product.

x	$f(x)$
-2	0.1
-1	0.1
0	0.2
$+1$	0.4
$+2$	0.2

can be *measured* to some degree of accuracy. In the case of discrete variables, since only a distinct number of values are assumed, *counts* can be made at each of these values. Examples of continuous variables include weight, length, velocity, rate of production, dosage of a drug, and the length of life of a given product.

However, it may be argued that in the real world all counted or measured data are discrete. For example, if we want to measure weight, and the measuring instrument only permits a determination to the nearest thousandth of a pound, then the resulting data will be discrete in units of thousandths of a pound. Despite this discreteness of data caused by limitations of the measuring instruments, it is nevertheless useful in many instances to use mathematical models which treat certain variables as being continuous in nature. Furthermore, although in the real world, measured data are discrete, in essence, the variable under measurement is often continuous. Thus, if we use a continuous mathematical model of heights of individuals, where the underlying data are measured or discrete, we may conceive of this model not as a convenient approximation, but rather as a model of reality which is more accurate than the discrete data from which it was derived.

On the other hand, we often find it convenient to convert a variable which is conceptually continuous into one which is discrete. Thus, in the case of heights of individuals, rather than using measurements along a continuous scale, we may set up classifications such as tall, medium, and short. In Example 2-3, gross national product was treated as a discrete random variable in five distinct categories. Conceptually, it may be viewed as a continuous variable and is often so treated in econometric models.

The rather philosophical point has sometimes been made that one indication of progress in science is the extent to which discrete variables can be converted into continuous variables. Thus, the physicist treats color in terms of the continuous variable of wavelengths rather than the discrete classification of names of colors. Measurement of oral temperature by means of a thermometer treats human body temperature as varying along a continuous scale (although the resultant measurements are discrete), rather than as discrete as when the temperature is judged by placing a hand on the forehead of another person and classifying his temperature as "normal," "high," or in some other category. However, in applied problems, where a probability model is used to represent a real world situation, we may work either in terms of discrete or continuous random variables, whichever appear to be most appropriate for the problem or decision-making situation in question. Only probability distributions of discrete random variables will be discussed in the remainder of this chapter.

Characteristics of Probability Distributions

In the three preceding examples, it was seen that the sum of the probabilities in each probability distribution was equal to one. It is possible to summarize the characteristics of probability distributions somewhat more formally.

Given a finite or countably infinite sample space, a probability distribution of a random variable X, whose value at x is $f(x)$, possesses the following properties:[1]

(1) $f(x) \geq 0$ for all real values of x

(2) $\sum_x f(x) = 1$

Property (1) simply states that the probability function, f, of the random variable X is defined for all real values of x. There is a finite or countably infinite number of values of x for which $f(x) > 0$. At all other values of x, a value of zero is assigned to $f(x)$. The second property states that the sum of the probabilities in a probability distribution is equal to one. The notation $\sum_x f(x)$ means "sum the values of $f(x)$ for all values that x takes on."

Other terms are in use for probability distributions. We have already

[1] A somewhat simplified notation is used here. A mathematically more elegant notation would represent the values that the random variable X could assume as $x_1, x_2, \ldots, x_n$ with associated probabilities $f(x_1), f(x_2), \ldots, f(x_n)$. Then the two properties would appear as

(1) $f(x_i) \geq 0$ for all i

(2) $\sum_{i=1}^{n} f(x_i) = 1$

If X takes on a countably infinite number of values, then the second property would appear as

$$\sum_{i=1}^{\infty} f(x_i) = 1$$

seen that "probability function" and "probability distribution" are employed synonymously. Probability distributions of *discrete* random variables are often referred to as "probability mass functions," since the probabilities are "massed" at distinct points along the (say) x axis. The corresponding term for *continuous* random variables is "probability density function." The abbreviated terms "mass function" and "density function" are frequently used. The term "frequency function" is often used synonymously with "probability distribution" or "probability function," but sometimes it denotes only probability distributions of discrete random variables. To avoid confusion, we will use "probability distribution" and "probability function" as generic terms, and "probability mass function" and "probability density function" in the case of discrete and continuous random variables, respectively.

Cumulative Distribution Functions

Frequently, we are interested in the probability that a random variable is equal to or less than some specified value or greater than a given value. The *cumulative distribution function* is particularly useful in this connection. We may define this function as follows:

Given a random variable X, the value of the cumulative distribution function at x, denoted $F(x)$, is *the probability that X takes on values less than or equal to x*. Hence,

(2.1) $$F(x) = P(X \leq x)$$

In the case of a discrete random variable, it is clear that

(2.2) $$F(c) = \sum_{x \leq c} f(x)$$

The symbology $\sum_{x \leq c} f(x)$ means "sum the values of $f(x)$ for all values of x less than or equal to c."

Example 2-4 We return to Example 2-1 involving the experiment of tossing a coin two times. The probability of obtaining zero or less heads is $F(0) = 1/4$; one or less heads, $F(1) = 3/4$; and two or less heads, $F(2) = 1$. In Table 2-7 are shown the probability mass function and cumulative distribution for the random variable "number of heads."

A graph of the cumulative distribution function is given in Figure 2-3, which is a so-called "step function." That is, the values change in discrete "steps" at the indicated integral values of the random variable, X. Thus, $F(x) = 0$ to the left of the point $x = 0$, but steps up to $F(x) = 1/4$ at $x = 0$, and so forth. A dot is shown at the left of each horizontal line segment to indicate where the probability is read for the integral values of x. The values of the cumulative distribution function are read at these points on the *upper* line segments at each integral value of x.

Table 2-7 Probability Mass Function and Cumulative Distribution Function for Number of Heads in the Two Tosses of a Coin Experiment.

x	$f(x)$	$F(x)$
0	1/4	1/4
1	1/2	3/4
2	1/4	1

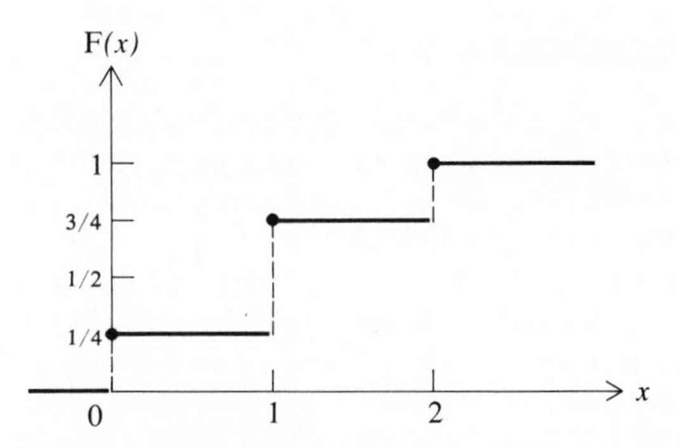

Figure 2-3 Graph of cumulative distribution function of number of heads in the two tosses of a coin experiment.

We note the following relations in this problem, which follow from the definition of a cumulative distribution function:

$$\begin{aligned} F(0) &= f(0) \\ F(1) &= f(0) + f(1) \\ F(2) &= f(0) + f(1) + f(2) \end{aligned}$$

The probabilities of more than zero, one, and two heads are given, respectively, by

$$\begin{aligned} 1 - F(0) &= 3/4 \\ 1 - F(1) &= 1/4 \\ 1 - F(2) &= 0 \end{aligned}$$

Example 2-5 Let us return to Example 2-3, which discussed an economist's subjective probability distribution of change in GNP. A few questions will illustrate some uses of the cumulative distribution function.

What was the probability assigned by the economist to

(a) the event that the change in GNP will not exceed an increase of 5%?

$$F(1) = f(-2) + f(-1) + f(0) + f(1) = 0.8$$
or
$$F(1) = 1 - f(2) = 1 - 0.2 = 0.8$$

(b) the event that GNP will not decline? The event "GNP will not decline" is the event "GNP will remain unchanged or will increase" and is the complement of the event "GNP will decrease." Thus, it is given by

$$1 - F(-1) = 1 - [f(-2) + f(-1)] = 1 - 0.2 = 0.8$$
or
$$1 - F(-1) = f(0) + f(1) + f(2) = 0.8$$

(c) a change in GNP of 5% or less?

$$f(-1) + f(0) + f(1) = 0.7$$
or
$$F(1) - F(-2) = 0.8 - 0.1 = 0.7$$

2.2 Probability Distributions of Discrete Random Variables

In many situations, it is useful to represent the probability distribution of a random variable by a general algebraic expression. Probability calculations can then be conveniently made by substituting appropriate values into the algebraic model. The mathematical expression is a compact form of summarizing the nature of the process that has generated the probability distribution. Thus, the statement that a particular probability distribution is appropriate in a given situation, contains a considerable amount of information concerning the nature of the underlying process thus described. In this section, the following probability distributions of discrete random variables will be discussed: the uniform, binomial, multinomial, hypergeometric, and Poisson distributions.

2.3 The Uniform Distribution

In certain situations, equal probabilities may be assigned to all of the possible values that a random variable may assume. Such a probability distribution is referred to as a *uniform distribution*. For example, suppose a fair die is rolled once. The probability is 1/6 that the die will show any given number on its uppermost face. The probability mass function in this case may be written as

$$f(x) = 1/6 \text{ for } x = 1, 2, \ldots, 6$$

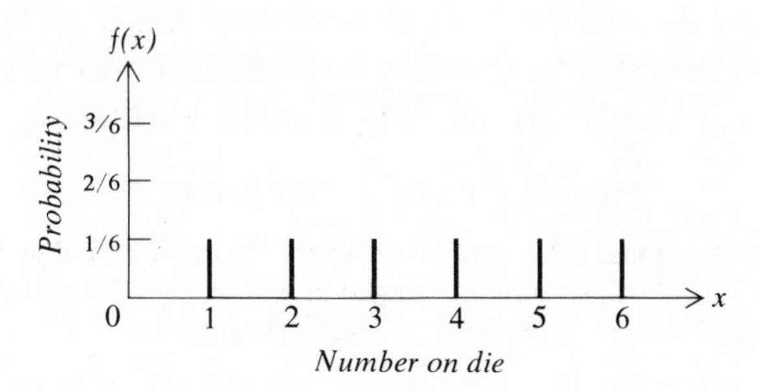

Figure 2-4 Graph of probability mass function of numbers obtained in a roll of a die.

A graph of this distribution is given in Figure 2-4.

As another illustration, let us consider the case of the Amgar Power Company, which produces electrical energy from geothermal steam fields. It takes about two years to build a production facility. In planning its production strategies, the company concludes that it is equally likely that demand two years hence will be from 80,000 to 120,000 kilowatts in increments of 10,000 kilowatts.

Therefore, the probability distribution established by Amgar Power Company for this future demand is

$$f(x) = 0.20 \text{ for } x = 80{,}000,\ 90{,}000, \ldots, 120{,}000$$

At a later point, we will examine how such information is used in decision making procedures which determine optimal courses of action.

2.4 The Binomial Distribution

The binomial distribution is undoubtedly the most widely applied probability distribution of a discrete random variable. It has been used to describe an impressive variety of processes in business and the social sciences as well as other areas. The type of process which gives rise to this distribution is usually referred to as a *Bernoulli trial* or as a *Bernoulli process*.[1] The mathematical model for a Bernoulli process is developed from a very specific set of assumptions involving the concept of a series of experimental trials.

Let us envision a process or experiment characterized by repeated trials. The trials take place under the following set of assumptions:

---

[1]Named after James Bernoulli (1654–1705), a member of a family of Swiss mathematicians and scientists, who did some of the early significant work on the binomial distribution.

(1) There are two mutually exclusive possible outcomes on each trial, which are referred to as "success" and "failure." In somewhat different language, the sample space of possible outcomes on each experimental trial is $S = \{\text{failure, success}\}$.

(2) The probability of a success, denoted p, remains constant from trial to trial. The probability of a failure, denoted q, is equal to $1 - p$.

(3) The trials are independent. That is, the outcomes on any given trial or sequence of trials do not affect the outcomes on subsequent trials.

The outcome on any specific trial is determined by chance. Such processes are referred to as "random processes" or "stochastic processes." Bernoulli trials are one example of such processes.

Our aim is to develop a formula for the probability of x successes in n trials of a Bernoulli process. We shall first take a simple specific case of coin tossing as an example of a series of Bernoulli trials. We calculate the probability of obtaining exactly two heads in five tosses. The resulting expression is then generalized.

Let us consider the experiment of tossing a fair coin five times. We may treat each toss as one "Bernoulli trial" or "one trial of a Bernoulli process." The possible outcomes on any particular trial are a head or a tail. Assume the appearance of a head to be a success. Of course, in these problems, the classification of one of the two possible outcomes as a "success" is completely arbitrary and there is no necessary implication of desirability or goodness involved. For example, we may choose to refer to the appearance of a defective item in a production process as a success and a non-defective item as a failure. Or, if a process of births is treated as a series of Bernoulli trials, the appearance of a female (male) may be classified as a success, and a male (female) a failure.

Returning to the five tosses of a fair coin, suppose that the sequence of outcomes is

$$H\ T\ H\ T\ T$$

H and T being used to denote head and tail, as usual. We now introduce a convenient coding device for outcomes on Bernoulli trials.

Let $x_i = 0$ if the outcome on the ith trial is a failure, and
$x_i = 1$ if the outcome on the ith trial is a success.

Then, the outcomes of the sequence of tosses given above may be written as

$$1\ 0\ 1\ 0\ 0$$

Since the probability of a success and a failure on a given trial are p and

q, respectively, the probability of this particular sequence of outcomes is, by the multiplication rule,

$$P(1, 0, 1, 0, 0) = p\,q\,p\,q\,q = q^3p^2$$

In the notation used for this probability, for simplicity, commas have been used to separate the outcomes of the successive trials, although actually we have here the joint probability of the intersection of the events which occurred on the five trials. This is the probability of obtaining the specific sequence of successes and failures, in the order in which they occurred. However, we are interested not in any specific order of results, but rather in the probability of obtaining a given number of successes in n trials. What then is the probability of obtaining exactly two successes in five Bernoulli trials? The following nine other sequences satisfy the condition of exactly two successes in five trials:

1 1 0 0 0
1 0 0 1 0
1 0 0 0 1
0 1 1 0 0
0 1 0 1 0
0 1 0 0 1
0 0 1 1 0
0 0 1 0 1
0 0 0 1 1

By the same reasoning earlier used, each of these sequences has the same probability, q^3p^2. We can obtain the number of such sequences from the formula for the permutation of n objects, x of which are of one type, and $n - x$ of another type. Thus, the number of distinguishable sequences in this problem is $P_m(5; 2, 3) = 10$. However, we indicated in Equation (1.17) that

$$P_m(n; x, n - x) = \binom{n}{x} = \frac{n!}{x!\,(n - x)!}$$

Thus,

$$P_m(5; 2, 3) = \binom{5}{2} = 10$$

and we may write

$$P(\text{exactly 2 successes}) = \binom{5}{2}q^3p^2$$

In the case of the fair coin example, we assign a probability of $1/2$ to p and $1/2$ to q. Hence,

$$P(\text{exactly 2 heads}) = \binom{5}{2}\left(\frac{1}{2}\right)^3\left(\frac{1}{2}\right)^2 = \frac{10}{32} = \frac{5}{16}$$

This result may be generalized to obtain the probability of (exactly)

x successes in n trials of a Bernoulli process. Let us assume $n - x$ failures occurred followed by x successes, in that order. We may then represent this sequence as

$$\underbrace{000\cdots0}_{\substack{n-x \\ \text{failures}}}\quad\underbrace{111\cdots1}_{\substack{x \\ \text{successes}}}$$

The probability of this particular sequence is $q^{n-x}p^x$. The number of possible sequences of n trials which would contain exactly x successes is $\binom{n}{x}$.[2] Therefore, the probability of obtaining x successes in n trials of a Bernoulli process is given by[3]

$$f(x) = \binom{n}{x} q^{n-x}p^x \text{ for } x = 0, 1, 2, \ldots, n \tag{2.3}$$

If we denote by X the random variable number of successes in these n trials, then clearly

$$f(x) = P(X = x)$$

The possible values of $f(x)$ are displayed in Table 2-8. The fact that this is a probability distribution is verified by noting that (1) $f(x) \geq 0$ for all real values of x and (2) $\sum_x f(x) = 1$. The first condition is seen to be true by noting that since p and n are non-negative numbers, $f(x)$ cannot be negative. The second condition is true because

$$\sum_x \binom{n}{x} q^{n-x}p^x = (q + p)^n = 1^n = 1$$

The second condition requires some explanation. The sum of the terms representing probabilities of numbers of successes is the familiar binomial expansion. The binomial $(q + p)^n$ is expanded as follows:

$$(q + p)^n = q^n + nq^{n-1}p^1 + \frac{n(n-1)}{2}q^{n-2}p^2 + \cdots + nq^1p^{n-1} + p^n$$

---

[2]Because the combination notation is universally used in the binomial probability distribution, that convention is followed here. Note, however, that conceptually, we have here the number of distinct permutations that can be formed of n objects, $n - x$ of which are of one type and x of the other. Since $P_m(n; n - x, x) = \binom{n}{x}$, the combination notation may be used instead of the conceptually more revealing one of permutations.

[3]The following method of writing the mathematical expression for such a probability distribution is often used:

$$\begin{aligned} f(x) &= \binom{n}{x} q^{n-x}p^x \text{ for } x = 0, 1, 2, \ldots, n \\ &= 0, \text{ elsewhere} \end{aligned}$$

In this and other places where it is clear that $f(x)$ is equal to zero for other than the specified values of the random variable, the notation on the last line will not be included.

Table 2-8 The Probability Distribution for the Random Variable Number of Successes in n Trials of a Bernoulli Process.

Number of Successes x	*Probability* $f(x)$
0	$\binom{n}{0}q^{n-0}p^0$
1	$\binom{n}{1}q^{n-1}p^1$
2	$\binom{n}{2}q^{n-2}p^2$
.	.
.	.
.	.
x	$\binom{n}{x}q^{n-x}p^x$
.	.
.	.
.	.
n	$\binom{n}{n}q^{n-n}p^n$
Total	1

It can be easily verified that term by term these values are identical with the respective probabilities of zero to n successes. That is,

$$f(0) = \binom{n}{0}q^{n-0}p^0 = q^n$$

$$f(1) = \binom{n}{1}q^{n-1}p^1 = nq^{n-1}p^1$$

$$f(2) = \binom{n}{2}q^{n-2}p^2 = \frac{n(n-1)}{2}q^{n-2}p^2$$

$$\vdots \qquad \vdots \qquad \vdots$$

$$f(n) = \binom{n}{n}q^{n-n}p^n = p^n$$

Therefore, the terms "binomial probability distribution" or simply "binomial distribution" are usually used to refer to the probability distribution resulting from a Bernoulli process. In summary, in problems, where the assumptions of a Bernoulli process are met, we can obtain the probabilities of zero, one, and so forth successes in n trials from the respective terms of the binomial expansion of $(q + p)^n$, where q and p denote the probabilities of failure and success on a single trial and n is the number of trials.

The binomial distribution has two parameters,[4] n and p. Each pair of values for these parameters establishes a different distribution. Thus, the binomial is, in fact, a family of probability distributions. Since computations become laborious for large values of n, it is advisable to make use of special tables. Selected values of the binomial cumulative distribution function are given in Appendix A, Table A-1. The values of

$$F(c) = \sum_{x \leq c} f(x) \quad \text{for } x = 0, 1, 2, \ldots, n$$

are shown in that table for $n = 2$ to $n = 20$ and $p = 0.05$ to $p = 0.50$ in multiples of 0.05. Values of $f(c)$, cumulative probabilities for p values greater than 0.50, and probabilities that x is greater than a given value or lies between two values can be obtained by appropriate manipulation of these tabulated values. Some of the examples which follow illustrate the use of the table. More extensive tables have been published by the National Bureau of Standards and Harvard University, but even such tables usually do not go beyond $n = 50$ or $n = 100$. For large values of n, approximations are available for the binomial distribution, and the exact values generally need not be determined.

It is important to note that in the case of the binomial distribution, as with any other mathematical model, the correspondence between the real world situation and the model must be carefully established. In many cases, the underlying assumptions of a Bernoulli process are obviously not met. For example, suppose that in a production process, items produced by a certain machine tool are tested as to whether they meet specifications. If the items are tested in the order in which they are produced, then the assumption of independence would doubtless be violated. That is, whether an item meets specifications would not be independent of whether the preceding item(s) did. If the machine tool had become subject to wear, it is quite likely that if it produced an item which did not meet specifications, the next item would fail to conform to specifications in a similar way. Thus, whether or not an item is defective would *depend* on the characteristics of preceding items. In the coin tossing illustration, on the other hand, we conceived of an experiment

---

[4]In this context, the term "parameters" refers to variables which are sufficient to specify a probability distribution. When particular values are assigned to the parameters of a probability function, a specific distribution in the family of possible distributions is defined. For example, $n = 10$, $p = 1/2$ specifies a particular binomial distribution; $n = 20$, $p = 1/2$ specifies another.

in which a head or tail on a particular toss did not affect the outcome on the next toss.

It can be seen from the assumptions underlying a Bernoulli process that the binomial distribution is applicable to the situations of *sampling from a finite universe with replacement* or *sampling from an infinite universe*, with or without replacement. In either of these cases, the probability of success may be viewed as remaining constant from trial to trial and the outcomes as independent among trials. If the population size is large relative to sample size, that is, if the sample constitutes only a small fraction of the population, and if p is neither very close to zero or one in value, the binomial distribution is often sufficiently accurate, even though sampling may be carried out from a finite universe without replacement. It is difficult to give universal rules of thumb on what constitutes a sufficiently large sample for this purpose. Some practitioners suggest a population size at least ten times the sample size. However, clearly, the purpose of the calculations determines the required degree of accuracy. Furthermore, in general, approximations are relatively closer for terms near the center of the distribution than in the tails and for sums of terms rather than for individual terms.

Example 2-4 The tossing of a fair coin five times was used earlier in this chapter as an example of a Bernoulli process; the probability of obtaining two heads (successes) was calculated. Compute the probabilities of all possible numbers of heads and thus establish the particular binomial distribution which is appropriate in this case.

Solution: This problem is an application of the binomial distribution for $p = 1/2$ and $n = 5$. Letting X represent the random variable "number of heads," the probability distribution is as follows:

x	$f(x)$
0	$\binom{5}{0}\left(\frac{1}{2}\right)^5\left(\frac{1}{2}\right)^0 = \frac{1}{32}$
1	$\binom{5}{1}\left(\frac{1}{2}\right)^4\left(\frac{1}{2}\right)^1 = \frac{5}{32}$
2	$\binom{5}{2}\left(\frac{1}{2}\right)^3\left(\frac{1}{2}\right)^2 = \frac{10}{32}$
3	$\binom{5}{3}\left(\frac{1}{2}\right)^2\left(\frac{1}{2}\right)^3 = \frac{10}{32}$
4	$\binom{5}{4}\left(\frac{1}{2}\right)^1\left(\frac{1}{2}\right)^4 = \frac{5}{32}$
5	$\binom{5}{5}\left(\frac{1}{2}\right)^0\left(\frac{1}{2}\right)^5 = \frac{1}{32}$
	1

Example 2-5 In Chapter 1, the probability of obtaining at least one six in two rolls of a die (or in one roll of two dice) was calculated in several different ways to illustrate basic principles. Solve the same problem using the binomial distribution.

Solution: We view the two rolls of the die as Bernoulli trials. Defining the appearance of a six as a success, $p = 1/6, q = 5/6$, and $n = 2$. It is instructive to examine the entire probability distribution.

x	$f(x)$
0	$\binom{2}{0}\left(\frac{5}{6}\right)^2\left(\frac{1}{6}\right)^0 = \left(\frac{5}{6}\right)^2$
1	$\binom{2}{1}\left(\frac{5}{6}\right)^1\left(\frac{1}{6}\right)^1 = 2\left(\frac{5}{6}\right)\left(\frac{1}{6}\right)$
2	$\binom{2}{2}\left(\frac{5}{6}\right)^0\left(\frac{1}{6}\right)^2 = \left(\frac{1}{6}\right)^2$
	1

The expressions at the right-hand side of the $f(x)$ column have been given in the form with which the student is probably most familiar for the terms in the expansion of $(5/6 + 1/6)^2$.

The required probability is

$$P\binom{\text{at least}}{\text{one six}} = f(1) + f(2) = 2(5/6)(1/6) + (1/6)^2 = 11/36$$

Example 2-6 An interesting correspondence took place in 1693 between Samuel Pepys, author of the famous *Diary*, and Isaac Newton, in which Pepys posed a probability problem to the eminent mathematician. The question as originally stated by Pepys was[5]:

"*A* has six dice in a box, with which he is to fling a six
B has in another box 12 dice, with which he is to fling two sixes
C has in another box 18 dice, with which he is to fling three sixes
(Question)—Whether *B* and *C* have not as easy a task as *A* at even luck?"

In rather flowery seventeenth century English, Newton replied and said essentially in modern parlance, "Sam, I do not understand your question." Newton asked whether individuals *A*, *B*, and *C* were to throw independently and whether the question pertained to the obtaining of *exactly* one, two, or three sixes or *at least* one, two, or three sixes.

After an exchange of letters, in which Pepys supplied little help in answering

---

[5]Schell, Emil D., "Samuel Pepys, Isaac Newton and Probability," *The American Statistician*, October, 1960, pp. 27–30.

these queries, Newton decided to frame the question himself. In modern language, Newton's wording would appear somewhat as follows:

"If A, B, and C toss dice independently, what are the probabilities that:

A will obtain at least one six in a roll of six dice?

B will obtain at least two sixes in a roll of twelve dice?

C will obtain at least three sixes in a roll of eighteen dice?"

Newton's reply to these questions involved some rather tortuous arithmetic to obtain the correct result. His work doubtless represented a very respectable intellectual feat, considering the infantile state of probability theory at that time. Today, virtually any beginning student of probability theory, standing on the shoulders of the giants who came before him, would immediately see the application of the binomial distribution to the problem. Let us denote by $P(A)$, $P(B)$, and $P(C)$, respectively, the probabilities that A, B, and C would obtain the specified events. Then,

$$P(A) = 1 - \binom{6}{0}(5/6)^6(1/6)^0 \approx 0.67$$

$$P(B) = 1 - \binom{12}{0}(5/6)^{12}(1/6)^0 - \binom{12}{1}(5/6)^{11}(1/6)^1 \approx 0.62$$

$$P(C) = 1 - \binom{18}{0}(5/6)^{18}(1/6)^0 - \binom{18}{1}(5/6)^{17}(1/6)^1 - \binom{18}{2}(5/6)^{16}(1/6)^2 \approx 0.60$$

Thus, $P(A) > P(B) > P(C)$.

Pepys honestly admitted that he did not understand Newton's calculations and furthermore that he didn't believe the answer. He argued that since B throws twice as many dice as A, why can't he simply be considered as two A's? Thus, he would have at least as great a probability of success as A. Of course, Pepys' question indicated that he was rather confused. There is no reason why the probability of at least two sixes in a roll of twelve dice should be twice the probability of at least one six in a roll of six dice, and, as seen by the above calculations, indeed it is not.

Example 2-7 A project manager has determined that a certain subcontractor fails to deliver certain standard orders on time in about 25% of the orders given to him. If this situation is viewed as a Bernoulli process, determine from Appendix A, Table A-1 the probabilities that in ten orders the subcontractor

(a) will fail to deliver three or less orders on time.
(b) will fail to deliver between three and five (inclusive) orders on time.
(c) will deliver three or more orders on time.
(d) will deliver at most eight orders on time.
(e) will fail to deliver exactly two orders on time.
(f) will fail to deliver seven or more orders on time.

Solution: Let $p = 0.25$ stand for the probability that an order will not be delivered on time. Then $q = 0.75$ and $n = 10$. X represents the number of orders not delivered on time. Note that a failure to deliver an order on time is considered a "success" in this problem despite the undesirability of this outcome.

(a) $P(X \leq 3) = F(3) = \sum_{x=0}^{3} \binom{10}{3} (.75)^{10-x} (.25)^x$. From Appendix A, Table A-1, $F(3) = 0.7759$

(b) The probability of obtaining three, four, or five successes is given by the difference between "five or less successes" and "two or less successes." Thus,

$$P(3 \leq X \leq 5) = F(5) - F(2) = 0.9803 - 0.5256 = 0.4547$$

(c) The event "three or more failures" is the same as the event "seven or less successes." Hence,

$$P(X \leq 7) = F(7) = 0.9996$$

(d) The event "at most eight failures" is the same as "eight or less failures" or "two or more successes."

$$P(2 \leq X \leq 10) = F(10) - F(1) = 1.0000 - 0.2440 = 0.7560$$

(e) The probability of "exactly two successes" is given by the difference between the probabilities of "two or less successes" and "one or less successes."

$$P(X = 2) = F(2) - F(1) = 0.5256 - 0.2440 = 0.2816$$

(f) "Seven or more successes" is the complement of the event "six or less successes." Therefore,

$$P(X \geq 7) = 1 - P(X \leq 6) = 1 - F(6) = 1 - 0.9965 = 0.0035$$

The binomial distribution is often very useful in decision-making situations in business. One area in which it has been very widely applied is in quality control. In acceptance sampling plans, inspection is carried out on the articles drawn in a sample, and lots or shipments are either accepted or rejected on the basis of the sample evidence. Such plans are widely used in industry for incoming materials, at various stages of manufacturing processes, for outgoing final product, and for inspection by purchasers of material shipped by suppliers. The following example illustrates the use of the binomial distribution in an acceptance sampling plan.

Example 2-8 Suppose a firm requires that shipments of certain electromechanical components which it purchases should contain 10% or fewer defective items. It decides to use the following sampling plan as the basis for accepting or rejecting incoming shipments. A sample of ten components will be drawn at random from the shipment and tested. If one or less components are defective, the shipment will be accepted. If more than one component is defective, the shipment will be rejected. Incoming shipments are sufficiently large to assume that the binomial distribution is applicable. What are the statistical implications of this sampling plan?

Solution: The probability that an incoming shipment will be accepted is the probability that the sample of ten components drawn from it will contain one or less defectives. If p is the proportion of defective components in the shipment, the probability of observing one or less defectives in a sample of ten is

$$P_a = \binom{10}{0}(1-p)^{10}p^0 + \binom{10}{1}(1-p)^9 p$$
$$= (1-p)^{10} + 10(1-p)^9 p$$

The symbol P_a is often used in quality control work to denote the probability of acceptance. In keeping with conventional practice, the characteristics of this sampling plan may be summarized as ($c = 1, n = 10$), where c is referred to as the "acceptance number," that is, if c or fewer defectives are observed in the sample of n items, the lot from which the sample was drawn is accepted. If more than c defectives are observed in the sample, the lot is rejected.

A graph of the acceptance probabilities for all possible lot (or shipment) proportion defectives, usually referred to as an operating characteristic curve, is shown in Figure 2-5. On the X axis are shown the possible proportion defectives contained in incoming lots. On the vertical axis are shown the probabilities that these incoming lots will be accepted when the given sampling plan is applied. These probabilities may be determined from Table A-1 of Appendix A for p values

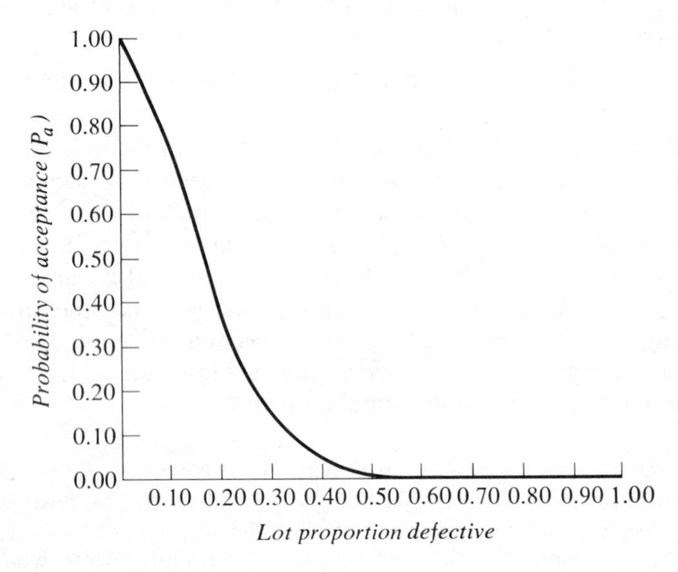

Figure 2-5 Graph of operating characteristic curve for acceptance sampling plan ($c = 1, n = 10$).

at intervals of 0.05. Probabilities for other p values may be calculated by use of logarithms. Four-place logarithms are given in Appendix A, Table A-4.

Some very interesting aspects of this sampling plan can be discerned. The purchaser wishes to accept shipments that contain 10% or fewer defective components. The sampling plan calls for 10% or fewer defectives in the sample in order for the shipment from which the sample was drawn to be accepted. Thus, without further analysis, the sampling plan might appear to be a very reasonable one. However, let us view the problem both from the standpoint of the producer of the components (the seller) and the consumer (the purchaser).

If the producer sends out a shipment which is just in conformity with requirements, that is, it contains exactly 10% defective items, the probability of acceptance is

$$P_a = (.9)^{10} + 10\,(.9)^9\,(.1)^1 = 0.74$$

Viewing this probability from the long run standpoint, if the producer sends out a stream of shipments each containing 10% defective components, 74% of these shipments will yield one or less defectives and will, therefore, be accepted. Thus, 26% of these lots will be rejected. This seems quite unfair to the producer. That is, just because of chance variations in the sampling process, these shipments containing 10% defectives, thus meeting requirements, are erroneously rejected over one-quarter of the time. The probability of a lot or shipment being rejected which is in conformity with requirements or specifications is referred to as the "producer's risk." The "producer's risk" for shipments containing less than 10% defectives is obviously less than 0.26, as can be seen from the graph of the operating characteristic curve.

It might be thought that since this sampling plan is unfair to the producer, it therefore operates to the benefit of the consumer. However, it is unfair to the consumer as well. For example, assume an incoming lot contains 15% defective components, and thus does not meet requirements. The probability that such a shipment will be accepted is 0.54 (Appendix A, Table A-1, $n = 10$, $p = 0.15$, $c = 1$). In other words, shipments that are 15% defective, and thus should be rejected, are accepted more than half the time. The probability of a shipment being accepted that does not meet requirements is referred to as the "consumer's risk." In this problem, the consumer's risk is very high for shipments which contain more than 10% defectives, but for which the percentage defective is nevertheless quite close to 10%. For example, the probability of acceptance of a shipment which has 12% defective components is 0.66.

Producer's risk and consumer's risk are examples of so-called Type I and Type II errors, respectively, which are discussed in detail in Chapter 7 under the subject of hypothesis testing. A Type I error involves the erroneous rejection of a true hypothesis, and a Type II error is the erroneous acceptance of a false hypothesis. As can be seen in the foregoing illustration, in an acceptance sampling context, a Type I error, the incorrect rejection of a true hypothesis takes the form of a producer's risk, or the incorrect rejection of a "good" lot, that is, one that meets specifications. Similarly, a Type II error, the incorrect acceptance of a false hypothesis takes the form of a consumer's risk, or the incorrect acceptance of a "bad" lot, that is, one that does not meet specifications.

In summary, this sampling plan does not appear to provide a satisfactory decision making procedure. The plan does not discriminate well between good and bad shipments. This stems primarily from the small sample size employed in the plan and the lack of symmetry in the binomial distribution when p differs from 0.50. Small samples from moderately defective lots ($p < .50$) tend to understate the proportion defective in these lots. For example, it was seen in the above illustration where $n = 10$ that shipments which contained 10% defective items produced samples with 10% or less defectives (one or fewer defectives) 74% of the time. Thus, such shipments produce samples which are as good as or better than the shipments about three-quarters of the time and samples which are worse only about one-quarter of the time. An opposite tendency is present when lots are severely defective ($p > 0.50$).

The above considerations highlight an important property of the binomial distribution. When p is equal to 0.50, the distribution is symmetrical. For example, see Figure 2-1 where $p = 0.50$ and $n = 2$. When p differs from 0.50, the distribution is asymmetrical ("skewed"). This property is illustrated in Figures 2-6 and 2-7 where the binomial distributions for $p = 0.1$, $n = 10$ and $p = 0.9$, $n = 10$, respectively, are plotted.

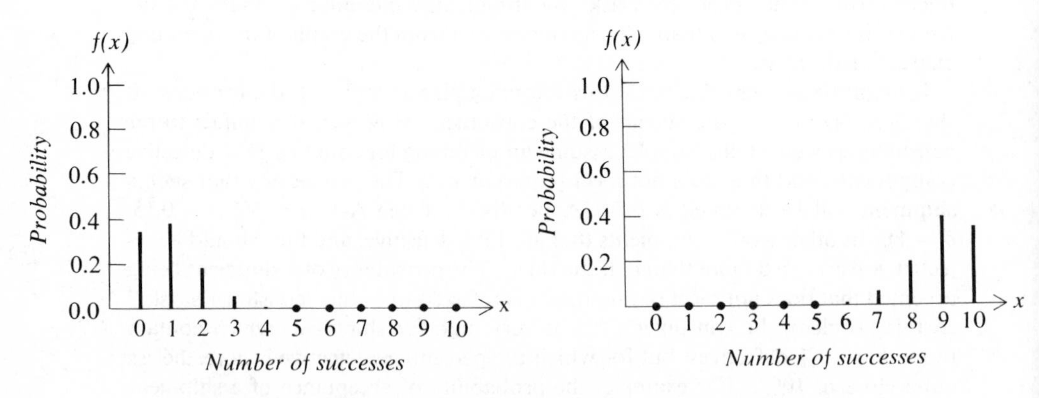

Figure 2-6 Graph of binomial distribution for $p = 0.1$, $n = 10$.

Figure 2-7 Graph of binomial distribution for $p = 0.9$, $n = 10$.

2.5 The Multinomial Distribution

In the case of the binomial distribution, there were two possible outcomes on each experimental trial. The multinomial distribution represents a straightforward generalization of the binomial for the situation where there are more than two possible outcomes on each trial.

The assumptions underlying the multinomial distribution may be stated in a completely analogous manner to those of the binomial distribution.

(1) There are k mutually exclusive possible outcomes on each trial, which may be referred to as $E_1, E_2, \ldots, E_k$. Therefore, the sample space of possible outcomes on each trial is $S = \{E_1, E_2, \ldots, E_k\}$.

(2) The probabilities of outcomes $E_1, E_2, \ldots, E_k$, denoted $p_1, p_2, \ldots, p_k$, respectively, remain constant from trial to trial.

(3) The trials are independent.

Under these assumptions, the probability that there will be x_1 occurrences of E_1, x_2 occurrences of $E_2, \ldots, x_k$ occurrences of E_k in n trials is given by

(2.4) $$f(x_1, x_2, \ldots, x_k) = \frac{n!}{x_1!\, x_2! \cdots x_k!} p_1^{x_1} p_2^{x_2} \ldots p_k^{x_k}$$

where $x_1 + x_2 + \cdots + x_k = n$ and $p_1 + p_2 + \cdots + p_k = 1$.

The expression $f(x_1\ x_2, \ldots, x_n)$ is the general term of the multinomial distribution

$$(p_1 + p_2 + \cdots + p_k)^n$$

Analogously, in the binomial distribution, the probability of x successes in n trials is given by

$$f(x) = \frac{n!}{x!\,(n-x)!} q^{n-x} p^x$$

which is the general term of $(q + p)^n$, or in terminology similar to that of the multinomial distribution, $(q + p)^n$ may be written

$$(p_1 + p_2)^n$$

where $p_1 + p_2 = 1$.

Example 2-9 If eight dice are rolled, what is the probability that there will be three ones, two twos, zero threes, two fours, zero fives, and one six?

Solution: Applying the multinomial distribution, this probability is given by

$$f(3, 2, 0, 2, 0, 1) = \frac{8!}{3!\, 2!\, 0!\, 2!\, 0!\, 1!} \left(\frac{1}{6}\right)^3 \left(\frac{1}{6}\right)^2 \left(\frac{1}{6}\right)^0 \left(\frac{1}{6}\right)^2 \left(\frac{1}{6}\right)^0 \left(\frac{1}{6}\right)^1$$

Example 2-10 The Taylor Distributing Company has classified its accounts receivable into three categories, A, B, C. There were 40% type A, 40% type B, and 20% type C accounts. A sample of three accounts was drawn at random from the entire list of accounts receivable. Give the probability distribution of the numbers of each type of account in the sample, assuming the multinomial distribution is applicable.

Solution: If x_1, x_2, and x_3 stand for the numbers of type A, B, and C accounts, then the appropriate multinomial distribution is

$$f(x_1, x_2, x_3) = \frac{3!}{x_1!\, x_2!\, x_3!} (0.40)^{x_1} (0.40)^{x_2} (0.20)^{x_3}$$

$$x_1 + x_2 + x_3 = 3 \quad \text{and} \quad x_i = 0, 1, 2, 3$$

This probability distribution is given in Table 2-9.

Table 2-9 Multinomial Distribution for Numbers of Type A, B, and C Accounts Receivable; $p_1 = 0.40$, $p_2 = 0.40$, $p_3 = 0.20$, and $n = 3$.

(x_1, x_2, x_3)	$f(x_1, x_2, x_3)$
3, 0, 0	0.064
0, 3, 0	0.064
0, 0, 3	0.008
2, 1, 0	0.192
2, 0, 1	0.096
1, 2, 0	0.192
1, 1, 1	0.192
1, 0, 2	0.048
0, 2, 1	0.096
0, 1, 2	0.048
	1.000

2.6 The Hypergeometric Distribution

In Section 2.4, the binomial distribution was discussed as the appropriate probability distribution for situations in which the underlying assumptions of a Bernoulli process were met. A major application of the binomial distribution was seen to be the computation of probabilities for the *sampling of finite universes with replacement*. In most practical situations, sampling is carried out *without replacement*. For example, if a sample of families is selected in a city in order to estimate the average income of all families in the city, sampling units are ordinarily not replaced prior to the selection of subsequent ones. That is, families are not replaced in the original population, and thus given an opportunity to appear more than once in the sample. In fact, such samples are usually drawn in a single operation, without there being any possibility of drawing the same family twice. Also, in a sample drawn from a production process, articles are generally not replaced and given an opportunity to reappear in the sample. Thus, both in the case of human populations or universes of physical objects, sampling is ordinarily carried out without replacement. In this section, we discuss the hypergeometric distribution as the appropriate model for the non-replacement sampling situation.

The following example illustrates the type of situation for which the hypergeometric distribution is appropriate. Assume a list of 1000 persons 950 of whom are adults, 50 of whom are children. Numbers from one to 1000 are assigned to these individuals. These numbers are printed on 1000 identical discs which are placed in a large bowl. A sample of five chips is drawn at random from the bowl *without replacement*. These five chips may be drawn simultaneously or successively. For ease of reference, we shall refer to this situation as the drawing of a random sample of five persons from the group of 1000, although, of course, the chips rather than persons were in fact sampled. The process of numbering chips to correspond to persons and the sampling of chips is simply a device to insure randomness in the process of sampling the population of 1000 persons.

What is the probability that none of the five persons in the sample is a child? An alternative way of wording the question is, "What is the probability that all five persons in the sample are adults?" The sample space of possible outcomes in this experiment is the total number of samples of five persons that can be drawn from the population of 1000 persons. This is the number of combinations that can be formed of 1000 objects taken five at a time, $\binom{1000}{5}$. We can compute the required probability by obtaining the ratio of the number of sample points favorable to the event "none of the five persons is a child" to the total number of points in the sample space. The number of ways five adults can be drawn from the 950 adults is $\binom{950}{5}$. The number of ways zero children can be selected from the 50 children is $\binom{50}{0}$. Therefore, the total number of ways of selecting five adults and 0 children from the population of 950 adults and 50 children is $\binom{950}{5}\binom{50}{0}$.

Hence, the probability of obtaining zero children (and five adults) in this sample of five persons is

$$\frac{\binom{950}{5}\binom{50}{0}}{\binom{1000}{5}}$$

Carrying out the arithmetic reveals a very interesting fact.

$$\frac{\binom{950}{5}\binom{50}{0}}{\binom{1000}{5}} = \frac{\dfrac{950!}{5!\,945!} \times \dfrac{50!}{0!\,50!}}{\dfrac{1000!}{5!\,995!}} = \frac{950 \times 949 \times 948 \times 947 \times 946}{1000 \times 999 \times 998 \times 997 \times 998} = 0.7734$$

Grouping the product obtained as a multiplication of five factors, we have

$$P(\text{zero children}) = \left(\frac{950}{1000}\right)\left(\frac{949}{999}\right)\left(\frac{948}{998}\right)\left(\frac{947}{997}\right)\left(\frac{946}{996}\right)$$

which is the result we would have arrived at if we had simply solved the original problem in terms of conditional probabilities. That is, the probability of obtaining an adult on the first draw is 950/1000; the probability of obtaining an adult on the second draw given that an adult was obtained on the first draw is 949/999, and so forth. Therefore, the joint probability of obtaining zero children (five adults) if the sampling is carried out without replacement is given by the multiplication of the five factors shown.

We can now state the general nature of this type of problem and the hypergeometric distribution as a solution to it. Suppose there is a population which contains N elements, X of which are of one type, termed "successes," $N - X$ of another type, denoted "failures." The corresponding terminology for a random sample of n elements drawn without replacement is that we require x successes and $n - x$ failures. The data of this general problem are tabulated below:

Population	*Required Sample*
X = number of successes[6]	x = number of successes
$N - X$ = number of failures	$n - x$ = number of failures
N = total number in population	n = total number in sample

Then, the *hypergeometric distribution* which gives the probability of x successes in a random sample of n elements drawn *without replacement* is

$$\textbf{(2.5)} \qquad f(x) = \frac{\binom{N-X}{n-x}\binom{X}{x}}{\binom{N}{n}} \quad \text{for } x = 0, 1, 2, \ldots, [n, X]$$

The symbol $[n, X]$ means the smaller of n or X. For example, if there had been only ten children in the population, (X), in the illustration given above and the sample size had been 50, (n), the largest value that the number of children in the sample, (x), could take on would be ten, (X). On the other hand, if X exceeded n, clearly x could be as large as n.

The hypergeometric distribution bears a very interesting relationship to the binomial distribution. Suppose, in the case of the population containing 950 adults and 50 children, we had been interested in the same probability of

[6]Note that in order to maintain parallel notation for the population and sample in this case, the symbol X does *not* denote the *random variable* for number of successes in the sample, but is instead the total number of successes in the population which is sampled.

obtaining zero children in a random sample of five persons if the sample were randomly drawn *with replacement*. Then, letting $q = 0.95$, $p = 0.05$ and $n = 5$ in the binomial distribution, we have

$$f(0) = \binom{5}{0}(0.95)^5(0.05)^0 = (0.95)^5 = 0.7738$$

Just as in the case of the hypergeometric distribution, where the required probability could have been computed by using the multiplication rule for *dependent* events, here in the case of the binomial distribution, the probability could have been computed by simply using the multiplication rule for *independent* events, thus,

$$P(\text{zero children}) = \left(\frac{950}{1000}\right)\left(\frac{950}{1000}\right)\left(\frac{950}{1000}\right)\left(\frac{950}{1000}\right)\left(\frac{950}{1000}\right)$$

It may be noted that the hypergeometric and binomial probability values are extremely close in this illustration, agreeing exactly in the first three decimal places. It can be shown that when N increases without limit, the hypergeometric distribution approaches the binomial distribution. Therefore, the binomial probabilities may be used as approximations to hypergeometric probabilities when n/N is small. A frequently used rule of thumb is that the population size should be at least ten times the sample size ($N > 10n$) for the approximations to be used. However, the governing considerations, as usual, include such matters as the purpose of the calculations, whether a sum of terms rather than a single term is being approximated, and whether terms near the center or tails of the distribution are involved.

Just as the multinomial distribution represents the generalization of the binomial distribution when there are more than two possible classifications of outcomes, the hypergeometric distribution can be similarly extended. No special name is given to the more general distribution; it also is referred to as the "hypergeometric distribution." Assume a population which contains N elements, X_1 of type one, X_2 of type two, . . . , X_k of type k, and that we require in a sample of n elements drawn without replacement that there be x_1 elements of type one, x_2 of type two, . . . , x_k of type k. Tabulating the data in an analogous fashion to the two outcome case, we have

Population (Number of Elements)	*Required Sample (Number of Elements)*
X_1 of type one	x_1 of type one
X_2 of type two	x_2 of type two
.	.
.	.
.	.
X_k of type k	x_k of type k

The *hypergeometric distribution*, which gives the probability of obtaining x_1 occurrences of type one, x_2 occurrences of type two, . . . , x_k occurrences of type k in a random sample of n elements drawn without replacement, is

$$\textbf{(2.6)} \quad f(x_1, x_2, \ldots, x_k) = \frac{\binom{X_1}{x_1}\binom{X_2}{x_2}\cdots\binom{X_k}{x_k}}{\binom{N}{n}} \quad \text{for } x_i = 0, 1, 2, \ldots, [n, X_i]$$

where

$$\sum_{i=1}^{k} X_i = N \quad \text{and} \quad \sum_{i=1}^{k} x_i = n$$

Example 2-11 The Humblest Oil Corporation has 100 service stations in a certain community which it has classified according to merit of geographic location as follows:

Merit of Location	*Number of Stations*
Excellent	22
Good	38
Fair	27
Poor	10
Disastrous	3
	100

The corporation has a computer program for drawing random samples (without replacement) of its service stations. In a random sample of 20 of these stations, what is the joint probability of obtaining six excellent, six good, four fair, three poor, and one disastrous stations according to the above location classification?

Solution:

$$f(6, 6, 4, 3, 1) = \frac{\binom{22}{6}\binom{38}{6}\binom{27}{4}\binom{10}{3}\binom{3}{1}}{\binom{100}{20}}$$

Example 2-12 What is the probability that in a hand of cards dealt from a well shuffled deck you will have either all spades or all clubs?

Solution: There are $\binom{52}{13}$ possible bridge hands. Using the hypergeometric distribution and the addition rule, the required probability is

$$\frac{\binom{13}{13}\binom{13}{0}\binom{13}{0}\binom{13}{0}}{\binom{52}{13}} + \frac{\binom{13}{0}\binom{13}{13}\binom{13}{0}\binom{13}{0}}{\binom{52}{13}} = \frac{1}{\binom{52}{13}} + \frac{1}{\binom{52}{13}} \approx \frac{2}{635 \text{ billion}}$$

If you are dealt such a hand, it is fair to say you have observed a rare event.

2.7 The Poisson Distribution

Another very useful probability function is the Poisson distribution, named for a Frenchman who developed it during the first half of the nineteenth century.[7] The distribution can be used in its own right and also for approximation of binomial probabilities. The former use is by far the more important one and in this context has had many fruitful applications in a wide variety of fields. We will first discuss the Poisson as a distribution in its own right and then as an approximation to the binomial distribution.

The Poisson as a Distribution in Its Own Right

The Poisson distribution has been usefully employed to describe the probability functions of phenomena such as product demand; demands for service; numbers of telephone calls that come through a switchboard; numbers of accidents; numbers of traffic arrivals such as trucks at terminals, airplanes at airports, ships at docks, and passenger cars at toll stations; and numbers of defects observed in various types of lengths, surfaces, or objects.

All of these illustrations have certain elements in common. The given occurrences can be described in terms of a discrete random variable, which takes on values 0, 1, 2, and so forth. For example, product demand can be characterized by 0, 1, 2, etc., units purchased in a specified time period; number of defects can be counted as 0, 1, 2, etc., in a specified length of electrical cable. The product demand example may be viewed in terms of a process which produces random occurrences in continuous time. The defects example pertains to random occurrences in a continuum of space. In cases such as the one in which numbers of defects are counted, the continuum may not only

---

[7]Siméon Denis Poisson (1781–1840), was particularly noted for his applications of mathematics to the fields of electrostatics and magnetism. He wrote treatises in probability, calculus of variations, Fourier's series, and other areas.

be one of length, but area or volume as well. Thus, there may be a count of the number of blemishes in areas of sheet-metal used for aircraft or the number of a certain type of microscopic particle in a unit of volume such as a cubic centimeter of a solution. In all of these cases, there is some rate in terms of number of occurrences per interval of time or space which characterizes the process producing the outcome.

Using as an example the occurrences of defects in a length of electrical cable, we can indicate the general nature of the process which produces a Poisson probability distribution. This process is described more formally in the sub-section entitled "The Poisson Process." The length of cable has some rate of defects per interval, say, two defects per meter. If the entire length of cable is subdivided into very small subintervals, say, of one millimeter each: (1) the probability that exactly one defect occurs in this subinterval is a very small number and is constant for each such subinterval, (2) the probability of two or more defects in a millimeter is so small it may be considered to be zero, (3) the number of defects that occurs in a millimeter does not depend on where that subinterval is located, and (4) the number of defects that occurs in a subinterval does not depend on the number of events in any other non-overlapping subinterval.

In the foregoing example, the subinterval was one of length. Analogous sets of conditions would characterize examples in which the subinterval is a unit of area, volume, or time.

The Nature of the Poisson Distribution

As indicated in the previous discussion, the Poisson distribution is concerned with occurrences that can be described by a discrete random variable. This random variable, denoted X, can take on values $x = 0, 1, 2, \ldots$, where the three dots mean "*ad infinitum.*" That is, the domain of the Poisson probability function consists of all non-negative integers. The probability of exactly x occurrences in the Poisson distribution is:

(2.7) $$f(x) = \frac{\mu^x e^{-\mu}}{x!} \quad \text{for } x = 0, 1, 2, \ldots$$

where μ is the mean number of occurrences per interval and $e = 2.71828\ldots$ (the base of the Naperian or natural logarithm system).

As can be seen from (2.7), the Poisson distribution has a single parameter symbolized by the Greek letter μ ("mu"). If we know the value of μ, we can write out the entire probability distribution. The parameter μ can be interpreted as the average number of occurrences per interval of time or space which characterizes the process producing the Poisson distribution. The average referred to here is the arithmetic mean, which is discussed extensively in Chapters 3 and 4. For our present discussion, it suffices to note that an arithmetic mean is the familiar average obtained by totalling the values of a set of observations and dividing by the number of observations. For example, the arithmetic mean of 3, 8, and 4 is $(3 + 8 + 4)/3 = 5$. Hence, in the case of the Poisson distribution, μ can be interpreted as the arithmetic mean of occurrences per interval or in other words, as an average rate of occurrences per interval. Thus, μ may represent an average of three units of demand per day, 5.3 demands for service per hour, 1.2 aircraft arrivals per five minutes, 1.5 defects per ten feet of electrical cable, and so forth.

In order to illustrate the way in which probabilities are calculated in the Poisson distribution, we consider the following example. A study revealed that the number of telephone calls per minute that come through a certain switchboard between 10:00 A.M. and 11:00 A.M. on business days is distributed according to the Poisson probability function with an average μ of 0.4 calls per minute. What is the probability distribution of the number of telephone calls per minute during the specified time period?

Let X represent the random variable "number of telephone calls per minute" during the given time period. Then, $\mu = 0.4$ calls per minute is the parameter of the Poisson probability distribution of this random variable. The probability that exactly zero calls will occur (come through the switchboard), per minute is given by substituting $x = 0$ in the Poisson probability function Equation (2.7). Hence,

$$P(X = 0) = f(0) = \frac{(0.4)^0 e^{-0.4}}{0!} \tag{2.8}$$

Since $(0.4)^0 = 1$ and $0! = 1$, expression (2.8) becomes simply

$$f(0) = e^{-0.4} = 0.670 \tag{2.9}$$

The value 0.670 for $f(0)$ can be found in Appendix A, Table A-9, where exponential functions of the form e^x and e^{-x} are tabulated for values of x from 0.00 to 6.00 at intervals of 0.10.

Continuing with the calculation of the Poisson probability distribution, we find the probability of exactly one call per minute by substituting $x = 1$ in (2.7). Hence, $f(1)$ is given by

$$P(X = 1) = f(1) = \frac{(0.4)^1 e^{-0.4}}{1!} = (.4)(.670) = 0.268 \tag{2.10}$$

In Table A-3, Appendix A, are listed values of the cumulative distribution function for the Poisson distribution. That is, values of $F(c) = \sum_{x=0}^{c} f(x)$, or the probabilities of c or fewer occurrences, are provided for selected values of the parameter μ. As in Table A-1 for the binomial cumulative distribution, probabilities such as $1 - F(c)$ or $a \leq f(x) \leq b$ can be obtained by appropriate manipulation of the tabulated values.

The use of Table A-3 will be illustrated in terms of the above example. To obtain the probability of zero calls per minute, using $c = 0$ and $\mu = 0.4$ in Table A-3, we find the value of $F(0) = 0.670$. Of course, this is also the value of $f(0)$, since the probability of zero or fewer occurrences equals the probability of zero occurrences. Therefore, as before, $f(0) = 0.670$.

We find the probability of exactly one telephone call per minute by subtracting the probability of zero calls from the probability of one or fewer calls, that is,

$$f(1) = F(1) - F(0) = 0.938 - 0.670 = 0.268$$

Similarly, using Table A-3, we find the values of $f(2)$, $f(3)$, and $f(4)$ to be

$$\begin{aligned} f(2) &= F(2) - F(1) = 0.992 - 0.938 = 0.054 \\ f(3) &= F(3) - F(2) = 0.999 - 0.992 = 0.007 \\ f(4) &= F(4) - F(3) = 1.000 - 0.999 = 0.001 \end{aligned}$$

Although, as indicated earlier, the random variable X in the Poisson distribution takes on the values 0, 1, 2, . . . , ad infinitum, $F(4) = 1.00$ in this problem. This means that the probabilities of 5, 6, . . . , occurrences are so small that they would appear as zeros in the first three decimal places.

The required probability distribution for this problem is given in Table 2-10. Several other illustrations of the use of the Poisson distribution are given in Examples 2-13, 2-14, and 2-15.

Table 2-10 Poisson Probability Distribution of the Number of Telephone Calls That Come Through a Certain Switchboard Between 10:00 A.M. and 11:00 A. M. on Business Days.

Number of Calls x	*Probability* $f(x)$
0	0.670
1	0.268
2	0.054
3	0.007
4	0.001
	1.000

Example 2-13 Airplanes have been observed to arrive on weekdays at a certain small airport at an average rate of three for the one-hour period 1:00 P.M. to 2:00 P.M. If these arrivals are distributed according to the Poisson probability distribution, what are the probabilities that

(a) exactly zero airplanes will arrive between 1:00 P.M. and 2:00 P.M. next Monday?
(b) either one or two airplanes will arrive between 1:00 P.M. and 2:00 P.M. next Monday?
(c) a total of exactly two airplanes will arrive between 1:00 and 2:00 P.M. during the next three weekdays?

Solution: In this problem, we may use the parameter $\mu = 3$ arrivals per day for the time period 1:00 P.M.–2:00 P.M. Let X represent a random variable denoting the number of arrivals during the specified time period. The mathematical solutions are given for (a), (b), and (c) to illustrate the theory involved. However, the answers may be determined by looking up values in Table A-3 of Appendix A as indicated.

(a) The random variable X follows the Poisson distribution with the parameter $\mu = 3$. Thus,

$$P(X = 0) = f(0) = \frac{3^0 e^{-3}}{0!} = 0.050$$

This value may be obtained from Appendix A, Table A-3, for $\mu = 3$, $c = 0$. We note that $f(0) = F(0)$.

(b) Since exactly one arrival and exactly two arrivals are mutually exclusive events, we have, by the addition rule,

$$f(1) + f(2) = \frac{3^1 e^{-3}}{1!} + \frac{3^2 e^{-3}}{2!} = 0.373$$

This value can be obtained from Appendix A, Table A-3, for $\mu = 3$. The required probability is $F(2) - F(0) = 0.423 - 0.050 = 0.373$.

(c) A total of exactly two arrivals in three weekdays during the time period 1:00 P.M.–2:00 P.M. can be obtained by having two arrivals on the first day, none on the second day and none on the third day during the specified one-hour period and so forth. The total number of ways in which the event in question can occur is shown in Table 2-11.

Table 2-11 Possible Ways of Obtaining a Total of Exactly Two Arrivals in Three Weekdays.

	Number of Arrivals	
Day 1	*Day 2*	*Day 3*
2	0	0
0	2	0
0	0	2
1	1	0
1	0	1
0	1	1

Let P_2 represent the required probability. Using the multiplication and addition rules, and again using the parameter μ = three arrivals per day during the period 1:00 P.M.–2:00 P.M., we have

$$\begin{aligned} P_2 &= 3\,[f(2)][f(0)]^2 + 3\,[f(1)]^2[f(0)] \\ &= 3\left(\frac{3^2e^{-3}}{2!}\right)\left(\frac{3^0e^{-3}}{0!}\right)^2 + 3\left(\frac{3^1e^{-3}}{1!}\right)^2\left(\frac{3^0e^{-3}}{0!}\right) \\ &= \frac{81}{2}e^{-9} = 0.005 \end{aligned}$$

The solution is greatly simplified if we change the time interval for which the parameter μ is stated. This has the effect of changing the random variable in the problem. Thus, if μ = three arrivals *per day* during the time period 1:00 P.M.–2:00 P.M., then μ = nine arrivals *per three days* during the same time period. The probability of exactly two arrivals in three weekdays during the given one-hour period can then be obtained by computing $P(X = 2)$, where X is a Poisson distributed random variable denoting the number of arrivals *per three days*. The required probability is, therefore, obtained by simply computing $f(2)$ in a Poisson distribution with the parameter $\mu = 9$.

$$P_2 = f(2) = \frac{9^2e^{-9}}{2!} = \frac{81}{2}e^{-9} = 0.005$$

This value can be obtained from Appendix A, Table A-3, for $\mu = 9$. The probability is given by $F(2) - F(1) = 0.006 - 0.001 = 0.005$.

This problem illustrates the point that considerable simplification of computations for Poisson processes can often be accomplished by convenient choices of parameters.

It is worthwhile to note a few points concerning the appropriateness of the Poisson distribution in the preceding example. It was stated at the beginning of the problem that the airplane arrivals were distributed according to the Poisson distribution. Whether it is appropriate to consider the past arrival distribution as a Poisson distribution during the specified time periods depends on the nature of the past data. Actual relative arrival frequencies can be tabulated and compared with the theoretical probabilities given by a Poisson distribution. Tests of "goodness of fit" for judging the closeness of actual and theoretical frequencies are discussed in Chapter 9.

In practical work, the question often arises whether a given mathematical model is likely to be applicable in a certain situation. This requires careful examination of whether the underlying assumptions of the model are apt to be fulfilled by the real world phenomena. For example, in this problem, suppose certain cargo deliveries are made either on Mondays or Tuesdays between 1:00 P.M. and 2:00 P.M. Assuming that if a delivery is made on Monday, it will not be made on Tuesday, the independence assumption (4) of a Poisson process is clearly violated. That is, the number of arrivals during the one-hour time period on Tuesday *depends* on the number of arrivals during the corresponding time period on Monday, and vice versa. Furthermore, if the nature of the aircraft arrivals is such that Monday and Tuesday always have more arrivals between 1:00 P.M. and 2:00 P.M. than do other weekdays, then assumption (3) is violated. That is, if we were to count arrivals for the one-hour period for a given day (or two days and so forth), then the number of occurrences obtained clearly would depend upon the day on which the count was begun. It is indeed a rare event when the assumptions of a probability distribution are perfectly met by a real world process. Experience in a given field aids considerably in making judgments as to whether so great a departure from assumptions has occurred that a model may no longer be applicable.[8] In the final analysis, actual comparison of the data generated by a process with the probabilities of the theoretical distribution is the best way of determining the appropriateness of the distribution. Of course, even if a given mathematical model (or other type of model) has provided a good description of past data, there is no guarantee that this state of affairs will continue into the future. The analyst must be alert to changes in the environment which would make the model inapplicable.

Example 2-14 A department store has determined in connection with its inventory control system that the demand for a certain housefurnishing item was Poisson distributed with the parameter $\mu = 4$ per day.

(a) Determine the probability distribution of the daily demand for this item.

(b) If the store stocks five of these items on a particular day, what is the probability that the demand will be greater than the supply?

Solution: Let X represent the random variable number of items sold per day.

(a) The probability distribution of X is

x	$f(x)$
0	0.018
1	0.074
2	0.146
3	0.195
4	0.196
5	0.156
6	0.104
7	0.060
8	0.030
9	0.013
10	0.005
11	0.002
.	.
.	.
.	.

[8] An appropriate thought here is perhaps contained in the anonymous bit of advice, "Good judgment comes from experience, and experience comes from poor judgment."

The sum of the probabilities for demand from zero through eleven units is 0.999. Therefore, the sum of the probabilities for twelve or more units is only 0.001.

The probabilities can be obtained from Appendix A, Table A-3, using the relationship $f(x) = F(x) - F(x - 1)$.

(b) The probability that demand will be greater than five units is the complement of the probability that it will be five units or less. Thus, from Appendix A, Table A-3, we have

$$P(x > 5) = 1 - F(5) = 1 - 0.785 = 0.215$$

Example 2-15 The number of entries made in each of six accounts receivable is distributed according to the Poisson probability distribution with parameter $\mu = 1$ per day. Entries in the accounts may be assumed to be independent. What is the probability that on a specified day

(a) none of the six accounts will receive any entries?
(b) each of the six accounts will receive at least one entry?
(c) exactly three accounts will receive no entries?

Solution: This problem illustrates a situation in which two different probability distributions may have to be used to provide a solution. In this case, the Poisson and binomial distributions are applicable.

(a) Since the number of entries in a given account is Poisson distributed with an average of one entry per day ($\mu = 1$), the probability that a given account will receive zero entries on a specified day is

$$P \text{ (no entry in a given account)} = f(0) = \frac{1^0 e^{-1}}{0!} = e^{-1}$$

Entries in different accounts are independent events. Therefore, by the multiplication rule, we have

$$P \text{ (no entry in all six accounts)} = (e^{-1})^6 = e^{-6} = 0.002$$

(b) The event that a given account will receive at least one entry is the complement of the event that the account receives no entries. Therefore,

$$P \text{(a given account receives at least one entry)} = 1 - e^{-1}$$

By the multiplication rule for independent events,

$$P \text{(at least one entry in each of the six accounts)} = (1 - e^{-1})^6 = 0.064$$

(c) Let p equal the probability that a given account receives no entries on a specified day. From part (a), $p = e^{-1}$. We may consider p to be the probability of success in a Bernoulli trial. Thus, $q = 1 - e^{-1}$, and $n = 3$ in this problem.

$$P \text{(no entries in exactly three accounts)} = \binom{6}{3} (1 - e^{-1})^3 (e^{-1})^3 = 0.252$$

The Poisson as an Approximation to the Binomial Distribution

The foregoing discussion concerned the use of the Poisson probability function as a distribution in its own right. We turn now to a consideration of the Poisson distribution as an approximation to the binomial distribution.

The Bernoulli process was seen to give rise to a two parameter probability function—the binomial distribution. Since computations involving the binomial distribution become quite tedious when n is large, it is useful to have a simple method of approximation. The Poisson distribution is particularly suitable as an approximation when n is large and p is small.

Assume in the expression for $f(x)$ of the binomial distribution that n is permitted to increase without bound and p approaches zero in such a way that np remains constant. Let us denote this constant value for np as μ. Under these assumptions, it can be shown that the binomial expression for $f(x)$ approaches the following value:

$$f(x) = \frac{\mu^x e^{-\mu}}{x!}$$

where $\mu = np$, and e is the base of the natural logarithm system. As can be seen from expression (2.7), the value approached by the binomial distribution under the given conditions is the Poisson distribution. Hence, the Poisson distribution can be used as an approximation to the binomial probability function. In this context, the Poisson distribution, similar to the binomial distribution, gives the probability of observing x successes in n trials of an experiment, where p is the probability of success on a single trial. That is, x, n, and p are interpreted in the same way as in the binomial distribution.

Because of the assumptions underlying the derivation of the Poisson distribution from the binomial distribution, the approximations to binomial probabilities are best when n is large and p is small. A frequently used rule of thumb is that the approximation is appropriate when $p \leq 0.05$ and $n \geq 20$. However, the Poisson distribution sometimes provides surprisingly close approximations even in cases where n is not large nor p very small. As an illustration of how these approximations may be carried out, we return to the problem of sampling the population consisting of 950 adults and 50 children. The probability of observing zero children in a random sample of five persons drawn with replacement was previously computed from the binomial distribution. We now compute the same probability using the Poisson distribution. Since n is only 5 in this problem, this is not a prototype situation for the use of the Poisson distribution for approximating binomial probabilities. Rather, it is an example of the surprisingly small errors that are sometimes observed in certain cases, even though n is small, and it is used here simply to carry forward the arithmetic for a familiar illustration.

The binomial parameters in this problem were $p = 0.05$ and $n = 5$. Therefore,

$$\mu = np = 5 \times 0.05 = 0.25$$

Thus, in the Poisson distribution, the probability of zero successes (children) is

$$f(0) = \frac{(0.25)^0 e^{-0.25}}{0!} = e^{-0.25}$$

From Appendix A, Table A-3, using $c = 0$ and $\mu = 0.25$, we find the value of $F(0)$, which in this case is equal to $f(0)$ (since the probability of zero or fewer successes equals the probability of zero successes). Therefore,

$$f(0) = 0.779$$

This figure is the same in the first two decimal places as the corresponding number obtained from the binomial probability.

However, the percentage errors would be much larger for the other terms of the binomial representing probabilities of one, two, and so forth, successes. It is recommended, therefore, that the Poisson approximations not be used unless the conditions for n and p in the aforementioned rule of thumb are met.

The parameter μ can be interpreted as the average number of successes per sample of size n. This can be seen from the fact that $\mu = np$. Since p is the probability of success per trial and n is the number of trials, multiplication of n by p gives the average number of successes per n trials. For example, to interpret this in terms of the foregoing problem, we take a long run relative frequency viewpoint. The population contains $p = 0.05$ children. A random sample of $n =$ five persons was drawn with replacement from this population. Suppose samples of size $n = 5$ were repeatedly drawn with replacement from the same population and the number of children was recorded for each sample. It can be proven mathematically, and it seems intuitively reasonable, that the *average proportion* of children per sample of five persons is equal to $p = 0.05$. Also, it follows that the *average number* of children per sample of five persons is equal to $np = 5(.05) = 0.25$ children. The average referred to here is the arithmetic mean, obtained by totalling the proportions or numbers of children for all samples and dividing by the number of samples.

The following example represents a more justifiable use of the Poisson approximation to binomial probabilities than the above illustration which involved a small sample size.

Example 2-16 An automatic machine produces washers, three percent of which are defective according to a severe set of specifications. If a sample of 100 washers is drawn at random from the production of this machine, what are the probabilities of observing

(a) exactly three defectives?

(b) between two and four defectives, inclusive?

It may be assumed that the universe of product sampled was sufficiently large to warrant binomial probability calculations.

Solution: (a) With the parameters $p = 0.03$ and $n = 100$, the probability of exactly three defectives, using the binomial distribution, is

$$f(3) = \binom{100}{3}(.97)^{97}(.03)^3 = 0.227$$

An approximation to this probability is given by the Poisson distribution with parameter

$$\mu = np = 100 \times 0.03 = 3$$

The Poisson probability of exactly three defectives is

$$f(3) = \frac{3^3 e^{-3}}{3!} = 0.224$$

which may be determined from Appendix A, Table A-3, using the relationship $f(3) = F(3) - F(2)$. Thus, the approximation is in error by about 1%

(b) From a sufficiently extensive table of values of the binomial distribution, it can be determined that

$$P(2 \leq x \leq 4) = \sum_{x=2}^{4} \binom{100}{x}(.97)^{100-x}(.03)^x = 0.623$$

The corresponding probability according to the Poisson distribution with parameter $\mu = 3$ can be determined from Appendix A, Table A-3, as $P(2 \leq x \leq 4) = F(4) - F(1) = 0.815 - 0.199 = 0.616$. Again, the approximation error is about 1%.

The Poisson Process

In the preceding discussion, the uses of the Poisson distribution both as a distribution in its own right and as an approximation to the binomial probability function were explained. We return now to a fuller and more complete discussion of the "Poisson process" which produces random occurrences in a continuum of time or space.

A Poisson process is characterized by a specific set of assumptions. It is easiest to explain these assumptions in terms of a continuum of time. In Figure 2-8, continuous time is represented by a horizontal axis labelled t, with measurement commencing at $t = 0$. Isolated events in time are represented

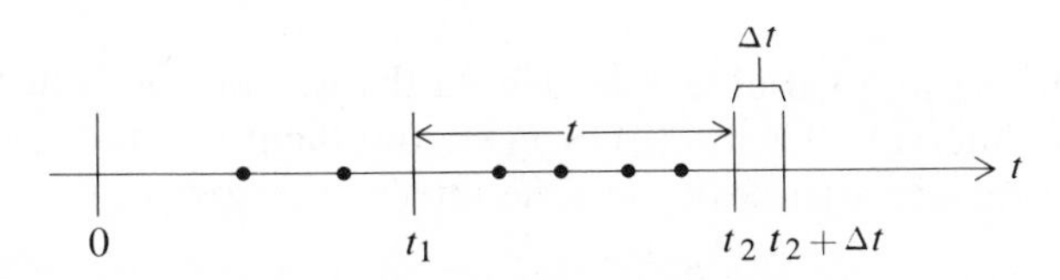

Figure 2-8 Graph of time scale and occurrences of events in a Poisson process.

by dots along this line. Two arbitrary time points, t_1 and t_2 and a theoretically infinitesimally small time interval, Δt, are shown. The time elapsed between t_1 and t_2 is denoted as t.

The following assumptions underlie the Poisson process:

(1) The probability that exactly one occurrence takes place in the interval Δt is given by $\lambda \Delta t$, where λ is some positive number.

(2) The probability of two or more events in the time interval Δt is so small that it may be considered equal to zero. Thus, the probability of exactly no occurrences in Δt is approximately $1 - \lambda \Delta t$.

(3) The probability that exactly n events will occur in the time interval t does not depend on the location of t_1.

(4) The probability that exactly n events will occur in the time interval t is independent of the number of events which occurred prior to t_1 and does not depend on the number of events which occurred in any subinterval from 0 to t_1.

Some comment on these assumptions is in order. In assumption (1), the parameter λ is often referred to as the "intensity" of the process. From assumptions (3) and (4) it is apparent that the probability of n events in a time interval t is the same, regardless of where the time interval is located. This is the same as saying the average number of events per unit time is constant. Hence, if we consider a time interval t which is divided into small units of time, and if there are λ events per unit time, the average number of events in time interval t is λt.

Assumption (3) is sometimes referred to as the "stationarity assumption." This means that the probability of a specified number of events in a given time interval does not depend on the location of that interval on the time scale. Assumption (4) is an "independence assumption"—the occurrences in any two non-overlapping intervals are independent events.

Let the probability that exactly n events will occur in a time interval t be denoted $P(n)$. It can be proved from the above assumptions that

(2.11)
$$P(n) = \frac{(\lambda t)^n e^{-\lambda t}}{n!} \quad \text{for } n = 0, 1, 2, \ldots$$

This expression is easily converted to the notation previously used in (2.7) by denoting the average number of occurrences per time interval t as

$$\mu = \lambda t$$

and using the random variable X to denote the number of occurrences in the specified time interval. Then, $P(n) = f(x)$, and the probability that a Poisson process will generate x successes in time interval t is given by

$$f(x) = \frac{\mu^x e^{-\mu}}{x!} \quad \text{for } x = 0, 1, 2, \ldots$$

Although the assumptions and Equation (2.11) were stated in terms of a process which produces random occurrences in continuous time, the entire discussion as indicated earlier, pertains equally well to random occurrences in a continuum of space. In all of these cases, the parameter μ is interpreted as the average number of occurrences (successes) per interval of time or space.

2.8 Joint Probability Distributions

The discussion to this point has been concerned with probability distributions of discrete random variables, considered one at a time. These are often referred to as *univariate probability distributions* or *univariate probability mass functions*. It may be recalled that in such distributions, probabilities are assigned to events in a sample space, where the sample points pertain to outcomes for a single random variable. In most realistic decision making situations more than one factor must be taken into account at a time. Frequently, the joint effects of several variables, some or all of which are interdependent, must be analyzed, often in terms of what their impact is upon some objective the decision maker is attempting to achieve.

In this section we consider the joint probability distributions of discrete random variables, or stated differently, probability distributions of two or more discrete random variables. Such functions are frequently referred to as *multivariate probability distributions*, the term *bivariate probability distribution* being used for the two variable case.

We return to the random experiment of rolling a die twice (or rolling two dice once) for a simple example of a bivariate probability distribution. Let X and Y represent two random variables denoting the outcomes on the first and second rolls of the die and $P(X = x \cap Y = y)$ the joint probability that X takes on the value x, and Y takes on the value y[9]. The bivariate probability distribution of X and Y is given in Table 2-12.

The values in the cells of Table 2-12 are the joint probabilities of the respective outcomes denoted by the column and row headings for X and Y. Thus the figure 1/36 in the upper left-hand corner of the table is the joint probability that the number one will be obtained on the first roll, followed by a one on the second roll, and so forth. Also displayed in this table are the separate univariate probability distributions of X and Y. These are generally referred to as *marginal distributions*, since they are found literally in the margins of the joint probability table. The marginal distribution of X consists of the values of X shown in the column headings and the column totals at the bottom;

---

[9]The notation $P(X = x \cap Y = y)$ should really read $P\{(X = x) \cap (Y = y)\}$. The simplified symbolism is in common use, and will be employed in this text.

Table 2-12 Bivariate Probability Distribution of Outcomes on Two Rolls of a Die.

y \ x	1	2	3	4	5	6	*Row Totals*
1	1/36	1/36	·	·	·	1/36	1/6
2	1/36	1/36	·	·	·	1/36	1/6
3	·	·	·	·	·	·	·
4	·	·	·	·	·	·	·
5	·	·	·	·	·	·	·
6	1/36	1/36	·	·	·	1/36	1/6
Column Totals	1/6	1/6	·	·	·	1/6	1

the marginal distribution of Y consists of the values of Y shown in the row headings and the row totals at the right-hand side of the table. In this particular case, the marginal probability distributions of X and Y are identical. They are shown in Table 2-13. Since the letter f is used in $f(x)$ to denote the marginal probability distribution of the random variable X, to avoid confusion, the letter g is used in $g(y)$ to symbolize the marginal probability distribution of the random variable Y.

Table 2-13 Marginal Probability Distribution of X.

x	1	2	3	4	5	6
$f(x)$	1/6	1/6	1/6	1/6	1/6	1/6

Marginal Probability Distribution of Y.

y	1	2	3	4	5	6
$g(y)$	1/6	1/6	1/6	1/6	1/6	1/6

The probability distribution of a single discrete random variable is graphed by displaying the values of the random variable along the horizontal axis and the corresponding probabilities along the vertical axis. In the case of a bivariate distribution, two axes are required for the values of the random variables and a third for the measurement of probabilities. Usually, the joint values of the two variables are depicted on a plane (the x, y plane) and the associated probabilities are read along an axis perpendicular to the plane. A graph of the probability distribution for two rolls of a die is shown in Figure 2-9.

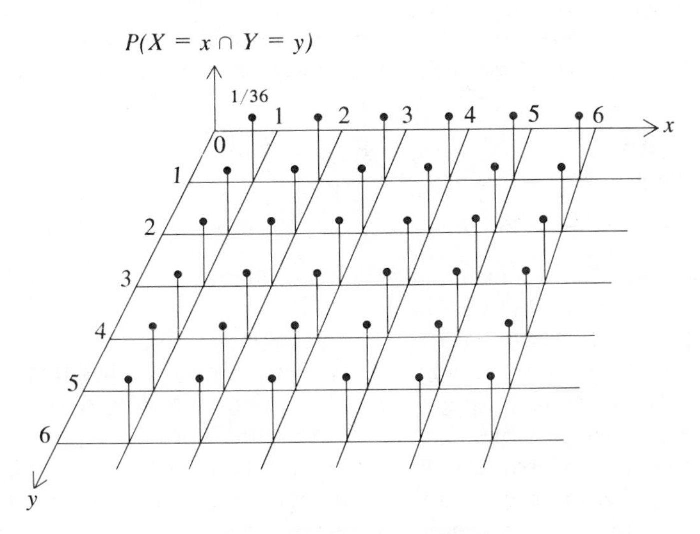

Figure 2-9 Graph of joint probability distribution for two rolls of a die.

Discrete Bivariate Distributions

Joint probability distributions for two discrete random variables can now be discussed somewhat more formally. In the one variable case, the symbol $f(x)$ is used to denote the probability that the random variable X takes on the value x; that is $f(x) = P(X = x)$. Analogously, in the bivariate case, $f(x, y)$ denotes the joint probability that X takes on the value x and Y assumes the value y; that is, $f(x, y) = P(X = x \cap Y = y)$. The conditions for an arbitrary function $f(x, y)$ to be a bivariate probability distribution are:

(1) $f(x, y) \geq 0$ for all real values of x and y

(2) $\sum_y \sum_x f(x, y) = 1$

We return briefly to the two rolls of a die illustration for an interpretation of these conditions. The first condition is met, since the probability of every pair of x and y values is defined and is non-negative for all real values of x and y. In fact, we note that for 36 pairs of x and y values $f(x, y) = 1/36$. For all other pairs, it is understood that $f(x, y) = 0$. The second condition, $\sum_y \sum_x f(x, y) = 1$, is analogous to the univariate condition that $\sum_x f(x) = 1$. The double summation $\sum_y \sum_x f(x, y)$ means that the values in the cells of the probability table are totalled first across rows (over the X variable) and then down columns (over the Y variable). Of course, the order of the summation does not matter. Totalling could proceed first down the columns of the table and then across rows, in which case the double sum would be written $\sum_x \sum_y f(x, y)$. In the die example, the sum of 36 terms of 1/36 each is equal to 1.

A succinct way of writing the joint probability distribution in the two rolls of a die problem is

$$f(x, y) = 1/36 \quad \text{for } x = 1, 2, \ldots, 6, y = 1, 2, \ldots, 6$$

Again, we note that we have simplified the notation here by omitting the statement that $f(x, y) = 0$ elsewhere.

Simple formulas can be given for probability distributions which are derivable from the joint distribution. It was observed earlier that marginal probability distributions for a given variable are obtained by summing across values of the other variable. Thus, for example, in the dice problem, if we want the marginal probability of obtaining a two on the first roll, that is, $P(X = 2)$, we sum the probabilities in the column $x = 2$ over all values of y and obtain $f(2) = 1/6$. In general, therefore, we have the following formulas for marginal probability distributions for the two discrete random variables X and Y:

Marginal Probability Distributions

(2.12)
$$f(x) = \sum_y f(x, y)$$

(2.13)
$$g(y) = \sum_x f(x, y)$$

These are general expressions for marginal distributions. In the illustration of obtaining the marginal probability of a two on the first roll of a die, one particular value of the distribution was calculated. Symbolically, we may write for that example

$$f(2) = \sum_{y=1}^{6} f(2, y) = f(2, 1) + f(2, 2) + \cdots + f(2, 6) = 1/6$$

Another important type of distribution obtainable from a joint probability function is the *conditional probability distribution.* It may be recalled that when the terminology of events was used, the probability of the intersection of two events was shown to be equal to the product of the unconditional probability of the first event and the conditional probability of the second event, given the first. Symbolically, if A_1 and A_2 are the two events, it was shown, (1.5), that

$$P(A_1 \cap A_2) = P(A_1)\,P(A_2 \mid A_1)$$

Using probability distribution notation, the analogous statement is

$$f(x, y) = f(x)\,g(y \mid x) \tag{2.14}$$

where $g(y \mid x)$ represents the probability that the discrete random variable Y equals y, given that X equals x. That is, $g(y \mid x) = P(Y = y \mid X = x)$. Solving Equation (2.14) for $g(y \mid x)$ gives the

Conditional Probability Distributions for Y

$$g(y \mid x) = \frac{f(x, y)}{f(x)}, \; f(x) > 0 \tag{2.15}$$

It is indicated in (2.15) that $f(x) > 0$, since division by zero is not a mathematically defined operation. Note that there is a separate conditional probability distribution of the random variable Y for each admissible value of X.

Returning to the terminology of events, if $P(A_2 \mid A_1) = P(A_2)$, then, as given in (1.7),

$$P(A_1 \cap A_2) = P(A_1)\,P(A_2)$$

and A_1 and A_2 are said to be *independent events.*

Thus, if $g(y \mid x) = g(y)$, then

$$f(x, y) = f(x)\,g(y) \tag{2.16}$$

and X and Y are said to be *independent random variables.*

In the case of events A_1 and A_2, if the order of the events is interchanged, the multiplication rule is written

$$P(A_1 \cap A_2) = P(A_2)\,P(A_1 \mid A_2)$$

Analogously, if the order of X and Y are interchanged, we may write

$$f(x, y) = g(y)\,f(x \mid y) \tag{2.17}$$

and we obtain the

Conditional Probability Distributions for X

$$f(x \mid y) = \frac{f(x, y)}{g(y)}, \; g(y) > 0$$

In the case of the dice illustration, clearly X, the outcome on the first roll, and Y, the outcome on the second roll, are independent. All of the conditional probabilities are equal to the corresponding marginal probabilities. For example, the conditional probability of obtaining a three on the second roll given that a two was obtained on the first roll is equal to the marginal probability of getting a three on the second roll. In symbols,

$$g(3 \mid 2) = \frac{f(2, 3)}{f(2)} = \frac{1/36}{1/6} = 1/6$$

and

$$g(3) = \sum_{x=1}^{6} f(x, 3) = 1/6$$

As indicated in Section 1.3, in a sense, every probability is conditional, in that it is dependent upon the sample space which has been defined. For example, what is the probability that an executive has an income before taxes of $25,000 or more? Obviously, the answers are different if the sample space in question is (a) the United States, (b) New York City, or (c) the steel industry. Some writers include S, the symbol for the sample space in the symbolism for probabilities, and thus even marginal probabilities appear as conditional upon the sample space defined. To avoid cluttering up the notation, throughout this text, it will be assumed that appropriate sample spaces have been defined, and the simpler notation, excluding the S, will be used.

Multivariate Probability Distributions

Joint probability distributions for three or more random variables can be represented by a straightforward generalization of the two variable case. Aside from increased computational effort, joint distributions for three or more variables do not pose new analytical problems. For completeness, the conditions for a function of n discrete random variables $X_1, X_2, \ldots, X_n$ to be a probability distribution are given here.

(1) $f(x_1, x_2, \ldots, x_n) \geq 0$ for all real values of $x_1, x_2, \ldots, x_n$

(2) $\sum_{x_n} \cdots \sum_{x_2} \sum_{x_1} f(x_1, x_2, \ldots, x_n) = 1$

Example 2-17 Table 2-14 presents the results of a sample survey of the employment status of the labor force by age in a certain community:

Table 2-14 Sample of the Labor Force in a Certain Community Classified by Age Group and Employment Status.

Employment Status	*Under 25*	*Age Group (in Years)* 25 and Under 45	*45 and Over*	*All Ages*
Unemployed	120	250	130	500
Employed	1880	4750	2870	9500
Labor Force	2000	5000	3000	10,000

(a) If a person is drawn at random from this labor force, what is
(1) the joint probability that he is under 25 years of age and unemployed?
(2) the marginal probability that he is under 25 years of age?
(3) the conditional probability that he is unemployed, given that he is under 25 years of age?
(b) What are the marginal probability distributions of age and employment status?
(c) What is the conditional probability distribution of age given that an individual in this labor force is unemployed?
(d) Is there evidence of dependence between employment status and age in this labor force?
(e) Retaining the same marginal totals, what would the numbers of persons in each cell of the table have to be for independence to exist between age and employment status?

Solution: In order to answer these questions, we convert the table of numbers of occurrences (absolute frequencies) to probabilities (relative frequencies), by dividing each number in Table 2-14 by 10,000, the total number of persons in the sample. These relative frequencies are given in Table 2-15.

Table 2-15 Relative Frequencies of Occurrence for a Sample of the Labor Force of a Certain Community Classified by Age Group and Employment Status.

Employment Status	*Under 25*	*Age Group (in Years)* 25 and Under 45	*45 and Over*	*All Ages*
Unemployed	0.012	0.025	0.013	0.050
Employed	0.188	0.475	0.287	0.950
Labor Force	0.200	0.500	0.300	1.000

Let X and Y be random variables representing age group and employment status, respectively, which take on values as follows:

Age Group	x	*Employment Status*	y
Under 25	0	Unemployed	0
25 and under 45	1	Employed	1
45 and over	2		

The joint probability distribution now appears as given in Table 2-16. The answers to the questions asked may now be given

(a)(1) P (under 25 ∩ unemployed) $= P(X = 0 \cap Y = 0) = f(0, 0) = 0.012$

(2) P (under 25) $= \mathrm{P}(X = 0) = f(0) = 0.200$

(3) P (unemployed | under 25) $= P(Y = 0 \mid X = 0) = g(0 \mid 0)$

$$= \frac{f(0, 0)}{f(0)} = \frac{0.012}{0.200} = 0.06$$

Table 2-16 Joint Probability Distribution Derived From Tables 2-14 and 2-15.

$y \backslash x$	0	1	2	$g(y)$
0	0.012	0.025	0.013	0.050
1	0.188	0.475	0.287	0.950
$f(x)$	0.200	0.500	0.300	1.000

(b) The marginal probability distributions given in the margins of the probability table are

x	$f(x)$
0	0.200
1	0.500
2	0.300
	1.000

y	$g(y)$
0	0.050
1	0.950
	1.000

(c) This conditional probability distribution is given by

$$P(X \mid Y = 0) = f(x \mid 0) = \frac{f(x, 0)}{g(0)}, \text{ where } x = 0, 1, 2$$

Thus, each of the joint probabilities in the first row of Table 2-15 or 2-16 is divided by the marginal probability or total of that row. That is, the joint probabilities of being unemployed and in the specified age groups are divided by the marginal probability of being unemployed. Therefore, the required conditional probability distribution is

x	$f(x \mid 0)$
0	$0.012/0.050 = 0.24$
1	$0.025/0.050 = 0.50$
2	$0.013/0.050 = 0.26$
	1.00

(d) Employment status and age are ***not independent*** random variables. That is, it is not true that $f(x, y) = f(x)\, g(y)$ for *all* values of x and y. For example,

$$f(0, 0) \neq f(0)\, g(0)$$

Numerically,

$$(0.012) \neq (0.200)(0.050)$$

It may be noted that, for age group "25 and under 45," ($x = 1$), the joint probabilities do factor into the product of the respective marginal probabilities. For example,

$$f(1, 0) = f(1)\, g(0)$$

or

$$0.025 = (0.500)(0.050)$$

However, this is not sufficient. The equality must hold for *all* values of X and Y for these variables to be considered independent.

Another way of indicating that age and employment status are not independent random variables is to note that the marginal distributions are not equal to the corresponding conditional distributions. Thus for age, the conditional and marginal probability distributions are

x	$f(x \mid 0)$	$f(x \mid 1)$	$f(x)$
0	0.240	0.198	0.200
1	0.500	0.500	0.500
2	0.260	0.302	0.300
	1.000	1.000	1.000

The specific nature of the dependence between age and employment status is that the percentage unemployed decreases with age. The unemployment rates are: under 25, 6.0%; 25 and under 45, 5.0%; 45 and over, 4.3%.

It is important to observe that the relationship between age and employment status depends to a certain extent on the arbitrary classifications used for the two variables. For example, if narrower age classifications had been used, it might have been found that the heaviest unemployment rates in the "under 25" group were among the teenagers and that although the unemployment rate was low for the "45 and over" group, the rate was quite high for persons 60 years and older. Similarly, if a narrower breakdown for employment had been used, such as including part-time employment and multiple job-holding classifications, greater insights might have been gained into the underlying relationships between age and employment status.

(e) As indicated in (d), if age and employment status were independent, the marginal probability distributions would be equal to the corresponding conditional probability distributions. One way of interpreting this in terms of the problem is to observe that since 0.05 of persons in all age groups were unemployed ($g(0) = 0.050$), then, under the assumption of independence, the same proportion of individuals in *each* age group would be unemployed. Thus, 5% of 2000, 5000, and 3000 persons would be unemployed in the three respective age groups. In terms of marginal totals, the arithmetic would be:

Age group	
Under 25:	$\frac{500}{10{,}000} \times 2000 = 100$ unemployed
25 and under 45:	$\frac{500}{10{,}000} \times 5000 = 250$ unemployed
45 and over:	$\frac{500}{10{,}000} \times 3000 = 150$ unemployed

Therefore, the numbers of persons in each cell of the table under the independence assumption are as given in Table 2-17.

Table 2-17 Numbers of Persons Under the Assumption of Independence Between Age and Employment Status.

Employment Status		*Age Group (in years)*		
	Under 25	*25 and Under 45*	*45 and Over*	*All Ages*
Unemployed	100	250	150	500
Employed	1900	4750	2850	9500
Labor Force	2000	5000	3000	10,000

Example 2-18 An operations research worker developed a mathematical expression for the joint probability distribution of two discrete economic variables as follows:

$$f(x, y) = (1/40)(x^2 + 3y),\ x = 0, 1, 2, 3 \qquad y = 0, 1$$

(a) Display the joint bivariate probability distribution of X and Y.

(b) Derive mathematical expressions for the marginal probability distributions of X and Y.

(c) Derive the conditional probability distributions of the Y variable in two different ways.

Solution: The joint bivariate probability distribution can be obtained by evaluating $f(x, y)$ for the admissible values of X and Y. For example, if $X = 0$ and $Y = 1$,

$$f(0, 1) = (1/40)(0 + 3(1)) = 3/40$$

(a) Joint bivariate probability distribution of

$$f(x, y) = (1/40)(x^2 + 3y),\ x = 0, 1, 2, 3,\ y = 0, 1$$

y \ x	0	1	2	3	$g(y)$
1	3/40	4/40	7/40	12/40	26/40
0	0/40	1/40	4/40	9/40	14/40
$f(x)$	3/40	5/40	11/40	21/40	1

(b)

$$\begin{aligned} f(x) &= \sum_{y=0}^{1} f(x, y) = \sum_{y=0}^{1} (1/40)(x^2 + 3y) \\ &= (1/40)(x^2 + 3(0)) + (1/40)(x^2 + 3(1)) \\ &= (1/40)(2x^2 + 3),\ x = 0, 1, 2, 3 \end{aligned}$$

Substitution of the admissible values, $x = 0, 1, 2, 3$, yields the marginal probability distribution of X as displayed in the first and last rows of the joint bivariate probability distribution.

$$\begin{aligned} g(y) &= \sum_{x=0}^{3} f(x, y) = \sum_{x=0}^{3} (1/40)(x^2 + 3y) = (1/40)(0 + 3y) + (1/40)(1 + 3y) \\ &+ (1/40)(4 + 3y) + (1/40)(9 + 3y) = (1/40)(14 + 12y),\ y = 0, 1 \end{aligned}$$

Substitution of the admissible values $y = 0, 1$ gives the marginal distribution of Y as shown in the first and last columns of the joint bivariate probability distribution.

(c) The conditional probability distributions of the Y variable can be obtained by dividing the joint probabilities by the appropriate marginal probabilities.

y	$g(y \mid 0)$	$g(y \mid 1)$	$g(y \mid 2)$	$g(y \mid 3)$
0	$\frac{0/40}{3/40} = 0$	$\frac{1/40}{5/40} = \frac{1}{5}$	$\frac{4/40}{11/40} = \frac{4}{11}$	$\frac{9/40}{21/40} = \frac{9}{21}$
1	$\frac{3/40}{3/40} = 1$	$\frac{4/40}{5/40} = \frac{4}{5}$	$\frac{7/40}{11/40} = \frac{7}{11}$	$\frac{12/40}{21/40} = \frac{12}{21}$

Another way of obtaining the conditional probability distributions is to derive an analytical expression for these distributions and then substitute numerical values. Thus,

$$g(y \mid x) = \frac{f(x, y)}{f(x)} = \frac{(1/40)(x^2 + 3y)}{(1/40)(2x^2 + 3)} = \frac{x^2 + 3y}{2x^2 + 3}$$

$$g(y \mid 0) = \frac{0 + 3y}{2(0) + 3} = y; \; y = 0, 1$$

$$g(y \mid 1) = \frac{1 + 3y}{2(1) + 3} = \frac{1 + 3y}{5}; \; y = 0, 1$$

$$g(y \mid 2) = \frac{2^2 + 3y}{2(4) + 3} = \frac{4 + 3y}{11}; \; y = 0, 1$$

$$g(y \mid 3) = \frac{3^2 + 3y}{2(9) + 3} = \frac{9 + 3y}{21}, \; y = 0, 1$$

Substitution of the admissible values for y yields the same probability distributions as the other method.

PROBLEMS

1. State whether the following random variables are discrete or continuous.
   (a) Strength of a steel beam in pounds per square inch.
   (b) The weight of a supposedly 16-ounce box of breakfast cereal.
   (c) X equals 0 if the weight of a supposedly 16-ounce box of cereal is less than 16 ounces, and 1 if 16 ounces or more.
   (d) The number of defective batteries in a lot of 1000.
2. Which of the following are valid probability functions?

   (a) $f(x) = \dfrac{x}{2} \qquad x = -1, 0, 1, 2$

   (b) $f(x) = \dfrac{5 - x}{14} \qquad x = 0, 1, 2, 3$

   (c) $f(x) = \dfrac{x^2 + 1}{60} \qquad x = 0, 1, \ldots, 5$

3. Find k such that the following are probability functions.
   (a) $\frac{k}{x^2}$ $\quad x = 1, 2, 3$
   (b) kx $\quad x = 1/4, 1/2, 3/4$
   (c) $\frac{k}{x}$ $\quad x = 1, 2, 3, 4$

4. A corporation economist, in building a model to predict a company's sales, developed the following categories of change:
   Sales down more than 3%
   Sales down 3% or less
   Sales unchanged
   Sales up 3% or less
   Sales up more than 3%

   Let X be a random variable associated with sales change and assign subjective probabilities to form a probability function. Then graph the probability function and the cumulative probability function.

5. The probability distribution of sales of a new drug during the first month is:

	x	$f(x)$
At least $15,000	2	.2
At least $10,000 but less than $15,000	1	.3
At least $5000 but less than $10,000	0	.4
Less than $5000	−1	.1

   Find and graph the cumulative distribution.

6. Let $X = 1$ if a band blade lasts less than three days
   $X = 2$ if a band blade lasts at least three days but less than seven days
   $X = 3$ if a band blade lasts at least seven days but less than 14 days
   $X = 4$ if a band blade lasts at least 14 days but less than 21 days

   and
   $$F(x) = x^2/16$$
   Find $f(x)$.

7. Let X = the number of minutes it takes to drain a soda filler of a particular flavor in order to change over to a new flavor. The probability distribution for X is:
   $f(x) = x/15 \qquad x = 1, 2, \ldots, 5$
   (a) Prove that $f(x)$ is a probability function.
   (b) What is the probability that it will take exactly three minutes to change over?
   (c) What is the probability that it will take at least two minutes but not more than four minutes?
   (d) Find the cumulative distribution (in table form).
   (e) What is the probability that it will at most take three minutes?
   (f) What is the probability that it will take more than two minutes?

8. Construct a situation in which the random variable can be said to have a uniform probability distribution.

9. In studying the past records of Southeast Airlines it has been found that the actual arrival of the scheduled 10:00 A.M. flight from Washington, D. C. due in Kennedy Airport, New York at 11:35 A.M. is uniformly distributed by minutes in the range of 11:15 A.M. to 12:35 P.M. Let $X = 1$ represent 11:15 A.M., $X = 2$, 11:16 A.M., and so forth.
   (a) Write out the mathematical expression for $f(x)$.
   (b) What is the probability the plane will be late?
   (c) What is the probability the plane will arrive after 12 noon?
   (d) What is the probability the plane will arrive on or after 12 noon?

10. A quality control inspector is concerned about the number of defectives being produced by three (identical) machines. It is assumed that "in the long run" all three produce the same proportion, p, of defectives. Ten items are made on each machine, representing independent trials.
    (a) What is the probability distribution of the number of defectives produced on machine one?
    (b) What is the probability distribution of the total number of defectives produced?

11. A certain delicate manufacturing process produces 20% defects. A new process is being tested. A sample of 100 items is produced. The new process will be installed if the sample yields 15 or fewer defectives. We assume that the production of 100 items corresponds to 100 independent trials of the process.
    Write a symbolic expression, specifying all numerical values involved, for the exact probability of 15 or fewer defectives, if the new process also produces 20% defectives on the average. Do not carry out the arithmetic.

12. A firm bills its accounts on a 1% discount for payment within ten days and full amount due after ten days. In the past, 30% of all invoices have been paid within ten days. If during the first week of July the firm sends out eight invoices, what is the probability that
    (a) No one takes the discount?
    (b) Everyone takes the discount?
    (c) Three take the discount?
    (d) At least three take the discount?

13. A certain portfolio consists of six stocks. The investor feels the probability that each stock will go down in price is .4 and that these price movements are independent. What is the probability that exactly three will decline? That three or more will decline? Does the assumption of independence here seem logical? If not, is the binomial distribution the appropriate probability distribution for this problem?

14. A certain type of plastic bag in the past has burst under a pressure of 15 pounds 20% of the time. If a prospective buyer tests five bags chosen at random, what is the probability that exactly one will burst?

15. A manufacturer of 60-second development film advertises that 95 out of 100 prints will develop. A person buys a roll of twelve prints and finds two do

not develop. If the manufacturer's claim is true, what is the probability that two or more prints will not develop?

16. An economist claims he can predict whether a company's gross sales over the given year will or will not increase, from the previous year's annual statement of the company. He is given ten companies' reports selected randomly and asked to predict whether each company's gross sales will increase or not. He predicts six correctly. If one were to guess randomly, what is the probability that he would make six or more correct predictions?

17. Two employees of a company were arguing over whether or not a large barrel of widgets had been produced by machine number one. The first employee, L. B. Jones, said he was sure it could not have come from that machine for he had taken a random sample of 20 widgets from the barrel and found them all perfect. Since it is known that machine number one produces 10% defective, Jones stated that he was almost certain that a random sample of 20 items would contain at least one defective in view of the fact that a binomial distribution was applicable. Do you agree with Jones? Use figures to support your position.

18. An oil exploration firm is formed with enough capital to finance ten ventures. The probability of any exploration being successful is .1. What are the firm's chances of
    (a) Exactly one successful exploration?
    (b) At least one successful exploration?
    (c) Going bankrupt in the ten ventures?

19. A fifteen question true-false examination is given in which each correct answer is worth 6-2/3 points. If a student guesses randomly on *each* question what is the probability that
    (a) He will score 60 or above if nothing is deducted for wrong answers?
    (b) He will score 60 or above if 6-2/3 points are deducted for each wrong answer? Possible scores range from 100 to −100.
    (c) He will score less than 0 if 6-2/3 points are deducted for each wrong answer?

20. A professional salesman feels that with an average quality product he should typically be able to make one sale out of every ten attempts; with an above average product the sales ratio should be one out of five and with a below average product, one out of every 20. In 18 attempts, he makes one sale. What is the probability of this event if
    (a) The product is below average in quality?
    (b) The product is average?
    (c) The product is above average?

21. A used car salesman claims that the odds are only 3:1 against his selling a car to any particular customer. If he attempts to sell automobiles to eight customers on a given day what is the probability that he will make at least one sale?

22. A physician knows from long experience that the probability is 0.25 that a

patient with a certain disease will recover. At the present time he has three patients with the disease. Find the probability that

(a) All three will recover.
(b) At least one will recover.

23. A small manufacturer of electrical wire submits a bid on each of four different government contracts. In the past this manufacturer's bid has been the low-bid (i.e., he was awarded the contract) 15% of the time. Assuming this relative frequency is a correct probability assignment and assuming independence, what is the probability that the firm will not obtain any of the four contracts?

24. A cashier at a checkout counter of a supermarket makes mistakes on the bills of 20% of the customers she checks out. Of the first five people she checked out, what is the probability that she made no mistakes?

25. A time study engineer is directed to establish time standards for the job of circuit assembler. One of the elements of the circuit assembler's work cycle consists in reaching into the conveyor belt of a continuously operating Biagara circuit former machine and recovering a handful of circuits (six are consistently removed by the operator, since the human hand easily grasps six but finds it impossible to grasp seven). If each housing requires five circuits, and the Biagara machine turns out 10% defectives, what is the probability that the worker will be required to perform this job element more than once in order to complete the assembly of one housing?

26. Graph the binomial distribution for $n = 11, p = .2, .5$, and $.8$. What can be said about skewness and the symmetry of the binomial distribution as the value of p departs from .5?

27. In the past, 10% of all printed circuits of a certain type were defective. If a person has ten printed circuits and requires seven good ones to complete the construction of a certain machine, what is the probability he will be able to complete construction?

28. An evenly balanced gear is covered by a safety guard over 7/8 of its surface. Near the circumference of the gear is a small opening used for the lubrication of internal feeding channels. What is the probability that the wheel can be lubricated in three or fewer stoppages of the machine, if it is not possible to control the final position of the gear when the machine is shut off, since the gear must coast to a stop?

29. A firm's office contains 20 typewriters. The probability that any one typewriter will not work on a given day is .05.

    (a) What is the probability that exactly one will not work on a given day?
    (b) What is the probability that at least two will not work on a given day?
    (c) Use the Poisson approximation to solve parts (a) and (b) and compare your results.

30. According to the manager of a certain motel, the probability that persons inquiring about a room will want to take a room at the motel is .1.

(a) If the motel has only two vacancies what is the *exact* probability that after 20 people inquire, the motel will still not be full?
(b) Use the Poisson approximation and compare your answer to that found in (a).

31. In reference to Problem 30, suppose the probability that someone would want to take a room was .05 instead of .1, what would be the exact probability and the Poisson approximation to the probability? Why do you think there is a difference between the "goodness" of the approximation between the two problems?

32. A company is trying to decide whether to use quality control plan A or B. Plan A reads "Inspect 20 out of every 1000 items, and accept the 1000 items if not more than five are defective." Plan B reads "Inspect 20 out of every 1000 items and accept the 1000 items if not more than seven are defective." Which plan has a higher consumer's risk? Producer's risk? What factors do you think might make one plan "better" than the other?

33. A manufacturer of suitcases ships the suitcases to its retail outlets in lots of 100. Five suitcases are inspected from each lot before shipment from the warehouse. If none are defective, the lot is shipped to the retailer; if one or more are defective, the lot is sent back to the factory. Production is large enough to assume that the binomial distribution is applicable. Graph the operating characteristic curve for this acceptance sampling plan.

34. Manufacturer A of breakfast cereals uses the following acceptance sampling plan: "From each case of 100 16-ounce boxes of cereal inspect ten and if more than two weigh less than 16 ounces, reject the case and sell at a discount." Manufacturer B of breakfast cereals uses the following acceptance sampling plan. From each case of 100 16-ounce boxes of cereal it inspects five, and if more than one weighs less than 16 ounces, the case is rejected and sold at a discount. Graph the operating characteristic curves and compare them.

35. A quality control system installed by Electronic Components, Inc. attempts to reduce loss due to production of defectives by categorizing components as good and defective, and if defective, as to the degree of defectiveness.

	X	$P(X)$
Let	1 = good	.75
	2 = defective; recheck	.10
	3 = defective but repairable	.03
	4 = defective, but has salvage value	.07
	5 = defective, no value	.05

(a) Assuming independence, if eight parts are selected and tested on a given morning, what is the probability of obtaining four good components and one of each type of defective?
(b) On a given morning, what is the probability that four out of the eight items tested are good?
(c) What is the name of the distribution used in (a) and that used in (b)?

36. A bin contains 1000 light bulbs, of which 100 are defective. You draw a sample of ten light bulbs without replacement. What probability distribution would yield the exact probability that you will draw at least eight good bulbs? Explain the reason for your answer. What discrete probability distribution can be used in this case to approximate the probability and why? What is the approximate probability?

37. Certain machine parts are shipped in lots of 20 items. Three parts are selected from each lot and tested. The lot is acceptable if no defectives are among the three tested. What is the probability that a lot containing four defectives will be accepted?

38. In a small Southern town there are 800 white citizens registered to vote and 400 non-white citizens registered. Trial juries of twelve people are supposedly selected at random from the list of registered voters with the possibility that a person can serve more than once. In the past month there have been five trials and the composition of each was eleven white citizens and one non-white citizen. What would be the probability of this occurrence if juries were actually selected at random?

39. A hardware store has 50 accounts receivable with open balances. Of these, ten have balances in excess of $25. An auditor selects at random ten accounts receivable for audit. What is the probability that exactly five of these ten accounts will have balances in excess of $25?

40. A retail hamburger company has 100 franchised stores. The stores are evaluated by sales volumes as follows:

Sales	*Number of Stores*
High	35
Medium	50
Low	15

As a sales promotion device, a clown will visit ten stores this month. A clown is the company's symbol. In order to avoid possible accusations of favoritism, management selected the ten stores by a random drawing.

(a) What is the probability that no low volume stores will be visited?

(b) What is the probability that three high volume, three low volume, and four medium volume stores will be visited?

41. The bylaws of the Majestic Rendering Company state that membership on committees of the board of trustees shall be determined by lottery, and that committees may make recommendations only when there exists unanimous agreement. A four-man committee on managerial appointments is to be selected from the full board of 30 members to recommend a new vice president of personnel. Eighteen members of the full board favor selection of Mr. Anderson, while twelve members favor Mr. Brown. Assuming no board member changes his mind, what is the probability that the committee will recommend the appointment of Mr. Brown?

42. The Slaughter Meat Company has 80 employees. Twenty are members of the International Slicers Union, 32 are members of the Brotherhood of Cleavers, and the remainder are non-union. Each day an employee is selected at random to work one hour overtime to clean the work area. In the past week, one member from each union and three non-union men worked overtime. The union stewards complained that the union men were being discriminated against. What is the probability of this particular selection? What is the probability of three or more non-union men being selected?

43. Five universities each send three representatives to a local convention. Five representatives are selected at random to serve as discussion leaders. What is the probability that
    (a) exactly one representative from University A is a discussion leader?
    (b) each university has a discussion leader?

44. A sales representative receives 100 leads a day, five of which are excellent sales prospects. He draws five at random to visit on a particular day from the 100 leads received that day. What is the exact probability that he does not draw any of the five excellent prospects? Calculate an approximate probability.

45. The number of car accidents between 5:00 P.M. and 6:00 P.M. on Friday on the Surekill Expressway is distributed according to the Poisson distribution with a mean of three. What is the probability that on a given Friday there will be no accidents during the indicated time interval?

46. It is estimated that the number of taxicabs waiting to pick up customers in front of the Reading Terminal follows a Poisson distribution with a mean of 3.2 cabs per unit of observed time. What is the probability that on a random observation more than five cabs will be waiting?

47. A certain production process produces 1% defective items. Use the Poisson approximation to answer the following questions. In a lot of 1000 what is the probability that
    (a) exactly ten are defective?
    (b) at most ten are defective?

48. The average number of thread defects in a standard bolt of cloth produced by Burmont Mills is six. What is the probability that
    (a) a bolt will have eight or more thread defects?
    (b) a bolt will have no thread defects?
    (c) that five bolts selected at random will all have eight or more thread defects?

49. A die casting machine operates in such a way that the number of defective castings produced per day has a Poisson distribution with a mean of four. What is the probability that on a given day three castings will be defective?

50. Suppose that the number of white blood cells per five cubic centimeters of undiluted blood counted under a microscope averages four. We have five test tubes, each containing a sample of five cubic centimeters of undiluted

blood. Find the probability that every one of the five test tubes will have at most five white blood cells.

51. The number of lathes breaking down between the hours 8:00 A.M. and 4:00 P.M. per weekday in the machine shop of Eastinghouse Corporation has a Poisson distribution with a mean equal to 6. What is the probability that more than ten machines will break down on any given weekday during the indicated time period?

52. In an industrial complex the average number of fatal accidents per month is one-half. The number of accidents per month is adequately described by a Poisson distribution. What is the probability that four months will pass without a fatal accident?

53. A certain department store's telephone sales department receives 2000 calls on a given day. The probability that any call will result in a sale is .01. Use the Poisson approximation to find the probability that the 2000 calls result in
(a) exactly 15 sales.
(b) more than 15 sales.

54. A check of 1000 50-pound crates of oranges from Ripe Groves revealed the following data:

Number of Spoiled Oranges in Crate	*Number of Crates*
0	450
1	370
2	130
3	40
4	10

It is hypothesized that the number of spoiled oranges per crate has a Poisson distribution with a mean of .8. Assuming this hypothesis is correct, calculate the probabilities of getting zero, one, two, three, and four spoiled oranges in a crate selected randomly and compare these results with the above data. Would you be apt to agree or disagree with the hypothesis?

55. The following probability distributions pertain to the number of freight and passenger ships arriving on a given day at a certain port.

x	$f(x)$	y	$g(y)$
0	.1	0	.3
1	.3	1	.3
2	.4	2	.2
3	.2	3	.2

X = number of freight ships
Y = number of passenger ships
If the number of freight ships has no effect on the number of passenger ships arriving, what is the joint distribution of X and Y?

56. An operations research analyst developed the following mathematical expression for the joint distribution of two discrete economic variables:

$$f(x, y) = \frac{1}{13}(x^2 + y^2) \qquad \begin{array}{l} x = 0, 1, 2 \\ y = 0, 1 \end{array}$$

(a) Display the joint bivariate probability distribution of X and Y.
(b) Derive mathematical expressions for the marginal probability distributions of X and Y.
(c) Derive the conditional probability distributions of the Y variable in two different ways.
(d) Are X and Y independent?

57. Let

$$x = \begin{cases} 1 \text{ if salesman is from New York office} \\ 2 \text{ if salesman is from Philadelphia office} \\ 3 \text{ if salesman is from Trenton office} \end{cases}$$

$$y = \begin{cases} 1 \text{ if salesman grosses under \$1000 per week for the year} \\ 2 \text{ if salesman grosses less than \$2000 but at least \$1000 per week} \\ \quad \text{for the year} \\ 3 \text{ if salesman grosses at least \$2000 per week for the year} \end{cases}$$

Given

$$f(x, y) = \frac{36}{121xy}$$

(a) Prove that $f(x, y)$ is a probability function and explain its meaning.
(b) Find the marginal distribution for X and explain its meaning.
(c) Find the marginal distribution for Y and explain its meaning.
(d) Find the conditional distribution for X given Y and explain its meaning.
(e) Are X and Y independent? Explain the significance of your answer.

58. The following table gives the results of a sample survey of 100 test communities on different types of packaging:

Type of Packaging

Sales	*Ordinary Bottle*	*Ordinary Cans*	*Flip-top Cans*	*Screw-top Bottle*	*Total*
Under 1000 cases	15	16	9	8	48
At least 1000 cases	10	9	16	17	52
Total	25	25	25	25	100

(a) If a test community is selected at random for further study, what is
  (1) the probability that it had sales of less than 1000 cases and was tested with ordinary cans?
  (2) the marginal probability that it was tested with ordinary cans?
  (3) the conditional probability that it had sales of less than 1000 cases given that it was tested with ordinary cans?
(b) What are the marginal probability distributions of sales and type of packaging?

(c) What is the conditional distribution of sales given that a screw-top bottle is used?
(d) Is there evidence of dependence between packaging and sales?
(e) Retaining the same marginal totals, what would the numbers in the cells of the table have to be for exact independence to exist between type of packaging and sales?

59. The research department of a certain corporation developed a mathematical expression for the joint distribution of two discrete management variables as follows:

$$f(x, y) = \frac{1}{58}(x^2 + 2y^2) \qquad \begin{matrix} x = 1, 2, 3 \\ y = 1, 2 \end{matrix}$$

(a) Display the joint bivariate probability distribution of X and Y.
(b) Derive mathematical expressions for the marginal probability distributions of X and Y.
(c) Derive mathematical expressions for the conditional probability distributions of the Y variable.
(d) Derive mathematical expressions for the conditional probability distributions of the X variable.

60. Random variables are defined for sales and advertising expenditures as follows:

$$X = \begin{cases} 0 \text{ if sales are less than \$10,000 per month} \\ 1 \text{ if sales are at least \$10,000, but less than \$20,000 per month} \\ 2 \text{ if sales are at least \$20,000 per month} \end{cases}$$

$$Y = \begin{cases} 0 \text{ if advertising expenditures are less than \$10,000 per year} \\ 1 \text{ if advertising expenditures are at least \$10,000 per year, but less than} \\ \quad \text{\$20,000 per year} \\ 2 \text{ if advertising expenditures are at least \$20,000 per year} \end{cases}$$

An operation research team has found the following joint probability relationship for a certain industry:

y \ x	0	1	2
0	.08	.07	.05
1	.11	.18	.12
2	.01	.07	.31

(a) What is $f(1, 2)$?
(b) Find the marginal probability distributions.
(c) Find $f(X/Y = 0), f(X/Y = 1), f(X/Y = 2)$
(d) Are X and Y independent? What does this answer tell you?

61. A poll of the employees of Bajo Inc. in reference to a particular labor proposal gave the following results:

Opinion	*Type of Position* Skilled	Unskilled	*Total*
For	105	75	180
Against	75	150	225
No opinion	20	75	95
Total	200	300	500

Does a person's type of position affect his opinion on the labor proposal? Interpret your answer.

62. An economic researcher hypothesizes the following joint probability distribution for two economic variables:

$$f(x, y) = \frac{(x + 1)(y + 1)^2}{30} \qquad \begin{array}{l} x = 0, 1, 2 \\ y = 0, 1 \end{array}$$

Does the hypothesized distribution imply that the two variables are independent? Prove your answer.

CHAPTER THREE

Empirical Frequency Distributions

The subjects we have discussed so far have fallen within the field of probability theory. In this and the next chapter we introduce some basic concepts of statistical methods. We begin with the notion of an *empirical* frequency distribution as a device for summarizing the variation in sets of numerical observations in convenient tabular form. We will then discuss descriptive measures of the characteristics of these distributions, such as averages and measures of dispersion. In Chapter 4 we discuss the same types of measures for *theoretical* frequency distributions. In accordance with general usage, we shall usually refer to "empirical frequency distributions" simply as "frequency distributions." The term "theoretical frequency distribution" is synonymous with "probability distribution."

The methods to be discussed in this chapter are useful for describing patterns of variation in data. Variation is a basic fact of life. As individuals, we differ in age, sex, height, weight, intelligence, in the quantities of the world's goods we possess, in the amount of our good or bad luck, and in a myriad of other characteristics. In the world of business, there are variations observed in the articles produced by manufacturing processes, in the yields of the economic factors of production, in production costs, financial costs, marketing costs, and so forth. Such variations occur both in data observed at a point in time as well as in data occurring over a period of time.

The term "cross-sectional data" refers to data observed at a point in time, whereas "time series data" pertains to sets of figures which vary over a period of time. Frequency distribution analysis is concerned with cross-sectional data. In particular, such analysis deals with data where the order in which the observations were recorded is of no importance, as for example, the ages of the present members of the labor force in the U. S., the present wage distribution of employees in the automobile industry, or the distribution of U. S. corporations by the amounts of their net worth on a given date. On the other hand, if we recorded quality control data of a manufactured product we would ordinarily be very much concerned with the order in which the articles were produced. For example, if a sudden run of defective articles were produced, we would be interested in knowing when this occurred and what was the general time pattern of production of defective and good articles. Similarly, in the study of economic growth, we might be interested in the variation over time of such data as real income per person or real gross national product per person. General methods of time series analysis are treated in Chapter 11.

3.1 Frequency Distributions

When we are confronted with large masses of ungrouped data, that is, listings of individual figures, it is extremely difficult to perceive the important characteristics and salient information they contain. To take an extreme example, if we merely had a listing of the ages of the approximately 200 million persons in the U. S., the tremendous quantity of figures would doubtless prevent us from being able to extract the significant aspects of the data. However, if a frequency distribution of the numbers were formed, many of these features would readily be discernible.

A frequency distribution or frequency table is simply a table in which the data are grouped into classes and the numbers of cases which fall in each class are recorded. The numbers in each class are referred to as "frequencies," hence the term "frequency distribution." When the numbers of items are expressed by their proportions in each class, the table is usually referred to as a "relative frequency distribution," or simply a "percentage distribution." Returning to the example on the ages of the U. S. population, if age classes were set up, say, in intervals of five years, for example, zero and under five, five and under ten, and so forth, and the proportions of the total population belonging to each class were specified, it would be relatively easy to extract the general characteristics of the age distribution of the population. If similar distributions were compared for, say, the years 1915 and 1965, a number of important features would be observable without any further statistical analysis. For example, the range of ages in both distributions would be clear at a glance. The higher proportions of young persons below the age of 20 and older persons over the age of 65 in 1965 than in 1915 would stand out. Also, smaller percentages of persons in the age categories 25 to 35 years would be observed for 1965, reflecting the decline in births during the Thirties and early Forties. In summary, through the simple device of constructing frequency distributions, some of the underlying characteristics of the nation's age composition for each year emerge from the data, and generalizations about age patterns are thus facilitated.

The way in which the classes of a frequency distribution are stated depends on the nature of the data. In all cases, we deal with objects having characteristics associated with them that can be counted or measured. Thus, individuals have characteristics such as color, nationality, and religion, and counts can be made of the numbers of persons that fall in each of the relevant categories. For example, if only two color classifications are used, white and non-white, the frequency distribution of color for residents of Vinetown on January 1, 1969 may be shown as in Table 3-1.

Table 3-1 Frequency Distribution of Residents of Vinetown on January 1, 1969 Classified by Color.

Color	*Number of Persons*
White	40,222
Non-White	10,310
Total	50,532

Characteristics, such as color, nationality and religion, that can be expressed in qualitative classifications or categories are often referred to as "attributes" or "categorical variables." When dealing with such cases, we make counts of the number of persons or objects that fall into each attribute classification. Using somewhat different language, it is clear that attributes or categorical variables can be classified as *discrete random variables*. It is always possible to encode the attribute classifications to make them numerical. Thus, for example, in the preceding illustration, "white" could have been denoted 0 and "non-white" 1. There also are certain cases where the data naturally fall into simple numerical classifications which may be viewed as values of a discrete random variable. For example, families may be grouped according to number of children; the classes would run 0, 1, 2, and so forth.

In the case of continuous random variables, data are obtained by numerical measurements rather than counting. When large numbers of measurements are made, it is convenient to use intervals or groupings of values and to tabulate the numbers or frequencies of occurrence in each class. In this connection, a few problems have to be resolved concerning the number of class intervals, the size of these intervals, and the manner in which class limits should be stated.

3.2 Construction of a Frequency Distribution

In order to illustrate the method of constructing a frequency distribution, let us consider the figures in Table 3-2, which represent the daily sales of 75 small retail establishments on a particular day. It is difficult to discern any particular pattern in these unarrayed and ungrouped data. However, when a frequency distribution is formed, the nature of the data clearly emerges.

The questions of how many classes to use and what the size of these classes should be are interrelated. Frequency distributions generally are constructed with about five to 20 classes. Wherever feasible, it is desirable to use class intervals of equal sizes because comparisons of frequencies among classes are facilitated and subsequent calculations from the distribution are simplified. However, this is not always a practical procedure. For example, in the case of data on annual incomes of families, in order to show the detail for the portion of the frequency distribution where the majority of incomes lie, class intervals of $1000 or $2000 may be used up to about $10,000; then intervals of $5000 may be used up to $25,000 and a final class of $25,000 and over may be shown for the relatively small numbers of families having these highest incomes. It is clear that if we tried to maintain equal size classes of (say) $1000 throughout the entire range of income, there would be too many classes. On the other hand, if much larger class intervals were used, too many families would be lumped together in the first one or two classes, and we would lose the information concerning how these incomes were distributed. The use of unequal class intervals and an open-ended interval for the highest class provides a simple way out of the dilemma. An open-ended class interval is one which contains only one specific limit and an "open" or unspecified

Table 3-2 Daily Sales of 75 Small Retail Establishments on a Particular Day.

$255.50	$373.25	$242.11	$256.61	$305.02
285.32	358.21	279.44	261.12	311.74
287.52	297.01	302.60	289.73	282.90
215.23	299.42	275.21	259.72	283.91
310.52	216.76	294.55	260.13	315.72
313.98	289.62	293.65	280.27	325.62
248.92	257.92	276.31	292.44	286.43
341.81	267.94	278.22	291.03	288.06
333.12	274.05	250.27	295.43	318.12
291.72	282.62	274.08	336.22	325.41
290.41	308.62	272.01	296.21	345.79
254.19	235.72	268.03	298.65	351.62
334.27	251.43	366.54	309.37	314.78
228.42	259.12	354.83	282.79	306.43
233.91	258.14	365.42	281.34	312.04

value at the upper or lower end, as for example *$25,000 and over* or *110 pounds and under*. The use of unequal class sizes and open-ended intervals generally becomes necessary in cases where most of the data are concentrated within a certain range, where gaps appear in which relatively few items are observed, and finally where there are a very few extremely large or extremely small values. Open-ended intervals are sometimes also used to retain confidentiality of information. For example, the identity of the small number of individuals or companies in the highest class may be general knowledge, and stating an upper limit for the class might be considered as excessively revealing.

Let us assume that for the list of sales figures shown in Table 3-2 we would like to set up a frequency distribution having about eight classes and that we wanted them to be of equal size. A simple formula to obtain an estimate of the appropriate size interval is

$$c = \frac{H - L}{k}$$

where c = the size of the class interval
H = the value of the highest item
L = the value of the lowest item
k = the number of classes

This formula for class interval size simply divides the total range of the data, that is, the difference between the values of the highest and lowest observations, by the number of classes. The resultant figure indicates how large the class interval would have to be in order that the entire range of the data may be covered in the desired number of classes. Some considerations involved in determining an appropriate number of classes are discussed in Section 3.4.

For the retail sales data, c = ($373.25 − $215.23)/8 = $19.76. Since it is desirable to have convenient sizes for class intervals, the $19.76 figure may be rounded to $20.00 and the distribution may be tentatively set up on that basis. The frequency distribution shown in Table 3-3 results from a tally of the number of items that fall in each $20.00 class interval.

Table 3-3 Frequency Distribution of Daily Sales of 75 Small Retail Establishments on a Particular Day.

Daily Sales	*Number of Establishments*
$215.00–234.99	4
235.00–254.99	6
255.00–274.99	13
275.00–294.99	22
295.00–314.99	15
315.00–334.99	6
335.00–354.99	5
355.00–374.99	4
Total	75

Some important features of these data are immediately discernible from the frequency distribution. The approximate value of the range, or difference between the values of the highest and lowest items, is revealed. Of course, since the identity of the individual items is lost in the grouping process, we cannot tell from the frequency table alone what the exact values of the highest and lowest items are. However, for all practical purposes, the range of the data is communicated by the table. Also, the frequency distribution gives at a glance some notion of how the elements are clustered. For example, more of the retail sales figures fall in the $275.00–294.99 interval than in any other single class. When the frequencies in the two classes immediately preceding and following the $275.00–294.99 grouping are added to the 22 which are in that interval, a total of 50 or 2/3 of the 75 retail establishments are accounted for. Furthermore, the distribution shows how the data are spread or dispersed throughout the range from the lowest to the highest value. We can determine by a quick perusal, whether the items are bunched near the center of the distribution or perhaps spread rather evenly throughout. Also, we can see whether the frequencies fall away rather symmetrically on either side of a class near the center of the distribution or whether there is a decided lack of such symmetry. We will now consider various statistical measures for describing these characteristics of a frequency distribution in a more exact manner, but much information can be gained by simply observing the distribution itself. It should be noted that there is no single perfect frequency distribution for a given set of data. Several alternative distributions with different class interval sizes and different highest and lowest values may be equally appropriate.

3.3 Class Limits

The way in which class limits of a frequency distribution are stated depends upon the nature of the data. Figures on ages are a good illustration of this point. Suppose that ages had been rounded according to the *last completed year*, that is, ages were recorded as of the last birthday. Then a clear and unambiguous way of stating the class limits is as given in Table 3-4.

Table 3-4 Ages of 100 Individuals.

Age (As of Last Birthday)	*Frequency*
15 and under 20	7
20 and under 25	28
25 and under 30	40
30 and under 35	19
35 and under 40	6
Total	100

Consider the first class interval in Table 3-4, "15 and under 20." Since ages have been recorded as of the last birthday, this class encompasses individuals who have reached at least their fifteenth birthday but not yet their twentieth birthday. If a person is 19.999 years of age, that is, a fraction of a day away from his twentieth birthday, he falls into the first class. However, upon attaining his twentieth birthday, he falls in the second class, "20 and under 25." Thus, these class intervals are five years in size. The midpoints of the classes, that is, values located halfway between the class limits, are respectively 17.5, 22.5, 27.5, 32.5, and 37.5. These values are used in computations of statistical measures for the distribution. It is to be noted that with class limits established and stated as above, the "stated limits" are in fact, the true boundaries or "real limits" of the classes. Of course, there are other ways of wording the limits such as "at least 15 but under 20," "15 to but not including 20," and so forth.

Suppose, on the other hand, that age data have been rounded to the nearest birthday. According to a widely accepted convention, the class limits would be stated as in Table 3-5. Despite the fact that the stated limits in each class are only four years apart, it is important to realize that the size of these class intervals is still five years. For example, since the ages are given as of the nearest birthday, everyone who is between 14.5 and 19.5 years of age falls in the class "15–19." Thus, when data recorded to the nearest unit are grouped

Table 3-5 Ages of 100 Individuals.

Age (As of Nearest Birthday)	*Frequency*
15–19	6
20–24	27
25–29	38
30–34	21
35–39	8
Total	100

into frequency distribution classes, the lower real limit or lower boundary of any given class lies one-half of a unit below the lower stated limit and the upper real limit or upper boundary lies one-half of a unit above the upper stated limit. The midpoints of the class intervals may be obtained by averaging the lower and upper real limits, or the lower and upper stated class limits. For example, the midpoint of the class "15–19" is 17, which is the same figure obtained by averaging 14.5 and 19.5.

In summary, when raw data are rounded to the last unit, the stated class limits and real class limits are identical. When raw data are measured to the nearest unit, the respective real limits are one-half of a unit removed from the corresponding stated limits. In the case of both types of data, midpoints of classes are located halfway between the stated limits, or equivalently, halfway between the real limits.

Sometimes, raw data are exact and have not been recorded either to the last unit or nearest unit. For example, weekly wages of a group of workers may be numbers such as $151.22, $153.75, $154.38, and so forth. Let us assume these data are exact to the penny, and that mills, or tenths of pennies, were not recorded nor used in the determination of the wage figures. Then the raw data are really discrete in units of cents. A conventional method of stating the class intervals in such a case is $150.00–159.99, $160.00–169.99, etc. The stated limits and real limits are the same, although even in this case, some people treat the real limits as being one-half cent removed from the stated limits. Regardless of the point of view adopted, however, the midpoint is halfway between $150.00 and $159.99 or $154.995, which for practical purposes, is $155.00. Similarly, the midpoint of the next class would be $165.00, etc.

One other widely used convention which unfortunately introduces a slight error is worthy of mention. Often, class limits are expressed as follows, even when data have been measured (say) to the nearest tenth of a unit: 20–24.9, 25–29.9, 30–34.9, etc. The purpose of this method of stating class limits is to avoid ambiguity in classifying observations such as 20, 25, 30, etc. In such cases, even though the observations have been rounded, the data are often treated as though they are measured along a continuum. Hence, midpoints are expressed as depending upon class limits along the scale, *rather than the numbers* which are placed in the classes. The scale is thought of as depicted in the diagram below:

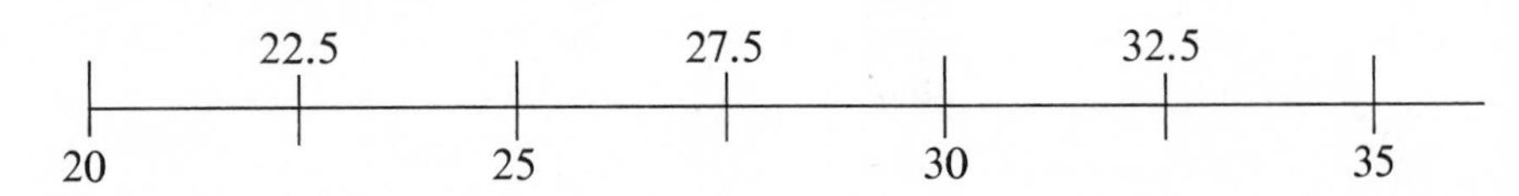

In such a continuous scale, the real upper limit of one class is also conceptually the lower limit of the next class. Hence, midpoints are taken as values halfway between the lower and upper real limits as shown, for example, 22.5, 27.5, and 32.5. However, it may be noted that if the data are *indeed rounded to the nearest tenth of a unit*, the real limits are 19.95, 24.95, 29.95, etc., and the midpoints should be 22.45, 27.45, etc.

Unfortunately, the more desirable conventions are not universally observed. Often, the situation requires one to use a frequency distribution constructed by others, and the nature of the raw data may not be clearly indicated. Needless to say, standards of good practice oblige the producer of the frequency distribution to indicate the nature of the underlying data. Let us assume that the raw data given in Table 3-2 are rounded to the nearest penny.

In all cases, the class intervals should be mutually exclusive and it should be clear into which class each item falls. If class limits are stated in the following manner: 30–40, 40–50, etc., it is not clear whether 40 belongs to the first class or the second.

3.4 Other Considerations in Constructing Frequency Distributions

A number of other points should be taken into account in the construction of a frequency distribution. If the data are such that there are concentrations of particular values, it is desirable to have these values at the midpoints of class intervals. For example, assume that data are collected on the amounts of the lunch checks in a students' dining room. Suppose these checks predominantly occur in multiples of five cents, although not exclusively so. If class intervals are then set up as $0.70–0.74, $0.75–0.79, etc., then a preponderance of items would be concentrated at the lower limits, which occur in multiples of five cents. In the calculation of certain statistical measures from the frequency distribution, the assumption is made that the midpoints of classes are average (arithmetic mean) values of the items in these classes. If, in fact, most of the items lie at the lower limits of the respective classes, a systematic error is introduced by this assumption, since the actual averages within classes will typically fall below the midpoints.

Another factor to be taken into account in constructing an empirical frequency distribution is the desirability of having a relatively smooth progression of frequencies. Many frequency distributions of business and economic data are characterized by having one class which contains more items than any other single class, and a more or less gradual dropping off of frequencies on either side of this class. Table 3-3 gives an example of such a distribution. As indicated in Section 3.2, the distribution may not be at all symmetrical. However, the important point concerning the smoothness of frequency progressions is that it is undesirable to have erratic increases and decreases of frequencies from class to class which tend to obscure the overall pattern. Erratic progressions of frequencies often arise from the use of class intervals that are too small. Increasing the size of class intervals usually results in a smoother progression of frequencies. However, wider classes reveal less detail than narrower classes. Thus, a compromise must be made in the construction of every frequency distribution. At one extreme, if we use class interval sizes of one unit each, every item of raw data is assigned to a separate class; at the other extreme, if we use only one class interval as wide as the range of the data, all items fall in the single class. Within the limits of these considerations, a great deal of freedom exists for the choice of an appropriate class interval size.

A final point to be considered in setting up a frequency distribution is that the numerical values of statistical measures computable from the grouped data should be close to the analogous values calculated from the ungrouped raw data. Since the user of the frequency distribution ordinarily does not have the raw data available, he makes his computations solely from the distribution. Clearly, if the values he calculates from the frequency distribution depart considerably from the corresponding measures computed from the ungrouped raw data, assuming no computational errors, then distortion of the statistical characteristics of the data has been introduced by the grouping process.

Let us take an example of how this comparison can be made. The arithmetic mean daily sales figure for the frequency distribution of the 75 small retail establishments given in Table 3-3 will be compared with the corresponding arithmetic mean computed from the raw data given in Table 3-2 from which the distribution was constructed. The arithmetic mean will be discussed more fully in Section 3.8.

The arithmetic mean of a set of values is obtained by summing the values of the items and dividing by the number of items. In symbols, if there are n values, $X_1, X_2, \ldots, X_n$, then the arithmetic mean of these values, denoted $\bar{X}$ is

(3.1)
$$\bar{X} = \frac{X_1 + X_2 + \cdots + X_n}{n} = \frac{\sum_{i=1}^{n} X_i}{n}$$

Since, in business and economic statistics, subscript indexes are often dropped, this formula may be written as:

(3.2)
$$\bar{X} = \frac{\Sigma X}{n}$$

The arithmetic mean of the ungrouped data on daily sales given in Table 3-2 is

$$\bar{X} = \frac{\$21{,}812.59}{75} = \$290.83$$

The arithmetic mean of the frequency distribution on daily sales in Table 3.3 is obtained in essentially the same manner. However, since the identity of individual items has been lost because of grouping, an estimate must be obtained for the total value of daily sales, ΣX, which appears in the numerator of the arithmetic mean calculation for the ungrouped data. This estimate is obtained by multiplying the midpoint of each class in the distribution by the frequency of that class and summing over all classes. In symbols, if X denotes the midpoint of a class and f the frequency, the arithmetic mean of a frequency distribution may be computed from the following formula:

(3.3)
$$\bar{X} = \frac{\Sigma fX}{n}$$

The computation of $\bar{X}$ for the frequency distribution shown in Table 3-3 is given in Table 3-6. The mean of $290.60 is quite close to $290.83, the mean of the original ungrouped data, differing from that figure by less than 0.1 percent. Thus, for practical purposes, no distortion has been introduced by the grouping process. Similar checks could be carried out for other statistical measures. No widely accepted rules of thumb exist for the desired degree of closeness between statistical measures computed from grouped and ungrouped data. The measures can be computed for alternative frequency distributions by experimenting with different size class intervals and different lower limits for the first class. In the final analysis, judgment plays a significant role in the decision as to what constitutes a suitable frequency distribution for a set of ungrouped figures. In general, several possible distributions may be appropriate for the same set of raw data.

Table 3-6 Computation of Arithmetic Mean of the Frequency Distribution Given in Table 3-3.

Daily Sales	*Number of Establishments* f	*Midpoints* X	fX
$215.00–234.99	4	$225.00	$ 900
235.00–254.99	6	245.00	1470
255.00–274.99	13	265.00	3445
275.00–294.99	22	285.00	6270
295.00–314.99	15	305.00	4575
315.00–334.99	6	325.00	1950
335.00–354.99	5	345.00	1725
355.00–374.99	4	365.00	1460
	$n = \Sigma f = 75$		$\Sigma fX = \$21,795$

$$\bar{X} = \frac{\Sigma fX}{n} = \frac{\$21,795}{75} = \$290.60$$

3.5 Graphic Presentation of Frequency Distributions

Graphs are often useful for presenting the salient features of a set of statistical data as contrasted with statistical tables, which show more specific detail. The use of graphs for displaying frequency distributions will be illustrated in terms of the data on retail sales shown in Table 3-3. One method is to represent the frequency of each class by a rectangle or bar. Such a chart is generally referred to as a *histogram*. A histogram for the frequency table given in Table 3-3 is shown in Figure 3.1. In agreement with the usual convention, values of the variable are depicted on the horizontal axis and frequencies of occurrence are shown on the vertical axis.

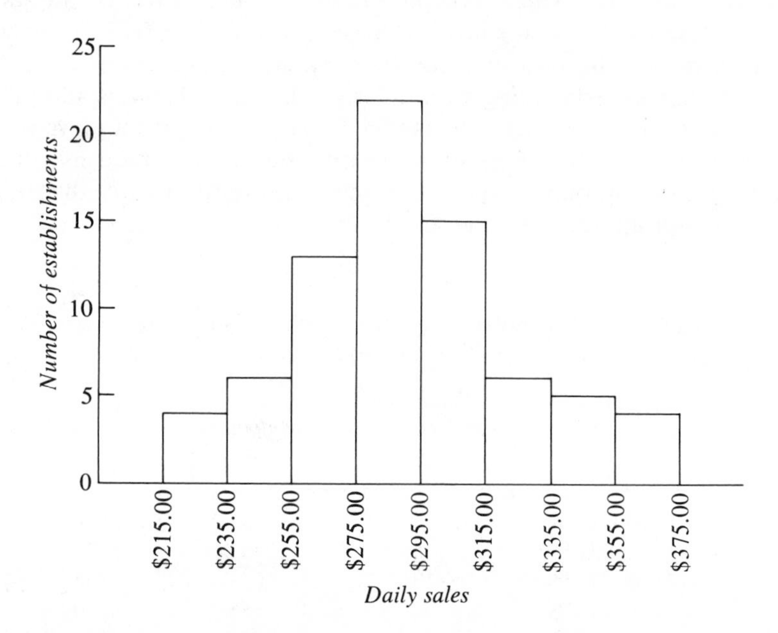

Figure 3-1 Histogram of frequency distribution of daily sales of 75 small retail establishments on a particular day from the data given in Table 3-3.

Class limits for the bars on a histogram may be shown either in terms of the real or apparent limits. Apparent limits have been used in Figure 3.1 with only the lower limit of each class being shown along the horizontal axis. Thus, a minor inaccuracy is present in the sense that the lower limit of each class appears visually as though it were also the upper limit of the preceding class. Of course, this would be strictly true only if the data were continuous. However, the minor error involved is justifiable on the basis of convenience of presentation.

An alternative method for the graphic presentation of a frequency distribution is the frequency polygon. In this type of graph, the frequency of each class is represented by a dot at the appropriate height plotted opposite the midpoint of each class. The dots are joined by line segments. Thus, a many-sided figure, or polygon is formed. Stated differently, a frequency polygon is the line graph obtained by joining the midpoints of the tops of the bars in a histogram. By convention, the polygon is closed at both ends of the distribution by line segments, drawn, respectively, from the dot representing the frequency in the lowest class to a point on the horizontal axis one-half of a class interval below the lower limit of the first class, and from the dot representing the frequency in the highest class to a point one-half of a class interval above the upper limit of the last class. A frequency polygon for the distribution given in Table 3-3 is shown in Figure 3.2. It is important to realize that the line segments are drawn merely for convenience in reading the graph. That is, the only points that are of significance are the plotted frequencies for the given midpoints. Interpolation for intermediate values between such points would be meaningless. Often the midpoints of classes are shown on the horizontal axis directly below the plotted points rather than class limits such as are shown in Figure 3.2.

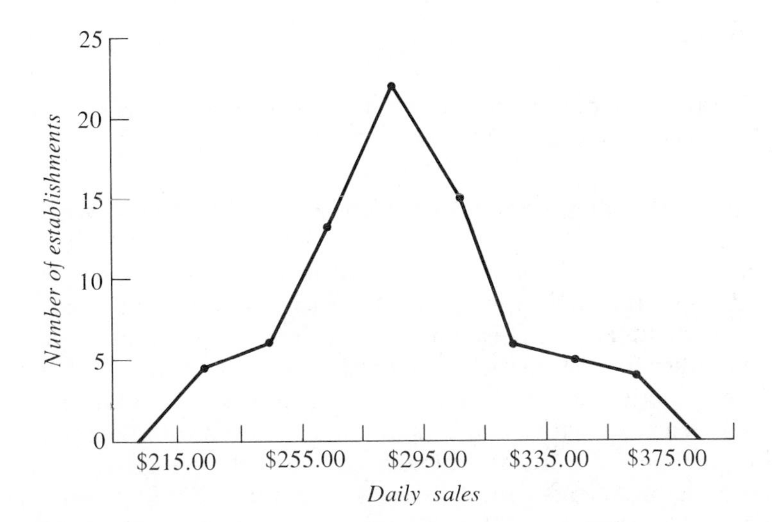

Figure 3-2 Frequency polygon for the distribution of daily sales of 75 small retail establishments on a particular day from the data given in Table 3-3.

If the class sizes in a frequency distribution were gradually reduced and the number of items were increased, the frequency polygon would approach a smooth curve more and more closely. Thus, as a limiting case, the variable of interest may be viewed as continuous rather than as occurring in discrete classes, and the polygon would assume the shape of a smooth curve. The frequency curve approached by the polygon for the retail daily sales data would thus appear as shown in Figure 3.3.

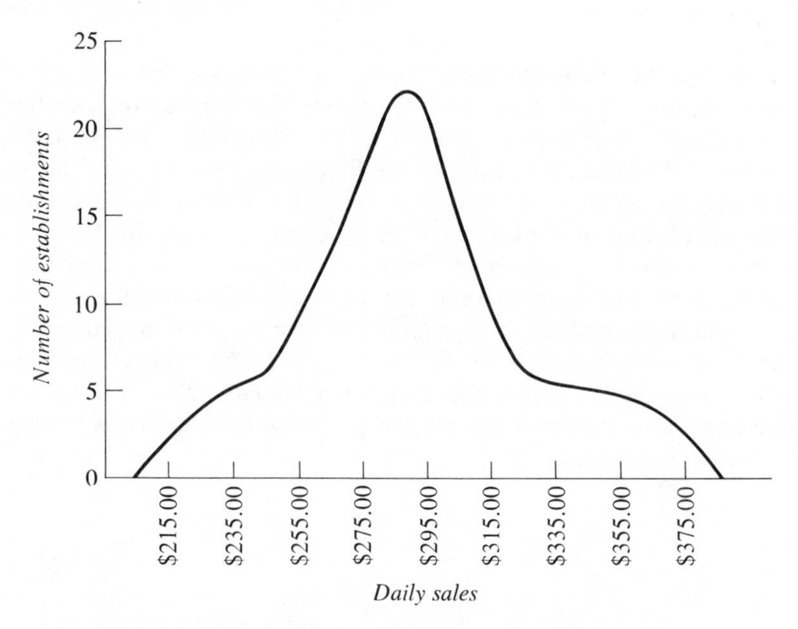

Figure 3-3 Frequency curve for the distribution of daily sales of 75 small retail establishments on a particular day from data given in Table 3-3.

3.6 Cumulative Frequency Distributions

Sometimes interest centers in the number of cases that lie below or above specified values rather than within intervals as shown in a frequency distribution. In such situations, it is convenient to use a *cumulative* frequency distribution rather than the usual frequency distribution. A so-called "less-than" cumulative distribution is shown for the daily sales distribution shown in Table 3-3. The cumulative numbers of establishments with sales less than the lower class limits of $215.00, $235.00, and so forth are given. Thus, there were zero establishments with daily sales of less than $215.00, four establishments with less than $235.00 of sales, $4 + 6 = 10$ establishments with less than $255.00, and so forth.

The graph of a cumulative frequency distribution is referred to as an *ogive*, (pronounced "ojive"). The ogive for the cumulative distribution shown in Table 3-7 is given in Figure 3.4. The plotted points represent the number of establishments having less than the amounts of sales shown on the horizontal axis directly below the points. The vertical coordinate of the last point represents the sum of the frequencies, in this case, 75. The S-shaped configuration depicted in Figure 3.4 is quite typical of the appearance of a "less than" ogive. A "more than" ogive for the daily sales distribution would have class limits reading "more than $215.00," "more than $235.00," and so forth. In this case a reversed S-shaped figure would have been obtained, sloping downward from the upper left to the lower right on the graph.

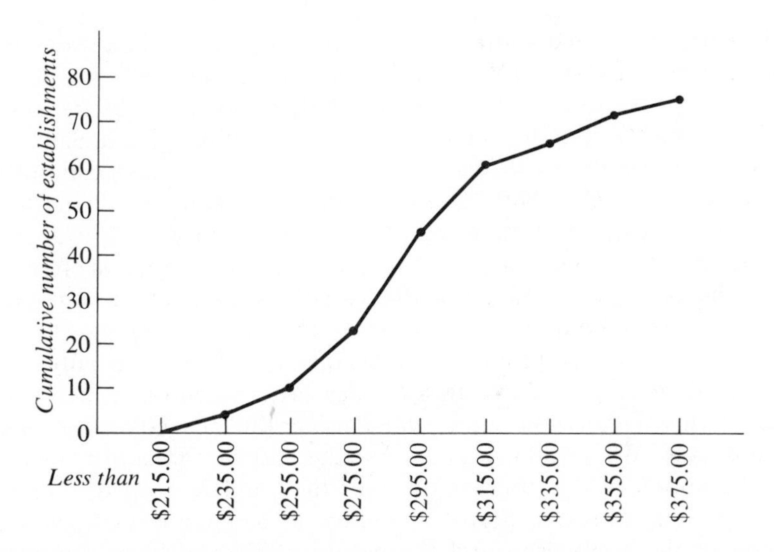

Figure 3-4 Ogive for the distribution of daily sales of 75 small retail establishments on a particular day.

Table 3-7 Cumulative Frequency Distribution of Daily Sales of 75 Small Retail Establishments on a Particular Day.

Daily Sales	*Number of Establishments*
Less than $215.00	0
Less than 235.00	4
Less than 255.00	10
Less than 275.00	23
Less than 295.00	45
Less than 315.00	60
Less than 335.00	66
Less than 355.00	71
Less than 375.00	75

3.7 Descriptive Measures for Frequency Distributions

It was indicated in Section 3.2 that once a frequency distribution is constructed from a set of figures, certain features of the data become readily apparent. For most purposes, it is necessary to have a more exact description of these characteristics than can be ascertained by a casual glance at the distribution. Thus, analytical measures are usually computed which describe such characteristics as the *central tendency, dispersion,* and *skewness* of the data. These measures constitute summary descriptions of the frequency distribution, which is itself a summarization of the set of original data.

Averages are the measures used to describe the characteristic of *central tendency* or *location* of data. One such measure has been referred to earlier in this chapter (Section 3.4), namely, the arithmetic mean. This is doubtless the most familiar of the averages; in fact, in common usage, it is often referred to as "the average." Also, in ordinary conversation or in newspapers or periodicals, we encounter such terms as the "average income," "average growth rate," "average profit rate," "average man," etc. Actually, there are several different types of averages or measures of central tendency implied in these terms. In this section, we will consider the most commonly employed and most generally useful averages. Averages attempt to convey in summary form the notion of "central location" or the "middle property" of a set of data. As we shall see, the type of average to be employed depends on the purpose of the application and the nature of the data being summarized.

Dispersion refers to the spread or variability in a set of data. One method of measuring this variability is in terms of the difference between the values of selected items in a distribution, such as the difference between the values of the highest and lowest items. Another more comprehensive method is in terms of some average of the deviations of all of the items from an average. Dispersion is, of course, a very important characteristic of data, in that interest frequently centers as much upon the uniformity or lack of uniformity in a set of data as upon their central tendency.

Skewness refers to the symmetry or lack of symmetry in the shape of a frequency distribution. This characteristic is of particular importance in connection with judging the typicality of certain measures of central tendency.

We begin the discussion of averages or measures of location by considering the most familiar one, the arithmetic mean.

3.8 The Arithmetic Mean

As indicated in the preceding section, there are a variety of ways of describing the central tendency or central location of a set of data, and probably the most widely used and most generally understood is the average known as the *arithmetic mean.* The arithmetic mean was briefly discussed in Section 3.4 in connection with the construction of frequency distributions. We now consider this average in greater detail. The arithmetic mean, often simply referred to as the *mean,* is the total of the values of a set of observations divided by their number. Thus, as indicated in (3.1) if $X_1, X_2, \ldots, X_n$ represent the values of n items or observations, the arithmetic mean of these items, denoted $\bar{X}$ is defined as

$$\bar{X} = \frac{X_1 + X_2 + \cdots + X_n}{n} = \frac{\sum_{i=1}^{n} X_i}{n}$$

If the subscripts are dropped, the formula becomes

$$\bar{X} = \frac{\Sigma X}{n}$$

For example, suppose the personal loan department of a bank made the following loans on a certain day: \$500, \$250, \$300, \$600, and \$425. Then the arithmetic mean loan is

$$\bar{X} = \frac{\$500 + \$250 + \$300 + \$600 + \$425}{5} = \frac{\$2075}{5} = \$415$$

The mean of \$415 may be thought of as the size of loan that would have been made if the total amount lent was \$2075 and all loans were the same size. That is, the mean is the value each item would have if they were all identical and the total value and number of items remained unchanged.

A brief note on symbolism is appropriate at this point. In keeping with standard statistical practice, the symbol $\bar{X}$ used in expressions (3.1), (3.2), (3.3), and in this section will be understood throughout this text to pertain to the mean of a *sample* of observations. The number of observations in the sample is denoted by a small letter n. A value such as $\bar{X}$, that is, a number computed from sample data, is referred to as a *statistic*. A statistic may be used as an estimate of an analogous population measure, known as a *parameter*. Thus, the sample mean $\bar{X}$ is a statistic which may be thought of as an estimate of the mean of the population from which the sample was drawn. The population mean is a parameter. It is conventional to denote population parameters by Greek letters and sample statistics by Roman letters. In keeping with this convention, a population mean will be denoted in this text by the Greek letter μ. The number of observations in a population is symbolized by the capital letter N. Therefore, the formula for the mean of a population is

$$\mu = \frac{X_1 + X_2 + \cdots + X_N}{N} = \frac{\sum_{i=1}^{N} X_i}{N}$$

If the subscripts are dropped, we have

$$\mu = \frac{\Sigma X}{N}$$

It may be noted that Greek letters are also used to denote parameters of probability distributions. Hence, for example, the symbol μ was used in Chapter 2 to represent the mean of the Poisson distribution.

Given a set of observations, we are often interested in the difference between these observations and the mean. Such differences are referred to as deviations. For example, in the case of the five loans given above, the deviation of the value of the first loan \$500 from the mean, \$415 is \$500 − \$415 = \$85. An important property of the arithmetic mean is that the sum of the deviations of a set of observations from their mean is equal to zero. It is very simple to prove this property. Let us denote the value of the ith item in a set of n observations as X_i, the mean as $\bar{X}$, and the deviation of the ith item from the mean as x_i. Then,

$$x_i = X_i - \bar{X}$$

the deviation of a particular observation from the mean of the observations.

Summing these deviations over all n observations gives

$$\sum_{i=1}^{n} x_i = \sum_{i=1}^{n} (X_i - \bar{X}) \tag{3.4}$$

By Rule 3 of Appendix B,

$$\sum_{i=1}^{n} (X_i - \bar{X}) = \sum_{i=1}^{n} X_i - \sum_{i=1}^{n} \bar{X}$$

By Rule 2 of Appendix B,

$$\sum_{i=1}^{n} \bar{X} = n\bar{X}$$

Therefore,

$$\sum_{i=1}^{n} x_i = \sum_{i=1}^{n} X_i - n\bar{X} = \sum_{i=1}^{n} X_i - \sum_{i=1}^{n} X_i = 0 \tag{3.5}$$

This simple property is the basis of many shortcut calculational formulas in statistics. It follows that if the sum of the deviations of a set of items from their mean is zero, then the similar sum from any other value in the series must not be zero. How this simple fact is utilized is illustrated below. Although the emphasis in this text is upon the use and interpretation of statistical methods rather than upon computational routines, shortcut methods of calculating the mean for both ungrouped and grouped data will be described because of the usefulness and importance of these methods. The basic procedures used in these shortcut methods are:

(1) *transforming* or *coding* the original data,
(2) carrying out calculations in terms of the coded data, and

(3) decoding the results by taking account of the transformation to yield the result which would have been obtained using the original observations.

In general, a coding procedure may involve subtracting a constant from each observation, adding a constant to each observation, or dividing or multiplying each observation by a constant, or some combination of two or more of the foregoing operations.

Shortcut Calculation of the Mean—Ungrouped Data

With simple sets of ungrouped data, such as the five personal loan figures discussed above, it is clear that shortcut methods of computing the mean are not particularly helpful. In fact, the methods are most useful in terms of time saving and simplication of arithmetic when dealing with grouped data. However, the techniques are clearly illustrated through the use of ungrouped data and simple generalizations are made for the case of data grouped by classes.

The shortcut method of computing the mean for ungrouped data involves:

(1) the selection of an assumed (or arbitrary) mean

(2) calculation of an average deviation from this assumed mean, and

(3) the addition of this average deviation as a correction factor to the assumed mean to obtain the true mean. This correction factor is positive if the assumed mean lies below the true mean, negative if it lies above.

We shall give the formula for the shortcut method of computing the mean for ungrouped data, apply the method to the five personal loan figures given in Section 3.8, and then derive the formula algebraically.

Shortcut Method for the Arithmetic Mean—Ungrouped Data:

(3.6)
$$\bar{X} = \bar{X}_a + \frac{\sum_{i=1}^{n} d_i}{n}$$

where $\bar{X}$ = the arithmetic mean
$\bar{X}_a$ = the assumed mean
$d_i = X_i - \bar{X}_a$ = the deviation of the ith observation from the assumed mean of the observations
n = the number of observations

If subscript indexes are dropped, the formula becomes

(3.7)
$$\bar{X} = \bar{X}_a + \frac{\Sigma d}{n}$$

The application of the formula to the five personal loan figures of Section 3.8 is illustrated in Table 3-8, with a choice of $400 as the assumed mean.

Table 3-8 Calculation of the Arithmetic Mean for Ungrouped Data by the Shortcut Method: Personal Loan Data.

Amount of Loans X	*Deviation from Assumed Mean* d
$500	+$100
250	− 150
300	− 100
600	+ 200
425	+ 25
	Σd = +$ 75

$$\bar{X} = \bar{X}_a + \frac{\Sigma d}{n} = \$400 + \$75/5 = \$400 + \$15 = \$415$$

Thus, the same answer of $415 is obtained as when the mean was directly computed from the definition. Of course, the shortcut method is perfectly general, and the mean may have been assumed to lie inside or outside the range of the given data. It is usually advisable to select as an assumed mean a simple round number close to the center of the range of the data. An interesting special case is the situation in which the assumed mean is zero. Then the deviations are the items themselves. The correction factor to be added to zero is thus the mean of the original data. In this sense, the arithmetic mean with which the student is familiar and has doubtless computed many times is a special case of the shortcut formula.

The derivation of the formula is quite simple. The deviation of an observation from the assumed mean may be expressed as a sum of two terms by subtracting and adding $\bar{X}$ as follows:

$$d_i = X_i - \bar{X}_a = (X_i - \bar{X}) + (\bar{X} - \bar{X}_a)$$

Taking the sum of the expressions on the left and right sides of this equation over all n observations gives

$$\sum_{i=1}^{n} d_i = \sum_{i=1}^{n} (X_i - \bar{X}) + \sum_{i=1}^{n} (\bar{X} - \bar{X}_a)$$

From Equation (3.5)

$$\sum_{i=1}^{n} (X_i - \bar{X}) = 0$$

Therefore,

$$\sum_{i=1}^{n} d_i = \sum_{i=1}^{n} (\bar{X} - \bar{X}_a) \tag{3.8}$$

By Rule 2, Appendix B, since $(\bar{X} - \bar{X}_a)$ is a constant,

$$\sum_{i=1}^{n} (\bar{X} - \bar{X}_a) = n(\bar{X} - \bar{X}_a) \tag{3.9}$$

Substituting (3.9) into (3.8) gives

$$\sum_{i=1}^{n} d_i = n(\bar{X} - \bar{X}_a) \tag{3.10}$$

Solving (3.10) for $\bar{X}$ yields the shortcut formula

$$\bar{X} = \bar{X}_a + \frac{\sum_{i=1}^{n} d_i}{n}$$

From this proof, we can now see the significance of the earlier statement that if the sum of the deviations from the true mean of a set of observations is zero, the sum of the deviations from any other value is not equal to zero. As shown in the proof, the arithmetic mean of the deviations from the assumed mean is the correction factor which must be added to the assumed mean to obtain the value of the true mean.

The use of coding will now be discussed for data grouped in the form of a frequency distribution.

Shortcut Method for the Arithmetic Mean—Grouped Data

In Section 3.4 it was indicated that the mean of a frequency distribution can be computed by a generalization of the definition for the mean of ungrouped data. That is, the mean of a frequency distribution was given in Equation (3.3) as

$$\bar{X} = \frac{\Sigma fX}{n}$$

where X represents the midpoint of a class, f the frequency in a class, and n the total number of observations. Using subscripts, this formula becomes

$$\bar{X} = \frac{\sum_{i=1}^{k} f_i X_i}{n} \tag{3.11}$$

where $\bar{X}$ = the arithmetic mean
f_i = the frequency of the ith class
X_i = the midpoint of the ith class
n = the number of observations
k = the number of classes in the frequency distribution.

Since the computation of the mean by this method was illustrated in Section 3.4, it will not be repeated here.

There are two different shortcut methods of calculating the mean from grouped data, one referred to simply as the "shortcut method," the other as the "step-deviation method." The shortcut method is a direct generalization of the corresponding method for ungrouped data. It utilizes deviations of midpoints of class intervals from the assumed mean. The step-deviation method proceeds one step further by expressing these deviations in terms of class interval units. In the case of frequency distributions with equal size class intervals, the step deviation method involves simpler arithmetic, and statistical measures such as the arithmetic mean and standard deviation can be computed more rapidly. Both of these methods are explained below in terms of an illustrative example for a frequency distribution of personal loans from a bank.

Shortcut Method for the Arithmetic Mean—Grouped Data

The shortcut method of calculating the arithmetic mean from grouped data is essentially the same as the corresponding method for ungrouped data. In the case of ungrouped data, deviations of the observations are taken from an assumed mean. The mean of these deviations is calculated and is added as a correction factor to the assumed mean. When the data are grouped in the form of a frequency distribution, deviations of the midpoints of classes are taken from an assumed mean. The mean of these deviations is obtained by weighting the deviations by the frequencies of the classes, adding these values and dividing by the total number of items in the distribution. Then, as in the case of ungrouped data, this average deviation is added as a correction factor to the assumed mean to obtain the true mean. The formula for the grouped data shortcut method is given in Equation (3.12). It can be seen that this formula differs from the corresponding one for ungrouped data only in the inclusion of a symbol for the multiplication of deviations by frequencies.

Shortcut Method for the Arithmetic Mean—Grouped Data

(3.12)
$$\bar{X} = \bar{X}_a + \frac{\sum_{i=1}^{k} f_i d_i}{n}$$

where $\bar{X}$ = the arithmetic mean
$\bar{X}_a$ = the assumed mean
f_i = the frequency of the ith class
$d_i = X_i - \bar{X}_a$ = the deviation of the midpoint of the ith class from the assumed mean
n = the number of observations
k = the number of classes in the frequency distribution

Dropping subscripts, the formula becomes

$$\bar{X} = \bar{X}_a + \frac{\Sigma fd}{n} \tag{3.13}$$

The application of the shortcut method is illustrated in Table 3-9 for a frequency distribution of personal loans. In this case, an assumed mean of \$500 was taken. The calculated correction factor of \$132, when added to this assumed mean, yields \$632 as the mean size of personal loans.

Table 3-9 Calculation of the Arithmetic Mean for Grouped Data by the Shortcut Method: Personal Loan Data.

Amount of Loans	*Number of Loans* f	d	fd
\$ 0 and under \$ 200	6	−\$400	−\$ 2,400
200 and under 400	18	− 200	− 3,600
400 and under 600	25	0	0
600 and under 800	20	+ 200	+ 4,000
800 and under 1000	17	+ 400	+ 6,800
1000 and under 1200	14	+ 600	+ 8,400
	100		+\$13,200

$\bar{X}_a = \$500$

$$\bar{X} = \bar{X}_a + \frac{\Sigma fd}{n} = \$500 + \frac{\$13{,}200}{100} = \$500 + \$132 = \$632$$

Step-Deviation Method for the Arithmetic Mean—Grouped Data

The step-deviation method follows the pattern of coding data by calculating deviations from an assumed mean, determining a correction factor and adding it to the assumed mean. However, in the step-deviation method, the coding procedure is carried one step further than in the shortcut method.

Deviations of midpoints of classes are taken from the assumed mean as in the shortcut method, but they are then divided by the class interval size to yield deviations stated in *class interval units.* In a frequency distribution with equal size class intervals, the relationship between deviations in class interval units and deviations in the shortcut method is expressed by the formula

(3.14)
$$d'_i = \frac{d_i}{c}$$

where $d'_i = \frac{d_i}{c} = \frac{X_i - \bar{X}_a}{c}$ = the deviation of the midpoint of the ith class from the assumed mean in class interval units.

$d_i = X_i - \bar{X}_a$ = the deviation of the midpoint of the ith class from the assumed mean.

c = the size of the class interval, that is, the difference between the value of the lower limit of a class and the lower limit of the immediately preceding class.

The formula for the step-deviation method is given in Equation (3.15). This formula differs from the shortcut method in the correction factor term. After the d' values are averaged, the result must be multiplied by the size of the class interval in order to return to the units of the original data.

Step-Deviation Method for the Arithmetic Mean—Grouped Data

(3.15)
$$\bar{X} = \bar{X}_a + \left(\frac{\sum_{i=1}^{k} f_i d'_i}{n}\right)c$$

where $\bar{X}$ = the arithmetic mean

$\bar{X}_a$ = the assumed mean

f_i = the frequency of the ith class

$d'_i = \frac{X_i - \bar{X}_a}{c}$ = the deviation of the midpoint of the ith class from the assumed mean in class interval units

n = the number of observations

k = the number of classes

c = the size of a class interval

If we dispense with subscripts, this formula becomes

(3.15′)
$$\bar{X} = \bar{X}_a + \left(\frac{\Sigma fd'}{n}\right)c$$

The step-deviation method is illustrated in Table 3-10 for the same frequency distribution of personal loans given in Table 3-9, taking again an assumed mean of \$500. This method results in simpler arithmetic than either the direct definitional formula or the shortcut method, particularly if the class intervals and frequencies involve a large number of digits.

Table 3-10 Calculation of the Arithmetic Mean for Grouped Data by the Step-Deviation Method: Personal Loan Data.

Amount of Loans	*Number of Loans* f	d'	fd'
\$ 0 and under \$ 200	6	−2	−12
200 and under 400	18	−1	−18
400 and under 600	25	0	0
600 and under 800	20	+1	+20
800 and under 1000	17	+2	+34
1000 and under 1200	14	+3	+42
	100		+66

$\bar{X}_a = \$500$

$$\bar{X} = \bar{X}_a + \left(\frac{\Sigma fd'}{n}\right)c = \$500 + \left(\frac{66}{100}\right)\$200 = \$632$$

3.9 The Weighted Arithmetic Mean

In averaging a set of observations, it is often necessary to compute a so-called "weighted average" in order to arrive at the desired measure of central location. A *weighted arithmetic mean* of a set of observations is calculated by multiplying (or weighting) each observation by its appropriate weight, totalling these products, and then dividing this total by the sum of the weights. Symbolically, the weighted arithmetic mean is given by the following formula:

$$\bar{X}_w = \frac{\Sigma wX}{\Sigma w} \qquad \textbf{(3.16)}$$

where $\bar{X}_w$ = the weighted arithmetic mean
X = the values of the observations
w = the weights applied to the X-values

An illustration of the use of a weighted arithmetic mean is given in the following simple example involving the averaging of profit-to-sales ratios.

Example 3-1 Suppose a company consisted of three divisions, all selling different product lines. The net profit-to-sales ratios for these divisions for the year 1955 were: Division A, 4%; Division B 5%, and Division C, 6%. What was the average (arithmetic mean) net profit-to-sales ratio?

Before attempting to answer this question a few comments are in order. First, it must be indicated that the question is incompletely specified and is, therefore, ambiguous. It is not clear whether each division is to be treated as a unit regardless of its size or whether the importance of each division is in some sense to be considered in computing the average.

Secondly, the data in this problem are referred to as *ratios* of net profit-to-sales. A moment's digression on this point may be useful. A ratio is simply an expression of a relationship between the values of two (or more) variables. This relationship may be expressed in various ways. A fraction may be used, in which the numerator gives the value of one variable and the denominator expresses the number of units of the second variable. For example, referring to the three divisions of the company mentioned earlier, if division A had net profits of $400,000 and sales of $10,000,000, the ratio of net profits-to-sales might be expressed as the fraction $400,000/$10,000,000. The result of the division of the two numbers in the fraction may be expressed as a decimal, for example, 0.04; or as a percentage, for example, 4%. In fact, we may express the relationship in terms of any number of units of the variable in the denominator of the ratio; for example, $4 net profit per $100 of sales, $0.40 net profit per $10 of sales, or $0.04 net profit per $1 of sales.

Let us now return to the question, "What is the average (arithmetic mean) net profit-to-sales ratio?" If this ambiguous question is rephrased to read "What is the arithmetic mean net profit-to-sales ratio per division (without regard to the sales size of these divisions)?" the answer is given by

$$\bar{X} = \frac{\Sigma X}{n} = \frac{.04 + .05 + .06}{3} = \frac{.15}{3} = .05 = 5\% \text{ per division}$$

where X = the values of the ratios for the three divisions
n = the number of divisions

The result ($\bar{X}$) may be referred to as an "unweighted arithmetic mean" of the three ratios. In this computation, the net profit-to-sales ratios were totalled, and the result divided by the number of divisions. Note, therefore, that the result is stated as 5% *per division*, because of the appearance of the three divisions in the denominator. It is clear that this calculation disregards differences that may exist in the amount of sales of the three divisions. However, this does not signify that the average is therefore meaningless. If interest is centered upon obtaining a "representative" or "typical" profit ratio, and there are no extreme values to distort the representativeness of the unweighted mean, then this type of computation is a valid one. Of course, if one were seeking a "typical" or "representative" figure, it would be desirable to have more than the three observations present in this illustration. Furthermore, other averages such as the median or mode may be preferable to the arithmetic mean in the determination of a typical or representative value. However, the point of this discussion is that the unweighted arithmetic mean is not a meaningless figure; indeed it is the correct answer to the question earlier posed, with or without the parenthetical phrase, "without regard to the sales size of these divisions."

On the other hand, suppose the question we want to answer is, "What is the arithmetic mean net profit-to-sales ratio *for the three divisions combined?*" Equivalently, the question may be worded, "What is the arithmetic mean net profit-to-sales ratio *for the company as a whole?*" The answer to these questions is a figure which relates total net profits to total sales for the three divisions combined. It is clear that if we have only the profit ratios for the three divisions, we do not have enough information to compute the required average. However, if we are given the dollar sales for each of the three divisions, that is, the denominators of the three net profit-to-sales ratios, then these figures can be used as "weights" in calculating the desired average. Specifically, the weighted arithmetic mean would be computed according to Equation (3.16), where X represents the net profit-to-sales ratios of the three divisions and w represents the sales of the three divisions. Assume that the sales were: Division A, \$10,000,000; Division B, \$10,000,000; and Division C, \$30,000,000. The computation of the required weighted average is given in Table 3.11.

It can be seen in Table 3-11 that when the net profit-to-sales ratio or percentage for each division is multiplied (weighted) by the sales of that division, the resultant figure is the net profit of the division. That is, (Net Profit/Sales) (Sales) = Net Profit. Thus, when the sum of the net profits (ΣwX) is divided by the sum of the sales figures (Σw), we have the desired average.

Table 3-11 Calculation of the Weighted Arithmetic Mean Net Profit-to-Sales Ratio for the Three Divisions of a Company, 1955.

Division	*Net Profit to Sales (percent)* X	*Sales* w	*Net Profit* wX
A	4	\$10,000,000	\$ 400,000
B	5	10,000,000	500,000
C	6	30,000,000	1,800,000
		\$50,000,000	\$2,700,000

$$\bar{X}_w = \frac{\$2,700,000}{\$50,000,000} = 5.4\%$$

That is, the weighted arithmetic mean of the three profit ratios can be seen to be nothing more than the figure that would have been obtained by totalling net profits for the company as a whole (the three divisions combined), and dividing that figure by total sales for the company as a whole. Since dollar sales are in the denominator of the $\Sigma wX/\Sigma w$ ratio, the answer of 5.4% may be interpreted in terms of dollars of profit per dollar of sales, that is, an average of $0.054 profit per dollar of sales.

The weights that were applied to the three profit-to-sales ratios in this problem were the actual dollar amounts of sales for the three divisions. That is, the weights used were the values of the denominators of the original ratios. An alternative procedure would be to weight the ratios by a *percentage breakdown* of the denominators, in this case, a percentage breakdown of total sales. For example, in the computation shown in Table 3-11, if instead of weights of $10,000,000, $10,000,000, and $30,000,000, weights of 20%, 20%, and 60% had been applied, the same answer of 5.4% would have resulted. Indeed, if any figures in the same proportions as $10,000,000, $10,000,000, and $30,000,000 had been employed, the same numerical answer would have resulted. It is clear that the reason the weighted arithmetic mean of 5.4% exceeded the unweighted mean of 5.0% was that in the weighted mean calculation, greater weight was applied to the 6% profit figure for Division *C* than to the corresponding 4% figure for Division *A*. This had the effect of pulling the weighted average up toward the 6% figure.

The net profit-to-sales ratio example given above is an illustration of the principle of weighting ratios by their denominators to obtain a weighted average. In certain special situations, as in the calculation of index numbers for measuring price movements, the weights need not represent the denominators of the ratios that are being combined. For example, assume that we have a number of "price relatives" for food commodities in 1965 relative to 1960. Thus, the price relative for bread is (say) the price of a loaf of bread in 1965 as a ratio of the price of the exact same loaf in 1960. In one method of index number construction such price relatives are weighted by the amounts of money that consumers spent on these commodities in the base period. However, whenever it is desired that the weighted mean represent the total of the numerators of the ratios being averaged divided by the total of the denominators, that is when a mean ratio for all companies *combined*, all cities *combined* etc. is wanted, then weights representing the denominators of the original ratios or a percentage breakdown of these denominators must be used.

Shifts in the pattern of weights applied to a set of ratios sometimes represent the key to a correct analysis of a given situation. For example, consider the following question, "Is it possible for each of the divisions in the calculation displayed in Table 3-11 to have increased its net profit-to-sales ratio from one year to another, while the net profit-to-sales percentage for the company as a whole has decreased?" A first superficial reaction might be to reason as follows: "If each of the three component divisions of a company had become more profitable, how could the company as a whole have been less profitable?" However, a moment's thought indicates that this result is possible, simply by having a greater relative weight applied to Division *A*'s profit percentage, and less relative weight to Division *C*'s percentage. The figures in Table 3-11 pertain to the year 1955. Let us assume that the divisional net profit-to-sales ratios and dollar volumes of sales in 1965 were as follows: Division *A*, 4.2% on $40,000,000 of sales; Division *B*,

5.2% on $30,000,000 of sales; and Division *C*, 6.2% on $30,000,000 of sales. Therefore, each division had increased its profit ratio from 1955 to 1965. The calculation of the weighted mean net profit-to-sales ratio for 1965 is given in Table 3-12.

Thus, although each division's profit rate increased from 1955 to 1965, the profit rate for the company as a whole dropped from 5.4% to 5.1%. It may be noted that the dollar amount of sales of each division remained the same (Division *C*) or increased (Divisions *A* and *B*). However, the relevant point in explaining the decrease from 5.4% to 5.1% is that in 1955 greater relative weight was applied to the highest profit rate (Division *C*, weight 60%), than to the lowest profit rate (Division *A*, weight 20%); whereas in 1965, greater relative weight was applied to the lowest profit rate (Division *A*, weight 40%) than to the highest profit rate (Division *C*, weight 30%).

Table 3-12 Calculation of the Weighted Arithmetic Mean Net Profit-to-Sales Ratio for the Three Divisions of a Company, 1965.

Division	*Net Profit to Sales (%)* X	*Sales* w	*Net Profit* wX
A	4.2	$ 40,000,000	$1,680,000
B	5.2	30,000,000	1,560,000
C	6.2	30,000,000	1,860,000
		$100,000,000	$5,100,000

$$\bar{X}_w = \frac{\$5,100,000}{\$100,000,000} = 5.1\%$$

Of course, if the lowest divisional profit rate in 1965 exceeded the highest divisional profit rate for 1955, that is, if there was no area of overlap in the range of the profit rates for the two years, then it would not have been possible for the profit rate for the company as a whole to have been lower in 1965 than in 1955. The important point here is to note that since the profit rate for the company as a whole is an *average* of the profit rates of the component divisions, it must fall within the range of the individual items which are averaged.

An interesting point arises from a consideration of the above illustration. That is, any ratio such as the net profit-to-sales percentage for the company as a whole may be thought of as the weighted average of the ratios of its possible components. Thus, for example, a death rate for a city may be conceived of as a weighted mean of the death rates for males and females, for whites and non-whites, for the various age classes, for the component religious groupings, etc. Therefore, in determining reasons for differences or changes in ratios of two totals such as net profits and sales, the crux of the analysis is frequently in terms of differences or changes in the relevant component ratios. As another illustration of this point, suppose one is faced with the information that the crude death rate (number of deaths divided by population) in St. Petersburg, Florida is greater than the corresponding death rate in Los Angeles, California. What accounts for this higher mortality rate? In an informal search for explanations, people acquainted with the two cities would probably agree that the reason for this difference in death rates is not to be found in the relative atmospheric purity of the two cities. That is, it is highly doubtful that the smog-laden atmosphere of Los Angeles is more conducive to lower mortality incidence than the atmosphere which envelops the Florida city. Of course, a large number of factors can be suggested which conceivably could be contributory elements. However, if one is given the further information that St. Petersburg is a favorite retirement city, the critical point is apparent. Because of the retirement phenomenon, St. Petersburg has a much higher proportion of persons in the "65 and older" age group than does Los Angeles. If we think of the crude death rate of a city as the weighted mean of the death rates for component age groupings of the population, then the high death rates of the older age groups get relatively greater weight in the case of St. Petersburg than in Los Angeles, and the lower death rates of the younger age groupings gets correspondingly less weight.

Example 3-2 Consider the following frequency distribution of current ratios (current assets divided by current liabilities), for 200 firms in the same industry:

Current Ratio (Current Assets ÷ Current Liabilities)	*Number of Firms*
.5 and under 1.0	8
1.0 and under 1.5	48
1.5 and under 2.0	60
2.0 and under 2.5	40
2.5 and under 3.0	28
3.0 and under 3.5	16
	200

Assume that total current liabilities for all 200 firms amounted to 600 million dollars. If the arithmetic mean of the above distribution were multiplied by 600 million dollars, would the result equal total current assets for the 200 firms? Why or why not?

The multiplication would *not* ordinarily equal total current assets for the 200 firms. It would be erroneous to argue that the reason is the assumption in the calculation of the arithmetic mean of the frequency distribution that the midpoints of the class intervals are the arithmetic means of the observations in these intervals. Even if this assumption were perfectly correct, the indicated multiplication would not yield total current assets for the 200 companies. The reason for this is that the arithmetic mean of the frequency distribution would have to be equal to the ratio of total current assets to total current liabilities for the 200 companies in order for the multiplication to yield the required result. The ratio of total current assets to total current liabilities for all companies combined is conceptually the weighted arithmetic mean of the current ratios of the 200 companies. However, it is important to realize that the arithmetic mean of the given frequency distribution is not an estimate of the weighted mean, but rather an estimate of the unweighted mean of the 200 current ratios. That is, it is merely an estimate of the figure which would have been obtained by dividing the sum of the original current ratios by 200, the number of firms. Of course, in certain very special cases in which the unweighted mean is equal to the weighted mean of the given ratios, the multiplication referred to earlier would yield total current assets, except for the discrepancy introduced by the assumption concerning midpoints of classes. For example, in the far-fetched situation where all 200 firms have the same current liabilities, the unweighted mean and the weighted mean of the current ratios would be equal. Another special situation would be one in which the weights (current liabilities), although different, just happen to be distributed in such a way that the unweighted mean and weighted mean were equal.

If the arithmetic mean of the frequency distribution of current ratios using number of firms as frequencies is an estimate of the unweighted mean, what type of frequencies would be required to obtain an estimate of the weighted mean of the current ratios?

It is clear from the earlier discussion that if we had the original 200 current ratios, that is, if we had the ungrouped data, the weighted mean could be obtained by applying the formula, $\bar{X}_w = \Sigma wX/\Sigma w$. In other words, we would multiply each ratio by the amount of current liabilities represented in that ratio (the denominator), add up these products, and divide by total current liabilities. However, if the ratios are grouped into class intervals, and the identity of the original data is thereby lost, we can obtain an estimate of the weighted mean by using as frequencies total current liabilities of the companies represented in each class. In this procedure, the midpoint assumption means using the midpoint as an estimate of the weighted mean current ratio for each class.

We summarize this discussion using symbols.

Let CA_i = the current assets of the ith firm

CL_i = the current liabilities of the ith firm

$\dfrac{CA_i}{CL_i}$ = the current ratio of the ith firm

Then, the mean of the distribution using number of firms as frequencies is an estimate of the unweighted mean

$$\bar{X} = \frac{\sum_{i=1}^{200} \frac{CA_i}{CL_i}}{200}$$

The mean of the distribution, using as frequencies total current liabilities of the companies represented in each class, is an estimate of the weighted mean

$$\bar{X}_w = \frac{\sum_{i=1}^{200} CA_i}{\sum_{i=1}^{200} CL_i}$$

3.10 The Median

The *median* is another well-known and widely used average. It has the connotation of the "middlemost" or "most central" value of a set of numbers. For ungrouped data, it is defined simply as the value of the central item when the data are arrayed by size. If there is an odd number of observations, the median is directly ascertainable. On the other hand, if there is an even number of items, there are two central values. In that case, by convention, a value halfway between these two central observations is designated as the median.

For example, suppose a test revealed that the lifetimes of five electronic devices in hours were 852, 931, 1010, 1027, and 1068. The median lifetime would be 1010 hours. If another device were tested, and its lifetime determined to be 1020 hours, the array would now read 852, 931, 1010, 1020, 1027, and 1068. The median would be a value halfway between 1010 and 1020, or 1015.

Another way of viewing the median is as a value below and above which lie an equal number of items. Thus, in the preceding illustration involving five observations, two lie above the median and two below. In the example involving six observations, three fall above and three fall below the median. Of course, in the case of an array with an even number of items, any value lying between the two central items may, strictly speaking, be referred to as a median. However, as indicated earlier, the convention is to use the midpoint between the two central items. In the case of tied values at the center of a set of observations, there may be, of course, no value such that equal numbers of items lie above and below it. Nevertheless, the central value, as defined in the preceding paragraph, is still designated as the median. For example, in the following array: 52, 60, 60, 60, 60, 61, 62, the number 60 is the median, although unequal numbers of items lie above and below this value.

In the case of a frequency distribution, since the identity of the original observations is not retained, the median, of necessity, is an estimated value.

Because in a frequency distribution the data are arranged in order of magnitude, frequencies can be cumulated to determine the class in which the median observation falls. It is then necessary to make some assumption about the way in which observations are distributed in that class. Conventionally the assumption is made that observations are equally spaced or evenly distributed throughout the class containing the median. The value of the median is then established by a linear interpolation. The procedure is illustrated for the distribution of personal loans previously given in Table 3-9. That distribution is shown again in Table 3-13. First, the calculation of the median is explained without the use of symbols. Then the procedure is generalized by stating it in the form of a formula.

In the distribution shown in Table 3-13, since there are 100 loans represented, the median lies between the 50th and 51st loans. Since 49 loans occur

Table 3-13 Calculation of the Median for a Frequency Distribution: Personal Loan Data.

	Number of Loans f	
\$ 0 and under \$ 200	6	
200 and under 400	18	
400 and under 600	25	
		$\Sigma f_p = 49$
600 and under 800	20	
800 and under 1000	17	
1000 and under 1200	14	
	100	

prior to the class "\$600 and under \$800," it is clear that the median is contained in that class. Assuming that the 20 loans are evenly distributed between \$600 and \$800, we can determine the median observation by interpolating 1/20 of the distance through this \$200 class. In summary, the median is calculated by adding 1/20 of \$200 to the \$600 lower limit of the class containing the median. That is,

$$Md = \$600 + \left(\frac{50 - 49}{20}\right)\$200 = \$600 + \left(\frac{1}{20}\right)\$200 = \$610.00$$

Thus, the formula for calculating the median of a frequency distribution is

$$Md = L_{Md} + \left(\frac{n/2 - \Sigma f_p}{f_{Md}}\right)c \tag{3.17}$$

where Md = the median
L_{Md} = the (real) lower limit of the class containing the median
n = the total number of observations in the distribution
Σf_p = the sum of the frequencies in classes preceding the one containing the median
f_{Md} = the frequency of the class containing the median
c = the size of the class interval

At this point, one may feel that we have located the value of the 50th observation rather than one falling midway between the 50th and 51st. However, the value determined is indeed one lying halfway between the 50th and 51st observations. This can be ascertained by examining the assumption of an even distribution of items within the class in which the median falls.

If there are 20 observations in the $200 class from $600 to $800, we may think of the class as being divided into 20 equal observation intervals of $10 each. This is depicted in Figure 3-5. The observations are assumed to be

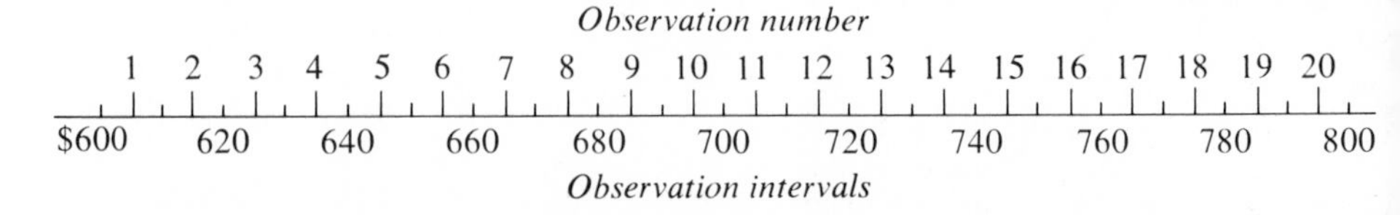

Figure 3-5 Diagram depicting the meaning of the assumption concerning an even distribution of observations.

located at the midpoints of these observation intervals. An interpolation of 1/20 through the class interval brings us to the end of the first observation interval, $610, which is seen to be a value halfway between the first and second observations. Since 49 frequencies preceded this class, the median of $610 is a value lying midway between the 50th and 51st observations.

3.11 Characteristics and Uses of the Arithmetic Mean and Median

The preceding sections have concentrated on the mechanics of calculating means and medians for ungrouped and grouped data. We now turn to the characteristics and uses of these averages.

The arithmetic mean is doubtless the most widely used and most familiar measure of central tendency. In fact, to many people it is "the average." The mean possesses the advantage of being a rigidly defined mathematical value and, therefore, is capable of algebraic manipulation. For example, the means of two related distributions can be combined by suitable weighting. Also if two of the quantities in the formula $\bar{X} = \Sigma X/n$ are known, the third can be obtained directly. Because of such mathematical properties, the arithmetic mean is used more often in advanced statistical techniques than any of the other averages.

The arithmetic mean is also important in connection with statistical inference as well as for descriptive purposes. For example, the arithmetic mean of a random sample of observations may be used to estimate the value of the corresponding arithmetic mean of the population from which the sample was drawn. Thus, the mean family income of a sample of families in the city of New York may be used as an estimate of the mean income of all families in that city. In this type of estimation, the mean is a more reliable estimator than other averages, such as the median, mode, or geometric mean. That is, the mean is less affected by sampling fluctuations than are the other measures of central tendency and it estimates the corresponding population figure more closely, on the average.

Another useful interpretation of the mean is as an estimated value for any item in a distribution. It may be recalled that the sum of the deviations of a set of observations from their mean is equal to zero. Therefore, if the mean of the set is used as an estimate of the value of any observation picked at random, and this procedure is repeated, then on the average, the mean amount of error, taking account of the sign of the error or deviation, is zero.

A disadvantage of the mean is its tendency to be distorted by extreme values at either end of a distribution. In general, it is pulled in the direction of these extremes. Another disadvantage is that the mean cannot be determined from a frequency distribution which has an open-ended interval such as "$25,000 and over," since the midpoint of such an interval is unknown. However, one way of handling this problem if the order of magnitude of the figures in the open-ended interval is known, is to make assumptions which will permit the calculation of the mean. What is needed is an estimate of the total value (fX) of observations in the open-ended interval. Therefore, an arithmetic mean for these items may be estimated and multiplied by the number of items in the interval to give an estimate of the total value. For example, in the above illustration of family incomes of $25,000 and over, an estimate is obtained of the total incomes falling in that interval.

Another approach to the problem of the open-ended interval is as follows. Assume a mean for this open-ended interval which you are certain represents a lower limit for the actual mean. Calculate the mean of the frequency distribution. Then repeat this procedure, assuming a mean for the open-ended interval which represents a probable upper limit. Thus, two means are obtained for the distribution which are quite likely to bracket the actual mean. The two means calculated under the different assumptions are apt to be quite close together particularly if the contribution of the open-ended interval to the ΣfX for the entire distribution is relatively small.

The median is also a very useful measure of central tendency. Its relative freedom from distortion by skewness in a distribution makes it a particularly desirable average for descriptive purposes. Thus, it is often used to convey the idea of a "typical" observation. It is primarily affected by the number of observations rather than their size. This can be seen by considering an array in which the median has been determined. Now assume that the largest item is increased (say) one hundredfold. The median remains unchanged. The arithmetic mean would, of course, be pulled toward the large extreme item.

The median may be interpreted as a "best estimate" value in a somewhat different sense than is the mean. As previously indicated, the mean is a best estimate in the sense that if observations are repeatedly drawn at random from the distribution and the mean is used to estimate each value, the mean amount of error or deviation is zero. On the other hand, if the same experiment of repeated estimation is used with the *median* being the estimated value, then the average (mean) amount of absolute error (that is, disregarding the sign of the error) is a *minimum*. In other words, the average deviation of the observations from the median is less than from any other value in the distribution. Thus, in an estimation situation, if one wants to minimize the average absolute amount of error and the sign of the error is not particularly important, then the median is preferable to the arithmetic mean.

Another merit of the median is the fact that it can ordinarily be computed for open-ended distributions. This is true, since the value of the median is determined solely from the interval in which it falls. It would virtually never fall into the open-ended interval, since that interval would have to contain more than one-half of the frequencies to include the median. It is extremely improbable that such a frequency distribution would ever be constructed. Therefore, for practical purposes, the median can be computed for open-ended distributions.

The major disadvantage of the median is that it is an average of position and hence is not a mathematical concept suitable for further algebraic treatment. For example, if one knows the medians of two distributions, there is no algebraic way of combining these two figures to obtain the median of the two combined distributions.

Also, the median tends to be a rather unstable value if the number of items is small. Furthermore, as observed earlier, the median is a less reliable average than the mean for estimation purposes, since it is more affected by sampling variations.

3.12 The Mode

Another average, which is conceptually very useful, but is often not explicitly calculated is the *mode*. In French, to be "in the mode" implies to be in fashion. The mode as a statistical average is the observation that occurs with the greatest frequency and thus is the most fashionable value. The mode is rarely determined for ungrouped data. The reason is that quite accidentally or haphazardly an item might occur more often than any others, yet may lie at the lower or upper end of the array of observations, and be a very unrepresentative figure. Therefore, a determination of the mode is generally not even attempted for ungrouped data.

In the case of data grouped into a frequency distribution, it is not possible to specify the observation that occurs most frequently, since the identity of the individual items is lost. However, since the data are grouped, we can determine the so-called "modal class," or the class which contains more observations than any other. Of course, class intervals should be of the same size when this determination is made. As a practical matter, most analysts of business and economic data do not usually proceed any further than the specification of a modal class in the measurement of a mode for an empirical frequency distribution. When the location of the modal class is considered along with the arithmetic mean and median, much useful information is generally conveyed not only about central tendency but also about the skewness of a frequency distribution.

Several interpolation formulas have been developed for determining the location of the mode within the modal class. These usually involve assumptions concerning the use of frequencies in the classes preceding and following the modal classes as weighting factors which tend to pull the mode up or down from the midpoint of the modal class. We shall not present any of these formulas here. However, it is useful to consider the meaning of such an estimated mode. Let us visualize the frequency polygon of a distribution and then the frequency curve approached as a limiting case by gradually reducing class size. In the limiting situation, the variable under study may be considered as continuous rather than as occurring in discrete classes. The mode may then be thought of as the value on the horizontal axis lying below the maximum point on the frequency curve (see Figure 3.6).

In the case of a frequency distribution, the mode has the connotation of typical or representative value. It specifies a location in the distribution at which there is a maximum of clustering. In this sense, it serves as a standard against which to judge the representativeness or typicality of other averages. If a frequency distribution is symmetrical, the mode, median, and mean

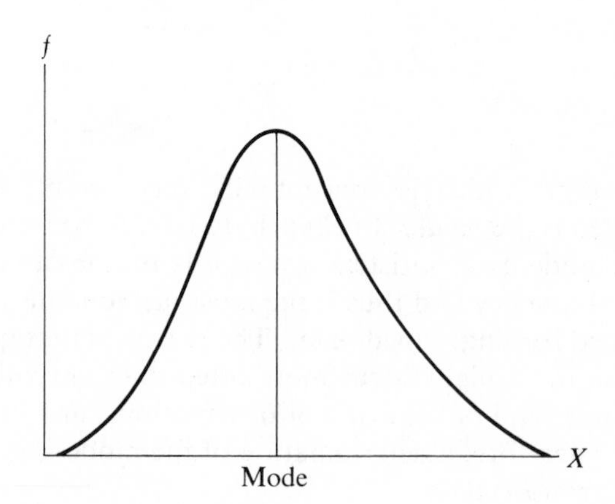

Figure 3-6 The location of the mode.[1]

coincide. As earlier noted, when there are extreme values in a distribution, the arithmetic mean is pulled in the direction of these extremes. Stated somewhat differently, in a skewed distribution, the mean is pulled away from the mode toward the extreme values. The median also tends to be pulled away from the mode in the direction of skewness, but is not affected as much as the mean. If the mean exceeds the mode, a distribution is said to have "positive skewness" or is "skewed to the right"; if the mean is less than the mode, the terms "negative skewness" and "skewed to the left" are used. The order in which the averages fall in these types of distributions is depicted in Figure 3.7.

Many distributions of economic data in the U. S. are found to be skewed to the right. Examples include the distributions of incomes of individuals,

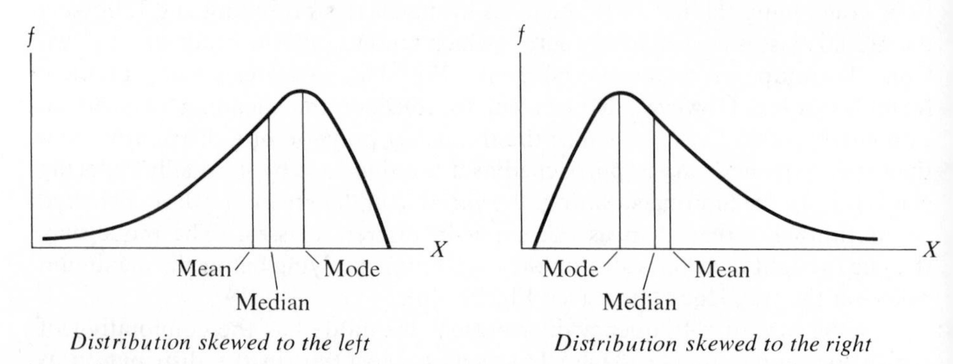

Figure 3-7 Skewed distributions depicting typical positions of averages.

[1]In this and subsequent graphs, the X on the horizontal axis denotes values of the observations and the f on the vertical axis denotes frequency of occurrence.

savings of individuals, corporate assets, sizes of farms, company sales within many industries, and so forth. In many of these instances, the arithmetic mean is pulled so far from the mode as to be a very unrepresentative figure.

Multimodal Distributions

If more than one mode appears, the frequency distribution is referred to as "multimodal"; if there are two modes it is referred to as "bimodal." Extreme care must be exercised in analyzing such distributions. As an example, consider a situation in which one wants to compare the arithmetic mean heights of undergraduate students at University *A* and University *B*. Assume that the mean calculated for University *A* exceeds that of University *B*. If one concludes from this finding that undergraduate students at University *A* are taller, on the average, than those at University *B*, without recognizing the fact that the height distribution for each of these universities is bimodal, serious errors of inference may be made. In order to illustrate the principle involved, let us assume that the mean height for undergraduate women is five feet, four inches and the mean height for undergraduate men is five feet, nine inches at each of these universities. Also, let us assume that the individual distributions of heights of women and men are symmetrical and that there are the same total number of undergraduate students on the two campuses. However, suppose that 70% of the undergraduates at University *A* are men, whereas only 50% of the undergraduates at University *B* are men. The frequency curves of the distributions of heights at the two universities would appear as indicated in Figure 3.8. Clearly, the mean height of students at University *A* exceeds that at University *B*. This is true simply because there

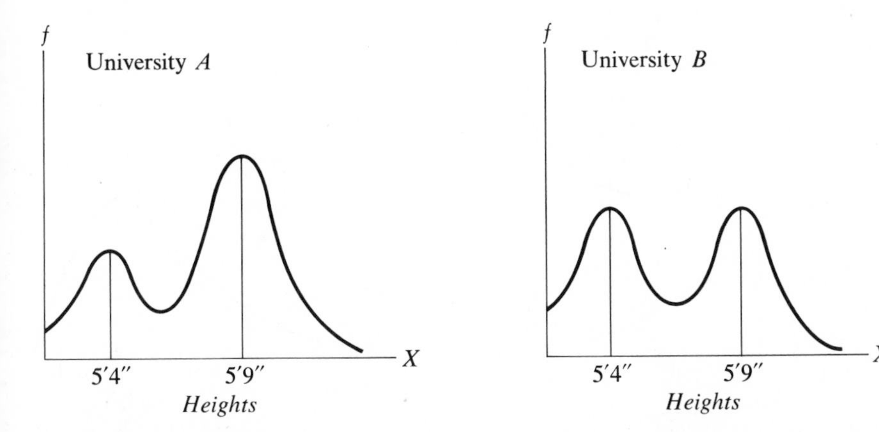

Figure 3-8 Bimodal frequency distributions: Heights of undergraduate students at two Universities.

is a higher percentage of men at University A. However, if one were ignorant of this fact, he might be tempted to infer reasons why students at University A are taller than at University B. The fact of the matter is that women at the two universities are equal in height. The same is true for the men. What is required here is to separate two height distributions on each campus, one for men and one for women. A comparison of the mean heights of women on two campuses would reveal their equality; similarly, for men.

The principle involved here is one of homogeneity of the basic data. The fact that a height distribution is bimodal suggests that there are two different "cause systems" present, and that two distinct distributions should be separated out. The data on heights may be said to be non-homogeneous with respect to the factor of sex. Other examples of bimodal distributions might include the merging of wage data for unskilled and skilled workers, data on dimensions of products from two different suppliers, and so forth.

If a basis for separating a bimodal distribution into two distributions cannot be found, then extreme care must be used in describing the data. In cases such as shown for University B in Figure 3.8, where the heights of the two modes are about equal, the arithmetic mean and median will probably fall between the modes and will be unrepresentative of the large concentrations of values lying at the modes below and above these averages.

3.13 Comparison of Averages

The differences between the values of the mean, median, and mode reveal useful information concerning the shape of a frequency distribution. If a unimodal distribution is perfectly symmetrical as shown in Figure 3.9a, the

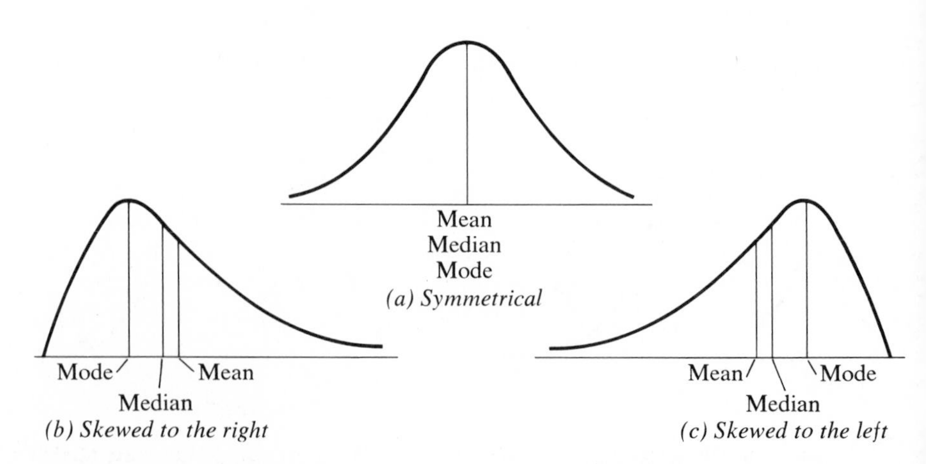

Figure 3-9 Location of averages in three types of distributions.

mean, median, and mode all coincide. On the other hand, if the distribution is not symmetrical (skewed), the median and mean tend to be pulled away from the mode in the direction of the skewness. The typical location of the three averages in moderately skewed distributions is shown in Figure 3.9b and c. In a distribution skewed to the right, as in Figure 3.9b, the mean and median tend to fall to the right of the mode, with the mean being pulled somewhat further than the median in the direction of the extreme values. This is true because the mean is more directly affected by the extreme values than is the median, which depends solely on the centrally located value. Skewness to the right is also referred to as "positive skewness," since the mean minus the median is a positive number. Measures of skewness have been developed which are functions of this difference[2] (or the difference between the mean and the mode), but such measures are infrequently calculated in business and economic applications. The frequency distribution of personal loan data shown in Tables 3-9 and 3-10 is an example of a distribution skewed to the right. If the midpoint of the modal class is taken as the mode, the respective values of the three averages are: mode = \$500, median = \$610, and mean = \$632.

Similarly, in the case of skewness to the left ("negative skewness"), the mean and median tend to fall to the left of the mode, with the mean falling somewhat further out in the left-hand tail of the distribution. Skewness to the right is much more frequently encountered in business and economic data, with many distributions displaying a concentration of frequencies at relatively low values, with a small number of extremely high values causing a tail extending to the right. This type of configuration tends to arise in the case of data, as for example, size of firm, company net worth, family income, etc., where there is a fixed or effective lower limit such as zero.

The major use of the mode is as a basis of comparison for other averages. Since the mode is a value at which there is maximum clustering of frequencies, the typicality of the median and mean is often judged in terms of how far these figures fall from the mode. However, typicality of averages also depends upon the spread or dispersion of items in a distribution. For example, it is possible to have a perfectly symmetrical distribution in which the mode, median, and mean coincide, yet none of these averages may be very typical because of the wide spread of the individual observations.

One of the main disadvantages of the mode is the fact that it is not a rigidly defined mathematical average for empirical frequency distributions. Thus, it tends to have little application in more advanced statistical techniques.

---

[2]One well-known measure is Pearson's Coefficient of Skewness, which is defined as

$$\text{Skewness} = \frac{3(\text{Mean} - \text{Median})}{\text{Standard deviation}}$$

3.14 The Geometric Mean

In business and economic problems, questions frequently arise concerning average percentage rates of change over time. Neither the mean, median, nor mode is the appropriate average to use in these instances. For example, consider the following time series for annual sales of the Gallison Equipment Company from 1960 through 1963:

	Sales
1960	$10,000,000
1961	8,000,000
1962	12,000,000
1963	15,000,000

What was the average percentage rate of change per year in sales? To answer this question, we must specify what we mean by the "average percentage rate of change per year." The most generally useful interpretation of this term is the constant percentage rate of change which if applied each year would take us from the first to the last figure. Hence, in the above illustration we would be interested in that constant yearly percentage rate of change which would be required to move from $10,000,000 of sales in 1961 to $15,000,000 in 1963. Clearly, none of the previously discussed averages provides the correct answer to this question. The correct answer can be obtained through the use of the geometric mean, or what amounts to the same thing, through the use of the familiar compound interest formula. In the discussion which follows, the geometric mean is defined, and the relationship between this average and compound interest calculations is indicated. Then these concepts are applied to the average percentage rate of change per year in sales of the Gallison Company.

The geometric mean, (G), of n numbers is defined as the nth root of the product of n numbers.

(3.18) $$G = \sqrt[n]{X_1 \cdot X_2 \cdots X_n}$$

This definition can be contrasted to that of the arithmetic mean, ($\bar{X}$), which is the sum of the n numbers divided by n,

$$\bar{X} = \frac{X_1 + X_2 + \cdots + X_n}{n}$$

In the calculation of the geometric mean, the n numbers are multiplied together, then the nth root is extracted, whereas in the case of the arithmetic mean, the n numbers are added together, and then this total is divided by n. It can be shown that the geometric mean of n numbers is always equal to or less than the arithmetic mean. When all of the numbers are identical, the two averages are equal. For example, consider the numbers 9 and 4. The arithmetic mean is $(9 + 4)/2 = 6\text{-}1/2$. The geometric mean is $\sqrt{9 \times 4} = 6$. On the other hand, if the two numbers are identical, such as 6 and 6, the arithmetic mean is $(6 + 6)/2 = 6$ and the geometric mean is $\sqrt{6 \times 6}$ or also 6.

In the preceding simple examples, because the geometric mean involved a number whose square root was an integer, its calculation was trivial. In more realistic problems, logarithms must be used in the calculation. Thus, to determine the geometric mean of n numbers, we first obtain the logarithm of the geometric mean by adding together the logs of the n numbers and dividing this total by n. Then the antilogarithm is taken to yield the geometric mean. Symbolically, we have

(3.19)
$$\log G = \frac{\Sigma \log X}{n}$$

and

(3.20)
$$G = \text{antilog}\,\frac{\Sigma \log X}{n}$$

For example, suppose the geometric mean of the three numbers 10, 12, and 16 is required. That is,

$$G = \sqrt[3]{10 \times 12 \times 16}$$

Taking logarithms gives

$$\log G = (1/3)(\log 10 + \log 12 + \log 16)$$

From Appendix Table A-4, we have

$$\begin{aligned} \log 10 &= 1.0000 \\ \log 12 &= 1.0792 \\ \log 16 &= \underline{1.2041} \\ & 3.2833 \end{aligned}$$

$$\begin{aligned} \log G &= (1/3)(3.2833) = 1.0944 \\ G &= \text{antilog } 1.0944 = 12.43 \end{aligned}$$

Now we turn to the compound interest formula as an application of the geometric mean. Suppose \$100 was invested at 5% compound interest for three years. Let

P_0 = original principal (\$100)
P_1 = amount accumulated at the end of the first year
P_2 = amount accumulated at the end of the second year
P_3 = amount accumulated at the end of the third year

Then,[3]

$$P_1 = \$100(1.05) = \$105.0$$
$$P_2 = \$100(1.05)(1.05) = 100(1.05)^2 = \$110.2$$
$$P_3 = \$100(1.05)(1.05)(1.05) = \$100(1.05)^3 = \$115.8$$

In general, if

P_0 = original principal
r = rate of interest expressed as a decimal
n = number of compounding periods
P_n = amount accumulated at the end of n periods

then,

(3.21) $$P_n = P_0(1 + r)^n$$

If interest is compounded at different rates in each time period, and if these successive rates are denoted $r_1, r_2, \ldots, r_n$, then the amount accumulated at the end of n periods with an original principal of P_0(dollars) is

(3.22) $$P_n = P_0(1 + r_1)(1 + r_2)\cdots(1 + r_n)$$

Thus, if interest were earned on \$100 at 3%, 5%, and 6%, respectively, in three periods, the amount accumulated at the end of the third period would be

$$P_3 = \$100(1.03)(1.05)(1.06) = \$114.6$$

The relationship between compound interest and the geometric mean can now be seen by a joint consideration of formulas (3.21) and (3.22). If the expression for P_n in (3.21) is substituted on the left-hand side of (3.22), we have, after cancellation of P_0,

$$(1 + r)^n = (1 + r_1)(1 + r_2)\cdots(1 + r_n)$$

Taking the nth root of both sides of this equation gives,

(3.23) $$(1 + r) = \sqrt[n]{(1 + r_1)(1 + r_2)\cdots(1 + r_n)}$$

Thus, $1 + r$ is seen to be the geometric mean of the $(1 + r_1), (1 + r_2), \ldots, (1 + r_n)$ values, since it is the nth root of the product of these n quantities. Let us interpret this result. The $r_1, r_2, \ldots, r_n$ figures may be thought of as percentage rates of change between the successive $P_0, P_1, \ldots, P_n$ values, whereas the $(1 + r_1), (1 + r_2), \ldots, (1 + r_n)$ figures represent the relative relationships between each P value and the preceding one. For example, in the preceding illustration, where $r_1 = 3\%$,

$$1 + r_1 = \frac{P_1}{P_0} = \frac{\$103}{\$100} = 1.03$$

---

[3]The accumulated amounts are shown to four significant digits because we used Table A-4 of Appendix A, which is a four place table of logarithms.

Therefore, the \$103 total at the end of the first period was 3% greater than \$100, and expressed as a relative rate of change, 103% of the \$100. The $1 + r$ value in equation (3.23) is thus the *average relative* and r is the *average percentage rate* of *change* per period in the time series of P values. If the $r_1, r_2, \ldots, r_n$ are not all equal, then the calculated r value can be interpreted as the rate that would have prevailed had there been a constant percentage rate of change from P_0 to P_n. Actually, this average percentage rate of change can be computed by using only the first and last figures, P_0 and P_n, respectively. An explicit formula for this average is given by solving Equation (3.21) for r,

(3.24)
$$r = \sqrt[n]{\frac{P_n}{P_0}} - 1$$

Another way of showing that the average percentage rate of change over time depends only on the first and last figures in a time series is to substitute P_1/P_0 for $1 + r_1$, P_2/P_1 for $1 + r_2$, etc., in Equation (3.23). All P values other than P_0 and P_n cancel as follows:

$$1 + r = \sqrt[n]{\frac{\not{P}_1}{P_0} \cdot \frac{\not{P}_2}{\not{P}_1} \cdot \frac{\not{P}_3}{\not{P}_2} \cdots \frac{\not{P}_{n-1}}{\not{P}_{n-2}} \cdot \frac{P_n}{\not{P}_{n-1}}} = \sqrt[n]{\frac{P_n}{P_0}}$$

These ideas will be illustrated by the case of the Gallison Equipment Company referred to earlier. The calculation is carried out for illustrative purposes in two different ways, first using Equation (3.23) and then Equation (3.24). The computation by Equation (3.23) is shown in Table 3-14. Sales for each year relative to the preceding year are obtained first. Then the geometric mean sales relative, $(1 + r)$, is derived by obtaining the logarithms of the sales relatives, the mean of these logarithms (0.0587) and the antilogarithm of that figure (1.145). Subtracting 1 gives an average rate of change of 14.5% per year.

The same figure is obtained by Equation (3.24), where $P_0 = \$10{,}000{,}000$, $P_n = \$15{,}000{,}000$, and $n = 3$.

$$r = \sqrt[3]{\frac{\$15{,}000{,}000}{\$10{,}000{,}000}} - 1$$
$$= \sqrt[3]{1.5} - 1 = 1.145 - 1 = .145$$
$$= 14.5\% \text{ per year}$$

Ordinarily, the first method, Equation (3.23) is not used, since the calculation by the simple form of the compound interest formula (3.24) is easier. However, it is important to keep in mind that the typicalness of the computed average can only be judged in terms of the individual percentage rates of change. For example, in the case of the sales data presented in Table 3-14, the annual rates of change were −20%, + 50%, and +25%. Clearly, the 14.5% average

Table 3-14 Calculation of Average Annual Percentage Rate of Change in Sales of the Gallison Equipment Company, 1960–1963.

Year	(1) *Sales*	(2) $(1 + r_i)$ *Sales Relatives*	(3) $(log(1 + r_i))$ *Logs of Sales Relatives*
1960	\$10,000,000		
1961	8,000,000	0.8000	9.9031 − 10
1962	12,000,000	1.5000	0.1761
1963	15,000,000	1.2500	0.0969
			10.1761 − 10

$$\log(1 + r) = \frac{\Sigma \log(1 + r_i)}{n} = \frac{0.1761}{3} = 0.0587$$

$$1 + r = 1.145$$

$$r = .145 = 14.5\% \text{ per year}$$

rate of change is not typical of such widely fluctuating figures. On the other hand, if sales had gone from \$10,000,000 to \$15,000,000 at relatively constant annual rates of change, the computed average would be a typical figure.

In interpreting an average, such as the 14.5% rate of change in the preceding example, it is important to note the effect of using only the terminal points, P_0 and P_n, of the time series. In dealing with economic data in which cycles appear, if the measurement of an average percentage rate of change is made by taking the first figure, P_0, at the trough of a business cycle, while the last point, P_n, is taken at the peak of a cycle, clearly a greater average rate will tend to be obtained than if the two selected terminal points were both midway between troughs and peaks of cycles. This effect is often well illustrated in national political campaigns when candidates refer to economic growth rates under earlier administrations. Often, grossly inconsistent statements seem to be made by opposing candidates about average growth rates, even when measured in terms of similar data such as real gross national product per capita. A closer examination generally reveals substantially different selections of terminal points over which rates of change are measured. In more careful analytical work, trend lines are often fitted to time series data, and average growth or decline rates may be determined from these trend lines. This certainly is a fairer procedure than measuring rates of change between arbitrarily selected terminal points. Trend lines are discussed in Chapter 11.

3.15 Dispersion — Distance Measures

Central tendency, as measured by the various averages discussed in the preceding sections, is an important descriptive characteristic of statistical data. However, although two sets of data may have similar averages, they may differ considerably with respect to the spread or dispersion of the individual observations. Measures of dispersion describe this variation in numerical observations.

There are two types of measures of dispersion. The first, which may be referred to as distance measures, describes the spread of data in terms of the distance between the values of selected observations. The very simplest of such measures is the range or the difference between the values of the highest and lowest items. For example, if the bids supplied by five contractors in competing for a particular contract are $102,435, $124,793, $137,605, $156,782, and $163,451, the range of these bids is $163,451 − $102,435 = $61,016. Such a measure of dispersion may be useful for obtaining a rough notion of the spread in a set of data, but is certainly inadequate for most analytical purposes. A disadvantage of the range is that it describes dispersion in terms of only two selected values in a set of observations. Thus, it ignores the nature of the variation among all other observations. Furthermore, the two numbers used, the highest and the lowest, are extreme rather than typical values.

Other distance measures of dispersion employ more typical values. For example, the *interquartile range* is the difference between the *third quartile* and the *first quartile* values. The third quartile is a figure such that three-quarters of the observations lie below this figure; the first quartile is a figure such that one-quarter of the observations lie below it. Thus, the distance between these two numbers measures the spread between the values which bound the middle 50% of the values in a distribution. However, a main disadvantage of such a measure is again that it does not describe the variation *among* the items between the middle 50% (nor among the lower and upper one-fourth of the values). The method of calculation of quartile values will not be explicitly discussed here, but for frequency distribution data, its calculation proceeds in a manner completely analogous to that of the median, which is itself, the *second quartile* value. That is, two-quarters of the observations in a distribution lie below the median, and two-quarters lie above it. Quartiles are special cases of general measures known as *fractiles,* which refer to values which exceed specified fractions of the data. Thus, the ninth decile exceeds 9/10 of the items, the ninety-ninth percentile exceeds 99/100 of the items in a distribution, etc. Clearly, many arbitrary distance measures of dispersion could be developed, but they are infrequently used in practical applications.

3.16 Dispersion — Average Deviation Methods

The most comprehensive descriptions of dispersion are those which are in terms of an *average deviation* from some measure of central tendency. One such measure is the *mean deviation* from the median. This measure is obtained by calculating the absolute deviation of each observation from the median, and then averaging these deviations by taking their arithmetic mean. The formula for the mean deviation may be written

$$\text{Mean deviation} = \frac{\Sigma\,|X - Md|}{n} \tag{3.25}$$

Absolute deviations, that is, deviations in which signs are ignored, are calculated, since the *amounts* of the differences of observations from the median rather than the *direction* of these differences is of main interest.

As an example, let us return to a set of observations referred to in Section 3.10, where the median was discussed. A test had revealed that the lifetimes of five electronic devices in hours were 852, 931, 1010, 1027, and 1068; the median was 1010 hours. The calculation of the mean deviation (62.4 hours) for these data is shown in Table 3-15. This figure may be interpreted as follows. On the average, the lifetimes of these five electronic devices deviated from the median figure (1010 hours) by 62.4 hours.

Table 3-15 Calculation of the Mean Deviation of the Lifetimes in Hours of Five Electronic Devices.

Lifetimes in Hours X	*Deviations from Median* $\lvert X - Md\rvert$
852	158
931	79
1010	0
1027	17
1068	58
	312

$$\text{Mean deviation} = \frac{\Sigma|X - Md|}{n} = \frac{312}{5} = 62.4 \text{ hours}$$

The mean deviation calculation above is often referred to as an *absolute measure* of dispersion, that is, it is in the units of the original data, namely, hours.

For purposes of comparing variation among different sets of data, a *relative measure* is required. This is essential whenever the sets of data to be compared are expressed in different units, or even when the data are in the same units but of different orders of magnitude. A relative measure of dispersion can be obtained by dividing the absolute measure by the average from which deviations were obtained. Thus, the relative mean deviation is defined by the formula

$$\text{(3.26)} \qquad \text{Relative mean deviation} = \frac{\text{Mean deviation}}{Md}$$

For the data in Table 3-15, the relative mean deviation is 62.4/1010 = 6.2%. This indicates that the five electronic devices deviated on the average, by 6.2% from the median lifetime. Since the relative mean deviation is expressed as a percentage, it is a figure which can be compared with a similar measure derived from another set of data. On the other hand, if the data in the other distribution were expressed in inches or pounds, the absolute measures of dispersion, that is, the mean deviations could not be compared because of the dissimilar units. Even if the data in another distribution were lifetime figures in hours, that is, in the same units as the data in Table 3-15, but if the median, were, say, of the order of magnitude of 10,000 hours, a relative measure would again be required for comparative purposes. For example, a mean deviation of 62.4 hours in a set of data whose median is 10,000 hours yields a relative mean deviation of 62.4/10,000 = 0.6%. This represents much less relative dispersion (or much greater relative uniformity) than the set of observations in Table 3-15, which has the same absolute measure of dispersion (62.4 hours), but a much smaller median. This point may be characterized by the following comparison. An inch added to the height of the Empire State Building would be of no consequence. On the other hand, an inch added to the end of Cleopatra's nose might very well have changed the course of history.

The mean deviation is sometimes measured from the mean rather than the median. It can be shown that the mean deviation is a minimum when measured from the median. That is, the mean deviation is less when the median is used as a base than if the measurement is taken around any other figure. This is often pointed to as a reason for preferring the use of the median as a base, but in actual application, the mean is used probably just as frequently as the median mainly because of the wide usage of the mean as a measure of central tendency.

The reason for the use of absolute deviations (ignoring signs) in the calculation of the mean deviation is perhaps most easily seen in the case where the mean is used as a base. If deviations from the mean were recorded with their correct signs, positive and negative, and then averaged, the answer would always be zero. This follows from the fact that the sum of the signed deviations from the mean is equal to zero. Thus, unless signs are ignored, the average of the deviations would not constitute a measure of variation.

The calculation of the mean deviation for data grouped in the form of a frequency distribution will not be illustrated here. However, the method is essentially the same as for ungrouped data. If the median is used as a base, absolute deviations of midpoints of classes are taken from the median; these deviations are multiplied by the frequencies of the respective classes and summed. A division of this total by the number of observations yields the mean deviation. The formula for the mean deviation of a frequency distribution may be written

$$\text{(3.27)} \qquad \text{Mean deviation} = \frac{\Sigma f\,|X - Md|}{n}$$

where f = frequency of a class
X = midpoint of a class interval
Md = median
n = total number of observations

The use of the mean deviation as a measure of dispersion is clearly preferable to a distance measure such as the range or the interquartile range, since the mean deviation takes into account every item in a set of data. Thus, the mean deviation is a more comprehensive measure of variation than is a distance measure. Nevertheless, the mean deviation suffers from a disadvantage, since algebraic positive and negative signs are ignored in its calculation. This causes the mean deviation to have poor combining properties insofar as its use in more advanced statistical techniques is concerned. Thus, the mean deviation has gained very limited acceptance as regards both theoretical and applied work in statistics. The variance and standard deviation, to be discussed next are doubtless the most widely used measures of dispersion.

The Variance and the Standard Deviation

The variance is a measure of variation which is expressed in terms of average deviation from some measure of central tendency. Specifically, the variance of a sample, denoted s^2, is the arithmetic mean of the squared deviations from the sample mean. In symbols, if $X_1, X_2, \ldots, X_n$ represent the values of n sample observations, the variance is defined by

$$\text{(3.28)} \quad s^2 = \frac{(X_1 - \bar{X})^2 + (X_2 - \bar{X})^2 + \cdots + (X_n - \bar{X})^2}{n} = \frac{\sum_{i=1}^{n} (X_i - \bar{X})^2}{n}$$

If subscripts are dropped, this becomes

(3.29)
$$s^2 = \frac{\Sigma (X - \bar{X})^2}{n}$$

Although the variance measures the extent of variation in the values of a set of observations, it is in units of squared deviations or squares of the units of the original numbers. In order to obtain a measure of *dispersion* in terms of the units of the original data, the square root of the variance is taken. The resulting measure is known as the standard deviation. Thus, the standard deviation of a sample of n observations is given by

(3.30)
$$s = \sqrt{\frac{\Sigma(X - \bar{X})^2}{n}}$$

By convention, the positive square root is used. The standard deviation is sometimes referred to as the root-mean-square deviation. That is, it is the square *root* of the *mean* of the *squared* deviations. It is clear that the standard deviation is a measure of the spread in a set of observations. If all of the values in a sample were identical, each deviation from the mean would be zero. In such a completely uniform distribution, the standard deviation would be equal to zero, its minimum value. On the other hand, as items are dispersed more and more widely from the mean, the standard deviation becomes larger and larger.

Many presentations of elementary statistics define the variance and standard deviation with a divisor of $n - 1$ rather than n. To understand the reason for this, let us return to the concepts of statistic and parameter referred to in Section 3.8. Using the convention that population parameters are designated by Greek letters, the sample statistic s^2 may be thought of as an estimate of the corresponding population parameter, σ^2. That is, σ^2 is the population variance or the mean of the squared deviations around the population mean, μ. It can be shown that s^2, when defined with an $n - 1$ divisor is a so-called "unbiased estimator" of the population parameter, σ^2. This means that if all possible samples of size n were drawn from the given population and the variances of these samples were averaged (arithmetic mean), this average would be equal to the population variance, σ^2. Thus, when the sample variance is defined with an $n - 1$ divisor, on the average, it correctly estimates the population variance.

For several reasons, in this chapter on descriptive statistics particularly, the definition with a divisor of n rather than $n - 1$ will be used. First of all, the criterion of unbiasedness is not the only way of judging goodness of estimation. Many situations may be cited in which to be correct on the average is not good enough. Other criteria of goodness of estimation can be suggested which may be equally important to lack of bias. Furthermore, a good estimation procedure would take into account the costs or seriousness of errors of estimate. From that point of view, a myopic concentration on unbiasedness in estimation is undesirable. Another point that is rarely mentioned in elementary treatments is the fact that although when a divisor of $n - 1$ is used, s^2 is an unbiased estimator of σ^2, s is not an unbiased estimator of σ. Thus, the division by $n - 1$ does not accomplish unbiased estimation of both the variance and standard deviation. A final point may be noted. If the sample size, n, is large, the difference between the numerical values obtained for s (or s^2), when using n rather than $n - 1$, is generally negligible. Thus, in many economic or business problems, where sample sizes may be rather large, the discussion as to whether to divide by n or $n - 1$ is in the nature of a tempest in a teapot. For purposes of this text, it is felt that a more integrated presentation of descriptive statistics is accomplished by defining the standard deviation with the divisor n rather than $n - 1$. The objective of this rather lengthy digression is to alert the reader to the alternative definitions he may encounter and to specify some of the issues involved.[4]

In the immediately ensuing paragraphs, we shall concentrate on methods of computing the standard deviation, both for ungrouped and grouped data. This will be followed by discussion of how this measure of dispersion is used. However, the major uses of the standard deviation are in connection with sampling theory and statistical inference, which are discussed in subsequent chapters.

The same illustrative personal loan data as were used in the case of the arithmetic mean will again be employed in the calculation of the standard deviation. Also, the shortcut methods of computation utilized for the mean will be discussed for the standard deviation.

In Section 3.8, the following ungrouped data were given as the amounts of loans made by the personal department of a bank on a certain day: \$500, \$250, \$300, \$600, and \$425. The arithmetic mean loan was \$415. The calculation of the standard deviation using the defining formula (3.30) is illustrated in Table 3-16.

The result \$128.1 for the standard deviation in Table 3-16 is an absolute measure of dispersion, which as previously indicated means that it is stated in the units of the original data. Whether this is a great deal or only a small amount of dispersion cannot be immediately determined. This sort of judgment is based on the particular type of data analyzed, for example, in this case, personal loan data. Furthermore, as we have already seen, relative

---

[4]One justification for the $n - 1$ divisor arises in connection with the concept of "degrees of freedom." Here, it is argued that since the sum of the deviations $X - \bar{X}$ must be equal to zero, only $n - 1$ of them are independent. That is, once $n - 1$ of the deviations are independently specified, the nth deviation is immediately determined by the constraint that the sum of the deviations must be equal to zero. Thus, it is said that there are $n - 1$ "degrees of freedom" present. In many areas of statistical methods, such as the analysis of variance and design of experiments, the concept of degrees of freedom is of considerable importance.

Table 3-16 Calculation of the Standard Deviation for Ungrouped Data by the Defining Formula: Personal Loan Data.

Amount of Loan X	*Deviation from Mean* $X - \bar{X}$	*Squared Deviations* $(X - \bar{X})^2$
\$500	\$500 − \$415 = \$85	7225
250	250 − 415 = −165	27225
300	300 − 415 = −115	13225
600	600 − 415 = 185	34225
425	425 − 415 = 10	100
$\bar{X}$ = \$415	\$0	82000

$$s = \sqrt{\frac{\Sigma(X - \bar{X})^2}{n}} = \sqrt{\frac{82{,}000}{5}} = \sqrt{16{,}400} = \$128.1$$

measures of dispersion are preferable to absolute measures for comparative purposes.

A warning must be expressed concerning the above standard deviation. Purposely, only a small number of observations were included to illustrate the method of calculation. However, in such cases, the choice of n or $n - 1$ in the denominator of the formula has a substantial effect on the results. For example, if $n - 1 = 4$ had been used as a divisor instead of five, the resultant standard deviation would have been \$143.2. As indicated earlier, the difference in results between the two methods of calculation tends to be smaller for larger sample sizes. It should be emphasized that we cannot say that either method is correct and the other is not. Both are measures of dispersion, and theoretically one could devise an infinite number of measures of this characteristic. However, the modern tendency is that if one wishes to estimate a population standard deviation, rather than to describe the dispersion in the sample itself, the divisor $n - 1$ is generally used.

Shortcut Calculation of the Standard Deviation—Ungrouped Data

The shortcut technique of computing the standard deviation for ungrouped data is a simple extension of the similar calculation for the arithmetic mean. Again, the method involves the selection of an assumed mean and calculation of deviations from this assumed mean. In terms of a tabular arrangement of computations, only one further column is required in the case of the standard deviation, namely, the squares of these deviations. Without proof, we state in (3.31) the formula for the shortcut method.

Shortcut Method for the Standard Deviation—Ungrouped Data

$$s = \sqrt{\frac{\Sigma d^2}{n} - \left(\frac{\Sigma d}{n}\right)^2} \tag{3.31}$$

The symbols have the same significance as in the case of the mean. That is, d represents a deviation from an assumed mean and n is the number of observations. The calculation may be viewed as follows. Under the radical sign, the first term is the mean of squared deviations from the assumed mean. The second term may be thought of as the square of a correction factor, which must be subtracted to adjust for the fact that deviations were not taken from the true mean. The correction factor is the average deviation from the assumed mean. The calculation of the standard deviation of the five personal loan figures by the shortcut method is illustrated in Table 3-17. As in the computation of the mean by the shortcut method, an assumed mean of $400 has been taken. Of course, the same result for the standard deviation, $128.1, is obtained as in the calculation by the defining formula, given in Table 3-16.

Table 3-17 Calculation of the Standard Deviation for Ungrouped Data by the Shortcut Method: Personal Loan Data.

Amount of Loans X	*Deviation from Assumed Mean* d	*Squared Deviations* d^2
$500	+$100	10,000
250	− 150	22,500
300	− 100	10,000
600	+ 200	40,000
425	+ 25	625
Total	+ 75	83,125

$$s = \sqrt{\frac{\Sigma d^2}{n} - \left(\frac{\Sigma d}{n}\right)^2} = \sqrt{\frac{83{,}125}{5} - \left(\frac{75}{5}\right)^2} = \sqrt{16{,}400} = \$128.1$$

A particularly useful form of the shortcut formula when computers or even desk calculating machines are used is the shortcut method using an assumed mean of zero. In this case, the deviations from the assumed mean are equal to the values of the original observations and the correction factor

$\Sigma d/n$ becomes the arithmetic mean $\bar{X} = \Sigma X/n$. The shortcut formula may then be written as in (3.32).

Shortcut Method for the Standard Deviation, Assumed Mean of Zero—Ungrouped Data

(3.32) $$s = \sqrt{\frac{\Sigma X^2}{n} - (\bar{X})^2}$$

The application of this formula to the personal loan data gives

$$s = \sqrt{\frac{943{,}125}{5} - (415)^2} = \$128.1$$

The advantage of this formula from the standpoint of computing equipment is that the calculation of deviations from a mean is not required. Since only values of observations and their squares are involved, they can be stored in the machine and cumulative sums can be obtained. This is a more efficient process for machine computation than the alternative methods involving calculations of deviations.

An important property of the variance and standard deviation is the fact that the sum of the squared deviations around the arithmetic mean is a minimum. That is, the mean of the squared deviations is a smaller figure when calculated around the arithmetic mean than around any other figure. This significant relationship, which helps account for the pervasive application of the variance and standard deviation in more advanced statistical techniques, can be used to help in remembering that the correction factor (squared) must be subtracted from the first term under the radical sign in formulas such as (3.31) and (3.32). Since the first terms involve sums of squared deviations around assumed means, these figures exceed the corresponding sum for the true standard deviation, and therefore a subtraction is required.

In the calculation of the standard deviation for data grouped into frequency distributions, it is merely necessary to adjust the foregoing formulas to take account of this grouping. The defining formula generalizes to

(3.33) $$s = \sqrt{\frac{\Sigma f(X - \bar{X})^2}{n}}$$

where, as usual for grouped data, X represents the midpoint of a class, f the frequency in a class, $\bar{X}$ the arithmetic mean, and n the total number of observations. The use of this formula is illustrated in Table 3-18 for the frequency distribution of personal loans previously shown in Table 3-10.

The saving in computational effort and time accomplished by the use of shortcut methods for frequency distribution data can be illustrated in the case of these personal loan data. Let us bypass the shortcut method and move directly to the step-deviation technique of calculating the standard deviation.

Table 3-18 Calculation of the Standard Deviation for Grouped Data by the Defining Formula: Personal Loan Data.

Amount of Loans	*Number of Loans* f	$(X - \bar{X})$	$(X - \bar{X})^2$	$f(X - \bar{X})^2$
\$ 0 and under \$ 200	6	\$ 100 − \$632 = −\$532	283,024	1,698,144
200 and under 400	18	300 − 632 = − 332	110,224	1,984,032
400 and under 600	25	500 − 632 = − 132	17,424	435,600
600 and under 800	20	700 − 632 = 68	4,624	92,480
800 and under 1000	17	900 − 632 = 268	71,824	1,221,008
1000 and under 1200	14	1100 − 632 = 468	219,024	3,066,336
	100			8,497,600

$$s = \sqrt{\frac{\Sigma f(X - \bar{X})^2}{n}} = \sqrt{\frac{8,497,600}{100}} = \$291.5$$

As in the case of the similar calculation for the arithmetic mean, the coding procedure involves taking deviations of midpoints of classes from an assumed mean and stating them in class interval units. Only one additional column of values, $f(d')^2$, is required to compute the standard deviation by the step-deviation method as compared to the corresponding arithmetic mean computation given in Table 3-10. The formula for the step-deviation method is given in Equation (3.34). All of the symbols have the same meaning as in Equation (3.15′) for the arithmetic mean and the computation in Table 3-10.

Step Deviation Method for the Standard Deviation—Grouped Data

$$s = c\sqrt{\frac{\Sigma f(d')^2}{n} - \left(\frac{\Sigma fd'}{n}\right)^2} \tag{3.34}$$

It may be noted that Equation (3.34) represents an adaptation of the shortcut formula for ungrouped data. The terms under the square root sign involve a multiplication by frequencies to account for grouping. Since the computation under the square root sign is in class interval units (because of the use of d' values), the result obtained after extracting the square root must be multiplied by the size of the class interval (c). The use of Equation (3.34) is illustrated in Table 3-19.

The same assumed mean, $\bar{X}_a$ = \$500 was used in Table 3-19 as in the calculation of the arithmetic mean given in Table 3-10. The advantage gained

Table 3-19 Calculation of the Standard Deviation for Grouped Data by the Step Deviation Method: Personal Loan Data.

Amount of Loans	*Number of Loans* f	d'	fd'	$f(d')^2$
\$ 0 and under \$ 200	6	−2	−12	24
200 and under 400	18	−1	−18	18
400 and under 600	25	0	0	0
600 and under 800	20	+1	+20	20
800 and under 1000	17	+2	+34	68
1000 and under 1200	14	+3	+42	126
	100		+66	256

$\bar{X}_a = \$500$

$$s = c\sqrt{\frac{\Sigma f(d')^2}{n} - \left(\frac{\Sigma fd'}{n}\right)^2} = \$200\sqrt{\frac{256}{100} - \left(\frac{66}{100}\right)^2} = \$200\ (1.457)$$

$s = \$291.4$

by using the step-deviation method, particularly in terms of the much smaller numbers involved, is evident in comparing Tables 3-18 and 3-19. The small discrepancy in the answer obtained for the standard deviation in these two computations is due to rounding error.

The step-deviation method is clearly the most desirable technique to use in the case of frequency distributions with equal size classes. If the classes are unequal in size, the shortcut formula $s = \sqrt{\frac{\Sigma fd^2}{n} - \left(\frac{\Sigma fd}{n}\right)^2}$ may be used, where d is a deviation of a class midpoint from an assumed mean in units of the original data.

Uses of the Standard Deviation

The standard deviation of a frequency distribution is very useful in describing the general characteristics of the data. For example, in the so-called "normal distribution," which is discussed extensively in Chapters 6, 7, and 8, the standard deviation is used in conjunction with the mean to indicate the percentage of items that fall within specified ranges. Hence, if a population is in the form of a normal distribution the following relationships apply:

$\mu \pm \sigma$ includes 68.3% of all of the items
$\mu \pm 2\sigma$ includes 95.5% of all of the items
$\mu \pm 3\sigma$ includes 99.7% of all of the items

For example, if a production process is known to produce items which have a mean length of $\mu = 10$ inches and a standard deviation of 1 inch, then we can infer that 68.3% of the items have lengths between $10 - 1 = 9$ inches and $10 + 1 = 11$ inches. About 95.5% have lengths between $10 - 2 = 8$ and $10 + 2 = 12$ inches, and 99.7% have lengths between $10 - 3 = 7$ and $10 + 3 = 13$ inches. Thus, in a normal distribution, a range of $\mu \pm 3\sigma$ includes virtually all the items in the distribution. As discussed in Chapter 6, the normal distribution is perfectly symmetrical. If the departure from a symmetrical distribution is not too great, the rough generalization that virtually all the items are included within a range from 3σ below the mean to 3σ above the mean still holds.

The standard deviation is also useful for describing how far individual items in a distribution depart from the mean of the distribution. Suppose the population of students who took a certain aptitude test displayed a mean score of $\mu = 100$ with a standard deviation of $\sigma = 20$. If a certain student obtained a score of 80 on the examination, his score can be described as lying one standard deviation below the mean. The terminology usually employed is that his *standard score* is -1, that is, if his examination score is denoted as X, then

$$\frac{X - \mu}{\sigma} = \frac{80 - 100}{20} = -1$$

The standard score of an observation is simply the number of standard deviations the observation lies below or above the mean of the distribution. Hence, in the example, the student's score deviates from the mean by -20 units, which is equal to -1 in terms of units of standard deviations away from the mean. If standard scores are computed from sample rather than universe data, the formula $(X - \bar{X})/s$ would be used instead.

Comparisons can thus be made for items in distributions which differ in order of magnitude or in the units employed. For example, if a student scores 120 on an examination in which the mean was $\mu = 150$ and $\sigma = 30$, his standard score is $(120 - 150)/30 = -1$. This standard score places the 120 at the same number of standard deviations below the mean as the 80 in the preceding example. Analogously, we could compare standard scores in a distribution of wages with comparable figures in a distribution of length of employment service, etc.

The standard deviation is doubtless the most widely used measure of dispersion, and considerable use is made of it in later chapters of this text.

3.17 Relative Dispersion — Coefficient of Variation

As observed earlier, the standard deviation is an absolute measure of dispersion, whereas for comparative purposes, a relative measure is required. Such a relative measure is obtained in the case of the standard deviation by expressing it as a percentage of the arithmetic mean. The resulting figure, referred to as the coefficient of variation, V, is defined symbolically in Equation (3.35).

(3.35)
$$V = \frac{s}{\bar{X}}$$

Thus, for the frequency distribution of personal loan data, whose standard deviation is \$291.5 with a mean of \$632, the coefficient of variation is

$$V = \frac{\$291.5}{\$632} = 46.1\%$$

Let us assume that the foregoing figures were observed for a sample of the loans of a finance company in a particular month. Six months later, a similar sample revealed a standard deviation of \$350 with an arithmetic mean of \$995. The coefficient of variation for this later set of loans is $V = s/\bar{X} = 350/995 = 35.2\%$. Therefore, the loans in the later period were relatively more uniform, or stated differently, displayed less relative variation than did the loans in the earlier period. It may be noted that the loans in the later period had the larger standard deviation, but because of the increase in the average size of loans, relative dispersion had decreased.

Both absolute and relative measures of dispersion are widely used in practical sampling problems. To give just one example, a question frequently arises as to the sample size required to yield an estimate of a universe parameter with a specified degree of precision. For example, the finance company referred to earlier may want to know how large a random sample of its loans it must study in order to estimate the average dollar size of all of its loans. If the company wants this estimate within a specified number of *dollars*, an absolute measure of dispersion is appropriate. On the other hand, if the company wants the estimate to be within a specified *percentage* of the true average figure, a relative measure of dispersion would be used.

3.18 Errors of Predictions

In this chapter, we have discussed descriptive measures for empirical frequency distributions, with emphasis on measures of central tendency and dispersion. Some interesting relationships between these two types of measures are observable when certain problems of prediction are considered. Suppose we want to guess or "predict" the value of an observation picked at random from a frequency distribution. Let us refer to the penalty of an incorrect prediction as the "cost of error." If there were a *fixed* cost of error on each prediction, no matter what the size of the error, we should guess the mode as the value of the random observation. This would give us the highest probability of guessing the *exact value* of the unknown observation. Assuming repeated trials of this prediction experiment we would thus minimize the average (arithmetic mean) cost of error.

Suppose, on the other hand, the cost of error varies directly with the size of error regardless of its sign, that is, whether the actual observation is above or below the predicted value. In this case, we would want a prediction which minimizes the average *absolute error*. The median would be the "best guess," since it minimizes average absolute deviations. The mean deviation about the median would be a measure of this minimum cost of error.

Finally, suppose the cost of error varies according to the square of the error. For example, an error of two units costs four times as much as an error of one unit. In this situation, the mean should be the predicted value, since the average of the squared deviations about it is less than around any other figure. Here the variance or standard deviation would represent a measure of this minimum error. Another point which we have observed previously for the mean is that the average amount of error, taking account of sign, would be zero.

A practical business application of these ideas is in the determination of the optimum size of inventory to be maintained. Let us assume a situation in which the cost of overstocking a unit (cost of overage) is equal to the cost of being short one unit (cost of underage). Further, it may be assumed that the cost of error varies directly with the absolute amount of error. For example, having two units in excess of demand costs twice as much as one unit. In this situation, the optimum stocking level is the median of the frequency distribution of numbers of units demanded.

PROBLEMS

1. The average size loan given by a certain bank during the year 1968 was $2500. Would you consider this a parameter or a statistic if:
   (a) you were interested in the average loan size between 1958 and 1968?
   (b) you were only interested in the average loan size for 1968?
2. Give a business or economic example of a case in which a computed average would be considered a parameter and a case in which the same average would be considered a statistic.

3. The commissions paid to nineteen Metroperidock insurance salesmen for a one-week period were:

$125.00	$138.50	$216.74	$ 82.75	$140.40
169.95	212.55	158.00	162.63	190.15
175.00	101.82	205.01	160.00	228.00
182.05	238.09	171.13	198.50	

Construct:

(a) a frequency table, assuming the data were recorded to the last cent,
(b) a histogram,
(c) a frequency polygon.

4. Daily sales of a small retail establishment are given below:

$ 97.60	$132.67
102.65	145.68
141.02	175.92
174.68	106.34
92.06	125.27
172.21	127.72
83.77	137.66
104.01	83.66
102.44	116.70
156.52	136.79
149.59	129.99
136.97	125.41
124.31	124.17
123.23	91.70
118.94	128.29

(a) Construct a frequency distribution for sales during the period, assuming the data were recorded to the last cent. Use about five classes.
(b) Sketch a frequency polygon showing the distribution in part (a).

5. The ages measured to the last birthday of the employees of Smith, Inc. are as follows:

22	21	26	42
28	39	20	32
35	45	49	28
31	49	42	30
31	39	37	36
33	47	38	48

Set up a frequency table having six classes and specify the midpoint of each class.

6. Compute the cumulative frequency distribution for the distribution given below.

Analysis of Ordinary Life Insurance

Policy Size	*Number of Ordinary Life Insurance Policies in Force per 1000 Policies, Oct. 1966*
Under $1,000	31
$1000 – 2499	181
2500 – 4999	108
5000 – 9999	213
10,000 or more	467

SOURCE: *Life Insurance Fact Book 1967*, Institute of Life Insurance, New York, 1967.

7. Comment critically on the following systems of designating class intervals:

(a)	(b)
83–102	83–under 102
102–121	103–under 121
etc.	122–under 141
	etc.

8. In reference to Problem 3, calculate:
   (a) the arithmetic mean from the original data.
   (b) the arithmetic mean using the frequency table.
   (c) Which is the true arithmetic mean value? Explain your answer.

9. The Center City Bank reports bad debt ratios (dollar losses to total dollar credit extended) of .04 for personal loans and .02 for industrial loans in 1966. For the same year, the Neighborhood Bank reports bad debt ratios of .05 for personal loans and .03 for industrial loans. Can one conclude from this that Center City's overall bad debt ratio is less than Neighborhood's? Justify.

10. Consider the following data:

Percentage of Civilian Labor Force Unemployed 1960
Delaware Valley Area, Pennsylvania

County	*% Unemployed*	*Civilian Labor Force*
Bucks	3.6	114,395
Delaware	3.8	214,758
Montgomery	2.5	206,324
Philadelphia	6.5	843,160

SOURCE: U. S. Population Census 1960, Pennsylvania.
   (a) What is the unweighted average of the percent unemployed per county?
   (b) What is the weighted average of the percent unemployed for the four counties combined?
   (c) Explain the reason for the difference in the figures obtained in parts (a) and (b).

11. A certain manufacturing firm has three plants. At plant one, the number of accidents for the month of January 1967 was 40; at plant two, 45; and at plant three, 65. The number of man-hours worked at each plant for the month was plant one, 20,000; plant two, 30,000; and plant three 60,000.
    (a) What was the number of accidents per thousand man-hours at each plant?
    (b) Calculate the appropriate average number of accidents per thousand man-hours worked for the firm as a whole.

12. A small clothing manufacturing company does both retail and wholesale trade. For the first quarter ending June 30, its profit per item from the retail trade was \$2.10 and from the wholesale trade was \$1.30. If wholesale business accounted for 80% of the company's business, what is the (arithmetic) mean profit per item for the quarter?

13. The XYZ Sales Company has three consumer divisions: mail order, telephone service, and direct sales. The accounting department reports the following information for the three divisions:

Division	*Dollar Profit per Order*	*Total Number of Orders*	*Total Dollar Sales*
Mail Order	\$ 6.00	600	\$30,000
Telephone Service	3.00	1000	30,000
Direct Sales	10.00	400	80,000

    (a) Find the overall dollar profit per order for the three divisions combined.
    (b) An executive of the company notes that he would prefer to have data on profit per dollar of sales for each division. Compute these profit-to-sales ratios.

14. According to the Department of Agriculture, in order to run a dairy farm with 35 milk cows in an optimal manner, hired labor is required in the following quantities for the year:

Permanent labor	864 hours
Spring seasonal labor	108 hours
Summer seasonal labor	75 hours
Fall seasonal labor	80 hours

    If permanent labor costs \$1.50 per hour, spring labor \$1.35 per hour, summer labor \$1.15 per hour, and fall labor \$1.25 per hour, what is the average cost per hour for labor over the year?

15. The number of resignations received by a certain firm per month during 1967 was:

    8, 3, 5, 3, 4, 3, 1, 0, 3, 4, 0, 7

    Calculate and interpret the arithmetic mean, mode, and median.

16. Compute the mean and median of both the raw data and the frequency distribution in Problem 4.

17. Explain or criticize the following statement.

    The frequency distributions of family income, size of business, and wages of skilled employees all tend to be skewed to the right.

18. Criticize or explain the following statements:
    (a) In a bimodal frequency distribution, the median gives a more typical value than the arithmetic mean and therefore should be used for comparison with other distributions.
    (b) Firm *A* manufactures a mass production assembly line product; 85% of its employees are unskilled and 15% semi-skilled workers. Firm *B* makes special electronic parts; 20% of its employees are unskilled, 45% semi-skilled, and the remainder are skilled workers. It is thus logical to assume that the standard deviation of Firm *A*'s wages is less than that of Firm *B*'s.

19. Blah Beer Company has two production lines working each day. The number of cases produced per eight-hour shift for a week during June 1968 was:

    Line 1: 6921, 8205, 6658, 6835, 7830, 6935, 7563, 6503, 6777, 6250, 6000, 6935, 7000, 7012, 5530

    Line 2: 5655, 6138, 7891, 7951, 6666, 7891, 7589, 7002, 8131, 7662, 7555, 8495, 7832, 7477, 6886

    Using six classes and "and under" upper class limits,
    (a) Construct and graph the frequency distribution for the number of cases produced per eight-hour shift.
    (b) Construct and graph the frequency distribution for each production line separately.
    (c) Which answer, that in part (a) or part (b), gives a better picture of the distribution of production of cases of beer at the Blah Beer Company? Explain your answer.

20. The following data represent chainstore prices paid by farmers for two products during the month of March 1965:

Composition Roofing Price Range Dollars/90-pound Roll	*Number of Reports*	*Douglas and Inland Firs, 2 × 4's Standard or Better Price Range Dollars/Thousand Pound-foot*	*Number of Reports*
\$2.35 and under \$2.75	1	76–85	2
2.75 and under 3.15	6	86–95	4
3.15 and under 3.55	33	96–105	8
3.55 and under 3.95	51	106–115	7
3.95 and under 4.35	121	116–125	28
4.35 and under 4.75	50	126–135	48
4.75 and under 5.15	44	136–145	57
5.15 and under 5.55	13	146–155	71
5.55 and under 5.95	5	156–165	45
		166–175	20
		176–185	1

SOURCE: Agricultural Handbook No. 326 U.S.D.A., Economics Research Service, Nov., 1966, Washington, D. C.

(a) Graph the frequency distribution for each product.
(b) Indicate the median and modal class for each distribution.
(c) Calculate and graph the cumulative distribution for each frequency distribution.
(d) Compare the skewness of the two distributions.

21. The following table presents a frequency distribution of the gross income of males in Philadelphia in 1960.
Frequency Distribution of Gross Income of Males, Philadelphia, Pa., 1960

Income	*Number of Males*
$999 or less	70,347
1000 –$1999	68,109
2000 – 2999	66,735
3000 – 3999	87,751
4000 – 4999	106,429
5000 – 5999	94,714
6000 – 6999	53,388
7000 – 9999	54,329
10,000 and over	25,447

SOURCE: U. S. Census of Population, 1960.

(a) Why do you think open-ended intervals are used in this distribution?
(b) Which is the modal class?
(c) Which is the median class?

22. In one distribution, the mean is larger than the median; in a second distribution, the mean and median are equal. Can one conclude from this that there is greater variability in the first distribution than in the second? Justify.

23. The following are the earnings for all employees of Company *A* for the week ended September 23, 1967:

Earnings for Week Ended 9/23/67	*Number of Employees with Given Earnings*
$ 87.50 and under $ 95.00	2
95.00 and under 102.50	7
102.50 and under 110.00	9
110.00 and under 117.50	14
117.50 and under 125.00	10
125.00 and under 132.50	6
132.50 and under 140.00	2
	50

(a) Compute the arithmetic mean of the above distribution.
(b) Is the answer in (a) the same as you would have obtained had you calculated the following ratio?

$$\frac{\text{Total earnings for all employees of Company } A \text{ during the week ended September 23, 1967}}{\text{Total number of employees of Company } A \text{ during the week ended September 23, 1967}}$$

Why or why not?

(c) Compute the median of the above distribution.
(d) In which direction are these data skewed?
(e) The comptroller of the company stated that the total payroll for the week ended September 23, 1967, was $5675.18. Do you have any reason to doubt this statement? Support your position very briefly.
(f) Would you say that the arithmetic mean that you computed in part (a) provides a satisfactory description of the typical earnings of these 50 employees in the week of September 23, 1967? Why or why not?

24. The distribution below gives the dollar cost per unit of output for 200 plants in the same industry:

Dollar Cost per Unit of Output	*Number of Plants*
$1.00 and under $1.02	6
1.02 and under 1.04	26
1.04 and under 1.06	52
1.06 and under 1.08	58
1.08 and under 1.10	39
1.10 and under 1.12	15
1.12 and under 1.14	3
1.14 and under 1.16	1
Total	200

(a) Calculate the arithmetic mean of the distribution given above.
(b) Calculate the median of the distribution.
(c) Is the answer in (a) the same as you would have obtained had you computed the following ratio?

$$\frac{\text{Total dollar cost for the 200 companies}}{\text{Total number of units of output of the 200 companies}}$$

(d) Would you be willing to say that 50% of the *units* produced cost less than your answer to part (b) above? Explain.
(e) These 200 plants employed an average of 100 employees each. The standard deviation of the number of employees was 20 employees. Relatively speaking, do these 200 plants vary more in their employment or in their cost per dollar of output? Demonstrate statistically.
(f) Suppose that the last class above had read "$1.14 and over." What effect, if any, would this have had on your calculation of the arithmetic mean and of the median? Briefly justify your answers.

25. Observe the following distribution of the monthly gross pay of all semi-skilled employees of the Norbert Manufacturing Company in November 1967.

Monthly Gross Pay	*Number of Semi-skilled Employees with Given Gross Pay*
\$242.50 and under \$282.50	8
282.50 and under 322.50	15
322.50 and under 362.50	27
362.50 and under 402.50	20
402.50 and under 442.50	14
442.50 and under 482.50	11
482.50 and under 522.50	5
	100

(a) Compute the arithmetic mean of the above distribution.
(b) Compute the median of the above distribution.
(c) In what direction, if any, are these data skewed?
(d) What is your best estimate of the total gross pay of these 100 semi-skilled employees during November 1967? Comment briefly on the accuracy of your estimate.
(e) Compute the standard deviation of the above distribution.
(f) A union official has stated that there was a fair amount of dispersion in the monthly gross pay of these employees of the Norbert Manufacturing Company in November 1967. Do you agree? Comment.

26. The following table shows the distribution of states by the percentage of the popular vote cast for President Kennedy in the 1960 election. (The popular vote does not include minor party candidates or unpledged electors' tickets; that is, it includes only the vote for Nixon and Kennedy.)

% of Two-Party Popular Vote for Kennedy	*Number of States*
35 and under 40	2
40 and under 45	9
45 and under 50	16
50 and under 55	17
55 and under 60	2
60 and under 65	4
Total	50

SOURCE: *New York Times*, November 18, 1960.

(a) 1. Compute the arithmetic mean of the above frequency distribution.
2. Compute the standard deviation.
(b) Does your answer in (a), part 1, give the percentage of the total U.S. two-party popular vote for Kennedy? Why or why not?

(c) In Nixon's 25 best states (the states which he carried with the larges percentages) his percentage of the popular vote exceeded what value

27. Sales of Tarn Engineering Inc. for the year ending June 30, 1950 were $845,600 Sales for the year ending June 30, 1960 were $1,255,100. What was th yearly average percentage rate of increase?

28. Observe the following time series:

Sales of a Product Produced by the ABC Company

Year	*Number of Units Sold*
1930	4000
1960	2000

(a) Compute the average percentage rate of change in number of units sol per five-year period.
(b) If this five-year rate of change persisted until the year 2000, what woul sales of this product by the ABC Company be in that year? Set up you answer in the form of an equation from which a numerical answer coul be obtained if desired. You need not do the arithmetic.

29. The following table shows the market value of production of the Treadmil Company from 1940 to 1965 at five-year intervals:

Year	*Production (in thousands of dollars)*
1940	$20,000
1945	15,114
1950	28,267
1955	32,467
1960	36,572
1965	25,000

(a) Compute the *mean amount* of change *per five-year period* in value of production.
(b) Compute the *average percentage rate* of change *per five-year* period in value of production.
(c) Is the average computed in part (b) typical? Why or why not?

30. (a) Mr. Compound wants to earn 9% on his investment each year. He wishes to invest $4000 and reinvest interest received each year. How many years must he follow his plan to double his money? Give your answer as a whole number.
(b) If he had followed his plan half as long, how much would he have?
(c) If he had followed his plan twice as long (as your answer to part (a)) how much would he have? Comment on the relationship among your answers.

31. The closing prices of two common stocks traded on the American Stock Exchange for a week in September 1968 were

	Highfly	*Stabil*
Monday	$28	$28
Tuesday	34	26
Wednesday	18	22
Thursday	20	24
Friday	25	25

(a) Compare the two stocks simply on the basis of measures of central tendency.
(b) Compute the standard deviation for each of the two stocks. What information do the standard deviations give concerning the price movements of the two stocks?

32. The standard deviation of sales for the past five years for Company A was $1405 and for Company B $18,580. Can we conclude that sales are more stable for Company A than B?
33. The following is a distribution of the weights of 100 draftees who reported to an army camp for basic training.

Weights (in pounds)	*Frequency*
120 and under 140	14
140 and under 160	20
160 and under 180	36
180 and under 200	18
200 and under 220	8
220 and under 240	4
	100

(a) Compute the standard deviation and the arithmetic mean of the distribution of weights.
(b) Can one compare the variability in the weights of the draftees with the variability in their heights? If "yes," how? If "no," why not?

34. The following is the distribution of amounts spent for research and development and for marketing for the year 1968 by seven drug firms and seven cosmetic firms:

Expenditures (in	*Drug Companies*		*Cosmetic Firms*	
millions of dollars)	*R and D*	*Marketing*	*R and D*	*Marketing*
0 and under 1	1	1	4	0
1 and under 2	1	3	2	1
2 and under 3	3	2	1	3
3 and under 4	2	1	0	3

(a) Compute the arithmetic mean for each type of company, for each type of expenditure.

(b) Compute the standard deviation for each type of company for each type of expenditure.
(c) Based on the given data, what conclusions can you draw concerning these expenditures?

35. The following is a distribution of lifetimes, in hours, of 100 vacuum tubes.

Lifetime (hours)	*Frequency*
100 and under 200	12
200 and under 300	28
300 and under 400	20
400 and under 500	18
500 and under 600	14
600 and under 700	8
	100

(a) Compute the mean and coefficient of variation for the distribution.
(b) Compute and *interpret* the median.
(c) Suppose that the last class had read "600 and over." What effect, if any, would this have had on your calculations in parts (a) and (b) above?

CHAPTER FOUR

Theoretical Frequency Distributions

Descriptive measures for empirical frequency distributions were discussed in Chapter 3. In this chapter, the corresponding techniques will be considered for theoretical frequency distributions, that is, for probability distributions. These descriptive measures for probability distributions are essential components of modern quantitative techniques employed as aids for decision-making under conditions of uncertainty. They are used in a wide variety of business, governmental, and military applications. In this chapter, only descriptive measures for probability distributions of discrete random variables will be discussed. Although the corresponding treatment for continuous variables is not conceptually more difficult, calculus methods are required.

In Chapter 2, we considered certain probability distributions for discrete random variables as the appropriate mathematical models for real-world situations under specific sets of assumptions. Sometimes, from the nature of a problem, it is relatively easy to specify a suitable probability model. In other situations, the appropriate model is suggested only after substantial numbers of observations have been taken and empirical frequency distributions have been constructed. However, whatever the method by which we arrive at probability distributions, it is essential to be able to capture their salient properties in a few summary measures. These measures are the subject of this chapter.

4.1 Mathematical Expectation

Mathematical expectation or expected value is a basic concept in the formulation of summary measures for probability distributions. It is a fundamental notion employed for hundreds of years in areas such as the insurance industry, and is currently used in many applications in fields such as operations research, management science, systems analysis, managerial economics, and so forth. Let us consider a very simple problem to obtain an understanding of this concept.

Suppose the following game of chance were proposed to you. A "fair" coin is tossed. If it lands "heads," you win \$10; if it lands "tails," you lose \$5. What would you be willing to pay for the privilege of playing this game?

On any particular toss, you will either win \$10 or lose \$5. However, let us think in terms of a repeated experiment, in which we toss the coin and play the game many times. Since the probability assigned to the event "head" is 1/2 and to the event "tail" 1/2, in the long run, you would win \$10 on one-half of the tosses and lose \$5 on one-half. Therefore the average winnings per toss would be obtained by weighting the outcome \$10 by 1/2 and $-$\$5 by 1/2 to yield a weighted mean of \$2.50 per toss. In terms of Equation (3.16), this weighted mean is

$$\bar{X}_w = \frac{\Sigma wX}{\Sigma w} = \frac{(\$10)(1/2) + (-\$5)(1/2)}{1/2 + 1/2} = \$2.50 \text{ per toss}$$

Of course, when the weights are probabilities in a probability distribution, as is true in this example, the sum of the weights is equal to one. Therefore, in such cases, the formula could be written without showing the division by the sum of the weights.

The average of \$2.50 per toss is referred to as the "expected value" or "mathematical expectation" of the winnings. Note that on a single toss, only two outcomes are possible, namely, win \$10 or lose \$5. If these two possible winnings are viewed as the possible values of a random variable, which occur with probabilities of 1/2 each, then the expected value of the random variable can be seen to be the mean of its probability distribution. More formally, if X is a discrete random variable which takes on the value x with probability $f(x)$, then the expected value of X, denoted $E(X)$, is

(4.1)
$$E(X) = \sum_x xf(x)$$

That is, to obtain the expected value of a discrete random variable, each value that the random variable can assume is multiplied by the probability of occurrence of that value, and then all of these products are totaled.

Applying this formula for the expected value to the problem of tossing a coin, we have

$$E(X) = x_1 f(x_1) + x_2 f(x_2)$$

$$E(X) = \$10(1/2) + (-\$5)(1/2) = \$2.50 \text{ per toss}$$

We see that this is the same calculation performed earlier using the weighted mean formula, except that the sum of the weights, which is equal to one, is not explicitly shown as a divisor.

A brief comment on the meaning of the "expected value of an act" is pertinent at this point. We have seen that the expected value of a random variable is a weighted average of the values that the random variable can assume, where the weights are the probabilities of these random variable values. Analogously in decision making under uncertainty, if a certain action is taken and a specific event occurs, there is a specified payoff. The expected value of an act is the weighted average of these payoffs, where the weights are the probabilities of occurrence of the various events.

Now, to return to the question, "What would you be willing to pay for the privilege of playing this game?" it would appear reasonable that you should be willing to pay up to \$2.50 on each toss. For the game to represent a perfectly "fair bet" from the probability standpoint, you should pay *exactly* \$2.50 per toss. Then, in the long run, you would come out even. However, if we question different people, they might be willing to pay various sums to play the game. The fact that different decisions would be made is not necessarily an irrational situation. It simply reflects the different attitudes of these people toward financial risks. A figure such as the \$2.50 mathematical expectation in the above problem is often referred to as an "expected monetary value." The point being made here is that it is not always appropriate to use expected monetary value as the basis for decision making. Actually, in many business situations expected monetary value does represent an appropriate guide toward decision. For the moment, however, let us expand on the unsuitability of expected *monetary* value as a criterion for action in certain situations. We will then return to the conditions under which expected monetary value is an appropriate criterion.

Suppose we restate the coin tossing problem in terms of choice among alternative courses of action. You are asked to choose between two options, A_1 and A_2. A_1 again involves the flipping of a fair coin only one time. If it lands heads, you win \$10; if tails, you lose \$5. A_2 involves a certainty of neither a gain nor a loss. Which do you prefer, A_1 or A_2? Let us assume you choose A_1. We have already seen that the expected monetary value of A_1 is \$2.50. The monetary value of option A_2 is \$0. The choice of A_1 certainly seems reasonable. We can also use the terminology that the expected value of alternative A_2 is \$0, since as we shall see at a later point in this chapter, the expected value of a constant is the constant itself (Equation (4.5)).

Now, let us change the problem by shifting the decimal place in the gains and losses of alternative A_1. You now are asked to choose between the following two options A_1 or A_2. A_1 again involves the flipping of a fair coin. If it lands heads, you win \$100,000; if tails, you lose \$50,000. A_2 is the certainty of neither a gain nor a loss. Which do you now prefer? The expected monetary value of A_1 is \$25,000. That is, $E(A_1) = \$100{,}000\ (1/2) + (-\$50{,}000)\ (1/2) = \$25{,}000$. The corresponding figure for A_2 is \$0. Despite the fact that alternative A_1 has the higher expected monetary value, it would not at all be unreasonable to prefer A_2.

A loss of \$50,000 might be so catastrophic to you as an individual that you simply would not be willing to risk such a calamity. In conclusion, you might be willing to use expected monetary value as a criterion for decision in the first pair of alternatives, where the amounts of money involved were quite small. On the other hand, you might not be willing to use this criterion for very large sums of money. In terms of business decision making, businessmen would be apt to stray from expected *monetary* value as a criterion of choice when the possible extreme outcomes of an alternative are either too disastrous or too beneficial.

A basic reason why expected monetary value can't be used as a decision criterion for the entire scale of values, running from very small to very large amounts is that it assumes a proportional relationship between money amounts and the *utility* of these amounts to the decision maker. If \$10 million is not considered ten times as useful or valuable as \$1 million, and twice as valuable as \$5 million, then the expected *monetary* value criterion is not valid. Stating it in a positive way, therefore, expected monetary value *is* a valid decision criterion in situations in which utility is proportional to monetary amounts. A second assumption for the appropriate use of the expected monetary value criterion is that the decision maker is neutral to risk. That is, the act of risk taking does not, in and of itself, add or subtract any utility from a given risk situation. If this assumption is not valid, and if the decision maker actively seeks out a gamble merely because he enjoys gambling or carefully avoids a gamble because he does not like gambling, then clearly he will not act on the basis of expected monetary value.

The discussion of utility functions in subsequent chapters on decision theory indicates how logically consistent decisions can be made regardless of whether the above two assumptions are met. In this chapter, we will assume we are dealing with situations in which expected monetary value is a valid guide, that is, the two assumptions hold. Thus, decision makers will act in such a way as to *maximize expected monetary value.* This means they will choose alternatives which maximize expected monetary profits or which minimize expected monetary costs.

Risk and Uncertainty

The concept of expected value is extremely useful in the analysis of decision making under conditions of *uncertainty.* It is helpful at this point to take a somewhat closer look at what is meant by uncertainty. Many writers make a sharp distinction between "risk" and "uncertainty." They reserve the term "risk" for situations in which *objective probabilities* are available, either on the basis of a priori reasoning or relevant past frequency of occurrence experience. Hence, a priori reasoning would provide the basis for the assignment of probabilities for tosses of coins, rolls of dice, etc., where a knowledge of the physical structure of the coins or dice suffices for the establishment of the probability values. Past relative frequencies of occurrence would provide these assignments in actuarial situations, as for example, in assessing probabilities of death, accidents, sickness, or fires for insurance purposes.

On the other hand, these writers use the term "uncertainty" to apply to situations in which *only subjective probabilities* can be applied to the events in question, and there is insufficient basis for a priori or past frequency of occurrence assignments.

Under this sort of distinction between risk and uncertainty, it is clear that most decisions in the business world or in the field of governmental affairs fall under the heading of uncertainty. However, modern usage tends to treat this distinction between risk and uncertainty as merely one of degree. At one extreme, there is the risk situation in which probabilities of events are treated as known; at the other extreme there is virtual or complete ignorance concerning events. The modern tendency in the analysis of decision making is to refer to this entire spectrum as one of uncertainty. Hence, the older idea of risk is included under uncertainty and this latter term tends to be the only one used whether probabilities are assigned either on an objective or subjective basis. In keeping with modern usage, uncertainty will be used in this broader sense throughout this text.[1]

---

[1]One special use of the term "risk" has arisen in Bayesian decision theory. Hence, in Chapter 15 of this text, risk is used to refer to expected opportunity losses, where the meaning is the loss of making an error times the probability of making the error.

Example 4-1 Suppose an insurance company offers a 45-year old man a \$1000 one-year term insurance policy for an annual premium of \$12. Assume that the number of deaths per 1000 is five for persons in this age group. What is the expected gain for the insurance company on a policy of this type?

We may think of this problem as representing a chance situation in which there are two possible outcomes, (1) the policy purchaser lives or (2) he dies during the year. Let X be a random variable denoting the dollar gain to the insurance company for these two outcomes. The probability that the man will live through the year is 0.995. In this case, the insurance company collects the premium of \$12. The probability that the policy purchaser will die during the year is 0.005. In this case, the company has collected a premium of \$12, but must pay the claim of \$1000, for a net gain of −\$988. Thus, X takes on the values \$12 and −\$988 with probabilities 0.995 and 0.005, respectively. The calculation of expected gain for the insurance company is displayed in Table 4-1.

Table 4-1 Calculation of Expected Gain for an Insurance Company on a One Year Term Policy.

Outcome	x	$f(x)$	$x\,f(x)$
Policy Holder Lives	\$12	.995	\$11.94
Policy Holder Dies	−988	.005	−4.94
		1.000	\$7.00

$$E(X) = \sum_x x\,f(x) = \$7.00$$

It may be noted that in setting a premium for this policy, the insurance company would have to take into account usual expenses of doing business as well as the expected gain calculation.

Example 4-2 A financial manager for a corporation is considering two competing investment proposals. For each of these proposals, he has carried out an analysis in which he has determined various net profit figures and has assigned subjective probabilities to the realization of these returns. For proposal A, his analysis shows net profits of \$20,000, \$30,000, or \$50,000 with probabilities 0.2, 0.4, and 0.4, respectively. For proposal B, he concludes there is a 50-50 chance of a successful investment, estimated as producing net profits of \$100,000 or an unsuccessful investment, estimated as a break-even situation involving \$0 of net profit. Assuming each proposal requires the same dollar investment, which is preferable from the standpoint of expected monetary return?

Denoting the expected net profit on these proposals as $E(A)$ and $E(B)$, we have

$$E(A) = (\$20{,}000)(0.2) + (\$30{,}000)(0.4) + (\$50{,}000)(0.4) = \$36{,}000$$

$$E(B) = (\$0)(0.5) + (\$100{,}000)(0.5) = \$50{,}000$$

The financial manager would maximize expected net profit by accepting proposal B.

4.2 Expected Value of a Function of a Random Variable

In many business problems, as we have already noted, we may be interested in the expected value of a random variable. On the other hand, interest may center upon the expected value of some function of this basic random variable. For example, if the random variable is denoted by X, we may be interested in values assumed by $2 + 3X$. If X can take on the values $x = 0$, 1, 2, then the random variable $2 + 3X$ takes on the values $2 + 3x = 2, 5, 8$. That is, $2 + 3(0) = 2$, $2 + 3(1) = 5$, and $2 + 3(2) = 8$. Now, let us assume that X has a uniform or rectangular probability distribution; that is, the three values of X are equally likely. Therefore, the probabilities that X assumes the values 0, 1, and 2 are each 1/3. Then the probabilities that $2 + 3X$ assumes the values 2, 5, and 8 are also each 1/3. This idea is displayed in Table 4-2.

Table 4-2 Probability Distributions of a Random Variable (X) and a Function of the Random Variable ($2 + 3X$).

x	$f(x)$	$2 + 3x$	$f(x)$
0	1/3	2	1/3
1	1/3	5	1/3
2	1/3	8	1/3
	1		1

By Equation (4.1), the expected value of the random variable X is $E(X) = (0)(1/3) + 1(1/3) + 2(1/3) = 1$. It certainly seems reasonable that the expected value of the random variable $2 + 3X$ is equal to $E(2 + 3X) = 2(1/3) + 5(1/3) + 8(1/3) = 5$. This idea can be summarized in general as follows: If X is a discrete random variable, whose probability at x is $f(x)$, and $g(X)$ is any function of X, then the expected value of $g(X)$, denoted $E[g(X)]$ is

(4.2)
$$E[g(X)] = \sum_{x} g(x) f(x)$$

Thus, in words, the definition given in Equation (4.2) says, "In order to obtain the expected value of a function of a random variable, multiply the values that the function can take on by the corresponding probabilities for the basic random variable, and sum all of these products."

Example 4-3 A production manager who was engaged in planning for the next fiscal year was considering the production costs for a possible contract with a customer for some complex assembly units. From discussions with sales representatives, he concluded that there was about a 50-50 chance that the customer would order exactly one unit. However, he felt certain that no more than three would be ordered, with two units more likely than three. Treating the number of units to be ordered as a random variable, he assessed the probability distribution as follows:

Number of Units Ordered x	*Probability* $f(x)$
1	0.5
2	0.3
3	0.2
	1.0

In considering production costs, the production manager concluded that there would be a \$40,000 variable cost per unit and \$20,000 fixed cost regardless of the number of units produced. Therefore, denoting total production cost as $g(x)$, he specified total production cost for this customer as a function of the number of units ordered as follows:

$$g(x) = \$20{,}000 + \$40{,}000x$$

What was the expected total production cost for this customer?

Total production cost is a function of the basic random variable "number of units produced." Expected total production cost is \$88,000 as shown in Table 4-3.

Table 4-3 Calculation of Expected Total Production Cost, $E[g(X)]$.

Number of Units Ordered x	*Total Production Cost* $g(x)$	*Probability* $f(x)$	$g(x)\,f(x)$
1	\$ 60,000	0.50	\$30,000
2	100,000	0.30	30,000
3	140,000	0.20	28,000
		1.00	\$88,000

$$E[g(X)] = \sum_x g(x)\,f(x) = \$88{,}000$$

Expectation of a Linear Function of a Random Variable

The preceding examples were illustrations of calculations of expected values of linear functions of a random variable. That is, the function $g(X)$ was of the form $g(X) = a + bX$, which plots as a straight line on the usual two-dimensional graph with rectangular coordinates. It is useful to obtain a general expression for such an expected value. In order to do this, we return to the definition of the expected value of a function of a random variable,

$$E[g(X)] = \sum_x g(x)f(x)$$

Substituting the linear expression for $g(X)$ into both sides of the definition gives

$$E[a + bX] = \sum_x (a + bx)f(x) \tag{4.3}$$

Multiplying the factors inside the summation sign on the right-hand side of (4.3), and using summation Rules 1 and 3 of Appendix B, we have

$$\sum_x (a + bx)f(x) = \sum_x af(x) + \sum_x bxf(x) = a\sum_x f(x) + b\sum_x xf(x) \tag{4.4}$$

Since $\sum_x f(x) = 1$ and $\sum_x xf(x) = E(X)$, we may state in conclusion that

$$E(a + bX) = a + bE(X) \tag{4.5}$$

Applying Equation (4.5) to the preceding example of a linear function, we have

$$E(2 + 3X) = 2 + 3\,E(X) = 2 + 3(1) = 5$$

The expected value of a function of a random variable obeys the same rules as are true for summations. For example, $\sum_{i=1}^{N} aX_i = a\sum_{i=1}^{N} X_i$, and as we have seen in Equation (4.5), $E(aX) = aE(X)$, where a is a constant.

Also, just as

$$\sum_{i=1}^{N} (X_i + Y_i) = \sum_{i=1}^{N} X_i + \sum_{i=1}^{N} Y_i$$

it can be shown that

$$E(X + Y) = E(X) + E(Y) \tag{4.6}$$

That is, the expectation of a sum of two random variables is equal to the sum of the expectations of these random variables. This rule generalizes as well to the case of N random variables, where N is any finite number.

Similarly, in expressions such as $E[(X - a)^2]$, the operation inside the brackets is carried out before the expected value is obtained. Therefore,

$$E[(X - a)^2] = E[X^2 - 2aX + a^2] = E(X^2) - 2aE(X) + a^2$$

A couple of interesting points may be noted about Equation (4.5). The expected value of a constant is the constant itself. That is, $E(a) = a$. Also, it is evident that the expected value of a linear function of a random variable depends only on the expected value of the random variable and on no other descriptive measure of the probability distribution of that random variable. Thus, $E(a + bX)$ is a function of $E(X)$ and not of the dispersion, skewness or other properties of the probability distribution of X. An application of this idea occurs in the discussion of utility functions on page 638. Let us consider the case of an individual for whom the utility of a sum of money may be expressed as a linear function of that sum of money. If a payoff in dollars is denoted by X and the utility of the payoff by $U(X)$, the utility function in this case may be written $U(X) = a + bX$. Then, the expected value of utility, or more succinctly, the "expected utility" of the payoff is $E[U(X)] = a + bE(X)$. Thus, the utility of the payoff is a function of $E(X)$, which is an expected monetary value. The importance of this relationship, from the viewpoint of decision theory is that *a decision maker who has a linear utility function* maximizes his expected utility by maximizing expected monetary values.

4.3 Moments of Probability Distributions

In Chapter 3, a system of descriptive measures for empirical frequency distributions was discussed. In this section, a similar set of descriptive measures for theoretical frequency distributions, or probability distributions, is explained. Just as frequency distributions may be characterized in terms of central tendency, dispersion, skewness, etc., probability distributions may be similarly described. *Moments of a probability distribution* are used for this purpose. They represent a convenient and unifying method for summarizing many of the most commonly used descriptive statistical measures of probability distributions and, as indicated in this section, of empirical frequency distributions as well.

If X is a discrete random variable with probability distribution $f(X)$, then the kth moment about the origin of the distribution is denoted by μ_k' and may be defined as the expected value of X^k. The origin of the distribution simply refers to the "zero point" or, in other words, if we think of the values of the random variable X as being displayed on a horizontal axis, the origin is the point at which $X = 0$. In symbols the definition of the kth moment around the origin is given by

Moments Around the Origin of a Probability Distribution

(4.7) $$\mu'_k = E(X^k) = \sum_x x^k f(x)$$

Thus, if $k = 1$, we have $\mu'_1 = E(X) = \sum_x x f(x)$. If $k = 2$, $\mu'_2 = E(X^2) = \sum_x x^2 f(x)$, and so forth. It can be seen that the first moment around the origin (that is, around zero), μ'_1, is simply the expected value of X, or the mean of the probability distribution, $f(X)$. Since this value is used so frequently as a measure of central tendency for theoretical frequency distributions, it is given a simpler special symbol, μ. Often, a subscript is attached to the symbol for the mean to denote the random variable whose mean is obtained. Thus, μ_X denotes the mean of the random variable X, μ_Y denotes the mean of Y, etc.[2]

For the description of characteristics of probability distributions other than central tendency, *moments around the mean* are generally used. These moments are defined in an analogous manner to moments around the origin, except that the mean, μ, is subtracted from values of the random variable X before the exponent k is applied. Therefore, moments around the mean of a probability distribution may be defined as follows. If X is a discrete random variable with probability distribution $f(X)$, then the *kth moment around the mean* of the distribution is denoted by μ_k and is the expected value of $(X - \mu)^k$. In mathematical terms, the definition is

Moments Around the Mean of a Probability Distribution

(4.8) $$\mu_k = E(X - \mu)^k = \sum_x (x - \mu)^k f(x)$$

If $k = 1$, we have $\mu_1 = E(X - \mu) = \sum_x (x - \mu) f(x)$, which is equal to zero. This is easily proven, since $E(X - \mu) = E(X) - E(\mu) = \mu - \mu = 0$. This is the same property that was observed for empirical frequency distributions in Chapter 3, where it was shown that the sum of the deviations from the arithmetic mean is equal to zero. If the sum of these deviations is equal to zero it follows, of course, that their mean is also equal to zero.

If $k = 2$, $\mu_2 = E(X - \mu)^2 = \sum_x (x - \mu)^2 f(x)$. This is the variance of a probability distribution. The special symbol σ^2 is used for μ_2, and σ is used for the standard deviation. Thus, $\sigma = \sqrt{\mu_2}$. As in the case of the mean, a subscript is often attached to the symbol for the standard deviation to indicate the appropriate random variable. Thus σ_X denotes the standard deviation of the random variable X, σ_Y denotes the standard deviation of Y, etc.

The first moment around the origin, μ, is a measure of central tendency,

---

[2]In this text, particularly in the chapters on statistical decision procedures, lower case subscripts are generally used. Thus, for example, the means of X and Y are denoted by μ_x and μ_y, respectively.

and the second moment around the mean, σ^2, is a measure of dispersion. It can be shown that the third moment around the mean, $\mu_3 = E(X - \mu)^3$ is a measure of skewness of the probability distribution. If the distribution is symmetrical, $\mu_3 = 0$. If the distribution is skewed to the right, μ_3 is positive; for skewness to the left, μ_3 is negative. Graphs depicting distributions which are skewed to the right and left were shown in Figure 3-7. As we have seen in the case of empirical frequency distributions, descriptive measures of characteristics other than central tendency and dispersion are infrequently calculated in applied work. Similarly, moments of higher order than the second are not often computed for probability distributions.

It is instructive to compare the form of descriptive measures for empirical and theoretical frequency distributions. For example, in Chapter 3, the mean of an empirical frequency distribution was defined as $\bar{X} = \left(\sum_{i=1}^{n} f_i X_i\right)/n$ or $\bar{X} = (\Sigma fX)/n$ after subscript indexes were dropped. The relative frequency f_i/n in that formula plays the same role as the probability $f(x)$ in the formula $\mu = \sum_x xf(x)$ in the case of a theoretical frequency distribution. It is, therefore, possible to summarize descriptive measures for empirical frequency functions in the form of moments, just as in the case of probability distributions. Thus, in an empirical frequency distribution, if the kth moment around the origin is denoted by m'_k, it may be defined as

(4.9)
$$m'_k = \frac{\sum_{i=1}^{h} f_i X_i^k}{n}$$

where f_i is the absolute frequency or number of occurrences in the ith class interval, X_i is the midpoint of the ith interval, n is the total of the absolute frequencies, and h is the number of class intervals.

If $k = 1$, then

$$m'_1 = \frac{\sum_{i=1}^{h} f_i X_i}{n}$$

which is the mean of the distribution. Thus, just as the special symbol μ is used for μ'_1, the symbol $\bar{X}$ is used for m'_1. The second empirical moment around the origin is

$$m'_2 = \frac{\sum_{i=1}^{h} f_i X_i^2}{n}, \text{ etc.}$$

The *kth moment around the mean* of an empirical frequency distribution is denoted m_k and is defined as

(4.10)
$$m_k = \frac{\sum_{i=1}^{h} f_i (X_i - \bar{X})^k}{n}$$

As previously observed, the first moment around the mean, denoted m_1, is equal to zero. If $k = 2$,

$$m_2 = \frac{\sum_{i=1}^{h} f_i(X - \bar{X})^2}{n}$$

which is the variance of the distribution. The special symbol s^2 is used for m_2, just as σ^2 is used for μ_2.

The symbolism for the first two moments of empirical and theoretical frequency distributions is summarized in Table 4-4.

Table 4-4 Symbolism for the First Two Moments of Empirical and Theoretical Frequency Distributions.

Around Origin		*Around Mean*	
Empirical	*Theoretical*	*Empirical*	*Theoretical*
$m_1' = \bar{X}$	$\mu_1' = \mu$	$m_1 = 0$	$\mu_1 = 0$
m_2'	μ_2'	$m_2 = s^2$	$\mu_2 = \sigma^2$

A particularly convenient computational formula for σ^2 is given by $\sigma^2 = \mu_2' - \mu^2$. This formula is completely analogous to the shortcut method of computing s^2 given by Equation (3.32). This method of computing σ^2 is usually a labor saver, since it is in terms of the first and second moments around the origin and thus does not involve calculations using deviations from the mean. The formula can easily be proven as follows:

$$\sigma^2 = \sum_x (x - \mu)^2 f(x)$$

Expanding the squared expression gives

$$\begin{aligned}\sigma^2 &= \sum_x (x^2 - 2x\mu + \mu^2) f(x) \\ &= \sum_x x^2 f(x) - 2\mu \sum_x x f(x) + \mu^2 \sum_x f(x) \\ &= \mu_2' - 2\mu^2 + \mu^2\end{aligned}$$

Adding the last two terms, we can write

(4.11) $$\sigma^2 = \mu_2' - \mu^2$$

Another way of writing Equation (4.11) is

(4.12) $$\sigma_X^2 = E(X^2) - [E(X)]^2$$

In words, the variance of the random variable X is equal to the expected value of X^2 minus the square of the expected value of X.

We illustrate the calculation of theoretical moments for a few simple probability distributions.

Example 4-3 If a true coin is tossed three times, what are the expected value and standard deviation of the number of heads?

The calculation of the expected value and standard deviation from the definitional formulas for moments is given in Table 4-5. Also shown is the alternative calculation for the standard deviation, following from (4.11).

Table 4-5 Expected Value and Standard Deviation of Number of Heads in Coin Tossing Problem.

Number of Heads x	*Probability* $f(x)$	$xf(x)$	$(x-\mu)$	$(x-\mu)^2$	$(x-\mu)^2 f(x)$
0	1/8	0	−3/2	9/4	9/32
1	3/8	3/8	−1/2	1/4	3/32
2	3/8	6/8	+1/2	1/4	3/32
3	1/8	3/8	+3/2	9/4	9/32
	1	1 1/2			24/32 = 3/4

$$\mu = E(X) = \sum_{x=0}^{3} xf(x) = 1\ 1/2 \text{ heads}$$

$$\sigma^2 = E(X-\mu)^2 = \sum_{x=0}^{3} (x-\mu)^2 f(x) = 3/4$$

$$\sigma = \sqrt{3/4} = 0.87 \text{ heads}$$

Alternative Calculation of the Standard Deviation

$$\mu_2' = E(X^2) = \sum_{x=0}^{3} x^2 f(x) = 0(1/8) + 1(3/8) + 4(3/8) + 9(1/8) = 24/8 = 3$$

$$\sigma^2 = \mu_2' - \mu^2 = 3 - (3/2)^2 = 3/4$$

$$\sigma = \sqrt{3/4} = 0.87 \text{ heads}$$

Example 4-4 Compute the mean and variance for the total obtained on the uppermost faces in a roll of two unbiased dice.

Let X denote the specified total on the two dice. Then,

$$\mu = \sum_{x=2}^{12} (x)f(x) = 2(1/36) + 3(2/36) + 4(3/36) + 5(4/36) + 6(5/36)$$
$$+ 7(6/36) + 8(5/36) + 9(4/36) + 10(3/36) + 11(2/36) + 12(1/36) = 7$$

$$\sigma^2 = \sum_{x=2}^{12} x^2 f(x) - \mu^2 = 4(1/36) + 9(2/36) + 16(3/36) + 25(4/36) + 36(5/36)$$
$$+ 49(6/36) + 64(5/36) + 81(4/36) + 100(3/36) + 121(2/36)$$
$$+ 144(1/36) - (7)^2 = 54\text{-}5/6 - 49 = 5\text{-}5/6$$

Example 4-5 The Amiable Conglomerate Company estimates the net profit on a new product it is launching to be \$3,000,000 during the first year if it is "successful," \$1,000,000 if it is "moderately successful," and a loss of \$1,000,000 if it is "unsuccessful." The firm assigns the following probabilities to first year prospects for this product: successful, 0.15; moderately successful, 0.60; and unsuccessful, 0.25. What are the expected value and standard deviation of first year net profit for this product?

Let X denote first year net profit in millions of dollars, then,

$E(X) = (3)(.15) + (1)(.60) + (-1)(.25) =$ \$0.80 million or \$800,000

$$\sigma_X^2 = E(X^2) - [E(X)]^2 = 9(.15) + 1(.60) + 1(.25) - (.80)^2 = 2.20 - .64 = 1.56$$

$$\sigma_X = \sqrt{1.56} \approx 1.25 \text{ million or } \$1{,}250{,}000$$

4.3 Variance of a Function of a Random Variable

In Section 4.2, the expected value of a function of a random variable was considered, with emphasis on expectations of linear functions. We now consider variances of a function of a random variable, with similar emphasis on linear functions. Just as in the case of the expected value, we can obtain the variance of a function of a random variable, without first having to find the probability distribution of the new function.

It was shown in Equation (4.5) that $E(a + bX) = a + b\,E(X)$, where X is a discrete random variable and a and b are constants. The variance of this linear function, denoted σ_{a+bX}^2, can be shown to be equal to σ_{bX}^2, which in turn, can be shown to be equal to $b^2\sigma_X^2$. A proof of this relationship follows. However, first, the matter of notation will be considered. The variance of the function $a + bX$, which we have denoted σ_{a+bX}^2 is sometimes symbolized as Var $(a + bX)$, $V(a + bX)$, or $\sigma^2(a + bX)$. We will use the notation in which the random variable whose variance is sought is shown as a subscript. This is a widespread convention, particularly in sampling theory and statistical inference. This convention will be used throughout the text. Analogously, $E(a + bX)$ can be written μ_{a+bX}. In the discussion of sampling theory and statistical inference in subsequent chapters, we will often employ the latter type of notation.

The proof that $\sigma_{a+bX}^2 = b^2\sigma_X^2$ follows directly from the definition of a variance. Let the expected value of X be denoted μ. Then, from (4.5), $\mu_{a+bX} = a + b\mu$. Therefore, the variance of $a + bX$ is

$$\sigma^2_{a+bX} = \sum_x [(a + bX) - (a + b\mu)]^2 f(x)$$

Carrying out the subtraction inside the brackets and factoring out the b we have

$$\sigma^2_{a+bX} = \sum_x [b(x - \mu)]^2 f(x) = b^2 \sum_x (x - \mu)^2 f(x) \tag{4.13}$$

The expression $\sum_x (x - \mu)^2 f(x)$ in (4.13) is simply σ^2_X. Therefore, we may write

$$\sigma^2_{a+bX} = b^2 \sigma^2_X \tag{4.14}$$

It can be seen from (4.14) that the variance of a constant is equal to zero. That is, $\sigma^2_a = 0$. In terms of statistical data, we can interpret this result as follows. If we have a series of numbers which are all identical, there is no variability among them and, therefore, their variance is zero. Also, we note from (4.14) that the variance of a constant times a random variable is equal to the square of the constant times the variance of the random variable.

Variance of a Sum or Difference of Independent Random Variables

If a and b are constants and X and Y are independent random variables, then it can be shown that

$$\sigma^2_{aX+bY} = a^2 \sigma^2_X + b^2 \sigma^2_Y \tag{4.15}$$

If a and b are each equal to one, we have

$$\sigma^2_{X+Y} = \sigma^2_X + \sigma^2_Y \tag{4.16}$$

Thus, the variance of the sum of two *independent* random variables is equal to the sum of the variances of these variables. Equation (4.16) generalizes as well to the case of N independent random variables, where N is any finite number.

If $a = 1$ and $b = -1$, Equation (4.15) becomes

$$\sigma^2_{X-Y} = (1)^2 \sigma^2_X + (-1)^2 \sigma^2_Y = \sigma^2_X + \sigma^2_Y \tag{4.17}$$

Equation (4.17) expresses a relationship that is very useful in sampling theory. In words, the variance of the difference between two *independent* random variables is equal to the sum of the variances of the random variables.

We have observed in Equation (4.6) that the expected value of the sum of two random variables is equal to the sum of the expected values of these variables. Note that there is no restriction of independence in the case of expected values. That is, *Equation (4.6) holds, whether or not X and Y are independent.* The analogous relationship to Equation (4.15) for expected values is

(4.18) $$E(aX + bY) = aE(X) + bE(Y)$$

or in subscript notation,

(4.19) $$\mu_{aX+bY} = a\mu_X + b\mu_Y$$

The following examples illustrate the principles developed for expected values and variances.

Example 4-6 What are the effects on the values of the arithmetic mean and variance of a family income distribution if \$1000 is added to every family's income? If every income is doubled?

Let X be a random variable denoting the value of a family's income prior to the increase; μ_X and σ_X^2 are the mean and variance of this variable, respectively. Then if \$1000 is added to each family's income, we have by (4.5) and (4.14), respectively,

$$\mu_{X+\$1000} = \mu_X + \$1000$$

and

$$\sigma_{X+\$1000}^2 = \sigma_X^2$$

Thus, the mean family income increases by \$1000, but the variance remains unchanged.

If every income is doubled, then we have again by (4.5) and (4.14)

$$\mu_{2X} = 2\mu_X$$

$$\sigma_{2X}^2 = (2)^2\sigma_X^2 = 4\sigma_X^2$$

In this case, the mean income is doubled, but the variance is quadrupled.

Example 4-7 A project manager is in charge of a project to develop a scale model of an assembly to be placed in production at a later time. He has scheduled the following three activities to complete this project: (1) Design of equipment, (2) Training of personnel, (3) Assembly of equipment. The times required to complete these activities are independent of one another. Based on this manager's experience with similar projects in the past, he estimates the means (expected times) and standard deviations of the completion times in weeks for each activity as follows:

Activity	*Expected Time*	*Standard Deviation*
1. Design of Equipment	10	4
2. Training of Personnel	5	1
3. Assembly of Equipment	6	3

Estimate the mean and standard deviations of completion times for the entire project.

Let μ_1, μ_2, μ_3 and σ_1, σ_2, and σ_3 denote the expected values and standard deviations of the completion times of the three activities, respectively. Then the expected value and standard deviation of completion time for the entire project are:

$$\mu = \mu_1 + \mu_2 + \mu_3 = 10 + 5 + 6 = 21 \text{ weeks}$$

$$\sigma = \sqrt{\sigma_1^2 + \sigma_2^2 + \sigma_3^2} = \sqrt{4^2 + 1^2 + 3^2} = 5.1 \text{ weeks}$$

Example 4-8 A certain technical institution of higher learning requires both a verbal and mathematical aptitude test for entering students. In a particular year, the mean grade of entering students in the mathematics examination was 700 with a standard deviation of 100. The mean grade in the verbal examination was 600 with a standard deviation of 80. In connection with its admissions procedure, this institution calculates a weighted score for these two examinations in which the grade on the mathematics examination is weighted twice as heavily as the grade on the verbal test. If X is the score on the mathematics test and Y the weight on the verbal test, then the total score, Z is

$$Z = 2X + Y$$

Two students are randomly drawn from the entering class. The grade on the mathematics examination, (X), is determined for the first student; the grade on the verbal examination, (Y), is determined for the second. If a total score, Z, is determined for these two test grades using the foregoing formula, what are the mean and standard deviation of the distribution of the Z score?

By Equation (4.18)

$$E(Z) = 2E(X) + E(Y)$$
$$= 2(700) + 600 = 2000$$

By Equation (4.15)

$$\sigma_Z^2 = 2^2\sigma_X^2 + \sigma_Y^2$$
$$= 4(100)^2 + (80)^2 = 46{,}400$$

$$\sigma_Z = \sqrt{46{,}400} = 215.4$$

It may be noted that the application of Equation (4.15) for the variance of the sum implied that X and Y are independent random variables. The values of X and Y in this example were determined for two different students. Hence the assumption of independence appears to be a reasonable one. That is, it is unlikely that the grade on the mathematics examination of the first randomly selected student, X, depends upon the grade on the verbal examination of the second student, Y, and vice versa. On the other hand, it is much more likely that the grades on the two examinations for the *same student* would indeed be dependent. Methods of measuring the nature of the dependence and the degree of correlation or association between two variables are discussed in Chapter 10.

PROBLEMS

1. Let $X = -\$.10$ if a customer does not purchase from a telephone sales promotion and $X = +\$1.89$ if a customer does purchase from a telephone sales promotion. If the probability that a customer will purchase is 5%, what is the expected value of X? Interpret the result.
2. Let $X = 0$ if an item is defective and $X = 1$ if an item is not defective. If 10% of all the items are defective, what is the expected value of X? Interpret the result.
3. The number of freighters entering Hyannis Harbor between 8:00 A.M. and 2:00 P.M. on Fridays has the following probability distribution:

x	$f(x)$
0	.10
1	.20
2	.40
3	.15
4	.10
5	.05

   What is the expected number of freighters entering this harbor during the specified time period on a given Friday?
4. The demand for a certain product is a discrete random variable and is uniformly distributed in the range of ten to 20 items per day. State the probability distribution of this random variable. What is the expected demand per day?
5. The probability distribution for the number of customers entering a certain restaurant between 2:00 P.M. and 3:00 P.M. on a weekday is as follows:

x	$f(x)$
10	.01
11	.05
12	.13
13	.18
14	.26
15	.18
16	.13
17	.05
18	.01

What is the expected number of customers entering the restaurant during the given time period?

6. Suppose an insurance company offers a particular home owner a \$15,000 one-year term fire insurance policy on his home. Assume that the number of fires per 1000 dwellings of this particular home safety classification is two. For simplicity, assume that any fire will cause damages of at least \$15,000. If the insurance company charges a premium of \$36 a year, what is its expected gain from this policy? Does this mean that the company will earn that many dollars on *this particular policy?*

7. In Problem 9 of Chapter 2, what is the expected time of arrival of the 10:00 A.M. flight from Washington, D.C.?

8. Do you agree or disagree with each of the following statements? Explain your decision.
   (a) The expected value is the most typical or most common outcome.
   (b) If the variance of a random variable X is 10, then the variance of the new random variable $X/10$ is 1.
   (c) The second moment about the mean must be less than the second moment about the origin.

9. A local retail clothing company has ten stores. The net profit per week from each store has an expected value of \$300 and a variance of \$2500. What is the expected total net profit per week and standard deviation for all ten stores? Assume independence.

10. A manager of a motel in Cape Cod, Mass., is contemplating the construction of a prefabricated temporary housekeeping addition to his motel. The unit has a life of three seasons. The cost of constructing the unit allocated over the three seasons (the motel is only open May 15–Oct. 15) is \$8 per day. The cost of servicing the unit when occupied is \$1 per day. The unit rents for \$20 per day. If the probability that the unit will be rented on any particular day is .7, what is the expected profit per day? What is the expected profit for the life of the unit? (There are 153 days in each season.)

11. An economist has hypothesized the following mathematical expression for the probability distribution of a certain discrete economic variable.

$$f(x) = \frac{x+1}{10} \qquad x = 0, 1, 2, 3$$

Calculate the variance of the probability distribution of this variable by two different methods.

12. An operations research analyst developed the following probability distribution for a discrete variable in a managerial application:

$$f(x) = \frac{1}{18}(x^3 + 3) \qquad x = 0, 1, 2$$

Calculate the variance by two different methods.

13. In the game of roulette there are 36 numbers, 1 through 36, and the numbers 0 and 00. On a spin of the roulette wheel it is equally probable that a ball will rest on any of the 38 numbers. A person places a $1 bet on the numbers 1–12. That is, if the ball rests on any one of these numbers, the person wins $2, whereas if it rests on any other number, he loses his $1. The person places the bet 1000 consecutive times. What is his expected profit (loss) on each roll? For the 1000 rolls?

14. A certain machinist produces five to eight finished pieces during an eight-hour shift. An efficiency expert wants to assess the value of this machinist, where value is defined as value added less the machinist's labor cost. The value added for the work the machinist does is estimated at $5 per item. The machinist earns $3.50 per hour. From past records, the machinist's output is computed to have the following probability distribution:

Number of Pieces Produced Per Eight Hours	*Probability*
5	.2
6	.4
7	.3
8	.1

What is the expected monetary value of the machinist to the company per eight-hour day?

15. You are offered the following game of chance. First you roll a die. You pay $2 times the number on the uppermost face of the die. Then you flip a coin twice and receive $6 for each head. What is the expected profit or loss from this game?

16. Mrs. Plywood, Inc. earns $1200 a year from each of its franchises for rights to its name. In 1968 the company had six franchises, and wanted to add five more by the beginning of 1969. The president of the company assigned the following probabilities to the numbers of new franchises that would be obtained by 1969:

Number of New Franchises by 1969	*Probability*
0	.01
1	.03
2	.06
3	.10
4	.30
5	.50

What are Mrs. Plywood, Inc.'s expected earnings from its franchising rights and the variance of these earnings for 1969?

17. A baker produces ten lemon pies each morning at a cost of $4. He sells each pie for $.69, and those not sold at the end of the day are thrown away.

Assume the following probability distribution for pie demand on any given day:

Number of Pies Demanded	*Probability*
0	.001
1	.010
2	.035
3	.050
4	.065
5	.100
6	.120
7	.340
8	.115
9	.050
10	.114

Find the expected profit per day.

18. Let X and Y be independent random variables with

$$E(X) = 15$$
$$E(Y) = 10$$
$$\sigma^2(X) = 20$$
$$\sigma^2(Y) = 25$$

Find

(a) $E(7X + 2)$
(b) $E(X^2)$
(c) $E[X - E(X)]$

19. In reference to Problem 7 of Chapter 2, what is the mean and standard deviation of the number of minutes it takes to drain a soda filler?

20. As manager of a certain company, you must decide which of two new products to introduce this year. Your research group has given you the following probability distributions for the net profits of each product:

Product A		*Product B*	
Net Profit	*f(Net Profit)*	*Net Profit*	*f(Net Profit)*
−\$2000	.2	\$ 0	.4
0	.3	+ 1000	.3
+ 2000	.3	+ 2000	.2
+ 5000	.2	+ 3000	.1

Find the expected net profit and standard deviation for each product. Which product would you introduce, assuming other factors to be equal?

21. You are considering buying one of two stocks, Volatile or Stability, both priced presently at \$46, for a one-month trading venture. The probability distribution for the closing prices of the two stocks (rounded to nearest \$1) one month hence is estimated as:

Volatile		*Stability*	
Price	*f(Price)*	*Price*	*f(Price)*
44	.1	44	.005
45	.1	45	.015
46	.1	46	.030
47	.1	47	.100
48	.1	48	.350
49	.1	49	.350
50	.1	50	.100
51	.1	51	.030
52	.1	52	.015
53	.1	53	.005

Find the expected values and risk of each stock. (Financial analysts often refer to variance as "risk.") Which stock would you purchase and why?

22. For each of the following probability functions find the first two moments about the mean and origin:
(a) $X/3$ $\qquad X = 1, 2$
(b) $12/25X$ $\qquad X = 1, 2, 3, 4$
(c) $2X/5$ $\qquad X = 1/4, 1/2, 3/4, 1$

23. A person has just hired a building contractor to build a house for him. The house will be built in four stages. First, the contractor lays the foundation; second, he builds the frame and exterior; third, he puts in the wiring, plumbing, and interior; and last, he does the landscaping. Each stage must be completed before the next step is started. In attempting to get an estimate of when the house will be totally completed, the purchaser is able to get the following information from the persons in charge of each stage:

Stage	*Expected Time of Completion of Stage (in Weeks)*	*Standard Deviation (in Weeks)*
I	6	3
II	8	2
III	5	2
IV	2	1

What is the expected value and standard deviation of completion time for the house, assuming the completion times of the stages are independent?

24. Let $x_i = 0$ if the ith person does not return a reply card.
$x_i = 1$ if the ith person does return a reply card.
If the probability that the ith person will return the reply card is p, what is the variance of x_i?

25. Let $x_i = 0$ if there is a failure on the ith trial of an experiment.
$x_i = 1$ if there is a success on the ith trial of an experiment.
The probability of success on the ith trial is p and is constant for all i. Also,

the trials are independent (i.e., a Bernoulli process). Let $X = \Sigma x_i$ (X is binomial). Interpret X and calculate its mean and variance.

26. If $E(X) = np$ and $\sigma^2(X) = npq$, what are the mean and variance of X/n?

27. Assume the average salary of a public school teacher in Southern cities of over 150,000 population is $6100 with a standard deviation of $300, while in midwestern cities of over 150,000 population it is $6550 with a standard deviation of $400. What is the mean and standard deviation of the difference between a public school teacher's salary in the two areas, assuming independence?

28. Let $X = 0$ if a motel's price is less than $10 for a double room.
$= 1$ if a motel's price is less than $20 but at least $10 for a double room.
$= 2$ if a motel's price is $20 or more for a double room.
Let X refer to motels in Cape Cod during the month of August. Let Y be a random variable which assumes values with the same meaning as X, but refers to motels in Miami during the month of August. The frequency distributions for X and Y are

x	$f(x)$	y	$f(y)$
0	.2	0	.4
1	.4	1	.3
2	.4	2	.3

(a) If a person telephones randomly a motel in Cape Cod during August, what is the expected price he will be quoted?
(b) If a person telephones randomly a motel in Miami during August, what is the expected price he will be quoted?
(c) What is the expected difference in prices he will be quoted between the two locations?

29. A life insurance sales office has 20 salesmen. The arithmetic mean commission per week for these salesmen is $115 with a variance of $100. Assuming independence,

(a) what are the mean and variance for the *total commissions* paid in this sales office?
(b) if every salesman is given a flat $10 a week raise what would be the new mean and variance of weekly commissions for these salesmen?
(c) if the commission rate is increased by 10%, what would be the new mean and variance of weekly commission for these salesmen?

30. A statistical research assistant has developed the following probability distributions for the number of books checked out during any given hour between 8:00 A.M. and 5:00 P.M., Monday through Friday, from the adults' and children's section of the Balmouth Library.

Number of Books Checked out from Children's Section	*Probability*	*Number of Books Checked out from Adults' Section*	*Probability*
10	.01	10	.01
11	.09	11	.02
12	.24	12	.08
13	.30	13	.14
14	.15	14	.20
15	.12	15	.31
16	.05	16	.12
17	.03	17	.10
18	.01	18	.02

(a) What are the mean and variance of the number of books checked out in the children's section?
(b) What are the mean and variance of the number of books checked out in the adults' section?
(c) What are the mean and variance of the difference between the number of books checked out in the two sections? (Assume independence.)

31. Notilt Ladder Company has a factory in Pennsylvania and a factory in Georgia. The average number of ladders sold per month at the Pennsylvania factory is 2840 with a standard deviation of 200, while at the Georgia factory the average number of ladders sold is 2590 with a standard deviation of 100.
(a) What are the mean and standard deviation of the total number of ladders sold per month by the two factories?
(b) What are the mean and standard deviation of the difference between the number of ladders sold per month by the two factories?

32. A certain small marina in Yarmouth can handle five boats on any given day. Two types of boats, sailboats and runabouts, use the marina. From historical data the number of demands on any given day for facilities for each type of boat is:

Sailboats		*Runabouts*	
Number of Demands	*Probability*	*Number of Demands*	*Probability*
0	.1	0	.2
1	.3	1	.4
2	.3	2	.3
3	.2	3	.1
4	.1		

What is the expected number of demands on facilities and the corresponding standard deviation from sailboats? From runabouts? What is the expected total number of demands and the respective standard deviation if the demands from these types of boats are independent?

33. The average yield in bushels of grain per acre in the James River Basin is dependent upon the amount of yearly rainfall according to the following relationship:

$$Y = 1.0 + 2R + e$$

where Y = average yield
R = total rainfall for the year
e = a random variable with mean zero and variance of 4.0, and is independent of R

If $E(R) = 6$ and $\sigma^2(R) = 2.25$, what is the expected value and variance of the average yield?

34. A familiar economic relationship is

$$C = a + bY + e$$

where a = a constant
Y = national income in current dollars
b = a constant (marginal propensity to consume)
C = national consumption measured in current dollars
e = a random variable with mean of zero and a variance of σ_e^2, and is independent of C.

If Y is $820 billion, b is .6, and a is $250 billion, what is the expected value of C and the variance of C?

35. You are in charge of drilling for an oil company. At a certain stage in drilling, a geologist estimates the following probabilities:

Outcome	*Oil*	*Gas*	*Water*
Probability	.2	.3	.5

If X, the value of the result at the first level is $5 million for oil, $1 million for gas, and $0 for water, what is the expected monetary value of X?

CHAPTER FIVE

Statistical Investigations and Sampling

5.1 Formulation of the Problem

A statistical investigation arises out of the need to solve some sort of a problem. Problems may be classified in a variety of ways. One such classification is (1) the problem of choosing among alternative courses of action and (2) the problem of informational reporting. The problems of managerial decision-making typically fall under category (1). For example, an industrial corporation wishes to choose a particular plant site from several alternatives. A financial vice-president wishes to decide among alternative methods of financing planned increases in productive capacity. An advertising manager must choose media in which he will advertise from many possible choices. Under heading (2), many illustrations can be given of statistical data which are collected for reporting purposes. For example, a trade association may report to its members on the characteristics of these companies. A research organization may publish data on the relationship between achievement of school children and the socio-economic characteristics of the parents of these children. An economist may report data on the frequency distribution of family incomes in a particular city. Even in the case of informational reporting, the data collected should have an ultimate decision-making purpose for someone.

Throughout our lives we are involved in answering questions and solving problems. For many of these questions and problems, careful, detailed investigations are simply inappropriate. For example, to answer the question, "What clothes should I wear today?" or the question, "What type of transportation should I take to get to a friend's house?" does not require painstaking, objective, scientific investigation. This book, on the other hand, is concerned with the investigation of problems which do require careful planning and an objective, scientific approach to arrive at meaningful solutions. Many of these problems arise as rather vague, original questions. These questions must be translated into a series of other questions, which then form the basis of the investigation. In most carefully planned investigations, the problem will be defined and re-defined many times. The purposes and importance of an investigation will determine the type of study to be conducted. There are instances, where the objectives are particularly hard to define because of the large number of uses that will be made of the data and the very large mass of research consumers who will utilize the results of the study. For example, the U. S. Bureau of the Census publishes a wide variety of data on population, housing, manufactures and retail trade. It cannot specify in advance the many uses that will be made of these data. However, in all studies it is critical to spell out as meticulously as possible the purposes and objectives of the investigation. All subsequent analysis and interpretation depend upon these objectives, and it is only by spelling out very carefully what these objectives are that we can know what questions have been answered by the inquiry.

5.2 Design of the Investigation

The types of investigations in which we are interested are what may be referred to as "controlled inquiries." We are all familiar with the scientist who controls variables in the laboratory. He exercises control by manipulating the things and the events that he is investigating. For example, a chemist may hold the temperature of a gas constant, while he varies pressure and observes changes in volume.

Observational Studies

In most statistical investigations in business and economics, it is not possible to manipulate people and events in the direct way that a physical scientist manipulates his experimental materials. For example, if we want to investigate the effect of income on a person's expenditure pattern, it would not be feasible for us to vary this individual's income. On the other hand, we can observe the different expenditure patterns of people who fall in different income groups, and therefore we can make statistical generalizations about how expenditures vary with differences in income. This would be an example of a so-called *observational* study. In this type of study the analyst essentially examines historical relationships that exist among variables of interest. By observing the important and relevant properties of the group under investigation, the study can be carried out in a controlled manner. For example, if we are interested in how family expenditures vary with family income and color, we can record data on family expenditures, family incomes and color, and then tabulate data on expenditures by income and color classifications, such as white, black. In this way, if we observe the differences in family expenditures for white and black families within the same income group, we have, in effect, "controlled" for the factor of income. That is, since the families observed are in the same income group, income cannot account for the differences in the expenditures observed.

If observational data represent historical relationships, it may be particularly difficult to ferret out causes and effects. For example, suppose we observe past data on the advertising expenses and sales of a particular company. Also, let us assume, that both of these series have been increasing over time. It may be quite incorrect to assume that it is the changes in advertising expenditures that have caused sales to increase. If a company's practice in the past had been to budget 3% of last year's sales for advertising expense, one may state that advertising expenses depend upon sales with a one year lag. In this situation sales might be increasing quite independently of changes in advertising expenses. Thus, one certainly would not be justified in concluding that changes in advertising expenses cause changes in sales. The point may also be made that many other factors, other than advertising, may have influenced changes in sales. If data were not available on these other factors, it would not be possible to determine cause and effect relationships from these past observational data. The specific difficulty in attempting to derive cause-effect relationships in mathematical terms from historical data is that the various environmental factors which are pertinent will not ordinarily have been controlled nor have remained stable.

Direct Experimentation Studies

Direct experimentation studies are being increasingly used in fields other than the physical sciences, where they have been traditionally employed. In such studies, the investigator directly controls or manipulates factors which affect a variable of interest. For example, a marketing experimenter may vary the amounts of direct mail exposure to a particular consumer audience. He may also use different types of periodical advertising and observe the effects upon some experimental group. Various combinations of these direct mail exposures and periodical advertising, as well as other types of promotional expenditures, such as sales force, may be used. Thus the investigator may be able to observe from his experiment that high levels of periodical advertising produce high sales effects only if there is a high concentration of sales force activity. These scientifically controlled experiments for generating statistical data to which only brief reference is being made here can be very efficiently utilized to reduce the effect of uncontrolled variations. The real importance of this type of planning or design is that it gives greater assurance that the statistical investigation will yield valid and useful results.

Ideal Research Design

An important concept of a statistical inquiry is that of the ideal research design. The investigators should think through at the design stage what the ideal research experiment would be without reference to the limitations of data available, or data that can be feasibly collected. Then, if compromises must be made because of the practicalities of the real world situation, the investigator will at least be completely aware of the specific compromises and expedients that have been employed. As an example, suppose we wanted to answer the question as to whether women or men are better automobile drivers. Clearly, it would be incorrect simply to obtain past data on the accident rates of men versus women. First of all, men drive under quite different conditions than do women. For example, driving may constitute a large proportion of the work that many men do. On the other hand, women may drive primarily in connection with their duties as housewives. The conditions of such driving differ considerably with respect to exposure to accident hazards. Many other reasons may be indicated for differences in accident rates between men and women apart from the essential driving ability of these two groups. Thus, as a first approximation to the ideal research design, perhaps we would like to have data for quite homogeneous groups of men and women, for example, women and men of essentially the same age, driving under essentially the same driving conditions, using the same types of automobiles. It may not be within the resources of a particular statistical investigation to gather data of this sort. However, once the ideal data required for a meaningful answer to the question have been thought through, the limitations of other somewhat more practical sets of data become apparent.

5.3 Construction of Methodology

An important phase of a statistical investigation is the construction of the conceptual or mathematical model to be used. A model is simply a representation of some aspect of the real world. Mechanical models are very profitably used in industry as well as other fields of endeavor. For example, airplane models may be tested in a wind tunnel. Ship models may be tested in experimental water basins. Various types of experiments may be carried out by varying certain factors and observing the effect of these variations on the mechanical models employed. Thus, we can manipulate and experiment upon the models and draw corresponding inferences about their real world counterparts. The advantages of this procedure are obvious as compared to attempting to manipulate an experiment using the real world counterparts, such as actual airplanes or ships after they are constructed. In statistical investigations, mathematical models are often used to state in mathematical terms the relationships among the relevant variables. These models are conceptual abstractions which attempt to describe, predict, and often to control real world phenomena. For example, the law of gravity describes and predicts the relationship between the distance an object falls and the time elapsed. This conceptual model describes the relationship between time and distance if an object is dropped in a vacuum. Such models can be tested by physical experimentation.

In well designed statistical investigations, the nature of the model or models to be employed should be carefully thought through in the planning phases of the study. In fact, the nature of these models provides the conceptual framework which dictates the type of the statistical data to be collected. Let us consider a few simple examples. Suppose a market research group wants to investigate the relationship between expenditures for a particular product and income and several other socio-economic variables. The investigators may want to use a mathematical model such as a regression equation, to be discussed later in this book, which states in mathematical form the relationship among the above variables. When the investigators determine the variables that are most logically related to the expenditures for the product, they also determine the types of data that will have to be collected in order to construct their model.

Even in the case of relatively simple informational reporting, there is a conceptual model involved. For example, suppose an agency wishes to determine the percentage of unemployed in a given community. Also, assume that the agency must gather the data by means of a sample survey of the labor force in this community. The ratio "proportion unemployed" is itself a model. It states a mathematical relationship between the numerator (number of persons unemployed) and the denominator (total number of persons in the labor force). The agency may wish to go further and state the range within which it is highly confident that the true percentage unemployed falls. In such a situation, as we shall see later when we study estimation of population values, there is an implicit model which is the probability distribution of a sample proportion.

Suppose a company wishes to establish a systematic procedure for accepting or rejecting shipments received from a particular supplier. Various types of models have been used to solve this sort of problem. The company may decide to accept or reject shipments on the basis of testing some hypothesis concerning the percentage of defective items observed. On the other hand, it may decide to base its acceptance procedure on the arithmetic mean value of some characteristic which is considered important. Other procedures are possible in which, for example, a formal decision model may be constructed. For these types of models, the probability distribution of the percentage defective produced by this company in the past may be required as well as data on the percentage of defective articles observed in a sample drawn from the particular incoming shipment in question. Obviously, the nature of the data to be observed and furthermore the nature of the analysis to be carried out will flow from the type of conceptual model used in the investigation.

5.4 Some Fundamental Concepts

Statistical Universe

In the problem formulation stage, it is necessary to define very carefully the relevant *statistical universe* of observations. The universe or *population*, consists of the total collection of items or elements that fall within the scope of a statistical investigation. The purpose of defining a statistical population is to provide very explicit limits for the data collection process and for the inferences and conclusions that may be drawn from the study. The items or elements which comprise the population may be individuals, families, employees, schools, corporations, and so forth. Time and space limitations must be specified, and it should be clear whether or not a particular element falls within or outside the universe.

In survey work, a listing of all the elements in the population is referred to as the *frame*, or *sampling frame*. A *census* is a survey which attempts to include every element in the universe. The word "attempts" is used here because often in surveys of very large populations, despite every effort to do so, complete coverage may not be effected. Thus, for example, the Bureau of the Census readily admits that its national "censuses" of population invariably result in underenumerations. Strictly speaking, any partial enumeration of a population constitutes a *sample*, but the term "census" is used as indicated here. In most practical applications, it is not even feasible to attempt a complete enumeration of a population, and, therefore, typically, only a sample of items is drawn. If the population is well defined in space and time, the problem of selecting a sample of elements from it is considerably simplified.

Let us illustrate some of the above ideas by means of a simple example. Suppose we draw a sample of 1000 families in a large city to estimate the arithmetic mean family income of all families in the city. The aggregate of all families in the city constitutes the universe and each family is an element of the universe. The income of the family is a characteristic of the unit. A listing of all families in the city would comprise a frame. If instead of drawing the sample of 1000 families, an attempt had been made to include all families in the city, a census would have been conducted. The definition of the universe would have to be specific as to the geographic boundaries that constitute the city and also the time period for which income would be observed. The terms "family" and "income" would also have to be rigidly specified. Of course, the precise definitions of all of these concepts would depend upon the underlying purposes of the investigation.

The terms *universe* and *sample* are relative. An aggregate of elements which constitutes a population for one purpose may merely be a sample for another. Thus, if one wants to determine the average weight of students in a particular classroom, the students in that room would represent the population. However, if one were to use the average weight of these students as an estimate of the corresponding average for all students in the school, then the students in the one room would be a sample of the larger population. The sample might not be a good one from a variety of viewpoints, but nevertheless, it is a sample.

If the number of elements in the population is fixed, that is, if it is possible to count them and come to an end, the population is said to be finite. Such universes may range from a very small to a very large number of elements. For example, a small population might consist of three balls in an urn; a large population might be the retail transactions which occur in a large city during a one-year period. A point of interest concerning these two examples is that the balls in an urn represent a fixed and unchanging population, whereas the retail transactions illustrate a dynamic population, which might differ considerably over time and space.

Infinite Populations

An *infinite population* is composed of an infinitely large number of items. Usually, such populations are conceptual constructs in which data are generated by processes which may be thought of as repeating indefinitely, such as the rolling of dice and the repeated measurement of weight of an object. Sometimes, the population that is sampled is finite, but is so large that it makes little practical difference if it is considered to be infinite. For example, suppose that a population consisting of 1,000,000 manufactured articles contains 10,000 defectives. Thus, 1% of the articles are defective. If two articles are randomly drawn from the lot in succession, without replacing the first article after it is drawn, the probability of obtaining two defectives is

$$\left(\frac{10{,}000}{1{,}000{,}000}\right)\cdot\left(\frac{9999}{999{,}999}\right)$$

For practical purposes, this product is equal to (0.01) (0.01) = 0.0001. If the population were considered to be infinite with 1% defective articles, the probability of obtaining two defectives is exactly 0.0001. Frequently, in such situations where a finite population is very large relative to sample size, it is simpler to treat this population as infinite. The important point here is that since a finite population is depleted by sampling without replacement whereas an infinite population is inexhaustible, then if the depletion causes the population to change only slightly, it may be simpler for computational purposes to consider the population as infinite.

Sometimes, an infinite population may be considered as being generated by a finite population which is repeated indefinitely into the future. For example, a company may draw a sample from a lot from a particular supplier in order to decide whether to purchase from this supplier in the future. Thus, the purchaser makes a decision concerning the *manufacturing process* which produces future lots. The particular lot which is sampled for test purposes is a finite population. The process which produces the particular lot may be viewed as an infinite population. Care must be exercised in such situations to insure that the manufacturing process is indeed a stable one, and may validly be viewed as a single universe. Future testing may in fact reveal differences of such a magnitude that the conceptual universe should be viewed as having changed.

Target Populations

Another useful concept is the *target population*, or the universe about which inferences are desired. Sometimes in statistical work, it is impractical or perhaps impossible to draw a sample directly from this *target population*, but it is possible to obtain a sample from a very closely related one. That is, the list of elements which constitutes the *frame* that is sampled may be related to but is definitely different from the list of elements which comprises the target population. For example, suppose it is desired to predict the winner in a forthcoming municipal election by means of a polling technique. The target population is the collection of individuals who will cast votes on election day. However, in the nature of the case, it is not possible to draw a sample directly from this population, since the specific individuals who will show up at the polls on election day are unknown. It may be possible to draw a sample from a closely related population, such as the eligible voting population. In this case, the list of eligible voters constitutes the sampling frame. The percentage of the eligible voting population which would vote for a given candidate may differ from the corresponding figure for the election day population. Furthermore, the percentage of the eligible voter population which would vote for a given candidate will probably change as the election date approaches. Thus, we have a situation in which the population that can be sampled changes over time, and is different from that about which inferences are to be made. In the case of election polling, the situation is further complicated by the fact that at the time the sample is taken, many individuals may not have made up their minds concerning the candidate for whom they will vote. Therefore, some assumption must be made about how these "undecideds" will break down as to voting preferences. "In depth" interviews are often used for this purpose in which questions are asked of the undecideds concerning the issues and individuals in the campaign to help determine for whom the respondents will probably vote. In carefully run election polls, numerous sample surveys are taken, spaced through time, including some investigations near the election date. Then, trends can be determined in voting composition, and inferences can be made from populations which are defined very close in time to election day. Some of the instances of incorrect predictions in national elections have resulted from failures to deal properly with the problem of undecideds and from cessations of sampling at time periods too far removed from election day. There is no easy answer to the question of how to adjust for the fact that the populations sampled are different from the election day population. For example one approach to the problem of non-voters is to conduct post-election surveys to determine the composition of the non-voting group and to estimate their probable voting pattern had they shown up at the polls. Historical information of this sort could conceivably be used to adjust polls of eligible voters. However, this is an expensive procedure, and the appropriate method of adjustment is fraught with problems.

In many statistical investigations, the target population coincides with a population that can be sampled. However, in any situation where one must sample a past statistical universe, and yet must make estimates for a future universe, the above mentioned problem of inference about the target universe is present.

Control Groups

Probably the most familiar setting involving the concept of a control group is the situation in which an experimental group is given some type of treatment. In order to determine the effect of the treatment, another group is included in the experiment, and is not given the treatment. These "no-treatment" cases are known as the "control group." The effect of the treatment can then be determined by comparing the relevant measures between the "treatment" and "control" groups. For example, in testing the effectiveness of an inoculation against a particular disease, the inoculation may be administered to a group of school children (the treatment group) and not given to another group of school children (the control group).[1] The effectiveness of the inoculation can then be determined by comparing the incidence of the relevant disease between the two groups. As a point in the design of the experiment, it may be noted that there should be no systematic difference between the two groups at the outset which would make one group more susceptible to the disease than the other. Therefore, such experiments are sometimes designed with so-called "matched-pairs," where pairs of persons having similar characteristics, one from the treatment and one from the control group, are drawn into the experiment. For example, if age and health are felt to have some effect on incidence of the disease, pairs of school children may be drawn who are similar with respect to these characteristics. The treatment is given to one child of a pair and not to the other. Since the children are of similar ages and have the same health backgrounds, these factors cannot explain the fact that one child contracts the disease, whereas the other does not. In the language of experimental design, age and general health conditions may be said to have been "designed out" of the experiment. Numerous other techniques are employed in experimental design to insure that treatment effects can be meaningfully discerned.

The concept of a control group is important in many statistical investigations in the areas of business and economics. In fact, in many instances, the results of an investigation may be uninterpretable unless one or more suitable control groups have been included in the study. It is a sad fact that often after statistical investigations are completed at considerable expense it is found that because of faulty design and inadequate planning, the results cannot be meaningfully interpreted or the data collected are inappropriate

---

[1]Difficult ethical questions arise in cases of this sort which involve human experimentation. If the inoculation is indeed effective, its use should clearly not be withheld from anyone who wants it. In cases of new treatments whose effectiveness is highly questionable, yet human experimentation appears necessary, the treatment group often is composed entirely of volunteers.

for testing the hypotheses in question. It is of paramount importance that during the planning stage, the investigators project themselves to the completion of the study. They should ask the question, "If the collected data show thus-and-so, what conclusions can we reach?" This simple yet critical procedure will often highlight difficulties connected with the study design.

A few examples follow to illustrate the use of control groups in statistical studies. Suppose a mail-order firm decided to conduct a study to determine the characteristics of its "high volume" customers. Its purpose is to determine the distinguishing characteristics of these heavy purchasers in order to direct future campaigns to non-customers who possess similar attributes. Assume that the firm decides to study all of its high volume customers. At the conclusion of the investigation, it will be able to make statements such as, "The income of high volume customers is so-many dollars." Or, it may calculate that $X\%$ of these heavy purchasers have a certain characteristic. Such population figures will be of virtually no use unless the company has an appropriate control group against which to compare these figures. The important point is for the company to be able to isolate the distinguishing characteristics of high volume purchasers. Thus, in studying its customers, the company should have separated them into two groups: "high volume" and "non-high volume." Now, if it studied both groups, it would be in a position to determine those properties which are different between the two groups. Thus, returning to the earlier statements, if the company found that the high volume and non-high volume customers had the *same* mean incomes and that in both groups $X\%$ possessed a certain characteristic, it could not use these properties to distinguish between the two groups. The properties that differed most between the two groups would obviously be the most useful ones for spelling out the distinguishing characteristics of high volume purchasers. In summary, the firm could have used the non-high volume customers as a control group against which to compare the properties of the high volume group, which in the terminology used earlier would represent the "treatment group."

A comment in the form of a warning is pertinent in the above illustration. Care must be used in the selection of the properties of the two groups to be observed. These properties should bear some logical relationship to the characteristic of high versus non-high volume purchases. Otherwise, the properties may be spurious indicators of the distinguishing characteristics between the two groups. For example, income level would be logically related to purchasing volume. If the high volume purchaser group had a substantially higher income than the non-high volume group, then income would evidently be a reasonable distinguishing characteristic. On the other hand, suppose the high volume purchaser group happened to have a higher percentage of persons who wore black shoes at the time of the survey than did the non-high purchaser group. This characteristic of shoe color would *not* seem to be logically related to volume of purchases. Hence, we would not be surprised if the relationship between shoe color and volume of purchases disappeared in subsequent investigations or even reversed itself.

A couple of comments may be made on the construction of control groups. If the treatment group is symbolized as A and the control group as B, then an alternative control group to the one used would have been the treatment and control groups combined, or $A + B$. Thus, in the above example, if the relevant data had been available for the "high volume" and "non-high volume" customers combined, or for all customers, this group could have constituted the control. For example, let us assume for simplicity that there were equal numbers of high volume and non-high volume customers. Suppose that 90% of high volume customers possessed characteristic X, whereas only 50% of non-high customers had this characteristic. The same information would be given by the statements that 90% of high volume customers possessed characteristic X, while 70% of *all* customers possessed this characteristic. The 70% figure, of course, is the weighted mean of 90% and 50%. With the knowledge of equal numbers of persons in the high volume and non-high volume groups, it can be inferred that 50% of non-high volume customers had the property in question. This point is of importance, because sometimes historical data may be available for an entire group $A + B$, whereas available resources may permit a study only of the treatment group A or what is the more usual case only a sample of this group. However, the use of a historical control group is a dangerous procedure, because systematic changes may have taken place in the treatment and control groups over time or in the surrounding conditions of the experiment. Therefore, the more scientifically desirable procedure is to design the treatment and control groups for the specific investigation in question.

The general objectives of an investigation determine the control groups to be used. Thus, for other purposes, individuals who are not customers of a firm, or customers who have not purchased a specific product could conceivably constitute appropriate control groups. It may also be noted that time considerations and available resources usually permit only samples to be drawn from the treatment and control populations rather than complete enumerations of these populations.

A couple of other brief examples of the use of control groups will be given here. A national commission wanted to investigate insurance conditions in cities in which civil disturbances in the form of riots had occurred. Specifically, one of the matters the commission wished to study was cancellation rates for burglary and fire and theft policies in sections of these cities ("riot areas") primarily affected by the riots. The main purpose was to determine whether individuals and businesses in these areas were having difficulty in retaining such policies because of cancellations by insurance companies. It became clear in the planning stages of the study that it would not be sufficient merely to measure cancellation rates in the cities in which riots had occurred, because it would not be possible in the absence of other information to judge whether these rates were low, average, or high. Therefore, a sample of individuals and businesses in cities which had not experienced riots was used as a control group. Another control group was established consisting of individuals and businesses in the "non-riot areas" of the cities that had experienced riots. Thus, the data on cancellation rates could be meaningfully interpreted. Comparisons were made between cities that had experienced riots and those which had not. Further comparisons were made between cancellation rates in riot areas and non-riot areas in cities where these disturbances had been present. The data disclosed that burglary and fire and theft insurance cancellation rates were higher in cities in which riots had occurred. Furthermore, within cities in which these civil disturbances had been present, cancellation rates were higher in sections where riots were experienced than in non-riot sections. It may be noted, parenthetically, that if the company policies on cancellations were known, and data were available in suitable form, the same information could have been obtained from the company records. However, such was not the case. Therefore, the aforementioned sample survey was required to obtain the indicated data.

Another illustration is the case of a company which wished to determine whether its labor costs were "out-of-line" with similar costs throughout the company's industry group. It obtained the ratio of labor costs to total operating costs for the company and compared these to a published distribution of such ratios for all firms in the industry for the same time period. In this situation, all firms in the industry constituted the control group. This is an illustration of a rather obvious need for and choice of a control group. In many situations, the need and choice are somewhat more subtle.

Types of Errors

The concept of error is a central one throughout all statistical work. Wherever we have measurement, inference, or decision making, the possibility of error is present. In this section, we shall concern ourselves with errors of measurement. The problems of errors of inference and decision making are treated in subsequent chapters.

It is useful to distinguish two different types of errors which may be present in statistical measurements, namely, *systematic errors* and *random errors*. Systematic errors, as the term implies, cause a measurement to be incorrect in some systematic way. If observations have arisen from a sample drawn from a statistical universe, systematic errors are of a type which persist even when the sample size is increased. They are errors involved in the procedures of a statistical investigation and may occur in the planning stages, or during or after the collection process. Another term, conventionally employed for systematic error, is *bias*. Among the causes of bias are faulty design of a questionnaire such as misleading or ambiguous questions, systematic mistakes in planning and carrying out the collection and processing of the data, nonresponse and refusals by respondents to provide information, and too great a discrepancy between the sampling frame and the target universe. As a generalization, these errors may be viewed as arising primarily from inaccuracies or deficiencies in the measuring instrument.

On the other hand, *random errors* or *sampling errors* may be viewed as arising from the operation of a large number of uncontrolled factors, conveniently subsumed under the term "chance." As an example of this type of error, if repeated random samples of the same size are drawn from a statistical universe (replacing each sample after it is drawn), a particular statistic, such as an arithmetic mean, will differ from sample to sample, even if the same definitions and procedures are used. These sample means tend to distribute themselves below and above the "true" population parameter (arithmetic mean), with small deviations between the statistic and the parameter occurring relatively frequently and large deviations occurring relatively infrequently. The word "true" has quotation marks around it, because it refers to the figure that would have been obtained through equal complete coverage of the universe, that is, a complete census using the same definitions and procedures as had been used in the samples. The difference between the mean of a particular sample and the population mean is said to be a *random error* or a *sampling error*, as it is termed in later chapters. The complete collection of factors which could explain why the sample mean differed from the population mean is unknown, but we can conveniently lump them all together and refer to the difference as a random or chance error. A random error is one which arises from differences between the outcomes of trials (or samples) and the corresponding universe values using the same measurement procedures and instruments. The sizes of the differences are indications of reliability or precision. Random errors decrease on the average as sample size is increased. It is precisely for this reason that we prefer a larger sample of observations to a smaller one, all other things being equal. That is, since sampling errors are on the average smaller for larger samples, the results are more reliable or more precise.

Systematic and random errors may occur in experiments where the variables are manipulated by the investigator or in survey work where observations are made on the elements of a population without any explicit attempt to manipulate directly the variables involved. A few examples will be given here of how bias or systematic error may be present in a statistical investigation. The problems of how random errors are measured and what constitute suitable models for the description of such errors represent central topics of statistical methods and are discussed extensively later in this chapter and Chapters 6 through 9.

Systematic Error—Biased Measurements

The possible presence of biased measurements in an experimental situation may be illustrated by a simple example. Suppose that a group of individuals measured the length of a 36-inch table top using the same yardstick. Let us further assume that the yardstick although calibrated as though it were 36 inches long was, in fact, 35 inches long, and this fact was unknown to the individuals making the measurements. There would then be a systematic error of one inch present in each of the measurements, and a statistic such as an arithmetic mean of the readings would reflect this bias. In this type of situation, the systematic error could be detected if another, correctly calibrated, yardstick were used as a standard against which to test the incorrect one. This is an important methodological point. Often, systematic error can be discovered through the use of an independent measuring instrument. Even if the independent instrument is inaccurate, a comparison of the two measuring instruments may give clues as to where the search for sources of bias should be made. The variation among the individual measurements using the incorrect yardstick would be a measure of what is usually referred to as "experimental error," that is, differences among individual observations that are not attributable to specific causes of variation. It may be noted that the observations may have been very precise, in the sense that each person's measurement was very close to that of every other person. Thus, the random error would be small, and there would be good repeatability because in repeating the experiment, each measurement would be close to preceding measurements. These random or chance errors may be assumed to be compensating in that some observations would tend to be too large and some too small. Since the table top is 36 inches long, the measurements would tend to cluster around a value about one inch greater than the true length of the table top. In summary, we have a model in which each individual measurement may be viewed as the sum of three components, (1) the true value, (2) systematic error, and (3) experimental error. This relationship is stated in equation form below

$$\text{(5.1)} \qquad \begin{matrix}\text{Individual}\\ \text{Measurement}\end{matrix} = \begin{matrix}\text{True}\\ \text{Value}\end{matrix} + \begin{matrix}\text{Systematic}\\ \text{Error}\end{matrix} + \begin{matrix}\text{Experimental}\\ \text{Error}\end{matrix}$$

Systematic Error—Literary Digest Poll

A classic case of the presence of systematic error in a survey sampling procedure is that of the *Literary Digest* presidential election of 1936. During the election campaign between Franklin D. Roosevelt and Alf Landon, the *Literary Digest* magazine sent questionnaire ballots to a very large list of persons whose names appeared in telephone directories and automobile registration lists. Over two million ballots, amounting to about one-fifth of the total number sent out, were returned by the respondents. On the basis of these replies, the *Literary Digest* erroneously predicted that Landon would be the next president of the U.S. The reasons why the results of this survey were so severely biased are rather clear. In 1936, during the great depression period, the presidential vote was cast largely along economic lines. The group of the electorate which did not own telephones or automobiles did not have an opportunity to be included in the sample. This group, which represented a lower economic level than possessors of telephones and automobiles, voted predominantly for Roosevelt, the Democratic candidate. A second reason stemmed from the non-response group, which represented about four-fifths of those polled. Typically, individuals of higher educational and higher economic status are more apt to respond to voluntary questionnaires than those with lower economic and educational status. Therefore, the non-response group doubtless contained a higher percentage of this lower status group than did the group which responded to the questionnaire. Again, this factor added a bias due to under-representation of Democratic votes. In summary, the sample used for prediction purposes contained a greater proportion of persons of higher socio-economic status than were present in the target population, namely, those who cast votes on election day. Since this factor of socio-economic status was related to the way people voted, a systematic overstatement of the Republican vote was present in the sample data.

A couple of methodological lessons can be derived from this example. First, as earlier noted, it is a dangerous procedure to sample a frame which differs considerably from the target population. Second, procedures must be established to deal with the problem of nonresponse in statistical surveys. Clearly, even if the proper target universe had been sampled in this case, the problem of nonresponse would still have to be properly handled.

Systematic Error—Method of Data Collection

Another example of bias will now be given which illustrates that the direction of systematic error may be associated with the nature of the agency which collects the data as well as the method by which the data are collected.

The alumni society of a large Eastern university decided to gather information from the graduates of that institution to determine a number of characteristics including their current economic status. One of the questions of interest was the amount of last year's gross income, suitably defined. A mail questionnaire was sent to a random sample of graduates, and the results were tabulated from the returns. When frequency distributions were made and averages were calculated by year of graduation, it became clear that the income figures were unusually high as compared to virtually any existing external data which could be examined. In other words, the income figures were clearly biased in an upward direction. It is fairly easy in this case to speculate on the causes of this upward systematic error. In this type of mail questionnaire, a higher non-response rate could be expected from those graduates whose incomes were relatively low than from those with higher incomes. That is, it appears reasonable that those with higher incomes would have a greater propensity to respond than others. Furthermore, if there were instances of misreporting of incomes, it is probable that these tended to be overstatements rather than understatements, because of the desire to appear relatively economically successful.

On the other hand, let us consider the same sort of data as reported to the Internal Revenue Service on annual income tax returns. Doubtless, it is safe to say that there is relatively little overstatement of gross incomes. Indeed, it seems reasonable to suppose that there is a downward bias in these data in the aggregate. It may be noted that since responses to the Internal Revenue Service are mandatory, the effect of nonresponse may be considered negligible. Thus, the interesting situation is presented here of the same type of data being gathered by two different agencies, one set being biased in an upward direction, the other in the opposite direction. Therefore, in using secondary statistical information, it is clear that informed critical judgment must be exercised to extract meaningful inferences. This judgment must include practical considerations such as methods of data collection and auspices under which studies are conducted. Of course, false reporting is not easily overcome, particularly in situations in which no independent objective data are available against which the reported information may be checked.

5.5 Fundamentals of Sampling

Purposes

Sampling is important in most applications of quantitative methods to managerial and other business problems. There are a wide variety of reasons why this is so. In certain instances, sampling may represent the only possible or practicable method to obtain the desired information. For example, in the case of processes, such as manufacturing, where the universe is conceptually infinite including all future as well as current production, it is not possible to accomplish a complete enumeration of the population. On the other hand, in destructive sampling of a finite population it is possible to effect a complete enumeration of the universe, but it would not be practical to do so. For example, if a military procurement agency wanted to test a shipment of bombs, it could detonate all of the bombs in a destructive testing procedure and obtain complete information concerning the quality of the shipment. However, since there would be no usable product remaining, a sampling procedure is clearly the only practical method by which to assess the quality of the shipment.

Sampling procedures are often employed for overall effectiveness, cost, timeliness, and other reasons. A complete census, although it does not have sampling error introduced by a partial enumeration of the universe, nevertheless often contains greater total error than does a sample survey. This is true because greater care can usually be exercised in a sample survey than in carrying out censuses. Errors in collection, classification, and processing of information may be considerably smaller in the case of sample surveys which can be carried out under far more carefully controlled conditions than substantially larger scale complete censuses. For example, response errors arising from lack of information, misunderstood questions, faulty recall, and other reasons may only be reduced by intensive and expensive interviewing and measurement methods, which may be feasible in the case of a sample but may be prohibitively costly for a complete enumeration.

An important characteristic of information obtained from samples is that although error may be present, it may be small enough for decision making purposes. An outstanding example of the recognition of this point is the virtually universal acceptance of quality assurance procedures based on sampling by American industry during and since World War II. During that war, military purchasing agencies introduced the policy that material need not be perfect, but must meet *acceptable quality levels*, often expressed in terms of a certain fraction defective. The government required vendors to introduce definite sampling inspection plans for the control of manufacturing processes and acceptance sampling plans for finished product. As the purchaser, the military also utilized such plans itself. Interestingly enough, American industry (and subsequently that of many other countries as well) discovered that the use of these sampling schemes rather than 100% inspection procedures with the latter's unrealistic and unrealizable goal of perfect quality, actually accomplished an improvement in average quality levels at reduced cost. The sampling plans can be designed to result in desired levels of producer's and consumer's risk. (See Example 2-8, Chapter 2.)

The employment of sampling rather than censuses for purposes of timeliness occurs in a wide variety of areas. A notable example is the wide array of government data on economic matters such as income, employment, and prices, which are collected on a sample basis at periodic intervals. Here, timeliness of publication of the results is of considerable importance. The more rapid collection and processing of data afforded by sampling procedures represents a telling advantage over corresponding census methods.

Random and Non-random Selection

Items can be selected from statistical universes in a variety of ways. It is useful to distinguish random from non-random methods of selection. In this book, attention is focused upon random or probability sampling, that is, sampling in which the probability of inclusion of every element in the universe is *known.* Non-random sampling methods are referred to as "judgment sampling," that is, selection methods in which judgment is exercised in deciding which elements of a universe to include in the sample. Such judgment samples may be drawn by choosing "typical" elements or groups of elements to represent the population. They may even involve random selection at one stage, but may allow the exercise of judgment in another. For example, areas may be selected at random in a given city, and interviewers may be instructed to obtain specified numbers of persons of given types within these areas, but may be permitted to make their own decisions as to which individuals are brought into the sample.

This book deals only with random or probability sampling methods rather than judgment sampling because of the clearcut superiority of probability selection techniques. The basic reason random sampling is preferable to judgment sampling is that in judgment selection there is no objective method of measuring the precision or reliability of estimates made from the sample. On the other hand, in random sampling, the precision with which estimates of population values can be made is obtainable from the sample itself. This is a very important advantage, since random sampling techniques thus provide an objective basis for measuring errors due to the sampling process and for stating the degree of confidence to be placed upon estimates of population values

Judgment samples can sometimes be usefully employed in the planning and design of probability samples. For example, where expert judgment is available, a pilot sample may be selected on a judgment basis in order to obtain information which will aid in the development of an appropriate sampling frame for a probability sample.

Simple Random Sampling

We have seen that a *random sample* or *probability sample* is a sample drawn in such a way that the probability of inclusion of every element in the population *is known.* There are a wide variety of types of such probability samples, particularly in the area of sample surveys. Experts in survey sampling have developed a large body of theory and practice aimed toward the optimal design of probability samples. This is a highly specialized area to which an entire course or two in a graduate program in statistics is often allocated. We will concentrate upon the simplest and most fundamental probability sampling method, namely, *simple random sampling*. The major body of statistical theory is based upon this method of sampling. Other sample designs are discussed in Chapter 8.

We first define a simple random sample for the case of a finite population of N elements. A simple random sample of n elements is a sample drawn in such a way that *every combination of n elements has an equal chance of being the sample selected.* Since most practical sampling situations involve sampling *without replacement*, it is useful to think of this type of sample as one in which each of the N population elements has an equal probability, $(1/N)$, of being the one selected on the first draw, each of the remaining $N - 1$, has an equal probability, $(1/N - 1)$, of being selected on the second draw, and so on until the nth sample item has been drawn. Since there are $\binom{N}{n}$ possible samples of n items, the probability that any sample of size n will be the one drawn is $1/\binom{N}{n}$.

This concept of a simple random sample may be illustrated by the following example. Let the population consist of three elements, A, B, and C. Thus, $N = 3$. Suppose, we wish to draw a simple random sample of two elements. Then, $n = 2$. Using formula (1.15) for the number of combinations that can be formed of N objects taken n at a time, we find the number of possible samples to be

$$\binom{3}{2} = \frac{3!}{2!1!} = 3$$

These three possible samples contain the following pairs of elements: (A, B), (A, C), and (B, C). The probability that any one of these three samples will be the one selected is $1/3$.

It is a property of simple random sampling that every element in the population has an equal probability of being included in the sample. However, many other sample designs possess this property as well, as for example, certain stratified sample and cluster sample procedures, discussed in Chapter 8.

Simple random samples were defined above for the case of sampling a finite population without replacement. If a finite population is sampled *with replacement*, the same element could appear more than once in the sample. Since, for practical purposes, this type of sampling is virtually never employed, it will not be discussed any further here.

On the other hand, simple random sampling of *infinite populations* is of importance, particularly in the context of sampling of processes. The following definition is often given corresponding to the one for finite populations. For an infinite population, a simple random sample is one in which *on every selection, each element of the population has an equal probability of being the one drawn.* This is difficult to visualize, in terms of actual sampling from a physical population. Therefore it is more satisfactory to take a more rigorous theoretical approach, and to use the language of random variables. Thus, we view the drawing of the sample as an experiment in which observations of values of a random variable are generated, and the successive sample observations or elements are the outcomes of trials of the experiment. Then, a simple random sample of n observations is defined by the presence of two conditions: (1) the n successive trials of the experiment are independent, and (2) the probability distribution of the random variable remains constant from trial to trial. In terms of sampling a physical population, we may interpret these as meaning: (1) the n successive sample observations are independent and, (2) the population composition remains constant from trial to trial.

To aid in the interpretation of the above definition, let us consider the case of drawing a simple random sample of n observations from a Bernoulli process. To make the illustration concrete, assume a situation in which a fair coin is tossed. A simple random sample of n observations would be the sample consisting of the outcomes on n *independent* tosses of the coin. Thus, if the number 0 denotes the appearance of a tail and 1 the appearance of a head, the following notation might designate a particular simple random sample of five observations, in which two tails and three heads were obtained in the indicated order: (0, 1, 0, 1, 1). In summary, we note that (1) the tosses of the coin were statistically independent, and (2) the probability of obtaining a head (or tail) remained constant from trial to trial. It is conventional to use an abbreviated method of referring to such a sample as "a sample of five independent observations from a Bernoulli process," or "a sample of five independent observations from a binomial distribution."

The term "random sample," although it properly refers to a sample drawn with known probabilities, is often used to mean "simple random sample." Therefore, we alert the student to this alternative usage.

Methods of Simple Random Sampling

Although it is easy to state the definition of a simple random sample, it is not always obvious how such a sample is to be drawn from an actual population.

Drawing Chips from a Bowl. We now restrict our attention to the most straightforward situation in which the population is finite and in which the elements are easily identified and can be numbered. For example, suppose there are 100 students in a college freshman class and we wish to draw a simple random sample of ten of these students without replacement. We could assign numbers from 1 to 100 to each of the students and place these numbers on physically similar disks (or balls, slips of paper, etc.), which could then be placed in a bowl. We shake the bowl to accomplish a thorough mixing of the disks and then proceed to draw the sample. The first disk is drawn, and we record the number written on it. We then shake the bowl again, draw the second disk, and record the result. The process is repeated until we have drawn ten numbers. The students corresponding to these ten numbers constitute the required simple random sample.

Tables of Random Numbers. If the population size is very large, the foregoing procedure can become quite unwieldy and time-consuming. Furthermore, it may introduce biases if the disks are not thoroughly mixed. Therefore, in recent years, there has been a marked tendency to use tables of random digits for the purpose of drawing such samples. These tables are useful for the selection of other types of probability samples, as well.

A table of random digits is simply a table of digits which have been generated by a random process. Usually, the digits are combined, for example, into groups of five digits each, for ease of use. Thus, a table of random digits could be generated by the process of drawing chips from a bowl similar to the one just described. The digits 0, 1, 2, . . . , 9 could be written on disks, the disks placed in a bowl, and then drawn, one at a time, *replacing the selected disk after each drawing.* Thus, on each selection, the population would consist of the ten digits. The recorded digits would constitute a particular sequence of random digits. These tables are now usually produced by a computer, which has been programmed to generate random sequences of digits.

We now illustrate the use of random digits using Table 5-1. Suppose there were 9241 undergraduates at a large university and we wished to draw a simple random sample of 300 of these students. Each of the 9241 students could be assigned a four-digit number, for convenience, say, from 0001 to 9241. This list of names and numbers would constitute the sampling frame. We now turn to a table of random digits in order to select a simple random sample of 300 such four-digit numbers. We may begin on any page in the table and proceed in any systematic manner to draw the sample. Assume we decided to use the first four columns of each group of five, beginning at the upper left and reading downward. The first five digits on the left-hand side designate the line number, so we ignore them. Starting with the second group of digits, we find the sequence 98389. Since we are using the first four digits, we have the number 9838. This exceeds the largest number in our population, 9241. Therefore, we ignore this number and read down to pick up the next four-digit number, 1724. This, then is the number of the first student in the sample. Reading down consecutively, we find 0128, 9818 (which we ignore), 5926, and so forth until 300 four-digit numbers between 0001 and 9241 have been specified. If any previously selected number is repeated, we simply ignore the repeated appearance, and continue. In this illustration, we read downward on the page. We could have read laterally, diagonally, or in any other systematic fashion. The important point is that each four-digit number has an equal probability of selection, regardless of what systematic method of drawing is used, and regardless of what numbers have already preceded.

Methods are available for drawing other types of samples than simple random samples, and even for situations where the elements have not been

Table 5-1 Random Digits.[a]

19300	98389	95130	36323	33381	98930	60278	33338	45778	86643	78214
19301	17245	58145	89635	19473	61690	33549	70476	35153	41736	96170
19302	01289	68740	70432	43824	98577	50959	36855	79112	01047	33005
19303	98182	43535	79938	72575	13602	44115	11316	55879	78224	96740
19304	59266	39490	21582	09389	93679	26320	51754	42930	93809	06815
19305	42162	43375	78976	89654	71446	77779	95460	41250	01551	42552
19306	50357	15046	27813	34984	32297	57063	65418	79579	23870	00982
19307	11326	67204	56708	28022	80243	51848	06119	59285	86325	02877
19308	55636	06783	60962	12436	75218	38374	43797	65961	52366	83357
19309	31149	06588	27838	17511	02935	69747	88322	70380	77368	04222
19310	25055	23402	60275	81173	21950	63463	09389	83095	90744	44178
19311	35150	34706	08126	35809	57489	51799	01665	13834	97714	55167
19312	61486	33467	28352	58951	70174	21360	99318	69504	65556	02724
19313	44444	86623	28371	23287	36548	30503	76550	24593	27517	63304
19314	14825	81523	62729	36417	67047	16506	76410	42372	55040	27431
19315	59079	46755	72348	69595	53408	92708	67110	68260	79820	91123
19316	48391	76486	60421	69414	37271	89276	07577	43880	08133	09898
19317	67072	33693	81976	68018	89363	39340	93294	82290	95922	96329
19318	86050	07331	89994	36265	62934	47361	25352	61467	51683	43833
19319	84426	40439	57595	37715	16639	06343	00144	98294	64512	19201
19320	41048	26126	02664	23909	50517	65201	07369	79308	79981	40286
19321	30335	84930	99485	68202	79272	91220	76515	23902	29430	42049
19322	33524	27659	20526	52412	86213	60767	70235	36975	28660	90993
19323	26764	20591	20308	75604	49285	46100	13120	18694	63017	85112
19324	85741	22843	16202	48470	97412	65416	36996	52391	81122	95157

[a]SOURCE: RAND Corporation, *A Million Random Digits with One Hundred Thousand Normal Deviates* (Glencoe, Ill.: Free Press, 1955), excerpt from page 387. Used by permission.

prelisted. Many of the tables include instructions on their use, and we will not pursue the subject any further here.

Other Sampling Designs

In this section, we have concentrated on simple random sampling. Most of the theoretical structure of statistical inference is based on this type of sampling. Thus, for example, the underlying theory of the next five chapters assumes this sampling procedure. However, in sample survey work, many other types of sample designs are used to accomplish increased sampling precision for fixed cost or to minimize cost for a fixed level of sampling precision. Even these designs, though, build on the foundation principles of simple random sampling. In order to appreciate how more sophisticated sample designs such as stratified random sampling and cluster sampling can represent an improvement over simple random sampling, the principles discussed in the next three chapters are essential. Therefore, in Section 8.6, we return to the subject of other sample designs, and we consider how such procedures accomplish their objectives.

PROBLEMS

1. The school board of Lower Fenwick wished to ascertain voter opinion concerning a special assessment to permit the expansion of school services. Lower Fenwick is an industrial community on the fringe of a metropolitan area and has a population of 25,000. There are 5000 pupils enrolled in the public schools of the community. The board selected a random sample of these children and sent questionnaires to their parents.
   (a) Identify the statistical universe from which the above sample was drawn.
   (b) Is the sample chosen a simple random sample of parents in Lower Fenwick? Of parents of public school children in Lower Fenwick? Why or why not?
   (c) If you had been asked to assist the board, would you have approved the universe it studied? Defend your position.
2. The personnel director of a large manufacturing concern wished to ascertain employee opinion with respect to the annual Christmas party. The party had been for employees only in the past, but the director thought husbands and wives of employees might be included for the next party. He selected a random sample of 100 persons who had attended the party the previous year and asked their opinions. Eighty wanted the party to be "employees only," but 20 requested that husbands and wives be included. Based on the results the director decided to continue the present practice. Briefly evaluate the procedure employed by the director, indicating any possible sources of bias and non-homogeneity.
3. Explain the difference between a sample and a census.
4. Give an example of a situation where the population of interest would be
   (a) an infinite population,
   (b) considered an infinite population yet in reality is a finite population,

(c) considered an infinite population, since it is generated by a finite population repeated indefinitely into the future,
(d) a target population.

5. If one is interested in the percentage of consumers in the New York City area who would wear a certain type of man's suit at varying prices, what is the population of interest? Is this a fixed and unchanging population or a dynamic population? Explain your answer.

6. What is the difference between an observational study and a "controlled inquiry"?

7. Discuss the need for and advantages of having a control group in a study, and give an example of a study in which use of a control group would be of value.

8. In each of the following situations, state whether a control group would be of use, and if so, what the control group would be.
   (a) You are interested in investigating the accident rates in low income urban areas and you have data on numbers of accidents occurring in low income urban areas, numbers of cars registered in low income areas, actual area of low income areas, and other similar information.
   (b) You are interested in evaluating the effectiveness of a new safety lighting program to be installed in your plant.

9. Distinguish clearly the difference between systematic errors and random errors. Explain which error will decrease with a larger sample size, which will not, and why. Which error can and should be eliminated?

10. State whether each of the following errors should be considered random, systematic, or both, and why:
    (a) In a study which attempted to estimate the percentage of students who smoke, the first 100 students who entered the student lounge, the only area in the building where students are permitted to smoke, were asked if they smoked. The study resulted in an overestimate of the true percentage.
    (b) In a study to estimate the average life of a certain type of vacuum tube, five tubes were purchased from five different stores, in five different wholesale sales regions. The average life of the five tubes tested was shorter than the "true" average life.
    (c) In a study to determine the true weight of a process which fills one pound cans, 50 cans were selected randomly and weighed on a scale that measured .1 ounce too heavy. The process in fact filled the cans on the average with one pound but the 50 cans averaged 1.06 pounds.
    (d) One thousand questionnaires were sent out asking the respondent to rate the community services of police, fire, garbage, and so forth as bad, adequate, or good. Of the 87 responses, a majority rated the services either bad or good. Yet the majority of the people in the community considered the services adequate.

11. Give at least three possible reasons why sampling procedures rather than a census may be desirable in certain situations.
12. State whether each of the following statements is true or false and explain your answer.
    (a) Judgment sampling is good, since we can get an objective measure of the random error.
    (b) Whenever costs permit a census to be taken, a census is always preferable to sampling.
    (c) Systematic errors can often be reduced by better procedures, while random errors can only be reduced by larger sample sizes.

CHAPTER SIX

Sampling Distributions

In Chapter 3, we examined the methods by which statistics such as the arithmetic mean and standard deviation are computed from the data contained in a sample. In Chapter 5, we discussed how simple random samples can be drawn from finite and infinite populations. We now consider how such statistics differ from sample to sample if repeated simple random samples of the same size are drawn from statistical populations. For example, a given statistic, such as a proportion or a mean, will vary from sample to sample. The probability distribution of such a statistic is referred to as a *sampling distribution.* Thus, we may have a sampling distribution of a proportion, a sampling distribution of a mean, etc. These sampling distributions are the basic concept of statistical inference and are of considerable importance in modern statistical decision theory as well. We commence our discussion by considering the sampling distributions of numbers of occurrences and proportions of occurrences.

6.1 Sampling Distribution of Number of Occurrences

We can illustrate the meaning and properties of the sampling distribution of a number of occurrences by means of a fictitious example. Let us assume a manufacturing process which produces articles of which 10% are defective and 90% are non-defective. We conceive of the production process as an infinite population. Thus, we may view the successive drawings of articles from the process as a series of Bernoulli trials. That is, the three requirements of a Bernoulli process may be interpreted in terms of this problem as follows: (1) there are two possible outcomes on each draw, defective or non-defective, (2) the probability of a defective (non-defective) remains constant from draw

Table 6-1 Probability Distribution of the Number of Defectives in a Simple Random Sample of Five Articles from a Process Containing 10% Defectives.

Number of Defectives x	*Probability* $f(x)$
0	$\binom{5}{0}(0.9)^5(0.1)^0 = 0.59$
1	$\binom{5}{1}(0.9)^4(0.1)^1 = 0.33$
2	$\binom{5}{2}(0.9)^3(0.1)^2 = 0.07$
3	$\binom{5}{3}(0.9)^2(0.1)^3 = 0.01$
4	$\binom{5}{4}(0.9)^1(0.1)^4 \approx 0.00$
5	$\binom{5}{5}(0.9)^0(0.1)^5 \approx 0.00$
	1.00

to draw, and (3) the draws are independent. We also note that the three conditions would hold equally as well if we assumed we were sampling a finite shipment of articles, but replaced each article prior to drawing the next one. In summary, the ensuing discussion pertains to the sampling of an *infinite universe* or a *finite universe with replacement.*

Returning to the production process, suppose we were to draw a simple random sample of five articles and note the number of defectives in the sample. This number is a random variable which can take on the values 0, 1, 2, 3, 4, or 5. Since we are dealing with a Bernoulli process, the probabilities of obtaining these numbers of defectives may be computed by means of a binomial distribution in which $p = 0.10$, $q = 0.90$, and $n = 5$. Therefore, the respective probabilities are given by the expansion of the binomial $(0.9 + 0.1)^5$. This probability distribution is shown in Table 6-1, using the same notation as in Chapter 2.

We now interpret the probability distribution given in Table 6-1 as a sampling distribution. The number of defectives observed in a sample of five articles is a sample statistic. Thus, Table 6-1 displays the probability distribution of this sample statistic. Let us take a long-run relative frequency of occurrence point of view and interpret the respective entries in Table 6-1 as follows. If we took repeated simple random samples of five articles each at random from a manufacturing process which produces 10% defective articles, in 59% of these samples we would observe zero defectives, in 33% we would find one defective, and so forth. In summary, the probability distribution may now be called a "sampling distribution of number of occurrences."

6.2 Sampling Distribution of a Proportion

Frequently, it is convenient to deal in terms of "proportion of occurrences" rather than "number of occurrences." We can convert the numbers to proportions of occurrences by dividing by sample size. The sample proportion, denoted $\bar{p}$ (pronounced p-bar) may be calculated from

$$\bar{p} = \frac{x}{n} \tag{6.1}$$

where x is the number of occurrences of interest and n is the sample size. In the above example, $\bar{p}$ takes on the possible values $0/5 = 0.00$, $1/5 = 0.20, \ldots,$ $5/5 = 1.00$ with the same probabilities as the corresponding numbers of defectives. We note in passing that the number of occurrences in a sample of size n is given by $x = n\bar{p}$. This may be seen by multiplying both sides of (6.1) by n. The sampling distribution of $\bar{p}$ is given in Table 6-2. In keeping with the usual convention, the probabilities are denoted $f(\bar{p})$.

Table 6-2 Sampling Distribution of the Proportion of Defectives in a Simple Random Sample of Five Articles from a Process Containing 10% Defectives.

Proportion of Defectives $\bar{p}$	*Probability* $f(\bar{p})$
0.00	0.59
0.20	0.33
0.40	0.07
0.60	0.01
0.80	0.00
1.00	0.00
	1.00

Properties of the Sampling Distribution of $\bar{p}$ and $n\bar{p}$

We turn now to the properties of the sampling distributions of number of occurrences, $n\bar{p}$, and proportion of occurrences, $\bar{p}$. The means and standard deviations of these distributions are of particular interest in statistical inference. The calculation of these two measures is given in Table 6-3 for number of

Table 6-3 Calculation of the Mean and Standard Deviation of the Sampling Distribution of the Number of Defectives in a Simple Random Sample of Five Articles from a Process Containing 10% Defectives.

$x = n\bar{p}$	$f(x)$	$xf(x)$	$x - \mu_x$	$(x - \mu_x)^2$	$(x - \mu_x)^2 f(x)$
0	0.59	0.00	−0.5	0.25	0.1475
1	0.33	0.33	+0.5	0.25	0.0825
2	0.07	0.14	+1.5	2.25	0.1575
3	0.01	0.03	+2.5	6.25	0.0625
4	0.00	0.00	+3.5	12.25	0.0000
5	0.00	0.00	+4.5	20.25	0.0000
	1.00	0.50			.4500

$$\mu_{n\bar{p}} = \mu_x = \Sigma x f(x) = 0.50 \text{ defectives}$$

$$\sigma_{n\bar{p}} = \sigma_x = \sqrt{\Sigma(x - \mu)^2 f(x)} = \sqrt{0.4500} = 0.67 \text{ defectives}$$

defectives and in Table 6-4 for proportion of defectives. The definitional formulas (4.1) and (4.8) were used for these computations. In actual applications, calculations such as those in Tables 6-3 and 6-4 are never made to obtain the mean and standard deviations of a binomial distribution. They are given here only to aid in the understanding of the meaning of sampling distributions. General formulas for these measures can be derived by substituting $\binom{n}{x} q^{n-x}p^x$ for $f(x)$ and $f(\bar{p})$ in the definitional formulas and performing appropriate manipulations. The results of these derivations are summarized in Table 6-5.

Table 6-4 Calculation of the Mean and Standard Deviation of the Sampling Distribution of the Proportion of Defectives in a Simple Random Sample of Five Articles from a Process Containing 10% Defectives.

$\frac{x}{n} = \bar{p}$	$f(\bar{p})$	$(\bar{p})f(\bar{p})$	$\bar{p} - \mu_{\bar{p}}$	$(\bar{p} - \mu_{\bar{p}})^2$	$(\bar{p} - \mu_{\bar{p}})^2 f(\bar{p})$
0.00	0.59	0.000	−.10	.01	.0059
0.20	0.33	0.066	+.10	.01	.0033
0.40	0.07	0.028	+.30	.09	.0063
0.60	0.01	0.006	+.50	.25	.0025
0.80	0.00	0.000	+.70	.49	.0000
1.00	0.00	0.000	+.90	.81	.0000
		0.100			.0180

$$\mu_{\bar{p}} = \mu_{x/n} = \Sigma\bar{p}f(\bar{p}) = 0.10 = 10\% \text{ defectives}$$

$$\sigma_{\bar{p}} = \sigma_{x/n} = \sqrt{\Sigma(\bar{p} - \mu_{\bar{p}})^2 f_{(\bar{p})}} = \sqrt{.0180} = .134 = 13.4\% \text{ defectives}$$

Table 6-5 Formulas for the Mean and Standard Deviation of a Binomial Distribution.

Random Variable	*Mean*	*Standard Deviations*
Number of Occurrences ($n\bar{p}$)	$\mu_{n\bar{p}} = np$	$\sigma_{n\bar{p}} = \sqrt{npq}$
Proportion of Occurrences ($\bar{p}$)	$\mu_{\bar{p}} = p$	$\sigma_{\bar{p}} = \sqrt{\frac{pq}{n}}$

Let us illustrate the use of the formulas given in Table 6-5 for the above distributions of number and proportion of defectives. Substituting $p = 0.10$, $q = 0.90$, and $n = 5$, we obtain

Number of Defectives

Mean: $\mu_{n\bar{p}} = np = 5 \times 0.1 = 0.50$ defectives

Standard Deviation: $\sigma_{n\bar{p}} = \sqrt{npq} = \sqrt{5 \times 0.1 \times 0.9} = \sqrt{0.45} = 0.67$ defectives

Proportion of Defectives

Mean: $\mu_{\bar{p}} = p = 0.10 = 10\%$ defectives

Standard Deviation: $\sigma_{\bar{p}} = \sqrt{\dfrac{pq}{n}} = \sqrt{\dfrac{0.10 \times 0.90}{5}} = \sqrt{0.0180} = .134$
$= 13.4\%$ defectives

Of course, these are the same results obtained in the longer calculations shown in Tables 6-3 and 6-4. Let us interpret these results in terms of the appropriate sampling distributions. The mean of the binomial distribution in this example, where the sample statistic is number of defectives $n\bar{p}$, is $\mu_{n\bar{p}} = n \cdot p = (5)(0.10) = 0.50$ defectives. This means that if repeated simple random samples of five articles each are drawn from a process containing 10% defectives, on the average, there will be one-half of a defective per sample. If, for example, $n = 200$ and $p = 0.10$, then, on the average, we would expect to obtain $(200)(0.10) = 20$ defectives per sample. The standard deviation, $\sigma_{n\bar{p}} = \sqrt{npq} = 0.67$ is a measure of the variation in number of defectives that is attributable to the chance effects of random sampling.

When the sample statistic is proportion of defectives $\bar{p}$, the mean $\mu_{\bar{p}} = p = 0.10$, or 10% defectives. This means that if we draw simple random samples of five articles each from a process which is 10% defective, on the

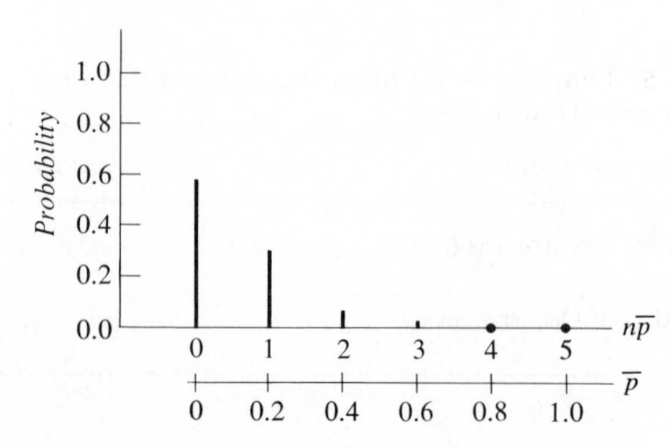

Figure 6-1 Graph of sampling distribution of $n\bar{p}$ and $\bar{p}$ for $p = 0.1, q = 0.9$, and $n = 5$.

average, we will observe 10% defectives in the samples. The standard deviation, $\sigma_{\bar{p}} = \sqrt{pq/n} = .134$ or 13.4%, is again a measure of variation attributable to the chance effects of sampling. Here the variation is in proportion of defectives.

The binomial distribution in this example is skewed as shown in Figure 6-1. Two horizontal scales are shown in this graph to depict corresponding values of $n\bar{p}$ and $\bar{p}$.

As we saw in Section 2.4, if p is less than 0.5, the distribution tails off to the right as in Figure 6-1. On the other hand, if p exceeds 0.5, the skewness is to the left. If p is held fixed, and the sample size n becomes larger, the sampling distributions of $n\bar{p}$ and $\bar{p}$ become more and more symmetrical. This is an important property of the binomial distribution from the standpoint of sampling theory and practice, which we examine further in Section 6.3.

6.3 Continuous Distributions

Thus far, in this text, we have dealt solely with probability distributions of *discrete* random variables. Probability distributions of continuous random variables are also of considerable importance in statistical theory. We turn therefore to an examination of such distributions, with particular emphasis on the meaning of their graphs. It is suggested that the reader review the definitions of discrete and continuous variables given in Section 2.1.

The binomial distribution which we have been discussing in this chapter is an example of a probability distribution of a *discrete* random variable. We have graphed such distributions by erecting ordinates at distinct values along the horizontal axis. To gain better insight into the meaning of a graph of the probability distribution of a continuous random variable, let us begin by graphing a binomial distribution as a *histogram*. We assume a situation in which a true coin is tossed two times and the random variable of interest is the number of heads obtained. We have previously seen that the probabilities of zero, one, and two heads are respectively 1/4, 1/2, and 1/4. This is an illustration of a binomial distribution in which $p = 1/2$ and $n = 2$. If the coin were tossed four times, the probabilities of 0, 1, 2, 3, and 4 heads are, respectively, 1/16, 4/16, 6/16, 4/16, and 1/16. This is a binomial distribution in which $p = 1/2$ and $n = 4$. Graphs of these distributions in the form of histograms are given in Figure 6-2. In these histograms, let us now interpret 0, 1, 2, 3, and 4 heads not as discrete values, but rather as midpoints of classes whose respective limits are $-1/2$ to 1/2, 1/2 to 1-1/2, 1-1/2 to 2-1/2, and so forth. The probabilities or relative frequencies associated with these classes are represented on the graph by the areas of the rectangles or bars. Thus, in the graph for $n = 4$, since the rectangle for the class interval 2-1/2 to 3-1/2

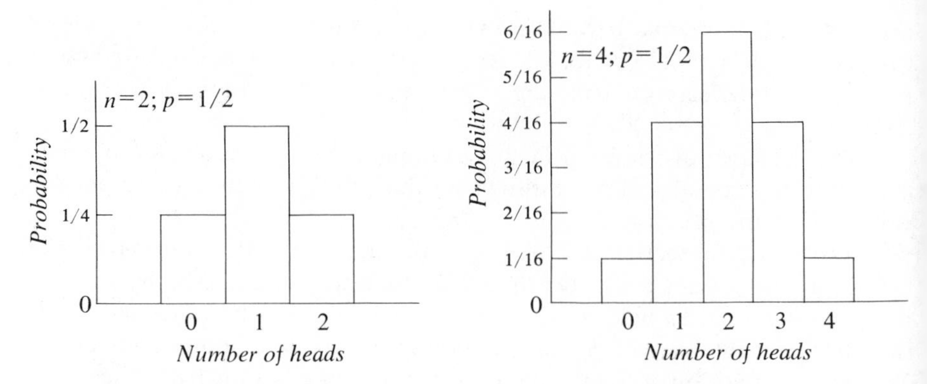

Figure 6-2 Histograms of the binomial distribution for $n = 2$, $p = 1/2$, and $n = 4$, $p = 1/2$.

has four times the area of that from 3-1/2 to 4-1/2, it represents four times the probability. If we were to represent the histogram for the case where $n = 4$ by means of a smooth continuous curve, the curve would pass through the rectangle for three heads as shown in Figure 6-3. It is clear that the shaded area under the curve for the class interval 2-1/2 to 3-1/2 is approximately equal to the area of the rectangle representing the probability of three heads, because the included area ABC is about equal to the excluded area CDE. In summary, in the approximation of a histogram by a smooth curve, the area under the curve bounded by the class limits for any given class represents the probability of occurrence of that class. In the foregoing illustration, if we had increased n greatly, say to 50 or 100 and decreased the width of the rectangles, we would see visually that the histogram appeared to approach

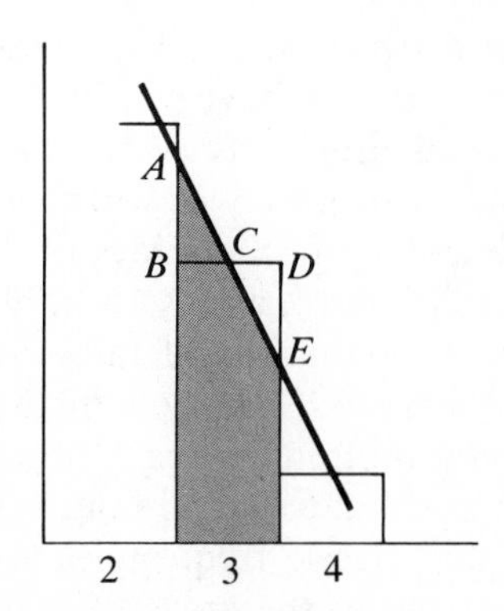

Figure 6-3 Approximation of a histogram by a continuous curve.

more and more closely a continuous curve. Since the total area of the rectangle in a histogram representing a probability distribution of a discrete random variable is equal to one, the total area under a continuous curve representing the probability distribution of a continuous random variable is correspondingly equal to one. The area under the curve lying between the two vertical lines erected at points a and b on the x axis represents the probability that the random variable X takes on values in the interval a to b.[1] This situation is depicted in Figure 6-4.

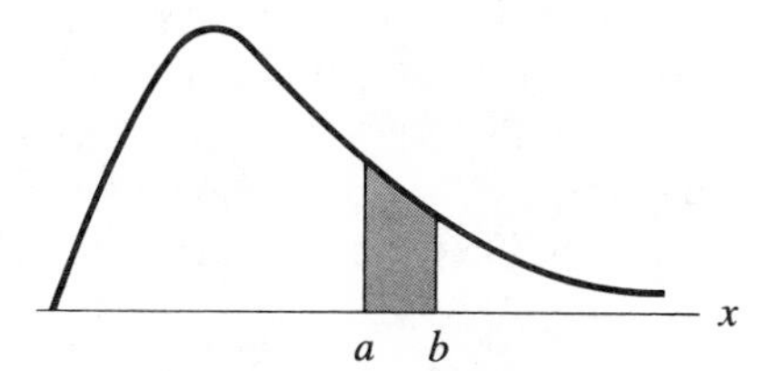

Figure 6-4 Graph of a continuous distribution: probability that the random variable X lies between a and b.

As indicated earlier, a probability for a discrete random variable is sometimes referred to as a "probability mass function." The corresponding term for a continuous random variable is "probability density function." In the continuous case, since there are an infinite number of points between a and b, the probability that X lies between a and b may be viewed as the sum of an infinite number of ordinates erected from a to b. Intuitively, this sum can be seen to be identical with the area bounded by the curve, the horizontal

---

[1]Let the value of the probability distribution of a random variable X at x be denoted $f(x)$. If X is discrete, the probability that X lies between a and b inclusive (in the closed interval (a, b)) is

$$P(a \leq X \leq b) = \sum_{x=a}^{b} f(x)$$

If X is continuous, the probability that X lies between a and b is

$$P(a \leq X \leq b) = \int_a^b f(x)\,dx$$

The reader acquainted with integral calculus can see that this definition in the continuous case is the counterpart of the summation in the discrete case. Also, it can be seen that the graphical interpretation of the probability in the continuous case is the area bounded by the curve whose value at x is $f(x)$, the X axis, and the ordinates at a and b. If the probability distribution is continuous at a and b, it makes no difference whether equal signs are shown for the interval from a to b.

axis, and the ordinates at a and b. In the discrete case, $f(x)$ denotes the probability that a random variable X takes on the value x. In the continuous case $f(x)$ cannot be interpreted as the probability of an event x, since there is an infinite number of x values. Thus, the probability of any one of them is considered to be equal to zero. In summary, for continuous random variables, probabilities can be interpreted graphically only in terms of *areas*.

By mathematics beyond the scope of this text, it can be shown that if, in the binomial distribution, p is held fixed while n is increased without limit, the binomial approaches a particular continuous distribution, referred to as the normal distribution, normal curve, or Gaussian distribution, after the mathematician and astronomer Karl Gauss. Although our illustration has been in terms of a case where $p = 1/2$, this is not a necessary condition for the proof. Even if the binomial is not symmetrical, that is, $p \neq 1/2$, the binomial still approaches the normal distribution as n increases. The shape of the normal curve is indicated in Figure 6-5. This discussion of the

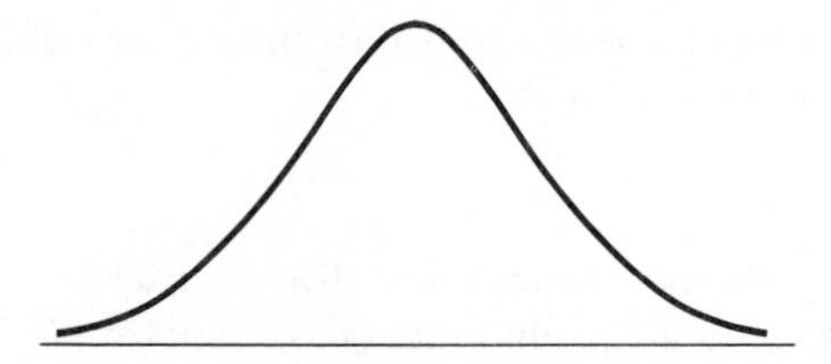

Figure 6-5 Graph of the normal distribution.

manner in which the binomial distribution approaches the normal distribution has been given here on a nonmathematical, intuitive basis in order to specify the general conclusions of these very important relationships. Actually, in the early mathematical derivations, the binomial variable was expressed in so-called "standard units" that is, $(x - \mu_{n\bar{p}})/\sigma_{n\bar{p}} = (x - np)/\sqrt{npq}$, and n was assumed to increase without limit. More modern proofs use other approaches to arrive at the same result.

A brief comment on "standard units" is useful at this point because such units are so widely employed, particularly in sampling theory and statistical inference. Standard units are merely an example of the previously mentioned "standard score" (see Section 3.16). The standard score is the deviation of a value from the mean of a frequency or probability distribution stated in units of the standard deviation. In general, it is of the form $(x - \mu)/\sigma$, where x denotes the value of the item, and μ and σ are the mean and standard deviation of the distribution, respectively. In the case of a binomially distributed random variable, as indicated in Table 6-5, the mean and standard deviation of X, the number of successes in n trials, are, respectively, np and $\sqrt{npq}$. Hence, the "standard score" or "standard unit" is $(x - np)/\sqrt{npq}$. Other terms that are also used to refer to standard scores or standard units include "standardized unit," "standardized form," and "standard form."

6.4 The Normal Distribution

The normal distribution plays a central role in statistical theory and practice, particularly in the area of statistical inference. Because of the relationship we observed in Section 6.3, the normal distribution is very useful as an approximation to the binomial in many instances where the latter distribution is the theoretically correct one. As we will see, calculations involving the normal curve are generally much easier than those involving the binomial distribution because of the simple compact form of tables of areas under the normal curve.

In addition to its use as an approximation to the binomial, the normal distribution is important in its own right in sampling applications. Before we consider such applications, let us examine the basic properties of the distribution.

Properties of the Normal Curve

Probability distributions of continuous random variables can be described by the same types of measures, such as means, medians, and standard deviations, as are used in the case of discrete random variables. One of the important characteristics of the normal curve is that we need only know the mean and standard deviation to be able to compute the entire distribution.

The normal probability distribution is defined by the equation

$$f(x) = \frac{1}{\sqrt{2\pi}\,\sigma} e^{-(1/2)[(x-\mu)/\sigma]^2} \tag{6.2}$$

In this equation, the mean and standard deviation, which determine the location and spread of the distribution, are denoted by μ and σ, respectively. These are said to be the two parameters of the normal distribution. This is analogous to the situation for the binomial distribution in which the parameters are n and p. π and e are simply constants which arise in the mathematical derivation and are approximately equal to 3.1416 and 2.7183, respectively. π is the familiar quantity which appears in numerous mathematical formulas, as for example, in the expression for the area of a circle, $A = \pi r^2$, where A denotes the area and r the radius. The constant e is the base of the natural logarithm system, as indicated in the discussion of the Poisson distribution (Section 2.7). Thus, for given values of μ and σ, if we substitute a value into Equation (6.2) for x, we can compute the corresponding value for $f(x)$. Following the usual convention, the values, x, of the random variable of interest are plotted along the horizontal axis, and the corresponding ordinates $f(x)$ along the vertical. In Figure 6-6 are shown three normal probability distributions, differing in their locations and spread. Thus, the mean of distribution (c) denoted μ_c is the largest, since the distribution lies farthest to the right on the x axis, and distribution (a) has the smallest mean, μ_a.

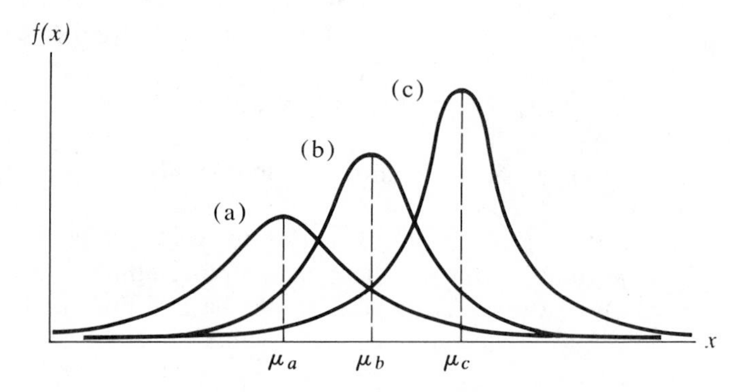

Figure 6-6 Three normal probability distributions.

On the other hand, the standard deviation of (c) is least, while that of (a) is greatest. Thus, the normal distribution defined by Equation (6.2) represents a family of distributions with the specific member of that family being determined by the values of the parameters μ and σ. Graphically, the normal curve is bell shaped and symmetrical around the ordinate erected at the mean, which lies at the center of the distribution. Recalling our previous interpretation of the graph of a continuous probability distribution, since one-half of the probability lies to the left (right) of the mean, the probability is 0.5 that a value of x will fall below (above) the mean. The values of x range from minus infinity to plus infinity. As we move further away from the mean, either to the right or left, the ordinates $f(x)$ get smaller and smaller. Thus, moving in either direction from the mean, the curve is *asymptotic* to the x axis; that is, the curve gets closer and closer to the horizontal axis, but never reaches it. However, for practical purposes, we rarely need to consider x values lying beyond three or four standard deviations from the mean, since virtually the entire area is included within this range. Stated differently, there is virtually no area in the tails of a normal distribution beyond 3 or 4 standard deviations from the mean.

Areas under the Normal Curve

We now turn to the use of the normal curve in terms of the areas under it. Although it was important to define the distribution as in Equation (6.2), in order to observe the relationship between x values and $f(x)$ values, in most applications in statistical inference we are not interested in the ordinates of the curve. Rather, since the normal curve is a continuous distribution and since the emphasis is upon its use as a probability distribution, we are interested in the areas under the curve.

It is convenient to use the term "normally distributed" for variables which have normal probability distributions, and we shall do so here. Of course, the term "normal" has no implication of quality, in the sense that other distributions are in some respect to be considered "abnormal." The term merely refers to probability distributions describable by Equation (6.2). Variables which are normally distributed occur in a variety of different units, such as dollars, pounds, inches, hours, etc. For convenience, it is useful to transform a normally distributed variable into such a form that a single table of areas under the normal curve would be applicable, regardless of the units of the original data. The transformation used for this purpose is that of the "standard unit" or, as it is often called in the case of a normal distribution, "standard normal deviate." As we noted earlier, to express an observation of a variable in standard units, we obtain the deviation of this observation from the mean of the distribution and then state this deviation in multiples of the standard deviation. For example, suppose a variable was normally distributed with mean 100 pounds, and standard deviation ten pounds. One observation of this variable is the value 120 pounds. What is this number in standardized units?

The deviation of 120 pounds from 100 pounds is +20 pounds, in units of the original data. Dividing +20 pounds by ten pounds, we obtain +2. Thus a deviation of +20 pounds from the mean lies two standard deviations above the mean if one standard deviation equals ten pounds.

Let us state this notion in general form. The number of standard units z for an observation x from a probability distribution is defined by

(6.3)
$$z = \frac{x - \mu}{\sigma}$$

where x = the value of the observation
μ = the mean of the distribution
σ = the standard deviation of the distribution.

Thus, in the illustration, $z = (120 - 100)/10 = 20/10 = +2$. The +2 indicates a value lying two standard deviations above the mean. If the observation had been 80, then $z = (80 - 100)/10 = -20/10 = -2$. The -2 denotes a value lying two standard deviations below the mean.

We now turn to another example to illustrate the use of a table of areas under the normal curve. Assume a manufacturing process which produces a certain electrical part, whose lifetime is normally distributed with an arithmetic mean of 1000 hours and a standard deviation of 200 hours. Before solving a number of probability problems concerning this distribution, we refer to Figure 6-7 to examine the relationship between values of the original variable (x values) and values in standard units (z values).

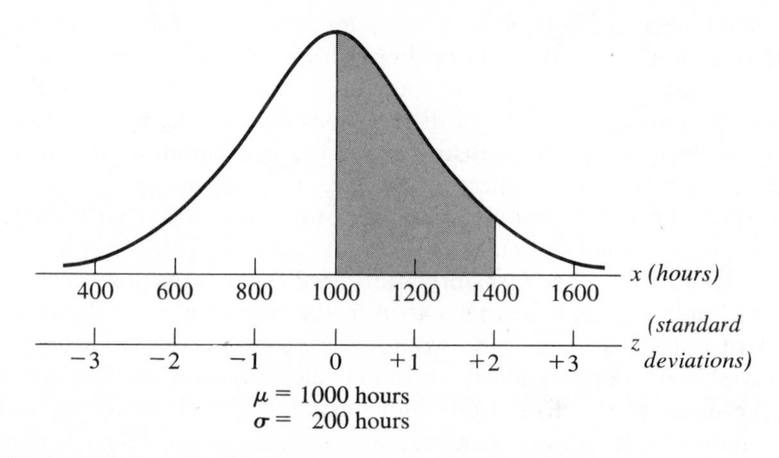

Figure 6-7 Relationship between x values and z values.

Suppose we wish to determine the proportion of parts produced by this process with lifetimes between 1000 and 1400 hours. This proportion or probability is indicated by the shaded area in Figure 6-7. We can obtain this value from Table A-5 of Appendix A, which gives areas under the normal curve lying between vertical lines erected at the mean and at specified points above the mean stated in multiples of standard deviations (z values). The left column of the table gives z values to one decimal place. The column headings give the second decimal place of the z value. The entries in the body of the table represent the area included between the vertical line at the mean and the line at the specified z value. Thus, returning to our example, the z value for 1400 hours is $z = (1400 - 1000)/200 = +2$. In Table A-5, we find the value 0.4772; hence, 47.72% of the area in a normal distribution lies between the mean and a value two standard deviations above the mean. In summary, 0.4772 is the proportion of parts produced by this process with lifetimes between 1000 and 1400 hours.

We now note a general point about the distribution of z values. Comparing the x scales and z scales in Figure 6-7, we see that for a value at the mean in the distribution of x, z is equal to zero. If an x value is at $\mu + \sigma$, that is, one standard deviation above the mean, $z = +1$, and so forth. Therefore, the probability distribution of z values, referred to as the "standard normal distribution" is simply a normal distribution with a mean of zero and a standard deviation equal to one.[2]

Example 6-1 What is the proportion of parts produced by the above process with lifetimes between 600 and 1400 hours?

First, we transform to deviations from the mean in units of the standard deviation.

$$\text{If } x = 1400;\ z = \frac{1400 - 1000}{200} = +2$$

$$\text{If } x = 600;\ z = \frac{600 - 1000}{200} = -2$$

Thus, we want to determine the area in a normal distribution that lies within two standard deviations of the mean. Table A-5 gives entries only for positive z values. However, since the normal distribution is symmetrical, the area between the mean and a value two standard deviations below the mean is the same as the area between the mean and a value two standard deviations above the mean. Hence, we double the area previously determined to obtain $2(0.4772) = 0.9544$ as the required area. In summary, about 95.5% of the parts produced by this process have lifetimes between 600 and 1400 hours. We also note the generalization that about 95.5% of the area in a normal distribution lies within two standard deviations of the mean. The required area is shown in Figure 6-8(a).

Example 6-2 What is the proportion of parts produced by this process with lifetimes between 1100 and 1350 hours?

Both 1100 and 1350 lie above the mean of 1000 hours. We can determine the required probability by obtaining (1) the area between the mean and 1350 and (2) the area between the mean and 1100, and then subtracting (2) from (1).

$$\text{If } x = 1350;\ z = \frac{1350 - 1000}{200} = \frac{350}{200} = 1.75$$

$$\text{If } x = 1100;\ z = \frac{1100 - 1000}{200} = \frac{100}{200} = 0.50$$

Table A-5 gives 0.4599 as the area corresponding to a z value of 1.75 and 0.1915 for 0.50. Subtracting 0.1915 from 0.4599 yields 0.2684 or 26.84% as the result. This area is shown in Figure 6-8(b).

---

[2]For the reader with a knowledge of calculus, we note that Table A-5 gives values of the integral $\int_0^z f(z)\,dz$, where $z = (x - \mu)/\sigma$ and $f(z) = (1/\sqrt{2\pi})\,e^{-z^2/2}$

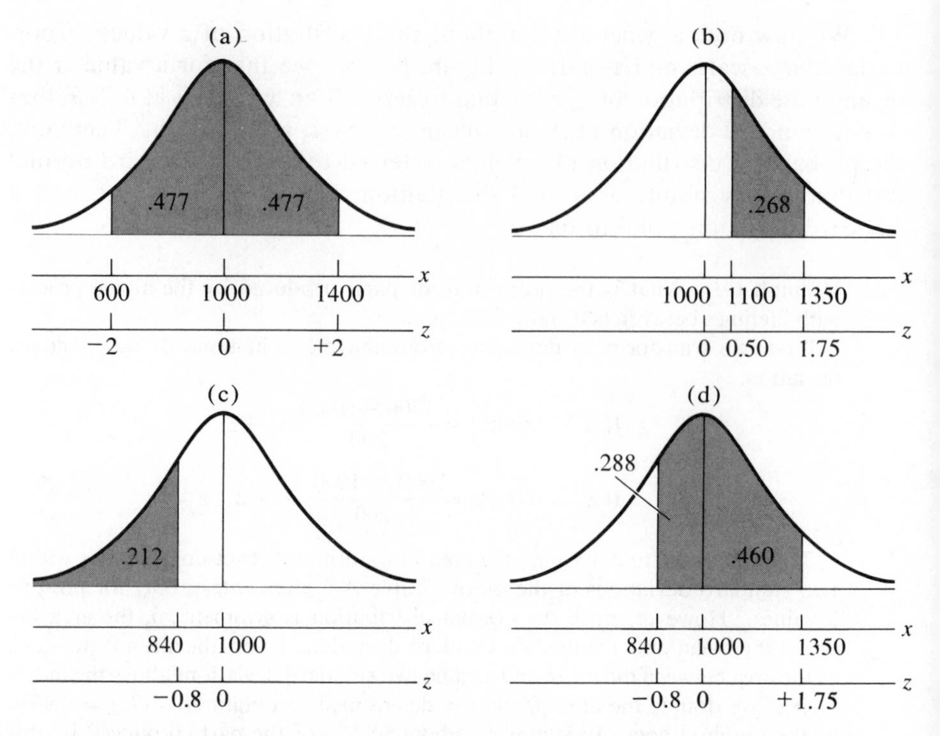

Figure 6-8 Areas corresponding to Examples 6-1–6-4.

Example 6-3 What is the proportion of parts produced by this process with lifetimes less than 840 hours?

The observation 840 hours lies below the mean. We solve this problem by determining the area between the mean and 840 and subtracting this value from 0.5000, which is the entire area to the left of the mean.

$$\text{If } x = 840; \; z = \frac{840 - 1000}{200} = -0.80$$

Since only positive z values are shown in Table A-5, we look up a z value of 0.80 and find 0.2881. This is also the area between the mean and a z value of -0.80. Subtracting 0.2881 from 0.5000 gives the desired result, 0.2119. The area corresponding to this probability is given in Figure 6-8(c).

Example 6-4 What is the proportion of parts produced by this process with lifetimes between 840 and 1350 hours?

Since 840 lies below the mean and 1350 lies above the mean, we determine (1) the area lying between 840 and the mean and (2) the area lying between 1350 and the mean, and add (1) to (2). The z values for 840 and 1350 were previously determined as -0.80 and $+1.75$, respectively, with corresponding areas of

0.2881 and 0.4599. Adding these two figures, we obtain 0.7480 as the proportion of parts with lifetimes between 840 and 1350 hours. The corresponding area is shown in Figure 6-8(d).

It was stated earlier that in the normal distribution, the range of the x variable extends from minus infinity to plus infinity. Yet, in the problems just considered, negative lifetimes were impossible. This illustrates the point that a variable may be said to be normally distributed provided that the normal curve constitutes a good fit to its empirical frequency distribution within a range of about three standard deviations from the mean. Since virtually all the area is included in this range, the situation in the tails of the distribution is considered negligible.

It is useful to note the percentages of area that lie within integral numbers of standard deviations from the mean of a normal distribution. These values have been tabulated in Table 6-5. Hence, as was observed in Example 6-1, about 95.5% of the area in a normal distribution lies within plus or minus two standard deviations from the mean. The reader should verify the other figures from Table A-5. Let us restate these probability figures in terms of

Table 6-5 Percentages of Area that Lie within Specified Intervals Around the Mean in a Normal Distribution.

Interval	*Area, %*
$\mu \pm \sigma$	68.3
$\mu \pm 2\sigma$	95.5
$\mu \pm 3\sigma$	99.7

rough statements of odds. Since about two-thirds of the area lies within one standard deviation, the odds are about two-to-one that in a normal distribution an observation will fall within that range. Correspondingly, the odds are about 95-to-5 or 19-to-1 for the two standard deviation range and 997 to 3 or about 332 to 1 for three standard deviations.

The Normal Curve as an Approximation to the Binomial

In Section 6.3, we indicated that the binomial distribution approaches the normal distribution when n becomes large. Therefore, the normal curve can be used as an approximation to the binomial distribution for the calculation of probabilities for which the binomial is the theoretically correct distribution. In general, the approximations are better when the value of p in the binomial is close to 1/2 than when p is close to 0 or 1, because for $p = 1/2$ the binomial distribution is symmetrical, and as we have seen the normal curve is a symmetrical distribution. However, the normal distribution often provides surprisingly good approximations even when p is not equal to 1/2 and even when n is not very large. We illustrate the use of the normal curve as an approximation to the binomial distribution by two examples. The first example illustrates the approximation of the probability of a single term in the binomial distribution by a normal curve calculation. The second example illustrates a corresponding calculation for a sum of terms in the binomial distribution.

Example 6-5 Assume a manufacturing process which, under particularly severe specifications, produces 25% defective items. What is the probability that a randomly drawn sample of 20 items will contain exactly four defectives?

Using Equation (2.3) for the binomial distribution in which $n = 20$, $p = .25$, and $q = .75$, we have $P(X = 4) = f(4) = \binom{20}{4}(.75)^{16}(.25)^4$. This probability is evaluated from Appendix A, Table A-1 as

$$P(X = 4) = F(4) - F(3) = .4148 - .2252 = .1896$$

In order to obtain the normal curve approximation to this probability of exactly four defectives, we set up a normal curve with the same mean and standard deviation as the given binomial distribution and find the area between 3.5 and 4.5 as shown in Figure 6-9. The reason for obtaining the area between

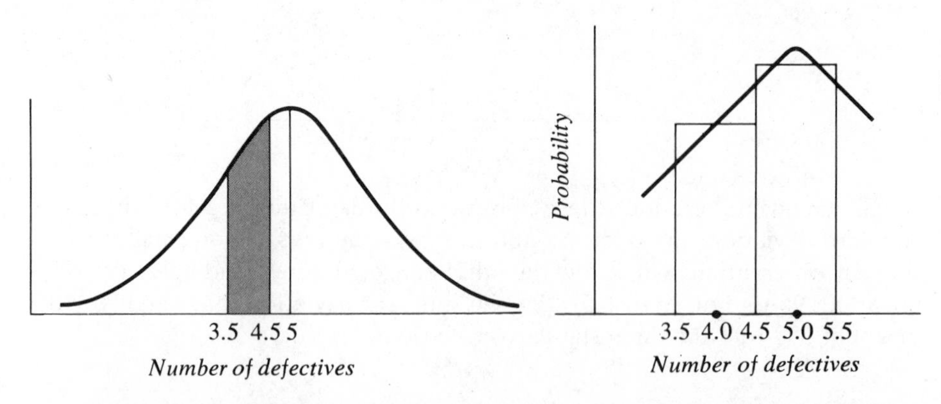

Figure 6-9 Area under the normal curve for the probability of obtaining exactly four defectives in a randomly drawn sample of 20 items from a process which produces 25% defectives.

Figure 6-10 Representation of a binomial as a histogram and the corresponding normal curve approximation.

3.5 and 4.5 is that the random variable in the binomial distribution is discrete, whereas in the case of the normal curve it is continuous. Hence, as shown in Figure 6-10, if the binomial probabilities are depicted graphically in the form of a histogram, the true probability of exactly four occurrences is given by the area of the rectangle erected at 4. To approximate this area by a corresponding area under the normal curve, four defectives can be treated as the midpoint of a class whose limits are 3.5 and 4.5. The mean and standard deviation of the binomial distribution in this problem are

$$\mu = np = (20)(.25) = 5$$

$$\sigma = \sqrt{npq} = \sqrt{(20)(.25)(.75)} = 1.94$$

Using these numbers as the mean and standard deviation of the approximating normal curve, we calculate the z values for 3.5 and 4.5 as follows

$$z_1 = \frac{3.5 - 5}{1.94} = -0.77$$

$$z_2 = \frac{4.5 - 5}{1.94} = -0.26$$

The areas for these z values are, respectively, .2794 and .1026; their difference yields the desired approximation, .2794 − .1026 = .1768. Hence, .1768 is the normal curve approximation to the true binomial probability of .1896.

Example 6-6 In the same manufacturing process as in Example 6-5, what is the probability that a randomly drawn sample of 20 items will contain four or more defectives?

Summing the appropriate terms in Equation (2.3), we find

$$P(X \geq 4) = \sum_{x=4}^{20} f(x) = \sum_{x=4}^{20} \binom{20}{x} (.75)^{20-x}(.25)^x$$

This probability is evaluated from Appendix A, Table A-5 as

$$P(X \geq 4) = 1 - F(3) = 1 - .2252 = .7748$$

The corresponding normal curve approximation is shown graphically in Figure 6-11. As indicated in the preceding problem, the z value for 3.5 is −0.77. Therefore, the desired area is .2794 + .5000 = .7794. The closeness of this approximation to the true binomial probability of .7748 illustrates the fact that normal curve approximations involving sums of terms usually are closer to the corresponding true probabilities than are approximations for individual terms in the binomial distribution.

The correction of one-half of a unit because the binomial distribution is discrete, whereas the normal curve is continuous, is always required when, as in Example 6-5, the probability of a *single term* in the binomial is desired. On the other hand, the correction is often dispensed with when *sums of terms* in the binomial are desired and the *sample size is large*. Thus, in

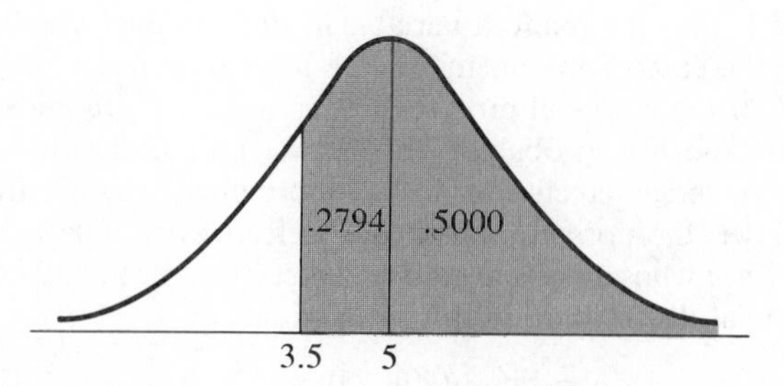

Figure 6-11 Area under the normal curve for the probability of obtaining four or more defectives in a randomly drawn sample of 20 items from a process which produces 25% defectives.

Example 6-6, if the sample size had been, say, 100, the z value for 4 rather than 3.5 may have been determined in calculating the probability of four or more defectives. Problems at the end of this chapter which require the use of this so-called "continuity correction" have been so indicated.

6.5 Sampling Distribution of the Mean

In Sections 6.1 and 6.2 we discussed sampling distributions of numbers of occurrences and percentages of occurrences. We now turn to another important probability distribution, namely, the sampling distribution of the arithmetic mean. For brevity, we shall use the term "the sampling distribution of the mean," or simply, "the sampling distribution of $\bar{x}$." To illustrate the nature of this distribution, let us return to the manufacturing process which produces electrical parts, whose lifetime is normally distributed with an arithmetic mean of 1000 hours and a standard deviation of 200 hours. We now interpret this process distribution as an infinite population from which simple random samples can be drawn. It is possible for us to draw a large number of such samples of a given size, for example, $n = 10$. We can compute the arithmetic mean lifetime of the ten parts in each sample. In accordance with our usual terminology, each such sample mean may be referred to as a "statistic." Since these statistics will usually differ from one another, we can construct a frequency distribution of these sample means. The universe mean of 1000 hours is the "parameter" around which these sample statistics will be distributed, with some sample means lying below 1000 and some lying above it. If we take any finite number of samples, the sampling distribution is referred to as an "empirical sampling distribution." On the other hand, if we conceive of the situation of drawing all possible samples of the given size, the sampling distribution is a "theoretical sampling distribution." It is such theoretical distributions upon which statistical inference is based, and on which we shall concentrate. These distributions are nothing more than probability distributions of the relevant statistics. In most practical situations, only one sample is drawn from a statistical population in order to test a hypothesis or to estimate the value of a parameter. The work implied in generating a sampling distribution by drawing repeated samples of the same size is virtually never carried out, except perhaps as an academic experiment. However, it is important for the reader to realize that the underlying theoretical structure for decisions based on single samples is the sampling distribution.

Sampling from Normal Populations

What are the salient characteristics of the sampling distribution of the mean, if samples of the same size are drawn from a population in which values are normally distributed? To obtain an answer to this question, let us begin by assuming that the sample size is five. Interpreting this in terms of our problem, let us assume that a random sample of five electrical parts is drawn from the above-mentioned population, and the mean lifetime of these five parts, denoted $\bar{x}_1$, is determined. Then, another sample of five parts is drawn, and the mean $\bar{x}_2$ is determined. Let us assume that the first mean was equal to 990 hours. Thus, it falls below the population mean. Assume the second mean was equal to 1022 hours. Hence it lies above the population mean. The theoretical frequency distribution of $\bar{x}$ values of all such simple random samples of five articles each would constitute the sampling distribution of the mean for samples of size five. Intuitively, we can see what some of the characteristics of such a distribution might be. A sample mean would be just as likely to lie above the population mean of 1000 hours as below it. Small deviations from 1000 hours would occur more frequently than large deviations. Furthermore, because of the effect of averaging, we would expect less dispersion or spread among these sample means than among the values of the individual items in the original population. That is, the standard deviation of the sampling distribution of the mean would be less than the standard deviation of the values of individual items in the population.

A couple of other characteristics of sampling distributions of the mean might be noted. If samples of size 50 rather than five had been drawn, another sampling distribution of the mean would be generated. Again we would expect the means of these samples to cluster around the population mean of 1000 hours. However, we would expect to find even less dispersion among these sample means than in the case of samples of size five. Thus, the standard deviation of the sampling distribution, which measures chance error inherent in the sampling process, would decrease with increasing sample size. Another characteristic of these sampling distributions which is not at all intuitively obvious but can be proved mathematically is that if the original population distribution is normal, sampling distributions of the mean will also be normal. Furthermore, as is indicated below, it can be shown that even if the population is non-normal, the sampling distribution of the mean is approximately normal for large sample sizes. A graphical presentation is given in Figure 6-12, which displays the relationships we have just discussed for the case of a normal population. For the population distribution, the horizontal axis represents values of individual items (x values). For the sampling distributions, the horizontal axis represents the means of samples of size five and 50. Since all three of the distributions are probability distributions of continuous random variables, the vertical axis pertains to probability densities.

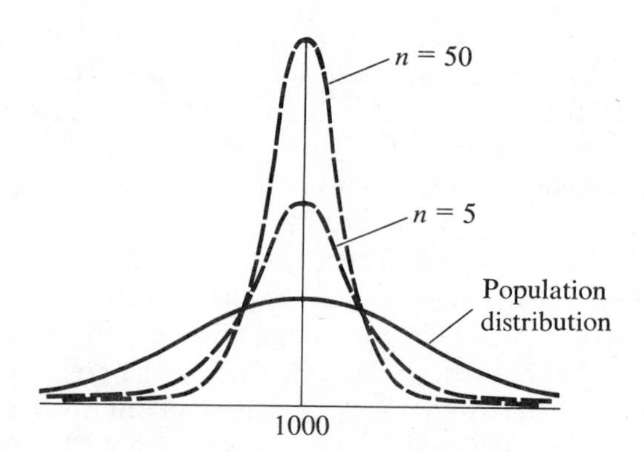

Figure 6-12 Relationship between a normal population distribution and normal sampling distributions of the mean for $n = 5$ and $n = 50$.

The foregoing material introduces the following theorem:

Theorem 6.1
If a random variable X is normally distributed with mean μ_x and standard deviation σ_x, then the mean $\bar{x}$ in a simple random sample of size n is also normally distributed with mean $\mu_{\bar{x}} = \mu_x$ and standard deviation

$$\sigma_{\bar{x}} = \frac{\sigma_x}{\sqrt{n}}$$

In this statement of the theorem, it may be noted that we have used somewhat more formal language than in the preceding discussion. Instead of saying that the values of individual items in a population are normally distributed, we refer to a normal distribution of the random variable X. The mean and standard deviation of the population are referred to as the mean and standard deviation of the random variable, μ_x and σ_x, respectively. It will be useful to employ this type of symbolism for population parameters throughout the remainder of this text. That is, the population variable is indicated as a subscript to the relevant parameters. A similar convention will be used for sampling distributions where the relevant statistic is also indicated as a subscript.

A very interesting aspect of the theorem is that the expected value (mean) of the sampling distribution of the mean, symbolized $\mu_{\bar{x}}$, is equal to the original population mean μ_x. This relationship is proved in Rule 13 of Appendix C for the even more general case of simple random samples of size n from *any* infinite population. The standard deviation of the sampling distribution of the mean, usually referred to as the "standard error of the mean" and denoted $\sigma_{\bar{x}}$, is given by

(6.4)
$$\sigma_{\bar{x}} = \frac{\sigma_x}{\sqrt{n}}$$

This relationship is proved in Rule 12 of Appendix C, again for the more general case of sampling from *any* infinite population.

Very important implications follow from Equation (6.4). We can think of any sample mean, $\bar{x}$, as an estimate of the population mean, μ_x. The difference between the statistic $\bar{x}$ and the parameter μ_x, $\bar{x} - \mu_x$ is referred to as a "sampling error." Thus, if $\bar{x}$ were exactly equal to μ_x and were used as an estimate of μ_x, there would be no sampling error. Therefore, $\sigma_{\bar{x}}$, which is a measure of the spread of the $\bar{x}$ values around μ_x, is a measure of *average sampling error*. That is, it measures the amount by which $\bar{x}$ can be expected to vary from sample to sample. Another interpretation is that $\sigma_{\bar{x}}$ is a measure of the *precision* with which μ_x can be estimated by using $\bar{x}$. Referring to Equation (6.4), we see that $\sigma_{\bar{x}}$ varies directly with the dispersion in the original population, σ_x, and inversely with the square root of the sample size, n. Thus, as might be expected, the greater the dispersion among the items in the original population, the greater the expected sampling error in using $\bar{x}$ as an estimate of μ_x. As population dispersion decreases, so does the expected sampling error. In the limiting case in which every item in the population has the same value, the population standard deviation is equal to zero; therefore, the standard error of the mean is also zero. This indicates that the mean of a sample from such a population would be a perfect estimate of the corresponding population mean, since there could be no sampling error. For example, if every item in the population were equal to 100 pounds, the population mean would be 100 pounds. Any sample would contain items all of which were equal to 100 pounds, and the sample mean would be 100 pounds. Thus, in this limiting case, the sample mean would estimate the population mean with perfect precision. As the population dispersion increases, estimation precision decreases.

The fact that the standard error of the mean varies inversely with the square root of sample size means that there is a certain type of diminishing returns in sampling effort. A quadrupling of sample size only halves the standard error of the mean; multiplying sample size by nine cuts the standard error to one-third its previous value.

Let us list in summary form the important properties of the sampling distribution of the mean. If the population is normally distributed with mean μ_x and a standard deviation σ_x, the sampling distribution of $\bar{x}$ has these properties:

(1) It has a mean equal to the population mean. That is, $\mu_{\bar{x}} = \mu_x$.
(2) It has a standard deviation equal to the population standard deviation divided by the square root of sample size. That is, $\sigma_{\bar{x}} = \sigma_x/\sqrt{n}$.
(3) It is normally distributed.

It is instructive to compare the properties of the population distribution and the sampling distribution of the mean in terms of a particular numerical example. We return to the normal population distribution of the lifetimes of electrical parts whose mean, μ_x, was equal to 1000 hours and whose standard deviation, σ_x, was equal to 200 hours. We saw from Example 6-1 that about 95.5% of the parts produced by the process fell within a range of two standard deviations from the population mean, or between 600 and 1400 hours. Symbolically,

$$\mu_x \pm 2\sigma_x = 1000 \pm 2(200) = 1000 \pm 400 = 600 \text{ to } 1400 \text{ hours}$$

This is a range that includes 95.5% of individual parts produced by the process. What is the corresponding range of two standard deviations on either side of the mean of the sampling distribution of the mean for samples of 100 articles?

The standard deviation of the sampling distribution of the mean is the standard error of the mean, $\sigma_{\bar{x}}$. Thus, we can write

$$\sigma_{\bar{x}} = \frac{\sigma_x}{\sqrt{n}} = \frac{200 \text{ hours}}{\sqrt{100}} = \frac{200 \text{ hours}}{10} = 20 \text{ hours}$$

Marking off a range of $2\sigma_{\bar{x}}$ on both sides of $\mu_{\bar{x}}$ enables us to write

$$\mu_{\bar{x}} \pm 2\sigma_{\bar{x}} = 1000 \pm 40 = 960 \text{ to } 1040 \text{ hours}$$

Let us interpret this range in a relative frequency of occurrence sense. If repeated simple random samples of 100 articles each were drawn from the population under discussion, about 95.5% of the *mean* lifetimes observed in these samples would lie between 960 to 1040 hours. Thus, although there is considerable variation among the lifetimes of the individual parts in the population, the odds are roughly 19 to 1 (95 to 5) that the mean lifetime of a sample of 100 parts would lie within 40 hours of the process average of 1000 hours. Using a range of three standard errors from the mean, the odds would be about 332 to 1 that the mean of a sample of 100 parts would lie within 60 hours of the process average of 1000 hours. We will see in some detail in later chapters how such information can be used in inference and decision making. However, a brief suggestion of one such use will be given at this point. Suppose that in connection with a quality control test, the firm producing these parts wished to determine whether the process average lifetime had changed from 1000 hours. If in a simple random sample of size 100, the mean lifetime differed by more than 60 hours from 1000, it could be concluded that it was extremely unlikely that such a sample had been drawn from a process whose mean was 1000 hours. Thus, assuming the sample was indeed a random one, it would appear reasonable to infer that the process average had changed. In actual process quality control programs, large numbers of smaller samples are tested over time rather than judgments being made on the basis of a single sample as large as 100 items. Nevertheless, the principles involved in inferences from the sample data are essentially the same.

Sampling from Non-Normal Populations

We concluded in the foregoing discussion that if a population is normally distributed, the sampling distribution of $\bar{x}$ is also normal. However, many population distributions of business and economic data are not normally distributed. What then is the nature of the sampling distribution of $\bar{x}$? It is a remarkable fact that for almost all types of population distributions, the sampling distribution of $\bar{x}$ is approximately normal for sufficiently large samples. Although a proof of the mathematical theorem on which this statement is based is beyond the scope of this text, we will give a simple illustration in terms of the rolling of dice and also report some rather famous experimental evidence to aid in making this relationship between populations and sampling distributions appear intuitively plausible.

Let us consider the experiment of rolling a fair die two times. In sampling terminology, we may view this as a situation in which a sample of two independent observations ($n = 2$) is drawn from a uniformly (rectangularly) distributed population. The population distribution is the following probability distribution for the numbers on the face of a single die:

x	1	2	3	4	5	6
$f(x)$	1/6	1/6	1/6	1/6	1/6	1/6

As we noted in Chapter 2, a graph of this distribution consists of six equal-sized ordinates plotted at the respective discrete outcomes 1, 2, . . . , 6. Also, in Chapter 2, we examined the probability distribution of the total of the numbers appearing on the faces of a die rolled two times (or two dice rolled once). However, since our focus is now upon the sampling distribution of $\bar{x}$, we are interested not in the *total* of the numbers, but rather their *arithmetic mean.* Hence, for example, if a two is obtained on the first roll, and a three on the second, the total is five, and the mean is 2-1/2. The probability of a total of five on two rolls of a die was earlier seen to be 4/36, and, therefore this is also the probability of a mean of 2-1/2. Hence, we can readily write down the sampling distribution of $\bar{x}$ for samples of size two. This distribution is given in Table 6-6 and its graph is displayed in Figure 6-13. As can be seen from the graph, the sampling distribution is symmetrical and unimodal despite

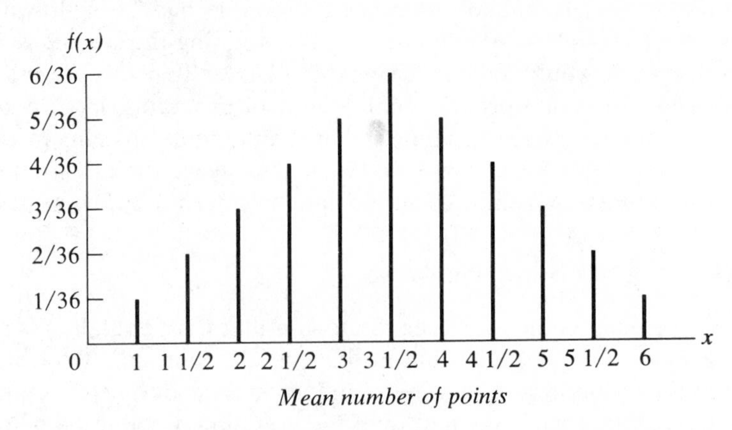

Figure 6-13 Graph of the sampling distribution of the mean for two rolls of a fair die.

Table 6-6 Sampling Distribution of the Mean for Two Rolls of a Fair Die.

Total	*Mean* $\bar{x}$	*Probability* $f(\bar{x})$
2	1	1/36
3	1-1/2	2/36
4	2	3/36
5	2-1/2	4/36
6	3	5/36
7	3-1/2	6/36
8	4	5/36
9	4-1/2	4/36
10	5	3/36
11	5-1/2	2/36
12	6	1/36
		1

the fact that the population is rectangularly distributed. As sample size (number of tosses) is increased, it can be shown that the distribution of $\bar{x}$ approaches the normal distribution. Thus, in summary, although the population distribution is rectangular, even for a sample size as small as $n = 2$, the sampling distribution of $\bar{x}$ has a single mode and is symmetrical, and this distribution approaches normality as the sample size becomes large.

Walter Shewhart, in a trail-blazing treatise on statistical methods applied to the control of quality of manufactured product, described a series of sampling experiments he conducted in order to observe the behavior of empirical sampling distributions of the mean.[3] He placed numbered chips in each of three bowls, the contents of one bowl representing a normal distribution as closely as could be obtained with discrete data, the second representing a discrete rectangular universe, and the third a discrete triangular universe. The triangular universe was simply a case where the lowest numbered chips occurred most frequently and the highest numbered chips occurred least frequently, with the shape of the ordinates of the frequency distribution appearing as a triangle. From each bowl he drew 1000 simple random samples of four chips (replacing each sample after drawing), and calculated the mean of each sample. For each of the three populations, he observed that the sampling distribution of $\bar{x}$ was very close to normal. It is indeed quite striking that even for samples as small as size four from populations very dissimilar to normal distributions, the sampling distributions of $\bar{x}$ were approximately normal!

This relationship between the shapes of the population distribution and the sampling distribution of the mean has been summarized in what is often referred to as the most important theorem of statistical inference, namely, the *central limit theorem*. The theorem is stated in terms of the z-variable for the sampling distribution of the mean and the approach of the distribution of this variable to the standard normal distribution. The formal statement of the theorem is as follows.

Central Limit Theorem

Theorem 6.2
If a random variable X, either discrete or continuous, has a mean μ_x and a finite standard deviation σ_x, then the probability distribution of $z = (\bar{x} - \mu_x)/\sigma_{\bar{x}}$ approaches the standard normal distribution as n increases without limit.

This theorem is indeed a very general one, since it makes no restrictions

---

[3] W. A. Shewhart, *Economic Control of Quality of Manufactured Product* (New York: D. Van Nostrand Company, Inc., 1931).

on the shape of the original population distribution. The requirement of a finite standard deviation is not a practical restriction at all, since virtually all distributions involved in real-world problems satisfy this condition. The reader should also note that the z-variable in this theorem is a transformation in terms of deviations of the sample mean $\bar{x}$ from the mean of the sampling distribution of means, μ_x, stated in multiples of the standard deviation of that distribution, $\sigma_{\bar{x}}$. Thus, it is exactly the same type of transformation as was used in the illustrative examples given earlier in this chapter. Here, however, all the values pertain to the sampling distribution of the mean.

It is useful at this point to restate Theorems 6.1 and 6.2 in the less formal context of sampling applications and to summarize the importance of the concepts involved.

Theorem 6.1 Restated
If a population distribution is normal, the sampling distribution of $\bar{x}$ is also normal for samples of all sizes.

Theorem 6.2 Restated
If a population distribution is non-normal, the sampling distribution of $\bar{x}$ may be considered to be approximately normal for a large sample.

The first of these theorems states that the sampling distribution of $\bar{x}$ will be *exactly* normal if the population is normal. The second, the central limit theorem, assures us that no matter what the shape of the population distribution is, the sampling distribution of $\bar{x}$ approaches normality as the sample size increases. The important point is that for a wide variety of population types, samples do not even have to be very large for the sampling distribution of $\bar{x}$ to be approximately normal. For example, only in the case of very highly skewed populations would the sampling distribution of $\bar{x}$ be appreciably skewed for samples larger than about 20. For most types of populations, the approach to normality is quite rapid as n increases.

Finite Population Multiplier

In our discussion of sampling distributions we have dealt with infinite populations. However, many of the populations in practical problems are finite, as for example, the employees in a given industry, the households in a city, and the counties in the U.S. Despite the fact that we may be dealing with finite populations, it turns out that as a practical matter formulas already obtained for infinite populations also can be applied in most cases to finite populations as well. In those cases where the results for infinite populations are not directly applicable, a simple correction factor applied to the formula for the standard deviation of the relevant sampling distribution is all that is required.

In simple random sampling from an *infinite population* we have seen that the sampling distribution of $\bar{x}$ has a mean $\mu_{\bar{x}}$ which is equal to the population mean, μ_x, and a standard deviation $\sigma_{\bar{x}}$ which is equal to $\sigma_x/\sqrt{n}$. The analogous situation in sampling from a *finite population* is that the mean $\mu_{\bar{x}}$ of the sampling distribution of $\bar{x}$ again is equal to the population mean, μ_x, but the standard deviation (standard error of the mean) is given by the following formula:

Standard Error of the Mean for Finite Populations

(6.5) $$\sigma_{\bar{x}} = \sqrt{\frac{N-n}{N-1}}\frac{\sigma_x}{\sqrt{n}}$$

where N is the number of elements in the population and n is the number in the sample. The quantity $\sqrt{(N-n)/(N-1)}$ is usually referred to as the "finite population correction" or the "finite correction factor." Thus, we see that in the case of a finite population, the standard error of the mean is equal to the finite population correction multiplied by $\sigma_x/\sqrt{n}$, which is the standard error of the mean in the infinite case. Since the finite population correction is approximately equal to one when the population size, N, is large relative to the sample size n, the standard error of the mean, $\sigma_{\bar{x}}$ in the case of sampling finite populations is for practical purposes equal to $\sigma_x/\sqrt{n}$ as in the case of an infinite population. Let us examine the last statement.

The factor $\sqrt{(N-n)/(N-1)}$ is approximately equal to $\sqrt{(N-n)/N}$ for large populations, since the subtraction of one in the denominator is negligible. We can now write

$$\sqrt{\frac{N-n}{N}} = \sqrt{1-\frac{n}{N}} = \sqrt{1-f}$$

where $f = n/N$ is referred to as the "sampling fraction," since it measures the fraction of the population contained in the sample. Thus, if the population size $N = 1000$, and the sample size $n = 10$, then $f = 10/1000 = 1/100$. In such a case, where the population is large relative to the sample, the finite population correction is very close to one. Here, for example, $\sqrt{1 - 1/100} \approx 1$, where the symbol $\approx$ denotes "approximately equal to." In summary, in this case, $\sigma_{\bar{x}}$ is practically equal to $\sigma_x/\sqrt{n}$.

Since it is often convenient to use the $\sqrt{1-f}$ form of the finite population correction, we restate the formula for $\sigma_{\bar{x}}$ as follows:

Standard Error of the Mean for Finite Populations

(6.6) $$\sigma_{\bar{x}} = \sqrt{1-f}\,\frac{\sigma_x}{\sqrt{n}} = \sqrt{1-\frac{n}{N}}\,\frac{\sigma_x}{\sqrt{n}}$$

where we have used the equal sign ($=$) rather than the approximately equal sign ($\approx$).

We now can see somewhat more clearly the relationship between sampling error, population size, and sample size. Returning to the case of the population of 1000 items, suppose that we had taken a sample of all 1000 items. Substituting into Equation (6.6), we have $\sigma_{\bar{x}} = \sqrt{1 - 1000/1000}\ \sigma_x/\sqrt{1000} = 0$. Thus, $\sigma_{\bar{x}}$, which measures expected sampling error, has a value of zero. This indicates that if we draw every item in the population into the sample, there is no sampling error involved in using the sample mean $\bar{x}$ as an estimate of the population mean, μ_x. As indicated previously, if the population is large relative to the sample, the finite population correction is approximately equal to one, and we have $\sigma_{\bar{x}} \approx \sigma_x/\sqrt{n}$. If the population is infinite, then $f = 0$ and we have again the formula for the case of an infinite population, $\sigma_{\bar{x}} = \sigma_x/\sqrt{n}$. A generally employed rule of thumb is that the formula $\sigma_{\bar{x}} = \sigma_x/\sqrt{n}$, which is strictly speaking applicable only for infinite populations, may be used whenever the size of the population is at least 20 times that of the sample, or in other words, whenever the sample represents 5% or less of the population.

A very striking implication of Equation (6.6) is that so long as the population is large relative to the sample, sampling precision becomes a function of sample size alone and does not depend on the relative proportion of the population sampled. Of course, it is assumed in this statement that the population standard deviation is constant. For example, let us assume a situation in which we draw a simple random sample of $n = 100$ from each of two populations. Each population has a standard deviation equal to 200 units ($\sigma_x = 200$). In order to observe the effect of increasing the number of elements in the population, we further assume the populations are of different sizes, namely, $N = 10{,}000$ and $N = 1{,}000{,}000$. The standard error of the mean for the population of 10,000 elements is by Equation (6.6)

$$\sigma_{\bar{x}} = \sqrt{1 - \frac{100}{10{,}000}}\ \frac{200}{\sqrt{100}} \approx \sqrt{1}\ \frac{200}{10} \approx 20$$

If the population is increased to 1,000,000, we have

$$\sigma_{\bar{x}} = \sqrt{1 - \frac{100}{1{,}000{,}000}}\ \frac{200}{\sqrt{100}} \approx \sqrt{1}\ \frac{200}{10} \approx 20$$

Thus, increasing the population from 10,000 to 1,000,000 has virtually no effect on the standard error of the mean, since the finite population correction is approximately equal to one in both instances. Indeed, if the population were increased to infinity, the same result would again be obtained for the standard error.

The finding that it is the absolute size of the sample and not the relative proportion of the population sampled that basically determines sampling precision is a point that many people find difficult to accept intuitively. In fact, prior to the introduction of statistical quality control procedures in American industry, arbitrary methods such as sampling 10% of the items of incoming shipments, regardless of shipment size, were quite common. There tended to be a vague feeling in these cases that approximately the same sampling precision was obtained by maintaining a constant sampling fraction. However, it is clear that widely different standard errors were the result because of large variations in the absolute sizes of the samples. The interesting principle that emerges from this discussion, for cases in which the populations are large relative to the samples, is that it is the absolute amount of work done (sample size), not the amount of work that might conceivably have been done (population size), that is important in determining sampling precision. The extent to which this finding can be applied to other areas of human activity we leave to the reader's judgment.

In our subsequent discussion of statistical inference we will be concerned with measures of sampling error for proportions as well as for means. Therefore, we note at this point the corresponding formula for the standard error of a proportion in sampling a finite population. In Section 6.1, it was indicated that the standard error of a proportion is given by $\sigma_{\bar{p}} = \sqrt{pq/n}$. Since our discussion referred to sampling as a *Bernoulli process*, it pertained to the sampling of an infinite population. The corresponding formula for the standard error of a proportion for a simple random sample of size n from a finite population is

Standard Error of a Proportion for Finite Populations

(6.7)
$$\sigma_{\bar{p}} = \sqrt{1-f}\sqrt{\frac{pq}{n}}$$

The same sorts of approximation considerations discussed in the case of the mean are pertinent here as well. Hence, if the size of the population is at least 20 times that of the sample, the formula for the case of infinite populations may be used.

It is of interest to note that the standard errors for means and proportions really are not two different formulas, but are merely two different forms of the same formula. The standard error of the mean for an infinite population is $\sigma_{\bar{x}} = \sigma_x/\sqrt{n}$. In the case of proportions, we may view the underlying binomial population as a *binary* population, that is, the random variable X takes on the value 0 if a failure occurs and 1 if a success occurs. The probability distribution of the population is then

x	$f(x)$
0	$q = 1 - p$
1	p

It can be shown that the mean, μ_x, of this population is equal to p and the standard deviation, σ_x, is equal to $\sqrt{pq}$. The proportion of successes in a sample of n independent observations, denoted $\bar{p}$, is nothing more than the mean of the zeros and ones denoting failures and successes. That is, $\bar{p}$ is the sum of the zeros and ones divided by n, or $\bar{p}$ = number of successes in sample/number of observations in sample. Therefore, by substitution into $\sigma_{\bar{x}} = \sigma_x/\sqrt{n}$, where $\bar{p} = \bar{x}$ and $\sigma_x = \sqrt{pq}$, we have

$$\sigma_{\bar{p}} = \frac{\sqrt{pq}}{\sqrt{n}} = \sqrt{\frac{pq}{n}}$$

The point of view that a proportion may be treated as a mean of binary values is often a very useful one.

Interpretation of a Sampling Distribution from a Finite Population

In this chapter we have discussed the meaning of sampling distributions from infinite populations. Although in the immediately preceding paragraphs we noted the effects of sample size and other factors on measures of sampling error in the case of a finite population, we have not explicitly discussed the meaning of the sampling distribution itself. Since most applied problems involve sampling from finite populations, it is important to investigate the nature of a sampling distribution in such cases. We turn, therefore, to a simple illustration which will be useful for that purpose.

Let us assume that we have a group of five families, and we have recorded the number of persons or "family size" in each. For purposes of our illustration, these five families constitute a population from which we may draw samples. Of course, with such a small population, we would not employ sampling techniques in a practical situation, but would simply obtain whatever information we desired, such as descriptive measures of family size, directly from the entire population. However, in order to gain insight into the nature of sampling distributions from finite populations, we will consider the implications of sampling this population of five elements. Our interest will be focused upon the sampling distribution of means. We begin by calculating the mean and standard deviation of the family size data for the population. These computations are shown in Table 6-7, using methods described in Chapter 3.

In order to examine a particular sampling distribution, let us deal with the case where $n = 2$. That is, we will construct the random sampling distribution for simple random samples of two families each. What would constitute a simple random sample in this situation? Since the population contains five elements and since in practical problems we ordinarily sample without replacement, a simple random sample of two of these elements is a sample drawn in such a way that every sample of two families each has an equal probability of being the sample selected. In this problem there are $\binom{5}{2} = 5!/2!3! = 10$

possible samples. These ten possible sample combinations are displayed in Table 6-8. We have noted previously that a simple random sample of n

Table 6-7 Computation of the Mean and Standard Deviation of Family Size for a Population of Five Families.

Family	*Family Size* x	x^2
A	4	16
B	3	9
C	2	4
D	5	25
E	7	49
	21	103

$$\mu_x = \frac{\Sigma x}{N} = \frac{21}{5} = 4.2 \text{ persons}$$

$$\sigma_x = \sqrt{\frac{\Sigma x^2}{N} - \mu_x^2} = \sqrt{\frac{103}{5} - (4.2)^2} = \sqrt{2.96} = 1.7 \text{ persons}$$

elements is a sample drawn in such a way that every combination of n elements has an equal chance of being the sample drawn. In general, in the sampling of a finite population containing N elements, there are $\binom{N}{n}$ such possible samples, and thus each sample has a probability equal to $1/\binom{N}{n}$ of being selected. In the case under discussion, each sample has a $1/10$ probability of selection.

By grouping the sample means and recording the probability of occurrence of each value, we can construct the sampling distribution of $\bar{x}$ for samples of size $n = 2$. This sampling distribution is shown in Table 6-9. Its mean and standard deviation, as determined by direct computation, are $\mu_{\bar{x}} = 4.2$ persons and $\sigma_{\bar{x}} = 1.1$ persons, respectively. It is a worthwhile exercise for the reader to check these calculations. We note that $\mu_{\bar{x}} = \mu_x = 4.2$ persons. Thus, in this particular example, we have illustrated the general principle that in the case of sampling a finite population without replacement, just as in the case of sampling an infinite population (or a finite population with replacement), the mean of the sampling distribution of means is equal to the population

Table 6-8 Possible Simple Random Samples of Two Families Each from a Population of Five Families.

Possible Samples of Two Families Each	*Sample Observations*	*Sample Mean* ($\bar{x}$)
A,B	4,3	3.5
A,C	4,2	3.0
A,D	4,5	4.5
A,E	4,7	5.5
B,C	3,2	2.5
B,D	3,5	4.0
B,E	3,7	5.0
C,D	2,5	3.5
C,E	2,7	4.5
D,E	5,7	6.0

Table 6-9 Sampling Distribution of the Mean for Simple Random Samples of Two Families Each from a Population of Five Families.

Sample Mean ($\bar{x}$)	*Probability* $f(\bar{x})$
2.5	1/10
3.0	1/10
3.5	2/10
4.0	1/10
4.5	2/10
5.0	1/10
5.5	1/10
6.0	1/10
	1

$\mu_{\bar{x}} = 4.2$ persons

$\sigma_{\bar{x}} = 1.1$ persons

mean. The standard error of the mean, $\sigma_{\bar{x}}$, can be calculated by Equation (6.5) as follows

$$\sigma_{\bar{x}} = \sqrt{\frac{5-2}{5-1}}\,\frac{1.7}{\sqrt{2}} = 1.1 \text{ persons}$$

which, of course, agrees with the value obtained by direct computation from the sampling distribution given in Table 6-9.

In summary, we can interpret the sampling distribution of $\bar{x}$ (or $\bar{p}$) in the case of sampling a finite population without replacement in the same way that we did for infinite populations. That is, it is the probability distribution of the means (or proportions) of all possible random samples of a given size. The theory summarized in Theorems 6.1 and 6.2 is used in applied problems for both finite and infinite populations. This is true despite the theoretical point that the normal distribution is continuous and a finite population necessarily implies discrete data. Thus, we may state again that in general *if a population is normally distributed, the sampling distribution of $\bar{x}$ is also normally distributed. If the population is not normally distributed, the sampling distribution of $\bar{x}$ is approximately normally distributed for sufficiently large samples.*

Information Derived from a Single Sample

In this chapter, we have examined the properties of sampling distributions and have emphasized that they are probability distributions of sample statistics for *all possible samples* of a given size. However, in practical problems, we generally have but a single random sample from the population in question. It is an extraordinary fact that it is possible to obtain information about the sampling distribution of all possible sample outcomes from a single sample, and thus it is possible to state the precision with which an estimate of a population parameter can be made from a statistic derived from a single sample. This fact has made possible the development of the theory and methods of statistical inference. Although the details of how this information is obtained from a single sample and how it is subsequently used represents the subject matter of the next two chapters, we note briefly at this point what the essential rationale is. We have seen how the essential descriptive measures of the sampling distribution of (say) $\bar{x}$ are related to population parameters. Thus, the mean of the sampling distribution of $\bar{x}$, $\mu_{\bar{x}}$, is equal to the mean of the population, μ_x. The standard deviation of the sampling distribution of $\bar{x}$, $\sigma_{\bar{x}}$, is a function of the population, σ_x. But we do not ordinarily know the mean and standard deviation of the population. As a matter of fact, if we had the necessary information to compute the parameters μ_x and σ_x, we would know the values of every item in the population and would have no need to resort to sampling. What we can do, however, is to use sample statistics from a single sample to estimate the corresponding population parameters. Thus, we use the sample mean $\bar{x}$ to estimate the population mean μ_x and the sample standard deviation s_x to estimate the population standard deviation, σ_x. These two estimates supply us with, respectively, (1) an estimate of a population parameter, and (2) a measure of the precision with which this parameter can be estimated. Point (1) is self evident. Point (2) requires a brief explanation. We have seen that the standard error of the mean $\sigma_{\bar{x}}$ is a measure of sampling error or estimate precision and that $\sigma_{\bar{x}}$ is equal to $\sigma_x/\sqrt{n}$. If we use the sample standard deviation s_x as a basis for estimating σ_x, then we have derived an estimate of $\sigma_{\bar{x}}$ from our sample. Thus, from our single sample, we have obtained estimates of $\mu_{\bar{x}}$ and $\sigma_{\bar{x}}$, two important parameters of the sampling distribution of $\bar{x}$, or the probability distribution of all possible means. In conclusion, we have both an estimate of a population parameter and a measure of the precision involved in estimating this population value.

Other Sampling Distributions

In this chapter, we have discussed sampling distributions of numbers and proportions of occurrences, and sampling distributions of the mean. In Chapter 2, we investigated other sampling distributions as well. For example, just as we were able to treat the binomial distribution as a sampling distribution of numbers of occurrences, under the appropriate conditions, the multinomial, Poisson, and hypergeometric distributions may similarly be used. However, it is frequently far simpler to use normal curve methods, based on the operation of the central limit theorem. There are a couple of other sampling distributions that we have not yet examined, which are of importance in elementary statistical inference. These are the Student *t*-distribution and the chi-square distribution. They will be discussed at the appropriate places in connection with statistical inference.

PROBLEMS

1. In an investment research office consisting of six employees, four favor purchasing a certain stock and two are against purchasing a certain stock. Three of these employees are selected at random to attend a meeting with a mutual fund which is a client. List all possible sets of three opinions which could be given to the client. Also assign the appropriate probabilities. Is this a sampling distribution?
2. The following are the weights of six corporate executives:

   182, 175, 185, 205, 195, 210 Mean: 192

   A committee of three is to be selected at random. List all possible weights of the persons on the committee, their mean weights, and compute the mean of these mean weights.
3. An inspector of a bottling company's assembly line draws ten filled bottles at random for inspection. If a bottle contains within a half of an ounce of the proper amount it is classified as good, otherwise it is classified as defective. List the possible numbers of defective bottles he could obtain in a sample, and

do the same for all possible proportions of defectives. Assuming the manufacturing process produces 10% defectives, assign probabilities to each of the possible sample outcomes. Calculate the mean and variance for the number of defectives and the proportion of defectives.

4. An investing service makes the following claim: "We can predict with 90% accuracy whether the Dow average will be higher or lower one week from the date of prediction." The investing service makes predictions about the Dow for a certain month (five predictions). Assume the reliability of any week's prediction is independent of any other week's prediction.
   (a) List all possible proportions of correct predictions the service can have for its five predictions.
   (b) Assign probabilities to each possible outcome under the assumption its claim is correct.
   (c) Assign probabilities to each possible action under the assumption the service makes random guesses (thus, the probability that any answer is correct is .5).

5. Suppose a population consists of six people, A, B, C, D, E, and F. A, B, C, and D do not smoke, while E and F do. We are interested in estimating the proportion of people who smoke based on a sample of three. List all possible samples of three persons each and the respective $\bar{p}$ values. Compute the mean $\bar{p}$.

6. Assume that 40% of the population favors a certain type of gun law legislation. If ten people selected at random are questioned in regard to the gun legislation, what would be the mean and standard deviation of the sampling distribution of the number favoring the gun legislation? If 20 people were questioned? If 40 people were questioned?

7. A political poll interviews 500 people at random and determines that X of them approve of candidate A.
   (a) What is the form of the distribution of X?
   (b) If 40% of the over-all population from whom the 500 were chosen favor A, what is the variance of X?

8. A certain manufacturing process turns out 10% defective items. If a sample of 20 items is drawn with replacement and inspected, what is the probability that
   (a) exactly two are defective?
   (b) more than two are defective?
   (c) less than two are defective?

   (Use normal approximation to the binomial with the continuity correction factor.)

9. According to a certain study, the probability that a reader will read any particular advertisement in Scanners Digest is .1. Multiroyal Inc., in a large advertising campaign, places 16 ads in Scanners Digest in two months. Use the normal approximation to the binomial with the continuity correction factor to find the approximate probability that a person reads

(a) exactly one Multiroyal ad.
(b) exactly two Multiroyal ads.
(c) exactly three Multiroyal ads.

10. What distinguishes a continuous variable from a discrete variable?

11. A statistical consultant has set up a mathematical discrimination model to classify an expensive solution as either "of workable quality" or "not of workable quality," on the basis of certain chemical tests. The model is such that it will classify a workable solution correctly 95% of the time. If 400 workable solutions are tested, what is the probability that more than 7% will be misclassified? (Use a normal distribution.)

12. The probability that any customer who enters the store will purchase a box of Alpine Milk is .2. If 2500 customers enter the store, what is the minimum number of boxes of Alpine Milk the store must have on hand if the probability that it will be out of stock is to be at most 1%?

13. A certain company has two production lines, A and B. Line A produces 10% defective items, while line B produces 15% defectives. A sample of 1000 items is drawn. What is the probability of getting 125 or more defectives if the sample comes from line A? What is the probability of getting 125 or less defectives if the sample comes from line B?

14. A mail-order magazine subscription company has found that 1% of all mail offerings are returned. If in a given week 10,000 offerings are mailed, what is the probability that
    (a) less than 1% will be returned?
    (b) between 70 and 151 are returned?
    (c) more than 109 are returned?
    (d) less than 85 are returned?

15. According to Pulp Paper Company, 5% of all trees cut down cannot be used in the processing of paper. If in a given period, 20 trees are cut down, find
    (a) the exact probability that two or more trees will not be usable for paper production, and
    (b) use the normal approximation to the binomial with the continuity correction factor to obtain an approximation to the exact probability.

16. Two companies, A and B, produce a certain type of specialized steel. Consider the thickness of this steel to be normally distributed. Roughly sketch the probability distributions for the two companies on the same graph for each of the following cases:
    (a) A has a mean of 3″ and a standard deviation of 1/2″,
    B has a mean of 3″ and a standard deviation of 1″.
    (b) A has a mean of 3″ and a standard deviation of 1″,
    B has a mean of 4″ and a standard deviation of 1″.

17. The weight of a pound can of cherry toffee produced by Kandy Corporation is normally distributed with a mean of one pound and a standard deviation of .04 pound. For each of the following probability questions, using

graphs with both X and Z axes, indicate the corresponding area under the normal curve. If a pound can of cherry toffee is bought at random from a store, what is the probability that it will weigh

(a) under a pound?
(b) at most .95 pound?
(c) more than 1.1 pounds?
(d) between .90 and 1.1 pounds?
(e) at least .90 pound?
(f) either more than 1.1 pounds or less than .90 pound?

18. Let X represent the strength of a certain type of hemlock beam produced by Outland Lumber Company. Assume X is normally distributed with a mean of 2000 psi and a standard deviation of 100 psi. A beam is drawn at random, and its strength is tested. Change each of the following probability statements made in terms of X into statements about the standardized variable Z. What is the probability that the tested strength will be
    (a) at least 2150 psi?
    (b) less than 1825 psi?
    (c) between 1875 psi and 2115 psi?
    (d) between 1795 psi and 1905 psi?
    (e) either more than 2250 psi or less than 1800 psi?
    (f) at most 2100 psi?

19. Let X, the time it takes for a new 3/8″ chain to break under 20,000 pounds of pressure, be normally distributed with a mean of 45 minutes and a standard deviation of ten minutes. State whether each of the following statements is true or false and why.
    (a) It is equally probable that the chain will break after 45 minutes as that it will break before 45 minutes.
    (b) It is more probable that the chain will break between 45 minutes and an hour's wear than between one-half hour and 45 minutes' wear.
    (c) It is equally probable that the chain will break before the first half hour's wear as it is that it will break after the first hour's wear.
    (d) A chain will break before the first hour and five minutes of wear 95.5% of the time.

20. The kilowatt demand at any given time on the Amgar Power Plant is normally distributed with a mean of 120,000 and a standard deviation of 10,000. If the plant can generate at most 150,000 kilowatts, what is the probability that at any given time there will be an overload?

21. You are employed by a firm that must buy ash-burners from a foreign distributor. It is known, that in the past, the mean life of the ash-burners received from this distributor has been 300 hours. It is further known that the standard deviation of the distribution of the life of ash-burners has been 30 hours. The foreign distributor has assured your management that manufacturing processes and quality control procedures have been stabilized to the point where there will be an average life of at least 300 hours. Acting on this statement, your management places an order for 100,000 ash-burners. A

sample of 900 is selected from the shipment and, the mean life is determined to be 303 hours.

(a) What fraction of the time would the mean life of ash-burners in a random sample of this size differ by as much or more than the difference that you have observed, through the operation of chance?
(b) Is the foreign distributor correct in asserting that the mean life of the ash-burners produced by him is at least 300 hours?

22. The scores on an achievement test given to 231,126 high school seniors are normally distributed about a mean of 500. The distribution has a standard deviation of 90.
    (a) What is the probability that an achievement score is less than 500?
    (b) What is the probability that an achievement score is between 320 and 680?
    (c) The probability is 0.85 that a score is more than what value?

23. The weight of a box of sugar toasted rice cereal is normally distributed with $\mu = 12$ ounces and $\sigma = .5$ ounce.
    If we pick a box at random, what is the probability that it weighs
    (a) less than 12 ounces?
    (b) less than 12.75 ounces?
    (c) between 11.5 and 12.75 ounces?

24. A group of children is under study, and the investigator has two measures of the aggression manifested by any child in a given aggression-inducing situation. He has a theory that the first measure of aggression X is a random variable which is normally distributed with mean 3 and variance 1. The second measure Y is thought to be also normally distributed, but with mean 2 and variance 4. Suppose that his theories are correct and further that X and Y are independent random variables.
    (a) Find $P(1.5 < X < 3.5)$.
    (b) Find $P(X < 2.5 \text{ and } 1.5 < Y < 3.5)$.
    (c) Find $P(X < 2.5 \text{ and } Y > 1.4)$.
    (d) Find $P(X < 2.5 \text{ or } Y > 1.4)$.
    (e) Between what two numbers do the middle 51% of the scores lie as measured by X, the first measure of aggression?

25. The weight of Greenbay Brand cookies packed in a box is normally distributed with $\mu = 12$ ounces and $\sigma = .5$ ounce.
    If an inspector from the FDA weighs one box chosen at random, what is the probability that the weight he observes is
    (a) more than 12 ounces?
    (b) more than 13 ounces?
    (c) between 11.5 and 13 ounces?

26. The mean income in Orange County, North Carolina is \$4271 with a standard deviation of \$1300.
    (a) Would it be correct to say approximately 99.7% of all people in Orange County earn between \$371 and \$8171? Why or why not?

(b) Would it be correct to say that if simple random samples of 100 inhabitants each were repeatedly drawn from Orange County, approximately 99.7% of the time the average income of the group would be between $3881 and $4661? Why or why not?

(c) Would it be correct to say that if simple random samples of 10,000 inhabitants each were repeatedly drawn from Orange County, approximately 99.7% of the time the average income of the group would be between $3310 and $5232? Why or why not?

(d) As an approximation, would (b) or (c) be more likely to be correct?

27. A specification calls for a drug to have a therapeutic effectiveness for a mean period of 50 hours. The standard deviation of the distribution of the period of effectiveness is known to be 16 hours. Shipments of the drug are to be accepted if the mean period lies between 48 and 52 hours in a sample of 64 items drawn at random.

    Suppose the actual mean period of effectiveness of the drug in a given shipment is 44 hours. What is the probability that the shipment will be accepted when, in fact, it should not be?

28. Given the following information about the present accounts receivable of a certain corporation: (1) Ten percent are accounts which have been classified as "dubious," as regards rapidity of collection. (2) The dollar sizes of accounts are normally distributed with an arithmetic mean of $10,000 and a standard deviation of $2000.

    (a) If five accounts are randomly selected, what is the probability of obtaining exactly one "dubious" account?

    (b) If one account is selected, what is the probability that it will be between $9000 and $12,000 in size?

    (c) If a random sample of 400 accounts is selected, what is the probability that the sample mean will be between $10,100 and $10,200 in size?

    (d) If one account is randomly selected, the probability is .15 that its size will exceed a certain dollar amount. What is this dollar amount?

29. (a) According to the 1967 County and City Statistical Abstract published by the Bureau of the Census, there are 257 automotive dealers in Delaware and the average sales per dealer is $502,680. Assume the standard deviation is $78,000. If 100 dealers are selected at random, what is the mean and standard deviation of the sampling distribution of the average sales per dealer for the 100 dealers?

    (b) According to the same source, there are 8216 dealers in California and the average sales per dealer is $626,540. Assume the standard deviation is also $78,000. If 100 California dealers are selected at random what is the mean and standard deviation of the sampling distribution of the average sales per dealer for the 100 dealers?

    (c) Compare the standard deviations of the sampling distributions in parts (a) and (b). Why are they different?

30. Do you agree or disagree with the following statements? Explain your answer.

(a) The probability that there are ten defectives in a sample of 200 is the same as that 95% are not defective in a sample of 200.
(b) In tossing a coin, the probability that the proportion of heads equals one-half goes to one as the number of tosses goes to infinity means that one could be reasonably certain that if one flipped a coin 10,000 billion times, one would get exactly 5000 billion heads.
(c) If the mean and variance of a variable are known, according to the Central Limit Theorem, we can use the normal distribution to approximate the probability that the variable will exceed some number.
(d) If X is normally distributed, the only information we need know about X to answer probability statements about it is its mean and standard deviation.
(e) The mean of a sample is always exactly normally distributed.

31. The average number of Xerox copies made per working day in a certain office is 356 with a standard deviation of 55. It costs the firm \$.03 a copy. During a working period of 121 days, what is the probability that the average cost per day is more than \$11.10?

32. What is the probability of drawing a simple random sample with a mean of 30 or more from a population with a mean of 28? The sample size is 100, the population variance is 81. Do we have to assume the population is normal?

33. The life of a Rollmore tire is normally distributed with a mean of 22,000 miles and a standard deviation of 1500 miles.

    (a) What is the probability that a tire will last at least 20,000 miles?
    (b) What is the probability that a tire will last more than 25,000 miles?
    (c) If a person buys four tires, what is the probability that the average life of the four tires exceeds 20,000 miles?

34. The sales on Monday of a particular small retail item are normally distributed with a mean of \$1852.75 and a standard deviation of \$285.15. Calculate the range in which 95.5% of the Monday sales figures will fall. If sales are averaged every month (arithmetic mean for four Mondays), calculate the range in which 95.5% of the average Monday sales would fall. If sales are averaged for the year (arithmetic mean for 52 Mondays), calculate the range in which 95.5% of the yearly average Monday sales would fall.

35. A clothing manufacturer has sales offices in Boston, New York, Washington, and Atlanta. Each office has 25 salesmen. The weekly sales for any salesman are normally distributed with a mean of \$1200 and a standard deviation of \$200. Within what range about the mean is the probability .997 that

    (a) a given salesman's weekly sales will fall?
    (b) the average weekly sales per salesman of the Atlanta sales office will fall?
    (c) the average weekly sales per salesman of the company will fall?

36. The mean income of a group of 35 participants in a marketing experiment is \$13,500 with a standard deviation of \$1000. The experiment is to be run in

groups of five. What is the mean income for the groups of five? What is the standard deviation?

37. The average salary of the presidents of 15 different small electronic controls companies is $28,875 with a standard deviation of $2865. A certain business magazine decides to make a study of the presidents of these 15 firms.
    (a) If five presidents are selected at random, what would be the mean and standard deviation of the average salary of the possible samples that could be drawn?
    (b) If ten presidents are selected at random, what would be the mean and standard deviation of the average salary of the possible samples that could be drawn?
    (c) If 15 presidents are selected at random what would be the mean and standard deviation of the average salary of the possible samples that could be drawn?

38. A large office building has five pay telephones in different locations throughout the building. The actual usage of each phone this month has been

Phone	*Number of calls from phone*
A	27
B	9
C	24
D	35
E	16

Two phones are to be picked at random for a study of the length of calls.
    (a) List all possible samples of two phones, and give the mean number of calls for each of these samples.
    (b) Calculate the mean and variance of the sampling distribution
        (1) from the table in part (a),
        (2) from the original data.

39. The average amount of leisure time spent reading per week by the 15 top executives of a certain company is 15 hours with a standard deviation of two hours. Five executives are selected at random and asked various questions, one of which is amount of leisure time per week spent on reading. The answers from the executives will constitute the "correct" standards for a test to be given to applicants for management positions. The mean of the five executives questioned was 14.8 hours with a standard deviation of 1.5 hours.
    (a) Are the 14.8 hours and 1.5 hours the mean and standard deviation of the sampling distribution of five executives selected from the population of 15?
    (b) Define the sampling distribution of the mean leisure reading time per week of five executives selected from the 15. Calculate the mean and standard deviation of the sampling distribution.

CHAPTER SEVEN

Statistical Decision Procedures— Hypothesis Testing

Most of the remainder of this text deals with methods by which rational decisions can be made when only *incomplete information* is available and *uncertainty* exists concerning outcomes which are critical to the success of these decisions. We have begun to see that probability theory is a useful and coherent framework for dealing with the problems of uncertainty. The field known as *statistical inference*, which is the subject matter of the next three chapters, uses this theory as a basis in making reasonable decisions from incomplete data. Statistical inference treats two different classes of problems: (1) hypothesis testing, which is discussed in this chapter and Chapter 9 and (2) estimation, which is examined in Chapter 8. In both cases, the particular problem at hand is structured in such a way that inferences about relevant population values can be made from sample data. The field of statistical inference has had a fruitful development since the latter half of the nineteenth century. Among the terms by which this body of theory and practice has come to be known is "classical statistics." A multitude of useful applications of these methods has been and continues to be made in a wide variety of fields. In Chapters 7 and 8, we examine the viewpoint and methodology of classical statistics. At the end of each of these chapters, we refer briefly to the contributions of "Bayesian decision theory," which represents an important extension of classical or traditional statistics. A fuller treatment of Bayesian decision theory is given in later chapters.

We turn now to the subject of hypothesis testing, one of the two basic subdivisions of classical statistical inference. A hypothesis in statistics is simply a quantitative statement about a population. At this point we will briefly and very informally summarize the rationale involved in testing such a hypothesis, and then proceed by explaining the details of these testing procedures in terms of illustrative examples.

7.1 Introduction — The Rationale of Hypothesis Testing

In order to gain some insight into the reasoning involved in statistical hypothesis testing, let us consider a hypothesis testing procedure which is nonstatistical, but with which we are all familiar. As it turns out, the basic process of inference involved is strikingly similar to that employed in statistical methodology.

Let us consider the process by which an accused individual is judged to be innocent or guilty in a court of law under our legal system. Under Anglo-Saxon law, the man before the bar is assumed to be innocent. The burden of proof of his guilt rests upon the prosecution. Using the language of hypothesis testing, let us say that we want to test a hypothesis, which we denote H_0, that the man before the bar is innocent. This means that an alternative hypothesis exists, H_1, that the defendant is guilty. The jury's job is to examine the evidence and to determine whether the prosecution has demonstrated that this evidence is inconsistent with the basic hypothesis H_0, of innocence. If the jurors decide the evidence is inconsistent with H_0, they reject that hypothesis, and therefore accept its alternative, H_1, that the defendant is guilty.

If we analyze the situation that results when the jury makes its decision, we find that there are the following four possibilities in terms of the basic hypothesis, H_0.

(1) The defendant is innocent (H_0 is true), and the jury finds that he is innocent (accepts H_0); hence the correct decision has been made.

(2) The defendant is innocent (H_0 is true), but the jury finds him guilty (rejects H_0); hence an error has been made.

(3) The defendant is guilty (H_0 is false), and the jury finds that he is guilty (rejects H_0); hence the correct decision has been made.

(4) The defendant is guilty (H_0 is false), but the jury finds him innocent (accepts H_0); hence an error has been made.[1]

In possibilities (1) and (3), we observe that the jury reaches the correct decision. On the other hand, in possibilities (2) and (4) it makes an error. Let us consider these errors in terms of conventional statistical terminology. In possibility (2), the hypothesis H_0 is erroneously rejected. This basic hypothesis, H_0, which is tested for possible rejection, is generally referred to in statistics as the "null hypothesis." Hypothesis H_1 is designated the "alternative hypothesis." To reject the null hypothesis when in fact it is true is referred to as a "Type I error." In possibility (4), the hypothesis H_0 is accepted in error. To accept the null hypothesis when it is false is termed a "Type II error." It may be noted that under our legal system the commission of the error designated as Type I has been considered to be far more serious than a Type II error. Thus, we feel that it is a more grievous mistake to convict an innocent man than to let a guilty man go free. Had we made H_0 the hypothesis that the defendant is guilty, the meaning of Type I and Type II errors would have been the reverse of the first formulation. What had previously been a Type I error would now become a Type II error, and Type II would now be Type I. In the statistical formulation of hypotheses, how we choose to exercise control over the two types of errors is a basic guide as to how we state the hypotheses to be treated. We will see in this chapter how this error control is carried out in hypothesis testing. The aforementioned situations are summarized in Table 7-1, where the column and row headings are stated in modern decision theory terminology and require a brief explanation. When hypothesis testing is viewed as a problem in decision making, there are two alternative actions that can be taken, "accept H_0, or "reject H_0." The two alternatives, truth or falsity of hypothesis H_0, are viewed as "states of nature" or "states of the world" which affect the "payoff" of the decision. The payoffs are listed in the table, and in the schematic presentation they are stated in terms of the correctness of the decision or the type of error incurred.

---

[1]This method of analyzing the consistency of evidence with the truth or falsity of a situation is not a recent one as can be seen by the following statement of the Greek philosopher Epictetus:

> "Appearances to the mind are of four kinds. Things either are what they appear to be; or they neither are, nor appear to be; or they are and do not appear to be; or they are not, and yet appear to be. Rightly to aim in all these cases is the wise man's task."
>
> Epictetus (*Discourses*)

Table 7-1 The Relationship Between Actions Concerning a Null Hypothesis and the Truth or Falsity of the Hypothesis.

Action Concerning Hypothesis H_0	*State of Nature*	
	H_0 is True	*H_0 is False*
Accept H_0	Correct decision	Type II error
Reject H_0	Type I error	Correct decision

We can see from the framework of the hypothesis testing problem that what we need is some criterion for the decision to either accept or reject the null hypothesis, H_0. Classical hypothesis testing attacks this problem by establishing decision rules based on data derived from simple random samples. The sample data are analogous to the evidence investigated by the jury. The decision procedure attempts to assess the risks of making incorrect decisions and in a sense, which we shall examine, to minimize them.

The Hypothesis Testing Procedure

As indicated at the beginning of this chapter, a statistical hypothesis is a quantitative statement about a population. A null hypothesis H_0 may simply be an assertion, for example, that a population mean μ_x is equal to a particular value, say $\mu_x = 30$. The alternative hypothesis H_1 may be that the population mean is equal to one or more other values than the one asserted under the null hypothesis, for example, $\mu_x \neq 30$. A simple random sample of size n is then drawn from the population in question and a sample statistic is observed in order to test the null hypothesis H_0 for possible rejection. In this example, the sample mean $\bar{x}$ would be observed. If the value of the sample statistic differs from the population value assumed under the null hypothesis H_0 by more than we would be willing to attribute to chance errors of sampling, we reject the null hypothesis H_0 and accept its alternative H_1. On the other hand, if the difference between the sample statistic and the population value assumed under H_0 is small enough to be attributed to chance sampling error, we do not reject the null hypothesis H_0. Actually this could mean either that we reserve judgment or accept H_0. In order to simplify the discussion, let us assume that failure to reject H_0 means we accept H_0. How do we know for what values of the sample statistic $\bar{x}$ we should accept H_0 and for what values we should reject H_0? The answer to this question is the essence of hypothesis testing.

The hypothesis testing procedure is simply a decision rule which specifies for every possible value of a statistic in a simple random sample of size n, whether the null hypothesis H_0 should be accepted or rejected. The set of possible values of the sample statistic is referred to as the *sample space*. Therefore, the test procedure divides (or partitions) the sample space into two mutually exclusive parts, called the acceptance region and the rejection region. The rejection region is also referred to as the critical region. We will illustrate the nature of the division of the sample space by means of two brief examples. Then the hypothesis testing procedure will be illustrated in considerable detail in Section 7.2.

Example 7-1 Suppose a construction company specified in purchasing rivets that in order to be acceptable the rivets must have a mean tensile strength of at least 40,000 pounds. Suppose the construction company decided to test the quality of a shipment by taking a simple random sample of n rivets drawn from the shipment. Assume further that the null hypothesis H_0 to be tested is "the true mean tensile strength of all rivets in the shipment is at least 40,000 pounds." The alternative hypothesis H_1 is "the true mean tensile strength of all rivets in the shipment is less than 40,000 pounds." These two hypotheses may be expressed mathematically as follows:

$$H_0: \mu_x \geq 40{,}000 \text{ pounds}$$

$$H_1: \mu_x < 40{,}000 \text{ pounds}$$

where μ_x denotes the mean tensile strength of the shipment.

In this case, the decision rule for accepting or rejecting H_0 would be as follows: If the mean of the sample, $\bar{x}$, is equal to or greater than some appropriate number c, accept H_0, and therefore accept the shipment. If the mean of the sample is less than c, reject H_0, and therefore reject the shipment. Hence the decision rule may be expressed as

(1) If $\bar{x} \geq c$, accept H_0

(2) If $\bar{x} < c$, reject H_0

This decision rule may be represented diagramatically by the following partition of the sample space of possible $\bar{x}$ values:

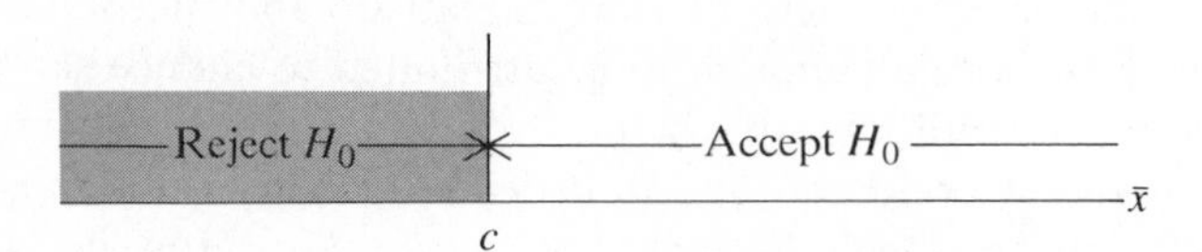

Example 7-2 An advertising agency developed a general theme for the commercials for a certain TV show based on the assumption that 50% ($p = .50$) of the viewers of this show were over 30 years of age. The agency conducted a survey of a simple random sample of n viewers of the show to determine whether there had been a change in this percentage. The agency constructed a test of the following hypotheses:

$$H_0: p = .50 \quad \text{over 30 years of age}$$

$$H_1: p \neq .50 \quad \text{over 30 years of age}$$

The decision rule constructed by the agency was that the hypothesis $H_0: p = .50$ was to be rejected if the sample percentage $\bar{p}$ of viewers over 30 years of age was less than some number c_1 (below .50), or was greater than some number c_2 (above .50). The decision rule may be written as

(1) If $c_1 \leq \bar{p} \leq c_2$, accept H_0

(2) If $\bar{p} < c_1$ or $\bar{p} > c_2$, reject H_0

The following is a schematic representation of the decision rule

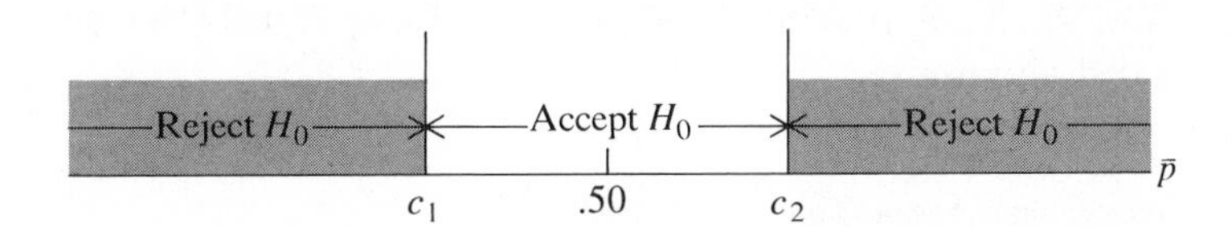

It may be noted that in the above example, as in Example 7-1, the sample space is divided into two mutually exclusive parts, an acceptance region and a rejection region. In Example 7-1, the rejection region is the set of points to the left of the number c. This type of hypothesis testing procedure is known as a "one-tailed test" or a "one-sided alternative." On the other hand, in Example 7-2, the rejection region consists of the set of points to the left of c_1 and to the right of c_2. This case is known as a "two-tailed test" or a "two-sided alternative."

We now turn to the application of some of these ideas. First, we will consider so-called "one-sample tests," which are tests of hypotheses based on data contained in a single sample. Tests involving means and proportions will be studied in that order. In the case of the one-sample tests discussed in Section 7.2, it is assumed that (1) the sample size is large ($n > 30$), (2) the sample size is decided upon before the test is conducted, and (3) the population standard deviation is known.

7.2 One-Sample Tests (Large Samples)

As a more detailed illustration of how the hypothesis testing procedure we have discussed can be used, we consider first an example of an acceptance sampling procedure. The Morgan Company, a manufacturer of space vehicle components, purchased regularly from a supplier a part which was required to withstand high temperatures. Specifications of mean heat resistance of at least 2250 degrees Fahrenheit (2250°F) were applied by the Morgan Company to shipments sent by this supplier. In the past, all of these shipments had met specifications. From long experience, it was found that the supplier's parts had a standard deviation of 300°F. That is, we will assume in this problem that it is *known* from experience that the standard deviation of heat resistance of parts in a shipment is 300°. In a simple random sample of 100 parts drawn from a *particular* shipment, a mean heat resistance of 2110°F was observed. Should the shipment be accepted, if it is desired that the risk of erroneously rejecting a shipment which meets specifications of a mean heat resistance of at least 2250°F be no more than 0.05? The shipment may be assumed to be very large relative to the sample of 100.

We proceed to convert this verbal problem to a hypothesis testing framework. As we shall see, this case is an example of a so-called "one-tailed test" or "one-sided alternative."

Tests Concerning a Mean: One-Tailed Test

We begin by stating in statistical terms the null hypothesis to be tested. Returning to our definition of a hypothesis as a quantitative statement about a population, we see that in this case the statement is about the mean of the population (shipment) sampled. It may be noted that an inference is to be drawn about the shipment, and the sample was randomly drawn from the shipment. Thus, the shipment is the population in this problem. The production process of the supplier may be viewed as a super population which gives rise to the shipment populations. Letting μ_x denote the mean heat resistance of the parts in this particular shipment population, we state the null and alternative hypothesis as follows:

$$H_0{:}\mu_x \geq 2250°\text{F}$$

$$H_1{:}\mu_x < 2250°\text{F}$$

where the standard deviation of the population is assumed to be known as 300°F. In symbols, $\sigma_x = 300°\text{F}$. The null hypothesis, H_0, states that the population mean heat resistance is equal to or greater than 2250°F (is at least

2250°F). If our decision procedure leads us to accept this hypothesis, our action will be to accept the shipment. On the other hand, if we reject H_0, we will reject the shipment. This means we accept the alternative hypothesis H_1, and therefore conclude that the shipment mean heat resistance is less than 2250°F. Our decision will be based on the data observed in the sample of 100 parts. Therefore, the question is simply this, "Are the sample data so inconsistent with the null hypothesis that we shall be forced to reject that hypothesis?" Before proceeding to examine this question more closely, we must discuss an important technical point. The null hypothesis must be stated in such a way that the probability of a Type I error can be calculated. This was the reason for including the equal sign in the statement of the null hypothesis, so that a particular value of μ_x was specified. The shipment mean heat resistance can exceed 2250°F in an infinite number of ways, one for each possible value of μ_x. Therefore, we cannot refer unambiguously to the probability of a Type I error in this situation as *the* probability of rejection of the null hypothesis when it is true. However, if we concentrate our attention on the single value, $\mu_x = 2250$°F, we can refer to a Type I error for that particular value, and we will be able to compute the probability of such an error once we have settled on a decision procedure. Hence, for the moment, we focus attention on the particular value $\mu_x = 2250$°F, for which the null hypothesis is true.

We now return to our question regarding the consistency of the sample data with the null hypothesis. In terms of this particular problem, we can word the question as follows, "Is a deviation as large as we have observed between a mean of 2110°F observed in a simple random sample of size 100 and a hypothesis population mean of 2250°F so great that we would be unwilling to attribute such a difference to chance errors of sampling a population with $\mu_x = 2250$°F?" If we conclude that the difference is so large it is improbable under the given hypothesis, then a "significant difference" is said to have been observed. An observed "significant difference" between a statistic and a parameter rejects the null hypothesis. Where the dividing line is set up for a "significant difference" versus a "non-significant difference" between the observed statistic and the parameter being tested depends on the risk we are willing to run of making a Type I error, that is, the risk of rejecting the null hypothesis when it is true. If we set up the dividing line at a point such that the probability is, say, 0.05 of erroneously rejecting the hypothesis that μ_x is equal to 2250°F, the test is said to have been conducted at the "5% significance level."

Conventional significance levels such as 0.05 and 0.01 are very frequently used in classical hypothesis testing, because of the desire to maintain a low probability of rejecting the null hypothesis H_0 when it is in fact true. These levels of significance are denoted by the Greek letter α. As we shall see, in one-tail test situations α represents the *maximum* probability of a Type I error. In two-tail test situations in which the null hypothesis consists of only one value of a population parameter, α represents *the* probability of a Type I error. We now summarize in Table 7-2 the alternative hypotheses tested in the present acceptance sampling problem and the possible actions concerning these hypotheses. After a brief discussion of the meaning of Type I and Type II errors in this problem, we proceed to an examination of how the hypothesis testing procedure is actually carried out quantitatively.

As we observe from Table 7-2, a Type I error in this problem, the incorrect rejection of the null hypothesis, takes the form of the rejection of a "good" shipment. The terms "good" and "bad" in this context will denote shipments which meet specifications or do not meet specifications, respectively.

Table 7-2 Acceptance Sampling Problem: Relationship Between Possible Actions and Hypotheses Concerning the Quality of a Shipment.

	State of Nature	
Action Concerning Hypothesis H_0	H_0: $\mu_x \geq 2250°F$ (*Shipment Meets Specifications*)	H_1: $\mu_x < 2250°F$ (*Shipment Does Not Meet Specifications*)
Accept H_0 (Accept Shipment)	No Error	Type II Error Acceptance of a Shipment Which Does Not Meet Specifications
Reject H_0 (Reject Shipment)	Type I Error Rejection of a Shipment Which Meets Specifications	No Error

As indicated in Example 2-8, this type of error has been termed in quality control work the "producer's risk." Thus, the producer runs the risk of having a good shipment rejected because the data in a sample drawn from the shipment were misleading due to chance sampling error. A Type II error in this problem takes the form of the acceptance of a bad shipment. This type of error is referred to as the "consumer's risk." That is, the consumer runs the risk of accepting a bad shipment. A realistic industrial sampling plan would require an equitable balancing of these types of risks, taking into account the costs involved to the producer and consumer. However, let us proceed with the present simplified problem, in order to study the classical hypothesis testing approach.

Decision Rules

We now turn to the question of how to establish a *decision rule* on which to base our acceptance or rejection of the shipment. As indicated earlier, an hypothesis testing decision rule is simply a procedure which specifies which action we should take for each possible sample outcome. Thus, we are interested in partitioning the sample space into a region in which we will reject the null hypothesis and a region in which we accept it. In every hypothesis testing problem, the partitioning of the appropriate sample space is accomplished from a consideration of the appropriate sampling distribution assuming the null hypothesis is true. This follows from the fact that specifying the probability of making Type I errors determines how the sample space will be partitioned. In this particular problem, since the question concerns *mean* heat resistance, the sampling distribution of means is the relevant distribution. We must use this distribution in our present problem to perform a test at the 5% significance level. Since our sample is large ($n = 100$), we use the central limit theorem and assume that the sampling distribution of means is normal. This distribution for samples of size 100 from a population, where $\mu_x = 2250°F$ (just in conformity with specifications) and $\sigma_x = 300°F$, is shown in Figure 7-1. As indicated in this graph, the shaded region represents 5% of the area under the normal curve. Referring to Table A-5, we see that 5% of the area in a normal distribution lies to the right of $z = +1.65$, and correspondingly the same percentage of area lies to the left of $z = -1.65$. This means that in our problem the dividing point below which we would reject a ship-

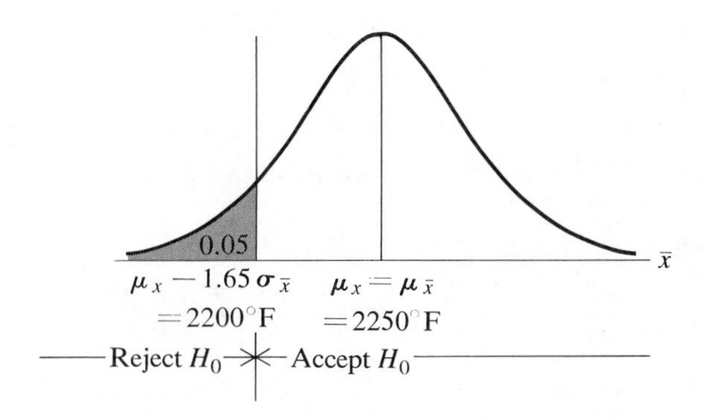

Figure 7-1 Sampling distribution of the mean showing regions of acceptance and rejection of H_0. Population parameters $\mu_x = 2250°F$, $\sigma_x = 300°F$ and sample size $n = 100$.

ment is a sample mean whose value is less than $\mu_x - 1.65\sigma_{\bar{x}}$. The standard error of the mean, $\sigma_{\bar{x}}$, is equal to

$$\sigma_{\bar{x}} = \frac{\sigma_x}{\sqrt{n}} = \frac{300°\text{F}}{\sqrt{100}} = 30°\text{F}$$

Thus, the critical value below which we would reject H_0 is

$$\mu_x - 1.65\sigma_{\bar{x}} = 2250°\text{F} - 1.65(30°\text{F}) \approx 2200°\text{F}$$

We can now see why this type of hypothesis testing situation is referred to as a one-tailed test or a one-sided alternative. Rejection of the null hypothesis takes place in only one tail of the sampling distribution.

In summary, the Morgan Company should proceed as follows in this problem. Upon drawing a single random sample of 100 parts from the incoming shipment and observing $\bar{x}$, the sample mean heat resistance, the Morgan Company should apply this decision rule

Decision Rule

(1) If $\bar{x} < 2200°\text{F}$, reject H_0 (reject the shipment)
(2) If $\bar{x} \geq 2200°\text{F}$, accept H_0 (accept the shipment)[2]

We can now answer the original question. Since the sample yielded a mean heat resistance of 2110°F, the null hypothesis H_0 should be rejected, and therefore the shipment should be rejected.

It is instructive to examine an alternative method of stating the decision rule. Instead of doing our work in terms of the original units, that is, in degrees, we could have calculated the z value in a standard normal distribution corresponding to an $\bar{x}$ value of 2110°F. If the z value lies to the left of the critical value of -1.65, H_0 is rejected; to the right of -1.65, H_0 is accepted. Thus the decision rule can be rephrased as follows:

Decision Rule

(1) If $z < -1.65$, reject H_0 (reject the shipment)
(2) If $z \geq -1.65$, accept H_0 (accept the shipment)

The arithmetic for the $\bar{x}$ value of 2110°F is

$$z = \frac{\bar{x} - \mu_x}{\sigma_{\bar{x}}} = \frac{2110°\text{F} - 2250°\text{F}}{30°\text{F}} = -\frac{140°\text{F}}{30°\text{F}} = -4.67$$

Therefore, since a mean of 2110°F falls 4.67 standard error units below the

[2]There is some ambiguity whether the equal sign should appear in the rejection or acceptance part of the decision rule. From the theoretical point of view, it is inconsequential. This follows from the fact that the normal curve is a continuous probability distribution. Thus, the probability of observing exactly $\bar{x} = 2200°\text{F}$ is zero.

mean of the sampling distribution and the dividing line is at 1.65 standard error units, the null hypothesis H_0 is rejected. This situation is displayed in Figure 7-2.

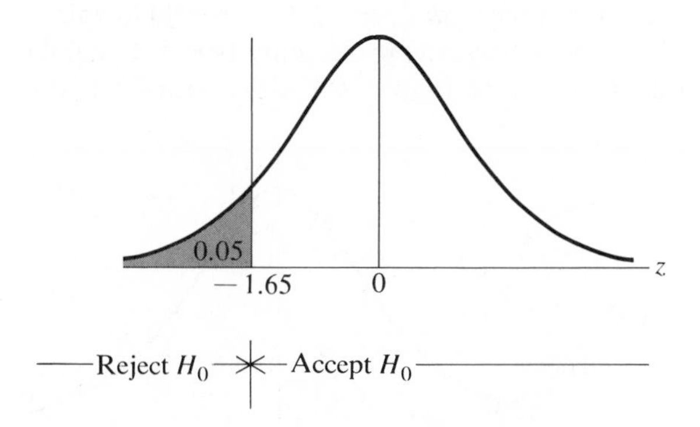

Figure 7-2 Standard normal curve for the acceptance sampling problem.

It is inconsequential which form of the decision rule we choose to use. However, as a practical matter, if the Morgan Company applied the same sampling procedure repeatedly to incoming shipments from its supplier, the first form would be simpler. Once the rule is established for a particular sample size, no further arithmetic is required. Nevertheless it is instructive in considering the rationale of the test to observe the implications of the computed z value. For example, in this case an $\bar{x}$ value of 2110°F corresponded to a z value of -4.67. Table A-5 does not give areas for z values above 4.0. Less than 0.0001 of the area in a normal distribution lies to the right of a z value of 4.0 or to the left of a z value of -4.0. Thus, interpreting the z value of -4.67 in terms of the acceptance sampling problem, if a sample of 100 parts was drawn at random from a shipment whose parts had a mean heat resistance of 2250°F and a standard deviation of 300°F, the probability of observing a sample mean of 2110°F or lower was less than 0.0001. Since this sample result is so unlikely under the hypothesis of a shipment mean of 2250°F, we reject that hypothesis.

Although we have solved the original problem of the Morgan Company, it is informative to continue the analysis. Having set up the sampling distribution of $\bar{x}$ to establish our decision rule, we can now obtain some insight into the meaning of the significance level $\alpha = 0.05$ in this type of one-tailed test problem. Suppose the supplier delivered a shipment which had a mean heat resistance in excess of 2250°F, say 2280°F. Note that 2280°F is a value that lies in the region $\mu_x \geq 2250$°F in which the null hypothesis is true. Retaining the same decision rule as originally set up, how does the probability of rejecting this shipment, that is, the probability of a Type I error, compare with the 0.05 figure for $\mu_x = 2250°F$? The appropriate sampling distribution for this case is one which is centered on $\mu_x = 2280°F$. The question is, "What is the probability that a simple random sample of 100 parts from a population with a mean heat resistance of $\mu_x = 2280°F$ would display a mean, $\bar{x}$, of less than 2200°F?" Obviously, the probability is less than 0.05 as can be seen from Figure 7-3. Comparing Figure 7-3 with Figure 7-1, the reader should

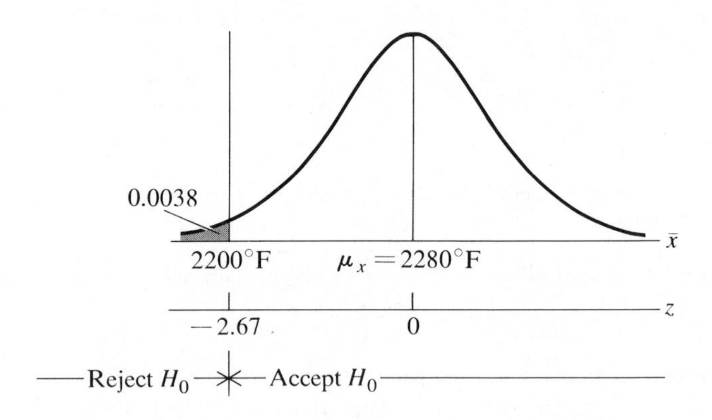

Figure 7-3 Sampling distribution of the mean; $\mu_x = 2280°F$, $n = 100$.

note that the dividing line for acceptance and rejection of H_0 remains constant at 2200°F once the decision rule has been established. With a population mean of 2280°F, the sampling distribution of $\bar{x}$ shifts to the right, leaving less than 0.05 of the area to the left of 2200°F. We can calculate the amount of this area in the usual way by determining the z value for 2200°F and looking up the corresponding area to its left in the standard normal distribution. We make the assumption that the population standard deviation σ_x remains unchanged and therefore the standard error of the mean $\sigma_{\bar{x}}$ also is unchanged.

$$z = \frac{2200°F - 2280°F}{30°F} = -\frac{80°F}{30°F} = -2.67$$

Referring to Table A-5, we find the figure 0.4962 corresponding to a z value of 2.67. By symmetry, then, 0.4962 of the area in a standard normal distribution lies between the mean and a value 2.67 standard deviations below the mean. The required area or probability is, therefore, $0.5000 - 0.4962 = 0.0038$, or less than 1%. The z scale and probability are shown in Figure 7-3. In summary, the probability of a Type I error if $\mu_x = 2280°F$ is 0.0038.

Generalizing from the preceding probability calculation, we can say that for values of μ_x greater than 2250°F, the probability of a Type I error is less than 0.05. The larger the value of μ_x, the lower is the probability of a Type I error. This makes sense in terms of the acceptance sampling problem. The greater the mean heat resistance in the incoming shipment, the lower is the probability that the shipment will be erroneously rejected. For $\mu_x = 2250°F$, the probability of a Type I error is 0.05. Now we can see the meaning of $\alpha = 0.05$, the significance level in this problem. It is the maximum probability of committing a Type I error. This sort of interpretation is typical for one-tailed tests. It is useful to observe the type of probability represented by this significance level. In terms of our present problem, it is the conditional probability of rejecting H_0, given that $\mu_x = 2250°F$. Expressing this symbolically, we can write

$$\alpha = 0.05 = P\,(\text{rejection of } H_0 \mid \mu_x = 2250°\text{F}) = P(\bar{x} < 2200°\text{F} \mid \mu_x = 2250°\text{F})$$

Summary of Procedure. In the discussion to this point, we have considered the situation in which the acceptance and rejection of the null hypothesis result in only two possible actions. Furthermore, we have concentrated on the determination of decision rules stemming from control of Type I errors without reference to the corresponding implications for Type II errors. While we shall deal with these matters subsequently, it is advisable not to clutter the present discussion with too many details. It is useful to summarize the hypothesis testing procedure discussed thus far as follows:

(1) A null hypothesis and its alternatives are drawn up. The null hypothesis is framed in such a way that we can compute the probability of a Type I error.

(2) A level of significance, α, is decided upon. This controls the risk of committing a Type I error.

(3) A decision rule is established by partitioning the relevant sample space into regions of acceptance and rejection of the null hypothesis. This partition is accomplished by a consideration of the relevant sampling distribution. The nature of the null hypothesis and the choice of α determine the partition.

(4) The decision rule is applied to the sample of size n. The null hypothesis is accepted or rejected. Rejection of the null hypothesis implies acceptance of the alternative.

Further Remarks. A number of points can be made concerning the statistical theory involved in the acceptance sampling problem of the foregoing discussion. First, the normal curve was used as the appropriate sampling distribution of the mean in that problem. If the population distribution of mean heat resistances is normal, the normal curve is the theoretically correct sampling distribution. It may be noted that no statement at all was made in the acceptance sampling problem about the population distribution. The normal curve was used for the sampling distribution of $\bar{x}$ under the central limit theorem argument that no matter what the shape of the population, the sampling distribution of $\bar{x}$ would be approximately normal for a sample as large as $n = 100$.

Second, no finite population correction was used in the calculation of the standard error of the mean in spite of the fact that the sample of 100 parts was drawn without replacement from a finite population. However, it was stated in the problem that the population size could be assumed to be very large relative to the sample size. Therefore, the finite correction factor may be assumed to be approximately equal to one in this case.

A third point is the fact that the standard deviation of the population, σ_x, was assumed to be known. We consider in Section 7.4 in the discussion of the t-distribution how to deal with the situation in which the population standard deviation is unknown.

Fourth, the nature of the z value computed in the problem is worth noting. In Chapter 6, when the idea of a standard score was discussed, it was in the context of a *population* which was normally distributed. In that case, $z = (x - \mu_x)/\sigma_x$ represented a deviation of the value of an individual item from the mean of the population expressed as a multiple of the population standard deviation. In our hypothesis testing problem, the z values were of the form $z = \bar{x} - \mu_x/\sigma_{\bar{x}}$. Such a z value represents a deviation of a sample mean from the mean of the sampling distributions of $\bar{x}$, stated in multiples of the standard deviation of that distribution, $\sigma_{\bar{x}}$. As we noted previously, the mean of the sampling distribution of $\bar{x}$, $\mu_{\bar{x}}$, is equal to the population mean, μ_x. Thus, we use μ_x and $\mu_{\bar{x}}$ interchangeably. As a generalization, in hypothesis testing problems, z values take the form

$$z = \frac{\text{statistic} - \text{parameter}}{\text{standard error}}$$

For example, in the hypothesis testing problem just discussed, the sample $\bar{x}$ value is the statistic; μ_x, the population mean, is the parameter; and $\sigma_{\bar{x}}$, the standard error of the mean, is the appropriate standard error.

Fifth, we note that the size of the sample, $n = 100$, was predetermined in our illustration. Thus, the case we discussed was one in which the sample was large, it was predetermined, and the construction of the decision rule was based on the control of only one type of incorrect decision, namely, Type I errors. The next section dealing with the power curve discusses the measurement of Type II errors for such a test.

Finally, it is important to realize that we did not *prove* that the null hypothesis was false, nor could sample evidence have proved that a null hypothesis is true. All that we can do is to discredit a null hypothesis or fail to discredit it on the basis of sample data. Actually, a single sample statistic such as $\bar{x}$ is consistent with an infinite number of hypotheses concerning μ_x.[3] From the standpoint of decision making and subsequent behavior, if sample data do not discredit a null hypothesis, we will act as though that hypothesis is true.

The Power Curve. The hypothesis testing procedure outlined thus far has concentrated upon the control of Type I errors. The question of how well this test controls Type II errors naturally arises. That is, when the null hypothesis is false, how frequently does the decision rule lead us to accept the null hypothesis erroneously? This question is answered by means of the *power curve* also called the *power function*, which can be computed from the information of the problem and the decision rule. The Greek letter β is used to denote the probability of a Type II error; thus, β represents the probability of accepting the null hypothesis when it is false. In the acceptance sampling problem, the null hypothesis H_0 is false for each value of μ_x in the alternative hypothesis $H_1: \mu_x < 2250°\text{F}$. Therefore, for each particular value of μ_x less than 2250°F, we can determine a β value. Actually, by convention, the power curve gives the complementary probability to β, that is, $1 - \beta$, for each value of the alternative hypothesis. Thus, it shows the probability of rejecting the null hypothesis for each value for which the null hypothesis is false, which, of course, represents in each case the probability of selecting the correct course of action. $1 - \beta$ is referred to as the "power of the test" against each particular value of the alternative hypothesis. For completeness, in a power curve, the probabilities of rejection are also shown for each value for which the null hypothesis is true. In summary, a power curve is a function which gives the probabilities of rejecting the null hypothesis H_0 for all possible values of the parameter tested. Therefore, it shows the ability of the decision rule to discriminate between true and false hypotheses.

The power curve for the acceptance sampling problem is shown in Figure 7-4. It has the typical reverse-S shape of a power curve in a one-tailed test with the rejection region in the left-hand tail. For a one-tailed test, with the rejection region in the right-hand tail, the curve would be S-shaped, or dropping from upper right to lower left on the graph. Rejection probabilities for the null hypothesis are shown on the vertical axis and possible values of the population parameter μ_x on the horizontal axis. Specifically, the figures plotted on the vertical axis are conditional probabilities of the form $P(\text{rejection of } H_0/\mu_x) = P(\bar{x} < 2200°\text{F}/\mu_x)$.

The nature of the power curve in Figure 7-4 can be obtained by considering a few of the plotted values. The value of $\alpha = 0.05$ is shown for

---

[3]Doubtless we have all had the disconcerting experience of observing a number of experts in disagreement after observing ostensibly the same basic set of data. Perhaps the reader shares the author's experience of finding it easiest to accept and reject hypotheses when there are no data available at all.

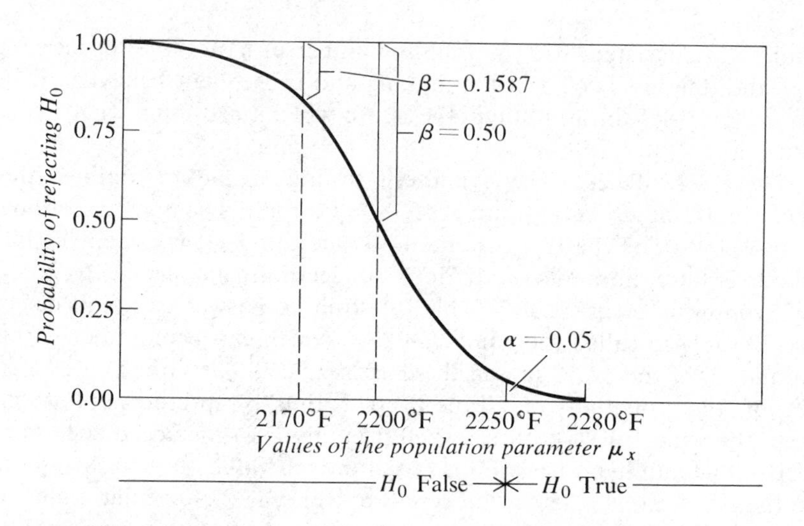

Figure 7-4 Power curve for the acceptance sampling problem.

$\mu_x = 2250°F$, indicating the significance level of the test. We can see that this is the maximum probability of erroneously rejecting the null hypothesis H_0, because the ordinates of the curve drop off to the right as μ_x increases in the region where H_0 is true. The heights of the ordinates of the power curve to the right of $\mu_x = 2250°F$ represent the probabilities of making Type I errors. One such ordinate was previously computed, namely, the probability of 0.0038 for $\mu_x = 2280°F$. The heights of the ordinates to the left of $\mu_x = 2250°F$ give the values of $1 - \beta$, or the probabilities of rejecting the null hypothesis when it is false. Therefore, the complementary distance from the curve to 1.0 are values of β, or probabilities of making Type II errors. Two such values are displayed in the graph; one for $\mu_x = 2200°F$ and one for $\mu_x = 2170°F$. We recall that our decision rule required the rejection of H_0 if the sample mean $\bar{x}$ was less than 2200°F. If the shipment or population mean is 2200°F, obviously the probability of observing $\bar{x}$ values less than 2200°F and, therefore, of rejecting H_0 is 0.50. This situation is displayed in Figure 7-5.

On the other hand, a bit of computation is required to obtain the β value for $\mu_x = 2170°F$. To compute this figure, we must refer to the sampling distribution of $\bar{x}$, given that $\mu_x = 2170°F$ and calculate the probability that a sample mean would lie in the rejection region, $\bar{x} < 2200°F$. The z value for 2200°F is

$$z = \frac{2200°F - 2170°F}{30°F} = +1.0$$

Thus, an $\bar{x}$ value of 2200°F lies one standard error unit of the mean to the

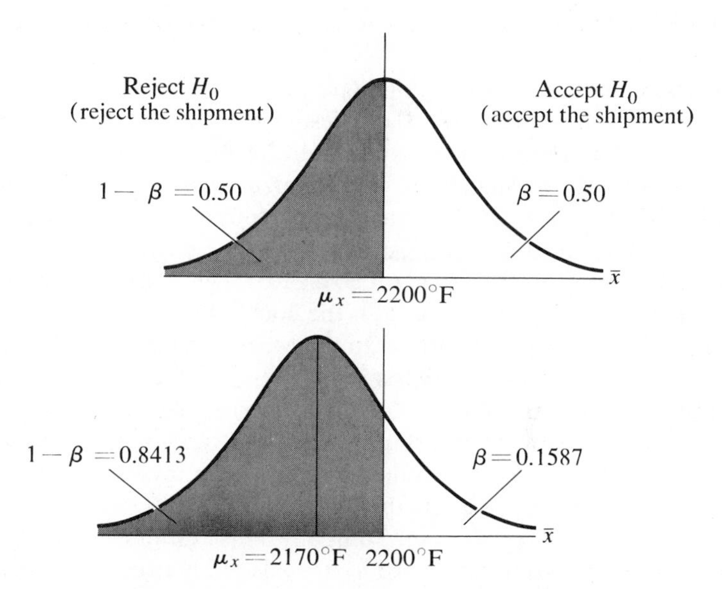

Figure 7-5 Graphs illustrating Type II error probabilities for μ_x = 2200°F and μ_x = 2170°F.

right of 2170°F. Referring to Table A-5, we find a figure of 0.3413, which when added to 0.5000, the area to the left of μ_x = 2170°F, gives 0.8413 as the probability of rejecting H_0. This is shown as the shaded region in Figure 7-5. Hence, the value of β for μ_x = 2170°F is 1.0000 − 0.8413 or 0.1587 or about 16%. Of course, this figure could have been obtained directly by subtracting 0.3413 from 0.5000, the area to the right of μ_x = 2170°F.

For values of μ_x slightly less than 2250°F, the probability of a Type II error is very high. In fact, as can be seen from Figure 7-4, these probabilities exceed 0.50 for μ_x values between 2200°F and 2250°F. This simply indicates that the power of the test is low when the values of μ_x under the null and alternative hypotheses are close together. For a fixed sample of size n, β can only be decreased by increasing α, and vice versa. If α is fixed, then as the sample size is increased, β is reduced for all values of the parameter in the region where H_0 is false. In this type of one-tailed test, the ideal power curve would be ⌐⌊ shaped with the vertical line occurring at μ_x = 2250°F. Thus, the probability of rejecting H_0 would always be equal to 1.0 when H_0 is false and 0.0 when H_0 is true. However, clearly this ideal curve is unattainable when sample data are used to test hypotheses, since sampling error will always be present. We may note that there is a trade-off relationship between Type I and Type II errors for a sample of fixed size, and thus under classical hypothesis tests such as the one discussed here, the level of significance should be decided by a consideration of the relative seriousness of the two types of errors.

The assessment or balancing of Type I and Type II errors requires some difficult comparisons on the part of the designer of the testing procedure. Let us view this matter solely from the standpoint of the Morgan Company in the present problem. For example, a Type I error means the rejection of the null hypothesis $H_0: \mu_x \geq 2250°F$ when that hypothesis is true. As indicated in Table 7-2, that means a rejection of a shipment that meets specifications. The economic costs of this type of error to the Morgan Company might involve the expense of delays caused in the company's manufacturing process by the unavailability of these parts. On the other hand, a Type II error means the acceptance of the null hypothesis when that hypothesis is false, i.e., when the alternative hypothesis $H_1: \mu_x < 2250°F$ is true. As shown in Table 7-2, this error means acceptance of a shipment which does not meet specifications. The cost of this type of error would be the expense involved in producing a product which might be defective because of the use of parts which do not meet specifications. The Morgan Company must establish the significance level for the test effecting an appropriate balance between the costs of the two types of errors. However, since classical hypothesis testing procedures do not explicitly include these economic costs as a formal part of the decision-making process, this is an extremely difficult balance to achieve. The Bayesian decision procedures discussed in Chapter 13 and succeeding chapters incorporate such costs in the formal structure of the decision analysis.

Operating Characteristic Curves. It is usual practice in industrial quality control to use operating characteristic curves, succinctly referred to as "O-C curves," rather than power curves to evaluate the discriminating power of a test. The O-C curve is simply the complement of the power curve. That is, the probability of acceptance rather than the probability of rejection of the null hypothesis is plotted on the vertical axis. The O-C curve corresponding to the power curve in Figure 7-4 is given in Figure 7-6.

Hence, for example, if the mean heat resistance of a shipment is 2170°F, the probability of accepting such a shipment by the aforementioned hypothesis testing procedure is given by the height of the O-C curve at $\mu_x = 2170°F$. As indicated in Figure 7.6, this is the probability of a Type II error for $\mu_x =$ 2170°F, or $\beta = 0.1587$. On the other hand, as shown in Figure 7-4, the height of the power curve for the same shipment mean, $\mu_x = 2170°F$, gives the probability of rejecting this shipment, or $1 - \beta$, which is equal to 0.8413. Thus $\beta = 0.1587$ is shown as the complementary distance from the power curve to the horizontal line at 1.00.

Test Concerning a Proportion: Two-Tailed Test

The preceding discussion dealt with a test of a hypothesis concerning a mean. We now turn to hypothesis testing for a proportion. As an illustration

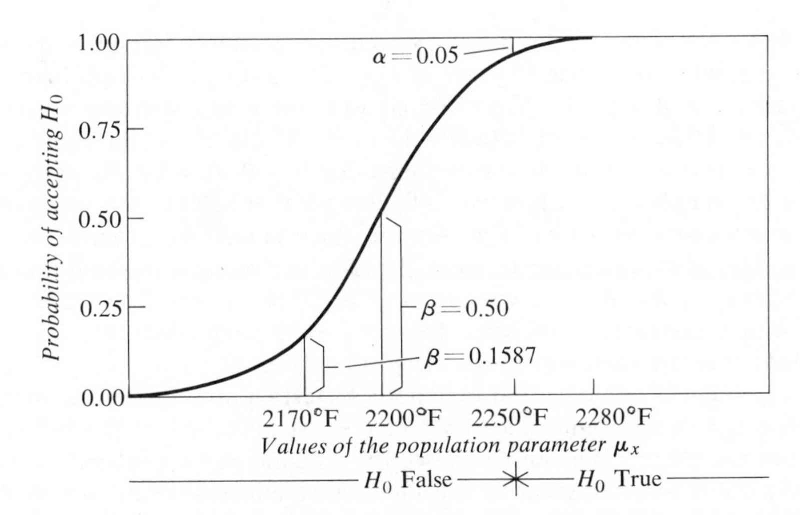

Figure 7-6 Operating characteristic curve for the acceptance sampling problem.

we return to Example 7-2. However, we will reword the problem in order to consider a test of a specific hypothesis. An advertising agency developed a general theme for the commercials for a certain TV show based on the assumption that 50% of the viewers of this show were over 30 years of age. The agency wished to change the general theme if the percentage had changed in either an upward or downward direction. If we use the symbol p to denote the proportion of *all* viewers of the TV show who were over 30 years of age, that is, p pertains to the *population* of viewers of the show, we can state the null and alternative hypotheses as follows:

$$H_0: p = .50 \text{ viewers over 30 years of age}$$

$$H_1: p \neq .50 \text{ viewers over 30 years of age}$$

Let us assume that the agency wished to run a risk of 5% of erroneously rejecting the null hypothesis of "no change," i.e., $H_0: p = .50$. That is, the agency decided to test the null hypothesis at the 5% significance level. Symbolically, this may be stated as $\alpha = 0.05$.

In order to test the hypothesis, the agency conducted a survey of a simple random sample of 400 viewers of the TV show. Of the 400 viewers, 210 were over 30 years of age and 190 were 30 years or less. What conclusion should be reached?

In this problem, as contrasted with the previous one, the null hypothesis concerns a single value of p, which is a hypothetical population parameter of 0.50. The alternative hypothesis includes all other possible values of p. The reason for setting up the hypotheses this way can be seen by reflecting on how the test will be conducted. The hypothesized parameter under the null hypothesis is $p = 0.50$. We have observed in a sample a certain proportion, denoted $\bar{p}$, who were over 30 years of age. The testing procedure involves a comparison of $\bar{p}$ with the hypothesized value of p to determine whether a significant difference exists between them. If $\bar{p}$ does not differ significantly from p, and we accept the null hypothesis that $p = 0.50$, what we really mean is that the sample is consistent with a hypothesis that half of the viewers of the TV show are over 30 years of age. On the other hand, if $\bar{p}$ is greater than 0.50 and a significant difference between $\bar{p}$ and p is observed, we will conclude that more than half of the viewers are over 30. If the observed $\bar{p}$ were less than 0.50, and a significant difference from $p = 0.50$ were observed, we would conclude that less than half of the viewers are over 30.

It is important to note that in hypothesis testing procedures, the alternate hypotheses and the significance level of the test must be selected before the data are examined. We can easily see the difficulty with a procedure which would permit the investigator to select α after examination of the sample data. It would always be possible to accept a null hypothesis simply by choosing a sufficiently small significance level, thereby setting up a large enough region of acceptance. Thus the first item in the order of events in the present problem is the usual one of setting up the competing hypotheses, stating the null hypothesis in such a way that a Type I error can be calculated. We have accomplished this by a single-valued null hypothesis, $H_0: p = 0.50$. Our next step is to set the significance level, which for illustrative purposes, we have taken as $\alpha = 0.05$.

We proceed with the test. The simple random sample of size 400 is drawn, the statistic $\bar{p}$ is observed and we can now establish the appropriate

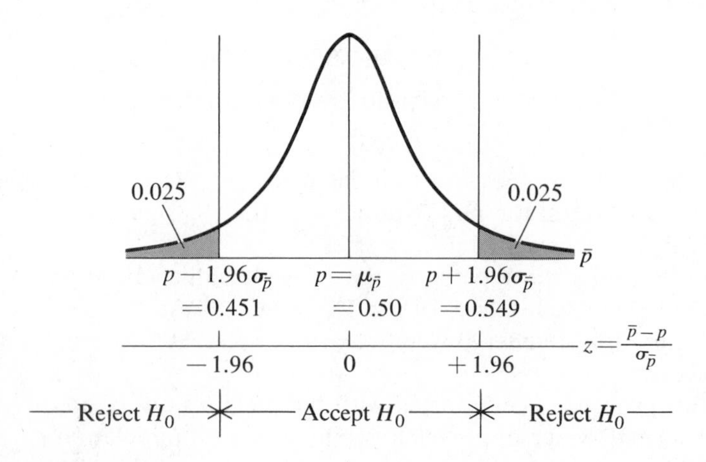

Figure 7-7 Sampling distribution of a proportion; $p = 0.50$, $n = 400$. Two-tailed test; $\alpha = 0.05$.

decision rule. Since the sample size is large, we can use the normal curve as an appropriate approximation for the sampling distribution of the percentage, $\bar{p}$. As in the preceding problem, for illustrative purposes, we will establish the decision rule in two different forms, first in terms of $\bar{p}$ values, then in terms of the corresponding z values in a standard normal distribution. The sampling distribution of $\bar{p}$ under the assumption that the null hypothesis is true has a mean of p and a standard deviation $\sigma_{\bar{p}} = \sqrt{pq/n}$. Again, we ignore the finite population correction, because the population is so large relative to the sample size. The sampling distribution of $\bar{p}$ for the present problem in which $p = 0.50$ is shown in Figure 7-7. Also shown is the horizontal axis of the corresponding standard normal distribution in terms of z values.

Since the null hypothesis will be rejected by an observation of a $\bar{p}$ value which lies significantly below or significantly above $p = 0.50$, we clearly are dealing with a two-tailed test. The critical regions (rejection regions) are displayed in Figure 7-7. The arithmetic involved in establishing regions of acceptance and rejection of H_0 is as follows. The standard error of $\bar{p}$ is given by

$$\sigma_{\bar{p}} = \sqrt{\frac{(0.50)(0.50)}{400}} = 0.025$$

Referring to Table A-5, in Appendix A we find that 2.5% of the area in a normal distribution lies to the right of $z = +1.96$, and therefore 2.5% also lies to the left of $z = -1.96$. Thus, we establish a significance level of 5% by marking off an acceptance range for H_0 of $p \pm 1.96\sigma_{\bar{p}}$. The calculation follows.

$$p + 1.96\sigma_{\bar{p}} = 0.50 + (1.96)(0.025) = 0.50 + 0.049 = 0.549$$

$$p - 1.96\sigma_{\bar{p}} = 0.50 - (1.96)(0.025) = 0.50 - 0.049 = 0.451$$

We now can state the decision rule. The agency draws a simple random sample of 400 viewers of the TV show, observes $\bar{p}$, the proportion in the sample who are over 30 years of age, and then applies the following decision rule:

Decision Rule

(1) If $\bar{p} < 0.451$ or $\bar{p} > 0.549$, reject H_0

(2) If $0.451 \leq \bar{p} \leq 0.549$, accept H_0

Again, for illustrative purposes, let us restate the decision rule in terms of z values.

Decision Rule

(1) If $z < -1.96$ or $z > +1.96$, reject H_0

(2) If $-1.96 \leq z \leq +1.96$, accept H_0

Applying this decision rule to the present problem, the observed sample $\bar{p}$ was

$$\bar{p} = \frac{210}{400} = 0.525$$

and

$$z = \frac{\bar{p} - p}{\sigma_{\bar{p}}} = \frac{0.525 - 0.500}{0.025} = +1.0$$

Therefore, the null hypothesis H_0 is accepted. This leads us to a rather negative conclusion. Had $\bar{p}$ fallen in the rejection region in the right-hand tail of the sampling distribution in Figure 7-7, that is, if $\bar{p}$ were greater than 0.549, we would conclude that more than half of the viewers of the TV show are over 30 years of age. If $\bar{p}$ had fallen in the left-hand tail of the rejection region, we would conclude that less than half are over 30. However, if as in this case, $\bar{p}$ lies in the acceptance region, we *cannot* conclude that more than half of the viewers are over 30 nor that less than half are over 30. The sample evidence is consistent with the hypothesis of a 50-50 split. Thus, acceptance of the null hypothesis means that on the basis of the available evidence we simply are not in a position to conclude that more than half nor less than half of the viewers are over 30 years of age. In some instances, the best course is to delay a terminal decision. Sequential decision procedures are often an appropriate technique to use where one of the alternatives is to delay the decision. A method of sequential decision making under uncertainty will be discussed in Bayesian decision theory.

Further Remarks. In this problem, the arithmetic was carried out in terms of proportion of successes. The calculations could have been made in terms of numbers of successes or percentage of successes rather than proportion. Of course, the conclusions are the same regardless of the method used. However, the arithmetic is often simpler if percentages are used. Hence, in this problem, if we work in percentage of successes, the standard error would be $\sqrt{(50)(50)/400} = \sqrt{6.25} = 2.5\%$. The fact that the standard error in this problem is 2.5% and the area in each of the rejection regions is also 2.5% is purely coincidental.

The hypothesized proportion in the null hypothesis in this problem was equal to 0.50. This stemmed from the fact that interest inhered in whether or not more than 50% of the viewers were over 30 years of age. On the other hand, if we wanted to test the assertion that the population proportion was 0.55, 0.60, or some other number, we would have used these figures as the respective hypothesized parameters. Also in this problem, the null hypothesis was single-valued ($p = 0.50$), while the alternative hypothesis was many-valued ($p \neq 0.50$). We saw that this resulted in a two-tailed test. Whether a test of a hypothesis is one-tailed or two-tailed depends on the question to be answered. For example, suppose the present problem had been framed as follows. An assertion had been made that at least 50% of the viewers of the TV show were over 30 years of age. Further assume that we wish to run a maximum risk of 5% of erroneously rejecting this assertion. The alternative hypotheses in this instance would be

$$H_0: p \geq 0.50$$

$$H_1: p < 0.50$$

This would involve a one-tailed test with a rejection region lying in the left-hand tail and containing 5% of the area under the normal curve. The critical region would be in the left tail, since only a significant difference for a $\bar{p}$ value lying below 0.50 could result in the rejection of the stated null hypothesis. The decision rule in terms of z values would be

Decision Rule

(1) If $z < -1.65$, reject H_0
(2) If $z \geq -1.65$, accept H_0

On the other hand, if the assertion had been that 50% or fewer of the viewers were over 30 years of age and if we wanted to run a maximum risk of 5% of erroneously rejecting this assertion, the rejection region would be the 5% area in the right tail. The corresponding alternative hypotheses and decision rule would be

$$H_0: p \leq 0.50$$

$$H_1: p > 0.50$$

Decision Rule

(1) If $z \geq +1.65$, reject H_0
(2) If $z < +1.65$, accept H_0

Standard normal distributions with the decision rules for these one-tailed tests are depicted in Figure 7-8. When tests are conducted for means or other statistical measures, they may also be either one- or two-tailed, depending upon the context of the problem.

The Power Curve. The power curve for the *two-tailed test* in the TV viewer problem just discussed is shown in Figure 7-9. The curve has the characteristic U-shape of a power function for a two-tailed test. The possible values of the parameter p are shown on the horizontal axis. The height of the ordinate at $p = 0.50$ is 0.05, which is the value of α. Since the only value of p for which the null hypothesis is true is 0.50, the ordinates at all other p values denote probabilities of rejecting the null hypothesis when it is false. The complements of these ordinates are equal to the values of β, or prob-

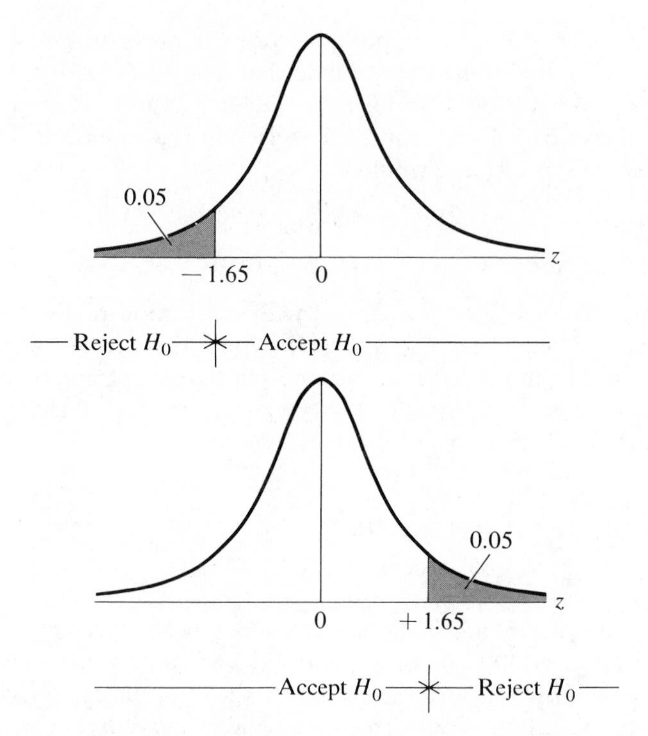

Figure 7-8 Standard normal distribution with decision rules in terms of z values: one-tailed tests ($\alpha = 0.05$).

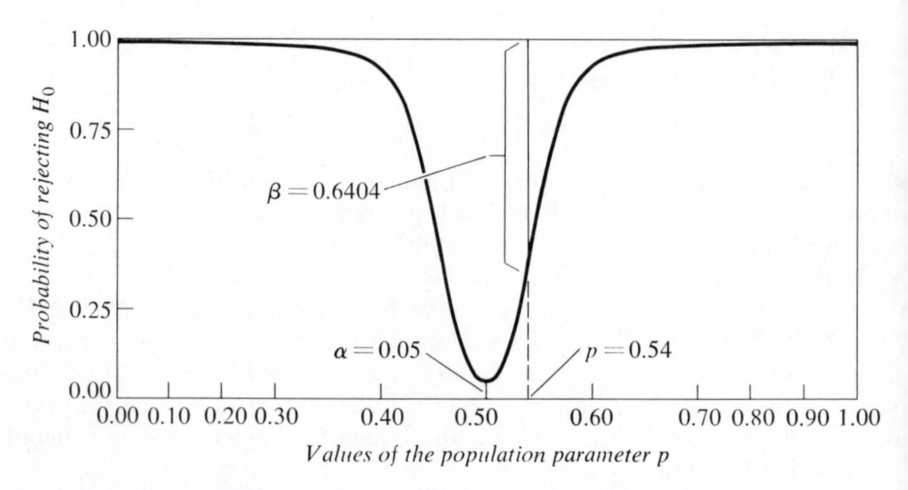

Figure 7-9 Power curve for the TV-viewer problem.

abilities of making Type II errors. One such β value is depicted on the graph for $p = 0.54$. We will carry out the calculation of this β value to illustrate the general method. Under the decision rule previously arrived at, the null hypothesis is accepted if $\bar{p}$ falls between 0.451 and 0.549. Thus, the question to be answered is, if a single random sample of size 400 is drawn from a population with parameter $p = 0.54$, what is the probability that the sample $\bar{p}$ will fall between 0.451 and 0.549? The areas corresponding to this probability are shown in Figure 7-10 in terms of the sampling distribution of $\bar{p}$

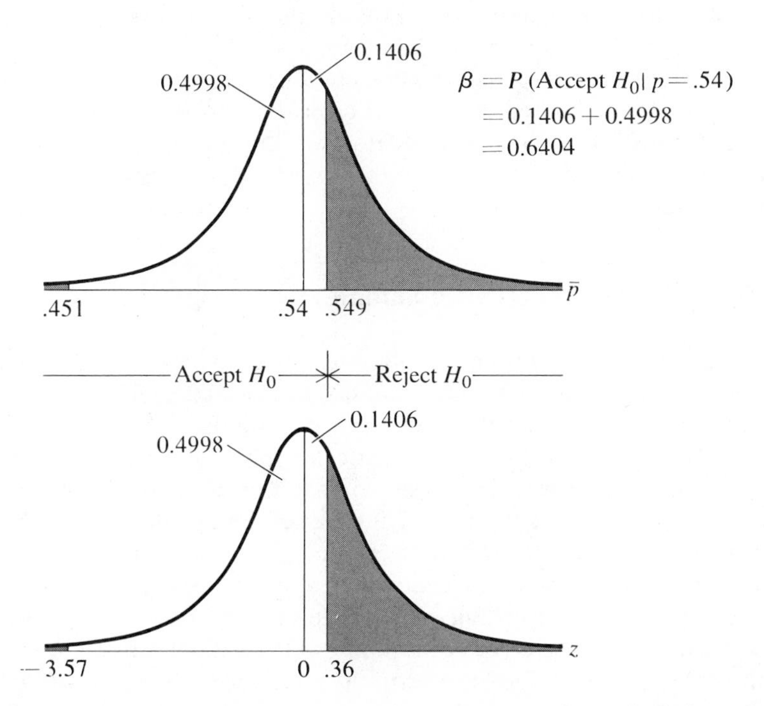

Figure 7-10 Sampling distributions of $\bar{p}$ and z, showing calculation of β for $p = .54$.

and the corresponding standard normal distribution of z values. The computations follow. The standard error of a proportion, $\sigma_{\bar{p}}$, if $p = 0.54$ is equal to $\sqrt{(0.54)(0.46)/400}$ or 0.0249.

$$\text{for } \bar{p}_1 = 0.451; \; z_1 = \frac{0.451 - 0.540}{0.0249} = -3.57$$

$$\text{for } \bar{p}_2 = 0.549; \; z_2 = \frac{0.549 - 0.540}{0.0249} = +0.36$$

From Table A-5 of Appendix A, we find that areas lying between the mean and these z values in a standard normal distribution are 0.4998 and 0.1406, respectively. Thus,

$$\beta = P(\text{Type II error} \mid p = 0.54) = 0.4998 + 0.1406 = 0.6404$$

This β value is shown in Figure 7-9. The power of the test for $p = 0.54$ is equal to $1 - \beta$, or 0.3596. By symmetry, the same β and $1 - \beta$ values would be obtained for $p = 0.46$, which is 4 percentage points below 50%, just as $p = 0.54$ is 4 percentage points above. Thus, although in this problem, we have controlled Type I errors at $\alpha = 0.05$, the test involves very high probabilities of Type II errors for values of p close to 0.50 and high probabilities for p values even a few percentage points removed from 0.50. In fact, it is only for p values as far removed as the critical points 0.451 and 0.549, that β values drop to 0.50 or less. The ideal power curve would be T-shaped with the rejection probability of H_0 equal to zero at $p = 0.50$ and equal to one for all other values of p.

7.3 Two-Sample Tests (Large Samples)

The discussion thus far has involved testing of hypotheses using data from a single random sample. Another important class of problems involves the question of whether statistics observed in two simple random samples differ significantly. Recalling that all statistical hypotheses are statements concerning population parameters, we see that this question implies a corresponding question about the underlying parameters in the populations from which the samples were drawn. For example, if the statistics observed in the two samples are arithmetic means, say $\bar{x}_1$ and $\bar{x}_2$, respectively, the question refers to whether we are willing to attribute the difference between these two sample means to chance errors of sampling. If on the basis of a test, we find that the difference is too large to attribute to chance errors, we will conclude that the populations from which the samples were drawn have *unequal means*. We illustrate these types of tests, first for differences between means and then for differences between proportions. In both cases, we assume as we did in Section 7.2 that we are dealing with large samples.

Test for Difference Between Means: Two-Tailed Test

A consulting firm conducting research for a client was asked to test whether the wage levels of unskilled workers in a certain industry was the same in two different geographical areas, referred to as Area A and Area B. The firm took simple random samples of these unskilled workers in the two areas and obtained the following sample data for weekly wages:

Area	*Mean*	*Standard Deviation*	*Size of Sample*
A	$\bar{x}_1 = \$90.01$	$s_1 = \$4.00$	$n_1 = 100$
B	$\bar{x}_2 = 85.21$	$s_2 = 4.50$	$n_2 = 200$

If the client wished to run a risk of 0.02 of incorrectly rejecting the hypothesis that the population means in these two areas were the same, what conclusion should be reached? Different sample sizes have been assumed in this problem in order to keep the example completely general. That is, the samples need not be of the same size.

Let us refer to the means and standard deviations of *all* unskilled workers in this industry in Areas *A* and *B*, respectively, as μ_1 and μ_2, and σ_1 and σ_2. These are the population parameters corresponding to the sample statistics $\bar{x}_1$ and $\bar{x}_2$, s_1 and s_2. The hypotheses to be tested are

$$H_0: \mu_1 - \mu_2 = 0$$

$$H_1: \mu_1 - \mu_2 \neq 0$$

That is, the null hypothesis asserts that the population parameters μ_1 and μ_2 are equal. As usual, the sample data must be compared to the hypothesis. We form the statistic $\bar{x}_1 - \bar{x}_2$, the difference between the sample means. If $\bar{x}_1 - \bar{x}_2$ differs significantly from zero, the hypothesized value for $\mu_1 - \mu_2$, we will reject the null hypothesis and conclude that the population parameters μ_1 and μ_2 are indeed different.

Since the risk of a Type I error has been set, we turn now to the determination of the decision rule which is based on the appropriate random sampling distribution. Let us examine a couple of the important characteristics of this distribution. The two random samples are independent, that is, the probabilities of selection of the elements in one sample are not affected by the selection of the other sample. Hence, $\bar{x}_1$ and $\bar{x}_2$ are independent random variables. We know from Rule 4 of Appendix C that the mean of the difference $\bar{x}_1 - \bar{x}_2$ is

(7.1) $$\mu_{\bar{x}_1 - \bar{x}_2} = \mu_{\bar{x}_1} - \mu_{\bar{x}_2}$$

We also know that $\mu_{\bar{x}_1} = \mu_1$ and $\mu_{\bar{x}_2} = \mu_2$, and since under the null hypothesis, $\mu_1 = \mu_2$, we have

$$\mu_{\bar{x}_1 - \bar{x}_2} = 0$$

By Rule 11 of Appendix C, the variance of the difference $\bar{x}_1 - \bar{x}_2$ is

(7.2) $$\sigma^2_{\bar{x}_1 - \bar{x}_2} = \sigma^2_{\bar{x}_1} + \sigma^2_{\bar{x}_2}$$

Equation (7.2) is an application of the principle that the variance of the difference between two *independent* random variables is equal to the sum of the variances of these variables. Taking the square root on both sides of (7.2), we obtain a formula for $\sigma_{\bar{x}_1-\bar{x}_2}$, the standard deviation of the difference $\bar{x}_1 - \bar{x}_2$,

(7.2a)
$$\sigma_{\bar{x}_1-\bar{x}_2} = \sqrt{\sigma^2_{\bar{x}1} + \sigma^2_{\bar{x}2}}$$

where $\sigma^2_{\bar{x}_1}$ and $\sigma^2_{\bar{x}_2}$ are simply the variances of the sampling distributions of $\bar{x}_1$ and $\bar{x}_2$. Since $\sigma^2_{\bar{x}_1} = \sigma^2_1/n_1$ and $\sigma^2_{\bar{x}_2} = \sigma^2_2/n_2$, Equation (7.2a) becomes

(7.3)
$$\sigma_{\bar{x}_1-\bar{x}_2} = \sqrt{\frac{\sigma^2_1}{n_1} + \frac{\sigma^2_2}{n_2}}$$

For the case under discussion, namely, that of large, independent samples, the sampling distribution of $\bar{x}_1 - \bar{x}_2$ is approximately normal by the central limit theorem argument. *In summary, therefore, if $\bar{x}_1$, and $\bar{x}_2$ are the means of two large, independent samples from populations with means μ_1 and μ_2 and standard deviations σ_1 and σ_2, respectively, and if we hypothesize that μ_1 and μ_2 are equal, then the sampling distribution of $\bar{x}_1 - \bar{x}_2$ may be approximated by a normal curve with mean $\mu_{\bar{x}_1-\bar{x}_2} = 0$ and standard deviation $\sigma_{\bar{x}_1-\bar{x}_2} = \sqrt{\sigma^2_1/n_1 + \sigma^2_2/n_2}$.* It is helpful to think of this sampling distribution as the frequency distribution that would be obtained by grouping the $\bar{x}_1 - \bar{x}_2$ values observed in repeated pairs of samples drawn independently from two populations having the same means.

The standard deviation $\sigma_{\bar{x}_1-\bar{x}_2}$ is referred to as the *standard error of the difference between two means*. We see from Equation (7.3) that we must know the population standard deviations in order to calculate this standard error. However, for *large samples*, we adopt the approximate procedure of using the sample standard deviations s_1 in place of σ_1 and s_2 for σ_2. The resulting estimated or approximate standard error is symbolized $s_{\bar{x}_1-\bar{x}_2}$ and may be written

(7.4)
$$s_{\bar{x}_1-\bar{x}_2} = \sqrt{\frac{s^2_1}{n_1} + \frac{s^2_2}{n_2}}$$

We can now proceed to establish the decision rule for the consulting firm problem. The test is clearly two-tailed because the hypothesis of equal population means would be rejected if $\bar{x}_1 - \bar{x}_2$ differed significantly from zero either by lying sufficiently far above or below it. The sampling distribution of $\bar{x}_1 - \bar{x}_2$ is shown in Figure 7-11. On the horizontal scale of the distribution is shown the difference between the sample means $\bar{x}_1 - \bar{x}_2$. As indicated, the mean of the distribution is equal to zero, or in other words, under the null hypothesis, the expected value of $\bar{x}_1 - \bar{x}_2$ is equal to zero. Another way of interpreting the zero is that under the null hypothesis $H_0: \mu_1 - \mu_2 = 0$, we have assumed that the mean wages of the populations of unskilled workers are the same in Area *A* and Area *B* for the industry in question. Since the significance level is 0.02, 1% of the area under the normal curve is shown in each tail. From Table A-5 of Appendix A we find that in a normal distribu-

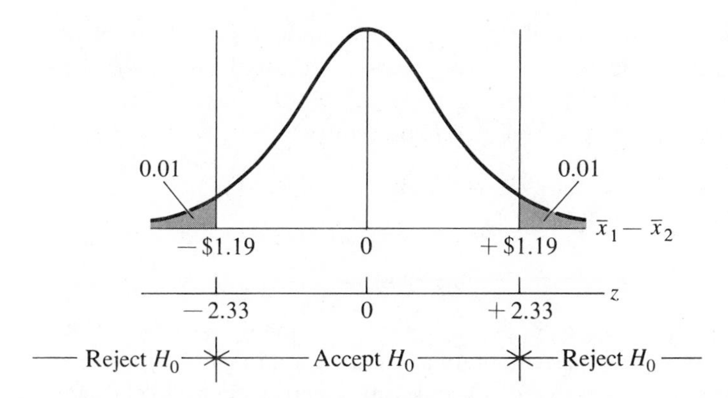

Figure 7-11 Sampling distribution of the difference between two means; two sided test; $\alpha = 0.02$.

tion 1% of the area lies to the right of $z = +2.33$ and by symmetry 1% lies to the left of $z = -2.33$. Thus, we would reject the null hypothesis if the sample difference $\bar{x}_1 - \bar{x}_2$ fell more than 2.33 standard errors from the expected value of zero. The estimated standard error of the difference between means, $s_{\bar{x}_1-\bar{x}_2}$ is by Equation 7.4 equal to

$$s_{\bar{x}_1-\bar{x}_2} = \sqrt{\frac{(\$4.00)^2}{100} + \frac{(\$4.50)^2}{200}} = \$0.51$$

and

$$2.33 s_{\bar{x}_1-\bar{x}_2} = (2.33)(0.51) = \$1.19$$

Thus, the decision rule may be stated as follows:

Decision Rule

(1) If $\bar{x}_1 - \bar{x}_2 < -\$1.19$ or $\bar{x}_1 - \bar{x}_2 > \$1.19$, reject H_0

(2) If $-\$1.19 \leq \bar{x}_1 - \bar{x}_2 \leq \$1.19$, accept H_0

In terms of z values, we have

Decision Rule

(1) If $z < -2.33$ or $z > +2.33$, reject H_0

(2) If $-2.33 \leq z \leq +2.33$, accept H_0

where

$$z = \frac{\bar{x}_1 - \bar{x}_2}{s_{\bar{x}_1-\bar{x}_2}}$$

The reader should note that this z value is in the usual form of the ratio (statistic − parameter)/standard error. The difference $\bar{x}_1 - \bar{x}_2$ is the statistic.

The parameter under test is $\mu_1 - \mu_2 = 0$, and thus need not be shown in the numerator of the ratio. As previously indicated, we have substituted an approximate standard error for the true standard error in the denominator.[4]

Using the decision rule in the present problem, we have

$$\bar{x}_1 - \bar{x}_2 = \$90.01 - \$85.21 = \$4.80$$

and

$$z = \frac{\bar{x}_1 - \bar{x}_2}{s_{\bar{x}_1 - \bar{x}_2}} = \frac{\$4.80}{\$0.51} = 9.4$$

Since \$4.80 far exceeds \$1.19 and correspondingly 9.4 far exceeds 2.33, the null hypothesis is rejected. Hence, it is extremely unlikely that these two samples were drawn from populations having the same mean. In terms of the problem, we conclude that the sample mean wages of unskilled workers in this industry *differed significantly* between Areas *A* and *B*. Since this discredits the hypothesis that the population mean wages are equal, we conclude that the population means *differ* between Area *A* and *B*. Note that it is incorrect to use the term "significant difference" when referring to the relationship between two population parameters, which in this case, are population means. Also, it is useful to observe that in this as in all other hypothesis testing situations, we are assuming random sampling. Obviously, if the samples were not randomly drawn from the two populations, the foregoing procedure and conclusion are invalid.

Test for Difference Between Proportions: Two-Tailed Test

Another important two-sample hypothesis testing case is one in which the observed statistics are proportions. The decision procedure is conceptually the same as in the case where the sample statistics are means; only the computational details differ. In order to illustrate the technique, let us consider the following example. Workers in the Stanley Morgan Company and Rock Hayden Company, two firms in the same industry, were asked whether they preferred to receive a specified package of increased fringe benefits or a specified increase in base pay. For brevity in this problem, we will refer to the companies as the S. M. Company and the R. H. Company and the proposed increases as "increased fringe benefits" and "increased base pay." In a simple random sample of 150 workers in the S. M. Company, 75 indicated

---

[4]Some authors reserve the symbol z for the case in which the population standard deviation is known and therefore the standard error in the denominator of the z ratio is also known. However, in this text, we use that symbol also in the case of large samples, where the population standard deviation is unknown and therefore an approximate standard error is substituted for a true standard error.

that they preferred increased base pay. In the R. H. Company, 103 out of a simple random sample of 200 preferred increased base pay. In each company, the sample was less than 5% of the total number of workers. It was desired to have a very low probability of erroneously rejecting the hypothesis of equal proportions in the two companies who preferred increased base pay. Therefore, a 1% level of significance was used for the test. Can it be concluded at the 1% level of significance that these two companies differed as regards the proportion of workers who preferred increased base pay?

Using the subscripts 1 and 2 to refer to the S. M. Company and R. H. Company, respectively, we can organize the sample data as follows:

S. M. Company	*R. H. Company*
$\bar{p}_1 = \frac{75}{150} = 0.50$	$\bar{p}_2 = \frac{103}{200} = 0.515$
$\bar{q}_1 = \frac{75}{150} = 0.50$	$\bar{q}_2 = \frac{97}{200} = 0.485$
$n_1 = 150$	$n_2 = 200$

where $\bar{p}_1$ and $\bar{q}_1$ refer to the sample proportions in the S. M. Company in favor of and not in favor of increased base pay, respectively. The sample size in the S. M. Company is denoted n_1. Corresponding notation is used for the R. H. Company. If we designate the population proportions in favor of increased pay in the two companies as p_1 and p_2, then in an analogous manner to the preceding problem, we set up the two alternative hypotheses

$$H_0: p_1 - p_2 = 0$$

$$H_1: p_1 - p_2 \neq 0$$

The underlying theory for the test is similar to that in the two-sample test for the difference between two means. If $\bar{p}_1$ and $\bar{p}_2$ are the observed sample proportions in large simple random samples drawn from populations with parameters p_1 and p_2, respectively, then the sampling distribution of the statistic $\bar{p}_1 - \bar{p}_2$ has a mean

(7.5) $$\mu_{\bar{p}_1-\bar{p}_2} = \mu_{\bar{p}_1} - \mu_{\bar{p}_2} = p_1 - p_2$$

and a standard deviation

(7.6) $$\sigma_{\bar{p}_1-\bar{p}_2} = \sqrt{\sigma^2_{\bar{p}_1} + \sigma^2_{\bar{p}_2}}$$

where $\sigma^2_{\bar{p}}$ and $\sigma^2_{\bar{p}_2}$ are the variances of the sampling distributions of $\bar{p}_1$ and $\bar{p}_2$. Under assumptions of a Bernoulli process, $\sigma^2_{\bar{p}_1} = p_1q_1/n_1$ and $\sigma^2_{\bar{p}_2} = p_2q_2/n_2$. Although the sampling was conducted without replacement, since each of the samples constituted only a small percentage of the corresponding population (less than 5%), the Bernoulli process assumption appears reasonable. Thus, Equation (7.6) becomes

(7.7)
$$\sigma_{\bar{p}_1-\bar{p}_2} = \sqrt{\frac{p_1q_1}{n_1} + \frac{p_2q_2}{n_2}}$$

If we hypothesize that $p_1 = p_2$, as in the null hypothesis in this problem, and if we refer to the common value of p_1 and p_2 as p, Equations (7.5) and (7.7) become

(7.8)
$$\mu_{\bar{p}_1-\bar{p}_2} = p - p = 0$$

and

(7.9)
$$\sigma_{\bar{p}_1-\bar{p}_2} = \sqrt{\frac{pq}{n_1} + \frac{pq}{n_2}} = \sqrt{pq\left(\frac{1}{n_1} + \frac{1}{n_2}\right)}$$

Since the common hypothesized proportion p under the null hypothesis is unknown, we estimate it for the hypothesis test by taking a weighted mean of the observed sample percentages. Referring to this "pooled estimator" as $\bar{p}$, we have

(7.10)
$$\bar{p} = \frac{n_1\bar{p}_1 + n_2\bar{p}_2}{n_1 + n_2}$$

The numerator of Equation (7.10) is simply the total number of "successes" in the two samples combined, and the denominator is the total number of observations in the two samples. The standard deviation in Equation (7.9), $\sigma_{\bar{p}_1-\bar{p}_2}$, is referred to as the *standard error of the difference between two proportions.* Substituting the "pooled estimator," $\bar{p}$ for p in (7.9), we have the following formula for the estimated or approximate standard error, $s_{\bar{p}_1-\bar{p}_2}$.

(7.11)
$$s_{\bar{p}_1-\bar{p}_2} = \sqrt{\bar{p}\bar{q}\left(\frac{1}{n_1} + \frac{1}{n_2}\right)}$$

We can now summarize these results. Let $\bar{p}_1$ and $\bar{p}_2$ be proportions of successes observed in two large, independent samples from populations with parameters p_1 and p_2. If we hypothesize that $p_1 = p_2 = p$, we obtain a pooled estimator $\bar{p}$ for p, where $\bar{p} = n_1\bar{p}_1 + n_2\bar{p}_2/n_1 + n_2$. *Then, the sampling distribution of* $\bar{p}_1 - \bar{p}_2$ *may be approximated by a normal curve with mean* $\mu_{\bar{p}_1-\bar{p}_2} = 0$ and estimated standard deviation,

$$s_{\bar{p}_1-\bar{p}_2} = \sqrt{\bar{p}\bar{q}\left(\frac{1}{n_1} + \frac{1}{n_2}\right)}$$

We may think of this sampling distribution as the frequency distribution of $\bar{p}_1 - \bar{p}_2$ values observed in repeated pairs of samples drawn independently from two populations having the same proportions.

Proceeding to the decision rule, we see again that the test is two-tailed because the hypothesis of equal population proportions would be rejected for $\bar{p}_1 - \bar{p}_2$ values which fall significantly above or below zero. The sampling distribution of $\bar{p}_1 - \bar{p}_2$ for the present problem is shown in Figure 7-12. Since

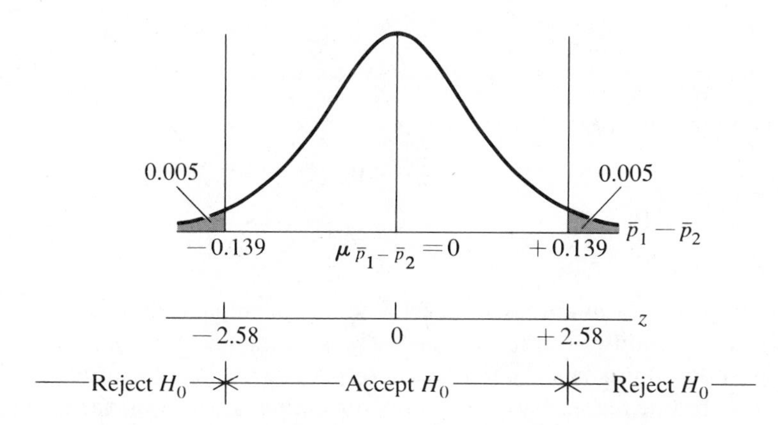

Figure 7-12 Sampling distribution of the difference between two proportions; two sided test; $\alpha = 0.01$.

the significance level $\alpha = 0.01$, $\alpha/2$ or 1/2 of 1% of the area under the normal distribution is shown in each tail. Referring to Table A-5, we find that 1/2 of 1% of the area in a normal curve lies above a z value of +2.58, and thus a similar percentage lies below $z = -2.58$. Hence, rejection of the null hypothesis $H_0: p_1 - p_2 = 0$ occurs if the sample difference $\bar{p}_1 - \bar{p}_2$ falls more than 2.58 standard error units from zero. By Equation (7.11), the standard error of the difference between proportions is

$$s_{\bar{p}_1-\bar{p}_2} = \sqrt{\bar{p}\bar{q}\left(\frac{1}{n_1} + \frac{1}{n_2}\right)} = \sqrt{(0.51)(0.49)\left(\frac{1}{150} + \frac{1}{200}\right)} = 0.054$$

where

$$\bar{p} = \frac{n_1\bar{p}_1 + n_2\bar{p}_2}{n_1 + n_2} = \frac{(150)(.50) + 200(.515)}{150 + 200} = \frac{75 + 103}{150 + 200} = 0.51$$

Hence,

$$2.58s_{\bar{p}_1-\bar{p}_2} = (2.58)(0.054) = 0.139$$

Therefore, the decision rule is the following

Decision Rule

(1) If $\bar{p}_1 - \bar{p}_2 < -0.139$ or $\bar{p}_1 - \bar{p}_2 > 0.139$, reject H_0

(2) If $-0.139 \leq \bar{p}_1 - \bar{p}_2 \leq 0.139$, accept H_0

In terms of z values, the rule is

Decision Rule

(1) If $z < -2.58$ or $z > +2.58$, reject H_0
(2) If $-2.58 \leq z \leq +2.58$, accept H_0

where

$$z = \frac{\bar{p}_1 - \bar{p}_2}{s_{\bar{p}_1 - \bar{p}_2}}$$

Applying this decision rule yields

$$\bar{p}_1 - \bar{p}_2 = 0.500 - 0.515 = -0.015$$

and

$$z = \frac{\bar{p}_1 - \bar{p}_2}{s_{\bar{p}_1 - \bar{p}_2}} = \frac{-0.015}{0.054} = -0.28$$

Thus, the null hypothesis is accepted. In summary, the sample proportions $\bar{p}_1$ and $\bar{p}_2$ did not differ significantly, and therefore we cannot conclude that the two companies differed with respect to the proportion of workers who preferred increased base pay. Our reasoning is based on the finding that assuming the population proportions were equal, a difference between the sample proportions of the size observed could not at all be considered unusual.

Test for Differences Between Proportions: One-Tailed Test

The two preceding examples illustrated two-tailed tests for cases where data are available for samples from two populations. Just as in the one-sample case, the question we wish to answer may give rise to a one-tailed test situation. In order to illustrate this point, let us examine the following problem.

Two competing drugs are available for treating a certain medical ailment. There are no apparent side-effects from administration of the first drug, whereas there are some definite side-effects of nausea and mild headaches in use of the second. A group of medical researchers has decided that it would nevertheless be willing to recommend use of the second drug in preference to the first if the proportion of cures effected by the second were higher than those by the first drug. The group felt that the potential benefits of achieving increased cures of the medical ailment would far outweigh the disadvantages of the possible accompanying side-effects. On the other hand, if the proportion of cures effected by the second drug was equal to or less than that of the first drug, the medical group would recommend use of the first. In terms of hypothesis testing, we can state the alternatives and consequent actions as

$$H_0: p_2 \leq p_1 \text{ (use the first drug)}$$

$$H_1: p_2 > p_1 \text{ (use the second drug)}$$

where p_1 and p_2 denote the population proportions of cures effected by the first and second drugs, respectively. Another way we may write these alternatives is

$$H_0: p_2 - p_1 \leq 0 \text{ (use the first drug)}$$

$$H_1: p_2 - p_1 > 0 \text{ (use the second drug)}$$

For purposes of comparison, we may note that the alternative hypotheses in the preceding problem, which was a two-tailed testing situation were

$$H_0: p_1 = p_2$$

$$H_1: p_1 \neq p_2$$

or in the alternative form in terms of differences

$$H_0: p_1 - p_2 = 0$$

$$H_1: p_1 - p_2 \neq 0$$

Clearly, the present problem involves a one-tailed test, in which we would reject the null hypothesis only if the sample difference, $\bar{p}_2 - \bar{p}_1$ differed significantly from zero and was a positive number.

The medical researchers used the drugs experimentally on two random samples of persons suffering from the ailment, administering the first drug to a group of 80 patients and the second drug to a group of 90 patients. By the end of the experimental period, 52 of those treated with the first drug were classified as "cured," whereas 63 of those treated with the second drug were so classified. The sample results may be summarized as follows:

First Drug	*Second Drug*
$\bar{p}_1 = \frac{52}{80} = 0.65$ cured	$\bar{p}_2 = \frac{63}{90} = 0.70$ cured
$\bar{q}_1 = \frac{28}{80} = 0.35$ not cured	$\bar{q}_2 = \frac{27}{90} = 0.30$ not cured
$n_1 = 80$	$n_2 = 90$

The pooled sample proportion cured is

$$\bar{p} = \frac{52 + 63}{80 + 90} = \frac{115}{170} = 0.676$$

and the estimated standard error of the difference between proportions is

$$s_{\bar{p}_2 - \bar{p}_1} = \sqrt{(0.676)(0.324)\left(\frac{1}{80} + \frac{1}{90}\right)} = 0.0719$$

Since the medical group wished to keep the probability low of erroneously adopting the second drug, it selected a 1% significance level for the test. One percent of the area under the normal curve lies to the right of $z = +2.33$. Therefore, the null hypothesis would be rejected if $\bar{p}_2 - \bar{p}_1$ falls at least 2.33 standard error units above zero. In terms of proportions,

$$2.33 s_{\bar{p}_2 - \bar{p}_1} = 2.33(0.0719) = 0.168$$

Hence, the decision rule is

Decision Rule

(1) If $\bar{p}_2 - \bar{p}_1 > 0.168$, reject H_0

(2) If $\bar{p}_2 - \bar{p}_1 \leq 0.168$, accept H_0

In terms of z values, the rule is

Decision Rule

(1) If $z > +2.33$, reject H_0

(2) If $z \leq +2.33$, accept H_0

where

$$z = \frac{\bar{p}_2 - \bar{p}_1}{s_{\bar{p}_2 - \bar{p}_1}}$$

In the present problem,

$$\bar{p}_2 - \bar{p}_1 = 0.70 - 0.65 = 0.05$$

and

$$z = \frac{0.70 - 0.65}{0.0719} = 0.70$$

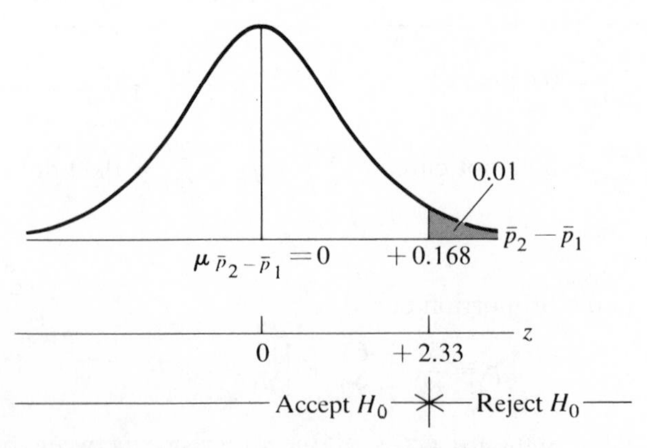

Figure 7-13 Sampling distribution of the difference between two proportions; one-sided test; $\alpha = 0.01$

Thus, the null hypothesis is accepted. On the basis of the sample data, we cannot conclude that the second drug accomplishes a greater proportion of cures than the first. The sampling distribution of $\bar{p}_2 - \bar{p}_1$ is given in Figure 7-13. The reader should note that it is immaterial whether the difference between proportions is stated as $\bar{p}_1 - \bar{p}_2$ or $\bar{p}_2 - \bar{p}_1$, but that care must be exercised concerning the correspondence between the way the hypothesis is stated, the sign of the difference between the sample proportions, and the tail of the sampling distribution in which rejection of the null hypothesis takes place.

7.4 The t-distribution — Population Standard Deviation(s) Unknown

The hypothesis testing methods we have discussed thus far are appropriate when the sample size is large. In this section, we concern ourselves with the case where the *sample size is small.* In order to understand the circumstances under which these small sample methods are suitable, it is useful to review some of the basic reasoning involved in large sample tests.

If a population is normally distributed, the sampling distribution of $\bar{x}$ is also normally distributed. If we form the statistic $z = (\bar{x} - \mu_x)/\sigma_{\bar{x}}$, then this statistic is normally distributed with a mean of zero and a standard deviation equal to one. As we have noted, the standard error of the mean in the denominator of the z statistic is related to the population standard deviation by the formula, $\sigma_{\bar{x}} = \sigma_x/\sqrt{n}$. Thus, the population standard deviation must be *known* to use the foregoing theory. In summary, if the population distribution is normal and the population standard deviation is *known,* the sampling distribution of $\bar{x}$ is also normal, and we can calculate the true standard error of the mean, $\sigma_{\bar{x}}$.

If the population distribution is non-normal, and the population standard deviation is *known,* the central limit theorem tells us that the sampling distribution of $\bar{x}$ approaches normality as the sample size is increased. Hence, for large samples, even when the population distribution is non-normal, the sampling distribution of $\bar{x}$ is approximately normally distributed. The same hypothesis testing methods are used as in the preceding case, where the population distribution was normal.

The case in which the distinction between large sample and small sample hypothesis testing methods becomes important is when the population standard deviation is *unknown* and therefore must be *estimated* from the sample observations. The main point is the fact that the ratio $(\bar{x} - \mu_x)/s_{\bar{x}}$ is not approximately normally distributed for all sample sizes, where $s_{\bar{x}}$ denotes an estimated standard error. As we have noted earlier, $s_{\bar{x}}$ is computed by the formula $s_{\bar{x}} = s_x/\sqrt{n}$, where s_x represents an estimate of the true population standard deviation. For large samples, the ratio $(\bar{x} - \mu_x)/s_{\bar{x}}$ is approximately a standard normal deviate, and we may use the methods discussed in Sections (7.2) and (7.3). However, since this statistic is not approximately normally distributed for small samples, the theoretically correct distribution, known as the t-distribution, must be used instead. Although the underlying mathematics involved in the derivation of the t-distribution is complex and properly beyond the scope of this text, it is possible to give an intuitive understanding of the nature of that distribution and its relationship to the normal curve.

The ratio $(\bar{x} - \mu_x)/s_{\bar{x}}$ is referred to as the "t-statistic." That is,

(7.12)
$$t = \frac{\bar{x} - \mu_x}{s_{\bar{x}}}$$

where

(7.13)
$$s_{\bar{x}} = \frac{s_x}{\sqrt{n}}$$

and

(7.14)
$$s_x = \sqrt{\frac{\Sigma\,(x - \bar{x})^2}{n - 1}}$$

Let us examine the t-statistic and its relationship to the standard normal deviate, $z = (\bar{x} - \mu_x)/\sigma_{\bar{x}}$. In the case of the standard normal deviate, since $\sigma_{\bar{x}} = \sigma_x/\sqrt{n}$, and σ_x is a constant, the z-statistic for any given sample size involves only one random variable, namely, $\bar{x}$, which appears in the numerator of the ratio. On the other hand, the t-statistic involves the ratio of two random variables. The random variable in the numerator is the same as in the case of the standard normal deviate. However, unlike the case of the z-statistic, the denominator is also a random variable. This follows from the fact that the population standard deviation σ_x is estimated by the use of s_x, which is computed by taking the sum of the squared deviations around the sample mean, $\bar{x}$. For any given sample of size n, the sample mean $\bar{x}$ is a random variable. In summary, since the t-statistic is a ratio of two random variables, its probability distribution has a different functional form from that of the z-statistic, which contains only one random variable.

One more comment is appropriate concerning the t-statistic before proceeding to the nature of the t-distribution. In the denominator of the t-statistic, s_x appears. It may be noted that an $n - 1$ is present in the formula for s_x, as contrasted to the n present in previous applications. We have earlier referred to s_x with an $n - 1$ in the denominator as an estimator of the population standard deviation. Its square, $s_x^2 = \Sigma(x - \bar{x})^2/(n - 1)$ is similarly an estimator of the population variance. It can be demonstrated mathematically that s_x^2 is an *unbiased* estimator of the population variance. That is, the expected value of this estimator is equal to the true population variance, σ_x^2.

Symbolically, $E(s_x^2) = \sigma_x^2$. Therefore, we can refer to s_x as the "square root of the unbiased estimator of the population variance." Strange as it may seem intuitively, although s_x^2 is an unbiased estimator of σ_x^2, s_x is not an unbiased estimator of σ_x. Nonetheless, s_x is a suitable estimator of σ_x and represents the correct mathematical functional form to appear in the denominator of the t-statistic. The number $n - 1$ is referred to as the *number of degrees of freedom.*

A brief explanation of the concept of "degrees of freedom" was given in Section 3.16, pp. 190. It is not feasible to give a single simple verbal explanation of this concept to cover all of its possible interpretations. For example, in Equation (7.15), ν can even be a non-integer, although for purposes of hypothesis testing, we are only interested in integral values. From a purely mathematical point of view, the number of degrees of freedom, ν, which appears in a probability distribution such as Equation (7.15) is simply a parameter of the distribution. However, in the present discussion, in which $s_x = \sqrt{\Sigma(x - \bar{x})^2/(n - 1)}$ is used as an estimator of the population standard deviation σ_x, the $n - 1$ may be interpreted as the number of independent deviations of the form $x - \bar{x}$ that are present in the calculation of s_x. Since the total of the deviations $\Sigma(x - \bar{x})$ for n observations is equal to zero, only $n - 1$ of them are independent. This means that if we were free to specify the deviations $x - \bar{x}$, we could designate only $n - 1$ of them independently. The nth one would be determined by the condition that the n deviations have to add up to zero. Therefore, in the estimation of a population standard deviation or a population variance, if an $n - 1$ divisor is used in the estimator, the terminology is that there are $n - 1$ degrees of freedoms present.

The t-distribution has been derived mathematically under the assumption of a normally distributed population. It has the following form

(7.15)
$$f(t) = c\left(1 + \frac{t^2}{\nu}\right)^{-(\nu+1)/2}$$

where

$t = \dfrac{\bar{x} - \mu_x}{s_{\bar{x}}}$, as previously defined

c = a constant required to make the area under the curve equal to unity

$\nu = n - 1$, the number of degrees of freedom

The variable t ranges from minus infinity to plus infinity. The constant c is actually a function of ν, (pronounced "nu"), so that for a particular value of ν, the distribution of $f(t)$ is completely specified. Thus, $f(t)$ is a family of functions, one for each value of ν. Just as is true of the standard normal distribution, the t-distribution is symmetrical and has a mean of zero. However, the variance of the t-distribution is greater than one, but approaches one

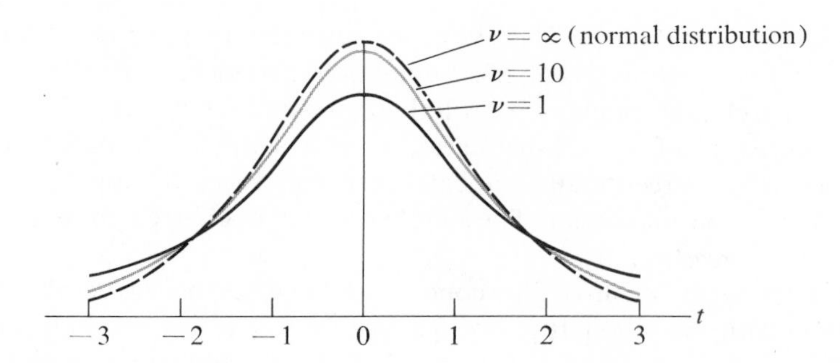

Figure 7-14 The t-distributions for $\nu = 1$ and $\nu = 10$ compared to the normal distribution ($\nu = \infty$).

as the number of degrees of freedom, and, therefore, the sample size, becomes large. Thus, the variance of the t-distribution approaches the variance of the standard normal distribution as the sample size increases. It can be demonstrated mathematically, that in the limit, that is, for an infinite number of degrees of freedom ($\nu = \infty$), the t-distribution and normal distribution are exactly equal. The approach to this limit is quite rapid. Hence, there is a widely practiced rule of thumb that samples of size $n > 30$ may be considered large and the standard normal distribution may appropriately be used as an approximation to the t-distribution, where the latter is the theoretically correct functional form. In Figure 7-14 are shown the graphs of several t-curves for different numbers of degrees of freedom. As can be seen from these graphs, the t-curves are lower at the mean and higher in the tails than is the standard normal distribution. As the number of degrees of freedom increases, the t-distribution rises at the mean and lowers at the tails until for an infinite number of degrees of freedom, it coincides with the normal distribution.

Further Remarks. Early work on the t-distribution was carried out by W. S. Gossett, who wrote under the pseudonym of "Student." Gossett was an employee of Guinness Brewery in Dublin. Since the brewery did not permit publication of research findings by its employees under their own names, Gossett adopted "Student" as a pen name. Consequently in addition to the term "t-distribution" used here, the distribution has come to be known as "Student's distribution" or "Student's t-distribution," and is so referred to in many other books and journals.

Hypothesis Testing Using the t-distribution: Small Samples, Population Standard Deviation(s) Unknown

Example 7-3 One Sample Test of a Hypothesis about the Mean: Two-Sided Test

The personnel department of a company developed an aptitude test for a certain type of semi-skilled worker. The developers of the test asserted a tentative hypothesis that the arithmetic mean grade obtained by this type of semi-skilled worker would be 100. It was agreed that this hypothesis would be subjected to a two-tailed test at the 5% level of significance. The aptitude test was given to a simple random sample of 16 of the semi-skilled workers with the following results:

$$\bar{x} = 94$$
$$s_x = 5 \text{ [computed by formula (7.14)]}$$
$$n = 16$$

The competing hypotheses are

$$H_0: \mu_x = 100$$
$$H_1: \mu_x \neq 100$$

To carry out the test, the following were calculated:

$$s_{\bar{x}} = \frac{s_x}{\sqrt{n}} = \frac{5}{\sqrt{16}} = 1.25$$

and

$$t = \frac{\bar{x} - \mu_x}{s_{\bar{x}}} = \frac{94 - 100}{1.25} = -4.80$$

Since the significance of this t value is judged from Table A-6 in Appendix A, a brief explanation of this table is required. In the table of areas under the normal curve, areas lying between the mean and specified z values were given. However, in the case of the t-distribution, since there is a different t curve for each sample size, no single table of areas can be given for all of these distributions. Therefore, for compactness, what is shown in a t table is the relationship between areas and t values for only a few "percentage points" in different t-distributions. Specifically, the entries in the body of the table are t values for selected areas in the two tails of the distribution combined. As an illustration of the use of the table, let us set up the areas of acceptance and rejection of the null hypothesis for the present problem. Since the sample size is 16, the number of degrees of freedom, $\nu = n - 1$, is $\nu = 16 - 1 = 15$. The heading of the left-hand column of Table A-6 shows the number of degrees of freedom. Hence, looking along the row for 15 degrees of freedom, under the column .05 we find the t value, 2.131. This means that in a t-distribution for 15 degrees of freedom, the probability is 5% that t is greater than 2.131 or is less than -2.131. Thus, in the present problem, at the 5% level of significance, the null hypothesis, $H_0: \mu_x = 100$ is rejected if a t value is observed which exceeds 2.131 or is less than -2.131. Since the computed t value in this problem is -4.80, the null hypothesis is rejected. In other words, we are unwilling to attribute the difference between our sample mean of 94 and the hypothesized population mean of 100 merely to chance errors of sampling. The t-distribution for this problem is shown in Figure 7-15.

A few remarks can be made about this problem. Since the computed t value of -4.80 was such a large negative number, the null hypothesis would have been rejected even at the 2% or 1% levels of significance (see Appendix Table A-6).

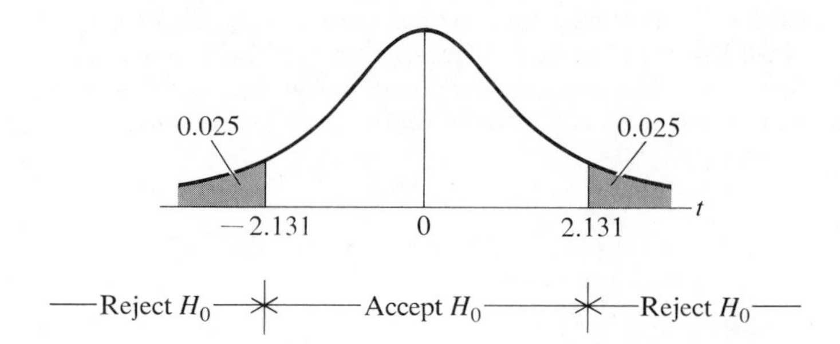

Figure 7-15 The t-distribution for 15 degrees of freedom.

Had the test been one-tailed at the 5% level of significance, we would have had to obtain the critical t value by looking under 0.10 in the caption of Table A-6, since the 0.10 figure is the combined area in both tails. Thus, for a one-tailed test at the 5% level of significance and a lower tail rejection region, in the present problem, the critical t value would have been -1.753.

It is interesting to compare these critical t values with analogous critical z values for the normal curve. From Table A-5 of Appendix A, we find that the critical z values at the 5% level of significance are -1.96 and 1.96 for a two-tailed test and -1.65 for a one-tailed test with a lower tail rejection region. As we have just seen, the corresponding figures for the critical t values in a test involving 15 degrees of freedom are -2.131 and 2.131 for a two-tailed test and -1.753 for a one-tailed test with a lower tail rejection region.

An underlying assumption is applying the t-test in this problem is that the population is closely approximated by a normal distribution. Since the population standard deviation, σ_x, is unknown, the t-distribution is the theoretically correct sampling distribution. However, if the sample size had been large, even with an unknown population standard deviation, the normal curve could have been used as an approximation to the t-distribution. As we saw in this problem, for 15 degrees of freedom, a combined total of 5% of the area in the t-distribution falls to the right of $t = +2.131$ and to the left of $t = -2.131$. The corresponding z values in the normal distribution are $+1.96$ and -1.96. As can be seen from Table A-6, the t value entry for 30 degrees of freedom is 2.042. The closeness of this figure to $+1.96$ gives rise to the usual rule of thumb of $n > 30$ as the arbitrary dividing line between large sample and small sample methods. We use this convenient rule in Chapters 7 and 8. However, what constitutes a suitable approximation really depends on the context of the particular problem. Furthermore, if the population is highly skewed, a sample size as large as 100 may be required for the assumption of a normal sampling distribution of $\bar{x}$ to be appropriate.

Example 7-4 Two-Sample Test for Means: Two-Sided Test Small simple random samples of the freshmen and senior classes were drawn at a large university and the amounts of money (excluding checks) that these individuals had on their persons were determined. The following statistics were calculated:

Freshmen	*Seniors*
$\bar{x}_1 = \$1.28$	$\bar{x}_2 = \$2.02$
$s_1 = 0.51$	$s_2 = 0.43$
$n_1 = 10$	$n_2 = 12$

As in Example 7-3, the sample standard deviations were computed by Equation (7.14) with an $n - 1$ divisor. At the 2% level of significance should we conclude that a significant difference was observed between the sample means?

The alternative hypotheses are

$$H_0: \mu_1 - \mu_2 = 0$$

$$H_1: \mu_1 - \mu_2 \neq 0$$

To test the null hypothesis, we use the t-statistic

$$t = \frac{(\bar{x}_1 - \bar{x}_2) - 0}{s_{\bar{x}_1 - \bar{x}_2}} = \frac{\bar{x}_1 - \bar{x}_2}{s_{\bar{x}_1 - \bar{x}_2}}$$

where $s_{\bar{x}_1 - \bar{x}_2}$ is the estimated standard error of the difference between two means.

Unlike the case of the large sample approach it is necessary here to assume equal population variances. An estimate of this common variance is obtained by pooling the two sample variances into a weighted average, using the numbers of degrees of freedom, $n_1 - 1$ and $n_2 - 1$, as weights. This pooled estimate of the common variance, which we will denote as s^2, is given by

$$s^2 = \frac{(n_1 - 1)s_1^2 + (n_2 - 1)s_2^2}{n_1 + n_2 - 2} \qquad (7.16)$$

Since the variance of the difference of two independent random variables is equal to the sum of the variances of these variables,

$$s_{\bar{x}_1 - \bar{x}_2}^2 = s_{\bar{x}_1}^2 + s_{\bar{x}_2}^2 = \frac{s^2}{n_1} + \frac{s^2}{n_2} \qquad (7.17)$$

The estimated standard error of the difference between two means is, therefore,

$$s_{\bar{x}_1 - \bar{x}_2} = \sqrt{\frac{s^2}{n_1} + \frac{s^2}{n_2}} = s\sqrt{\frac{1}{n_1} + \frac{1}{n_2}} \qquad (7.18)$$

A number of alternative mathematical expressions are possible for Equation (7.18), but because of its similarity in form to previously used standard error formulas, we shall use it in this form.

We now proceed to work out the present problem. Substitution into (7.16) gives

$$s^2 = \frac{9(0.51)^2 + 11(0.43)^2}{10 + 12 - 2} = .2187$$

and

$$s = \sqrt{.2187} = \$0.47$$

The estimated standard error is

$$s_{\bar{x}-\bar{x}_2} = (\$0.47)\sqrt{\frac{1}{10} + \frac{1}{12}} = \$0.20$$

Thus, the t value is

$$t = \frac{\$2.02 - \$1.28}{\$0.20} = \frac{\$0.74}{\$0.20} = 3.70$$

The number of degrees of freedom in this problem is $n_1 + n_2 - 2$, that is, $10 + 12 - 2 = 20$. One way of viewing the number of degrees of freedom in this case is as follows. In the one-sample case, where the sample standard deviation was used as an estimate of the population standard deviation, there was a loss of one degree of freedom; hence, the number of degrees of freedom was $n - 1$. In the two-sample case, each of the sample variances was used in the pooled estimate of population variance; hence, two degrees of freedom were lost, and the number of degrees of freedom is $n_1 + n_2 - 2$.

The critical t value at the 2% significance level for 20 degrees of freedom is 2.528 (See Appendix Table A-6). Since the observed t value, 3.70, exceeds this critical t value, the null hypothesis is rejected and we conclude on the basis of the sample data that the population means are indeed different. In terms of the problem, we are unwilling to attribute the difference in average amount of pocket money between freshmen and seniors at this university to chance errors of sampling. On the basis of this test we tentatively conclude that seniors at the university carry more pocket change than do freshmen.

7.5 The Design of a Test to Control Both Type I and Type II Errors

Relatively simple hypothesis testing situations have been considered thus far in this chapter in order to convey the basic principles of classical hypothesis testing. These tests have assumed that sample size had been determined in advance, and that only the risks of Type I errors were to be controlled formally by the decision procedure. Of course, it was assumed that a power curve would be constructed in all cases, and, therefore, an examination of the risks of Type II errors would be made for parameter values not included in the null hypothesis. However, nothing was included in the formal testing procedure to control the level of risk of a Type II error for any specific parameter value.

In this section, we consider a method of controlling the levels of both Type I and Type II errors simultaneously in the same test. In the previous tests, which controlled only Type I errors, that is, where only α was specified, one point on the power curve was determined. In a test designed to control both Type I and Type II errors, two specific points on the power curve are determined.

As an illustration, we return to the one-tailed test problem concerning a mean discussed in Section 7.2. That problem involved an acceptance sampling procedure in which $\alpha = 0.05$ for an incoming shipment whose parts have a mean heat resistance of 2250°F. This is what we referred to in quality control terminology as the "producer's risk." Thus, the producer ran a risk of a 5% probability of having a "good" shipment with $\mu_x = 2250$°F rejected in error. As we saw in that problem, this was the maximum risk of rejecting a good shipment, because the probability of erroneous rejection dropped below 5% for shipments with means in excess of 2250°F. Suppose now it was agreed to fix the "consumer's risk" or β at 0.03 for a shipment whose mean was 2150°F. That is, a shipment whose parts have a mean heat resistance of 2150°F does not meet specifications. We want the probability of erroneously accepting such a shipment to be 0.03. We continue to assume that the population standard deviation is $\sigma_x = 300$°F. The solution to this problem involves the determination of the *sample size* required to give the desired levels of control of both types of errors. After the required sample size has been calculated, the appropriate decision rule can be specified. We assume the sample size will be large enough to use normal sampling distributions for $\bar{x}$.

Figure 7-16 gives a graphic representation of the error controls specified in the preceding paragraph, and related information. In the original solution to this problem involving control of only Type I errors, the critical value for

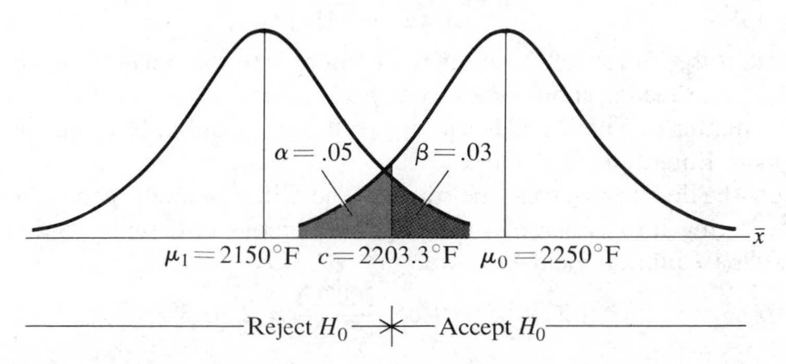

Figure 7-16 Acceptance sampling problem: control of Type I and Type II errors.

acceptance and rejection of the null hypothesis $H_0: \mu_x \geq 2250°F$ was 2200°F. The critical value is now unknown and will have to be evaluated. Let us denote the new critical value as C, the mean of 2250°F under the null hypothesis as μ_0, and the mean of 2150°F under the alternative hypothesis as μ_1. The area in the left tail of the sampling distribution of $\bar{x}$ when $\mu_0 = 2250°F$ is shown as .05 denoting the Type I error. As previously determined, this means that the critical point C lies 1.65 standard error units to the left of the mean $\mu_0 = 2250°F$. Therefore, we denote $z_0 = 1.65$. Under the alternative $\mu_1 = 2150°F$, we wanted the probability that an $\bar{x}$ value would lie in the acceptance region to be .03. From Table A-5, we ascertain that .03 of the area in a normal sampling distribution lies to the right of a value 1.88 standard error units above the mean. Hence, we denote $z_1 = 1.88$. Therefore, we can write the following relationships for the critical point C

$$C = \mu_0 - z_0\sigma_{\bar{x}} = \mu_0 - z_0 \frac{\sigma_x}{\sqrt{n}} \tag{7.19}$$

$$C = \mu_1 + z_1\sigma_{\bar{x}} = \mu_1 + z_1 \frac{\sigma_x}{\sqrt{n}} \tag{7.20}$$

Substituting the numerical values for this problem, we obtain

$$C = 2250°F - 1.65 \frac{(300°F)}{\sqrt{n}}$$

$$C = 2150°F + 1.88 \frac{(300°F)}{\sqrt{n}}$$

Setting the right-hand sides of these two equations equal to one another yields the solution $n = 112$ (to the nearest integer).

Therefore, a simple random sample of 112 parts from the incoming shipment would be required in order to obtain the desired levels of error control.

We can express this required sample size by solving the simultaneous Equations (7.19) and (7.20). The solution is

$$n = \left[\frac{(z_0 + z_1)\sigma_x}{(\mu_0 - \mu_1)}\right]^2 \tag{7.21}$$

Of course, substitution of the numerical values into this formula again yields $n = 112$. The reader should note that both z_0 and z_1 are taken as positive, since the matter of whether C lies above or below μ_0 and μ_1 is taken care of by the signs in Equations (7.19) and (7.20).

Now the decision rule can be stated. The critical value C can be obtained by substituting into either of the two simultaneous equations. Substituting into the first equation yields[5]

$$C = 2250°F - 1.65 \frac{(300°F)}{\sqrt{112}} = 2203.3°F$$

[5]Actually the values $C = 2203.2°F$ and $2203.4°F$ are obtained from the first and second equations, respectively. The discrepancy is due to rounding off n to an integral value.

Hence, the required decision rule is

Decision Rule

(1) If $\bar{x} < 2203.3°F$, reject H_0 (reject the shipment)

(2) If $\bar{x} \geq 2203.3°F$, accept H_0 (accept the shipment)

The power curve for this test is displayed in Figure 7-17.

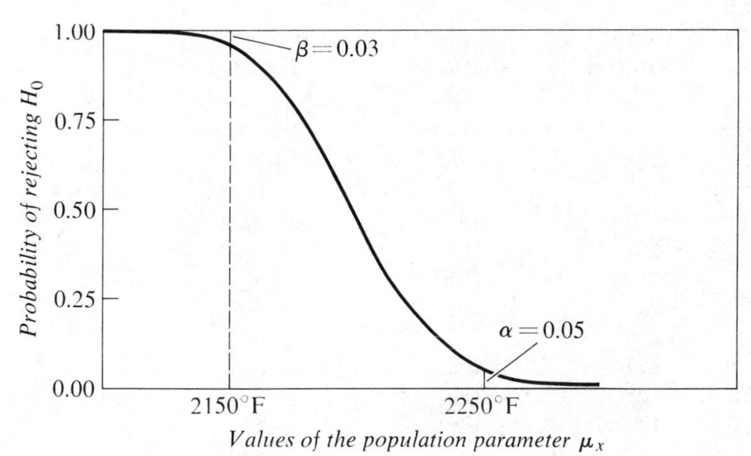

Figure 7-17 Power curve to control both Type I and Type II errors.

This one-tailed test for a mean was given as an illustration of the general method for controlling the levels of both Type I and Type II errors. Analogous tests can be constructed for proportions and for two-tailed tests as well.

In the type of quality control problem described here, the producer and the consumer might negotiate to determine the levels of Type I and Type II risks that they would agree to tolerate. These tolerable risks in turn would primarily depend upon the costs involved to the producer and consumer of these two types of errors.

7.6 The t-test for Paired Observations

In the two-sample tests considered thus far, the two samples had to be considered as independent. That is, it was necessary that the values of observations in one sample be *independent* of the values in the other. Situations arise in practice in which this condition does not hold. In fact, the two samples may consist of pairs of observations made on the same individual, object, or more generally, on the same selected population elements. Clearly, the independence condition is violated in these cases.

As a concrete example, let us consider a case in which a group of ten men was given a special diet, and it was desired to test weight loss in pounds at the end of a two-week period. The observed data are shown in Table 7-3.

Table 7-3 Weights Before and After a Special Diet for a Simple Random Sample of Ten Men.

Man	*Weight Before Diet* X_1	*Weight After Diet* X_2	*Difference in Weight* $d = X_2 - X_1$	$d - \bar{d}$	$(d - \bar{d})^2$
1	181	178	−3	+1	1
2	172	172	0	+4	16
3	190	185	−5	−1	1
4	187	184	−3	+1	1
5	210	201	−9	−5	25
6	202	201	−1	+3	9
7	166	160	−6	−2	4
8	173	168	−5	−1	1
9	183	180	−3	+1	1
10	184	179	−5	−1	1
			−40		60

$$\bar{d} = \frac{-40}{10} = -4 \text{ pounds}$$

$$s_d = \sqrt{\frac{\Sigma (d - \bar{d})^2}{n - 1}} = \sqrt{\frac{60}{10 - 1}} = 2.58 \text{ pounds}$$

$$s_{\bar{d}} = \frac{s_d}{\sqrt{n}} = \frac{2.58}{\sqrt{10}} = 0.82 \text{ pound}$$

As indicated in Table 7-3, X_1 denotes the weight before the diet and X_2 the weight after the diet. It would be incorrect to run a t-test to determine whether there is a significant difference between the mean of the X_1's and the mean of the X_2's, because of the non-independence of the two samples. Specifically, the first individual's weight after the test is certainly not independent of his weight before the test, etc. Each of the d values, $d = X_2 - X_1$ represents a difference between two observations on the same individual. The assumption is made that the subtraction of one value from the other removes the effect of factors other than that of the diet. Another way of stating this is that we assume that these other factors affect each member of any pair of X_1 and X_2 values in the same way.

We can state the alternative hypotheses to be tested as

$$H_0: \mu_2 - \mu_1 = 0$$

$$H_1: \mu_2 - \mu_1 < 0$$

where μ_1 and μ_2 are the population mean weights before and after the diet, respectively. Let us assume the test is to be carried out with $\alpha = 0.05$. The null hypothesis states that there is no difference between mean weight after the diet and mean weight before diet, whereas the alternative hypothesis says that mean weight after the diet is less than mean weight before. We can visualize the situation as one in which, if the null hypothesis is true, there is a population of numbers representing differences in the weight after the diet and before the diet, and the mean of these numbers is zero. Hence, we wish to test the hypothesis that our simple random sample of $d = X_2 - X_1$ values comes from this universe. The procedure used in these cases is to obtain the mean of the sample differences and to test whether this average, $\bar{d}$, differs significantly from zero. The estimated standard error of $\bar{d}$, denoted $s_{\bar{d}}$ is given by

$$s_{\bar{d}} = \frac{s_d}{\sqrt{n}}$$

where

$$s_d = \sqrt{\frac{\Sigma (d - \bar{d})^2}{n - 1}}$$

In this problem, $s_{\bar{d}} = 0.82$ pounds, as shown in the calculations in Table 7-3. Assuming the population of differences (d values) is normally distributed, the ratio $(\bar{d} - 0)/s_{\bar{d}}$ is a t-variate. Hence,

$$t = \frac{\bar{d} - 0}{s_{\bar{d}}} = \frac{-4}{0.82} = -4.88$$

The number of degrees of freedom is $n - 1$, where n is the number of d values. Hence, in this problem, $n - 1 = 10 - 1 = 9$. The test is one-tailed, because only a $\bar{d}$ value which is negative and significantly different from zero could result in acceptance of the alternative hypothesis, $H_1: \mu_2 - \mu_1 < 0$. Since $\alpha = 0.05$ and the test is one-tailed, we enter Table A-6 under the heading 0.10. The critical t value for 9 degrees of freedom is 1.833, which for our purposes is interpreted as -1.833. Since the observed t value of -4.88 is less than (lies to the left of) this critical point, the null hypothesis is rejected and we accept its alternative. Therefore, on the basis of this experiment, we conclude that the special diet does result in an average weight loss over a two-week period.

This method of pairing observations is also used to reduce the effect of extraneous factors which could cause a significant difference in means, whereas the factor whose effect we are really interested in may not have resulted in such a difference if this effect had been isolated.

For example, if medical experimenters wanted to test two different treatments to judge which was better, they might administer one treatment to one group of persons and the other treatment to a second group. Suppose on the basis of the usual significance test for means, it is concluded that one treatment is better than the other. Let us also assume that the group which received the supposedly better treatment was much younger and much healthier at the beginning of the experiment than the other group and that these factors could have an effect on the reaction to the treatments. Then clearly, the relative effectiveness of the two treatments would be obscured.

On the other hand, assume individuals were selected in pairs in which both members were about the same age and in about the same health condition. If the first treatment is given to one member of a pair, and the second treatment to the other, and then a difference measure is calculated for the effect of treatment, neither age nor health condition would affect this measurement. Ideally, we would like to select pairs which are identical in all characteristics other than the factor whose effects we are attempting to measure. Obviously, as a practical matter this is impossible, but the guiding principle is clear. Once differences are taken between members of each pair, the t-test proceeds exactly as in the preceding example. It may be noted that in the weight example, the differences were measured on the same individual, whereas in the present illustration the differences are derived from the two members of each pair.

The method of paired observations is a very useful technique. As compared to the standard two-sample t-test, in addition to the advantage that we do not have to assume the two samples are independent, we also need not assume that the variances of the two samples are equal.

7.7 Summary and Looking Ahead

In this chapter, we have considered some classical hypothesis testing techniques. These tests represent only a few of the simplest methods. All of the cases we have discussed thus far have involved only one or two samples. Methods are available for testing hypotheses concerning three or more samples. The cases we have dealt with thus far have tested only one parameter of a probability distribution. Techniques are available for testing whether an entire frequency distribution is in conformity with a theoretical model, such as a specified probability distribution. The tests we have considered involved a terminal decision on the basis of the sample evidence. That is, a decision concerning acceptance or rejection of hypotheses was reached on the basis of the evidence contained in one or two samples. Sequential decision procedures are available which permit postponement of decision pending further sample evidence. Some of the broader decision procedures are discussed in subsequent chapters.

Although classical hypothesis testing techniques of the type discussed in this chapter have been widely applied in a great many fields, it would be incorrect to infer that their use is non-controversial and that they can simply be employed in a mechanistic way. At this point, it suffices to indicate that the methods discussed are admittedly incomplete, and that Bayesian decision theory addresses itself to the required completion. Thus, for example, in hypothesis testing, the establishing of significance levels such as 0.05 or 0.01 inevitably appears to be a rather arbitrary procedure, despite the fact that the relative seriousness of Type I and Type II errors is supposed to be considered in designing a test. Although costs of Type I and Type II errors can theoretically be considered in the classical formulation, as a matter of actual practice, they are rarely included explicitly in the analysis. In Bayesian decision theory, the costs of Type I and Type II errors, as well as the payoffs of correct decisions, become an explicit part of the formal analysis.

Also, in classical hypothesis testing, decisions are reached solely on the basis of the present sample information without reference to any prior knowledge which may exist concerning the hypothesis under test. On the other hand, Bayesian decision theory provides a method for combining prior knowledge with current sample information for decision-making purposes. These Bayesian decision theory methods are discussed in subsequent chapters.

PROBLEMS

1. Distinguish between a parameter and a statistic.
2. Support or criticize the following statement: "A statistician will make very few errors if he sets a significance level (α) very low, say .001; hence he should always do so."
3. State null and alternative hypotheses in the following situations.
   (a) A production manager wishes to test his company's production process against the *Consumer Reports* claim that the company fills its 10-ounce packages with less than 10 ounces.
   (b) Jones Political Advisory Service claims the percentage of people who favor the Republican presidential candidate in New York State is the same as that in New Jersey.

(c) A safety engineer doubts a glove manufacturer's claim that the average width of a certain welding glove is 2″.
(d) The proportion of defective items produced by a certain process is reported to be at most 10%. A prospective buyer wishes to test this claim, using a low probability of rejecting the claim erroneously.

4. Agree with or criticize the following statements.
   (a) It is more important to control the Type I error than the Type II error, hence we should design our tests on the basis of controlling the Type I error.
   (b) It is impossible to control both the Type I and Type II errors, since to decrease one increases the other.
   (c) A Type I error occurs when we reject the null hypothesis incorrectly, and a Type II error occurs when we accept the null hypothesis incorrectly.
   (d) Once a decision has been made, we must then consider the possibility that Type I and Type II errors may both have occurred.
5. Distinguish briefly between
   (a) One-sided alternative and two-sided alternative.
   (b) Type I and Type II errors.
6. What is meant by "the power of a test"? How is the power related to the Type II error?
7. There are three alternative hypotheses, H_1, H_2, H_3, and one null hypothesis H_0. R is the rejection region of a test statistic X.
   $\text{Prob}\{X \in R \mid H_0\} = .07$
   $\text{Prob}\{X \in R \mid H_1\} = .62$
   $\text{Prob}\{X \in R \mid H_2\} = .88$
   $\text{Prob}\{X \in R \mid H_3\} = .95$
   (a) Find the power of this test for H_1.
   (b) Find the probability of a Type I error.
   (c) Find the β level for H_3.

   NOTE: $\text{Prob}\{X \in R \mid H_0\}$ is read "the probability that the test statistic X lies in the rejection region R given H_0."
8. Agree with or criticize the following statements.
   (a) The power curve has the probability of rejecting the null hypothesis on one axis and the possible values of the parameter on the other, while the operating characteristic curve has the possible values of the statistics on one axis and the probability of rejecting the null hypothesis on the other.
   (b) Beta represents the probability that one will reject the null hypothesis incorrectly.
   (c) If a person has a choice of applying two different tests to decide between a certain null hypothesis and alternative hypothesis, both tests having the same α level, then he should use the one with the smaller β error.
9. The Consumer Fraud Council claims that Skimpy Foods Inc. does not put the required weight of peanut butter in its 10-ounce bottle. A sample of 400 jars

is selected and weighed. From past experience it is known that the population standard deviation of weights of peanut butter in bottles is .3 ounce. Set up a test such that if the firm on the average places the required weight in its 10-ounce bottle, one would accuse it unjustly only once in 100 times.

10. A manufacturer claims that the average life of a certain type of transistor is at least 150 hours. It is known that the standard deviation of this type of transistor is 20 hours. A consumer wishes to test the manufacturer's claim, and accordingly tests 100 transistors. State the null hypothesis, the alternative hypothesis, and the decision rule. Use an α level of 5%, i.e., set up a rule which would reject the manufacturer's claim only 5% of the time if it is true.

11. Under the standard manufacturing process, the breaking strength of nylon thread is a random variable with mean 100 and standard deviation 5 (in appropriate units). A new cheaper process is tested; a sample of 25 threads is drawn, with the results: $\bar{x} = 98.4$, $s = 5.5$. Your assistant has noted that, "Since the sample mean is not significantly less than 100 at $\alpha = .05$, we have strong evidence that we should install the new process."
    (a) Is his statement about significance correct?
    (b) Suppose that a true mean strength of 98 would lead to large savings relative to process costs. Should this affect your assistant's conclusion? Why?

12. The specifications for a component of a spacecraft provide that the mean length should not be less than 101 millimeters with a standard deviation of 9 millimeters. A random sample of 100 components drawn from a very large shipment yields a mean length of 99 millimeters and a standard deviation of 9 millimeters.
    (a) If you were the manufacturer of spacecraft who had received this shipment would you accept it? Justify the position you take.
    (b) What is the probability of making a Type II error for a shipment of components having a mean length of 98 millimeters and a standard deviation of 9 millimeters if your action limit had been set at 99.04 millimeters?

13. A company has purchased a large quantity of steel wire. The supplier claims that the wire has a mean tensile strength of 80 pounds or more. The company tests a sample of 64 pieces of the wire and finds $\bar{x} = 79.1$ and $s = 5.6$ pounds.
    (a) Should the company dispute the claimed mean tensile strength on the basis of this evidence? Use a one-tailed test with a .05 level of significance. For this problem state *carefully*
        (1) The null and alternative hypotheses; and
        (2) the decision rule employed.
    (b) Assume that a subsequent shipment has in fact $\mu = 78.5$ and $\sigma = 5.6$. The company, unaware of this, employs the same test procedures and decision rule as above. What is the probability that the company will arrive at the correct decision with respect to the claimed mean tensile strength of 80 pounds?

14. A housing survey is to be undertaken to determine whether the proportion of substandard dwelling units in a certain large city has changed. At the time of the last Census of Housing, 10% of the dwelling units were classified as substandard. It is desired to maintain a .01 risk of erroneously concluding that a change has occurred in the proportion of substandard dwelling units. A simple random sample of 900 dwelling units is contemplated.
    (a) Set up the decision rule for this test.
    (b) Calculate the probability of concluding that no change has taken place if the true population proportion of substandard dwelling units is 11%. What type of error is involved in this erroneous conclusion of "no change"?

15. In a certain year the arithmetic mean interest rate on loans to all large retailers (i.e., those with assets of $5,000,000 or more) was 6.0% and the standard deviation was 0.2 percentage points. Two years later, a simple random sample of 100 loans to large retailers yielded an arithmetic mean interest rate of 6.015%.
    (a) Would you be willing to conclude that there has been a change in the average level of interest rates for large retailers? Assume you are willing to run a 5% risk of concluding there has been a change when in fact there has been no change.
    (b) Using the same decision making rule as in (a) above, what is the probability of making a Type II error if the average interest rate for all large retailers was 6.01%?
    (c) Explain *specifically in terms of this problem* the meaning of the probability you computed in (b).

16. Last year a wholesale distributor found that the mean sales per invoice was $60, the standard deviation $20. This year, a random sample of 400 invoices is to be drawn in order to test the hypothesis that the mean sale per invoice has not changed. It is assumed that σ will not change. The acceptance region of the test is agreed to be

    If $58.72 < \bar{x} < 61.28$, accept H_0

    (a) Suppose that in fact this year, $\mu = 61$. What is the probability of accepting H_0?
    (b) Calculate and explain the meaning of the power of the test when $\mu = 61$.
    (c) What level of significance was used for the test?

17. A package-filling device is set to fill cereal boxes with a (mean) weight of 20.10 ounces of cereal per box. The standard deviation of the amount actually put into the box is known to be .5 ounces. It is suspected that the package-filler is overfilling the box, thereby increasing materials costs. A random sample of 25 filled boxes is taken and weighed (net—not including container weight), yielding a mean weight of 20.251 ounces.
    (a) Show that 20.251 is not significantly larger than 20.10 at the $\alpha = .05$ level (i.e., that the hypothesis $\mu \leq 20.10$ cannot be rejected at this α level).

(b) Criticize the statement: "Since the excess is not statistically significant, we can be very confident that the mean fill is quite close to 20.10 ounces."

18. Assume that the Food and Drug Administration limits the caffein content of cola drinks to 1.2 grains per 12-ounce bottle and that the actual caffein content of cola drinks is normally distributed and varies from .55 to .85 grains per 12-ounce bottle. Let us estimate μ and σ as follows:

$$\mu = .70 \text{ grains} \qquad \sigma = .05 \text{ grains}$$

Suppose the FDA adopts an inspection plan that calls for rejection of a production lot if the population mean caffein content exceeds .7 grains per bottle. Then what values of sample means will lead to rejection of a lot? Assume sample size is 225 and that improper rejection of lots should not occur more than 1% of the time.

19. A standard intelligence test has been given for several years with an average score of 80 and a standard deviation of 7. A group of 25 students is taught with special emphasis on reading skill. If the 25 students obtained a mean grade of 83 on the exam, is there reason to believe that the special emphasis on reading skill increases the results of the test? Use $\alpha = .05$.

20. A certain manufacturing process produces a mean of 12.3 units per hour with a variance of 2.0. A new process is suggested which is expensive to install but would be worthwhile if production could be increased to an average of greater than 13.0 units per hour. In order to decide whether or not to make the change, 14 new machines are tested. They produce a mean of 13.3 units per hour. Should the new machines be purchased? Assume the process variance remains unchanged and use a 25% level of significance.

21. A certain type of hormone to be injected into hens is said to result in a mean increase in the weight of eggs of .3 ounce. A simple random sample of 9 eggs has an arithmetic mean of .4 ounce above the preinjection mean and a value of s ($n - 1$ denominator) equal to .2. Is this enough reason to accept the statement that the mean increase is .3 ounce? Use a two-tailed test with $\alpha = .05$.

22. Given the following hypothesis test:

H_0 : The percentage of defective items (p_0) produced by a certain process is 10.

H_1 : The percentage of defective items (p_0) produced by a certain process is greater than 10.

$$\alpha = .05$$

and the decision rule

If the number of defectives is greater than 14, reject H_0. Otherwise, accept H_0. The sample size is 100.

Comment on the following statements

(a) If p_0 is really greater than .1, the probability we will reject H_0 is .05.
(b) If 17 are defective, this proves that p_0 is greater than .1.
(c) The above is a one-tail test, because if p_0 is actually less than .1, we would make the same decisions as if it were actually .1.

(d) If 16 are found defective, the probability that we would make a Type II error is

$$P\left(z < \frac{14 - 16}{\sqrt{(100)(.16)(.84)}}\right)$$

23. Suppose you are responsible for the quality control of a certain part bought from a supplier. Inspection tests destroy the part, so you must use sampling. A 5% defective rate is tolerable but in your sample of 100 from a lot of 10,000, eight parts are defective. Is this sufficient evidence that the lot of parts has too many defectives?

24. The Constitution of the United States requires a two-thirds majority of both House and Senate to override a presidential veto. In a given situation the necessary majority was secured in the Senate, but the action of the House was uncertain. Prior to the actual vote in the House, newspaper reporters took a straw vote of 90 representatives, and 57 of them indicated their intention to vote to override the veto. If you wished to be wrong no more than five times in 100, assuming that the sample was taken at random, would you conclude that the veto would not be overridden? Justify your conclusion statistically.

25. The following test is set up: A sample of 100 people is selected at random and each is asked if he likes a certain new product. The company conducting the survey feels that it is necessary that at least 10% of all consumers like the product in order for the firm to continue marketing it. The firm therefore decides that if four or fewer people respond favorably, it will stop marketing the product. State in words the nature of the Type I error involved here. How large is it?

26. A bank in a growing metropolitan area determined that a new office located in the suburbs should be opened if more than 30% of the depositors using the city office would do business at the new branch. From the bank's list of active accounts, 400 depositors were selected at random. Of these, 144 indicated that they would patronize the new office if it were established. What is the probability of drawing a random sample, size 400, with 144 or more potential new office patrons, from a statistical universe which in reality contains exactly 30% potential patrons? The bank concluded that more than 30% of its depositors were potential users of the proposed new office. Should it have so concluded, in your opinion? Use $\alpha = .05$.

    In answering the above question, you had to locate a sample statistic (proportion) on a random sampling distribution. Draw a rough sketch of this distribution, showing

    (a) The value and location of the hypothesized parameter.
    (b) The value and location of the statistic.
    (c) The vertical and horizontal scale descriptions.
    (d) The portion of the distribution corresponding to the probability computed above.

27. Each person in a random sample of 400 registered voters in a community is asked whether or not he will vote for candidate A. Two-hundred and five

reply "yes" and 195 reply "no." (Assume no undecided votes.) Let p be the proportion of voters supporting candidate A and consider the hypotheses

$$H_0: p \leq 0.5$$

$$H_1: p > 0.5$$

The following test is proposed: Reject H_0 if $\bar{p}$, the sample proportion, is greater than 0.54.

(a) What is the greatest probability of a Type I error with the above test?
(b) Design a test such that α is at most 0.01, and test the hypothesis with the given sample data.
(c) If you are concerned about the probability of a Type II error and you want to minimize β, which of the two tests would you use? Why?

28. You are employed by a firm that must buy parts from a foreign distributor. It is known that in the past, 20% of the parts secured from this foreign distributor have proved to be defective. The foreign distributor has assured your management, however, that the manufacturing processes and quality control have improved to the point where less than 20% of any shipment will be defective. Acting on this statement, your management places an order for 100,000 parts. A sample of 400 is selected at random from the shipment and 68 are found to be defective.
(a) Is the foreign distributor correct in asserting that less than 20% of the shipment will be defective? Use $\alpha = .10$.
(b) What proportion of the time would the percentage defective in a random sample of this size differ due to chance from a universe value of 20% by as much or more than the difference that you have observed?

29. A manufacturer claims that a customer will find that no more than 8% of parts in a given shipment are defective. The customer decides to test this claim against the alternative hypothesis, $p > .08$, using a .05 probability of rejecting this claim when it is true.
(a) The customer randomly selects 200 parts from the shipment and finds 28 defectives. What conclusion should be reached?
(b) Might the decision you reached in part (a) be in error? If yes, what type of error would this represent? If no, explain.

30. The thickness of a certain type of welding glove should be 2″. If it is too thin it doesn't provide enough protection, and if it is too thick it is too bulky to use.
(a) A safety engineer obtains a sample of 100 gloves from Safety Mitt Manufacturer and finds the average thickness is 1.97″. If the standard deviation of the thickness of a glove is .1″, should the engineer conclude that the average thickness of Safety Mitt's gloves is 2″? Use an α level of .05.
(b) Safety Mitt Manufacturer claims that 95% of all its gloves meet the specification for thickness. In a sample of 100 gloves, seven do not meet the specifications. Based on this information should one agree with Safety Mitt's claim? Use an α level of .05 with the manufacturer's claim as the null hypothesis.

31. The major oil companies report that last year 5% of all credit charges for gasoline, car repairs and parts were never collected and must be written off as bad debts. Recently the oil companies have installed a central computer credit check system. That is, for any credit purchase of over $10, the local gas station must call the central computer center where after a computer search of the customer's payment record, the purchase is given a credit acceptance number or refused credit. Any sales of over $10 not given a credit acceptance number will not be honored as a credit sale by the oil company. To see if the system is effective, 1000 credit charges which were accepted were selected at random and it was found that 36 charges were uncollected and written off as bad debts. Do you think the system is effective? Use a 5% significance level.

32. In reference to the credit check system described in the last problem, a national all-purpose credit check system has also installed the same system, except that purchases must be at least $15 before a credit acceptance number is required. Prior to the installation of the system, 5% of all credit charges were uncollectable. A random sample of 1000 accounts drawn after the system was installed revealed 38 bad accounts.
    (a) Do you think this system is effective? Use a .05 significance level.
    (b) Would you agree with the assertion that the system using a $15 limit for the national credit card company is less effective than that for the oil companies in Problem 31? Use a .05 significance level.

33. Match the correct test statistics with the following four null hypotheses.

(a) $H_0 : \mu = \mu_0$	(1) $\bar{p}$
(b) $H_0 : p = p_0$	(2) $\bar{x}_1 - \bar{x}_2$
(c) $H_0 : \mu_1 = \mu_2$	(3) $\bar{x}_1 + \bar{x}_2$
(d) $H_0 : p_1 = p_2$	(4) $\bar{p}_1 - \bar{p}_2$
	(5) $\bar{x}$
	(6) $\bar{p}/\bar{x}$

In each of the following problems (Problems 34–35), a statistical procedure has been misused. Describe very specifically the incorrect procedures.

34. NASA is worried about the reliability of a certain magneto relay used in the Surveyor spacecraft. According to specifications, these relays should have an average life of 111 hours before failure. A sample of 25 switches is tested, and the average switch life is 108.8 hours with a standard deviation of 5. Since the z statistic,

$$z = \frac{108.8 - 111}{5/\sqrt{25}}$$

is less than 1.96, assuming that the average life is normally distributed, it was concluded at a 5% α level that the switches did not meet specifications.

35. The quality control department suspects that the proportion of defective fuses produced by a production line is above the acceptable 3% level. One-hundred fuses are chosen at random. If X is the number of defective fuses,

and it is assumed that X is normal with $\mu = 3$ and $\sigma^2 = 2.91$, the procedure to be used is to stop the production line unless $3 - 1.65\sqrt{2.91} \leq X \leq 3 + 1.65\sqrt{2.91}$. Use $\alpha = .05$.

36. A potential buyer of light bulbs bought 50 bulbs of each of two brands. Upon testing these bulbs, he found that brand A had a mean life of 1282 hours with a standard deviation of 80 hours, whereas brand B had a mean life of 1208 hours, with a standard deviation of 94 hours. The buyer wishes to be wrong no more than five times in 100 in saying that the two brands differ with respect to mean life. What should he conclude?

    In answering the above question, you had to locate a sample statistic (proportion) on a random sampling distribution. Draw a rough sketch of this distribution, showing

    (a) The value and location of the hypothesized parameter.
    (b) The value and location of the statistic.
    (c) The vertical and horizontal scale descriptions.
    (d) The portion of the distribution corresponding to the probability represented by $\alpha = .05$.
    (e) The point to the right of which lies 2 1/2% of the area under the distribution curve.

37. (a) In a simple random sample of 400 students in collegiate schools of business, 176 favored the addition of more required mathematics courses to be taken in the freshman and sophomore years. In a simple random sample of 400 students in liberal arts colleges, 144 favored the addition of more required mathematics courses. Do you believe there is a real difference in the attitude of the two groups? Justify your answer statistically and indicate the level of significance that you use.
    (b) Explain specifically the meaning of a Type I error in terms of this particular problem.

38. Two astronomers recorded observations on a certain star. The twelve observations obtained by the first astronomer have a mean reading of 1.20. The eight observations obtained by the second astronomer have a mean of 1.15. Past experience has indicated that each astronomer obtains readings with a variance of about .40. Does the difference between the two results seem reasonable?

39. A bank is considering opening a new branch in one of two neighborhoods. It desires to open the branch in the neighborhood having the higher mean income. From census records, two random samples of size 100 are selected, and the following information is obtained.

$$\begin{aligned} n_1 &= 100 & n_2 &= 100 \\ \bar{x}_1 &= \$6000 & \bar{x}_2 &= \$6100 \\ s_1 &= \$300 & s_2 &= \$400 \end{aligned}$$

The first neighborhood also has several small business firms that might be attracted to a new branch. Therefore, the bank wishes to avoid the error of concluding that the second neighborhood has a higher mean income when the true state of nature is that the first neighborhood has a mean income equal to or higher than the second neighborhood. Set up the appropriate null hypothesis and test at the 5% level of significance.

40. Suppose that in a simple random sample of 400 people from one city, 188 preferred a particular brand of soap to all others, and in a similar sample of 500 people from another city, 210 preferred the same product. Is there reason to doubt the hypothesis that equal proportions of persons in the two cities preferred this brand of soap at the 5% level of significance?

41. A railroad company installed two sets of 50 red oak ties each. The two sets were treated with creosote by two different processes. After a number of years of service, it was found that 22 ties in the first set and 18 ties in the second set were still in good condition. Is one justified in claiming that there is a real difference between the preserving properties of the two processes if he is willing to be wrong in so concluding no more than 5% of the time?

42. Prior to certain technological developments, the average breaking strength of a certain paper bag was 9 pounds, with a standard deviation of 1.5 pounds. With the new technological developments, the average breaking strength of the bag is now 10 pounds, with the same standard deviation. A jobber offers a lot of 1000 bags at a reduced rate, but does not know whether the bags were produced using the new or old technology. A sample of 25 bags was tested.
    (a) Set up a test such that the probability is .10 of concluding that the bags average 10 pounds in breaking strength when the true average is 9 pounds.
    (b) What is the probability of a Type II error for the test in part (a)? Define explicitly the meaning of a Type II error in this problem.

43. Ten people were placed on a special diet. The results after ten weeks were as follows

Person	*Weight in Pounds before Diet*	*Weight in Pounds after Diet*
A	210	209
B	200	194
C	185	184
D	174	174
E	193	191
F	190	190
G	184	180
H	225	225
I	240	237
J	215	212

Using the paired t-test, would you say the diet was successful? Use $\alpha = .05$.

44. A research group ran the following experiment in order to determine whether monetary incentives would increase learning. It selected a random sample

of students at a university. For some semesters, the students were given X dollars per week if their work met a certain standard, and no compensation otherwise. During other semesters, no rewards were given at all, regardless of performance. The results of eight students were as follows:

Student	*Average during Terms with Payment*	*Average during Terms without Payment*
1	3.10	3.00
2	2.95	2.45
3	2.00	2.25
4	1.95	1.95
5	3.80	3.75
6	2.43	2.65
7	2.65	2.55
8	2.40	2.20

Using a paired t-test, would you conclude that monetary incentives increase grade average? A higher grade average represents a better average. Use $\alpha = .05$.

CHAPTER EIGHT

Statistical Decision Procedures —Estimation

In Chapter 7, we discussed statistical hypothesis testing, which involved the acceptance or rejection of hypotheses concerning population parameters based on sample data. As indicated earlier, statistical estimation, or as it is more briefly termed, estimation, is the other major area classified under the heading of statistical inference. The subject of estimation is concerned with the methods by which population characteristics are estimated from sample information. Hence, both of the subclassifications of statistical inference deal with the question of how inferences can be made about population characteristics from information contained in samples.

8.1 Point and Interval Estimation

The basic reasons for the need to estimate population parameters from sample data are that it is ordinarily too expensive or it is simply infeasible to enumerate complete populations to obtain the required information. In the case of finite populations, the cost of complete censuses may be prohibitive and in the case of infinite populations, complete enumerations are impossible. Statistical estimation procedures provide us with the means of obtaining estimates of population parameters with desired degrees of precision. Numerous business and economic examples can be given of the need to obtain estimates of pertinent population parameters. A marketing organization may be interested in estimates of average income and other socio-economic characteristics of the consumers in a metropolitan area, a retail chain may wish an estimate of the average number of pedestrians per day who pass a certain corner, a production department may desire an estimate of the percentage of defective articles produced by a new production process, and a finance department may desire an estimate of average interest rates on mortgages in a certain section of the country. Undoubtedly, in all of these cases, exact accuracy is not required, and estimates derived from sample data would probably provide appropriate information to meet the demands of the practical situation.

Two different types of estimates of population parameters are of interest, *point estimates* and *interval estimates*. A point estimate is a single number which is used as an estimate of the unknown population parameter. For example, the arithmetic mean income of a sample of families in a metropolitan area may constitute a point estimate of the corresponding population mean for all families in that metropolitan area. The percentage of defectives observed in a sample may be used as an estimate of the corresponding unknown percentage of defectives in a shipment from which the sample was drawn.

A distinction can be made between an *estimate* and an *estimator*. Let us return to the illustration of estimating the population figure for arithmetic mean income of all families in a metropolitan area from the corresponding sample mean. The numerical value of the sample mean is said to be an *estimate* of the population mean figure. On the other hand, the statistical measure used, that is, the method of estimation is referred to as an *estimator*. For example, the sample mean, $\bar{x}$, is an estimator of the population mean. When a specific number is calculated for the sample mean, say $8000, that number is an *estimate* of the population mean figure. Of course, there would be other possible estimators of a population mean. For example, we could use a sample median, *Md*, the mean of the lowest and highest values in the sample observations, usually referred to as the mid-range, or some other method. In order to judge whether one estimator is better than another, it is necessary to have some criteria of "goodness of estimation." We discuss some possible criteria in the next section.

For most practical purposes, it would not suffice to have merely a single point estimate of a population parameter. Any single point estimate will be either right or wrong. It would certainly seem to be extremely useful, and perhaps even necessary, to have in addition to a point estimate, some notion of the degree of error that might be involved in using this estimate. *Interval estimation* is useful in this connection. Roughly speaking, an interval estimate of a population parameter is a statement of two values between which it is estimated that the parameter lies. Thus, an interval estimate in the example of the population arithmetic mean income of families in a metropolitan area might be $7100 to $8900. An interval estimate for the percentage of defectives in a shipment might be 3% to 5%. We may have a great deal of confidence or very little confidence that the population parameter is included in the range of the interval estimate. Therefore, it would seem necessary to attach some sort of probabilistic statement to the interval. From the viewpoint of classical statistical inference, this cannot be an objective relative frequency of occurrence probability statement, since there is no repetitive process which generates different values of the population parameter. If the population is constant, the population parameter is a single unknown number which either is or is not included in the estimated interval.

The procedure used to handle this problem is "confidence interval estimation." The confidence interval is an interval estimate of the population parameter. A confidence coefficient, for example, 90 or 95% is attached to this interval to indicate the degree of confidence to be placed upon the estimated interval. As we shall see in our subsequent discussion of this type of estimation, the confidence coefficient is not a probability that the population parameter lies in the stated interval. Instead, a coefficient, such as the above 90 or 95%, represents the proportion of such intervals which would be expected to include the population parameter, if a great many such intervals were calculated using the same estimation procedure. Confidence interval estimates which represent rather narrow intervals with high confidence coefficients attached are the most useful kind. For example, it is more useful for subsequent action to have a confidence interval statement that the percentage of defectives in a shipment lies in the interval 4 to 5% with a confidence coefficient of 99% than one which states that the interval is 1.5 to 7.5% with a confidence coefficient of 50%. The second interval statement clearly has two difficulties connected with it. Since the interval is wide, the information conveyed about the population parameter is vague, and since the confidence coefficient is low, there is excessive uncertainty even about this broad interval.

8.2 Criteria of Goodness of Estimation

Numerous criteria have been developed by which to judge the goodness of point estimators of population parameters. A rigorous discussion of these criteria requires some complex mathematics which falls outside the scope of this text. However, it is possible to gain an appreciation of the nature of these criteria in an intuitive non-rigorous way, and it is this sort of presentation we shall develop here.

Let us return to our illustration of estimating the arithmetic mean income of families in a metropolitan area. This arithmetic mean, assuming suitable definitions of income, family, and metropolitan area, is an unknown population parameter, which we designate as μ_x. If we took a simple random sample of families from this population, and calculated the arithmetic mean, $\bar{x}$, the median, Md, and the mid-range $(x_{\max} + x_{\min})/2$, where $x_{\max}$ and $x_{\min}$ are the largest and smallest sample observations, respectively, which method would be the best estimator of the population mean? Probably, your answer would be that the sample mean, $\bar{x}$, is the best estimator. In fact, if this question had not been raised, it might not even have occurred to you to use any statistic other than the sample mean as an estimator of the population mean. However, why do you think the sample mean represents the best estimator? It may not be very easy to articulate your answer to that question. As it turns out, the sample mean is preferable to the other candidate estimators by the generally utilized criteria of goodness of estimation of classical statistical inference. Let us examine the nature of a few of these criteria.

The following four criteria may be used to judge the goodness of a point estimator:

(1) Unbiasedness
(2) Consistency
(3) Efficiency
(4) Sufficiency

Unbiasedness. An estimator, such as a sample arithmetic mean, is a random variable. That is, if we take a simple random sample of n elements from a population, the sample mean may take on different values, depending upon which population elements are drawn into the sample. As is proven in Rule 13 of Appendix C, the expected value of this sample mean is the population mean. In symbols, we have

$$E(\bar{x}) = \mu_x \quad \textbf{(8.1)}$$

where $\bar{x}$ = the sample mean
μ_x = the population mean

If the expected value of a sample statistic is equal to the population parameter for which the statistic is an estimator, the statistic (or estimator) is said to be unbiased. Using more general terminology, if θ is a parameter to be estimated, and $\hat{\theta}$ is a sample statistic used to estimate θ, then $\hat{\theta}$ is said to be an *unbiased estimator* of θ if

(8.2) $$E(\hat{\theta}) = \theta$$

The bias of an estimator is, therefore, defined as

(8.3) $$\text{Bias} = E(\hat{\theta}) - \theta$$

Hence, if we say that a given estimator is unbiased, we are simply saying that this method of estimation is correct, *on the average*. In a relative frequency sense, if the method is employed repeatedly, the average of all estimates obtained from this estimator is equal to the value of the population parameter. In any given sample, the numerical estimate obtained from this estimator may not be equal to the parameter value; indeed, it is extremely unlikely that these two figures will be equal. Clearly, if an estimator is unbiased, this does not guarantee useful individual estimates. The differences of these individual estimates from the value of the population parameter may represent large errors. The fact that the bias or the long-run average of these errors is zero taken by itself may be of little practical importance. Furthermore, if two estimators are unbiased, we require additional criteria in order to evaluate preferredness.

The sample variance s_x^2 is a biased estimator of the variance of an infinite population σ_x^2, where

$$s_x^2 = \frac{\sum_{i=1}^{n} (x_i - \bar{x})^2}{n}$$

and

$$\sigma_x^2 = E(X - \mu_x)^2$$

for an infinite population, and all of the symbols on the right-hand sides have their usual meanings.

As proved in Rule 14 of Appendix C, an unbiased estimator can be constructed by multiplying the sample variance by the factor $n/(n-1)$. We then have

(8.4) $$E\left[\left(\frac{n}{n-1}\right)s_x^2\right] = E\left[\frac{\sum_{i=1}^{n} (x_i - \bar{x})^2}{n-1}\right] = \sigma_x^2$$

As can be seen from the nature of the proof in Rule 14,

$$\frac{\sum_{i=1}^{n} (x_i - \bar{x})^2}{n - 1}$$

is not an unbiased estimate of the variance of a finite population. Also, as previously noted, both

$$\sqrt{\frac{\sum_{i=1}^{n} (x_i - \bar{x})^2}{n}} \quad \text{and} \quad \sqrt{\frac{\sum_{i=1}^{n} (x_i - \bar{x})^2}{n - 1}}$$

are biased estimators of the population standard deviation.

Consistency. It is clear from the preceding discussion that knowing only that an estimator is unbiased gives us little information as to the goodness of that method of estimation. It would seem that closeness of the estimator to the parameter is of importance. Both the concepts of consistency and efficiency deal with this property of closeness. Consider the sample mean $\bar{x}$ as an estimator of the population parameter μ_x for an infinite population. What happens to the possible values of $\bar{x}$ as the sample size n increases? On an intuitive basis, we would certainly expect $\bar{x}$ to lie closer to μ_x, in some sense, as n becomes larger and larger. Loosely speaking, if an estimator, say $\hat{\theta}$, approaches the parameter θ closer and closer as the sample size n increases, $\hat{\theta}$ is said to be a *consistent estimator* of θ. Somewhat more rigorously, the estimator $\hat{\theta}$ is said to be a consistent estimator of θ if, as n approaches infinity, the probability approaches 1 that $\hat{\theta}$ will differ from the parameter θ by no more than an arbitrary constant.

Interpreting this idea of consistency in terms of sampling, it means that the sampling distribution of the estimator becomes more and more "tightly packed" around the population parameter as the sample size increases. Figure 8-1 illustrates this concept for the sample mean as an estimator of μ_x, the mean of an infinite population. The graph represents the respective sampling distributions of the sample mean, $\bar{x}$, for three independent samples of different sizes drawn from the same population; n_3 is a larger sample size than n_2, which is larger than n_1. We know from the relationship $\sigma_{\bar{x}}^2 = \sigma_x^2/n$ that the sampling variance $\sigma_{\bar{x}}^2$ decreases as n increases. It may be noted that all three sampling distributions center on the population parameter μ_x, since $\bar{x}$ is an unbiased estimator of μ_x.

Figure 8-2 illustrates the concept of consistency for $\hat{\theta}$, a *biased estimator* of a parameter θ. As in Figure 8-1 the graph represents the respective sampling distributions of an estimator, in this case, $\hat{\theta}$, for three independent samples of different sizes drawn from the same population, with $n_3 > n_2 > n_1$. None of the distributions are centered on the population parameter θ. Since $E(\hat{\theta})$ is not equal to θ, $\hat{\theta}$ is a biased estimator of θ. However, as can be seen from

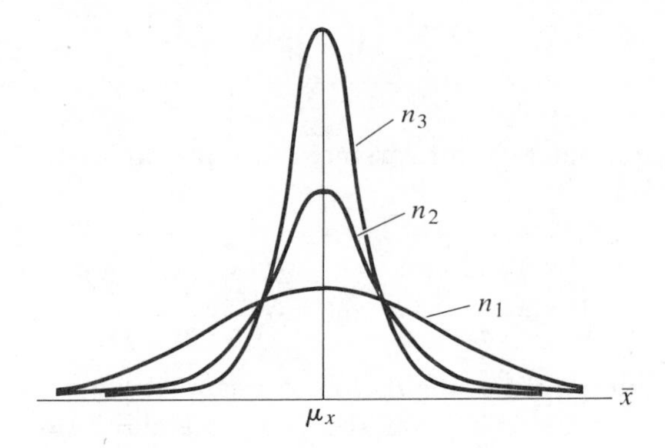

Figure 8-1 Sampling distributions of $\bar{x}$ as the sample size increases; $n_3 > n_2 > n_1$.

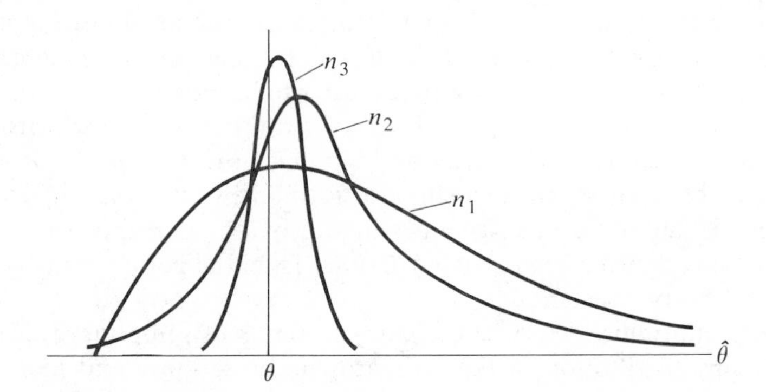

Figure 8-2 Sampling distributions of an estimator $\hat{\theta}$ which is biased but consistent; $n_3 > n_2 > n_1$.

the graph, as the sample size increases, the sampling distribution becomes increasingly "tightly packed" around $\hat{\theta}$.

Efficiency. The concept of efficiency refers to the sampling variability of an estimator. If two competing estimators are both unbiased, the one with the smaller variance (for a given sample size) is said to be relatively more efficient. Somewhat more formally, if $\hat{\theta}_1$ and $\hat{\theta}_2$ are two unbiased estimators of θ, their relative efficiency is measured by the ratio

(8.5) $$\frac{\sigma^2_{\hat{\theta}_2}}{\sigma^2_{\hat{\theta}_1}}$$

where $\sigma^2_{\hat{\theta}_1}$ is the smaller variance.

Let us consider as an example the situation of a simple random sample of size n drawn from a normal population with mean μ_x and variance σ_x^2. Suppose we want to consider the relative efficiency of the sample mean $\bar{x}$ and the sample median Md as estimators of the population mean, μ_x. Both estimators are unbiased. We know that the variance of the sample mean $\bar{x}$ is $\sigma_{\bar{x}}^2 = \sigma_x^2/n$. The variance of the sample median Md is approximately $\sigma_{Md}^2 = 1.57\sigma_x^2/n$. Therefore, the relative efficiency of $\bar{x}$ with respect to Md is $\sigma_{Md}^2/\sigma_{\bar{x}}^2 = (1.57\sigma_x^2/n)/(\sigma_x^2/n) = 1.57$. Interpreting this result in terms of sample sizes, if the sample median Md were used as an estimator of μ_x, the mean of a normal population, a sample size 57% larger would be required than if the sample mean $\bar{x}$ were used. Stated differently, the required sample size for the sample median would be 157% of that for the sample mean. Figure 8-3 shows the sampling distributions for these two estimators.

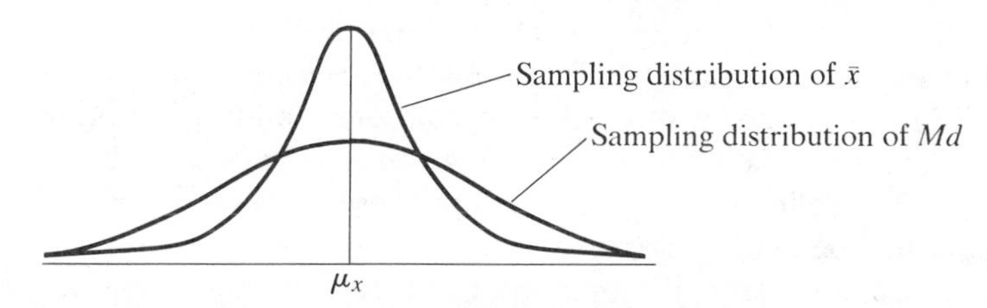

Figure 8-3 Sampling distributions of the mean ($\bar{x}$) and median (Md) for the same sample size.

Sufficiency. A sufficient statistic is an estimator that summarizes from the sample data all the information contained in these data, and no other estimator can provide additional information. We mention the criterion of sufficiency here for completeness, but we will not discuss its mathematical statement or methods for establishing sufficiency, since they fall outside the scope of this text. However, we may note that the two estimators that we will discuss in this chapter in connection with interval estimation, (1) $\bar{p}$, the sample proportion, and (2) $\bar{x}$, the sample mean, are both sufficient statistics.

8.3 Maximum Likelihood Estimation

Although the major focus of this chapter is upon interval rather than point estimation, we will discuss briefly one very important technique of point estimation, the *maximum likelihood* method. It was devised by the eminent British statistician, R. A. Fisher in the early 1920's. The method of maximum likelihood estimation is of importance because it is so widely applied and because maximum likelihood estimators ordinarily possess many desirable properties such as those discussed in Section (8.2). For example, it can be shown that if a sufficient estimator of a parameter exists, that estimator is the maximum likelihood one. Also, in addition to possessing the property of sufficiency, a maximum likelihood estimator under very general conditions, is both efficient[1] and consistent. Furthermore, as sample size becomes large, the sampling distribution of a maximum likelihood estimator approaches a normal distribution centered on the population parameter being estimated. However, maximum likelihood estimators are not necessarily unbiased. One notable example is the sample variance,

$$s_x^2 = \frac{\sum_{i=1}^{n} (x_i - \bar{x})^2}{n}$$

which is the maximum likelihood estimator of σ_x^2, but as indicated earlier, s_x^2 is not an unbiased estimator of σ_x^2.

An intuitive understanding of the maximum likelihood principle can be obtained from a simple example. Suppose there is a coin which is known to be "unfair," in the following way. Assume the probability of obtaining a head when the coin is tossed is either 1/4 or 3/4. This unknown probability will be denoted p. The following experiment is proposed. Three independent tosses of the coin will be made. From the sample outcomes, you are to guess what the correct value is for p, 1/4 or 3/4. The tosses are made, and a head, tail, and head are obtained in that order. If you had to decide now, solely on the basis of this information, what would you guess to be the value of p? Doubtless you would guess 3/4 to be a more "likely" value of p than 1/4. This is in agreement with the maximum likelihood estimate for p in this situation. Let us examine how such an estimate might be constructed.

Suppose we let $P(H, T, H; p = 3/4)$ and $P(H, T, H; p = 1/4)$ denote the joint probabilities of observing the sample outcomes "head," "tail," and "head," in that order, given that the probabilities of a head on a single trial are, respectively, 3/4 and 1/4. Then, by the multiplication rule

$$P(H, T, H; p = 3/4) = (3/4)(1/4)(3/4) = 9/64$$

and

$$P(H, T, H; p = 1/4) = (1/4)(3/4)(1/4) = 3/64$$

Thus, since the probability of having obtained the observed set of sample outcomes is 9/64 if the coin with $p = 3/4$ were tossed and only 3/64 if the

[1]In this context, an "efficient estimator" is one which possesses "minimum limiting variance" as compared to the "limiting variance" of all other possible estimators as sample size approaches infinity. The limiting variance is the value approached by the variance as sample size becomes large. Efficient estimators are also asymptotically unbiased; that is, in the limit, as sample size goes to infinity, such estimators are unbiased.

coin with $p = 1/4$ were tossed, it seems reasonable to decide on the basis of this evidence that $\hat{p} = 3/4$ rather than $\hat{p} = 1/4$. The symbol $\hat{p}$ is used to denote an estimate (or estimator) as contrasted to the symbol p, which denotes the true value of the population parameter being estimated. This estimate of $\hat{p} = 3/4$ is referred to as the maximum likelihood estimate, using the following rationale. The joint probability of obtaining a specific sequence of sample outcomes is referred to as a "likelihood function." The value of the estimator (in this example, estimate), for which the likelihood function is a maximum is referred to as the maximum likelihood estimator. Roughly speaking, we can say that $\hat{p} = 3/4$ gives a better explanation of the sample data than does $\hat{p} = 1/4$, in the sense that the probability of having observed these data under the hypothesis that $\hat{p} = 3/4$ is higher than under $\hat{p} = 1/4$.

Unfortunately but not unsurprisingly, maximum likelihood estimates or estimators are usually not so easily derivable as in the preceding example. In this example, the likelihood function was discrete and the parameter to be estimated could only take on two possible values. In many of the most interesting and important cases, the likelihood function is continuous and the parameter to be estimated may take on any value in a continuous interval. Calculus methods are usually required to maximize the function.[2]

Two problems of point estimation which are of interest in this chapter are the estimation of a population proportion and a population mean. In the case of estimating a population proportion, if we assume a Bernoulli process as generating n sample observations, it can be shown that the observed sample proportion of successes, $\bar{p}$ = number of successes/number of sample observations is the maximum likelihood estimator of the population parameter, p. In the case of estimating a population mean, denoted μ_x, it can be shown that $\bar{x}$, the mean of a simple random sample of n observations is the maximum likelihood estimator. These estimators are also unbiased, efficient, and consistent. Thus, it is not surprising that in many applications

---

[2]In contrast to the intuitive treatment given in this section, a brief mathematical explanation of maximum likelihood estimation would proceed as follows. Let X be a random variable, either discrete or continuous, with probability distribution $f(x; \theta)$, where θ denotes a parameter to be estimated. Assume n independent observations are drawn from $f(x; \theta)$. Denote the n random variables represented by these observations as $x_1, x_2, \ldots, x_n$. Then the likelihood function of these n random variables and the parameter θ is defined as

$$L(x_1, x_2, \ldots, x_n; \theta) = f(x_1; \theta) f(x_2; \theta) \cdots f(x_n; \theta)$$

where

$L(x_1, x_2, \ldots, x_n; \theta)$ = the likelihood function
$f(x_i; \theta)$ = the probability or probability density of the observation x_i given θ

Then the maximum likelihood estimator of θ, denoted $\hat{\theta}$, is the estimator which maximizes the likelihood function $L(x_1, x_2, \ldots, x_n; \theta)$ with respect to θ.

If the likelihood function is continuous and a maximum exists, conventional calculus methods of differentiation can often be used to determine that maximum.

of statistical methods, sample proportions or sample means are used as the "best" point estimators of the corresponding population parameters. Perhaps the reader has had occasion to calculate a sample proportion or mean, and may have intuitively used such a figure as an estimator of the corresponding population parameter.

8.4 Confidence Interval Estimation

As indicated in Section 8.1, for most practical purposes, it is not sufficient merely to have a single point estimate of a population parameter. Since the single estimate is undoubtedly incorrect, in the sense that it is extremely improbable that the estimate is exactly equal to the value of the population parameter, it is necessary to have an estimation procedure which gives some measure of the degree of precision involved. The standard procedure in classical statistical inference for this purpose is confidence interval estimation. The rationale of this type of estimation will be explained in terms of an illustrative example in which a population mean is the parameter to be estimated.

Interval Estimation of a Mean–Rationale

Suppose a manufacturer has a very large production run of a certain brand of tires and he is interested in obtaining an estimate of their arithmetic mean lifetime by drawing a simple random sample of 100 tires and subjecting them to a forced life test. Let us assume that from long past experience in manufacturing this brand of tires, it is known that the population standard deviation for a production run is $\sigma_x = 3000$ miles. Of course, ordinarily the standard deviation of a population is not known exactly and must be estimated from a sample, just as are the mean, proportion, or other parameters. However, for purposes of explaining the reasoning involved in confidence interval estimation, let us assume in this case that the population standard deviation is indeed known. The sample of 100 tires is drawn and a mean lifetime of 22,500 miles is observed. Thus, we denote $\bar{x} = 22{,}500$ miles. We shall address ourselves to the reasoning involved in answering the question, "What is the 95% confidence interval estimate for the mean lifetime of all tires in the production run?" Our answer in this case will be obtained by calculating the interval $\bar{x} - 1.96\sigma_{\bar{x}}$ to $\bar{x} + 1.96\sigma_{\bar{x}}$ and attaching a suitable probabilistic interpretation to this range. The nature of this interpretation will be deferred until the explanation of the example.

Just as in hypothesis testing, the procedure in confidence interval estimation is based on the concept of the sampling distribution. In this example, since we are dealing with the estimation of a mean, the appropriate distribution is the sampling distribution of the mean. We will review some fundamentals of this distribution in the case of the present example to lay the foundation for confidence interval estimation. Figure 8-4 shows the sampling distribution of the mean for simple random samples of size $n = 100$ from a population with an unknown mean, denoted μ_x and a standard deviation, $\sigma_x = 3000$ miles. It is assumed that the sample is large enough so that by the central limit theorem the sampling distribution may be assumed to be normal, even if the population is non-normal. The standard error of the mean, which is the standard deviation of this sampling distribution, is approximately equal to $\sigma_{\bar{x}} = \sigma_x/\sqrt{n}$. Strictly speaking, the finite population correction should be shown in this formula, but we will assume the population is so large relative to sample size that for practical purposes the correction factor is equal to one. The mean of the sampling distribution, $\mu_{\bar{x}}$, is equal to the population mean, μ_x.

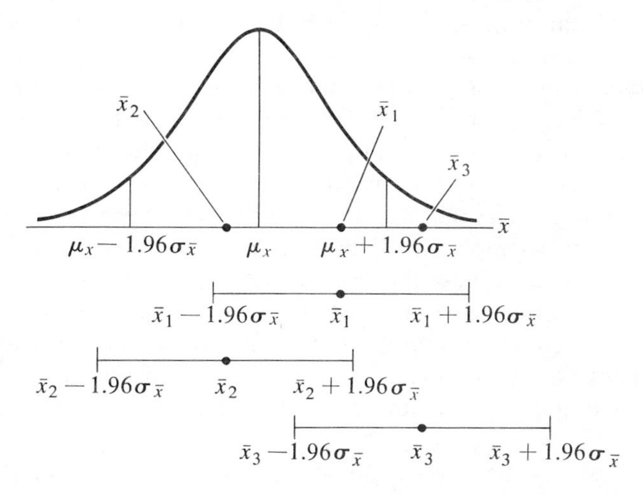

Figure 8-4 Sampling distribution of the mean and confidence interval estimates for three illustrative samples.

In our work with the normal sampling distribution of the mean, we have learned how to make conditional probability statements about sample means, given the value of the population mean. Thus, for example, in terms of the data of this problem, we can state that in drawing a simple random sample of 100 tires from the production run, the probability is 95% that the sample mean, $\bar{x}$, will lie within 1.96 standard error units of the mean of the sampling distribution, or from $\mu_x - 1.96\sigma_{\bar{x}}$ to $\mu_x + 1.96\sigma_{\bar{x}}$. This range is indicated on the horizontal axis of the sampling distribution in Figure 8-4. For emphasis, the ordinates have been shown at the end points of this range. As usual, we determine the 1.96 figure from Table A-5 of Appendix A, where we find that

47.5% of the area in a normal distribution is included between the mean of a normal distribution and a value 1.96 standard deviations to the right of the mean, and by symmetry 95% of the area is included in a range of plus or minus 1.96 standard deviations from the mean. Using a relative frequency interpretation, 95% of the $\bar{x}$ values of samples of size 100 would lie in this range if repeated samples were drawn from the given population. This is a *deductive statement*, in the sense that we deduce the probability from our knowledge of the population. Another way of looking at it is that this statement is a conditional probability about sample statistics, given the value of a population parameter. However, that type of statement is not what is needed in estimating a population parameter. What is required is an *inductive statement*, in which we make an inference about a population parameter from a knowledge of a sample statistic. Confidence interval estimation is a procedure for making such an inductive statement. It must be emphasized that the deductive statement about the range $\mu_x \pm 1.96\sigma_{\bar{x}}$ is *not* a confidence interval statement, *it is a probability statement.*

How then might we construct the desired type of inductive statement about the population parameter? To obtain an answer to this question, let us consider again the repeated simple random samples of size 100 from the population represented by the production run of tires. Let us assume our first sample yields a mean that exceeds $\mu_{\bar{x}}$ but falls between μ_x and $\mu_x + 1.96\sigma_{\bar{x}}$. The position of this sample mean, denoted $\bar{x}_1$ is shown on the horizontal axis of Figure 8-4. Suppose now that we set up an interval from $\bar{x}_1 - 1.96\sigma_{\bar{x}}$ to $\bar{x}_1 + 1.96\sigma_{\bar{x}}$. This interval is shown immediately below the graph in Figure 8-4. As can be seen in the figure, this interval, which may be written as $\bar{x}_1 \pm 1.96\sigma_{\bar{x}}$, includes the population parameter, μ_x. Of course, this simply follows from the fact that $\bar{x}_1$ fell within $1.96\sigma_{\bar{x}}$ from the mean of the sampling distribution, μ_x.

Now, let us assume our second sample from the same population yields the mean, $\bar{x}_2$, which lies on the horizontal axis to the left of μ_x, but again at a distance less than $1.96\sigma_{\bar{x}}$ away from μ_x. Again we set up an interval of the sample mean plus or minus $1.96\sigma_{\bar{x}}$, or from $\bar{x}_2 - 1.96\sigma_{\bar{x}}$ to $\bar{x}_2 + 1.96\sigma_{\bar{x}}$. This interval, shown below the graph in Figure 8-4, includes the population mean, μ_x.

Finally, suppose a third sample is drawn from the same population, with the mean $\bar{x}_3$, shown on the horizontal axis of Figure 8-4. This sample mean lies to the right of μ_x, but at a distance *greater than* $1.96\sigma_{\bar{x}}$ above μ_x. Now, when we set up the range $\bar{x}_3 - 1.96\sigma_{\bar{x}}$ to $\bar{x}_3 + 1.96\sigma_{\bar{x}}$, this interval does *not* include μ_x.

We can imagine a continuation of this sampling procedure, and we can assert that 95% of the intervals of the type $\bar{x} \pm 1.96\sigma_{\bar{x}}$ include the population parameter, μ_x. Now, we can get to the crux of confidence interval estimation. In the problem originally posed, as in most practical applications, only one sample was drawn from the population, not repeated samples. On the basis of the single sample, we were required to estimate the population parameter. The procedure is simply to establish the interval $\bar{x} \pm 1.96\sigma_{\bar{x}}$ and attach a suitable statement to it. The interval itself is referred to as a "confidence interval." Thus, for example, in our original problem the required confidence interval is

$$\bar{x} \pm 1.96\,\sigma_{\bar{x}} = \bar{x} \pm 1.96\,\frac{\sigma_x}{\sqrt{n}} = 22{,}500 \pm 1.96\,\frac{3000}{\sqrt{100}}$$
$$= 22{,}500 \pm 588 = 21{,}912 \text{ to } 23{,}088 \text{ miles}$$

We must be very careful how we interpret this confidence interval. It is incorrect to make a probability statement about this *specific* interval. For example, it is incorrect to state that the probability is 95% that the mean lifetime, μ_x, of all tires, falls in this interval. The population mean is not a random variable. Hence probability statements cannot be made about it. The unknown population mean, μ_x, either lies in the interval or it does not. We must return to the line of argument used in explaining the method and indicate that the intervals of the form $\bar{x} \pm 1.96\sigma_{\bar{x}}$ constitute the values of the random variable, not μ_x. Thus, the interpretation is that if repeated simple random samples of the same size were drawn from this population, and the interval $\bar{x} \pm 1.96\sigma_{\bar{x}}$ were constructed from each of them, then 95% of the statements that the interval contains the population mean, μ_x, would be correct. Another way of putting it is that in 95 samples out of 100 the mean, μ_x, would lie within intervals constructed by this procedure. The 95% figure is referred to as a "confidence coefficient" to distinguish it from the type of probability that is calculated when deductive statements are made about sample values from known population parameters.

Interval Estimation—Interpretation and Use

Despite the above interpretation of the meaning of a confidence interval, where the probability pertains to the estimation procedure rather than to the specific interval constructed from a single sample, the fact remains that the investigator ordinarily must make an inference on the basis of the single sample he has drawn. He will not draw the repeated samples implied by the interpretational statement. For example, in the tire illustration, an inference is required about the production run based on the particular sample of 100 tires in hand. If the confidence coefficient attached to the interval estimate is high, then the investigator will behave as though the interval estimate is correct. In the tire example, an interval estimate of a mean lifetime of 21,912 to 23,088 miles was obtained. This interval may or may not encompass the actual value of the population parameter, μ_x. However, since 95% of intervals so constructed would include the value of the mean lifetime, μ_x, of all tires in the production run, we will behave as though this particular interval does include the actual value. As noted earlier, it is desirable to obtain a relatively narrow interval with a high confidence coefficient associated with it. One without the other is not particularly useful. Thus, for example, in estimation of a proportion, say, the proportion of persons in the labor force who are unemployed, we can assert even without sample data that the percentage lies somewhere between 0 and 100% with a confidence coefficient of 100%. Obviously, this statement is neither very profound nor very useful because the interval is too wide. On the other hand, even if the interval is very narrow, but has a low associated coefficient, say 10%, the statement would again have little practical utility.

Confidence coefficients such as 0.90, 0.95, and 0.99 and two- or three-sigma limits, such as 0.955 or 0.997 are conventionally used. For a fixed confidence coefficient and population standard deviation, the only way to narrow a confidence interval and thus increase the precision of the statement is to increase the sample size. This is readily apparent from the way the confidence interval was constructed in the tire example. We computed $\bar{x} \pm 1.96\sigma_{\bar{x}}$, where $\sigma_{\bar{x}} = \sigma_x/\sqrt{n}$. If the 1.96 figure and σ_x remain constant, we can decrease the width of the interval only by increasing the sample size, n, since $\sigma_{\bar{x}}$ is inversely related to $\sqrt{n}$. Thus the marginal benefit of increased precision must be measured against the increase cost of sampling. In Section 8.5, we discuss a method of determining the sample size required for a specified degree of precision.

The type of confidence interval that we have been discussing is referred to as "two-sided" because two figures are given for the end-points or "confidence limits" of the interval. Sometimes, we are interested in a"one-sided interval," in which only one end-point of the interval is given, and a corresponding modification is made in interpretation. For example, in the tire illustration, the two-sided 95% confidence interval 21,912 to 23,088 miles was constructed. Using a long-run relative frequency interpretation, we may say that in repeated sampling, intervals constructed by the same procedure would include the population mean μ_x in 95% of the cases. Thus, in 2.5% of the samples μ_x would lie above the upper confidence limit and in 2.5% of the samples μ_x would lie below the lower limit. Now, suppose we had not been interested in an estimate of the mean lifetime of tires in the production run lying between two numbers, but rather in its being *at least* a given figure. Then, if we simply used the lower limit of 21,912 miles which was calculated by $\bar{x} - 1.96\sigma_{\bar{x}}$ and dispensed with the upper limit, we would have a one-sided interval with a confidence coefficient of 97.5%. Here, we are estimating that the population mean lifetime is at least 21,912 miles. Again using a long-run relative frequency interpretation, we would state that 97.5% of the intervals so constructed would include the population mean, and in 2.5% of the cases, the population mean would be expected to lie below the lower end-point of the interval. Clearly, in this type of "one-sided interval," where only a lower limit is stated, the unstated upper limit is really plus infinity. If only the upper limit were stated, the unstated lower limit would really be minus infinity.

In summary, if a two-sided confidence interval is desired, we construct the range $\bar{x} \pm z\sigma_{\bar{x}}$, where the multiple z is determined from the confidence coefficient. If a one-sided confidence interval is desired, we construct the range $\bar{x} - z\sigma_{\bar{x}}$ or $\bar{x} + z\sigma_{\bar{x}}$, whichever is appropriate. Just as is true for two-sided intervals, in the case of a one-sided interval, z, the multiple of standard errors is determined by the desired confidence coefficient. For example, in the illustration of the one-sided interval using the lower limit of 21,912 miles, if we had desired a confidence coefficient of 95% rather than 97.5%, we would have used a z value of 1.65 rather than 1.96; for a confidence coefficient of 99%, we would have used a z value of 2.33, etc.

One final point may be made before turning to confidence interval estimation of different types of population parameters. Ordinarily, as was indicated in the tire example, σ_x, the standard deviation of the population, is unknown. Therefore, it is not possible to calculate $\sigma_{\bar{x}}$, the standard error of the mean. However, just as in hypothesis testing, we can estimate the standard deviation of the population from a sample and use this figure to calculate an estimated standard error of the mean. We shall use this estimation technique in the examples which follow.

Interval Estimation of a Mean—Large Sample

We will use illustrative examples to discuss confidence interval estimation, and will concentrate first of all on situations where the sample size is large. The discussion will focus, in turn, on interval estimation of a mean, a proportion, the difference between means, and the difference between proportions. Then, a brief treatment of corresponding estimation procedures for small samples will be given.

As our first illustration of interval estimation of a mean from a large sample, let us consider the following problem.

Example 8-1 A group of students working on a summer project with a social agency took a simple random sample of 120 families in a well-defined "poverty area" of a large city in order to determine the mean annual family income of this area. The sample results were $\bar{x} = \$2810$, $s_x = \$780$, and $n = 120$, where the formula $s_x = \sqrt{[\Sigma(x - \bar{x})^2]/n}$ was used to calculate the sample standard deviation. What would be the 99% confidence interval for the mean income of all families in this poverty area?

The only way this problem differs from the foregoing illustration of the mean lifetime of tires is that the population standard deviation is unknown. Since the sample is so large, the usual procedure in this case is simply to use the sample standard deviation as an estimate of the corresponding population standard deviation. For small samples, this would not be an appropriate procedure. As previously noted, when the population standard deviation is unknown, the appropriate estimator of σ_x is $\sqrt{[\Sigma(x - \bar{x})^2]/(n - 1)}$, even though it is not unbiased. However, we repeat that when n is as large as in this problem, the difference between having an n or $n - 1$ in the denominator is of no practical consequence. Using s_x as an estimator of σ_x, we can compute an estimated standard error of the mean $s_{\bar{x}}$ as in Chapter 7. We have

$$s_{\bar{x}} = \frac{s_x}{\sqrt{n}} = \frac{\$780}{\sqrt{120}} = \$71.23$$

Hence, we may use $s_{\bar{x}}$ as an estimator of $\sigma_{\bar{x}}$, and because n is large we invoke the central limit theorem to argue that the sampling distribution of $\bar{x}$ is approximately normal. Again, we have assumed the finite population correction to be equal to one. The confidence interval, in general, is given by

(8.6) $$\bar{x} \pm z s_{\bar{x}}$$

where $s_{\bar{x}}$ now replaces $\sigma_{\bar{x}}$, which was used when the population standard deviation was known. For a 99% confidence coefficient in a two-sided interval, $z = 2.58$. Therefore, the required interval is

$$\$2810 \pm 2.58(\$71.23) = \$2810 \pm \$183.77$$

or the population mean is roughly between \$2626 and \$2994 with a 99% confidence coefficient. The same sort of interpretation given earlier for confidence intervals again applies here.

Interval Estimation of a Proportion—Large Sample

To illustrate confidence interval estimation for a proportion, we shall make similar assumptions as in the preceding example. In the following example, we assume a large simple random sample drawn from a population which is very large compared to the sample size.

Example 8-2 In the town of Smallsville, a simple random sample of 800 automobile owners revealed that 480 would like to see the size of automobiles reduced. What are the 95.5% confidence limits for the proportion of all automobile owners in Smallsville who would like to see car size reduced?

In this problem we desire a confidence interval estimate for p, a population proportion. We have obtained the sample statistic $\bar{p} = 480/800 = 0.60$, which is the sample proportion who wish to see car size reduced. For large sample sizes and for p values not too close to zero or 1.00, the sampling distribution of $\bar{p}$ may be approximated by a normal distribution with mean $\mu_{\bar{p}} = p$ and $\sigma_{\bar{p}} = \sqrt{pq/n}$. Here we encounter the same type of problem as in interval estimation of the mean. The formula for the exact standard error of a proportion, $\sigma_{\bar{p}} = \sqrt{pq/n}$, requires the values of the unknown population parameters p and q. Hence, we resort to a similar type of estimation as used in the case of the mean. If we substitute the sample statistics $\bar{p}$ and $\bar{q}$ for the parameters p and q in the formula for $\sigma_{\bar{p}}$, we can calculate an estimated standard error of a proportion $s_{\bar{p}} = \sqrt{\bar{p}\bar{q}/n}$. Using the same type of reasoning as in the case of interval estimation of the mean, we can state a two-sided confidence interval estimate for a population proportion as

(8.7) $$\bar{p} \pm zs_{\bar{p}}$$

In this problem, $z = 2$, since the confidence coefficient is 95.5% for a two-sided interval. Hence, substituting into Equation (8.7), we have as our interval estimate of the proportion of all automobile owners in Smallsville who would like to see car size reduced

$$0.60 \pm 2\sqrt{\frac{0.60 \times 0.40}{800}} = 0.60 \pm 0.0346$$

or the population proportion is estimated to be included in the interval 0.5654 to 0.6346, or roughly between 56.5% and 63.5% with a 95.5% confidence coefficient.

It may be noted that in this problem the inference about the parameter p pertains only to the population of all automobile owners in Smallsville, since that was the statistical universe that was sampled. It would not be valid to use the computed interval estimate as pertaining to a different population such as all automobile owners in the United States. For a valid inference about this national population, the universe originally sampled should have been national in scope.

Interval Estimate of the Difference Between Two Means—Large Samples

The foregoing examples of *estimation* of a population mean and proportion based on a single sample correspond to *hypothesis testing* of a population mean and proportion based on a single sample. Similarly, we will now examine interval estimation of the difference between means and the difference between percentages based on data obtained from two independent large samples. These cases parallel the two similar hypothesis testing situations discussed in Chapter 7. We examine first an example of confidence interval estimation of the difference between two population means.

Example 8-3 A large department store chain was interested in the difference between the average dollar amount of its delinquent charge accounts in the Northeastern and Western regions of the country for a certain year. The store took two independent simple random samples of these delinquent charge accounts, one from each region. The mean and standard deviation of the dollar amounts of these delinquent accounts were calculated to the nearest dollar with the following results.

Sample 1	*Sample 2*
$\bar{x}_1 = \$\ 65$	$\bar{x}_2 = \$\ 76$
$s_1 = \$\ 22$	$s_2 = \$\ 25$
$n_1 = \ 100$	$n_2 = \ 100$

The Northeastern region is denoted as 2 and the Western region as 1.

The analysts decided to establish 99.7% confidence limits for $\mu_2 - \mu_1$, where μ_2 and μ_1 denote the respective population mean sizes of delinquent accounts. Of course, a point estimate of $\mu_2 - \mu_1$ is given by $\bar{x}_2 - \bar{x}_1$. The required theory for the interval estimate is based on the fact that the sampling distribution of $\bar{x}_2 - \bar{x}_1$ for two large independent samples is exactly normal, if the population of differences is normal, with mean and standard deviation

$$\mu_{\bar{x}_2 - \bar{x}_1} = \mu_2 - \mu_1$$

and

$$\sigma_{\bar{x}_2 - \bar{x}_1} = \sqrt{\frac{\sigma_1^2}{n_1} + \frac{\sigma_2^2}{n_2}}$$

where all of the symbols have the same meaning as in Chapter 7.

Since the population standard deviations, σ_1 and σ_2 are unknown, and since the sample sizes are large, as in Chapter 7, the sample standard deviations may be substituted into the formula for $\sigma_{\bar{x}_2 - \bar{x}_1}$ to give an estimated standard error of the difference between two means,

$$s_{\bar{x}_2 - \bar{x}_1} = \sqrt{\frac{s_1^2}{n_1} + \frac{s_2^2}{n_2}}$$

As usual with problems of this type the population of differences may not be normal, and the population standard deviations are unknown. However, since the samples are large, we can use the central limit theorem to assert that the sampling distribution of $\bar{x}_2 - \bar{x}_1$ is approximately normal. The required confidence limits are given by

(8.8) $$(\bar{x}_2 - \bar{x}_1) \pm z s_{\bar{x}_2 - \bar{x}_1}$$

The calculation for $s_{\bar{x}_2 - \bar{x}_1}$ in this problem is

$$s_{\bar{x}_2 - \bar{x}_1} = \sqrt{\frac{(22)^2}{100} + \frac{(25)^2}{100}} = \$3.33$$

Since a 99.7% confidence interval is desired, the value of z is 3. Therefore, substituting into (8.8) gives

$$(\$76 - \$65) \pm 3(\$3.33) = \$11 \pm \$9.99$$

Hence, to the nearest dollar, confidence limits for $\bar{x}_2 - \bar{x}_1$ are \$1 and \$21. It is a worthwhile exercise for the reader to attempt to express in words specifically what this confidence interval means.

Interval Estimation of the Difference Between Two Proportions—Large Samples

The procedure for constructing a confidence interval estimate for the difference between two proportions is essentially the same as the corresponding technique for means.

Example 8-4 A credit reference service investigated two simple random samples of customers who applied for charge accounts in two different department stores. The service was interested in the proportion of applicants in each store who had annual incomes exceeding $10,000. It was decided to establish 90% confidence limits for the difference $p_1 - p_2$, where p_1 and p_2 represent the population proportions of applicants in each store whose incomes exceeded $10,000.

The sample data were

Store 1	*Store 2*
$\bar{p}_1 = 0.50$	$\bar{p}_2 = 0.18$
$\bar{q}_1 = 0.50$	$\bar{q}_2 = 0.82$
$n_1 = 150$	$n_2 = 160$

where these symbols have their conventional meanings. As in the preceding problem, we can start with a point estimate, where $\bar{p}_1 - \bar{p}_2$ is the obvious point estimate of $p_1 - p_2$. The sampling distribution of $\bar{p}_1 - \bar{p}_2$ may be assumed to be approximately normal with mean and standard deviation:

$$\mu_{\bar{p}_1 - \bar{p}_2} = p_1 - p_2$$

and

$$\sigma_{\bar{p}_1 - \bar{p}_2} = \sqrt{\frac{p_1 q_1}{n_1} + \frac{p_2 q_2}{n_2}}$$

Since the population proportions p_1 and p_2 are unknown, and since the sample sizes are large, $\bar{p}_1$ and $\bar{p}_2$ may be substituted for p_1 and p_2 to obtain the estimated standard error of the difference between percentages.

$$s_{\bar{p}_1 - \bar{p}_2} = \sqrt{\frac{\bar{p}_1 \bar{q}_1}{n_1} + \frac{\bar{p}_2 \bar{q}_2}{n_2}} \qquad (8.9)$$

It may be noted from Equation (8.9) that in confidence interval estimation the estimated standard error of the difference between percentages does not involve the pooling of the sample percentages $\bar{p}_1$ and $\bar{p}_2$. This follows from the fact that in contrast with the hypothesis testing illustration in Chapter 7, in estimation we do not establish a null hypothesis that the population parameters p_1 and p_2 are equal.

As in the procedure for differences between means, we use the argument that by the central limit theorem the sampling distribution of $\bar{p}_1 - \bar{p}_2$ is approximately normal, and establish confidence limits of

(8.10) $$(\bar{p}_1 - \bar{p}_2) \pm z s_{\bar{p}_1 - \bar{p}_2}$$

In this problem the value of $s_{\bar{p}_1 - \bar{p}_2}$ is

$$s_{\bar{p}_1 - \bar{p}_2} = \sqrt{\frac{(0.50)(0.50)}{150} + \frac{(0.18)(0.82)}{160}} = 0.051$$

and since a 90% confidence coefficient is desired, $z = 1.65$. Therefore, the required confidence interval for the difference in the proportion of the applicants in the two stores whose incomes exceeded $10,000 is

$$(0.50 - 0.18) \pm 1.65(0.051) = 0.32 \pm 0.084$$

The confidence limits are 0.236 and 0.404.

Interval Estimation—Small Samples

In the preceding examples, which involved large simple random samples, the sampling distributions were assumed to be normal. Using the statistic $\bar{x}$ as an example, the underlying theory was that $(\bar{x} - \mu_x)/\sigma_{\bar{x}}$ is exactly normally distributed if the population is normal and approaches a normal distribution for non-normal populations as the sample size becomes large. In the cases of large samples where the population standard deviation was unknown, we substituted the sample standard deviation instead in computing an estimated standard error of the mean and still used the normal distribution for establishing confidence intervals.

Just as in the case of hypothesis testing, if the population standard deviation, σ_x, is unknown, and if the sample size is small ($n \leq 30$), this procedure is not appropriate, and the theoretically correct t-distribution should be used instead. As was indicated in Chapter 7, the statistic $(\bar{x} - \mu_x)/s_{\bar{x}}$ is distributed according to the t-distribution with $n - 1$ degrees of freedom, where the estimated standard error of the mean $s_{\bar{x}}$ is equal to $s_x/\sqrt{n}$ and s_x is given by $\sqrt{\Sigma(x - \bar{x})^2/(n - 1)}$. We will illustrate the procedure for confidence interval estimation for the mean using the t-distribution.

Interval Estimate of a Mean–t-distribution

As an example of a confidence interval estimate of a mean for a small sample where the population standard deviation is unknown, let us return to the same problem setting as in the corresponding large sample case.

Example 8-5 Assume that a simple random sample of nine automobile tires was drawn from a large production run of a certain brand of tires. The mean and standard deviation of the lifetime of the tires in the sample were $\bar{x} = 22{,}010$

miles and $s_x = 2520$ miles, where s_x was computed by the formula $s_x = \sqrt{\Sigma(x - \bar{x})^2/(n-1)}$. The population standard deviation is unknown. What are the 95% confidence limits for the mean lifetime for all tires in this production run?

By the same type of reasoning as we used in the case of the normal sampling distribution for means, confidence limits for the population mean, using the t-distribution are given by

(8.11) $$\bar{x} \pm ts_{\bar{x}}$$

where t is determined for $n - 1$ degrees of freedom.

In this problem, the estimated standard error of the mean is

$$s_{\bar{x}} = \frac{2520}{\sqrt{9}} = 840 \text{ miles}$$

The number of degrees of freedom, which is one less than the sample size is $9 - 1 = 8$. Referring to Table A-6 of Appendix A under column .05 for eight degrees of freedom, we find $t = 2.306$. We recall that .05 is the correct column heading in the t-table because that means that an area of .025 is found in each tail of the distribution. Thus, the probability is 0.05 that $(\bar{x} - \mu_x)/s_{\bar{x}}$ lies below -2.306 or above $+2.306$, and 0.95 that it lies between those two values.

In this problem, substituting into Equation (8.11), we obtain the following 95% confidence limits

$$22{,}010 \pm 2.306(840) = 22{,}010 \pm 1937.04$$

Hence, to the nearest mile, the confidence limits for the estimate of the mean lifetime of all tires in the production run are 20,073 and 23,947 miles. The interpretation of the meaning of this interval and the associated confidence coefficient is the same as in the large-sample, normal distribution case.

Interval Estimate of the Differences Between Two Means–*t*-distribution

In the discussion of large sample methods, an approximate technique was given for confidence interval estimation of the difference between two means. More exact methods using the t-distribution exist, which strictly speaking are applicable whenever the population standard deviations are unknown, and should be used for small samples. Use of the t-distribution for the sampling distribution of the difference between means involves the assumption that the population distributions are normal. Since we do not want to bog down the present discussion of estimation with a proliferation of formulas, suffice it to say that t-distribution methods exist for confidence interval estimation of $\bar{x}_2 - \bar{x}_1$ when the population variances are unknown. Two different problems are ordinarily distinguished, one in which the population variances are assumed to be equal, $\sigma_1^2 = \sigma_2^2$, and the second in which the population variances are assumed to be unequal, $\sigma_1^2 \neq \sigma_2^2$.

8.5 Determination of Sample Size

In all of the examples thus far, the sample size n was given. However, an important question is how large should a sample be in a specific situation? If a sample is used which is larger than necessary, resources are wasted; if the sample is smaller than required, the objectives of the analysis may not be achieved.

Sample Size for Estimation of a Proportion.

Statistical inference provides the following type of answer to the size question. Let us assume an investigator desires to estimate a certain population parameter and he wants to know how large a simple random sample is required. We will make the assumption that the population is very large relative to the prospective sample size. He must answer two questions in order to specify the required sample size. (1) "What degree of precision is desired?" and (2) "How probable do you want it to be that the desired precision will be obtained?" Clearly, the greater the degree of desired precision, the larger will be the necessary sample size. Also, the greater the probability specified for obtaining the desired precision, the larger will be the required sample size. We will use illustrative examples to indicate the technique of sample size determination for estimation of a population proportion and a population mean, respectively.

Example 8-6 Let us assume a situation in which we would like to conduct a poll among eligible voters in a city in order to determine the percentage who intend to vote for the Democratic candidate in an ensuing election. We specify that we want the probability to be 95.5% that we will estimate the percentage that will vote Democratic within ± 1 percentage point. What is the required sample size?

An answer will be given to the question by first indicating the rationale of the procedure. Then this rationale will be condensed into a simple summary formula. The statement of the question states a relationship between sampling error that we are willing to tolerate, and the probability of obtaining this level of precision. In this problem, we have required that $2\sigma_{\bar{p}}$ be equal to 0.01. As can be seen in Figure 8-5, this means that we are willing to have a probability of 95.5% that our sample percentage $\bar{p}$ will fall within 0.01 of the true but unknown population proportion p. We may now write

$$2\sigma_{\bar{p}} = 0.01$$

or

$$2\sqrt{\frac{pq}{n}} = 0.01$$

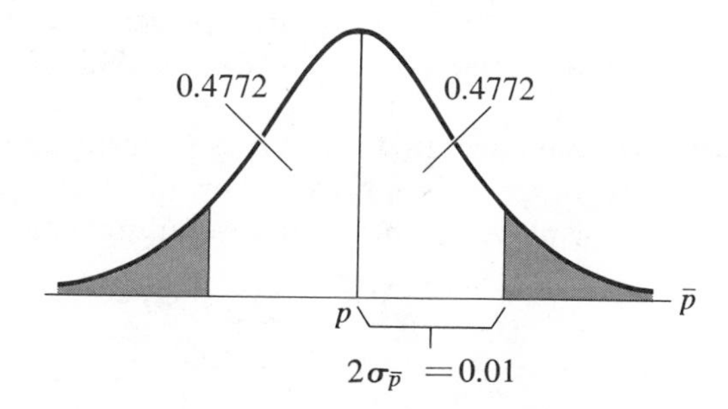

Figure 8-5 Sampling distribution of a proportion showing the relationship between error and probability of obtaining this degree of precision.

and

$$\sqrt{\frac{pq}{n}} = 0.005$$

In all of our previous problems, the sample size n was known, but here n is the unknown for which we must solve. However, it appears as though there are too many unknowns; the population parameters p and q as well as n. What we must do is estimate or guess values for p and q and then we can solve for n. Suppose we wanted to make a very conservative estimate for n. What should we guess as a value for p? In this context, by a conservative estimate we mean an estimate made in such a way as to insure that the sample size will be large enough to deliver the precision desired. In this problem, the "most conservative" estimate for n is given by assuming $p = 0.50$ and $q = 0.50$. This follows from the fact that the product pq is larger for $p = 0.50$ and $q = 0.50$ than for any other two possible p and q values, where $p + q = 1$. Thus, the largest or "most conservative" value of n is determined by substituting $p = q = 0.50$ as follows:

$$0.005 = \sqrt{\frac{0.50 \times 0.50}{n}}$$

Squaring both sides gives

$$0.000025 = \frac{0.50 \times 0.50}{n}$$

and

$$n = \frac{0.50 \times 0.50}{0.000025} = 10{,}000$$

Hence, to achieve the desired degree of precision, a simple random sample of 10,000 eligible voters would be required. Of course, the large size of this sample is attributable to the high degree of precision specified. If $\sigma_{\bar{p}}$ were doubled from 0.005 to 0.01, the required sample size would be cut down to one-fourth of

10,000 or 2500. This stems from the fact that the standard error varies inversely with the square root of sample size.

The arithmetic is simplified in these types of problems if we work in whole numbers of percentage points rather than decimals. For example, in whole numbers of percentage points, the preceding calculation becomes

$$1/2 = \sqrt{\frac{50 \times 50}{n}}$$

$$(1/2)^2 = 1/4 = \frac{50 \times 50}{n}$$

$$n = \frac{50 \times 50}{1/4} = 10{,}000$$

In this election problem, we assumed $p = q = 0.50$, although less conservative estimates are possible if it is believed that $p \neq 0.50$. In problems involving proportions, we would use whatever past knowledge we have to estimate p. For example, suppose we wanted to determine the sample size to estimate an unemployment rate, and we knew from past experience that for the community of interest the proportion of the labor force which was unemployed was somewhere between 0.05 and 0.10. We then would assume $p = 0.10$, since this would give us a more conservative estimate (larger sample size) than assuming $p = 0.05$ or any value between 0.05 and 0.10. The point is that if we assumed, say, $p = 0.07$, and in fact the true value of p was 0.09, the sample size we determined from a calculation involving $p = 0.07$ would not be large enough to give us the specified precision. On the other hand, assuming $p = 0.10$ assures us of obtaining the desired degree of precision regardless of what the true value of p is, in the range 0.05 to 0.10.

We can summarize this calculation for sample size by noting that we started with the statement that

$$e = z\sigma_{\bar{p}}$$

where e is the specified sampling error and z is the multiple of standard errors corresponding to the specified probability of obtaining this precision. Then, for infinite populations, or large populations relative to sample size, we have

(8.12)
$$n = \frac{z^2(p)(1 - p)}{e^2}$$

Hence, in the preceding voting problem, applying Equation (8.12) yields

$$n = \frac{(2)^2(0.50)(0.50)}{(0.01)^2} = 10{,}000$$

Sample Size for Estimation of a Mean

By an analogous calculation the required sample size for estimation of a mean can be determined. Suppose we wanted to estimate the arithmetic mean hourly wage rate for a group of skilled workers in a certain industry. Let us further assume that from prior studies we estimate that the population standard deviation of the hourly wage rates of these workers is about \$0.15. How large a sample size would be required to yield a probability of 99.7% that we will estimate the mean wage rate of these workers within ±\$0.03?

Since the 99.7% probability corresponds to a 3-standard error level, we can write

$$3\sigma_{\bar{x}} = \$0.03$$

or

$$\frac{3\sigma_x}{\sqrt{n}} = \$0.03$$

and

$$\frac{\sigma_x}{\sqrt{n}} = \$0.01$$

The population standard deviation, σ_x, is known from past experience. Hence, substituting $\sigma_x = \$0.15$ gives

$$\frac{\$0.15}{\sqrt{n}} = \$0.01$$

and

$$\sqrt{n} = \frac{\$0.15}{\$0.01} = 15$$

Squaring both sides yields the solution

$$n = (15)^2 = 225$$

Therefore, a simple random sample of 225 of these workers would be required. In summary, if we calculate $\bar{x}$ for the hourly wage rates of a simple random sample of 225 of these workers, we can estimate the mean wage rate of all skilled workers in this industry within \$0.03 with a probability of 99.7%.

In the problem discussed above, it was assumed that an estimate of the population standard deviation was available from prior studies. For example, this type of situation may arise in the case of governmental agencies which conduct repeated surveys of wage rates, population, and the like. If the population standard deviations (or estimated population standard deviations) in these past studies were not erratic or excessively unstable, they would provide useful bases for estimating σ_x values in the above sample size computational procedures.

Of course, sometimes an estimate of the population standard deviation may not be available from past experience. It may be possible, however, to get a rough estimate of σ_x if there is at least some knowledge of the total range of the basic random variable in the population. For example, suppose we know that the difference between the highest and lowest paid workers is about \$1.20. In a normal distribution a range of three standard deviations either side of the mean includes virtually the entire distribution. Thus, a range of $6\sigma_x$ includes almost all the frequencies, and we may state

$$6\sigma_x \approx \$1.20$$

or

$$\sigma_x \approx \$0.20$$

Of course, the population distribution is probably non-normal and \$1.20 may not be exact, either. Consequently, the estimate of σ_x may be quite rough. Nevertheless, we may be able to obtain a reasonably good approximate estimate of the required sample size in a situation where in the absence of this "guestimating" procedure, we may be at a loss for a notion of a suitable number of elements for the sample.

Expressing the technique of sample size estimation in the form of an equation yields

(8.13)
$$n = \frac{z^2\sigma_x^2}{e^2}$$

For the problem given earlier, this gives

$$n = \frac{(3)^2(\$0.15)^2}{(\$0.03)^2} = 225$$

In Equations (8.12) and (8.13), it is assumed that the n value which is determined is sufficiently large for the assumption of a normal sampling distribution to be appropriate and that the populations are large relative to this sample size.

8.6 Other Probability Sample Designs

In the discussion of statistical estimation up to this point, all estimates of population parameters were based on data obtained from *simple random samples.* However, in many practical situations, other sample designs may be preferable to simple random sampling in the sense that they may achieve greater precision of estimation at the same cost, or the same precision at lower cost. The subject of sample survey theory and methods is very specialized, and matters such as the selection of the optimal type of sample design and estimation method require a high level of expertise. "Do-it-yourself" methods are inadvisable for most sample surveys, and the obtaining of the advice or active involvement of a knowledgeable sampling specialist in the planning and implementation of such projects is usually a wise procedure. In this section, a brief discussion is presented of some alternative sample designs to simple random sampling and some of the issues involved in their selection and use. Stratified random sampling, cluster sampling, and systematic sampling are discussed in that order.

Stratified Random Sampling

In stratified random sampling, the population is classified into mutually exclusive subgroups or *strata*, and probability samples are drawn independently from each of these strata. The samples from each stratum may be obtained by simple random sampling, or some other form of probability sampling such as cluster sampling or systematic sampling. Sample statistics from each stratum may be combined to yield an overall estimate of a population parameter, or they may be compared with one another to reveal between-strata differences which may be of interest. In the ensuing discussion, we shall concentrate on the case where the objective is to obtain an overall estimate of a population parameter by combining the results from the individual strata.

The basic purpose of stratification as compared to simple random sampling is to obtain a reduction in sampling error or, synonymously, an increase in precision. This reduction in sampling error is accomplished by grouping together into strata elements which are more alike with respect to the characteristic under investigation than are the elements in the population as a whole. Stratification is most effective when the elements within strata are as homogeneous as possible, as regards the property to be studied, and, when the differences of elements among strata are as great as possible. We can see how this principle of stratification might operate to reduce sampling error in terms of a simple example.

Suppose we had a list of the incomes of every household in a city and we were interested in estimating the arithmetic mean household income by sampling. Let us compare on an intuitive basis the sizes of sampling error that might be expected in using first simple random sampling and then stratified random sampling. If we draw a simple random sample of n households, the sampling error in using the sample mean $\bar{x}$ to estimate the population mean household income μ_x may be measured by the standard error of the mean $\sigma_{\bar{x}}$. As we have seen earlier, $\sigma_{\bar{x}}$ is the standard deviation of the sampling distribution of $\bar{x}$, and thus measures the variation in $\bar{x}$ values that occurs in drawing repeated simple random samples of the same size from the given population. If we assume considerable variation among the incomes of the households in the population, ranging from very poor families to very rich families, there would correspondingly be a good deal of variation in the sampling distribution of $\bar{x}$. We can see this from the formula $\sigma_{\bar{x}} = \sigma_x/\sqrt{n}$, for infinite populations, in which $\sigma_{\bar{x}}$ varies directly with σ_x; that is, sampling error varies directly with population dispersion. Any particular simple random sample might be very unrepresentative of the population in the sense that it might contain a disproportionate number of high income or low income families. Indeed, *any* group of n households constitutes a possible sample. Therefore, we are not surprised that there would be considerable variation in the sample mean $\bar{x}$ from sample to sample.

On the other hand, suppose the households are grouped into, say, ten strata by income, ranging from the households with highest incomes in stratum one to the families with lowest income in stratum ten. Again, we wish to draw a sample of n households. If we draw a simple random sample of households within every stratum, and calculate the sample mean income for each of these strata, these means would contain relatively little sampling error. This follows from the fact that a sample drawn in, say, stratum ten can only contain low income households among which there is comparatively little variation in income. By properly weighting these $\bar{x}$ values (the weights are the population numbers of households in each stratum), we get a combined estimate of $\bar{x}$ which tends to be closer to the population mean μ_x, on the average, than under simple random sampling. Intuitively, we can see why this might be so. In simple random sampling, samples which contain only high income families or only low income families are possible, thus making for large variations in $\bar{x}$ from sample to sample. In stratified sampling, since at least some sampling must be carried out in each stratum, every sample of size n would contain representation from all ten strata. Thus, the variation in $\bar{x}$ values from sample to sample would tend to be less than in simple random sampling. If we imagine the extreme case in which all households within each stratum have the same income, we would need to draw only one household in each stratum and ascertain its income. This figure would be a perfect estimate of the mean income for households in the stratum. Then if we properly weighted these ten means to obtain a combined sample $\bar{x}$ value, this figure for $\bar{x}$ would be a perfect estimate of μ_x. That is, there would be no sampling error associated with this sample mean. On the other hand, there *would be* sampling error associated with the sample mean of a simple random sample of ten households from this population.

The foregoing example illustrated the simplest type of stratified sampling, namely stratified simple random sampling. This sample design involves the setting up of strata and then the drawing of simple random samples within each stratum. The example is unrealistic because the characteristic used to stratify the population, namely household income, was also the characteristic we were interested in estimating. Ordinarily, this is not the case. In the usual situation, the property used for stratification is different from the one to be estimated. As might be expected, greater precision is accomplished when the characteristic used for stratifying the population is related to the characteristic to be estimated than when this relationship does not exist. Hence, for example, in studying consumer spending, we might stratify consumer units by characteristics such as disposable income, size of consumer unit, and geographic location. In estimating unemployment of the labor force, we might stratify by race, sex, education, and age. In investigating profit margins of firms, we might stratify by type of industry and size of firm.

Estimation of a Population Mean and Its Precision

In order to illustrate the method of estimating a population mean and its precision in stratified simple random sampling, let us consider a simple example.

Example 8-7 It was desired to estimate average household durables expenditures in a city and also the comparison of such spending for white and non-white families. For this purpose the city's households were stratified by white and non-white race groups and simple random samples were drawn within each stratum. Before proceeding to the calculations, we introduce the required notation. Capital letters denote populations; small letters denote samples. The population is subdivided into L strata.

Let N = the total number of elements in the population and
N_i = the number of elements in the ith stratum.

Hence,

$$N = \sum_{i=1}^{L} N_i$$

Similarly, let

n = the total number of elements in the sample and
n_i = the number of elements in the sample from the ith stratum.

Hence,

$$n = \sum_{i=1}^{L} n_i$$

Let $\bar{x}_i$ = the sample mean in the ith stratum and
s_i = the sample standard deviation in the ith stratum.

(We assume here that the sample standard deviation was determined using a denominator of $n_i - 1$.) Data for this example are shown in Table 8-1.

Table 8-1 Stratified Sample of Households in a City.

Stratum (i)	*Number of Elements in Population* (N_i)	*Number of Elements in Sample* (n_i)	*Sample Mean Household Durables Expenditures* ($\bar{x}_i$)	*Sample Standard Deviation* (s_i)
$i = 1$ (White)	90,000	900	\$250	\$100
$i = 2$ (Non-white)	10,000	100	200	50
$L = 2$	$N = 100{,}000$	$n = 1000$		

An estimator of the population mean household durables expenditures is given by a weighted average of the strata sample means, where the population numbers of elements in each stratum are the weights. The reader should be able to prove easily that this is an unbiased estimator of the population mean. Symbolizing this estimator of the overall mean as $\bar{x}_{st}$, where the subscript "st" denotes "stratified," we have

(8.14) $$\bar{x}_{st} = \frac{N_1\bar{x}_1 + N_2\bar{x}_2 + \cdots + N_L\bar{x}_L}{N} = \sum_{i=1}^{L}\left(\frac{N_i}{N}\right)\bar{x}_i$$

Substituting into Equation (8.14) gives as an estimate of mean household durables expenditures in this city

$$\bar{x}_{st} = \left(\frac{90{,}000}{100{,}000}\right)\$250 + \left(\frac{10{,}000}{100{,}000}\right)\$200 = \$245$$

In order to obtain a measure of the precision of this estimate, we can calculate the variance of the sampling distribution of $\bar{x}_{st}$. In the usual way, let us denote the variance of $\bar{x}_{st}$ as $\sigma^2_{\bar{x}_{st}}$. Denoting the variance of a stratum sample mean as $\sigma^2_{\bar{x}_i}$, the variance of the overall mean is a weighted average of these $\sigma^2_{\bar{x}_i}$'s, where the weights are the squares of the relative number of elements in each stratum. This is easy to show by the additive property of variances for independent samples. The variance of $(N_i/N)\bar{x}_i$ is $(N_i/N)^2\sigma^2_{\bar{x}_i}$. We want the variance of $\bar{x}_{st}$ as defined in Equation (8.14). Since the strata sample means are independent, by Rule 9 of Appendix C, we have

(8.15) $$\sigma^2_{\bar{x}_{st}} = \sum_{i=1}^{L}\left(\frac{N_i}{N}\right)^2\sigma^2_{\bar{x}_i}$$

A point estimator of $\sigma^2_{\bar{x}_{st}}$ is given by

(8.16) $$s^2_{\bar{x}_{st}} = \sum_{i=1}^{L}\left(\frac{N_i}{N}\right)^2 s^2_{\bar{x}_i}$$

where

(8.17) $$s^2_{\bar{x}_i} = \frac{s_i^2}{n_i}\left(1 - \frac{n_i}{N_i}\right)$$

The variance $s^2_{\bar{x}i}$ is simply the variance of the stratum mean $\bar{x}_i$, which can be recognized as being in the familiar form of the variance of a sample mean, with $1 - n_i/N_i$ as the finite population correction. In this problem, these variances of strata means are

$$s^2_{\bar{x}1} = \frac{(\$100)^2}{900}\left(1 - \frac{900}{90{,}000}\right) = \$11$$

and

$$s^2_{\bar{x}2} = \frac{(\$50)^2}{100}\left(1 - \frac{100}{10{,}000}\right) = \$24.75$$

Substituting into (8.16) gives

$$s^2_{\bar{x}st} = \left(\frac{90{,}000}{100{,}000}\right)^2 (\$11) + \left(\frac{10{,}000}{100{,}000}\right)^2 (\$24.75) = \$9.16$$

The square root of $s^2_{\bar{x}st}$, denoted $s_{\bar{x}_{st}}$ is the standard error of $\bar{x}_{st}$. In this problem

$$s_{\bar{x}_{st}} = \sqrt{\$9.16} = \$3.03$$

The above results can be used according to the usual techniques of statistical inference. If we focus on the difference between average expenditures for whites and non-whites, we can test for the significance of the difference between the sample strata means $\bar{x}_1$ and $\bar{x}_2$. On the other hand, if interest mainly centers on the estimate of the overall population mean, we can use the overall sample mean $\bar{x}_{st}$ and its estimated standard error $s_{\bar{x}_{st}}$ to obtain confidence limits under the usual assumptions. For example, assuming the sampling distribution of $\bar{x}_{st}$ is approximately normal in this problem, 95.5% confidence limits for the population mean are given by

$$\bar{x}_{st} \pm 2s_{\bar{x}_{st}} = \$245 \pm 2(\$3.03)$$

Hence, the confidence limits are \$238.94 and \$251.06.

Methods of Allocation of the Sample to Strata

In the example given in Table 8-1, 1/10 of the population elements in each stratum were drawn into the sample. This is an example of *proportionate stratified sampling*, where the sampling fraction n_i/N_i is the same for all strata. This method of allocating a sample to strata is frequently used, mainly on grounds of convenience and simplicity. A proportionate stratified sample is an example of a "self-weighting sample," that is, a sample in which weights are not necessary. For example, in our illustration, if we merge the sample data from the two strata, the sample mean of the 1000 families is equal to the previously calculated weighted mean, $\bar{x}_{st}$. This follows from the fact that the sample numbers n_1 and n_2 are in the same proportion as the population numbers N_1 and N_2. Another way of looking at this relationship is that in proportionate stratified sampling as in simple random sampling, every element in the population has the same probability of inclusion in the sample. This self-weighting characteristic simplifies computational procedures, particularly when computing equipment is used and when a large number of population characteristics must be estimated from the same sample.

As might be expected, a proportionate allocation of a sample to strata does not generally yield optimum results in terms of minimizing sampling error. In a proportionate stratified sample, the sample is allocated to strata based only on the population number of elements in each stratum. If the cost of sampling elements from each of the strata is the same, better results can be obtained by also taking into account the variability in each stratum. The term *optimum allocation* is used to denote a method of allocating a fixed number of sample elements to strata in an optimal manner. In terms of minimizing sampling error for a given size sample, optimum allocation is obtained in stratified simple random sampling by assigning sample elements to strata in proportion to both the number of population elements and the standard deviation of the elements within each stratum. That is,

$$\frac{n_i}{n} = \frac{N_i\sigma_i}{\Sigma N_i\sigma_i}$$

Hence, the formula for accomplishing this optimum allocation is

$$n_i = n\left(\frac{N_i\sigma_i}{\Sigma N_i\sigma_i}\right) \tag{8.18}$$

We note from Equation (8.18) that in order to determine the exactly correct optimum allocation, we would have to know the standard deviations of each stratum, prior to drawing the sample. For planning purposes, estimates are made of the σ_i, based on prior information or in the case of large scale surveys, based on small pilot studies. For the purpose of illustrating the method of calculation of the optimum allocation, in our problem, we shall use the figures for s_i given in Table 8-1 as our estimates of the σ_i. Taking a sample of 1000 households as before, the data are $n = 1000$, $N_1 = 90{,}000$, $N_2 = 10{,}000$, $\sigma_1 = \$100$, $\sigma_2 = \$50$. Therefore, we obtain

For stratum 1: $N_1\sigma_1 = (90{,}000)(\$100) = \$9{,}000{,}000$
For stratum 2: $N_2\sigma_2 = (10{,}000)(\$\ 50) = 500{,}000$
$\Sigma N_i\sigma_i = \$9{,}500{,}000$

Thus, the optimal numbers of households to draw from each stratum are

$$n_1 = 1000\left(\frac{\$9{,}000{,}000}{\$9{,}500{,}000}\right) = 947.4$$

$$n_2 = 1000\left(\frac{\$500{,}000}{\$9{,}500{,}000}\right) = 52.6$$

Rounding off these results, the optimum allocation requires that 947 white households (stratum 1) and 53 non-white households (stratum 2) be drawn into the sample. The fact that a sample of only 53 non-white families is called for under optimum allocation as compared to 100 such families under proportionate stratified sampling is a result of the smaller variability of household durables expenditures in the non-white than in the white stratum.

It is important to realize that the optimum allocation we have just calculated pertains to the minimizing of sampling error in an estimate of the overall population mean. Obviously, if we also wanted to study the non-white households cross-classified by a number of other characteristics, a sample of only 53 households would doubtless be too small. For example, such a sample size might provide only a very small number of households for each of the cells in the cross-classification, *and in some cells, no households at all.* Also, if we wanted to make certain comparisons for non-white and white families, we might be moved in the direction of increasing the sample from the non-white stratum. Clearly, when there is a multiplicity of purposes to be accomplished in a survey, a principle of optimum allocation of sample size such as the one we have just discussed is apt to represent only one of the relevant considerations. Furthermore, if the cost of obtaining elements in the sample differs among strata, this should also be taken into account. Formulas and techniques are provided in the specialized literature of sample survey methods for accomplishing a minimum cost allocation of the sample among strata.

In many applications stratification is particularly useful in achieving gains in precision as compared to simple random sampling when there are extreme items which can be grouped into separate strata. In these cases, very high sampling fractions, often 100%, are used in the stratum or strata containing the extreme items. For example, consider an industry in which there are 3000 firms but the largest ten firms account for 90% of total sales. Suppose we wish to run a survey to estimate total sales and other characteristics of the industry. If we establish a separate stratum for the ten largest firms, and use a 100% sampling fraction in that stratum, that is, include all ten firms, then obviously we will account for 90% of total sales with no sampling error. No matter how we subdivide the remaining 2990 firms, sampling error has been restricted to that part of the population which accounts for only 10% of total sales. Even if only a relatively small sample is taken of the 2990 firms, total sales for the industry can be estimated with very high precision. This method of reducing sampling error is of considerable practical importance for economic data of various types, particularly for firms or individuals because of the high degree of concentration which often exists. For example, in many industries, the frequency distributions of firms by measures of size such as sales, income, assets, and net worth are highly skewed to the right. The establishment of separate strata for the large firms in the right-hand tails of these distributions is a very useful procedure.

Cluster Sampling

Cluster sampling is a technique in which the population is subdivided into groups or "clusters" and then a probability sample of these clusters is drawn and studied. For example, to carry forward the illustration considered under stratified sampling, suppose we wished to conduct a survey of expenditures on durable goods for all households in the United States. We might draw a simple random sample of counties in the United States, and then draw a simple random sample of households within the sampled counties. The counties are referred to as clusters, since for purposes of analysis the households are conceived of as being clustered into county units.

The example in the preceding paragraph illustrates so-called *multi-stage sampling*, since the sampling is carried on in more than one stage. In sample survey terminology, the counties are referred to as *primary sampling units* (p.s.u.'s), since they are the units sampled at the first stage. If a complete enumeration were attempted for all households in the sampled counties, the sample design would be referred to as *single-stage sampling*. However, since we assumed that a sample of households would be drawn within each primary sampling unit, this is an illustration of *two-stage sampling*. The units whose characteristics we wish to measure or count are referred to as *elementary units*. Thus, in our example, if we are interested in measuring household durables expenditures, a household represents the elementary unit. If, on the other hand, we were interested in characteristics of individuals within these households, an individual would constitute the elementary unit. A household would then represent a cluster of individuals. In summary, the county would be the primary sampling unit, the household the secondary sampling unit, and the individual the third stage sampling unit and also the elementary unit.

The primary purpose of cluster sampling is to accomplish cost savings in sample design. The nature of this cost reduction is most easily seen by a comparison of simple random sampling with cluster sampling. Let us continue with our illustration and assume that we want to determine average household durables expenditures in the United States. If we wished to draw a simple random sample of households, we would need an appropriate sampling frame, that is, a complete listing of all households from which to draw the sample. Suffice it to say that no such list exists, and it would clearly be prohibitively expensive to attempt to compile such a list. However, even if such a list were available, and numbers were assigned to each family, and tables of random numbers or some other technique were used to draw a simple random sample, serious practical difficulties would be encountered. The sampled households would be scattered throughout the United States, and the costs of travel, interviewing, and supervision of enumerators would be extremely high. On the other hand, if some form of multi-stage sampling were employed, then the successive stages of sampling might include clusters such as counties, political subdivisions within counties, and blocks. Listings usually exist at all of these stages. Listings of households for only the sampled blocks would then have to be constructed. The cost savings of listing, travel, interviewing, and supervision are obvious as compared to simple random sampling.

As regards sampling error, we have seen that in stratified sampling it is desirable to minimize differences among elements within strata and to maximize differences among strata. Exactly the opposite principle applies in the case of cluster sampling, where sampling error is reduced when the units within clusters are as heterogeneous as possible. As an intuitive extreme example, if a single cluster duplicates all of the heterogeneity which exists in the population, we would only need to draw this one cluster into our sample to have a good description of the population. However, in numerous instances, the clusters which can be set up conveniently may be quite homogeneous. In such cases, it is not ordinarily feasible to attempt to construct clusters which are heterogeneous, and the population must be dealt with in its existent form. For example, in our illustration of estimation of household durables expenditures, blocks within a city might constitute the clusters at one stage of a multi-stage sample design. Since high income households tend to be concentrated in certain blocks and low income households in others, the block clusters would be quite homogeneous. Nevertheless, it may be preferable and more feasible to use these blocks as clusters than to set up clusters consisting, say, of groups of blocks. The alternative of subsampling within blocks and including many more blocks throughout the city may also turn out to be infeasible from the cost standpoint. Frequently, a larger number of elementary units is required in cluster sampling to achieve the same precision as in simple random sampling. However, in terms of the cost required to obtain a fixed level of precision, a well designed cluster sample may be far superior to a simple random sample.

Systematic Sampling

In many practical applications, systematic sampling is used instead of simple random sampling as a method of obtaining random selection. In a systematic sample, every kth element is drawn from a population listed or arranged in some specific order. The starting point is selected at random from the first k elements. Thus, suppose the publisher of a news periodical has an alphabetically arranged list of subscribers, and he desires to send a questionnaire to a systematic sample of every fourth name on the list. In this case, $k = 4$. A starting point is selected at random from the first four names.

Assume the starting point is 2, that is, the second name. Then every fourth name is selected after number 2. Therefore, the systematic sample consists of names numbered 2, 6, 10, 14, and so forth.

If the elements of a population occur in random order, then the results of systematic sampling are very similar to those of simple random sampling. The assumption that the population is in random order can ordinarily be made in the case of populations of physical objects which have been thoroughly mixed. As an example, consider the output of a production process, where the order of production of the items has not been kept, and therefore these items do not appear to occur in any discernible sequence. In practice, even if the population is arranged in some particular sequence, but this order appears to be unrelated to the characteristic under investigation, the assumption of random order occurrence is frequently made. Thus, in the case of the alphabetical list of subscribers to the news periodical, if the questionnaire pertained to economic characteristics of these subscribers, it seems reasonable to assume that alphabetical order of names and economic characteristics are unrelated.

In certain instances, the results of systematic sampling may be very similar to those of proportionate stratified sampling. As an example, imagine a file of cards on which individual incomes have been recorded and these cards have been arranged in order from lowest to highest income. Further, let us assume that three strata of income have been defined: low, medium, and high. If we take a systematic sample of every tenth card, we will have a sample which includes 10% of the cards from each income stratum. Thus, in effect, we have a proportionate stratified sample, with a systematic sample within each stratum.

In many situations, systematic sampling may result in gains in precision over simple random sampling because of a stratification which exists in some spatial arrangement. For example, if a systematic sample is drawn of blocks in a city arranged in geographical order, the sample will include blocks from every area of the city. Again, the results of such sampling will be quite similar to those of stratified sampling, where the strata would be geographic areas consisting of collections of blocks. By comparison, a simple random sample of blocks would tend to result in greater sampling errors because of the wide differences in characteristics among blocks. Indeed, as we have seen, this tendency toward large sampling errors in simple random sampling when there is a great deal of population variation is the basic reason for introducing stratification into a sample design.

On the other hand, systematic sampling may under certain circumstances give poorer results than simple random sampling or alternative random selection techniques. Systematic sampling should be avoided whenever there is a periodic or cyclical variation in the order of the population elements. For example, suppose there are five workers denoted 1, 2, 3, 4, 5 who produce the same article and their work is placed along a moving production line in the order 1, 2, 3, 4, 5. We wish to draw a sample of the total production of the five men. Obviously, if we take a systematic sample of every fifth article, we will observe the work of only one worker.

Similarly, if we imagine city blocks, all of which have exactly 25 houses, if we start on a corner house and take a systematic sample of every 25th house, we will have a sample of only corner houses. If the characteristic of interest is the market value, and corner houses typically have higher market values than others on the block, we clearly would have a non-representative sample of the population of houses. Unless the investigator has specific knowledge that these types of periodicities do not exist, it is rather dangerous to use systematic sampling, especially since the sample itself may not give any indication of the existence of such periodicity.

Even in an alphabetical listing such as the periodical subscriber list referred to earlier, there may be subtle hidden cycles or periodic patterns. Various ethnic or national groups have last names in which certain first letters are used more frequently than others, and there are other letter sequencing patterns as well. Such factors can conceivably give rise to definite over-representation of certain groups in a systematic sample drawn from the alphabetical list if the hidden cycles happen to coincide with the sampling interval. There is a variation of systematic sampling which provides a safeguard against such hidden periodic patterns. Let us suppose we want to select every tenth name on an alphabetical list. We could select a name at random from the first ten names, another at random from the next ten names, and so forth. Thus, since there would be no fixed sampling interval, periodic patterns would not systematically affect the nature of the sample drawn.

Various combinations of the probability sampling techniques discussed in this chapter are often used in sample surveys. For example, in a national survey, counties may be stratified by one or more criteria, a sample may be drawn of townships within counties, blocks within townships, and dwelling units within blocks. If, for example, simple random sampling were used to select counties or townships, and systematic sampling were employed to select blocks or dwelling units, this sample design would represent a combination of stratification, clustering, simple random sampling, and systematic sampling. The effective design of such surveys requires expertise derived from experience as well as a thorough understanding of the necessary sampling methods and theory.

Other Sampling Methods

The sampling techniques discussed thus far in this chapter by no means exhaust the possible methods. A couple of others, *replicated sampling* and *sequential sampling* will be briefly mentioned here. *Replicated sampling*, or as it is sometimes called *interpenetrating sampling* involves the drawing and processing of two or more independent subsamples from the same population. The technique usually establishes "zones" within the population which contain equal numbers of sampling units. Then independent samples are drawn from these zones for subsample 1, subsample 2, etc. A major advantage of this type of sampling is the ease with which measures of sampling precision and systematic errors may be derived.

Another type of sampling which is particularly widely employed in industrial quality control work is *sequential sampling*. The method can be illustrated in terms of sampling of a shipment of manufactured parts by a purchasing company. We have discussed previously how such a purchaser may establish decision rules for acceptance or rejection of a shipment based on data obtained from a single sample drawn at random from the shipment. This method may be referred to as *single sampling*. *Double sampling* is a method which permits the decision to be deferred until the second sample has been drawn. If the first sample is very good, the shipment is accepted; if the first sample is very poor, the shipment is rejected. On the other hand, if the first sample is neither good nor bad enough to determine acceptance or rejection, another sample is drawn and the decision is made on the basis of the data from the first and second samples combined. *Sequential sampling* is simply an extension of the double sampling technique. It permits the decision to be deferred until any number of samples have been drawn. Sequential sampling usually requires a smaller average sample size than single sampling for the same quality protection (same levels of risks of Type I and Type II errors). Furthermore, very good or very poor shipments tend to be accepted or rejected very quickly on the basis of relatively small size samples. Increased attention in terms of larger sample sizes is reserved for shipments of intermediate quality, where the decisions are more difficult to make. Sequential sampling techniques are not restricted to quality control work, but are useful in a wide variety of business and other applications.

8.7 The Balancing of Errors

Our discussion of errors involved in estimation of population values from sample data has concentrated mainly upon random errors which measure the effects of chance in the sampling process. These errors are measured by the variance of the sampling distribution of the statistic around the expected value of the statistic. In the case of unbiased estimators, this expected value is equal to the true value of the parameter being estimated. However, for biased estimators, there is another component of error, namely, bias, which is the expected value of the statistic minus the true value of the parameter. A measure which summarizes these two sources of error is known as mean-squared error. *Mean-squared error* is defined as the expected value of the squares of the deviations of the sample estimates from the true value of the parameter. Thus, if we denote the true value of the population parameter as θ and the estimator as $\hat{\theta}$, then the mean-squared error is defined as $E[(\hat{\theta} - \theta)^2]$. Mean-squared error can be written

$$E[(\hat{\theta} - \theta)^2] = E[\hat{\theta} - E(\hat{\theta})]^2 + [E(\hat{\theta}) - \theta]^2 \tag{8.19}$$

where

$E[(\hat{\theta} - \theta)^2]$ = the mean-squared error

$E[\hat{\theta} - E(\hat{\theta})]^2$ = the variance of the estimator or the square of the standard error of the estimator

$E(\hat{\theta}) - \theta$ = the bias

In words, Equation (8.19) may be stated as

$$\text{Mean-Squared Error} = (\text{Standard Error})^2 + (\text{Bias})^2 \tag{8.20}$$

We can think of this relationship in terms of the right triangle depicted in Figure 8-6. If the estimator is unbiased, the mean-squared error is simply equal to the variance of the estimator. Fortunately, for biased estimators which are consistent, the bias tends to decrease rather rapidly as sample size increases and in most practical situations may be considered negligible.

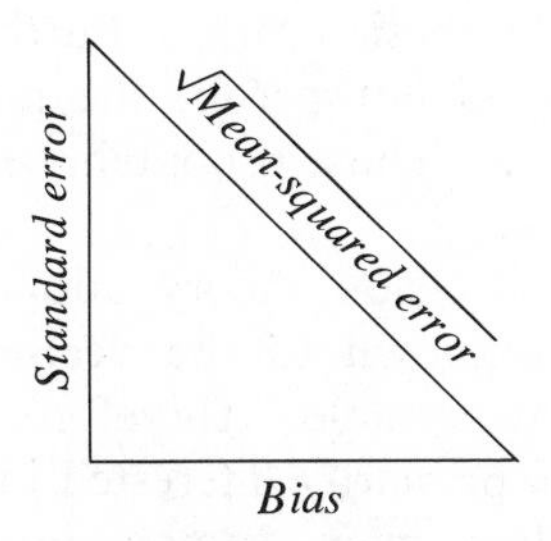

Figure 8-6 A right triangle showing the relationship between mean-squared error, the standard error of the estimator, and bias.

An analogous relationship for the total error of a survey can be specified. If we think of the *total error* of a survey as the hypotenuse of the triangle, the legs may be viewed as *random errors* and *systematic errors*. The random errors arising from sampling can be decreased by increasing sample size. However, if the systematic errors arising from non-sampling sources are very large compared to sampling error, increasing the sample size may not be a wise use of resources. Indeed, the total error may in fact be decreased by decreasing the sample size and using the resources saved to cut down on non-sampling errors. As an example, in a series of consumer expenditure surveys, it was found that when the sample data on consumption of alcoholic beverages reported by respondents were blown up to population figures, they consistently amounted to only about one-half of corresponding population totals, which were very accurately determined from distillers' alcoholic tax figures. It was evident that the respondents were under-reporting their spending on alcoholic beverages. Clearly, in this type of situation, in future similar surveys, increasing the sample size would be rather irrelevant. Instead efforts to decrease the response errors would have a greater payoff in terms of decreasing total error.

8.8 The Bayesian Decision Theory Approach

At the conclusion of Chapter 7, there was a brief discussion of Bayesian decision theory as a logical extension of classical statistics in the area of hypothesis testing. Bayesian decision theory provides a similar extension of confidence interval estimation. Classical methods permit estimation of population parameters based solely on data provided by the present sample. On the other hand, Bayesian decision theory furnishes a method for combining prior knowledge with current sample data for estimation purposes. Classical methods establish interval estimates of parameters with specified confidence coefficients. These confidence levels are rather arbitrary just as are significance levels in hypothesis testing. Further, as we have seen in this chapter, classical methods do not permit the population parameter to be treated as a random variable. Thus, probability statements cannot be made concerning these parameters.

In contrast, Bayesian decision theory adopts the viewpoint that if a population parameter is unknown to the decision maker, he may validly consider it to be a random variable. Therefore, he may make probability statements of the type he is primarily interested in—conditional probabilities concerning population values given sample information. Finally, payoff functions are formally incorporated in Bayesian analysis. Hence, the nature of the losses involved in incorrect estimation are made an explicit part of the formal problem structure. As in the case of the extension of hypothesis testing, decision theory extensions of classical estimation methods are discussed in subsequent chapters.

PROBLEMS

1. The parameter θ can take on one of three values, $\theta_1, \theta_2, \theta_3$. Event X occurs and has conditional probabilities of occurrence as follows

   $P(X \mid \theta_1) = .35 \qquad P(X \mid \theta_2) = .75 \qquad P(X \mid \theta_3) = .10$

   What is the maximum likelihood estimate of θ?

2. The parameter θ can take on one of three values, θ_1, θ_2, θ_3. An experiment is to be run to help estimate θ. The conditional probabilities of the experimental outcomes, given θ, are as follows

Experimental outcome	*True value of* θ θ_1	θ_2	θ_3
A	.8	.4	.02
B	.1	.5	.08
C	.1	.1	.90

If one were to use maximum likelihood estimation, what would be the best estimate of θ given that the experiment resulted in
   (a) *A?*
   (b) *B?*
   (c) *C?*

3. The following results were obtained from a random sample of 400 shoppers in the Philadelphia area. (1) 60% preferred sales help when purchasing clothing. (2) 30 of the 100 shoppers from New Jersey thought store *A* had the lowest prices in town. (3) 80 of the 300 shoppers from Philadelphia thought store *A* had the lowest prices in town.
   (a) Within what range would you be 90% confident of estimating the true percentage of shoppers who prefer sales help when purchasing clothing?
   (b) Construct a 95% confidence interval for the true difference in opinion of the New Jersey and Philadelphia shoppers about store *A*'s prices. Interpret the interval in light of the problem.

4. A simple random sample of 400 firms within a particular industry yielded an arithmetic mean number of employees of 232 and a standard deviation of 40.
   (a) Establish a 92% confidence interval for the population mean.
   (b) Precisely what is the meaning of a 92% confidence interval?

5. In a simple random sample of 1000 stockholders of Atlas Credit Corporation, 600 were in favor of a new issue of bonds (with attached warrants for purchase of common stock) while 400 were against it. Construct a 95.5% confidence interval for the actual proportion of all stockholders who are in favor of the new issue.

6. A random sample is to be selected to estimate the proportion of smokers at the University. The range of the estimate is to be kept within 5% with a confidence level of 95.5%. How large a simple random sample is required?

7. An efficiency expert made 100 random observations on a typist. In 56 of these observations he found that she was "idle."
   (a) Construct a 98% confidence interval for the proportion of the time that the typist is idle.
   (b) It was decided that the confidence interval in (a) was too large. If a new sample were drawn, how large would it have to be in order to estimate

the true proportion idle within three percentage points, with a confidence coefficient of 98%?

8. Explain the following statement in detail. The standard deviation of $\bar{x}$ ($\sigma_{\bar{x}}$) is both a measure of statistical error in point estimation and a measure of dispersion of a distribution.

9. Of the 14,714 families in West Falls, a random sample of 217 families was taken in order to determine the mean family income in this depressed area. A 95% confidence interval ($3812 to $4116) was established on the basis of the sample results.

   Using only the above information, which of the following statements are valid?

   (a) Of all possible samples of size 217 drawn from this population, 95% of the sample means will fall in the interval.
   (b) Of all possible samples of size 217 drawn from this population, 95% of the universe means will fall in the interval.
   (c) Of all possible samples of size 217 drawn from this population, 95% of the confidence intervals established by the above method will contain the universe mean.
   (d) 95% of families in West Falls have means between $3812 and $4116.
   (e) Using the above method, exactly 95% of the intervals so established will contain the sample mean $\bar{X}$.
   (f) We do not know whether the universe mean is in the interval $3812 to $4116.

10. A large shipment of electronic parts is received from Bell Supply House. The parts are to be sampled in order to estimate the proportion of unusable parts.

    (a) How large a random sample should be drawn if the objective is to estimate the proportion of unusable parts within four percentage points with 95% confidence? Assume the proportion of unusable parts is estimated from past experience to be .20 or less.
    (b) Assume, without regard to your answer in (a), that the receiving company drew a random sample of 300 parts and found 36 to be unusable. Construct a 99% confidence interval for the proportion of unusable parts in the shipment.

11. State whether each of the following statements is about an estimator or an estimate, and if about an estimate whether a point or interval estimate.

    (a) The sales manager feels that November's sales should be between $15,000 and $20,000.
    (b) In a discussion on the well-being of the community, the political party in power tends to refer to the *mean income* of a sample from the community, while the opposing party tends to refer to the *median income* of the sample.
    (c) A quality control engineer claims, after examining a proportion of items shipped, that 12% of all items in the shipment are defective.

12. List four criteria used to judge goodness of a point estimate, and discuss the meaning of each.

13. State whether each of the following statements is true or false, and explain your answer.

    (a) If a statistic is an unbiased estimator of a parameter, then that is the "best" estimator of the parameter.

    (b) It is better to calculate the sample variance as $\Sigma (X - \bar{X})^2/(n - 1)$ than as $\Sigma (X - \bar{X})^2/n$ if an unbiased estimator of σ^2 is desired.

14. A 95% confidence interval for the percentage of defective items in a lot of 100 is 10–12%. Is it correct therefore to say that the probability is 95% that either ten, eleven, or twelve items in the lot are defective?

15. The mean life of 100 light bulbs selected at random from a shipment of 50,000 bulbs was 1000 hours with a standard deviation of 200 hours. What are the 95% confidence limits for the true mean life of all 50,000 bulbs?

16. In a random sample of 100 families in city A, an arithmetic mean income of \$7000 and a standard deviation of \$1000 was observed. In a random sample of 400 families in city B, an arithmetic mean of \$7200 and a standard deviation of \$1000 was observed.

    (a) Construct a 95% confidence interval for the mean income of all families in city A.

    (b) Construct a 95% confidence interval for the mean income of all families in city B.

    (c) Construct a 95% confidence interval for the difference in mean incomes between the two cities.

17. A simple random sample of 100 students at a university yielded an arithmetic mean income of \$550, a modal income of \$600 and a standard deviation of \$120. Suppose you desire to estimate the mean income of the students of the university with 97% confidence. What would your interval estimate be?

18. A purchasing agent receives 50 pieces of grade B wire from both Barkoff and Estro Wire Companies. A laboratory examination yields the following results: (1) The average tensile strength of the Barkoff sample was 95.8 pounds and had a standard deviation of 5 pounds. (2) 24% of the Barkoff wire had a tensile strength of at least 100 pounds. (3) The average tensile strength of the Estro sample was 99 pounds and had a standard deviation of 5 pounds. (4) 20% of the Estro wire had a tensile strength of at least 100 pounds.

    (a) Construct a 90% confidence interval for the difference between the average tensile strengths of the wire of the two companies.

    (b) Construct a 90% confidence interval for the true difference between the proportion of wire of the two companies which has a tensile strength of at least 100 pounds.

    (c) Interpret your results in (a) and (b).

19. A statistician wishes to determine the average hourly earnings for employees in a given occupation in a particular state. He runs a pilot study and finds

that the point estimate for the mean is \$3.40 and the point estimate for the standard deviation is \$.25. He then specified that when he takes his random sample he wants to be 95.5% confident that the maximum error of his estimate will not exceed \$.02.

(a) Discuss the sense in which he will have 95.5% confidence in his estimate.
(b) What size sample should he take?

20. A manufacturer wishes to estimate the percentage defectives produced by one of his machines on the basis of a random sample of its output. He wishes to be 95% confident that his estimate lies within two percentage points of the true percentage defective. He feels quite certain that the machine does not produce more than 10% defectives. How large a simple random sample should the manufacturer take?

21. A sample of 25 college students selected at random from a normal population yielded a mean expenditure of \$250 per student with a standard deviation of \$40 ($n - 1$ denominator) for the week of Spring vacation.
    (a) Construct a 90% confidence interval for the true mean expenditure of all students in the population for the week of Spring vacation.
    (b) What would be the effect, if any, on the width of the interval in (a) if (1) a higher level of confidence were used and (2) the interval had been based on a sample of 50 students, assuming the same level of confidence, and same standard deviation?

22. State whether the following statements are true or false and explain your answers.
    (a) For small samples, if one is estimating the mean one uses $\bar{x} \pm ts_{\bar{x}}$.
    (b) The more efficient of two unbiased estimators has the narrower confidence interval.
    (c) If a 95% interval for the average strength of a certain type of wood beam is 12 to 15 psi, then one can conclude that in a sample of 100 wood beams 95 would have a strength between 12 and 15 psi.

23. One can always decrease the width of a confidence interval by increasing the sample size. Why then does one not always determine the desired width and then sample accordingly?

24. What are two ways of decreasing the width of a confidence interval for μ, given the best point estimator of a sample is $\bar{x}$?

25. The Minerva Plastic Company wishes to select a random sample of plastic bars from its production process in order to estimate the mean length of bars produced by the process. On the basis of past experience, it is estimated that the process standard deviation is about 0.5 foot. How large a sample is needed to estimate the process mean within ± 0.1 foot, with a 95.5% confidence coefficient?

26. Pollster *A* estimates the percentage of people who are going to vote "yes" on a bond issue as 50% on the basis of a sample of 100 people. Pollster *B* estimates the same percentage as 55% on the basis of a sample of 100 different

randomly selected people. Is there reason to say the difference between the pollsters' results is due to the different methods they use?

27. In evaluating various pension plans and funding methods, the Davis Corp. must determine the mean age of its work force. Since the company has several thousand employees, a sample must be taken. Estimate how large the sample should be in order for the appropriate interval estimate of the mean age to be no more than two years wide with 90% confidence, if the standard deviation of ages is taken to be ten years.

28. An estimate of p, the proportion of dwelling places that are vacant in a large city, is desired. How large a sample should be drawn if it is desired to estimate p within .02 with .98 confidence? It may be assumed that $.01 \leq p \leq .10$.

29. As a part of the planning for the marketing of a new product in a community of 400,000 families, a random sample of 100 families was chosen in order to estimate the community's income level. The mean family income in the sample was \$9000 and the standard deviation of the income distribution was \$200.
    (a) Construct a 95.5% confidence interval for the mean income of all families in the community.
    (b) If the range you obtained in (a) is larger than you are willing to accept, in what ways can you narrow it?

30. A manufacturer of electronic components has a contract with NASA to provide components for weather satellites. The contract specifies that the components must have a mean life of two months when operating under extreme temperature conditions. To determine whether its components meet the requirements specified, the company decides to test several of the components, selected randomly. Since the testing process consists of subjecting the components (which are very costly) to extreme conditions which destroy the components, it has been decided to take a small random sample. The results of this sample were

$$n = 9$$
$$\bar{x} = 9 \text{ weeks}$$
$$s = 4 \text{ weeks } (n - 1 \text{ divisor})$$

    (a) State a 95% confidence interval for μ, the population mean life of the components.
    (b) Can the company be sure that there is no more than a 5% chance that the population mean, μ, lies outside this interval? Explain.

31. A random sample of 100 invoices was drawn from some 10,000 invoices of the Ace Company. For the sample, the mean dollar sales per invoice was \$23.50 and the standard deviation was \$6.00.
    (a) Construct a .99 confidence interval for the mean.
    (b) Give the verbal meaning of your result in part (a).

32. A random sample of 1000 voters was selected from the Patriotic Party's

registration lists in a certain city. The individuals in the sample were asked which of two candidates they preferred. The following sample results were recorded.

Candidates	*No. of Individuals*
Hawk	900
Dove	100
Total	1000

(a) Estimate the percentage p of Doves in the population from which your sample was drawn.
(b) Set up a 95% confidence interval for the true percentage of Doves in the entire registration list.
(c) Interpret your answer to part (b) in terms of this problem.

CHAPTER NINE

Chi-Square Tests and Analysis of Variance

In Chapter 7, procedures were discussed for testing hypotheses using data obtained from a single simple random sample or from two such samples. For example, tests as to whether two population proportions or two population means were equal were considered. Obvious generalizations of such techniques are tests for the equality of more than two proportions or more than two means. The two topics discussed in this chapter supply these generalizations. *Chi-square tests* provide the basis for testing whether more than two population proportions may be considered to be equal; the *analysis of variance* tests whether more than two population means may be considered to be equal.

Chi-square tests will be discussed first in this

chapter, with the topics of (1) tests of goodness of fit and (2) tests of independence being considered in that order. Tests of goodness of fit provide a means for deciding whether a particular probability distribution is appropriate to use, based on a sample of observations. Tests of independence constitute a method for deciding whether the hypothesis of independence between different classificatory variables is tenable. It is this latter type of procedure which provides a test for the equality of more than two population proportions. Both types of chi-square tests furnish a conclusion on whether a set of observed frequencies differs so greatly from a set of theoretical frequencies that the hypothesis under which the theoretical frequencies were derived should be rejected.

9.1 Goodness of Fit Tests

One of the major problems in the application of the theory of probability, statistics, and mathematical models in general is that the real world phenomena to which they are applied usually depart somewhat from the assumptions embodied in the theory or models. For example, let us consider use of the binomial probability distribution in a particular problem. As indicated in Section 2.4, two of the assumptions involved in the derivation of the binomial distribution as the probability distribution for a Bernoulli process are

(1) The probability of a success, p, remains constant from trial to trial.
(2) The trials are independent.

Let us consider whether these assumptions are met in the following problem.

A firm bills its accounts on a 2% discount basis for payment within ten days and full amount due for payment after ten days. In the past, 40% of all invoices have been paid within ten days. In a particular week, the firm sends out 20 invoices. Is the binomial distribution appropriate for computing the probabilities that 0, 1, 2, . . . , 20 firms will take the discount for payment within ten days?

Considering the possible use of the binomial distribution, we can let $p = .40$ represent the probability that a firm will take the discount and $n = 20$ firms the number of trials. Does it seem logical to assume constancy of p, that is, the probability of taking the discount is .40 for each firm? Past relative frequency data of the taking of discounts for *each* firm could be brought to bear on the answer to this question. In most practical situations of this sort we would probably find that the practices of individual firms vary widely, with some firms virtually always taking discounts, some firms virtually never taking discounts, and most firms falling somewhere between these two extremes.

Does the assumption of independence seem tenable in this problem; that is, does it seem logical that whether one firm takes the discount is independent of whether another firm takes the discount? Probably not, since, general monetary conditions doubtlessly affect many of the firms in a similar way. For example, when money is "tight" and it is difficult for many firms to acquire adequate amounts of working capital, the fact that one firm does not take the discount is related to rather than *independent* of whether other firms have taken it. Also, there may be traditional practices in certain industries concerning the taking or not taking of discounts and other factors which would interfere with the independence assumption.

How great a departure from the assumptions underlying a probability distribution, or more generally, from the assumptions embodied in any theory or mathematical model, can be tolerated before we should conclude the distribution, theory, or model is no longer applicable? This is a very complex question and not one that can be readily answered by any simple universally applicable rule. The purpose of chi-square "goodness of fit" tests is to provide one type of answer to the preceding question by comparing *observed frequencies* with *theoretical* or *expected frequencies* derived under specified probability distributions or hypotheses.

The sequence of steps in performing goodness of fit tests is very similar to previously discussed hypothesis testing procedures. The following are the steps in testing for goodness of fit:

(1) A null and alternative hypothesis are established, and a significance level is selected for rejection of the null hypothesis.

(2) A random sample of observations is drawn from a relevant statistical population.

(3) A set of expected or theoretical frequencies is derived under the assumption that the null hypothesis is true. This generally takes the form of assuming that a particular probability distribution is applicable to the statistical population under consideration.

(4) The observed frequencies are compared to the expected, or theoretical, frequencies.

(5) If the aggregate discrepancy between the observed and theoretical frequencies is too great to attribute to chance fluctuations at the selected significance level, the null hypothesis is rejected.

We will illustrate goodness of fit tests and discuss some of the underlying theory for an example involving a uniform probability distribution (See Section 2.3). Additional examples follow.

Suppose a consumer research firm wished to determine whether or not there was a real preference in taste by coffee drinkers in a certain metropolitan area among five brands of coffee. The firm took a simple random sample of

1000 coffee consumers in the area and conducted the following experiment. Each consumer was given five cups of coffee, one of each brand A, B, C, D, and E without identification of the individual brands. The cups were presented to each consumer in a random order determined by sequential selection from five paper slips, each containing one of the letters A, B, C, D, and E. In Table 9-1 are shown the numbers of coffee consumers who stated that they liked the indicated brands best.

Table 9-1 Number of Coffee Consumers in a Certain Metropolitan Area Who Most Preferred the Specified Brand of Coffee.

Brand Preference	*Number of Consumers*
A	210
B	312
C	170
D	85
E	223
	1000

Denoting the true proportions of preference for each brand as p_A, p_B, p_C, p_D, and p_E, we can state the null and alternative hypotheses as follows:

$$H_0\colon p_A = p_B = p_C = p_D = p_E = 0.20$$

$$H_1\colon \text{The } p\text{'s are not all equal}$$

That is, if in the population from which the sample was drawn there were no differences in preference among the five brands, 20% of coffee drinkers would prefer each brand. An equivalent way of stating these hypotheses is

H_0: The probability distribution is uniform

H_1: The probability distribution is not uniform

In other words, the question we are raising is, "Should we consider the sample of 1000 coffee drinkers to be a random sample from a population in which the proportions who prefer each of the five brands are equal?" Of course, this hypothesis is only one of many that could conceivably be formulated. One of the strengths of the goodness of fit test discussed in this section is that it permits a variety of different hypotheses to be raised and tested.

If the null hypothesis of no difference in preference were true, the *expected* or *theoretical* number of the 1000 coffee drinkers in the sample who would prefer each brand would be $.20 \times 1000 = 200$. Hence, the expected frequency that corresponds to each of the observed frequencies in Table 9-1 is 200. We can now compare the set of observed frequencies with the set of theoretical frequencies derived under the assumption that the null hypothesis is true. The test statistic that is computed to make this comparison is known as chi-square, denoted χ^2. The computed value of χ^2 is

(9.1)
$$\chi^2 = \sum \frac{(f_0 - f_t)^2}{f_t}$$

where f_0 = an observed frequency
f_t = a theoretical (or expected) frequency.

As we can see from Equation (9.1), if every observed frequency is exactly equal to the corresponding theoretical frequency, the computed value of χ^2 is zero. This is the smallest value χ^2 can have. The larger the discrepancies between the observed and theoretical frequencies, the larger is χ^2.

The computed value of χ^2 is a random variable which takes on different values from sample to sample. That is, χ^2 has a sampling distribution just as do the other test statistics discussed in Chapter 7. We wish to answer the question "Is the computed value of χ^2 so large that we are required to reject the null hypothesis?" In other words, "Are the aggregate discrepancies between the observed frequencies, f_0, and theoretical frequencies, f_t, so large that we are unwilling to attribute them to chance, and have to reject the null hypothesis?" The calculation of χ^2 for the present problem is shown in Table 9-2.

The Chi-Square Distribution

Before we can answer the questions raised in the preceding paragraph, we must digress for a discussion of the appropriate sampling distribution. Then we will complete the solution to the coffee tasting problem. It can be shown that for large sample sizes the sampling (probability) distribution of χ^2 can be closely approximated by the chi-square distribution whose probability function is

(9.2)
$$f(\chi^2) = c(\chi^2)^{(\nu/2)-1}e^{-\chi^2/2}$$

where e = 2.71828 . . .
ν = number of degrees of freedom
c = a constant depending only on ν.

The chi-square distribution has only one parameter, ν, the number of degrees of freedom. This is similar to the case of the t-distribution (Equation (7.15)). Hence, $f(\chi^2)$ is a family of distributions, one for each value of ν.

Table 9-2 Calculation of Chi-Square Statistic for the Coffee Tasting Problem.

Brand Preference	(1) Observed Frequency f_0	(2) Theoretical (Expected) Frequency f_t	(3) $(f_0 - f_t)$	(4) $(f_0 - f_t)^2$	(5) $\frac{\text{Column (4)}}{\text{Column (2)}}$ $\frac{(f_0 - f_t)^2}{f_t}$
A	210	200	10	100	0.5
B	312	200	112	12,544	62.7
C	170	200	−30	900	4.5
D	85	200	−115	13,225	66.1
E	223	200	23	529	2.6
Total	1000	1000	0		136.4

$$\chi^2 = \sum \frac{(f_0 - f_t)^2}{f_t} = 136.4$$

χ^2 is a continuous random variable equal to or greater than zero. For a small value of ν, the distribution is skewed to the right. As ν increases, the distribution rapidly becomes symmetrical. In fact, for large values of ν, the chi-square distribution is closely approximated by the normal curve. Figure 9-1 depicts the chi-square distribution for 1, 5, and 10 degrees of freedom.

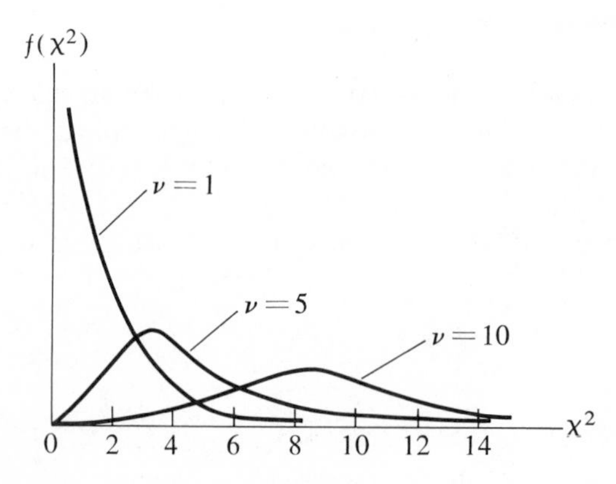

Figure 9-1 Chi-square distribution for 1, 5, and 10 degrees of freedom.

Since the chi-square distribution is a probability distribution, the area under the curve for each value of ν equals one. Because there is a separate distribution for each value of ν, it is not practical to construct a detailed table of areas. Therefore, for compactness, what is generally shown in a chi-square table is the relationship between areas and χ^2 values for only a few "percentage points" in different chi-square distributions.

In Table A-7 of Appendix A are shown χ^2 values that correspond to selected areas in the right-hand tail of the chi-square distribution. These tabulations are shown separately for the number of degrees of freedom listed in the left-hand column. The χ^2 values are shown in the body of the table and the corresponding areas are shown in the column headings.

As an illustration of the use of the chi-square table, let us assume a random variable having a chi-square distribution with 8 degrees of freedom. In Table A-7 we find a χ^2 value of 15.507 corresponding to an area of .05 in the right-hand tail. The relationships described in this illustrative problem are shown in Figure 9-2. Hence, if the random variable has a chi-square

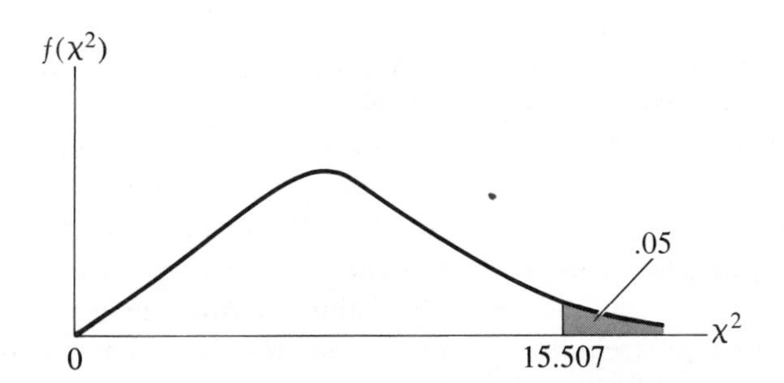

Figure 9-2 Chi-square distribution for eight degrees of freedom.

distribution with 8 degrees of freedom, the probability that χ^2 is greater than 15.507 is .05. Let us give the corresponding interpretation in a hypothesis testing context. If the null hypothesis being tested is true, the probability of observing a χ^2 figure greater than 15.507 because of chance variation is equal to .05. Therefore, for example, if the null hypothesis were tested at the .05 level of significance, and we calculated $\chi^2 = 16$, we would reject the null hypothesis, because so large a χ^2 value would occur less than five times in 100 if the null hypothesis were true. We now turn to a discussion of the rules for determining the number of degrees of freedom involved in a chi-square test.

Number of Degrees of Freedom

As we have seen earlier, a chi-square goodness of fit test involves a comparison of a set of observed frequencies denoted f_0, with a set of theoretical frequencies denoted f_t. Let the symbol k equal the number of classes for which these comparisons are made. For example, in the coffee tasting problem $k = 5$ because these are five classes for which we computed relative deviations of the form $(f_0 - f_t)^2/f_t$. To determine the number of degrees of freedom, we must reduce k by 1 for each restriction imposed. In the coffee tasting example, the number of degrees of freedom is equal to $\nu = k - 1 = 5 - 1 = 4$. The rationale for this computation follows.

In the calculations shown in Table 9-2, there are five classes for which f_0 and f_t values are to be compared. Hence, we start with $k = 5$ degrees of freedom. However, we have forced the total of the theoretical frequencies, Σf_t, to be equal to the total of the observed frequencies, Σf_0; that is, 1000. Therefore, we have reduced the number of degrees of freedom by 1 and there are now only 4 degrees of freedom. That is, once the total of the theoretical frequencies is fixed, only four of the f_t values may be freely assigned to the classes. When these four have been assigned, the fifth class is immediately determined because the theoretical frequencies must total 1000.

If any further restrictions are imposed in the calculation of the theoretical frequencies, the number of degrees of freedom is reduced by one for each such restriction. Hence, for example, if a sample statistic, such as the sample mean $\bar{x}$, is used as an estimate of an unknown population parameter, such as the population mean, μ, there would be a reduction of one degree of freedom. In summary, we can state the following rules for determining ν, the number of degrees of freedom is a chi-square test in which k classes of observed and theoretical frequencies are compared.

(1) If the only restriction is $\Sigma f_t = \Sigma f_0$, the number of degrees of freedom is $\nu = k - 1$.

(2) If in addition to the above restriction m parameters are replaced by sample estimates, the number of degrees of freedom is $\nu = k - 1 - m$.

Decision Procedure

We can now return to the coffee tasting example to perform the goodness of fit test. Let us assume we wish to test the null hypothesis at the .05 level of significance. Since the number of degrees of freedom is 4, we find the critical value of χ^2, which we denote as $\chi^2_{.05}$, to be 9.488 (Table A-7, Appendix A). This means that if the null hypothesis is true, the probability of observing a χ^2 value greater than 9.488 is .05. Specifically in terms of the problem, this means that if there were no difference in preference among brands, an aggregate discrepancy between the observed and theoretical frequencies larger than a χ^2 value of 9.488 would occur only five times in 100. We can state the decision rule for this problem in which $\chi^2_{.05} = 9.488$ as follows:

(1) If $\chi^2 > 9.488$, reject H_0
(2) If $\chi^2 \leq 9.488$, accept H_0

Since the computed χ^2 value in this problem is 136.4 and is thus very much larger than the critical $\chi^2_{.05}$ value of 9.488, we reject the null hypothesis. A reference to Table A-7 of Appendix A indicates that the computed χ^2 value of 136.4 is also far in excess of the critical value at the .01 level, namely, $\chi^2_{.01} = 13.277$. Hence, although in this problem we tested the null hypothesis H_0 at the .05 level of significance, we would have rejected H_0 at the .01 level as well. Therefore, our conclusion is that there exist real differences in consumer preference among the brands of coffee involved in the experiment. In statistical terms, we cannot consider the 1000 coffee drinkers in the experiment to be a simple random sample from a population having equal proportions who prefer each of the five brands. In terms of goodness of fit, we reject the null hypothesis that the probability distribution is uniform. Hence we conclude that the uniform distribution is decidedly not a "good fit" to the sample data.

We now turn to a number of other examples of chi-square goodness of fit tests.

Example 9-1 A management scientist was developing an inventory control system for a manufacturer of a diversified product line. He wanted to determine whether the Poisson distribution was an appropriate model for the demand for a particular product. He obtained the frequency distribution of the number of units of this product demanded per day for the past 200 business days. That distribution is shown in Columns (1) and (2) of Table 9.3. The mean number of units per day is shown at the bottom of the table, obtained by dividing the total of Column (3) by the total of Column (2).

Using the mean of this sample of observations, $\bar{x} = 3$ units of demand per day, as an estimate of the parameter μ of the corresponding theoretical Poisson distribution, the analyst calculated the Poisson probability distribution shown in the first two columns of Table 9-4. Multiplying the probabilities in Column (2) by 200 days, he obtained the theoretical or expected frequencies if the demand were distributed according to the Poisson distribution. These theoretical frequencies are shown in Column (3). For example, if the *probability* that 0 units would be demanded is 0.050, then in 200 days the *expected number* of days in which 0 units would be demanded is $.050 \times 200 = 10.0$ days. This figure of 10.0 days is the first entry in Column (3). The analyst was now able to apply a χ^2 goodness of fit test using the actual number of days in Column (2) of Table 9-3 as the observed frequencies, f_0, and the expected number of days in Column

Table 9-3 Number of Units of a Particular Product Demanded per Day for the Past 200 Business Days.

(1) *Number of Units Demanded Per Day* x	(2) *Observed Number of Days* f_0	(3) *Column* (1) × *Column* (2) $f_0 x$
0	11	0
1	28	28
2	43	86
3	47	141
4	32	128
5	28	140
6	7	42
7	0	0
8	2	16
9	1	9
10	1	10
Total	200	600

$$\bar{x} = \frac{600}{200} = 3 \text{ units per day}$$

(3) of Table 9-4 as the theoretical frequencies, f_t. These two sets of frequencies are shown, in Columns (2) and (3), respectively, of Table 9-5, where the calculation of the χ^2 value is carried out. The hypothesis under test in this problem may be stated as follows:

H_0: The population probability distribution is Poisson with $\mu = 3$

Assume we wish to test the null hypothesis at the .05 level of significance.

As can be seen in Table 9-5, the last four classes of Tables 9-3 and 9-4 for 7, 8, 9, and 10 units of demand have been combined into one class entitled "7 or more." Both f_0 and f_t values have been cumulated for the four classes and a single relative deviation of the form $(f_0 - f_t)^2/f_t$ has been calculated for the combined class. There are now eight classes, $k = 8$, in Table 9-5 for which the χ^2 value has been computed and from which the number of degrees of freedom will be determined. The reason for this combination of classes will be explained at the completion of the problem.

Let us now compute the number of degrees of freedom in the test. As indicated in the earlier discussion, the number of degrees of freedom is given by $\nu = k - 1 - m$, where m is the number of parameters that have been replaced by sample estimates. Since the sample mean $\bar{x}$ was used as the estimate of the

Table 9-4 Theoretical Distribution of Demand Assuming a Poisson Distribution.

(1) Number of Units Demanded per Day x	(2) Probability $f(x)$	(3) Column (2) × 200 Expected Number of Days f_t
0	0.050	10.0
1	0.149	29.8
2	0.224	44.8
3	0.224	44.8
4	0.168	33.6
5	0.101	20.2
6	0.050	10.0
7	0.022	4.4
8	0.008	1.6
9	0.003	0.6
10	0.001	0.2
Total	1.000	200.0

Table 9-5 Calculation of the Chi-Square Statistic for the Demand Distribution Problem.

(1) Number of Units Demanded per Day	(2) Observed Number of Days f_0	(3) Theoretical Number of Days f_t	(4) $f_0 - f_t$	(5) $(f_0 - f_t)^2$	(6) $\frac{(f_0 - f_t)^2}{f_t}$
0	11	10.0	1.0	1.00	.10
1	28	29.8	−1.8	3.24	.11
2	43	44.8	−1.8	3.24	.07
3	47	44.8	2.2	4.84	.11
4	32	33.6	−1.6	2.56	.08
5	28	20.2	7.8	60.84	3.01
6	7	10.0	−3.0	9.00	.90
7 or more	4	6.8	−2.8	7.84	1.15
Total	200	200.0	0		5.53

$$\chi^2 = \sum \frac{(f_0 - f_t)^2}{f_t} = 5.53$$

parameter μ in the Poisson distribution, $m = 1$. Hence the number of degrees of freedom is $\nu = 8 - 1 - 1 = 6$. For 6 degrees of freedom, the critical value of χ^2 at the .05 level of significance is $\chi^2_{.05} = 12.592$ (Table A-7, Appendix A). Therefore, since the observed χ^2 value of 5.53 is less than 12.592, we accept the null hypothesis.

In other words, the aggregate discrepancy between the observed and theoretical frequencies is sufficiently small for us to *conclude that the Poisson distribution* with $\mu = 3$ *is a good fit*. Based on this result, an appropriate amount of stock to carry can be determined.

In the foregoing example, the hypothesis that the population probability distribution is Poisson was tested and accepted. The same procedure could be used to test whether a shift away from a previously established Poisson distribution has occurred.

Rule Concerning Size of Theoretical Frequencies

As indicated earlier, for large sample sizes, the probability function of the computed χ^2 values can be closely approximated by the chi-square distribution given in Equation (9.2), which is the distribution of a *continuous* random variable. However, there are only a finite number of possible combinations of f_t values, and hence only a finite number of computed χ^2 values. Thus, a computed χ^2 figure is one value of a *discrete* random variable. If the sample size is large, the approximation of the probability distribution of this discrete random variable by the continuous chi-square distribution will be a good one. This is analogous to the situation in approximating the binomial distribution, which is discrete, by the normal curve, which is continuous (See Section 6.3).

When the expected frequencies, that is, the f_t values, are small, the approximation discussed in the preceding paragraph is inadequate. A frequently used rule is that each f_t value should be equal to or greater than 5. This is the reason the classes for 7, 8, 9, and 10 units of demand were combined in Example 9-1 in the computation of the χ^2 value. As shown in Table 9-5, the computed f_t value for the combined class is equal to 6.8, which satisfies the commonly used rule of thumb for a minimum expected frequency.

Example 9-2 A research investigator suspected that large families tended to have more male children than would be expected on the basis of chance. He felt that the distribution of male children was not purely random and that "male birth" events were not independent. He had data for a simple random sample of 320 families with five children each, which had been drawn for another purpose. However, he decided to conduct a partial test of his theory using these data. The frequency distribution of numbers of male children in these 320 families is shown in Table 9-6. The investigator decided to treat the birth process as a Bernoulli process and, therefore, to fit a binomial distribution to these data assuming male and female births were equally likely. He stated the null hypothesis to be tested as follows:

H_0: The population probability distribution is binomial with $p = 1/2$

The investigator decided to use a .05 level of significance, since that was the conventional level ordinarily used by other researchers in his field. He would rather have used a .01 level because he wanted to maintain a very low risk of rejecting his null hypothesis if it was indeed true. However, he felt that the practice of using .05 levels of significance in his field was so ingrained that he had no alternative.

Letting p = the probability of a male birth = 1/2
q = the probability of a female birth = 1/2
and n = number of children per family = 5,

he computed the binomial probability distribution given in Columns (1) and (2) of Table 9-7. The probabilities are the respective terms of the binomial

Table 9-6 Number of Male Children in a Sample of Families with Five Children Each.

(1) *Number of Male Children* x	(2) *Observed Number of Families* f_o
0	12
1	42
2	92
3	108
4	46
5	20
	320

Table 9-7 Calculation of Expected Frequencies in the Number of Male Children Problem: Assumed Binomial Distribution with $p = 0.50$ and $n = 5$.

(1) *Number of Male Births* x	(2) *Probability* $f(x)$	(3) *Column* (2) × 320 *Expected Number of Families* f_t
0	1/32	10
1	5/32	50
2	10/32	100
3	10/32	100
4	5/32	50
5	1/32	10
Total	1	320

$(1/2 + 1/2)^5$. Multiplying the probabilities in Column (2) by 320 families, the research investigator determined the expected number of families with zero, one, two, three, four, and five male children each. These theoretical frequencies are shown in Column (3) of Table 9-7. He then proceeded with the chi-square test of goodness of fit using the observed frequencies in Column (2) of Table 9-6 as the f_0 values and the expected frequencies in Column (3) of Table 9-7 as the f_t values. Table 9-8 shows the calculation of the χ^2 value.

The number of degrees of freedom is $\nu = k - 1 = 6 - 1 = 5$, since the only restriction was the requirement that $\Sigma f_t = \Sigma f_0$. No parameters were replaced by sample estimates. The critical value of χ^2 for 5 degrees of freedom is $\chi^2_{.05} = 11.070$. Hence, since the computed χ^2 value of 13.28 is greater than the critical value of 11.07, the null hypothesis was rejected *at the .05 level of significance.* The investigator was required to conclude that the binomial with $p = \frac{1}{2}$ was not a good fit. However, he was rather disgruntled because the critical level of χ^2 at the .01 level for 5 degrees of freedom is 15.086. Since the computed value of χ^2 was 13.28, which is less than 15.086, the null hypothesis would have been accepted at the .01 level. The researcher, being a very human fellow, continued to tell his friends about his theory. Furthermore, he concluded that his colleagues were rather unscientific chaps who went about uncritically using the .05 level to test their hypotheses.

Example 9-3 Another investigator who read the report described in Example 9-2 decided to test the binomial distribution for goodness of fit in a slightly different way using the same set of data. He felt that equal likelihood of male and female births should not be assumed for the population from which the sample was drawn. Therefore, he ran a chi-square goodness of fit test in which he

Table 9-8 Calculation of the Chi-Square Statistic for the Number of Male Children Problem: Fit of Binomial with $p = .50$ and $n = 5$.

Number of Male Children	*Observed Number of Families* f_0	*Theoretical Number of Families* f_t	$f_0 - f_t$	$(f_0 - f_t)^2$	$(f_0 - f_t)^2/f_t$
0	12	10	2	4	0.40
1	42	50	−8	64	1.28
2	92	100	−8	64	0.64
3	108	100	8	64	0.64
4	46	50	−4	16	0.32
5	20	10	10	100	10.00
	320	320	0		13.28

$$\chi^2 = \sum \frac{(f_0 - f_t)^2}{f_t} = 13.28$$

imposed the restriction of substituting the observed sample proportion of male children, denoted $\bar{p}$, for the binomial parameter p. To obtain $\bar{p}$, he calculated $\bar{x}$, the mean number of male children per family and divided that figure by 5. These computations are shown in Columns (1), (2), and (3) and at the bottom of Table 9-9. As indicated in Table 9-9, the sample proportion of male children was $\bar{p} = .522$. Using this $\bar{p}$ value as an estimate of the parameter p in a Bernoulli process, the investigator computed the binomial probabilities shown in Column 4 of Table 9-9. These probabilities are the terms in the binomial $(.478 + .522)^5$. Multiplying these probabilities by 320 yielded the expected frequencies shown in Column (5) of Table 9-9.

The researcher then proceeded with the chi-square test of goodness of fit given in Table 9-10, using the observed frequencies of Column (2) of Table 9-9 as the f_0 values and the expected frequencies of Column (5) of Table 9-9 as the f_t values. As shown in Table 9-10, the calculated χ^2 value was equal to 8.99. The number of degrees of freedom was determined from the formula $\nu = k - 1 - m$ as $\nu = 6 - 1 - 1 = 4$. As in Example 9-2, the number of classes was $k = 6$, but $m = 1$ because one sample estimate, namely $\bar{p}$, was used to replace a population parameter, in this case, p. Hence, there was a loss of an additional degree of freedom as compared to Example 9-2 in which this restriction was not imposed. This investigator, just as his colleague in the preceding example, tested for goodness of fit at the .05 level of significance. For 4 degrees of freedom, the critical χ^2 value is $\chi^2_{.05} = 9.488$. Since the computed χ^2 value was only 8.99, the null hypothesis that the population probability distribution is binomial (with $p = .522$) was accepted. Hence, this researcher concluded that the evidence represented by the observed frequency distribution on number of male children in the sample families was consistent with the hypothesis of the operation of a Bernoulli process for births, but he was quite skeptical concerning the idea that male and female births were equally likely.

The preceding example illustrates the tentative nature of conclusions drawn from hypothesis testing procedures. It is conceivable that the same sample evidence could have led to the acceptance of both hypotheses or the rejection of both hypotheses even at the same significance level. In the example, the sample evidence led to the rejection of one hypothesis and the acceptance of the other.

However, there is a subtle point involved in the example. The second investigator established the hypothesis he wished to test after he had examined the sample data. Under classical hypothesis testing procedures, the hypothesis should be set up before the data are gathered. But in actual practice, many hypotheses are tested after sample evidence is obtained, sometimes by persons who had control over collection and tabulation of the data, and sometimes by others. Classical statistics does not provide separate techniques for testing hypotheses before and after sample evidence has been collected. On the other hand, Bayesian decision theory, discussed in Chapters 13 through 17, provides different techniques for decision making prior to the obtaining of sample data and for the incorporation of sample information with prior knowledge.

Table 9-9 Calculation of Expected Frequencies in the Number of Male Children Problem: Assumed Binomial with $p = 0.522$ and $n = 5$.

(1) Number of Male Children x	(2) Observed Number of Families f_0	(3) Column (1) × Column (2) f_0x	(4) Probability $f(x)$	(5) Column (4) × 320 Expected Number of Families f_t
0	12	0	.025	8.00
1	42	42	.136	43.52
2	92	184	.298	95.36
3	108	324	.324	103.68
4	46	184	.178	56.96
5	20	100	.039	12.48
Total	320	834	1.000	320.00

$$\bar{x} = \frac{834}{320} = 2.61 \text{ male children per family}$$

$$\bar{p} = \frac{2.61}{5} = 0.522$$

Table 9-10 Calculation of the Chi-Square Statistic for the Number of Male Children Problem: Fit of Binomial with $p = .522$ and $n = 5$.

(1) Number of Male Children	(2) Observed Number of Families f_0	(3) Theoretical Number of Families f_t	(4) $f_0 - f_t$	(5) $(f_0 - f_t)^2$	(6) $(f_0 - f_t)^2/f_t$
0	12	8.00	4.00	16.00	2.00
1	42	43.52	−1.52	2.31	.05
2	92	95.36	−3.36	11.29	.12
3	108	103.68	4.32	18.66	.18
4	46	56.96	−10.96	120.12	2.11
5	20	12.48	7.52	56.55	4.53
Total	320	320.00	0		8.99

$$\chi^2 = \sum \frac{(f_0 - f_t)^2}{f_t} = 8.99$$

9.2 Tests of Independence

Another important application of the chi-square distribution is in testing for the independence of two variables on the basis of sample data. The general nature of the test is best explained in terms of a specific example.

Suppose a savings bank in a metropolitan area was required to move to a new location because of an urban redevelopment program. Two different locations, denoted A and B, were under consideration. The bank wished to determine whether the *preference* for the two locations by present depositors was *independent* of *depositor size*, as measured by account balances. All depositors with balances under a certain figure were classified as "small"; all others were classified as "large." The bank took a simple random sample of 1000 of its depositors and asked them which of the two locations they preferred. The results are shown in Table 9-11. This type of table, which has one basis of classification across the columns, in this case location preference, and another along the rows, in this case size of account, is known as a *contingency table*. If the table has two rows and two columns, as in Table 9-11, it is referred to as a *two-by-two* (often written 2×2) *contingency table*. In general, in an $r \times k$ contingency table, where r denotes the number of rows and k the number of columns, there are $r \times k$ cells. For example, in the 2×2 table under discussion there are $2 \times 2 = 4$ cells for which there are observed frequencies. In a 3×2 contingency table, there are $3 \times 2 = 6$ cells. The chi-square test consists in calculating expected frequencies under the hypothesis of independence, and comparing the observed and expected frequencies as in the goodness of fit tests.

The competing hypotheses under test in this problem may be stated as follows:

H_0: Preference for location is independent of depositor account size

H_1: Preference for location is not independent of depositor account size

Calculation of Theoretical or Expected Frequencies

As in the preceding chi-square tests, the theoretical frequencies are obtained by assuming the null hypothesis is true. We observe from Table 9-11 that 400 out of the 1000, or 40% of the depositors in the sample prefer location A. If the null hypothesis, H_0, is true, that is, if preference for location is independent of depositor size, then 400/1000 of the 360 small depositors and 400/1000 of the 640 large depositors would prefer location A.

Thus, the expected number of small depositors who prefer location A is

$$\frac{400}{1000} \times 360 = 144$$

Table 9-11 A Simple Random Sample of 1000 Depositors Classified by Preference for Bank Locations A and B and Size of Account.

Size of Account	(B_1) *Prefer Location A*	(B_2) *Prefer Location B*	*Total*
(A_1) *Small*	120	240	360
(A_2) *Large*	280	360	640
Total	400	600	1000

This is the expected frequency that corresponds to 120, the observed number of small depositors who prefer location A.

Similarly, the expected number of large depositors who prefer location A is

$$\frac{400}{1000} \times 640 = 256$$

The other two expected frequencies obtained by analogous reasoning are

$$\frac{600}{1000} \times 360 = 216$$

$$\frac{600}{1000} \times 640 = 384$$

In general, the theoretical or expected frequency for a cell in the ith row and jth column is calculated as follows:

$$(f_t)_{ij} = \frac{(\Sigma \text{ Row } i)(\Sigma \text{ Column } j)}{\text{Grand Total}} \tag{9.3}$$

where $(f_t)_{ij}$ = the theoretical (expected) frequency for a cell in the ith row and jth column

Σ Row i = the total of the frequencies in the ith row

Σ Column j = the total of the frequencies in the jth column

Grand Total = the total of all of the frequencies in the table.

For example, the theoretical frequency of 144 obtained for the cell in the first row and first column of Table 9-11, whose rationale of calculation was just explained, is computed by Equation (9.3) as

$$(f_t)_{11} = \frac{(360)(400)}{1000} = 144$$

In order to keep the notation uncluttered, we drop the subscripts denoting rows and columns for f_t values in the subsequent discussion.

The expected frequencies for the present problem are shown in Table 9-12. Because of the method of calculating the expected frequencies, the totals in the margins of the table are the same as the totals in the margins of the table of observed frequencies (Table 9-11). It is important to note that the method of computing the expected frequencies under the null hypothesis of independence is simply an application of the multiplication rule for independent events, given in Equation (1.7). For example, in Table 9-11, the "small" accounts and "prefer location A" categories have been denoted A_1 and B_1, respectively. Under independence, $P(A_1 \cap B_1) = P(A_1)P(B_1)$. The marginal probabilities $P(A_1)$ and $P(B_1)$ are given by

$$P(A_1) = \frac{360}{1000} = .36$$

$$P(B_1) = \frac{400}{1000} = .40$$

Hence, assuming independence, the joint probability of A_1 and B_1 is

$$P(A_1 \cap B_1) = P(A_1)\,P(B_1) = (.36)(.40) = .144$$

Multiplying this joint probability by the total frequency 1000, we obtain the expected frequency previously derived for the upper left-hand cell

$$.144 \times 1000 = 144$$

Table 9-12 Expected Frequencies for the Problem on the Relationship Between Preference for Bank Location and Depositor's Size of Account.

Size of Account	(B_1) *Prefer Location A*	(B_2) *Prefer Location B*	*Total*
(A_1) *Small*	144	216	360
(A_2) *Large*	256	384	640
Total	400	600	1000

Similar calculations would yield the other expected frequencies.

The Chi-Square Test

Denoting the observed frequencies in Table 9-11 as f_0 and the theoretical or expected frequencies in Table 9-12 as f_t, we calculate the χ^2 statistic in the usual way, as shown in Table 9-13. The cells have been denoted by intersection symbolism, namely, $A_1 \cap B_1$, $A_1 \cap B_2$, $A_2 \cap B_1$, and $A_2 \cap B_2$. As shown in the table, the calculated value of χ^2 is 10.42.

The number of degrees of freedom in a 2×2 contingency table is equal to 1. This can be rationalized intuitively as follows. When only one of the expected frequencies has been determined, all of the others are fixed by the marginal totals. For example, in the present problem, when the expected frequency of 144 for the "small and prefer location *A*" cell, $A_1 \cap B_1$, has been determined, the expected frequency for the "small and prefer location *B*" cell, $A_1 \cap B_2$, must be equal to 216 in order to add to 360 for the "small," A_1, marginal total. All other expected frequencies are similarly constrained by the marginal totals. Hence, we may say that there is only one degree of freedom in establishing the theoretical frequencies. *In general, in a contingency table containing r rows and k columns, there are* $(r - 1)(k - 1)$ *degrees of freedom*. Thus, in a 2×2 table, the number of degrees of freedom is $(2 - 1) \times (2 - 1) = 1$; in a 3×2 table, the number of degrees of freedom is $(3 - 1) \times (2 - 1) = 2$, etc.

Referring to Table A-7 in Appendix A, we find critical chi-square values of $\chi^2_{.05} = 3.841$ at the 5% level of significance and $\chi^2_{.01} = 6.635$ at the 1% level of significance. Since the computed chi-square value of 10.42 exceeds both of these figures, we would reject the null hypothesis of independence at

Table 9-13 Calculation of the χ^2 Statistic for the Bank Location Preference Problem.

Cell	*Observed Number of Depositors* f_0	*Expected Number of Depositors* f_t	$f_0 - f_t$	$(f_0 - f_t)^2$	$(f_0 - f_t)^2/f_t$
$A_1 \cap B_1$	120	144	−24	576	4.00
$A_1 \cap B_2$	240	216	24	576	2.67
$A_2 \cap B_1$	280	256	24	576	2.25
$A_2 \cap B_2$	360	384	−24	576	1.50
Total	1000	1000	0		10.42

$$\chi^2 = \sum \frac{(f_0 - f_t)^2}{f_t} = 10.42$$

both significance levels. Therefore, we conclude on the basis of the sample data that preference for the new location of the bank *is not* independent of depositor's account size. A glance at Table 9-11 tells us what the nature of the relationship between these two variables is. The proportion of small depositors who prefer location A is $120/360 = .33$, whereas the proportion of large depositors who prefer that location is $280/640 = .44$. Hence, there is a greater preference for location A among large depositors than among small ones.

Further Comments

We have seen how the chi-square test for independence in contingency tables is a means of determining whether or not a relationship exists between two bases of classification, or, in other words, whether a relationship exists between two variables. The variables are stated in the form of *qualitative categories* such as "small accounts," "large accounts," "prefer location A," and "prefer location B," and statistical observations are classified within these categories. Although this type of tabulation provides a basis for testing whether there is a dependence between the two classificatory variables, it does not yield a method for estimating the values of one variable from known or assumed values of the other. In the next chapter, where *regression* and *correlation analysis* is discussed, methods for providing such estimates are indicated. In that type of analysis the variables are stated in the form of numerical values rather than in qualitative categories. For example, instead of accounts being classified as "large" or "small," the account sizes could be stated in actual numbers of dollars. Instead of preference being stated as "prefer location A" or "prefer location B," the desirability of location B over A could be stated on a numerical scale running, say, from -100 to $+100$. Then, regression analysis provides a method for estimating a numerical value for how much more or less location B is preferred to location A by a depositor with an account of a specific dollar amount. Regression analysis, in particular, provides a very powerful tool for stating in explicit mathematical form the nature of the relationship that exists between two or more variables.

However, as was noted in the paragraph preceding the last one, at least some indication may be obtained of the nature of the relationship between the two variables in a contingency table by computing proportions such as .33 small depositors versus .44 large depositors who prefer location A. Also, if we consider proportions such as these, a fruitful line of reasoning can be developed. Hence, suppose we let $\bar{p}_1 = .33$ and $\bar{p}_2 = .44$. It can be shown by statistical theory that a test for the significance of the difference between these two proportions by the methods of Section 7.3 using Equation (7.11) to calculate the estimated standard error of the difference is algebraically identical to the chi-square test. This means that the test of the hypotheses of independence and non-independence which was carried out in the foregoing example is identical to the test of the following hypotheses

$$H_0: p_1 = p_2$$
$$H_1: p_1 \neq p_2$$

where p_1 denotes the *population proportion* of small depositors who prefer location A, and p_2 denotes the *population proportion* of large depositors who prefer location A.

9.3 *k* Sample Tests for Proportions

A powerful generalization develops from the preceding discussion. Just as we test whether two proportions are equal by a chi-square test applied to data in a 2×2 contingency table, we can test whether three or more proportions are equal in a contingency table that has three or more rows or columns. Thus, in the bank location problem, there were two categories of depositors account sizes, "small" and "large." If there were k categories, we could test the null hypothesis

$$H_0: p_1 = p_2 = \cdots = p_k$$

against the alternative that the p values are not all equal. Example 9-4 which follows is an illustration of this principle for a 3×3 contingency table.

Example 9-4 Table 9-14 summarizes the automobile and telephone ownership of a simple random sample of 10,000 families in a certain area. We will test the null hypothesis that telephone and automobile ownership are independent. As indicated by the preceding discussion, this is equivalent to testing the null hypothesis

$$H_0: p_1 = p_2 = p_3$$

Table 9-14 A Simple Random Sample of 10,000 Families Classified by Number of Automobiles and Telephones Owned.

Number of Telephones	*Number of Automobiles Owned*			
	Zero	*One*	*Two*	*Total*
Zero	1000	900	100	2000
One	1500	2600	500	4600
Two or more	500	2500	400	3400
Total	3000	6000	1000	10,000

against the alternative that the p's are not all equal. Here p_1 denotes the population proportion of non-automobile owning families who do not have telephones. The same proportions for "one-car" and "two-car" families are denoted by p_2 and p_3, respectively. This null hypothesis implies similar null hypotheses of equal population proportions of zero, one, and two automobile owning families who have one telephone, and two or more telephones. Let us assume it is desired to maintain a low risk of erroneously rejecting the null hypothesis of independence, and, therefore, the test is to be carried out at the .01 level of significance.

The procedure for finding the expected frequencies is the same as in the case of the 2×2 contingency table except that there are more categories involved. Hence, we observe from the marginal totals on the right-hand side of Table 9-14 that 2000/10,000, or 20%, of the families have no telephone. Therefore, under the null hypothesis, .20 of the families owning zero, one, and two automobiles, respectively, would have no telephone. Thus, the expected number of the 3000 non-automobile owning families that have no telephone is

$$\frac{2000}{10{,}000} \times 3000 = 600$$

This expected frequency corresponds to the figure 1000 in the upper left-hand cell of Table 9-14.

The expected number of the 6000 "one-car" families who have no telephone is

$$\frac{2000}{10{,}000} \times 6000 = 1200$$

This figure corresponds to the 900 shown in the first row. All other expected frequencies are obtained in similar fashion. The expected frequencies are presented in Table 9-15. Again denoting the observed frequencies as f_0 and the expected frequencies as f_t, we have shown the calculation of the χ^2 statistic in Table 9-16. No cell designations are indicated, but of course every f_0 value is compared with the corresponding f_t figure. As shown at the bottom of the table, the computed value of χ^2 is equal to 794.3. The number of degrees of freedom is $(r - 1)(k - 1)$ or $(3 - 1)(3 - 1) = 4$. In Table A-7 of Appendix A we find a critical value at the .01 level of significance of $\chi^2_{.01} = 13.277$. Since the computed χ^2 so greatly exceeds this critical value, the null hypothesis of independence between telephone and automobile ownership is emphatically rejected. Equivalently, we have rejected the null hypothesis, $H_0: p_1 = p_2 = p_3$, where as stated earlier, p_1, p_2, and p_3 denote the population proportions of zero-, one-, and two-car families, respectively, who do not have telephones. Reference to Table 9-14 makes it obvious why the null hypothesis was rejected. Of the 3000 families who did not own automobiles, $1000/3000 = .33$ did not own a telephone. Let $\bar{p}_1 = .33$. The corresponding proportions of one- and two-car owning families who did not own telephones were $\bar{p}_2 = 900/6000 = .15$ and $\bar{p}_3 = 100/1000 = .10$. Hence, we have concluded that it is highly unlikely that these three statistics represent samples drawn from populations that have the same proportions ($p_1 = p_2 = p_3$). Clearly, the proportion of non-telephone owning families declines as automobile ownership increases. The data suggest a strong relationship between the ownership of telephones and automobiles for the families studied.

Selected Comments

Since the sampling distribution of the χ^2 statistic, $\chi^2 = \Sigma(f_o - f_t)^2/f_t$ is only an approximation to the theoretical distribution defined in Equation (9.2), the sample size must be large for a good approximation to be obtained. As in the goodness of fit tests, in contingency tables, cells with frequencies of less than 5 should be combined.

Furthermore, in 2 × 2 tables, that is, when there is one degree of freedom, an adjustment known as Yate's correction for continuity may be used. This correction is introduced because the theoretical chi-square distribution is

Table 9-15 Expected Frequencies for the Problem on the Relationship Between Telephone and Automobile Ownership.

Number of Telephones	*Number of Automobiles Owned* Zero	One	Two	Total
Zero	600	1200	200	2000
One	1380	2760	460	4600
Two or More	1020	2040	340	3400
Total	3000	6000	1000	10,000

Table 9-16 Calculation of the χ^2 Statistic for the Telephone and Automobile Ownership Problem.

Observed Number of Families f_o	*Expected Number of Families* f_t	$f_o - f_t$	$(f_o - f_t)^2$	$(f_o - f_t)^2/f_t$
1000	600	400	160,000	266.7
1500	1380	120	14,400	10.4
500	1020	−520	270,400	265.1
900	1200	−300	90,000	75.0
2600	2760	−160	25,600	9.3
2500	2040	460	211,600	103.7
100	200	−100	10,000	50.0
500	460	40	1600	3.5
400	340	60	3600	10.6
Total 10,000	10,000	0		$\chi^2 = 794.3$

continuous, whereas the tabulated values in Table A-7, Appendix A are based on the distribution of the discrete χ^2 statistic of Equation (9.1). The correction is applied by computing the following χ^2 statistic

$$\chi^2 = \sum \frac{(|f_0 - f_t| - 1/2)^2}{f_t} \tag{9.4}$$

In this correction, 1/2 is subtracted from the absolute value of the difference between f_0 and f_t before squaring. The effect is to reduce the calculated value of χ^2 as compared to the corresponding calculation by Equation (9.1) without the correction.[1] In an example such as the one just discussed, where the expected frequencies are large, the effect of this correction is clearly unimportant, but it may be of greater significance for smaller samples.

As we have seen, in both chi-square goodness of fit tests and tests of independence, the null hypothesis is rejected when large enough values of χ^2 are observed. Some investigators have raised the question as to whether the null hypothesis should also be rejected when the computed value of χ^2 is too low, that is, too close to zero. This is a situation in which the observed frequencies, f_0, all appear to *agree too well* with the theoretical frequencies, f_t. The recommended course of action is to examine the data very closely to see whether errors have been made in the recording of the data. Perhaps the data rather than the null hypothesis should be rejected. An experience of one researcher is relevant to this point. He was analyzing some data on oral temperatures and found that a disturbingly large number of the recorded temperatures were equal to the "normal" figure of 98.6°F. He suspected these data as being "too good to be true." Upon investigation, he found that the temperatures were recorded by relatively untrained nurses aides. Several of them had misread temperatures by recording the number to which the arrow on the thermometer pointed, namely, 98.6°! Clearly, this was a case where the data rather than an investigator's null hypothesis should be rejected.

9.4 Analysis of Variance: *k* Sample Tests for Means

In Section 9.3, we saw that the *chi-square test is a generalization of the two sample test for proportions* and enables us to test for the significance of the difference among $k,(k > 2)$, sample proportions. Conceptually, this represents a test of whether the k samples can be treated as having been drawn from the same population, or in other words, from populations having the same proportions. Similarly, in this section we consider a very ingenuous technique known as the *analysis of variance*, which is a *generalization of the two sample*

[1]For a more complete discussion of Sheppard's correction, see Snedecor, George W., *Statistical Methods*, 5th ed., The Iowa State University Press, 1956.

test for means and enables us to test for the significance of the difference among k,($k > 2$), sample means. Analogously to the case of the chi-square test, this represents a test of whether the k samples can be considered as having been drawn from the same population, or more precisely, from populations having the same means.

A central point to realize is that although the analysis of variance is literally a technique which analyzes or tests variances, by doing so, it provides us with a test for the significance of the difference among *means*. The rationale by which a test of variances is in fact a test for means will be explained shortly.

As an example, we consider a problem in which it is desired to test whether three methods of teaching a basic statistics course differ in effectiveness. It has been agreed that student grades on a final examination covering the work of the entire course will be used as the measure of effectiveness. The three methods of teaching are

Method 1: The lecturer does not work out nor assign problems.
Method 2: The lecturer works out and assigns problems.
Method 3: The lecturer works out and assigns problems. Students are also required to construct and solve their own problems.

The same professor taught three different sections of students, using one of the 3 methods in each class. All of the students were sophomores at the same university and were randomly assigned to the 3 sections. In order to explain the principles of the analysis without cumbersome computational detail, we will assume there were only 12 students in the experiment, 4 in each of the 3 different sections. Of course, in actual practice, a substantially larger number of observations would be required to furnish convincing results. The final examination was graded on the basis of 25 as the maximum score and 0 as the minimum score. The final examination grades of the 12 students in the 3 sections are given in Table 9-17. As shown in the table, the mean grades for students taught by methods 1, 2, and 3 were 17, 20, and 23, respectively, and the overall average of the 12 students, referred to as the "grand mean," was 20. It may be noted that the grand mean of 20 is the same figure as would be obtained by adding up all 12 grades and dividing by 12.

Notation

At this point we introduce some useful notation. In Table 9-17, there are 4 rows and 3 columns. As in the discussion of chi-square tests for contingency tests, let r represent the number of rows and k the numbers of columns. Hence, there is a total of $r \times k$ observations in the table, in this case $4 \times 3 = 12$. Let X_{ij} be the score of the ith student taught by the jth method, where

$i = 1, 2, 3, 4$ and $j = 1, 2, 3$. Thus, for example, X_{12} denotes the score of student 1 taught by method 2 and is equal to 19; $X_{23} = 21$, etc. In this problem, the different methods of instruction are indicated in the columns of the table, and interest centers on the differences among the scores in the 3 columns. This is typical of the so-called "one factor (or one-way) analysis of variance," in which an attempt is made to assess the effect of only one factor (in this case, instructional method), on the observations. In the present problem, there are 3 columns. Hence, we denote the values in the columns as X_{i1}, X_{i2}, and

Table 9-17 Final Examination Grades of Twelve Students Taught by Three Different Methods.

	Teaching Method		
Student	1	2	3
1	16	19	24
2	21	20	21
3	18	21	22
4	13	20	25
Total	68	80	92
Mean	17	20	23

$$\text{Grand mean} = \frac{17 + 20 + 23}{3} = 20$$

X_{i3}, and the totals of these columns as $\sum_i X_{i1}$, $\sum_i X_{i2}$, and $\sum_i X_{i3}$. The subscript i under the summation signs indicates that the total of each of the columns is obtained by summing the entries over the rows. Adopting a simplified notation, we will refer to the means of the 3 columns as $\bar{X}_1$, $\bar{X}_2$, and $\bar{X}_3$, or in general, $\bar{X}_j$. Finally, we denote the grand mean as $\bar{\bar{X}}$ (pronounced "X double-bar"), where $\bar{\bar{X}}$ is the mean of all $r \times k$ observations. Since each column in our example contains the same number of observations, $\bar{\bar{X}}$ can be obtained by taking the mean of the 3 sample means $\bar{X}_1$, $\bar{X}_2$, and $\bar{X}_3$. This notation is summarized in Table 9-18. It is suggested that the reader study Table 9-18 carefully and compare the notation with the corresponding entries in Table 9-17.

The Hypothesis to be Tested

As indicated earlier, in the illustrative problem we are considering we want to test whether the effectiveness of the three methods of teaching a basic statistics course differ. We have calculated the following mean final examination scores of students taught by the 3 methods $\bar{X}_1 = 17$, $\bar{X}_2, = 20$, and $\bar{X}_3 = 23$. The statistical question is "Can the 3 samples represented by these 3 means be considered as having been drawn from populations having the same mean?" Denoting the population means corresponding to $\bar{X}_1$, $\bar{X}_2$, and $\bar{X}_3$ as μ_1, μ_2, and μ_3, respectively, we can state the null hypothesis as

$$H_0: \mu_1 = \mu_2 = \mu_3$$

This hypothesis is to be tested against the alternative

$$H_1: \text{The three means } \mu_1, \mu_2, \text{ and } \mu_3 \text{ are not all equal}$$

Hence, what we wish to determine is whether the differences among the sample means $\bar{X}_1$, $\bar{X}_2$, and $\bar{X}_3$ are too great to attribute to the chance errors of drawing samples from populations having the same means. If we do decide that the sample means differ significantly, our substantive conclusion is that the teaching methods differ in effectiveness.

Table 9-18 Notation Corresponding to the Data of Table 9-17.

i	X_{i1}	X_{i2}	X_{i3}
1	X_{11}	X_{12}	X_{13}
2	X_{21}	X_{22}	X_{23}
3	X_{31}	X_{32}	X_{33}
4	X_{41}	X_{42}	X_{43}
Total	$\sum_i X_{i1}$	$\sum_i X_{i2}$	$\sum_i X_{i3}$
Mean	$\bar{X}_1$	$\bar{X}_2$	$\bar{X}_3$
		$\bar{\bar{X}} = \dfrac{\bar{X}_1 + \bar{X}_2 + \bar{X}_3}{3}$	

Although we will specify the assumptions underlying the test procedure at the end of the problem, we indicate one of them at this point, namely, the assumption that the variances of the 3 populations are all equal. Therefore, rewording the hypothesis slightly, we can state that we wish to test whether our 3 samples were drawn from populations having the same means and variances.

Decomposition of Total Variation

Before discussing the procedures involved in the analysis of variance, we consider the general rationale underlying the test. If the null hypothesis that the 3 population means, μ_1, μ_2, and μ_3, are equal is true, then both the variation among the sample means $\bar{X}_1$, $\bar{X}_2$, and $\bar{X}_3$ and the variation within the 3 groups reflect chance errors of the sampling process. The first of these types of variation is conventionally referred to as "variation between the k means," "between-group variation," or "between-column variation," despite the English barbarism involved in using the word "between" rather than "among" when there are more than 2 groups present. The second type is referred to as "within-group variation" or "within-column variation." Between-column variation refers to variation of the sample means $\bar{X}_1$, $\bar{X}_2$, and $\bar{X}_3$ around the grand mean, $\bar{\bar{X}}$. On the other hand, within-column variation refers to the differences of the individual observations within each column from their respective means $\bar{X}_1$, $\bar{X}_2$, and $\bar{X}_3$.

Under the null hypothesis that the population means are equal, the between-column variation and the within-column variation would be expected not to differ significantly from one another, since they both reflect the same type of chance sampling errors. On the other hand, if the null hypothesis is false, and the population column means are indeed different, then the between-column variation should significantly exceed the within-column variation. This follows from the fact that the between-column variation would now be produced by the inherent differences among the column means as well as by chance sampling error. On the other hand, the within-column variation would still reflect chance sampling errors only. *Hence, a comparison of between-column variation and within-column variation yields information concerning differences among the column means.* This is the central insight provided by the analysis of variance technique.

The term "variation" is used in statistics in a very specific way to refer to a sum of squared deviations and is often referred to simply as a "sum of squares." When a measure of variation is divided by an appropriate number of degrees of freedom, as we have seen earlier in this text, it is referred to as a "variance." In the analysis of variance, such a variance is referred to as a "mean square." For example, the variation of a set of sample observations, denoted X, around their mean $\bar{X}$ is $\Sigma(X - \bar{X})^2$. Dividing this sum of squares by the number of degrees of freedom $n - 1$, where n is the number of observations, we obtain, $[\Sigma(X - \bar{X})^2]/(n - 1)$, the sample variance,[2] which as indicated in Section (7.4) is an unbiased estimator of the population variance. This sample variance can also be referred to as a "mean square."

We now proceed with the analysis of variance by calculating the between-column variation and within-column variation for our illustrative problem.

---

[2]As was stated in Chapter 3, both $[\Sigma(X - \bar{X})^2]/n$ and $[\Sigma(X - \bar{X})^2]/(n - 1)$ are known as the "sample variance."

Between-Column Variation

As indicated earlier, the between-column variation, or between-column sum of squares, measures the variation among the sample column means. It is calculated as follows:

$$\text{(9.5)} \qquad \begin{matrix}\text{Between-column}\\ \text{sum of squares}\end{matrix} = \sum_j r(\bar{X}_j - \bar{\bar{X}})^2$$

where r = number of rows (sample size involved in the calculation of each column mean)

$\bar{X}_j$ = the mean of the jth column

$\bar{\bar{X}}$ = the grand mean

$\sum_j$ = means that the summation is taken over all columns.

As indicated in Equation (9.5), the between-column sum of squares is calculated by the following steps:

(1) Compute the deviation of each column mean from the grand mean.

(2) Square the deviations obtained in Step (1).

(3) Weight each deviation by the sample size involved in calculating the respective mean. In the illustrative example all sample sizes are the same and are equal simply to the number of rows, $r = 4$.

(4) Sum up over all columns the products obtained in Step (3).

Table 9-19 Calculation of the Between-Column Sum of Squares for the Teaching Methods Problem.

$$(\bar{X}_1 - \bar{\bar{X}})^2 = (17 - 20)^2 = 9$$
$$(\bar{X}_2 - \bar{\bar{X}})^2 = (20 - 20)^2 = 0$$
$$(\bar{X}_3 - \bar{\bar{X}})^2 = (23 - 20)^2 = 9$$
$$\sum_j r(\bar{X}_j - \bar{\bar{X}})^2 = 4(9) + 4(0) + 4(9) = 72$$

The calculation of the between-column sum of squares for the numerical example involving three different teaching methods is given in Table 9-19. As indicated in the table, the between-column variation is equal to 72.

Within-Column Variation

The within-column sum of squares is a summary measure of the random errors of the individual observations around their column means. The formula for its computation is

(9.6)

$$\text{Within-column sum of squares} = \sum_j \sum_i (X_{ij} - \bar{X}_j)^2$$

where X_{ij} = the value of the observation in the ith row and jth column

$\bar{X}_j$ = the mean of the jth column

$\sum_j \sum_i$ means that the squared deviations are first summed over all sample observations within a given column, then summed over all columns.

As indicated in Equation (9.6), the within-column sum of squares is calculated as follows:

(1) Calculate the deviation of each observation from its column mean.
(2) Square the deviations obtained in Step (1).
(3) Add the squared deviations within each column.
(4) Sum over all columns the figures obtained in Step (3).

The computation of the within-column variation for the teaching methods problem is given in Table 9-20.

Total Variation

The between-column variation and within-column variation represent the two components of the total variation in the overall set of experimental data. The total variation or total sum of squares is calculated by adding up the squared deviations of all of the individual observations from the grand mean $\bar{\bar{X}}$. Hence, the formula for the total sum of squares is

(9.7)

$$\text{Total sum of squares} = \sum_j \sum_i (X_{ij} - \bar{\bar{X}})^2$$

The total sum of squares is computed by the following steps:

(1) Calculate the deviation of each observation from the grand mean.
(2) Square the deviations obtained in Step (1).
(3) Add the squared deviations over all rows and columns.

The total sum of squares or total variation of the 12 observations in the teaching methods problem is $(16 - 20)^2 + (21 - 20)^2 + \cdots + (25 - 20)^2 = 118$. Referring back to the results obtained in Table 9-19 and 9-20, we see that the total sum of squares, 118, is equal to the sum of the between-column sum of squares, 72, and the within-column sum of squares, 46. In general, the following relationship holds:

(9.8) Total variation = Between-column variation + Within-column variation

Although, as we have indicated earlier, the test of the null hypothesis in a one-factor analysis of variance involves only the between-column variation and the within-column variation, it is useful to calculate also the total variation. This computation is helpful as a check procedure, and is instructive in indicating the relationship between total variation and its components.

Short-Cut Computational Formulas

The formulas we have given for calculating the between-column sum of squares (9.5), within-column sum of squares (9.6), and the total sum of squares (9.7) are the clearest ones for revealing the rationale of the analysis of variance procedure. However, the following short-cut computational formulas are often used to calculate these sums of squares.

(9.9) Between-column sum of squares $= \dfrac{\sum_j T_j^2}{r} - C$

(9.10) Within-column sum of squares $= \sum_j \sum_i X_{ij}^2 - \sum_j \dfrac{T_j^2}{r}$

(9.11) Total sum of squares $= \sum_j \sum_i X_{ij}^2 - C$

where C, the so-called "correction term" is given by

(9.12) $$C = \frac{T^2}{kr}$$

and where T_j is the total of the r observations in the jth column and T is the grand total of all kr observations, that is,

(9.13) $$T = \sum_j \sum_i X_{ij}$$

and all other terms are as previously defined.

These formulas are especially useful when the column means and grand mean are not integers. The short-cut formulas not only save time and computational labor, but are also more accurate because of avoidance of rounding problems which usually occur with the use of Equations (9.5), (9.6), and (9.7).

Table 9-20 Calculation of the Within-Column Sum of Squares for the Teaching Methods Problem.

i	$(X_{i1} - \bar{X}_1)$	$(X_{i1} - \bar{X}_1)^2$	$(X_{i2} - \bar{X}_2)$	$(X_{i2} - \bar{X}_2)^2$	$(X_{i3} - \bar{X}_3)$	$(X_{i3} - \bar{X}_3)^2$
1	$(16 - 17) = -1$	1	$(19 - 20) = -1$	1	$(24 - 23) = 1$	1
2	$(21 - 17) = 4$	16	$(20 - 20) = 0$	0	$(21 - 23) = -2$	4
3	$(18 - 17) = 1$	1	$(21 - 20) = 1$	1	$(22 - 23) = -1$	1
4	$(13 - 17) = -4$	16	$(20 - 20) = 0$	0	$(25 - 23) = 2$	4
		34		2		10

$$\sum_j \sum_i (X_{ij} - \bar{X}_j)^2 = 34 + 2 + 10 = 46$$

Number of Degrees of Freedom

Although the preceding discussion has been in terms of *variation* or *sums of squares* rather than *variance*, the actual test of the null hypothesis in the analysis of variance involves a comparison of the *between-column variance* with the *within-column variance*, or in equivalent terminology, a comparison of the *between-column mean square* with the *within-column mean square*. Hence, the next step in our procedure is to determine the number of degrees of freedom associated with each of the measures of variation. As stated earlier in this section, if a measure of variation, that is, a sum of squares, is divided by the appropriate number of degrees of freedom, the resulting measure is a variance, that is, a mean square.

The number of degrees of freedom associated with the between-column sum of squares is $k - 1$. We can see the reason for this by applying the same general principles indicated earlier for determining number of degrees of freedom in t-tests and chi-square tests. Since there are k columns, or k group means, there are k sums of squares involved in measuring the variation of these column means around the grand mean. Because the sample grand mean is only an estimate of the unknown population mean, we lose one degree of freedom. An alternative view is that the between-column sums of squares is composed of k squared deviations of the form $(\bar{X}_j - \bar{\bar{X}})^2$. If $k - 1$ of these $\bar{X}_j$ values are assigned arbitrarily, then the kth value is determined to arrive at the figure for the between-column sum of square. Hence, there are $k - 1$ degrees of freedom present.

The number of degrees of freedom in our illustrative example which has three different teaching methods, that is, three columns, is $k - 1 = 3 - 1 = 2$.

The number of degrees of freedom associated with the within-column variation is $rk - k = k(r - 1)$. This may be reasoned as follows. There are a total of rk observations. In determining the within-column variation, the squared deviations within each column were taken around the column mean. There are k column means, each of which is an estimate of the true unknown population column mean. Hence there is a loss of k degrees of freedom, and k must be subtracted from rk, the total number of observations.

Alternatively, there are r squared deviations in each column taken around the column mean and a total sum of squares for the column. $r - 1$ of the sums of squares can be assigned arbitrarily, and the last becomes fixed in order for the sum to equal the column sum. Since there are k columns, we have $k(r - 1)$ degrees of freedom.

In the illustrative problem, the number of degrees of freedom associated with the within-column sum of squares is $k(r - 1) = 3(4 - 1) = 9$.

The number of degrees of freedom associated with the total variation is equal to $rk - 1$. There are rk squared deviations taken from the sample grand mean, $\bar{\bar{X}}$. Since $\bar{\bar{X}}$ is an estimate of the true but unknown population mean, there is a loss of one degree of freedom. Alternatively, in the determination of the total sum of squares, there are rk squared deviations. $rk - 1$ of them may be arbitrarily assigned, but the last one is then constrained in order for the sum to be equal to the total sum of squares.

In the illustrative example, the number of degrees of freedom associated with the total variation is $rk - 1 = (4)(3) - 1 = 11$.

Just as the between-column and the within-column variation sum to the total variation, the numbers of degrees of freedom associated with the between- and within-column variations add to the number associated with the total variation. In symbols,

$$rk - 1 = (k - 1) + (rk - k) \tag{9.14}$$

In the illustrative problem, the numerical values corresponding to Equation (9.14) are $11 = 2 + 9$.

The Analysis of Variance Table

An analysis of variance table for the teaching methods problem is given in Table 9-21. The calculations at the bottom of the table will be described presently. The table is in the standard form ordinarily employed to summarize the results of an analysis of variance. In Columns (1), (2), and (3), respectively, are listed the possible sources of variation, the sum of squares for each of these sources, and the number of degrees of freedom associated with each of the sums of squares. We again note that both sums of squares and numbers of degrees of freedom are additive, that is, these figures for between-column and within-column sources of variation add to the corresponding figure for total variation. Dividing the sums of squares in Column (2) by the numbers of degrees of freedom in Column (3) yields the between-column and within-column variances shown in Column (4). As indicated earlier, another name for a sum of squares divided by the appropriate number of degrees of freedom is a "mean square," and it is conventional to use this term in an analysis of variance table. Thus, in our illustrative problem, the between-column mean square denoted MS_b, is equal to $72/2 = 36$. The within-column mean square, denoted MS_w, is equal to $46/9 = 5.11$. The test of the null hypothesis that the population column means are equal is carried out by a comparison of MS_b to MS_w.

Table 9-22 gives the general format of a one-factor analysis of variance table. The sums of squares have been denoted as follows:

$$SS_b = \text{Between-Column Sum of Squares}$$

$$SS_w = \text{Within-Column Sum of Squares}$$

$$SS_t = \text{Total Sum of Squares}$$

Other notation is as previously defined or as given in the next subsection.

Table 9-21 Analysis of Variance Table for the Teaching Methods Problem.

(1) *Source of Variation*	(2) *Sum of Squares*	(3) *Degrees of Freedom*	(4) *Mean Square*
Between columns	72	2	36
Within columns	46	9	5.11
Total	118	11	

$$F(2, 9) = \frac{36}{5.11} = 7.05$$

$$F_{.05}(2, 9) = 4.26$$

Since $7.05 > 4.26$, reject H_0.

Table 9-22 General Format of a One-Factor Analysis of Variance Table.

(1) *Source of Variation*	(2) *Sum of Squares*	(3) *Degrees of Freedom*	(4) *Mean Square*
Between columns	SS_b	$\nu_1 = k - 1$	$MS_b = SS_b/(k-1)$
Within columns	SS_w	$\nu_2 = k(r-1)$	$MS_w = SS_w/k(r-1)$
Total	SS_t	$rk - 1$	

$$F(\nu_1, \nu_2) = \frac{MS_b}{MS_w}$$

The *F*-Test and *F*-Distribution

The comparison of the between-column mean square to the within-column mean square is made by computing their ratio, referred to as *F*. Hence, *F* is given by

$$F = \frac{MS_b}{MS_w} \tag{9.15}$$

In the *F*-ratio, the between-column variance is always placed in the numerator and the within-column variance in the denominator. Under the null hypothesis that the population column means are equal, the *F*-ratio would tend to be equal to 1. On the other hand, if the population column means do indeed differ, then the between-column mean square, MS_b, will tend to exceed the within-column mean square, MS_w, and the F-ratio will then be greater than 1. In terms of our illustrative problem concerning different teaching methods, if F is large, we will reject the null hypothesis that the population mean examination scores are all equal, that is, we will reject $H_0: \mu_1 = \mu_2 = \mu_3$. On the other hand, if F is close to 1, we will accept the null hypothesis. The answer to how large the test-statistic F must be in order to reject the null hypothesis is given by reference to the probability distribution of the F random variable. This distribution is complex, and its mathematical expression will be given here for reference only. Fortunately, critical values of the F-ratio have been tabulated for frequently used significance levels analogous to the case of the chi-square distribution. The probability density function of F is

(9.16)
$$f(F) = cF^{(\nu_1/2)-1}\left(1 + \frac{\nu_1 F}{\nu_2}\right)^{-(\nu_1+\nu_2)/2}$$

where ν_1 = the number of degrees of freedom of the numerator of F
ν_2 = the number of degrees of freedom of the denominator of F
c = a constant depending only on ν_1 and ν_2.

The underlying assumptions are that two random samples are drawn from normally distributed populations with equal variances σ_1^2 and σ_2^2. The term "homoscedasticity" is used in this and other statistical tests for the assumption of equal variances. Unbiased estimators $\hat{\sigma}_1^2$ and $\hat{\sigma}_2^2$ of the population variances are constructed from the sample, and

$$F = \frac{\hat{\sigma}_1^2}{\hat{\sigma}_2^2}$$

Similar to the distributions of t and χ^2, the F-distribution is actually a family of distributions. Each pair of values of ν_1 and ν_2 specifies a different distribution. F is a continuous random variable which ranges from zero to infinity. Since the variances in both the numerator and denominator of the F-ratio are squared quantities, F cannot take on negative values. The F-distribution has a single mode, and although the specific distribution depends on the value of ν_1 and ν_2, its shape is generally assymmetrical and skewed to the right. The distribution tends towards symmetry as ν_1 and ν_2 increase. We will use the notation $F(\nu_1, \nu_2)$ to denote the F-ratio defined in Equation (9.15), that is, $F = MS_b/MS_w$, where the numerator and denominator are between-column mean squares and within-column mean squares with ν_1 and ν_2 degrees of freedom, respectively. Table A-8 of Appendix A presents the critical values of the F-distribution for two selected significance levels, $\alpha = .05$ and $\alpha = .01$. In this table, ν_1 values are listed across the columns and ν_2 down the rows. There are two entries in the table corresponding to every pair of

ν_1, ν_2 values. The upper figure in light-face type is an F value that corresponds to an area of .05 in the right-hand tail of the F-distribution with ν_1 and ν_2 degrees of freedom. That is, it is an F value that would be exceeded only five times in 100 if the null hypothesis under test were true. The lower figure in boldface type is an F value corresponding to a .01 area in the right-hand tail.

We will illustrate the use of the F-table in terms of the teaching methods problem. Assuming we wish to test the null hypothesis $H_0: \mu_1 = \mu_2 = \mu_3$ at the .05 level of significance, we find in Table A-8 of Appendix A that for $\nu_1 = 2$ and $\nu_2 = 9$ degrees of freedom an F value of 4.26 would be exceeded 5% of the time, if the null hypothesis were true. As indicated at the bottom of Table 9-21, we denote this critical value as $F_{.05}(2,9) = 4.26$. This relationship is depicted in Figure 9-3. Again referring to Table 9-21, since the computed value of the F-ratio of the between-column mean square to the within-column mean square is 7.05, and therefore greater than the critical value of 4.26, we reject the null hypothesis. Hence, we conclude that the column

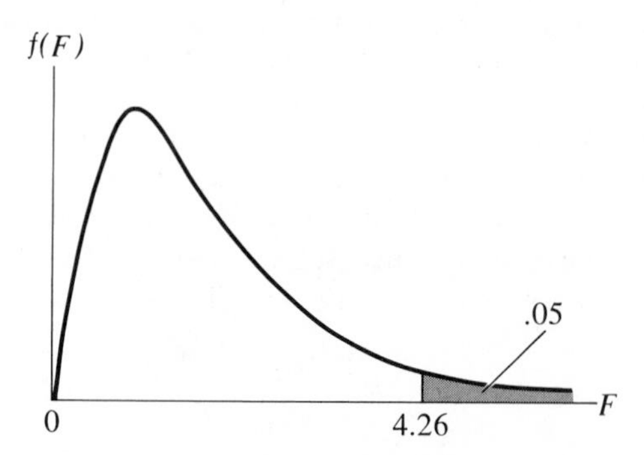

Figure 9-3 The F-distribution for the teaching methods problem indicating the critical F value at the 5% level of significance.

means, that is, the sample mean final examination scores in classes taught by the three teaching methods, differ significantly. The inference about the corresponding population means is that they are not all the same. Referring back to Table 9-17, we see that average grades under Method 3 exceed those under Method 2, which are higher than those under Method 1. Hence, based on these data, our inference is that the three teaching methods are not equally effective, and there is evidence that Method 3 is the most and Method 1 the least effective.

The foregoing example was used to illustrate the rationale involved in the analysis of variance, the statistical technique employed, and the nature of the conclusions that can be drawn. However, it is pertinent to repeat the caveat that the sample sizes in this illustration are doubtless too small for safe conclusions to be drawn about differences in effectiveness of different teaching methods. After all, only four observations were made under each teaching method. Although the risk of a Type I error was controlled at .05 in this problem, the risk of Type II errors may be intolerably high. Suffice it to say that larger sample sizes are generally required, and that similar to the hypothesis testing cases for one or two sample means discussed in Chapter 7, methods are available for controlling Type II errors at specified levels in hypothesis testing for more than two sample means. The interpretation of a Type I error specifically in terms of the present problem is the erroneous rejection of the null hypothesis that all three teaching methods are equally effective. A Type II error is the acceptance of the null hypothesis when in fact the effectiveness of all three teaching methods is not the same.

Further Remarks

It is useful to consider the nature of the conclusion and interpretation arrived at in the teaching methods example. We performed a one-factor analysis of variance, and because the ratio of the between-column variance to the within-column variance differed significantly from one, we rejected the hypothesis that the three population mean examination scores were the same. This conclusion was based on the reasoning that the within-column variance was a measure of random error and under the null hypothesis of equal population column means, the between-column variance also is a measure of random error. Rejection of this null hypothesis leads us to the conclusion that the variation among column means is in excess of these random or chance errors of sampling. Hence, we conclude that the population column means are indeed different. Since these mean examination scores constitute our measure of teaching effectiveness, we conclude that the three teaching methods are equally effective.

However, let us be somewhat more critical and look at the possible interpretations of our findings more closely. We assumed that the *same teacher* taught three different sections of the basic statistics course by the three specified methods, 1, 2, and 3. Method 3 appeared most effective and Method 1 least effective. Suppose that the instructor teaches these three sections on the same days and that the class taught by Method 3 is given early in the morning. Hence, the teacher is fresh, wide awake, and enthusiastic. On the other hand, the classes taught by Methods 2 and 1 are in the middle of the day and late afternoon, respectively. Let us assume that by late afternoon, the instructor is tired, sleepy, and rather unenthusiastic. Then, it is possible that the differences in teaching effectiveness are not really attributable solely to the different teaching methods, but rather to some unknown mixture of the difference in teaching methods and the aforementioned factors associated with the time of day. We can think of experimental designs which would tend to counteract or "control" the effect of the time of day factor. Suppose the class taught by Method 1 meets one-third of the time in the early morning, one-third of the time in the middle of the day, and one-third of the time late in the afternoon; similarly for classes taught by Methods 2 and 3. Then time of class meeting would not be a factor which differentially affects teacher effectiveness. Of course, practical scheduling considerations might militate against this type of staggered class arrangement. However, the principle of explicitly controlling for the effects of the time of day factor is clear.

If the staggered scheduling system was not feasible, another approach suggests itself. If it were suspected a priori, that Method 3 is the most effective and Method 1 the least effective teaching method, the class taught by Method 3 might be scheduled late in the afternoon, and the one by Method 1 early in the morning. Then if results were obtained such as in the illustrative example, we would have even greater confidence that Method 3 was most effective and Method 1 least effective. This conclusion follows from the fact that Method 3 had to overcome the disadvantage of time of day, whereas Method 1 had a time of day advantage. However, this type of experimental design is dangerous because the effect of the factor of teaching effectiveness might be blurred by the oppositely acting effect of the time of day factor. When two factors are acting together this way and we have no way of segregating their separate effects, "confounding" is said to be present. That is, the effects of teaching method and time of day are confounded.

Other factors certainly occur to the reader which might have affected the differences among the column means, that is, the mean examination scores. For example, suppose the four students taught by Method 3 had greater aptitude for statistics than those taught by Method 2, who in turn had better aptitude than those taught by Method 1. Then, clearly again we have confounded effects. The rejection of the null hypothesis of equal population mean examination scores might not reflect differences in teaching effectiveness nearly so much as differences in student aptitudes. Of course, it was assumed in our example that the students taught by the three methods were randomly selected from a sophomore class, but nevertheless the hypothesized differences in aptitude might have been present. Again, the point is that in a more elaborate experimental design, we might attempt to control explicitly for the effect of student aptitude.

The preceding discussion applies equally well to the hypothesis testing methods considered earlier, as for example in Chapter 7, because we might have had only two teaching methods to compare rather than three. In summary, mechanistic or rote application of statistical techniques such as hypothesis testing methods must be guarded against. In this text, we consider the general principles involved in some of the simpler, basic procedures. More refined and sophisticated techniques may very well be required in particular instances. For example, we have considered only one-factor analysis of variance. More elaborate experimental designs which attempt to control and test for the effects of more factors are available, and considerable expertise is often required for their proper application.

One of the points we have attempted to convey in the preceding discussion is that statistical results are virtually always consistent with more than one interpretation. Naïve leaping to conclusions must be guarded against and careful consideration must be given to alternative interpretations and explanations. We conclude this chapter with two anonymous humorous stories which are relevant to the point that alternative interpretations and explanations of experimental results are often possible.

An investigator wished to determine the differential effects involved in the intake of various types of mixed drinks. Therefore, he had subjects drink substantial quantities of scotch and water, bourbon and water, and rye and water. All of the subjects became intoxicated. The investigator concluded that since water was the one factor common to all of these drinks, the imbibing of water makes people drunk.

The heroine of our second story is a grammar school teacher, who wished to explain the harmful effects of drinking liquor to her class of eight-year-olds. She placed two glass jars of worms on her desk. Into the first jar, she poured some water. The worms continued to move about, and did not appear to have been adversely affected at all by the contact with the water. Then she poured a bottle of whiskey into the second jar. The worms became still and appeared to have been mortally stricken.

The teacher then called upon a student and asked, "Johnny, what is the lesson to be learned from this experiment?" Johnny, looking very thoughtful, replied, "I guess it proves that it is good to drink whiskey because it will kill any worms you may have in your body."

PROBLEMS

1. A set of five coins was tossed 1000 times. The number of times that 0, 1, 2, 3, 4, and 5 heads were obtained is shown in the table.

Number of Heads	*Number of Tosses*
0	36
1	138
2	348
3	287
4	165
5	26
Total	1000

Determine whether the binomial distribution is a good fit to these data. Assume the probability of a head is .5 and use a .05 level of significance.

2. Over a 100-day period, during a certain minute between 2:00 P.M. and 4:00 P.M., the number of phone calls coming into the switchboard of a company was as follows:

Number of Calls	0	1	2	3	4	5	6	7	8
Observed Number of Days	5	7	30	40	7	4	5	1	1

Fit a Poisson distribution to these data, and use a chi-square test to determine the "goodness of fit." Use a .01 level of significance.

3. A consumer research firm wished to determine whether there was a real difference in preference in taste by iced tea drinkers in a certain metropolitan area among five brands of iced tea. The firm took a simple random sample of 100 iced tea consumers in the area and conducted the following experiment. Five glasses of iced tea were given to each consumer. The glasses were marked *A, B, C, D,* and *E* and contained one of each of the individual brands. The cups were presented to each consumer in a random order determined by sequential selection from five paper slips, each containing one of the letters *A, B, C, D,* and *E.* In the table are shown the numbers of iced tea consumers who stated that they most preferred the indicated brands.

Brand Preferred	*Number of Consumers*
A	27
B	16
C	22
D	18
E	17

Using a chi-square test determine whether the null hypothesis

H_0: the probability distribution is uniform

should be rejected. Test using both a .05 and a .01 level of significance.

4. In 20 Confederate army corps units from 1860 to 1864, the number of deaths per army corp per year resulting from measles was as follows:

Number of Deaths	0	1	2	3	4	5	6	7
Number of Corps-Years	22	28	35	8	7	0	0	0

Use a chi-square test to determine whether the Poisson distribution is a "good fit."

5. A box contains a large number of balls of four different colors: black, white, red, and green. A sample of 16 balls drawn at random from the box revealed three black, seven white, four red, and two green balls. Test the hypothesis that the box contains equal proportions of differently colored balls, using a .05 level of significance.

6. A student looking for an easy statistics teacher was told by the departmental office that all three statistics teachers passed the same proportions of students. The student did some research and came up with the following results:

Student Performance	*Professor A*	*Professor B*	*Professor C*	*Total*
Number passed	42	43	38	123
Number failed	8	5	14	27
Total	50	48	52	150

Should the student believe what the office told him? Use a .05 level of significance.

7. A subscription service stated that preferences for different national magazines were independent of geographical location. A survey was taken in which 300 persons randomly chosen from three areas were given a choice among three different magazines. Each person expressed his or her favorite. The following results were obtained:

Region	*Magazine X*	*Magazine Y*	*Magazine Z*	*Total*
New England	75	50	175	300
Northeastern	120	85	95	300
Southern	105	110	85	300
Total	300	245	355	900

Would you agree with the subscription service's assertion? Use a .05 level of significance.

8. A professor wished to know if grades in a basic economics course were independent of the students' year in school. He obtained the grades of all the students taking the course in the first semester and set up the following table. Test the hypothesis that performance in the course is independent of the students' year in school using a .05 and .01 significance level. A grade in the course ranges from zero to 100.

	Grade Ranges			
Year	80–100	60–79	0–59	*Total*
Sophomore	20	46	14	80
Junior	15	40	5	60
Senior	10	20	10	40
Total	45	106	29	180

9. The editor of *The Star* did research to determine whether social class has any effect on newspaper buying. He took a poll of 150 people from each of three social classes and ascertained whether they read his paper or its competitor, *The Press*. The results were

Social Class	*The Star*	*The Press*	*Total*
Lower	80	70	150
Middle	90	60	150
Upper	50	100	150
Total	220	230	450

Test the hypothesis that choice of newspaper and social class are independent. Use a .05 level of significance.

10. Components are supplied to a television manufacturer by two subcontractors. Each component is tested with respect to five characteristics before it is accepted by the manufacturer. Records have been kept for one month on the numbers of different types of defects for each contractor, and from these the following table has been constructed.

	Type of Defect					
Supplier	*A*	*B*	*C*	*D*	*E*	*Total*
1	70	10	10	30	0	120
2	10	10	20	20	20	80
Total	80	20	30	50	20	200

Would you conclude that type of defect and supplier are independent? Use $\alpha = .01$.

11. The following table shows the location of a random sample of 200 members of a trade association by city type and geographic region.

	Geographic Region				
City Type	(B_1) *Eastern*	(B_2) *Southern*	(B_3) *Midwestern*	(B_4) *Far Western*	*Total*
(A_1) Large	35	10	25	25	95
(A_2) Small	15	10	15	15	55
(A_3) Suburb	25	5	10	10	50
Total	75	25	50	50	200

Are city type and geographical region independent? Test at the .05 level of significance.

12. Seven samples each of size ten are drawn from a normally distributed population. The means and variances of the seven samples are shown in the following table:

Sample	*Mean*	*Variance*
A	50	5
B	47	4
C	52	5
D	51	6
E	49	5
F	52	4
G	49	6

Would you conclude that the seven samples were drawn *randomly* from the same population? Is your conclusion the same for both the .05 and .01 levels of significance? Assume the variances were computed using an $n - 1$ denominator.

13. A manufacturer has a choice of three subcontractors from whom to buy parts. The manufacturer, before deciding from whom he will buy, purchases five batches from each subcontractor. There are the same number in each batch. The number of defectives per batch is given in the following table:

Batch	*Subcontractor A*	*Subcontractor B*	*Subcontractor C*
1	35	15	25
2	25	20	40
3	30	25	40
4	35	15	35
5	20	30	30

Would you conclude that there is no real difference among these subcontractors in the average number of defectives produced per batch? Use a .05 level of significance.

14. The manufacturer of a new product wished to select the best advertising display for his product. Because he had a choice of five different displays he randomly selected 25 different stores and placed each type of display in five stores. The following figures are the average amount and variance per display in terms of dozens sold during the first six months.

Type of Display	*Mean*	*Variance*
1	78	9
2	76	7
3	77	8
4	74	8
5	76	10

Can the manufacturer assume that it doesn't matter which display he uses? Assume the variances were computed using an $n - 1$ denominator and use a .01 level of significance.

15. As head of a department of a consumer research organization, you have the responsibility for testing and comparing lifetimes of light bulbs for four brands of bulbs. Suppose you test the lifetime of three bulbs of each of the four brands. Your test data are as follows, each entry representing the lifetime of a bulb, measured in hundreds of hours.

Brand

A	*B*	*C*	*D*
20	25	24	23
19	23	20	20
21	21	22	20

Can we assume that the mean lifetimes of the four brands are equal?

CHAPTER TEN

Regression and Correlation Analysis

10.1 Introduction

In Chapter 9, we discussed the use of the chi-square test in contingency tables as a means of determining whether or not a relationship exists between two bases of classification, or more briefly, whether a relationship exists between two variables. However, the contingency table represents a method of classifying statistical observations into qualitative categories, and like other methods of tabulation, it does not provide a basis for estimating the values of one variable from a knowledge of the other. We now turn to a very widely used body of statistical methods, namely, *regression* and *correlation analysis*, which enables us to deal with variables which are stated in terms of numerical values rather than in qualitative categories. Furthermore, these methods provide the bases for estimating the values of one variable from known or assumed values of one or more other variables and for measuring the strength of the relationships among the variables.

Equations are used in mathematics to express the relationships among variables. In fields such as geometry or trigonometry, these mathematical functions or equations express the *exact relationships* that are present among the variables of interest. Thus, the equation $A = s^2$ describes the relationship between s, the length of the side of a square, and A, the area of the square. The equation $A = ab/2$ expresses the relationship between b, the length of any side of a triangle, a, the altitude or perpendicular distance to that side from the angle opposite it, and A, the area of the triangle. If we substitute numerical values for the variables on the right-hand sides of these equations, we can calculate the *exact values* of the quantities on the left-hand sides. In the social sciences and in fields such as business and governmental administration, exact relationships are not generally observed among variables, but rather *statistical relationships* prevail. That is, certain average relationships may be observed among variables but these average relationships do not provide a basis for perfect predictions. For example, if we know how much money a corporation spends on television advertising, we cannot make an exact prediction of the amount of sales this promotional expenditure will generate. If we know a family's net income, we cannot make an exact forecast of the amount of money that family saves. On the other hand, we can measure statistically how sales vary, on the average, with differences in television advertising, or how family savings vary, on the average, with differences in income. Also, we can determine the amount of dispersion that exists around these average relationships. On the basis of these relationships, we may be able to estimate the values of the variables of interest closely enough for decision-making purposes. The techniques of regression and correlation analysis are important statistical tools used to accomplish this measurement and estimation process.

The term "regression analysis" refers to the methods by which estimates are made of the values of a variable from a knowledge of the values of one or more other variables, and to the measurement of the errors involved in this estimation process. The term "correlation analysis" refers to methods for measuring the strength of the association (correlation) among these variables.

We begin by discussing the case of a *two variable linear regression and correlation analysis*. The term "linear" means that an equation of a straight line of the form $Y = a + bX$, where a and b are numbers, is used to describe the average relationship that exists between the two variables and to carry out the estimation process. The factor whose values we wish to estimate is referred to as the *dependent variable* and is denoted by the symbol Y. The factor from which these estimates are made is called the *independent variable* and is denoted by X. The terms "dependent" and "independent" do not imply that there is necessarily any cause and effect relationship between the variables. What is meant is simply that estimates of values of the dependent variable Y may be obtained for given values of the independent variable X from a mathematical function involving X and Y. In that sense, the values of Y are dependent upon the values of X. The X variable may or may not be *causing* changes in the Y variable. Hence, if we are estimating sales of a product from figures on advertising expenditures, sales is the dependent variable and advertising expenditures is the independent variable. There may or may not be a causal connection between these two factors in the sense that changes in advertising expenditures cause changes in sales. In fact, in certain situations, the cause-effect relation may be just the opposite of what appears to be the obvious one. For example, suppose a company budgets a product's advertising expenditures for the next year as a flat percentage of the sales of that product during the preceding year. Then advertising expenditures are more directly dependent on sales (with a one year lag) than vice versa.

Let us consider some illustrative cases of variables which it is reasonable to assume are related to one another, that is, correlated. If suitable data were available, we might attempt to construct an equation which would permit us to estimate the values of one variable from the values of the other. We shall assume that the first named factor in each pair is the variable to be estimated, that is, the dependent variable, and the second one is independent. Consumption expenditures might be estimated from a knowledge of income; investment in telephone equipment from expenditures on new construction; personal net savings from disposable income; commercial bank discount rates from Federal Reserve Bank discount rates; and success in college from scholastic aptitude test scores.

Of course, additional definitions are required to attach meaning to the estimation problems listed above. Thus, in the illustration of consumption expenditures and income, we would have to specify whose expenditures and whose income are involved. If we wanted to estimate family consumption expenditures from family income, the family would be said to be the "unit of association." The estimating equation would be constructed from data representing observations of these two variables for individual families. We would have to define the variables more specifically. For example, we might be interested in estimates of annual family consumption expenditures from annual family net income, where again these terms would require precise definitions.

In each of these examples, it is possible to specify other independent variables which might be included to aid in obtaining good estimates of the dependent variable. Hence, in estimating a family's consumption expenditures, we might wish to use knowledge of the size of the family in addition to information on the family's income. This would be an illustration of *multiple regression analysis*, where two independent variables, that is, (1) family income and (2) family size are used to obtain estimates of a dependent variable. As indicated earlier in this chapter, we first consider two-variable problems involving a dependent factor and only one independent factor before turning to a brief discussion of multiple regression problems involving more than one independent variable.

10.2 Scatter Diagrams

In studying the relationship between two variables, it is advisable as a first step to plot the data on a graph. This allows visual examination of the extent to which the variables are related and aids in choosing the type of model that would be appropriate for estimation. The chart used for this purpose is known as a *scatter diagram*, which is a graph on which each plotted point represents an observed pair of values of the dependent and independent variables. We will illustrate this by plotting a scatter diagram for the data given in Table 10-1. These figures represent observations for a sample of ten families of annual expenditures on consumer durables, which we shall treat as the dependent variable, Y, or the factor to be estimated, and annual net income, X, which is the independent variable or the factor from which the estimates are to be made. We shall assume that the ten families constitute a simple random sample of families with $10,000 or less of annual net income in a metropolitan area in 1967. Although only ten families are doubtless too small a sample from which to draw very useful conclusions that would apply to all such families in a metropolitan area, we shall use such a small sample in order to require only a very modest amount of arithmetic. Furthermore, we have assumed relatively low incomes in order to simplify the numerical work.

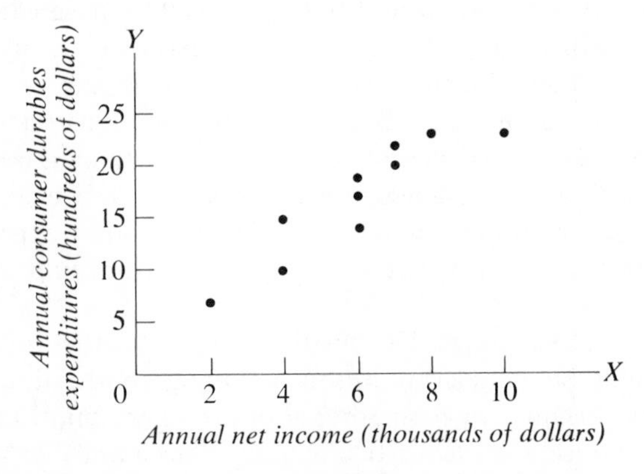

Figure 10-1 A scatter diagram of annual consumer durables expenditures and annual net income of a sample of 10 families in a metropolitan area in 1967.

Figure 10-1 presents the data of Table 10-1 on a scatter diagram. On the Y axis are plotted the figures on consumer durables expenditures and on the X axis annual net income. This follows the standard convention of plotting

Table 10-1 Annual Consumer Durables Expenditures and Annual Net Income of a Sample of Ten Families in a Metropolitan Area in 1967.

Family	*Annual Consumer Durables Expenditures (hundreds of dollars)* Y	*Annual Net Income (thousands of dollars)* X
A	23	10
B	7	2
C	15	4
D	17	6
E	23	8
F	22	7
G	10	4
H	14	6
I	20	7
J	19	6

SOURCE: Hypothetical data.

the dependent variable along the Y axis and the independent variable along the X axis. The pair of observations for each family determines one point on the scatter diagram. Thus, for family A, a point is plotted corresponding to $X = 10$ along the horizontal axis and $Y = 23$ along the vertical axis; for family B, a point is plotted corresponding to $X = 2$ and $Y = 7$, and so forth. An examination of the scatter diagram gives some useful indications of the nature and strength of the relationship between the two variables. For example, depending upon whether the Y values tend to increase as the values of X increase, or to decrease as the values of X increase, there is said to be a *direct* or *inverse* relationship, respectively, between the two variables. The configuration on Figure 10-1 indicates a general tendency for the points to run from the lower left to the upper right-hand side of the graph. Hence, as we move from low to higher income families, consumer durables expenditures tend to increase. This is an example of a *direct relationship* between the two variables. On the other hand, if the scatter of points runs from the upper left to the lower right, that is, if the Y variable tends to decrease as X increases, there is said to be an *inverse relationship* between the variables. Also, an examination of the scatter diagram gives an indication of whether a straight line appears to be an adequate description of the average relationship between the two variables. If a straight line is used to describe the average relationship between Y and X, a *linear relationship* is said to be present. On the other hand, if the points on the scatter diagram appear to fall along a curved line rather than a straight line, a *curvilinear relationship* is said to exist. Figure 10-2 presents illustrative combinations of the foregoing types of relationships. Parts (a), (b), (c), and (d) of Figure 10-2 show respectively direct linear, inverse linear, direct curvilinear, and inverse curvilinear relationships. As can be seen, the points tend to follow: in (a) a straight line sloping upward, in (b) a straight line sloping downward, in (c) a curved line sloping upward, and in (d) a curved line sloping downward. Of course, the relationships are not always so obvious. In (e) the points appear to follow a horizontal straight line. Such a case depicts a situation of "no correlation" between the X and Y variables, or no evident relationship, since the horizontal line implies no change in Y, on the average, as X increases. In (f) the points follow a straight line sloping upward as in (a), but there is a much wider scatter of points around the line than in (a). In our present problem, we shall assume that a visual examination of Figure 10-1 suggests that there is a direct linear relationship between the two variables. In the next section, we shall discuss the purposes of regression and correlation analysis and the types of procedures used to accomplish these objectives. Then these procedures will be illustrated in terms of the data of Table 10-1.

10.3 Purposes of Regression and Correlation Analysis

What does a regression and correlation analysis attempt to accomplish in studying the relationship between two variables, such as expenditures on consumer durables and net annual income of families? We will concentrate on these basic goals, which emphasize the relationships contained in the particular sample under study. At a later point we will consider other objectives involving statistical inference, that is, inferences concerning the population from which the sample was drawn.

The first two objectives and the statistical procedures involved in their accomplishment fall under the heading of *regression analysis*, while the third objective and related procedures are classified as *correlation analysis*. The objectives are stated below and the statistical measures used to achieve these objectives are named. However, the mathematical definitions of these measures are postponed until the discussion of their use in the illustrative problem involving family expenditures on consumer durables and family income.

1. *The first purpose of regression analysis is to provide estimates of values of the dependent variable from values of the independent variable.* The device

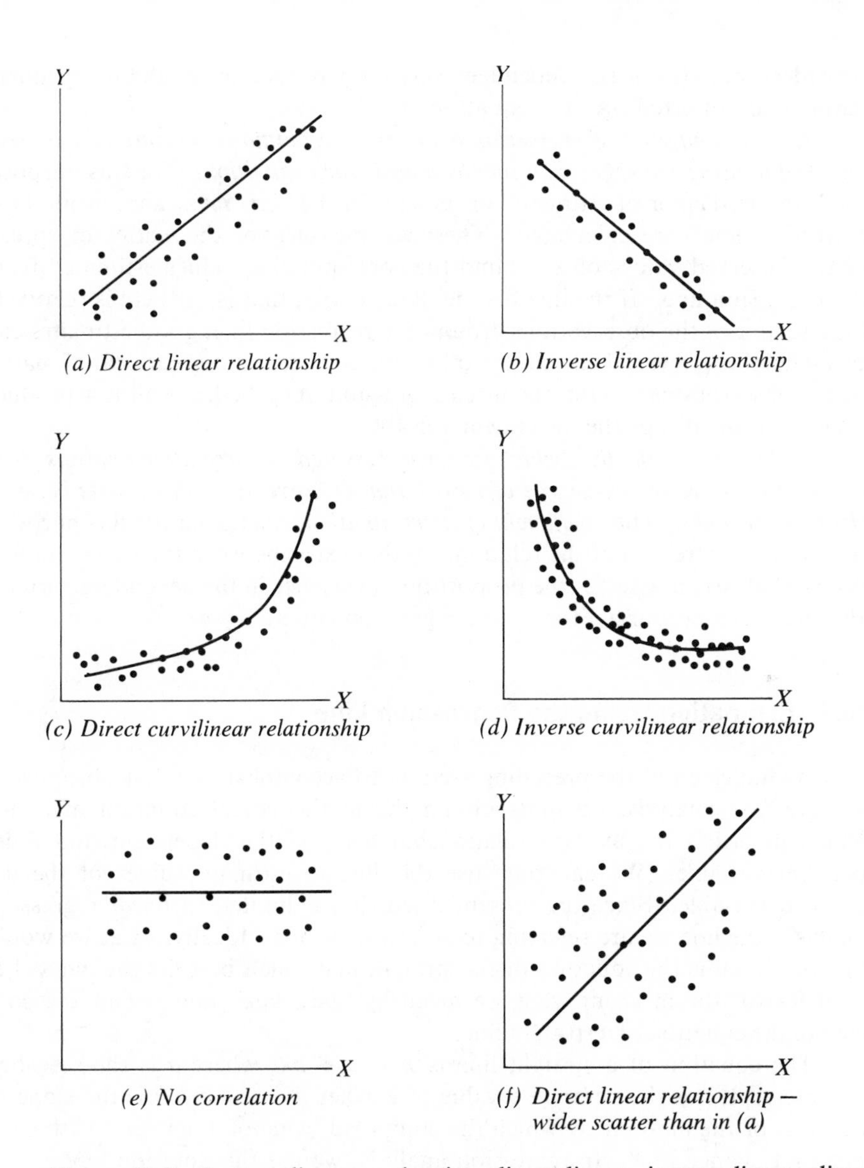

Figure 10-2 Scatter diagrams showing direct linear, inverse linear, direct curvilinear, and inverse curvilinear relationships.

used to accomplish this estimation procedure is the *regression line*, which is a line fitted to the data by a method to be subsequently described. The regression line describes the average relationship existing between the X and Y variables. Somewhat more precisely, it is a line which displays mean values of Y for given values of X. The equation of this line, known as the *regression equation*, provides estimates of the dependent variable when values of the independent variable are inserted into the equation.

2. *A second goal of regression analysis is to obtain a measure of the error involved in using the regression line as a basis for estimation.* For this purpose, the "standard error of estimate" or its square, the "error variance around the regression line" are calculated. These are measures of the scatter or spread of the observed values of Y around the corresponding values estimated from the regression line. If the line fits the data closely, that is, if there is relatively little scatter of the observations around the regression line, good estimates can be made of the Y variable. On the other hand, if there is a great deal of scatter of the observations around the fitted regression line, the line will not produce accurate estimates of the dependent variable.

3. *The third objective, which we have classified as correlation analysis, is to obtain a measure of the degree of association or correlation that exists between the two variables.* The *coefficient of determination*, calculated for this purpose, measures the strength of the relationship that exists between the two variables. As we shall see, it assesses the proportion of variance in the dependent variable that has been accounted for by the regression equation.

10.4 Estimation Using the Regression Line

As indicated in the preceding section, to accomplish the first objective of a regression analysis, we must obtain the mathematical equation of a line which describes the average relationship between the dependent and independent variable. We can then use this line to estimate values of the dependent variable. Since the present discussion is limited to *linear* regression analysis, the line we are referring to is a straight line. Ideally, what we would like to obtain is the equation of the straight line which best fits the data. Let us defer for the moment what we mean by "best fits," and review the concept of the equation of a straight line.

The equation of a straight line is $Y = a + bX$, where a is the so-called "Y intercept," or the computed value of Y when $X = 0$, and b is the slope of the line, or the amount by which the computed value of Y changes with each one unit change in X. In regression analysis, we use the notation

$$\bar{Y}_X = a + bX \qquad \textbf{(10.1)}$$

for the equation of the regression line. It is indicated at a later point why the particular symbol $\bar{Y}_X$ is used to denote a computed value of the dependent variable. It suffices at present to note that it is useful to use the different symbols $\bar{Y}_X$, which denotes a *computed* value of the dependent variable, and Y, which denotes an *observed* value. For example, in the data given in Table 10-1, the observed value of Y for family A is 23, or \$2300 annual consumer durables expenditures. The observed X value is ten, indicating that this family's net income is \$10,000. When we obtain a regression equation, we may wish to estimate annual consumer durables expenditures for a family with an annual net income of \$10,000. By substituting $X = 10$ into the regression equation, we can obtain the required estimate. Since the computed figure will in general be different from the observed value $Y = 23$, it is useful to have a separate symbol, such as $\bar{Y}_X$, to denote this estimated or computed value of the dependent variable.

Let us review by means of a simple illustration the relationship between the equation $\bar{Y}_X = a + bX$ and the straight line which represents the graph of the equation. Suppose the equation is

$$\bar{Y}_X = 2 + 3X \tag{10.2}$$

Thus, $a = 2$ and $b = 3$. If we substitute a value of X into this equation, we can obtain the corresponding computed value of $\bar{Y}_X$. Each pair of X and $\bar{Y}_X$ values represents a single point. Although only two points are required to determine a straight line, several pairs of X and $\bar{Y}_X$ values are shown for the line $\bar{Y}_X = 2 + 3X$ in Table 10-2. The graph corresponding to this line is shown in Figure 10-3. As can be seen on the graph, since the a value in the equation of the line is equal to two, the line intersects the Y axis at a height of two units. Also, since the b value, or slope of the line, is equal to three, we note in Table 10-2 that the $\bar{Y}_X$ values increase by three units each time X increases by one unit. This is shown graphically in Figure 10-3 as a rise of three units in the line when X increases by one unit.

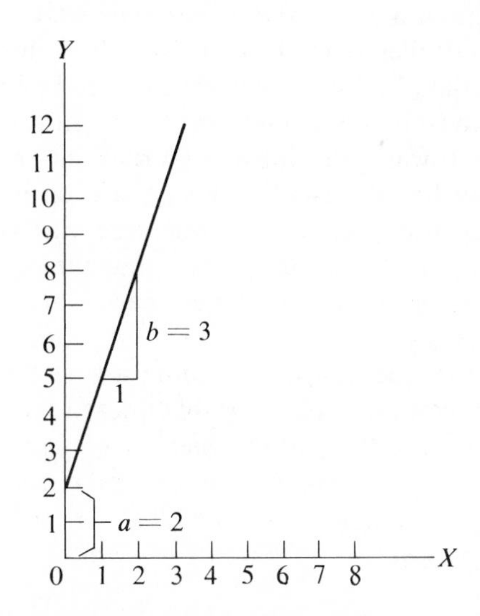

Figure 10-3 Graph of the line $\bar{Y}_X = 2 + 3X$.

Table 10-2 Calculation of Pairs of X and $\bar{Y}_X$ Values for the Line $\bar{Y}_X = 2 + 3X$.

X	$\bar{Y}_X$
0	2
1	5
2	8
3	11
4	14

The terms "regression line" and "regression equation" for the estimating line and equation stem from the pioneer work in regression and correlation analysis of the British biologist, Sir Francis Galton in the nineteenth century. In the course of his investigations of hereditary traits, one of Galton's studies concerned the relationship between heights of fathers and heights of their sons. He found that a direct relationship existed; that is, short fathers tended to have short sons, and tall fathers tended to have tall sons. Thus, on a scatter diagram, where Y = height of son and X = height of father, the points tended to fall along a line from lower left to upper right. However, in the data he examined, Galton found a regression toward an average height or as he termed it, a "regression to mediocrity." That is, there was a tendency for tall fathers to have sons who were shorter than they were, and for short fathers to have sons who were taller than their fathers. The lines which were fitted to the scatter diagram data in this early work came to be known as "regression lines" and the equations of these lines as "regression equations." Galton found this regression toward the mean in other natural characteristics as well. The terminology has persisted. Thus, these terms for the estimating line and estimating equation are used in the wide variety of fields in which regression analysis is applied, despite the fact that the original implication of a regression toward an average is not necessarily present in terms of the phenomena under investigation.

We now turn to the question of obtaining a best fitting line to the data plotted on a scatter diagram in a two-variable linear regression problem. The fitting procedure to be discussed is the *method of least squares*, which undoubtedly is the most widely applied curve fitting technique in statistics.

The Method of Least Squares

In order to establish a best fitting line to a set of data on a scatter diagram, we must have criteria concerning what constitutes *goodness of fit*. A number of such criteria, which might at first thought seem reasonable, turn out to be unsuitable. For example, we might entertain the idea of fitting a straight line to the data in such a way that one-half of the points fall above the line and one-half below. However, this requirement can be easily dismissed, since such a line may represent a quite poor fit to the data if, say, the points which fall above the line lie very close to it whereas the points below deviate considerably from it.

Another criterion that is suggested by the difficulty with the preceding one is to balance *the deviations* of the points above and below the line rather than to balance simply *the number* of points above and below the fitted line. The deviations referred to here are $Y - \overline{Y}_X$ values, that is, differences between observed values of the dependent variable and the corresponding estimated points on the regression line at the same X values. Thus, it should be noted that the deviations are measured parallel to the Y axis and are in the units of the Y variable. If a point lies above the fitted line, $Y - \overline{Y}_X$ is a positive value; if the point is below the line, $Y - \overline{Y}_X$ is negative. We can state this criterion of balancing deviations in terms of making the total of the positive and negative deviations equal to zero, or, in symbols, $\Sigma(Y - \overline{Y}_X) = 0$. However, it turns out that *any* straight line that passes through the point made by the mean of the X's and mean of the Y's $(\overline{X}, \overline{Y})$ satisfies this criterion. Therefore, the requirement that $\Sigma(Y - \overline{Y}_X) = 0$ is not sufficient to specify a good fitting line. Figure 10-4 shows a line with a positive slope (b is a positive number) which passes through $(\overline{X}, \overline{Y})$ and thus meets the aforementioned criterion. Also shown is a line with negative slope which passes through the same point, and therefore it too satisfies the same criterion. The negatively inclined line is clearly a poor fit to the points on the indicated scatter diagram. The graph also shows the relationship between Y, $\overline{Y}_X$, and $Y - \overline{Y}_X$ for a fitted line whose equation is $\overline{Y}_X = a + bX$.

It would seem reasonable to apply a criterion of goodness of fit which

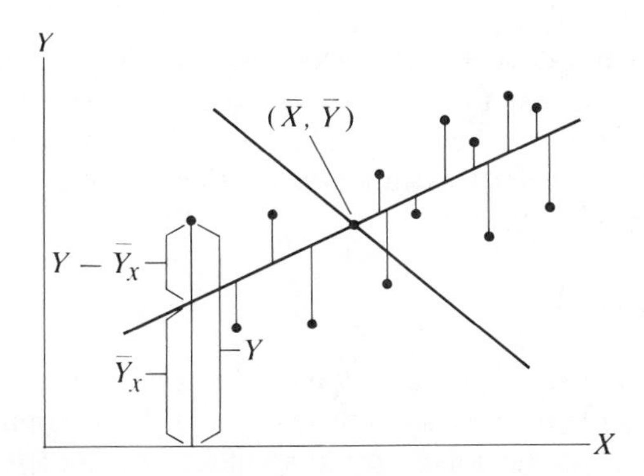

Figure 10-4 Two straight lines passing through the point $(\overline{X}, \overline{Y})$.

specifies that the fitted line should in some sense make the deviations of the observed points from the estimating line as small as possible. Hence, we might establish the requirement that the line that is fitted to the data must minimize the *absolute* deviations of the points from the line. Symbolically, this requires a straight line which minimizes $\Sigma |Y - \bar{Y}_X|$. We previously encountered absolute deviations in the calculation of the mean deviation in Section 3.16. Just as in the case of the mean deviation, the use of absolute values and the consequent disregarding of algebraic positive and negative signs in applying the criterion of minimizing $\Sigma |Y - \bar{Y}_X|$ turns out to be a disadvantage. The *method of least squares* discussed below, which minimizes the sum of the squared deviations $\Sigma(Y - \bar{Y}_X)^2$, is mathematically simpler and produces an estimating line which has many desirable properties.

Let us now consider the most generally applied curve fitting technique in regression analysis, namely, the method of least squares. This method imposes the requirement that the *sum of the squares* of the deviations of the observed values of the dependent variable from the corresponding computed values on the regression line must be a minimum. Thus, if a straight line is fitted to a set of data by the method of least squares it is a "best fit" in the sense that the sum of the squared deviations, $\Sigma(Y - \bar{Y}_X)^2$, is less than it would be for any other possible straight line. Another useful characteristic of the least squares straight line is that it passes through the point of means, $(\bar{X}, \bar{Y})$, and therefore, as explained earlier, makes the total of the positive and negative deviations equal to zero. In summary, the least squares straight lines possesses the following mathematical properties:

(10.3) $$\Sigma(Y - \bar{Y}_X)^2 \text{ is a minimum}$$

(10.4) $$\Sigma(Y - \bar{Y}_X) = 0$$

Nonlinear functions can also be fitted to a set of data by the method of least squares. In all cases, the least squares line has the characteristic that the sum of the squared deviations from it is less than from any other line of the same mathematical classification. For example, we may denote a second degree parabola by the equation $\bar{Y}_X = a + bX + cX^2$. If such a curved line is fitted by the method of least squares, then the sum of the squared deviations from it will be less than from any other second degree parabola that could have been fitted to the data. The nature of a second degree parabola and the method of fitting this function by the method of least squares are discussed in Section 10.10.

The Normal Equations

By using calculus methods to apply the condition that the sum of the squared deviations from a straight line must be a minimum, two equations are derived which can be solved for the values of a and b in the regression equation[1] $\bar{Y}_X = a + bX$. These equations are conventionally referred to as the "normal equations," although this expression is a misnomer in the sense that normal does not imply "normal curve" nor does it imply a "normal" versus "abnormal" situation. The normal equations are simply the mathematical expressions from which one obtains the values of the constants a and b in the least squares regression equation. The two normal equations, from which we determine the values of a and b in the fitted regression line are

$$\Sigma Y = na + b\Sigma X \tag{10.9}$$

$$\Sigma XY = a\Sigma X + b\Sigma X^2 \tag{10.10}$$

The procedure for using these normal equations will now be discussed in terms of our illustrative problem.

Computational Procedure

Let us return to the problem involving the sample of ten families and assume that after examination of the scatter diagram in Figure 10-1, we *decide* to fit a *straight line* to the data. We now can use the normal equations for this purpose. The various quantities n, ΣY, ΣX, ΣXY, and ΣX^2, required in these two equations are determined from the original observations, where n is the number of pairs of X and Y values, in this case, ten. For our illustration, the computation of the required totals is shown in Table 10-3. Although ΣY^2 is not needed for the normal equations, its calculation is also shown.

---

[1]The method of deriving the normal equations is as follows. We denote the sum of squared deviations which must be minimized as some function of the unknown quantities a and b. Thus, let

$$F(a, b) = \Sigma(Y - \bar{Y}_X)^2 \tag{10.5}$$

Substituting the right-hand expression for $\bar{Y}_X = a + bX$ into (10.5) gives

$$F(a, b) = \Sigma(Y - a - bX)^2 \tag{10.6}$$

We impose the condition of minimizing $F(a, b)$ by obtaining its partial derivatives with respect to a and b and setting them equal to zero. Thus,

$$\frac{\partial F(a, b)}{\partial a} = -2\Sigma(Y - a - bX) = 0 \tag{10.7}$$

$$\frac{\partial F(a, b)}{\partial b} = -2\Sigma(Y - a - bX)(X) = 0 \tag{10.8}$$

Solving Equations (10.7) and (10.8) yields the two normal equations

$$\Sigma Y = na + b\Sigma X$$

$$\Sigma XY = a\Sigma X + b\Sigma X^2$$

A check that the second derivatives of $F(a, b)$ are positive reveals that a minimum has been found.

This figure is useful for calculating the standard error of estimate to be discussed shortly. Substituting the quantities $n = 10$, $\Sigma Y = 170$, $\Sigma X = 60$, $\Sigma XY = 1122$, and $\Sigma X^2 = 406$ from Table 10-3 into the normal equations (10.9) and (10.10) gives

$$170 = 10a + 60b \tag{10.11}$$

$$1122 = 60a + 406b \tag{10.12}$$

We now have two equations in two unknowns, the a and b values for which we wish to solve. Probably the simplest way to evaluate these two simultaneous equations is to multiply the first equation by six and then subtract the

Table 10-3 Computations Required for a Regression and Correlation Analysis for the Data Shown in Table 10-1.

Family	Y	X	XY	X^2	Y^2
A	23	10	230	100	529
B	7	2	14	4	49
C	15	4	60	16	225
D	17	6	102	36	289
E	23	8	184	64	529
F	22	7	154	49	484
G	10	4	40	16	100
H	14	6	84	36	196
I	20	7	140	49	400
J	19	6	114	36	361
	170	60	1122	406	3162

first equation from the second in order to eliminate a. Carrying out this procedure yields a solution for the constant b as follows:

$$\begin{aligned} 1020 &= 60a + 360b \\ 1122 &= 60a + 406b \\ \hline 102 &= \qquad\quad 46b \end{aligned}$$

$$b = \frac{102}{46} = 2.22$$

We can now substitute this value of b into either equation to obtain the value for a. Substituting into Equation (10.11) gives

$$170 = 10a + 60(2.22)$$

$$10a = 170 - 133.2 = 36.8$$

$$a = \frac{36.8}{10} = 3.68$$

Hence, the least squares regression line is

$$\bar{Y}_X = 3.68 + 2.22X \tag{10.13}$$

If a family drawn from the same population had an annual net income of \$8000 in 1967, its estimated annual consumer durables expenditures from Equation (10.13) would be

$$\bar{Y}_X = 3.68 + 2.22(8) = 21.44 \text{ (hundred dollars)}$$

By plotting the point thus determined ($X = 8$, $\bar{Y}_X = 21.44$) and one other point, or by plotting any two points derived from the regression equation, we can plot the regression line. The line is shown in Figure 10-5, page 479.

Now that the regression line has been determined, we can verify the fact that it goes through the point of means, $(\bar{X}, \bar{Y})$. The arithmetic mean annual net income of the families in the sample is

$$\bar{X} = \frac{\Sigma X}{n} = \frac{60}{10} = 6 \text{ (thousand dollars)}$$

Substituting this value into the regression equation (10.13) for X, we obtain

$$\bar{Y}_X = 3.68 + 2.22(6) = 17 \text{ (hundred dollars)}$$

Thus, the regression line passes through the point (6, 17). Since $\bar{Y} = \Sigma Y/n = 170/10 = 17$ (hundred dollars), we have confirmed that the estimating line contains the point $(\bar{X}, \bar{Y})$. Thus, in addition to satisfying the least squares criterion, the fitted line is one for which $\Sigma(Y - \bar{Y}_X) = 0$.

An alternative simpler procedure to the use of the normal equations (10.9) and (10.10) for determining the values of a and b is given by obtaining a general solution for a and b from these equations. This solution is

$$a = \bar{Y} - b\bar{X} \tag{10.14}$$

$$b = \frac{\Sigma XY - n\bar{X}\bar{Y}}{\Sigma X^2 - n\bar{X}^2} \tag{10.15}$$

Substituting into these two expressions, we again obtain the previously determined values for a and b.

$$a = 17 - 2.22(6) = 17 - 13.32 = 3.68$$

$$b = \frac{1122 - 10(6)(17)}{406 - 10(6)^2} = \frac{102}{46} = 2.22$$

Interpretation of the Regression Equation

It is important to note that to this point no assumptions have been made concerning the probability distributions of the X and Y variables. Thus, only relatively limited interpretations of our results can be given. The fitting process we have carried out may be interpreted as one which has minimized the sum of the squared errors of estimation for the sample of ten families that we have studied. However, since we assumed the sample was *randomly* drawn from families with \$10,000 or less of annual net income in a metropolitan area in 1967, we may say that our computed regression equation should provide about as accurate estimates for such families not in the sample as for those which happened to be selected.

We now proceed with an interpretation of the regression equation (10.13) in terms of the problem. The b value in this equation is often referred to as the "regression coefficient" or "slope coefficient." The figure of $b = 2.22$ indicates first of all that the slope of the regression line is positive. Thus, as income increases, estimated consumer durables expenditures increase. Taking into account the units in which the X and Y variables are stated, $b = 2.22$ means that for two families whose 1967 annual net income differs by \$1000, the *estimated difference* in their annual consumer durables expenditures is \$222. This is an interpretation in terms of the *regression line*. If we think of the figure $b = 2.22$ in terms of the *sample studied*, we can say that for two families whose 1967 annual net income differed by \$1000, *on the average*, their annual consumer durables expenditures differed by \$222.

It is important to realize that the figure of $b = 2.22$ does not mean that as a family's income increases *over time* by \$1000, its estimated consumer durables expenditures would increase by \$222. This would constitute a *time series* interpretation as opposed to a *cross-sectional* interpretation; the latter is the valid one. A time series interpretation pertains to the way that Y *changes over time* as X changes over time, whereas a cross-sectional interpretation refers to the way that Y values *differ* per unit difference in the X values. Let us examine this point more closely.

Suppose the original set of observations had been figures for a particular family's annual consumer durables expenditures and annual net income by years. That is, the original data might have been arranged as follows:

	Annual Consumer Durables Expenditures Y	*Annual Net Income* X
1950	—	—
1951	—	—
1952	—	—
.	.	.
.	.	.
.	.	.

The data for both the expenditures and income variables are given in the form of a *time series*, that is, in a series of figures arranged chronologically.

In this case, the so-called "unit of association" for these variables is the year; that is, each pair of X and Y values is observed for a unit of a year. The unit of association is the element which binds or links the paired values of X and Y. Income and consumer durables expenditures are assumed to be related for the same year. If we plot the data on a scatter diagram, each point represents an observation of X and Y for the same year. Now, if a regression equation were fit to these data by the procedures we have just discussed, a *time series interpretation* would be the correct one. That is, we could refer to an estimated increase in consumer durables expenditures over time which is associated with a one unit increase in income for the family studied. Usually, time series data pertain to aggregates such as all families in a city, a region, or the country as a whole, as opposed to the single family in our present illustration.

On the other hand, in the case of the data for the sample of 10 families shown in Table 10-1 for which we computed a regression equation, the unit of association is *the family*. Hence, each pair of X and Y values refers to a different family. Such figures are referred to as *cross-sectional* data, in the sense that we have studied a cross section of families. Time is only incidental in these observations. The important fact is that each point on the scatter diagram pertains to a family; each pair of X and Y values is plotted for a family. In this case, as we have seen, the interpretation of the slope of the regression line is in terms of the estimated *difference between families* in the value of the dependent variable per unit *difference between families* in the independent variable. This point concerning the distinction between cross-sectional and time series interpretations is important, particularly in dealing with certain types of economic data. There is a sizable literature in economic research of studies which use cross-sectional data and time series data separately and of other studies which relate the measures derived from the two different types of data. In particular, many regression analyses have been carried out to obtain empirical measures of the consumption function, an important concept in the theories of John Maynard Keynes, the famous British economist (1883–1946). The consumption function is a function which relates consumption expenditures to income (both factors variously defined). In this connection, it is interesting to note that the regression coefficient or b value determined in our preceding illustrative problem may be interpreted as a cross-sectional estimate of what the economist refers to as the "marginal propensity to consume." That is, it is an estimate of the ratio of the difference in dollars of a particular type of consumption expenditures to a difference of a dollar in income. For example, adjusting the $b = 2.22$ value for the nature of the units of the X and Y variables, we have a "marginal propensity to consume coefficient" of 0.222. This means, of course, an estimated difference of \$0.222 in annual consumer durables expenditures for two families whose annual net income differed by \$1.00.

10.5 The Standard Error of Estimate

Now that we have seen how the regression equation accomplishes estimation, the first purpose of regression and correlation analysis referred to in Section 10.3, we can turn to the second objective, that of obtaining a measure of the error involved in using the regression line for estimation. In order to obtain an appreciation of the rationale implied in the type of measure used for this purpose, let us consider, first of all, how we might proceed in an estimation situation, where data for only one variable are available, rather than for two variables as in the case of regression analysis. Suppose you were asked the following question, "What is your best single estimate of the expenditures on consumer durables in 1967 by a family selected at random in the metropolitan area we have been discussing?" Let us assume that data concerning consumer durables expenditures for all families in the metropolitan area (population parameters) were available, and that information on other variables was not. A good choice for an estimating device would be the population arithmetic mean figure for consumer durables expenditures. Retaining the designation of the consumer durables spending variable as Y, let us denote the population mean of this variable as μ_Y. This estimator, μ_Y, would be a good one in at least two ways. First of all, if we think of repeating this experiment indefinitely (sampling with replacement), on the average our estimate would be correct. That is, the sum of the positive and negative deviations from the mean of the figures for the individual families in this population is equal to zero; in symbols, $\Sigma(Y - \mu_Y) = 0$. Secondly, the mean possesses the least squares property that $\Sigma(Y - \mu_Y)^2$ is a minimum. This means that the sum of the squared deviations of individual values from the population mean is less than from any other single figure. However, how precise an estimate would μ_Y be? The answer to this question depends on how much dispersion or scatter there is in the figures for the individual families around the mean, μ_Y.

Let us consider an extremely improbable special case. We assume that every family in the population spent exactly the same amount on consumer durables. In this case, the population standard deviation, denoted σ_Y, would be equal to zero. Hence, if we were to use μ_Y as an estimate of the expenditures on consumer durables for a family selected at random, our estimate would be perfectly correct. Furthermore, if we repeated the estimation experiment, every time we selected an additional family, we would predict its expenditures exactly.

On the other hand, suppose we assume the more realistic situation in which there is some dispersion in the family consumer durables expenditures figures around the mean. In that case μ_Y would still be used to estimate consumer durables spending for a family selected at random from the population. Clearly, however, if there is considerable dispersion among families in such expenditures, μ_Y might be quite erroneous as an estimate for a particular family. We can indicate the closeness of our estimating procedure in a manner with which we are familiar. *Let us make the assumption that family consumer durables expenditures are normally distributed with mean μ_Y and standard deviation σ_Y.* Then we can make statements of the following sort. The best single estimate for consumer durables expenditures for a family selected at random is μ_Y. The odds are about 2 to 1 that the family's figure will be found in the range $\mu_Y \pm \sigma_Y$ (probability 68.3%); the odds are about 19 to 1 that the figure is in the range $\mu_Y \pm 2\sigma_Y$ (probability 95.5%); and about 332 to 1 that it is in the range $\mu_Y \pm 3\sigma_Y$ (probability 99.7%). It may be noted that these are probability statements rather than confidence interval estimates. This follows from the fact that the population parameters and population distribution are known, and a probability statement is made concerning the value of an element selected at random from that population. If the standard deviation σ_Y were quite small, the ranges given above would be narrow, and relatively useful statements for estimation purposes could be made, whereas if σ_Y were large, the opposite situation would prevail.

The usefulness of the regression line for estimation purposes can be indicated by an analogous procedure to the one just discussed. If there is a great deal of scatter of the observed Y values around the line, estimates of Y values based on computed $\bar{Y}_X$ values on the regression line will not be very close. On the other hand, if every point falls on the regression line, insofar as the sample of observations studied is concerned, perfect estimates of the Y values can be made from the fitted regression line. As indicated in Section 10.3, the measure of the dispersion of the observations around the computed regression line is referred to as the *standard error of estimate.* Just as the standard deviation is a measure of the scatter of observations in a frequency distribution around the mean of that distribution, the standard error of estimate is a measure of the scatter of the observed values of Y around the corresponding computed $\bar{Y}_X$ values on the regression line. Since the regression line we have discussed is based on a sample of observations, we will use notation appropriate to a sample. Hence, the standard error of estimate is defined as

(10.16)
$$s_{Y.X} = \sqrt{\frac{\Sigma(Y - \bar{Y}_X)^2}{n}}$$

where, as previously, n is the size of sample.

It is useful to consider the nature of the notation for the standard error of estimate, $s_{Y.X}$. In the discussion of dispersion in Section 3.16, the symbol s was used to denote the standard deviation of a sample of observations. The use of the letter s in $s_{Y.X}$ is analogous since, as explained in the preceding paragraph, $s_{Y.X}$ is also a measure of dispersion computed from a sample. However, since both the variables Y and X are present in a two-variable regression and correlation analysis, subscript notation is required to distinguish among the various possible dispersion measures. Hence, as we have seen, the notation for the standard error of estimate, where Y and X are respectively the dependent and independent variables, is $s_{Y.X}$. On the other hand, if X were treated as the dependent and Y as the independent variable and a regression line of the form $\bar{X}_Y = a' + b'Y$ were computed, the standard error of estimate around this regression line would be denoted $s_{X.Y}$. In summary, the letter to the left of the period in the subscript is the dependent variable, the letter to the right denotes the independent variable. Subscripts are also required to distinguish standard deviations around the means of the two variables. Thus, s_Y denotes the standard deviation of the Y values of a sample around the mean $\bar{Y}$, and s_X denotes the standard deviation of the X values around their mean, $\bar{X}$.

In a realistic problem containing large numbers of observations, the computation of the standard error of estimate using Equation (10.16) clearly involves a great deal of arithmetic. The calculation of $\bar{Y}_X$ for each X value in the sample is required, and then the arithmetic implied by the formula must be carried out. It is useful to have a short-cut method which only involves quantities already computed. It is easy to see how such a short-cut formula might be derived. Since $\bar{Y}_X = a + bX$, the right-hand expression in this equation can be substituted for $\bar{Y}_X$ in Equation (10.16) and the resulting expression may be simplified in various ways. A convenient form of such a short-cut formula is given by

(10.17)
$$s_{Y.X} = \sqrt{\frac{\Sigma Y^2 - a\Sigma Y - b\Sigma XY}{n}}$$

All quantities required by Equation (10.17) were calculated for our illustrative problem in Table 10-3, or were computed in obtaining the constants of the regression line. Hence, the standard error of estimate for these data is

$$s_{Y.X} = \sqrt{\frac{3162 - (3.68)(170) - (2.22)(1122)}{10}} = 2.13 \text{ (hundreds of dollars)}$$

The standard error of estimate has been indicated in Figure 10-5. The scatter diagram for the illustrative problem, the regression line, and bands of one and two $s_{Y.X}$ in width have been shown above and below the regression line.

To this point, no probability assumptions have been introduced for either the dependent or independent variables. However, in order to use a measure of dispersion around the regression line as an indicator of error of estimation, some assumption concerning the distribution of points around the regression line is necessary. Assuming that the observed Y values are normally dis-

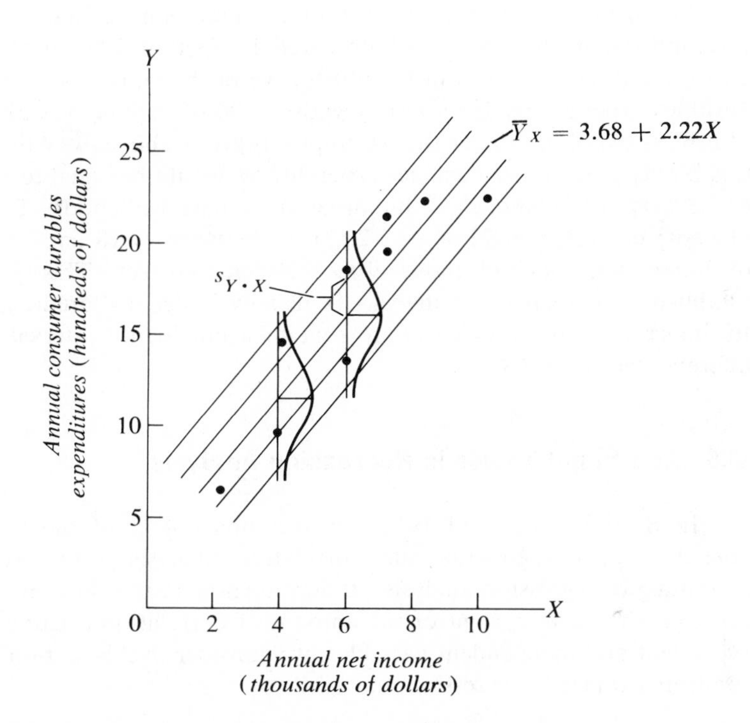

Figure 10-5 Least squares regression line and bands at distances of one and two standard errors of estimate.

tributed around the regression line, about 68% of the points will fall within the band made by parallel lines a distance of one standard error of estimate above and below the regression line, about 95.5% of the points will fall within a distance of two standard errors of estimate, and so forth. A visual interpretation of the normality assumption is given in Figure 10-5. Let us be somewhat more specific concerning this assumption. For every given X value, we can compute $\overline{Y}_X$, our estimate of the value of the dependent variable. We are assuming that

(1) The actual Y values for the given X form a normal distribution with $\overline{Y}_X$ as the arithmetic mean of the distribution and $s_{Y.X}$ as the standard deviation of the distribution.

(2) At every value of X, the dispersion of the Y values around the computed $\overline{Y}_X$ is the same; that is, it is assumed that the probability distributions of the Y values have the same standard deviation, $s_{Y.X}$, for every X value within the observed range.

The use of the standard error of estimate can be illustrated using the assumptions given above. We observed in Section 10.4 that using the regression equation derived in the illustrative problem, the estimated consumer durables expenditures for a family with $8000 of income was 21.44 hundreds of dollars ($2144). Under the assumptions given above, we can estimate that this family's consumer durables expenditures would be equal to 21.44 ± 2.13, or between 19.31 and 23.57 hundreds of dollars with odds of about 2 to 1 of being correct, and $21.44 \pm 2(2.13)$, or between 17.18 and 25.70 hundreds of dollars, with odds of about 19 to 1 of being correct. This type of interval established in regression analysis is usually called a "prediction interval." Its interpretation is analogous to that of a confidence interval in statistical inference (see page 376).

10.6 Additional Topics in Regression Analysis

In Sections 10.4 and 10.5, the accomplishment of the first two basic objectives of a regression and correlation analysis were discussed—both pertaining to regression analysis. Before turning to the third objective, which concerns the measurement of the amount of correlation existing between the dependent and independent variables, we consider in this section a number of additional topics in regression analysis.

Conditional Probability Distributions

As indicated in the preceding section, the use of the least squares method of fitting a regression line to a sample of observations does not permit any inferences about the population which was sampled unless some probabilistic assumptions are made. In the discussion of the standard error of estimate, it was indicated how an assumption could be made concerning normal distributions of Y values around the $\bar{Y}_X$ values of the regression line for given values of X. Then, a method for making an interval estimate for a family's consumer durables expenditures was given using this probability assumption and the computed standard error. That procedure provides an approximate prediction interval, and the method is a rough one appropriate only in the case of *large samples*. Again, we repeat the caveat that the sample in the illustrative problem is small ($n = 10$) and is used here merely for convenience of exposition and arithmetic. We turn now to more exact estimation procedures, and to more formal ways of expressing probabilistic assumptions underlying regression analysis.

In Chapter 2, a brief introduction was given to the concept of conditional probability distributions. Such distributions were considered for the case of discrete random variables. The patterns of variation around regression lines may be considered in terms of conditional probability distributions. For this purpose, some new notation will be introduced. Let us consider a regression model in which Y is a random variable and X is a controlled or predetermined variable. This situation may be viewed as one in which observations of the dependent variable Y are made for certain selected values of X. For example, in terms of our illustrative problem, we may select families which have certain specified annual incomes, say \$4000, \$6000, \$8000, etc., and observe what their expenditures were on consumer durables. If we now focus on families at a given income level, say \$6000, the probability distribution of the Y variable, consumer durables expenditures, is a *conditional probability distribution* of Y for $X = 6$ (thousand dollars). Such a conditional probability distribution may be denoted in the usual way as $f(Y \mid X = 6)$ or, in general, $f(Y \mid X)$. This distribution has a mean, which may be denoted $\mu_{Y.X}$ and a standard deviation which may be denoted $\sigma_{Y.X}$. The mathematical functional relationship between the values of X and the means of the conditional probability distributions of Y is called a *regression curve*. If we consider a model in which this mathematical relationship between $\mu_{Y.X}$ and X is a straight line, the regression curve is called a *regression line*. The equation of the regression line in this model may be written as

$$\mu_{Y.X} = \alpha + \beta X \tag{10.18}$$

In this model of a linear regression function, α and β are population parameters which must be *estimated* from sample data. It can be shown that, under the following assumptions, maximum likelihood estimates of α and β are exactly equal to the values obtained for a and b in the equation $\bar{Y}_X = a + bX$ by applying the method of least squares as in Section 10.4.[2]

(1) The deviations from the regression line, $Y - \mu_{Y.X}$, are independent of one another. This implies that successive deviations are unrelated.

(2) X is a controlled or predetermined random variable.

(3) The conditional probability distribution of Y given X, $f(Y \mid X)$, is normal. Its mean $\mu_{Y.X}$ is a linear function of X. Its standard deviation $\sigma_{Y.X}$ is independent of X.

A few comments about these assumptions are in order. The first assumption is generally more valid for cross-sectional data than for time series data. In the case of time series, since there usually is a definite dependence between successive observations, the deviations of these observations around a fitted line are ordinarily also not independent. This is seen most clearly in the case of time series data which move in cycles around a fitted trend line. Observa-

[2]It can also be shown mathematically that if the joint distribution of X and Y is the bivariate normal probability distribution, the method of maximum likelihood estimation for α and β is again the same as the method of least squares.

tions in portions of a cycle which lie above the trend line will have positive deviations, while those below will have negative deviations. Hence, deviations for successive observations tend to be related.

The operational importance of the second assumption is that the X variable is assumed to be known exactly, while Y is considered to be a random variable.

The third assumption concerning normality is not required for inferences about the constants in the regression equation. However, it is necessary for probability statements about the dependent variable Y. The statement that $\sigma_{Y.X}$ is independent of X is the assumption of uniform scatter around the regression line. That is, it is assumed that the standard deviations of the conditional distributions of Y are equal for every value of X.

We can now relate this discussion of the theoretical regression model to the preceding computation of the regression equation and standard error of estimate for a sample of n observations. The regression line $\mu_{Y.X} = \alpha + \beta X$ may be considered as the population or "true" relationship. If a value of X is substituted into this equation, the computed $\mu_{Y.X}$ figure is the mean of the conditional probability distribution of Y given X. In the regression equation $\bar{Y}_X = a + bX$, a is an estimate of the population parameter α, and b is an estimate of β. In fact, it can be shown that a and b are unbiased estimators of α and β and possess many other desirable properties as well.[3] We now can see why the notation $\bar{Y}_X$ was used in the regression equation $\bar{Y}_X = a + bX$, calculated from sample data. Previously, the bar over a letter representing a variable meant the mean of that variable; thus $\bar{Y}_X$ implies the mean of the conditional distribution of Y given X, or briefly, the *conditional mean*.

The standard deviation of the conditional probability distribution of Y given X, $\sigma_{Y.X}$, may similarly be considered as the population or "true" value of the standard error of estimate. It may be noted that in regression analysis we have used the following notation:

$$E(Y \mid X) = \mu_{Y.X}$$

$$\sigma(Y \mid X) = \sigma_{Y.X}$$

Although a and b, as calculated by the method of least squares, were unbiased estimators of α and β, the sample standard error of estimate $s_{Y.X} = \sqrt{\Sigma(Y - \bar{Y}_X)^2/n}$ is not an unbiased estimator of $\sigma_{Y.X}$. In the next paragraph, the problem of estimating a population measure of dispersion around the regression line is discussed.

[3]These estimators are also minimum-variance, consistent, asymptotically efficient, and sufficient.

Point Estimates of Dispersion Around the Regression Line

In Chapter 3 we saw that the sample variance,

$$s^2 = \frac{\Sigma(X - \bar{X})^2}{n}$$

is a biased estimator of the population variance, σ^2. On the other hand, the corresponding estimator with an $n - 1$ divisor rather than n, is an unbiased estimator. Using the symbol $\hat{\sigma}^2$ for the latter estimator, we may write

(10.19) $$\hat{\sigma}^2 = \frac{\Sigma(X - \bar{X})^2}{n - 1}$$

is an unbiased estimator of the population variance σ^2, or synonymously,

(10.20) $$E(\hat{\sigma}^2) = \sigma^2$$

An analogous situation pertains in the case of a measure of dispersion around the regression line. Let us refer to the square of the standard error of estimate as the variance around the regression line. This variance, the square of the expression $s_{Y.X}$ defined in (10.16),

$$s^2_{Y.X} = \frac{\Sigma(Y - \bar{Y}_X)^2}{n}$$

is a biased estimator of $\sigma^2_{Y.X}$, the variance around the population regression line. It can be shown that the use of a divisor of $n - 2$ rather than n produces an unbiased estimator of $\sigma^2_{Y.X}$. Denoting this estimator as $\hat{\sigma}^2_{Y.X}$, we may state that

(10.21) $$\hat{\sigma}^2_{Y.X} = \frac{\Sigma(Y - \bar{Y}_X)^2}{n - 2}$$

is an unbiased estimator of the population variance around the regression line, or synonymously,

(10.22) $$E(\hat{\sigma}^2_{Y.X}) = \sigma^2_{Y.X}$$

It is customary to use the square root of (10.21) as the estimator of the population standard error of estimate. This estimated standard error of estimate is defined by

(10.23) $$\hat{\sigma}_{Y.X} = \sqrt{\frac{\Sigma(Y - \bar{Y}_X)^2}{n - 2}}$$

The $n - 2$ in Equations (10.21) and (10.23) represents the number of degrees of freedom around the regression line. In general, the divisor is $n - k$, where k is the number of parameters in the regression equation which were estimated from the n observations. As we have seen, in the case of a linear regression function, the parameters to be estimated are α and β; thus $k = 2$. We now discuss how these ideas may be used in obtaining prediction intervals in regression analysis.

Prediction Intervals in Regression Analysis

We learned in Section 10.5 how to use the sample standard error of estimate to set up a prediction interval for a family's consumer durables expenditures, Y, given its net income, X. In the preceding paragraph, we considered how to adjust the standard of estimate for the number of degrees of freedom to obtain an estimate of the scatter of observations around the "true" or population regression line. We will now use this "adjusted" or estimated standard error of estimate to obtain a prediction interval for a Y value.

Returning to our illustrative problem, suppose we wish to establish a prediction interval for annual consumer durables expenditures for a family with a net annual income of \$8000. The family is assumed to have been drawn from the same universe as the sample of ten families used to establish the regression line. Further, let us assume that we would like to attach a 95% confidence coefficient to the prediction interval. Since the estimated standard error of estimate $\hat{\sigma}_{Y.X}$, which we will use, is an *estimated* standard deviation, the t-distribution is the appropriate one for establishing the required interval. That is, the present example is a case in which the sample size is small ($n \leq 30$), and $\hat{\sigma}_{Y.X}$ is an *estimate* of a population standard deviation rather than a *known* population standard deviation. Our point estimate is again obtained from the previously computed regression line, $\bar{Y}_X = 3.68 + 2.22X$. Hence, our best single estimate of this family's annual consumer durables expenditures is 21.44 hundred dollars. The estimated standard error of estimate can be obtained from the following short-cut formula analogous to (10.17), but containing the adjustment for degrees of freedom:

$$\textbf{(10.24)} \qquad \hat{\sigma}_{Y.X} = \sqrt{\frac{\Sigma Y^2 - a\Sigma Y - b\Sigma XY}{n - 2}}$$

Substituting into (10.24) gives

$$\hat{\sigma}_{Y.X} = \sqrt{\frac{3162 - (3.68)(170) - (2.22)(1122)}{8}} = 2.39 \text{ (hundreds of dollars)}$$

From Table A-6, for 8 degrees of freedom, the value of t is 2.306. Thus, the prediction interval is

$$\begin{aligned} \bar{Y}_X \pm t\hat{\sigma}_{Y.X} &= 21.44 \pm 2.306(2.39) \\ &= 21.44 \pm 5.51 \text{ (hundreds of dollars)} \end{aligned}$$

Hence, the prediction interval is \$1593 to \$2695, with an associated confidence coefficient of 95%. If n is large, the prediction intervals obtained by the two methods discussed to this point

(10.25) $$\bar{Y}_X \pm z s_{Y.X}$$

and

(10.26) $$\bar{Y}_X \pm t\hat{\sigma}_{Y.X}$$

would for practical purposes be about equal. That is, if n is large, the z values for the normal distribution are quite close to the corresponding t values in the t-distribution for the same confidence coefficient, and also $s_{Y.X}$, which is computed with a divisor of n, will be approximately equal to $\hat{\sigma}_{Y.X}$, which is computed with a divisor of $n - 2$.

Another possible approach to prediction interval estimation is to calculate $\hat{\sigma}_{Y.X}$, but to assume a normal distribution. This approach leads to the prediction interval

(10.27) $$\bar{Y}_X \pm z\hat{\sigma}_{Y.X}$$

For large sample sizes, prediction intervals established by (10.27) would be approximately the same as intervals constructed by (10.25) and (10.26).

Prediction Interval for a Conditional Mean

The cases of prediction interval estimation discussed thus far pertained to estimating a *specific* Y value. We were interested in a range within which we were highly confident that consumer durables expenditures fell for *a particular family* with a given annual net income. On the other hand, there are occasions when one wishes to establish a prediction interval for the *conditional mean of* Y rather than for a particular observation. Here, the question concerns establishing a range within which one is confident that the true *average value* of Y is included for a given value of X. In this case the appropriate prediction interval is

(10.28) $$\bar{Y}_X \pm t\frac{\hat{\sigma}_{Y.X}}{\sqrt{n}}$$

If the sample size is large, an approximate prediction interval is given by

(10.29) $$\bar{Y}_X \pm z\frac{s_{Y.X}}{\sqrt{n}}$$

Analogously to the prediction interval given in (10.27), if the sample size is large, another approximate prediction interval for the conditional mean of Y is given by

(10.30) $$\bar{Y}_X \pm z\frac{\hat{\sigma}_{Y.X}}{\sqrt{n}}$$

For large sample sizes, the prediction intervals given by (10.28), (10.29) and (10.30) would all be approximately equal.

Suppose we were interested in a prediction interval with a 95% confidence coefficient for *average* annual consumer durables expenditures for families whose annual net income is \$8000. Based on the regression line established for the sample of ten families, the best single estimate we can make of the conditional mean is again $\bar{Y}_X = 21.44$ hundreds of dollars. Using the t-distribution, we compute the following prediction interval:

$$21.44 \pm (2.306)\frac{2.39}{\sqrt{10}} = 21.44 \pm 1.74 \text{ (hundreds of dollars)}$$

Hence, the average consumer durables expenditures for families with \$8000 annual net income is predicted to be included in the range of \$1970 to \$2318.

It may be noted that this range is much narrower than the previously computed corresponding prediction interval of \$1593 to \$2695 for a particular family. However, this is reasonable because in the prediction interval for a particular family, it is the scatter of values for individual families around the regression line which is relevant. On the other hand, the establishment of the prediction interval for the conditional mean is analogous to setting up a confidence interval for a population mean, where the measure of precision is given by the standard error of the mean, $\sigma_{\bar{X}} = \sigma_X/\sqrt{n}$.

The usefulness of any prediction interval depends on the purposes for which it is to be used. For example, for long-range planning purposes, relatively wide limits may be appropriate and useful. On the other hand, for short-term operational decision making, narrower and therefore more precise intervals may be required.

Two other types of prediction intervals will be discussed after hypothesis testing in regression and correlation analysis have been considered.

10.7 Correlation Analysis—Measures of Association

In the preceding two sections, regression analysis was discussed, with emphasis on estimation and measures of error in the estimation process. We now turn to correlation analysis, in which the basic objective is to obtain a measure of the degree of association that exists between two variables. In this analysis, interest centers on the strength of the relationship between the variables, or in other words, how well the variables are correlated. Because of the difference in the objectives of regression and correlation analysis, it is not surprising that the underlying mathematical models of the two types of analysis differ.

In the model discussed in regression analysis, X was considered to be a controlled or predetermined variable fixed at specific values, whereas Y was a random variable. In our illustrative problem, for the data to be strictly in conformity with this model, families with certain predetermined annual net incomes, X, would have to be selected and observations would have to be made of their annual consumer durables expenditures, Y. The observed Y values for each given value of X are then interpreted as sample observations from a conditional probability distribution.

On the other hand, in the correlation analysis model *both* X and Y are considered to be random variables. In the most commonly used variant of this model, X and Y are assumed to be jointly normally distributed; the conditional probability distributions of Y are normal with the same standard deviation; similarly for X. Thus, for example, suppose we were interested in using the correlation analysis model to estimate weights, designated as Y, from a knowledge of height, denoted X, for adult males in a large city. If we drew a simple random sample of n adult males, and observed the weights and heights of persons in the sample, then both X and Y would be random variables. It would be assumed that heights are normally distributed and for each height, for example, 5′8″, 5′9″, etc., weights are also normally distributed. The standard deviations of these weight distributions are assumed to be equal. With these assumptions, all previously computed measures discussed under regression analysis are still valid. That is, assuming a linear relationship between height and average weight, the least squares regression line and the standard error of estimate would be computed in the same way as explained in Sections 10.4 and 10.5. However, the symmetrical assumptions for X and Y imply that Y may be as validly employed in estimating X as vice versa. That is, the analyst could construct a least squares regression equation in which X is a function of Y, and thus he could estimate height from a knowledge of weight. Insofar as the mathematics of the model is concerned, since both X and Y are random variables, and since it is equally meaningful to talk about the probability distributions of X for given values of Y as the probability distribution of Y for given values of X, the procedures of estimating Y from X and X from Y are equally valid.

On the other hand, if the regression model had been used, and if in the original data collection process, weights had been observed only for male adults of certain *predetermined* heights, then only weight would have been a random variable. In that case, it would not have been valid to construct a regression equation in which X was taken as a function of Y; hence it would not have been valid to attempt to estimate height from a knowledge of weight. Of course, even in the case of the correlation analysis model, good judgment, knowledge of the substantive field being studied, and the purpose of the study should aid in determining the estimation procedure. Hence, even if one had a simple random sample of male adults in which height and weight would *both* be random variables, since it seems more logical to say that weight depends upon height than vice versa, and if we assume the analyst's primary interest was in estimating weight from a knowledge of a person's height, then it would be patently silly to construct the other regression line which would make estimates of a person's height based on his weight.

Returning to our illustrative problem, the correlation model requires that the sample of ten families be drawn at random from the population and that observations of X and Y be made for each family. There must be no preselection of certain values for the X variable, annual income. We assume such a model in the discussion which follows.

The Coefficient of Determination

In Section 10.5, we considered how estimates of a particular value of the dependent variable Y might be made, first using the mean of the Y's as an estimating device, then using a regression equation which relates Y to a second variable X. The precision of estimation in these two cases was indicated by the scatter of points around the mean of the Y's and the scatter of points around the regression line, respectively. A measure of the amount of correlation that exists between Y and X can be developed in terms of the relative variation of points around the regression line and variance around the mean of the Y variable.

In order to present the rationale of this measure of strength of the relationship between Y and X, we will consider two extreme cases, (1) zero linear correlation and (2) perfect direct linear correlation. The term "linear" indicates that a straight line has been fitted to the X and Y values and the term "direct" indicates that the line is inclined from the lower left to the upper right-hand side of a scatter diagram.

Two sets of data are presented in Table 10-4 labeled (a) and (b). They are shown in the form of scatter diagrams in Figure 10-6. As we shall see, the data in (a) and (b) illustrate the case of zero linear correlation and perfect direct linear correlation, respectively. In the discussion which follows, we will assume the collections of units of association shown in (a) and (b) represent simple random samples from their respective universes. Therefore, we employ notation pertinent to the samples. An analogous argument can be presented assuming the observations represent population data. The notation would change correspondingly. The calculations given below the scatter diagrams in Figure 10-6 will be explained in terms of the data displayed in the charts.

Case (a) represents a situation in which $\bar{Y}$, the mean of the Y values, coincides with a least squares regression line fitted to these data. Even without doing the arithmetic, we can see why this is so. The slope of the regression line is equal to zero, because the same Y values are observed for each value of X. Thus, the regression line would coincide with the mean of the Y values, balancing deviations above and below the regression line. Another way of

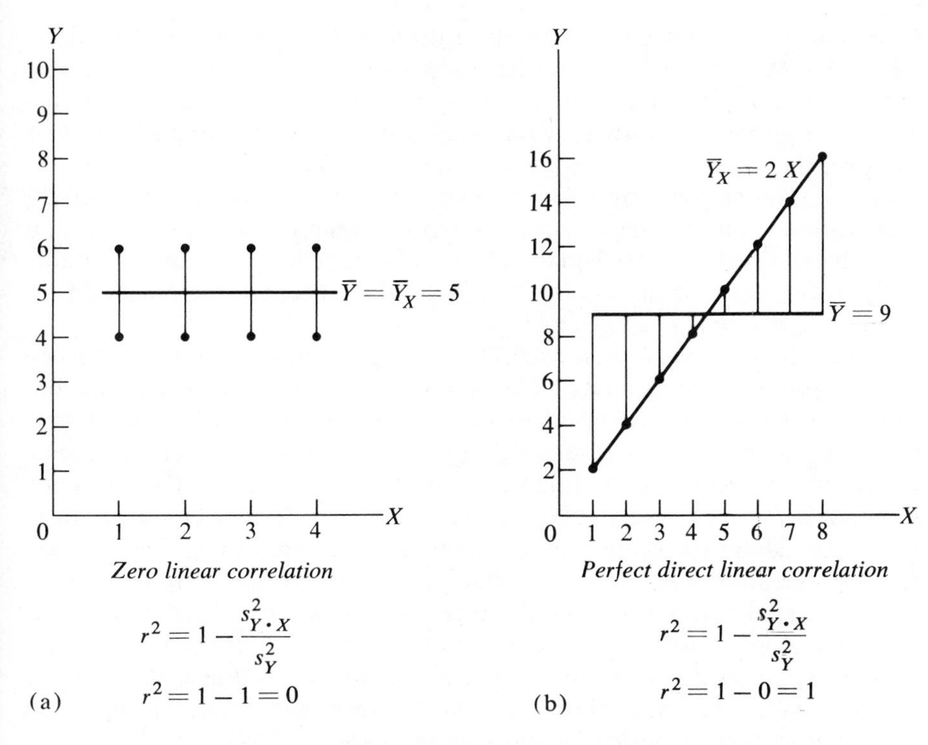

Figure 10-6 Scatter diagrams representing zero linear correlation and perfect direct linear correlation.

Table 10-4 Two Sets of Data Displaying (a) Zero Linear Correlation and (b) Perfect Direct Linear Correlation.

(a)			(b)		
Unit of Association	X	Y	*Unit of Association*	X	Y
A	1	4	*A*	1	2
B	1	6	*B*	2	4
C	2	4	*C*	3	6
D	2	6	*D*	4	8
E	3	4	*E*	5	10
F	3	6	*F*	6	12
G	4	4	*G*	7	14
H	4	6	*H*	8	16
		$\bar{Y} = 5$			$\bar{Y} = 9$

observing this relationship is in terms of the first of the two normal equations, $\Sigma Y = na + b\Sigma X$. Since $b = 0$, we have $a = \Sigma Y/n = \bar{Y}$. Hence, the regression line has a Y intercept equal to $\bar{Y}$. Since it is also a horizontal line, the regression line coincides with $\bar{Y}$. From the point of view of estimation of the Y variable, the regression line represents no improvement over the mean of the Y values. This can be shown by a comparison of $s^2_{Y.X}$, the variance around the regression line and s^2_Y, the variance around the mean of the Y values. In this case, the two variances are equal. These variances may be interpreted graphically as the mean of the squares of the distances between the points on the scatter diagram and $\bar{Y}$, shown in Figure 10-6(a).

Now, let us consider case (b). This is a situation in which the regression line is a perfect fit to the data. The regression equation is a very simple one which can be determined by inspection. The Y intercept is equal to zero, since the line passes through the origin (0, 0). The slope is equal to 2, because for every unit increase in X, Y increases by two units. Hence, the regression equation is $\bar{Y}_X = 2X$, and all of the data points lie on the regression line. Insofar as the data in the sample are concerned, perfect predictions are provided by this regression line. Given a value of X, the corresponding value of Y can be correctly estimated from the regression equation indicating a perfect linear relationship between the two variables. Again, a comparison can be made of $s^2_{Y.X}$ and s^2_Y. Since all points lie on the regression line, the variance around the line, $s^2_{Y.X}$, is equal to zero. On the other hand, the variance around the mean, s^2_Y, is some positive number, in this case, 21.

The relationship between the variances around the regression line and mean can be summarized in a single measure to indicate the degree of association between X and Y. The most commonly used measure for this purpose is the *sample coefficient of determination*,[4] defined as follows:

(10.31)
$$r^2 = 1 - \frac{s^2_{Y.X}}{s^2_Y}$$

As we shall see from the subsequent discussion, r^2 may be interpreted as the percentage of variance in the dependent variable Y that has been accounted for or "explained" by the relationship between Y and X expressed in the regression line. Hence, it is a measure of the degree of association or correlation between X and Y.

As indicated in Figure 10-6(a), when there is no linear correlation between X and Y, the sample coefficient of determination, r^2, is equal to zero. This follows from the fact that, since $s^2_{Y.X}$ and s^2_Y are equal, the ratio $s^2_{Y.X}/s^2_Y$

[4]A formula which adjusts for degrees of freedom is $r_c^2 = 1 - (1 - r^2)(n - 1)/(n - 2)$ where r_c^2 denotes the adjusted coefficient of determination. This formula can be derived by using in Equation (10.31) variances around the regression line and mean which have been adjusted for degrees of freedom.

equals one. Hence r^2 equals zero, because the computation of the coefficient of determination requires subtraction of this ratio from one.

On the other hand, as indicated in Figure 10-6(b), when there is perfect linear correlation between X and Y, the sample coefficient of determination, r^2, is equal to one. In this case, the variance around the regression line is equal to zero, while the variance around the mean is some positive number. Thus, the ratio $s_{Y.X}^2/s_Y^2$ equals zero. Hence, r^2 equals one when the value of this ratio is subtracted from one.

In realistic problems, r^2 falls somewhere between the two limits, zero and one. A value close to zero suggests that there is not much linear correlation between X and Y; a value close to one connotes a strong linear relationship between X and Y.

Interpretation of the Coefficient of Determination

The measure r^2 has been referred to as the *sample* coefficient of determination. This is because it pertains only to the sample of n observations studied. The corresponding *population coefficient of determination* is defined as

$$\rho^2 = 1 - \frac{\sigma_{Y.X}^2}{\sigma_Y^2} \tag{10.32}$$

The use of the symbol ρ^2 (rho squared) adheres to the usual convention of employing a Greek letter for a population parameter corresponding to the same letter in our alphabet which denotes a sample statistic. In the definition of ρ^2, $\sigma_{Y.X}^2$ is the variance around the population regression line $\mu_{Y.X} = \alpha + \beta X$ and σ_Y^2 is the variance around the population mean of the Y's, μ_Y. In probabilistic terms, $\sigma_{Y.X}^2$ is the variance of the conditional probability distributions of Y, whereas σ_Y^2 is the variance of the marginal (unconditional) distribution of Y.

It is useful to consider in more detail the specific interpretations that may be made of coefficients of determination. For convenience, only the sample coefficient r^2 will be discussed, but the corresponding meanings for ρ^2 are obvious.

An important interpretation of r^2 may be made in terms of variance in the dependent variable Y, which has been explained by the regression line, $\bar{Y}_X = a + bX$. The problem of estimation is conceived of in terms of "explaining" or accounting for the variation in the dependent variable Y. In this context s_Y^2 represents the total variance in Y. A regression line is fitted to the data and a certain portion of the variance in Y is explained or accounted for. However, if the regression line does not pass through all of the points on the scatter diagram, the deviations $(Y - \bar{Y}_X)$ represent variation that still has not been accounted for. The mean of the squares of these deviations, $s_{Y.X}^2$, represents this "unexplained variance." The ratio, $s_{Y.X}^2/s_Y^2$ is the proportion of total variance in Y, which remains unexplained by the regression equation; correspondingly $1 - s^2_{Y.X}/s^2_Y$ represents *the proportion of total variance in Y that has been explained by the regression equation.* These ideas may be summarized as follows:

$$r^2 = 1 - \frac{s^2_{Y.X}}{s^2_Y} = 1 - \frac{\text{unexplained variance in } Y}{\text{total variance in } Y}$$

Collecting terms over a common denominator, we obtain

(10.33) $$r^2 = \frac{\text{total variance in } Y - \text{unexplained variance in } Y}{\text{total variance in } Y}$$

$$r^2 = \frac{\text{explained variance in } Y}{\text{total variance in } Y}$$

A simple numerical example helps to clarify these relationships. Let $s^2_Y = 10$ and $s^2_{Y.X} = 4$. Thus, $r^2 = 1 - 4/10 = 6/10 = 60\%$. In this problem there are ten units of total variance in Y to be accounted for. After fitting the regression line, the residual variance or unexplained variance amounts to four units. Hence 60% of the total variance in the dependent variable is explained by the relationship between Y and X expressed in the regression line.

Figure 10-7 is a graphical interpretation of total and unexplained variance. In the scatter diagram in part (a), deviations of the observations are shown from the mean of the Y's. The total variance in Y is the mean of

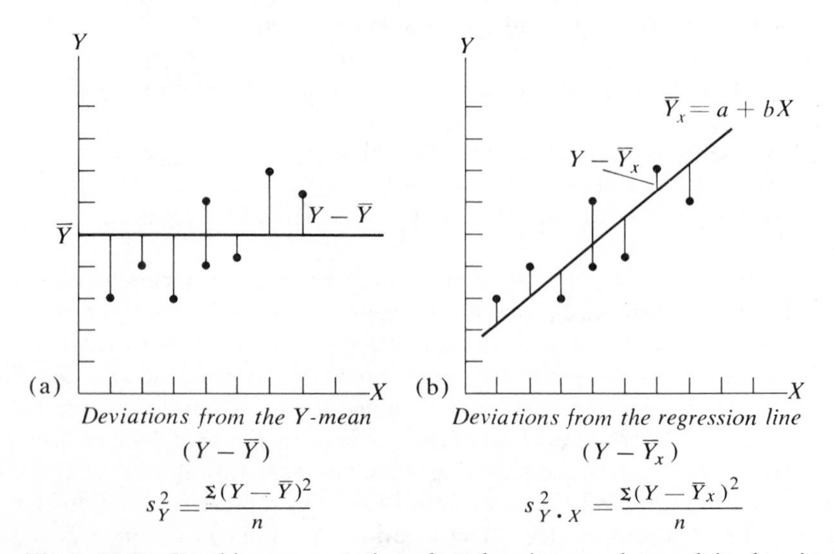

Figure 10-7 Graphic representation of total variance and unexplained variance.

the squares of these deviations. In the scatter diagram in part (b), deviations of the observations are shown from the regression line. The unexplained variance is the mean of the squares of these deviations.

Another interpretation of the value of r^2, with a slightly different emphasis is in terms of reduction in variance in the dependent variable. The total variance in Y, s_Y^2 is referred to as the "error variance around the mean of the Y's" to place stress on the error involved in using $\bar{Y}$ as a basis for estimating Y values. Similarly, $s_{Y.X}^2$ is called the "error variance around the regression line" as an indicator of the error involved in using the regression line for estimating Y values. Hence, r^2 measures the *proportionate reduction in error variance if the regression line rather than the mean of the Y's is used to estimate values of the Y variable.* Referring to the preceding simple numerical example, if $s_Y^2 = 10$ is the error variance around $\bar{Y}$, and if $s_{Y.X}^2 = 4$ is the error variance around $\bar{Y}_X = a + bX$, then there is a 60% reduction in error variance in going from 10 to 4.

The reduction in error variance may be construed as a measure of the improvement in using the regression line rather than the mean of the dependent variable for estimation. Thus, in the case of no linear correlation, there is no improvement. In the case of perfect linear correlation, there is a 100% improvement, because in using the regression line, error variance is reduced to zero.

It is useful to note the relationship which must exist between the two error variances. At worst, if there is no linear correlation between X and Y, no reduction in error variance will be accomplished in using the regression line rather than the mean of the Y's for estimation. In all other cases, there will be a reduction. Therefore, in symbols, we have the relationship

(10.34) $$s_Y^2 \geq s_{Y.X}^2$$

Calculation of the Sample Coefficient of Determination

Let us return to the problem involving the relationship between consumer durables expenditures and net income for our sample of ten families. In order to obtain the value of r^2 in that problem, we must only evaluate s_Y^2 because $s_{Y.X}^2$ has already been calculated. (Actually, $s_{Y.X}$, the square root of $s_{Y.X}^2$, was computed in Section 10.5.) The value of s_Y^2 for the ten values of Y given in Table 10-3 is 27.20. Since $s_{Y.X}^2 = 4.56$, the sample coefficient of determination is

$$r^2 = 1 - \frac{4.56}{27.20} = 1 - 0.168 = 0.832$$

Thus, for our sample of ten families, about 83% of the variance in annual consumer durables expenditures was *explained* by the regression equation, which related such expenditures to annual net income.

The computation of r^2 just carried out did not involve much arithmetic because there were only ten families in the sample. However, just as in the case of the standard error of estimate, it is useful to have shorter methods of calculation for situations in which there are large numbers of observations. These short-cut formulas are particularly helpful when computations are carried out by hand or on a calculating machine, but even when computers are used, they represent more efficient methods of computation. Such a formula, which only involves quantities already calculated, is

$$r^2 = \frac{a\Sigma Y + b\Sigma XY - n\bar{Y}^2}{\Sigma Y^2 - n\bar{Y}^2} \tag{10.35}$$

Substituting into Equation (10.35), we obtain

$$r^2 = \frac{(3.68)(170) + (2.22)(1122) - 10(17)^2}{(3162) - 10(17)^2} = 0.832$$

The Coefficient of Correlation

A widely used measure of the degree of association between two variables is the coefficient of correlation, which is nothing more than the square root of the coefficient of determination. Thus, the population and sample coefficients of correlation are, respectively,

$$\rho = \sqrt{\rho^2} \tag{10.36}$$

and

$$r = \sqrt{r^2} \tag{10.37}$$

Again, for convenience, our discussion will relate only to the sample value.

The algebraic sign attached to r is the same as that of the regression coefficient, b. Thus, if the slope of the regression line, b, is positive, then r is given a plus sign also; if b is negative, r is given a minus sign.[5] Hence, r ranges from a value of minus one to plus one. A figure of $r = -1$ indicates a perfect inverse linear relationship; $r = +1$ indicates a perfect direct linear relationship, and $r = 0$ indicates no linear relationship.

Illustrative scatter diagrams for the cases of $r = +1$ ($r^2 = 1$) and $r = 0$ ($r^2 = 0$) were given in Figure 10-6. A corresponding scatter diagram for $r = -1$, the case of perfect inverse linear correlation, is shown in Figure 10-8. As indicated in the graph, the slope of the regression line is negative and every point falls on the line. Thus, for example, if the slope of the regression line, b, were equal to -2, this would mean that with each increase of one

---

[5]An interesting relationship between r and b is given by $b = rs_Y/s_X$. Since s_Y and s_X are positive numbers, b has the same sign as r. We can also note that when the regression line is horizontal, $b = 0$ and also $r = 0$.

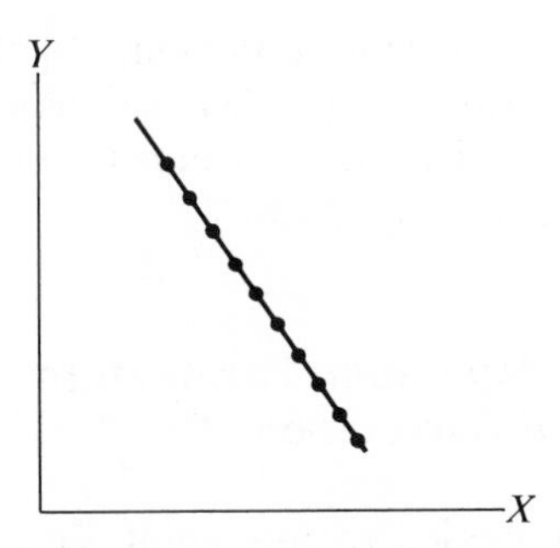

Figure 10-8 Scatter diagram representing perfect inverse linear correlation, $r = -1$.

unit in X, Y would decrease by two units. Since all points fall on the regression line in the case of perfect inverse correlation, $s^2_{Y.X} = 0$. Therefore, substituting into Equation (10.31) to compute the sample coefficient of determination, we have $r^2 = 1 - 0 = 1$. Taking the square root, we obtain $r = \sqrt{1} = \pm 1$. However, since the b value is negative, that is, X and Y are inversely correlated, we assign the negative sign to r, and $r = -1$.

In our illustrative problem,

$$r = \sqrt{0.832} = 0.912$$

The sign is positive because b was positive, indicating a direct relationship between consumer durables expenditures and net income.

Despite the rather common use of the coefficient of correlation, it is preferable for interpretation purposes to use the coefficient of determination. As we have seen, r^2, the coefficient of determination, can be interpreted as a proportion or a percentage figure. When the square root of a percentage is taken, the specific meaning becomes obscure. Furthermore, since r^2 is a decimal value (unless it is equal to zero or one), its square root, or r, is a larger number. Thus, the use of r values to indicate the degree of correlation between two variables tends to give the impression of a stronger relationship than is actually present. For example, an r value of $+0.7$ or -0.7, seems to represent a reasonably high degree of association. However, since $r^2 = 0.49$, less than one-half of the total variance in Y has been explained by the regression equation.

It is useful to observe that the values of r and r^2 do not depend on the units in which X and Y are stated nor on which of these variables is selected as the dependent or independent variable. Whether a value of r or r^2 may be considered high depends somewhat upon the specific field of application. With some types of data it is relatively unusual to find r values in excess of about 0.80. On the other hand, particularly in the case of time series data,

r values in excess of 0.90 are quite common. In the following section, we consider the matter of determining whether the observed degree of correlation in a sample is sufficiently large to justify a conclusion that correlation between X and Y actually exists in the population.

10.8 Inference about Population Parameters in Regression and Correlation

In the procedures discussed to this point, computation and interpretation of *sample* measures have been emphasized. However, as we know from our study of statistical inference, sample statistics ordinarily differ from corresponding population parameters because of chance errors of sampling. Therefore, it is useful to have a procedure which provides protection against the possible error of concluding from a sample that an association exists between two variables, while actually no such relationship exists in the population from which the sample was drawn. A hypothesis testing technique, such as those discussed in Chapter 7, can be employed for this purpose.

Inference about the Population Correlation Coefficient ρ

Let us assume the situation implied by the correlation model in which we take a simple random sample of n units from a population and make paired observations of X and Y for each unit. The sample correlation coefficient, r, as defined in Equation (10.37), is calculated. The procedure involves a test of the hypothesis that the population correlation coefficient, ρ, is equal to zero in the universe from which the sample was drawn. In keeping with the language used in Chapter 7, we wish to test the null hypothesis that $\rho = 0$ versus the alternative $\rho \neq 0$. Symbolically, we may write

$$H_0 : \rho = 0$$

$$H_1 : \rho \neq 0$$

If the computed r values in successive samples of the same size from the hypothesized population were distributed normally around $\rho = 0$, we would only have to know the standard error of r to perform the usual test involving the normal distribution. Although r values are not normally distributed, a similar procedure is provided by the statistic

$$t = \frac{r}{\sqrt{(1 - r^2)/(n - 2)}} \tag{10.38}$$

which has a t-distribution for $n - 2$ degrees of freedom. It may be noted that despite the previous explanation that r^2 is easier to interpret than r, the hypothesis testing procedure is in terms of r rather than r^2 values. The reason is that under the null hypothesis, $H_0: \rho = 0$, the sampling distribution of r leads to the t-statistic, which is a well-known distribution and is relatively easy to work with. On the other hand, under the same hypothesis of no correlation in the universe, r^2 values, which range from zero to one, would not even be symmetrically distributed, and the sampling distribution would be more difficult to deal with. The expression on the right in (10.38) is given in many references by its equivalent form, $r\sqrt{n-2}/\sqrt{1-r^2}$. Suppose we wish to test the hypothesis that $\rho = 0$ at the 5% level of significance for our illustrative problem involving ten families. Since $r = 0.912$ and $n = 10$, substitution into (10.38) yields

$$t = \frac{0.912}{\sqrt{\dfrac{1 - 0.832}{10 - 2}}} = 6.3$$

Referring to Table A-6, we find a critical t value of 2.306 at the 5% level of significance, for eight degrees of freedom. Therefore, the decision rule is

(1) If $-2.306 \leq t \leq 2.306$, accept H_0
(2) If $t < -2.306$ or $t > 2.306$, reject H_0

Since our computed t value is 6.3, far in excess of the critical value, we conclude that the sample r value differs significantly from zero. We reject the hypothesis that $\rho = 0$, and we conclude that there is a positive linear relationship between annual consumer durables expenditures and annual net income in the population from which our sample was drawn. Since the critical t value is 3.355 at the 1% level of significance (the smallest level shown in Table A-6), it is extremely unlikely that an r value as high as 0.91 would have been observed in a sample of ten items drawn from a population in which X and Y were uncorrelated.

A few comments may be made concerning this hypothesis testing procedure. First of all, this technique is valid only for a hypothesized universe value of $\rho = 0$. Other procedures must be used for assumed universe correlation coefficients other than zero.[6]

Second, only Type I errors are controlled by this testing procedure. That is, when the significance level is set at, say, 5%, the test provides a 5% risk of incorrectly rejecting the null hypothesis of no correlation. No attempt is made to fix the risks of Type II errors at specific levels.

---

[6]Fisher's z-transformation may be used when ρ is hypothesized to be nonzero. In this procedure, a change of variable is made from the sample r to a statistic z, defined as $z = \frac{1}{2}\log_e[(1 + r)/(1 - r)]$. This statistic is approximately normally distributed with mean $\mu_z = \frac{1}{2}\log_e[(1 + \rho)/(1 - \rho)]$ and standard deviation $\sigma_z = 1/\sqrt{n-3}$.

Third, even though the sample r value is significant according to this test, the amount of correlation may not be considered to be substantively important. For example, in a large sample, a quite low r value may be found to be significantly different from zero. However, since relatively little correlation has been found between the two variables, we may be unwilling to use the relationship observed between X and Y for decision making purposes. Furthermore, prediction intervals based on the use of the applicable standard errors of estimate may be too wide to be of practical use.

Fourth, the distributions of t values computed by Equation (10.38), as in previously discussed cases, approach the normal distribution as sample size increases. Hence, for large sample sizes, the t value is approximately equal to z in the standard normal distribution, and critical values applicable to the normal distribution may be used instead. For example, in the preceding illustration, in which the critical t value was 2.306 for eight degrees of freedom at the 5% level of significance, the corresponding critical z value would be 1.96 at the same significance level.

Inference about the Population Regression Coefficient β

In many cases, a great deal of interest is centered on the value of b, the slope of the regression line computed from a sample. For example, it was indicated earlier that in the case of our illustrative problem in which Y is an expenditures variable and X is an income variable, the regression coefficient b can be interpreted as a "marginal propensity to consume coefficient." If the variables in the illustrative problem had been stated in terms of logarithms and a regression equation of the form $\log \bar{Y}_X = a + b \log X$ had been fitted to the data, then the b value could be interpreted as a so-called "income elasticity coefficient," or the percentage change in expenditures per 1% change in income. Statistical inference procedures involving either hypothesis testing or confidence interval estimation are often useful for answering questions concerning the size of the population regression coefficient, β.

In order to illustrate the hypothesis testing procedure for a regression coefficient, let us return to the data in our problem, in which $b = 2.22$. We recall that the interpretation of this figure was that for two families whose annual net income in 1967 differed by \$1000, their estimated difference in annual consumer durables expenditures was \$222. Suppose that on the basis of similar studies in the same metropolitan area, it had been concluded that in previous years a valid assumption for the true population regression coefficient was $\beta = 2$. Can we conclude that the marginal propensity to consume coefficient had changed?

To answer this question, we use a familiar hypothesis testing procedure. We establish the following null and alternative hypotheses:

$$H_0: \beta = 2$$

$$H_1: \beta \neq 2$$

Assume that we were willing to run a 5% risk of erroneously rejecting the null hypothesis that $\beta = 2$. The procedure involves a t-test, in which the estimated standard error of the regression coefficient, denoted $\hat{\sigma}_b$ is given by

(10.39) $$\hat{\sigma}_b = \frac{\hat{\sigma}_{Y.X}}{\sqrt{\Sigma(X - \bar{X})^2}}$$

Hence, $\hat{\sigma}_b$, the estimated standard deviation of the sampling distribution of b values, is a function of the scatter of points around the regression line and the dispersion of the X values around their mean. The t-statistic computed in the usual way is given by

(10.40) $$t = \frac{b - \beta}{\hat{\sigma}_b}$$

We calculate $\hat{\sigma}_b$ according to (10.39) as follows

$$\hat{\sigma}_b = \frac{2.39}{\sqrt{46}} = 0.35$$

Substituting this value for $\hat{\sigma}_b$ into (10.40) gives

$$t = \frac{2.22 - 2}{0.35} = 0.63$$

Since the same level of significance and the same number of degrees of freedom are involved as in the preceding test for the significance of r, the decision rule is identical. With a critical t value of 2.306 at the 5% level of significance, we cannot reject the null hypothesis that $\beta = 2$. Hence, we cannot conclude that the coefficient for the marginal propensity to consume durables has changed for families in the given metropolitan area.

The corresponding confidence interval procedure involves setting up the interval

(10.41) $$b \pm t\hat{\sigma}_b$$

In this problem the 95% confidence interval for β is

$$2.22 \pm (2.306)(0.35) = 2.22 \pm 0.81$$

Therefore, we can assert that the population β figure is included in the interval 1.41 to 3.03 with an associated confidence coefficient of 95%.

It may be noted that we have used the t-distribution in both the hypothesis testing and confidence interval procedures just discussed. As in previous examples, we may note that normal curve procedures can be used for large sample sizes. Hence, for two-tailed hypothesis testing at the 5% level of significance and for 95% confidence interval estimation, the 2.306 t value given in the preceding examples would be replaced by a normal curve z value of 1.96.

10.9 Other Interval Estimates

As we have seen, even if the true regression line $\mu_{Y.X} = \alpha + \beta X$ were available, there would still be errors of estimation because of the variation of points around this regression line. The previously discussed standard errors of estimate measure this type of variation. There is another component of variation which arises from sampling errors that are present in the sample regression line. A sample regression line can differ from the true population regression line because of differences in the height of the line and the slope of the line. One form in which the sample regression equation can be expressed is

$$\bar{Y}_X = \bar{Y} + b(X - \bar{X}) \tag{10.42}$$

This regression equation expresses $\bar{Y}_X$, the estimated value of the dependent variable, as a function of $X - \bar{X}$, a deviation of an observation X from the mean $\bar{X}$. For a particular X value, the same estimate $\bar{Y}_X$ would be obtained from $\bar{Y}_X = a + bX$ or from the regression equation given in (10.42).

When we view the problem of estimating a conditional mean according to (10.42), we see that the variance of $\bar{Y}_X$ is composed of two parts, the variance in the average height of the regression line ($\bar{Y}$), and the variance in the slope of the line (b). Using Rule 9 of Appendix C, we have

$$\begin{aligned}\sigma^2(\bar{Y}_X) &= \sigma^2[\bar{Y} + b(X - \bar{X})] \\ &= \sigma^2(\bar{Y}) + \sigma^2[b(X - \bar{X})] \\ &= \sigma^2(\bar{Y}) + (X - \bar{X})^2\sigma^2(b)\end{aligned}$$

Hence,

$$\sigma^2(\bar{Y}_X) = \frac{\sigma^2_{Y.X}}{n} + (X - \bar{X})^2 \frac{\sigma^2_{Y.X}}{\Sigma(X - \bar{X})^2} \tag{10.43}$$

Substituting the more familiar subscript notation $\sigma^2_{\bar{Y}_X}$ for $\sigma^2(\bar{Y}_X)$ on the left-hand side of (10.43) and the estimator $\hat{\sigma}^2_{Y.X}$ for $\sigma^2_{Y.X}$ on the right-hand side, we obtain

$$\hat{\sigma}^2_{\bar{Y}_X} = \frac{\hat{\sigma}^2_{Y.X}}{n} + (X - \bar{X})^2 \frac{\hat{\sigma}^2_{Y.X}}{\Sigma(X - \bar{X})^2} \tag{10.44}$$

Taking the square root of (10.44) leads to the following formula for the stimated standard error of the conditional mean.

$$\hat{\sigma}_{\bar{Y}_X} = \hat{\sigma}_{Y.X}\sqrt{\frac{1}{n} + \frac{(X - \bar{X})^2}{\Sigma(X - \bar{X})^2}} \tag{10.45}$$

This expression for the estimated standard error of the conditional mean an be used to establish a confidence interval for the conditional mean, often alled a "confidence interval for the regression line." Such an interval is ndicated in the hypothetical scatter diagram in Figure 10-9. As can be seen

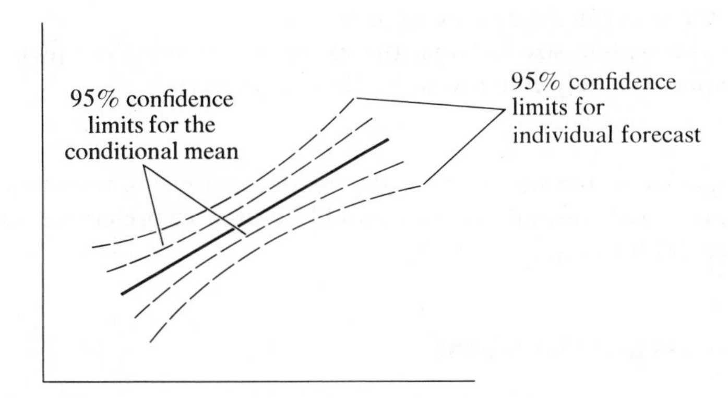

Figure 10-9 Confidence intervals for the conditional mean (regression line) and individual forecast (new observation).

n the diagram and as is evident from expression (10.45), the further the X alues lie from their mean, the greater is the width of the confidence band. f the sample size is large, $\Sigma(X - \bar{X})^2$ will be large compared to $(X - \bar{X})^2$ for ny particular X value. Hence, the estimated standard error for the conitional mean will be

$$\hat{\sigma}_{\bar{Y}_X} \approx \frac{\hat{\sigma}_{Y.X}}{\sqrt{n}} \tag{10.46}$$

which is the same formula as given in (10.28), the previously discussed method or establishing confidence limits for a conditional mean. The more exact tandard error formula (10.45) should be used in preference to (10.46) when he sample size is small.

An analogous situation arises in the case of estimating an individual value f the dependent variable Y. In this situation, we have a new observation rawn from the same population as the sample from which we constructed the egression line. We observe the X value for this new observation, and we wish to predict the value of Y. Let us denote this estimated value of Y as Y_{new} an its standard error as $\hat{\sigma}_{\text{new}}$. By a proof similar to that given for $\hat{\sigma}_{\bar{Y}_X}$, it can b shown that

$$\hat{\sigma}_{\text{new}} = \hat{\sigma}_{Y.X}\sqrt{1 + \frac{1}{n} + \frac{(X - \bar{X})^2}{\Sigma(X - \bar{X})^2}} \tag{10.47}$$

The estimate of Y_{new} is usually referred to as an "individual forecast and $\hat{\sigma}_{\text{new}}$ is called the "standard error of forecast." The expression for th standard error of forecast can be used to set up confidence limits for individu forecasts in the usual way. Such an interval is indicated in Figure 10-9. A in the case of confidence limits for the regression line, the further the X valu is from $\bar{X}$, the wider is the confidence interval.

When the sample size is large, the expression under the square root i (10.47) is approximately equal to one. Hence, we have

$$\hat{\sigma}_{\text{new}} \approx \hat{\sigma}_{Y.X} \tag{10.48}$$

Analogously to the situation for the conditional mean, for small sampl sizes the more exact formula (10.47) should be used in preference to the ex pression (10.24) for $\hat{\sigma}_{Y.X}$.

10.10 Caveats and Limitations

Regression and correlation analysis are very useful and widely applie techniques. However, it is important to understand the limitations of th methods and to interpret the results with care.

Cause and Effect Relationships

In correlation analysis, the value of the coefficient of determination, r^2, i calculated. This statistic measures the degree of association between tw variables. Neither this quantity nor any other statistical technique whic measures or expresses the relationship among variables can prove that on variable is the cause and one or more other variables are the *effects*. Indeec there has been philosophical speculation and debate through the centurie as to the meaning of cause and effect, and as to whether such a relationshi can ever be demonstrated by experimental methods. In any event, a measur such as r^2 does not prove the existence of a cause-effect relationship betwee two variables X and Y.

In a situation in which a high value of r^2 is obtained, X may be producin variations in Y, or third and fourth (etc.) variables W and Z may be producin variations in both X and Y. There have been numerous examples giver frequently humorous in nature, to demonstrate the pitfalls in attempting t draw cause-effect conclusions in such cases. For example, if the average salaries of ministers are associated with the average price of a bottle of scotch whiskey over time, that is, time represents the unit of association, a high degree of correlation between these two variables will probably be observed. Doubtless, we would be reluctant to conclude that it is fluctuations in ministers' salaries which cause the variations in the price of a bottle of scotch, or vice versa. This is a case where a third variable, which we may conveniently designate as the general level of economic activity, operates to produce variations in both of the aforementioned variables. From the economic standpoint, salaries of ministers represent the price paid for a particular type of labor; the cost of a bottle of scotch is also a price. When the general level of economic activity is high, both of these prices will tend to be high. When the general level of economic activity is low, as in periods of recession or depression, both of these prices tend to be lower than they were during more prosperous times. Thus, the high degree of correlation between the two variables of interest is produced by a third variable (and possibly others); certainly neither of the two variables is *causing* the variations in the other.

Furthermore, it is important to keep in mind the problem of sampling error. As we have seen, it is conceivable that in a particular sample a high degree of correlation, either direct or inverse, may be observed, while in fact there is no correlation (or very little correlation) between the two variables in the population.

Finally, in applying critical judgment to the evaluation of observed relationships, one must be on guard against "nonsense correlations" where no meaningful unit of association is present. For example, suppose we record in a column labeled X the distance from the ground of the skirt hemlines of the first 100 women who pass a particular street corner. In a column labeled Y, we record 100 observations of the heights of the Himalaya mountains along a certain latitude at five mile intervals. It is possible that a high r^2 value might be obtained for these data. Clearly the result is nonsensical because there is no meaningful unit or entity through which these data are related. In the illustrative example used in this chapter, expenditures and income were observed for the same family. We have seen that the unit of association might be a time period or some other entity. There must be a reasonable link between the variables studied which is embodied in the unit of association.

Extrapolation Beyond the Range of Observed Data

In regression analysis, an estimating equation is established on the basis of a particular set of observations. A great deal of care must be exercised in making predictions of values of the dependent variable based on values of the independent variable outside the range of the observed data. Such predictions are referred to as *extrapolations*. For example, in the illustrative problem considered in this chapter, a regression line was computed for families whos annual net income ranged from \$2000 to \$10,000. It would be extremel unwise to make a prediction of consumer durables expenditures for a famil with an annual net income of \$25,000 using the computed regression lin To do so would imply that the straight-line relationship could be projecte up to a value of \$25,000 for the independent variable. Clearly, in the absenc of other information, we simply do not know whether the same function form of the estimating equation is valid outside the range of the observed dat In fact, in certain cases, unreasonable or even impossible values may resu from such extrapolations. For example, suppose a regression with a negativ slope had been computed relating the percentage of defective articles produce Y with the number of weeks of on-the-job training received X, by a group c workers. An extrapolation for a large enough number of weeks of trainin would produce a negative value for the percentage of articles produced, whic is an impossible result. Clearly in this case, although the computed estimatin equation may be a good description of the relation between X and Y withi the range of the observed data, an equation with different parameters c even a completely different functional form is required outside this rang Without a specific investigation, one simply does not know what the ap propriate estimating device is outside the range of observed data. The maxin "To know how many teeth a horse has, you must open his mouth and cou his teeth," is relevant here.

However, there are instances where the exigencies of a situation requi an estimate, and obtaining additional data is either impractical or impossibl Extrapolations and alternative methods of prediction have to be engaged i but the limitations and risks involved must be kept constantly in mind.

Other Regression Models

To this point, we have considered only one particular form of the re gression model, namely, that of a straight-line equation relating the depende variable Y to the independent variable X. Sometimes theoretical considera tions indicate that this is the form of model required. On the other hand, linear model is often used because either the theoretical form of the relatio ship is unknown and a linear equation appears to be adequate, or the theore ical form is known, but is rather complex, and a linear equation may provi a sufficiently good approximation. In all cases, the determination of the mo appropriate regression model should be the result of a combination of the retical reasoning, practical considerations, and careful scrutiny of the availab data.

Often, the straight-line model $\bar{Y}_X = a + bX$ is not an adequate descri tion of the relationship between the two variables. In some situations, mode which involve transformations of one or both of the variables may provi better fits to the data. For example, if the dependent variable Y is transformed to a new variable log Y, a regression equation of the form

(10.49)
$$\log \bar{Y}_X = a + bX$$

may yield a better fit. Insofar as arithmetic is concerned, log Y is substituted for Y everywhere that Y appeared previously in the formulas. However, care must be used in the interpretation. The antilogarithm of $\log \bar{Y}_X$ must be taken to provide an estimate of the dependent variable Y for a given value of X. Also, it must be recognized that different assumptions are involved in this model as compared to the model $\bar{Y}_X = a + bX$. It is now assumed that log Y rather than Y is a normally distributed random variable.

Possible transformations include the use of square roots, reciprocals, logarithms, etc., of one or both of the variables. As an example of one such useful transformation, if a straight-line equation is fitted to the logarithms of both variables, the model is of the form

(10.50)
$$\log \bar{Y}_X = a + b \log X$$

The regression coefficient, b, in this model has an interesting interpretation, if as in the illustrative example used in this chapter, Y is a consumption variable and X is income. It was previously indicated that for such variables the regression coefficient b in the model $\bar{Y}_X = a + bX$ can be interpreted as a marginal propensity-to-consume coefficient. That is, it estimates the dollar change in consumption per dollar change in income. Analogously, in the model $\log \bar{Y}_X = a + b \log X$, the coefficient b can be interpreted as an income elasticity of consumption coefficient. That is, it estimates the *percentage change* in consumption per *one percent change* in income. Of course, fitting an equation of the form $\log \bar{Y}_X = a + b \log X$ in the illustration under discussion implies that the income elasticity of consumption is constant over the range of income observed. Similarly, a model of the form $\bar{Y}_X = a + bX$ implies that the marginal propensity to consume is constant over the range of observed income. In fact, according to Keynesian economic theory, the marginal propensity to consume (for total consumption expenditures) decreases with increasing income. The fitting of such regression models clearly cannot merely be a mechanistic procedure, but must involve a combination of knowledge of the field of application, good judgment, and experimentation.

In some applications, a curvilinear regression function may be more appropriate than a linear one. Polynomial functions are particularly convenient to fit by the method of least squares. The straight-line regression equation of the form $\bar{Y}_X = a + bX$ is a polynomial of the first degree, since X is raised to the first power. A second degree polynomial would involve a regression function of the form

(10.51)
$$\bar{Y}_X = a + bX + cX^2$$

in which the highest power to which X is raised is two. This is the equation of a second degree parabola, which has the characteristic that one change in direction can take place in Y as X increases, whereas in the case of a straight line, no changes in direction can take place. A third degree polynomial permits two changes in direction, etc. In the straight-line function, the amount of change in $\bar{Y}_X$ is constant per unit change in X. In the second degree parabola, the amounts of change in $\bar{Y}_X$ may decrease or increase per unit change in X, depending upon the shape of the function. Figure (10.10) shows two scatter diagrams for situations in which a second degree regression function of the form of (10.51) may provide a good fit. The probable shape of the regression function has been indicated. Analogous situations could be portrayed for cases of inverse relationships between X and Y.

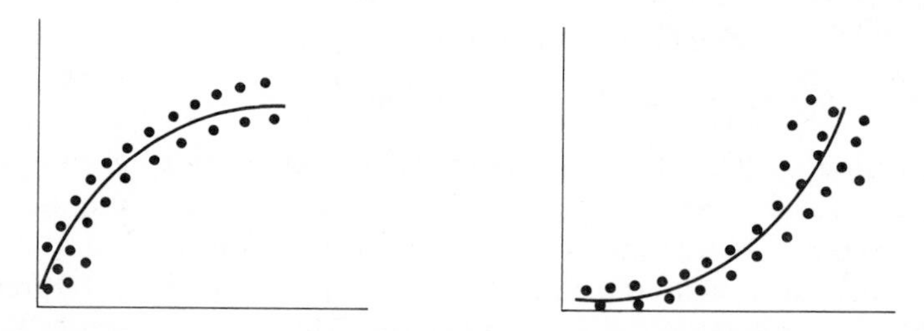

Figure 10-10 Scatter diagrams for two situations in which a second degree polynomial regression function might be appropriate.

In the case of the straight-line regression function, the application of the method of least squares leads to two normal equations which have to be solved for a and b. Analogously, to obtain the values of a, b, and c in the second degree polynomial function, the following three simultaneous equations must be solved.

$$\begin{aligned} \Sigma Y &= na + b\Sigma X + c\Sigma X^2 \\ \Sigma XY &= a\Sigma X + b\Sigma X^2 + c\Sigma X^3 \\ \Sigma X^2 Y &= a\Sigma X^2 + b\Sigma X^3 + c\Sigma X^4 \end{aligned} \tag{10.52}$$

Various computer programs have been developed which solve such normal equation systems, provide for transformations of the variables, and calculate all of the regression and correlation measures we have discussed (as well as others). In applied problems in which large quantities of data are present and considerable experimentation with the form of the regression model is required, or more complex models than those discussed to this point are appropriate, the use of modern computers may be the only feasible method of implementation.

Thus far, the discussion in this chapter has been limited to two-variable regression and correlation analysis. In many problems, the inclusion of more than one independent variable in a regression model may be required to provide useful estimates of the dependent variable. Suppose, for example, that in the illustration involving family consumer durables expenditures and family income, poor predictions were made based on the single independent variable income. Other factors such as family size, age of the head of the family, number of employed persons in the family, etc., might be considered as possible additional independent variables to aid in the estimation of consumer durables expenditures. When two or more independent variables are utilized, the problem is referred to as a *multiple regression and correlation analysis*. Although a detailed discussion of this type of analysis is beyond the scope of this text, a brief description of the technique is included in the next section. Following that description, a case study is given which illustrates the use of such a model in a marketing managerial problem.

10.11 Multiple Regression and Correlation Analysis

Multiple regression analysis represents a logical extension of two-variable regression analysis. Instead of a single independent variable, two or more independent variables are used to estimate the values of a dependent variable. However, the fundamental concepts in the analysis remain the same. Thus, just as in the analysis involving the dependent and only one independent variable, there are the following three general purposes of multiple regression and correlation analysis:

(1) To derive an equation which provides estimates of the dependent variable from values of the two or more independent variables.

(2) To obtain a measure of the error involved in using this regression equation as a basis for estimation.

(3) To obtain a measure of the proportion of variance in the dependent variable accounted for or "explained by" the independent variables.

The first purpose is accomplished by deriving an appropriate regression equation by the method of least squares. The second purpose is achieved through the calculation of a standard error of estimate, which just as in two-variable analysis, is simply the standard deviation of the observed values of the dependent variable around the estimated values computed from the regression equation. The third purpose is accomplished by computing the multiple coefficient of determination, which is analogous to the coefficient of determination in the two-variable case, and as indicated in (3) above, measures the proportion of variance in the dependent variable explained by the independent variables.

As an example, let us return to our illustrative problem in which family consumer durables expenditures were estimated from family net income, both variables being stated on an annual basis. As indicated in the preceding section, the use of additional variables to income might be considered to obtain improvement in the prediction of the dependent variable. Let us assume that family size is selected as a second independent variable. Estimates of consumer durable expenditures may now be made from the following multiple regression equation:

(10.53) $$Y_c = a + b_1X_1 + b_2X_2$$

where

Y_c = family consumer durables expenditures (estimated)

X_1 = family net income

X_2 = family size

and a, b_1, and b_2 are numerical constants which must be determined from the data in a manner analogous to that of the two-variable case. For simplicity, we have assumed a linear regression function. We have slightly changed the notation for the estimated value of the dependent variable from $\overline{Y}_X$, used in the simple two-variable analysis, to Y_c. The subscript c stands for "computed." This is a simplified notation which we will use in the following discussion. Some systems of notation attempt to indicate in the subscript of the dependent variable all of the independent variables in the regression equation. In this type of notation, the dependent variable is usually designated as X_1 and the independent variables as $X_2, X_3, \ldots, X_k$. For example, a linear regression equation of the type given in (10.53) involving two independent variables would be written

(10.54) $$X_{1.23} = a + b_2 X_2 + b_3 X_3$$

The symbol $X_{1.23}$ denotes a value of the dependent variable X_1 estimated from the independent variables X_2 and X_3. Sometimes a more elaborate system of notation is also used which designates the constants a, b, and b_2, as $a_{1.23}$, $b_{12.3}$, and $b_{13.2}$, respectively.

As an illustrative example, we will carry out a multiple regression and correlation analysis, fitting the linear regression equation (10.53) to data for the indicated variables. The basic data for family consumer durables expenditures, family income, and family size are shown in the first three columns of Table 10-4. The data for the first two of these variables are the same as those given in Table 10-1 for the two-variable problem previously solved.

The data on family size represent the total number of persons in each of the families in the sample.

The Multiple Regression Equation

We begin the analysis by using the method of least squares to obtain the best fitting three-variable linear regression equation of the form given in (10.53). In the two-variable regression problem, the method of least squares was used to obtain the best fitting straight line. In the present problem, the analogous geometric interpretation is that the method of least squares is used to obtain the best fitting plane. In a three-variable regression problem, the points can be plotted in three dimensions, along the X_1, X_2, and Y axes analogous to the case of a two-variable problem, in which the points are plotted in two dimensions along an X and Y axis. The best fitting plane would pass through the points as shown in Figure 10-11, with some falling above and some below the plane in such a way that $\Sigma(Y - Y_c)^2$ is a minimum. Whereas in our previous illustration involving two variables, two normal equations resulted from the minimization procedure, now three normal equations must be solved to determine the values of a, b_1, and b_2:[7]

(10.55)
$$\begin{aligned} \Sigma Y &= na + b_1\Sigma X_1 + b_2\Sigma X_2 \\ \Sigma X_1 Y &= a\Sigma X_1 + b_1\Sigma X_1^2 + b_2\Sigma X_1X_2 \\ \Sigma X_2 Y &= a\Sigma X_2 + b_1\Sigma X_1X_2 + b_2\Sigma X_2^2 \end{aligned}$$

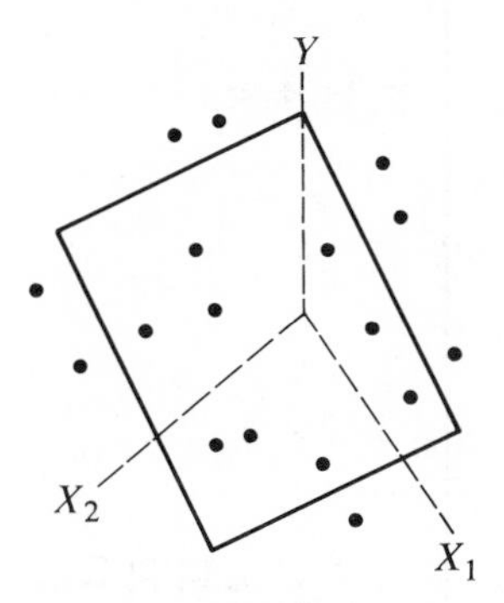

Figure 10-11 Graph of a multiple regression plane for data on the variables Y, X_1, and X_2.

---

[7]In a manner similar to that of the two-variable case, a function of the form

$$F(a, b_1, b_2) = \Sigma(Y - Y_c)^2 = \Sigma(Y - a - b_1X_1 - b_2X_2)^2$$

is set up. This function is minimized by the standard calculus method of taking its partial derivatives with respect to a, b_1, and b_2 and equating these derivatives to zero. This procedure results in the three normal equations of (10.55).

Table 10-4 Computations for Linear Multiple Regression Analysis: Expenditures on Consumer Durables, Y, Family Income, X_1, and Family Size, X_2.

Family	*Expenditures on Consumer Durables (hundreds of dollars)* Y	*Income (thousands of dollars)* X_1	*Family Size* X_2	X_1Y	X_2Y	X_1X_2	Y^2	X_1^2	X_2^2
A	23	10	7	230	161	70	529	100	49
B	7	2	3	14	21	6	49	4	9
C	15	4	2	60	30	8	225	16	4
D	17	6	4	102	68	24	289	36	16
E	23	8	6	184	138	48	529	64	36
F	22	7	5	154	110	35	484	49	25
G	10	4	3	40	30	12	100	16	9
H	14	6	3	84	42	18	196	36	9
I	20	7	4	140	80	28	400	49	16
J	19	6	3	114	57	18	361	36	9
Total	170	60	40	1122	737	267	3162	406	182
Mean	17	6	4						

The calculations of the required sums are shown in Table 10-4. Substituting into the normal equations (10.55) gives

$$170 = 10a + 60b_1 + 40b_2$$
$$1122 = 60a + 406b_1 + 267b_2$$
$$737 = 40a + 267b_1 + 182b_2$$

Solving these three equations simultaneously, we obtain the following values for a, b_1, and b_2:

$$a = 3.92$$
$$b_1 = 2.50$$
$$b_2 = -0.48$$

The calculations to obtain the values of a, b_1, and b_2 can be simplified somewhat by transforming (or coding) the original data into deviations from the mean of each of the variables. This procedure reduces by one the number of simultaneous equations which must be solved. Denoting the deviations from the means of Y, X_1, and X_2 by lower case letters y, x_1, and x_2, respectively, we have

(10.56)
$$y = Y - \bar{Y}$$
$$x_1 = X_1 - \bar{X}_1$$
$$x_2 = X_2 - \bar{X}_2$$

When expressed in terms of deviations the second and third normal equations of (10.55) become

(10.57)
$$\Sigma x_1 y = b_1 \Sigma x_1^2 + b_2 \Sigma x_1 x_2$$
$$\Sigma x_2 y = b_1 \Sigma x_1 x_2 + b_2 \Sigma x_2^2$$

These equations contain fewer terms than their counterparts in (10.55), since Σx_1 and Σx_2 are equal to zero. This is an example of the property that the sum of the deviations of any set of observations from their mean is equal to zero.

The required computations for the lower case normal equations of (10.57) are given in Table 10-5. Short-cut methods have been utilized to calculate the sums of squares and cross products. Substituting the numerical results of Table 10-5 into the normal equations (10.57) yields

$$102 = 46b_1 + 27b_2$$
$$57 = 27b_1 + 22b_2$$

Table 10-5 Calculations in Terms of Deviations from Means for the Multiple Regression Between Expenditures on Consumer Durables, Y, Family Income, X_1, and Family Size, X_2.

$$\Sigma y^2 = \Sigma Y^2 - (\bar{Y})(\Sigma Y) = 3162 - (17)(170) = 272$$
$$\Sigma x_1^2 = \Sigma X_1^2 - (\bar{X}_1)(\Sigma X_1) = 406 - (6)(60) = 46$$
$$\Sigma x_2^2 = \Sigma X_2^2 - (\bar{X}_2)(\Sigma X_2) = 182 - (4)(40) = 22$$
$$\Sigma x_1 y = \Sigma X_1 Y - (\bar{X}_1)(\Sigma Y) = 1122 - 1020 = 102$$
$$\Sigma x_2 y = \Sigma X_2 Y - (\bar{X}_2)(\Sigma Y) = 737 - (4)(170) = 57$$
$$\Sigma x_1 x_2 = \Sigma X_1 X_2 - (\bar{X}_1)(\Sigma X_2) = 267 - (6)(40) = 27$$

Solving these simultaneous equations for b_1 and b_2, we again find

$$b_1 = 2.50$$
$$b_2 = -0.48$$

The value of the constant a is obtained from

(10.58) $$a = \bar{Y} - b_1\bar{X}_1 - b_2\bar{X}_2$$

Substituting the values of $\bar{Y}$, $\bar{X}_1$, and $\bar{X}_2$ from Table 10-4 we find that a is the same value previously computed.

$$a = 17 - (2.5)(6) - (-0.48)(4) = 3.92$$

When large numbers of variables and observations are present, even calculations involving deviations from means are apt to be too laborious. The utilization of electronic computing equipment in such cases may represent the only feasible alternative.

The multiple regression equation may now be written as

(10.59) $$Y_c = 3.92 + 2.50X_1 - 0.48X_2$$

Let us illustrate the use of this equation for estimation. Suppose we want to estimate consumer durables expenditures for a family from the same population as the sample studied. The family's income is \$6000 and there are four persons in the family. Substituting $X_1 = 6$ and $X_2 = 4$ yields the following estimated expenditures on consumer durables:

$$\begin{aligned} Y_c &= 3.92 + 2.50(6) - 0.48(4) \\ &= 17.00 \text{ (hundreds of dollars)} \\ &= 1700 \text{ (dollars)} \end{aligned}$$

In two-variable analysis, we discussed the interpretation of the constants a and b in the regression equation. Let us consider the analogous interpretation of the constants a, b_1, and b_2 in the multiple regression equation. The constant a is again the Y intercept. However, now it is interpreted as the value of Y_c when X_1 and X_2 are both equal to zero. The b values are referred to in multiple regression analysis as *net regression coefficients*. The b_1 coefficient measures the change in Y_c per unit change in X_1 when X_2 is held fixed, and b_2 measures the change in Y_c per unit change in X_2 when X_1 is held fixed.[8]

Hence, in the present problem, the b_1 value of 2.50 indicates that if a family has an income which is \$1000 greater than another's (a one unit change in X_1) and *the families are of the same size* (X_2 is held constant), then the estimated expenditures on consumer durables of the higher income family exceed those of the other by 2.5 hundreds of dollars or \$250. Similarly, the b_2 value of -0.48 means that if a family has one person more than another (a one unit change in X_2) and *the families have the same income* (X_1 is held constant), then the estimated expenditures of the larger family are less than those of the smaller by \$48.

Two properties of these net regression coefficients are worth noting. The b_1 value of 2.50 hundreds of dollars implies that an increment of one unit in X_1, or a \$1000 increment in income, occasions an increase of \$250 in Y_c, estimated expenditures of consumer durables, regardless of the size of the family (for families of the sizes studied). Hence, an increase of \$1000 in income adds \$250 to estimated consumer durables expenditures, regardless of whether there are two or six people in the family. An analogous interpretation holds for b_2. These interpretations follow from the fact that a *linear* multiple regression equation was used, and are embodied in the assumption of linearity. It is advisable to plot scatter diagrams of the dependent variable against each of the independent variables at the beginning of the problem to examine the actual nature of the relationships which appear to be present.

A second property of regression coefficients is apparent from a comparison of the b value of 2.22 in the simple regression equation (10.13), $\bar{Y}_X = 3.68 + 2.22X$, previously obtained when family income, X, was the only independent variable, with the b_1 value of 2.50, the net regression coefficient of income in the multiple regression equation $Y_c = 3.92 + 2.50X_1 - 0.48\ X_2$, when the family size variable is included in the regression equation. The coefficient $b = 2.22$ in the simple two-variable regression equation makes no explicit allowance for family size. The net regression coefficient $b_1 = 2.50$, on the

---

[8]In terms of calculus, b_1 and b_2 are the partial derivatives of Y_c with respect to X_1 and X_2, respectively, that is,

$$\frac{\partial Y_c}{\partial X_1} = b_1 \text{ and } \frac{\partial Y_c}{\partial X_2} = b_2$$

other hand, "nets out" the effect of family size. A net regression coefficient may in general be greater or less than the corresponding regression coefficient in a two-variable analysis.

In this problem, the families with larger incomes were also the ones with larger family sizes. The positive correlation between income and family size is indicated by the correlation coefficient $r = .85$, as shown in Table 10-6. The foregoing pattern exemplifies an important characteristic of regression coefficients, regardless of the number of independent variables which have been included in the study. That is, a regression coefficient for any specific independent variable, for example, income, measures not only the effect on the dependent variable of income, but also the effect which is attributable to any other independent variables which happen to be correlated with it but have not been explicitly included in the analysis. This is true for both two-variable and multiple regression analyses.

When independent variables are highly correlated, rather odd results may be obtained in a multiple regression analysis. For example, a regression coefficient which is positive (negative) in sign in a two-variable regression equation may change to a negative (positive) sign for the same independent variable in a multiple regression equation containing other independent variables which are highly intercorrelated with the one in question. For example, in this problem, the dependent variable, consumer durables expenditures, Y, is positively correlated with family size, X_2, as indicated by the correlation coefficient of $+.74$ (Table 10-6). Hence, the regression coefficient for family size would also be positive in sign. However, as we have seen, the net regression coefficient for family size, b_2, in the three-variable regression equation is equal to $-.48$ and is thus negative in sign.

It will be shown in the discussion of statistical inference in multiple regression that the regression coefficients for highly intercorrelated independent variables tend to be unreliable. The importance of this is that when independent variables are highly intercorrelated, it is extremely difficult to separate out the individual influences of each variable. This can be seen by considering an extreme case. Suppose a two-variable regression and correlation analysis is carried out between a dependent variable, denoted Y, and an independent variable, denoted X_1. Further, let us assume that we introduce another independent variable X_2, which has perfect positive correlation with X_1, that is, the correlation coefficient between X_1 and X_2 is $+1$. We now conduct a three-variable regression and correlation analysis. It is clear that X_2 cannot account for or explain any additional variance in the dependent variable Y after X_1 has been taken into account. The same argument could be made if X_1 were introduced after X_2. As indicated in the ensuing discussion of statistical inference in multiple regression, the net regression coefficients, b_1 and b_2, in cases of high intercorrelation between X_1 and X_2 will tend not to differ significantly from zero. Yet, if separate two-variable analyses had been run between Y and X_1 and Y and X_2, the individual regression coefficients might have differed significantly from zero. There is a great deal of concern in fields such as econometrics and applied statistics with this problem of intercorrelation among independent variables, often referred to as *multicollinearity*. One of the simplest solutions to the problem of two highly correlated independent variables is merely to discard one of the variables.

The illustration in this section used only two independent variables. The general form of the linear multiple regression function for k independent variables $X_1, X_2, \ldots, X_k$ is

(10.60)
$$Y_c = a + b_1X_1 + b_2X_2 + \cdots + b_kX_k$$

The linear function which is fitted to data for two variables is referred to as a straight-line, for three variables a plane, for four or more variables, a *hyperplane*. Although we cannot visualize a hyperplane, its linear characteristics are analogous to those of the linear functions of two or three variables. With the use of electronic computers it is possible to test and include large numbers of independent variables in a multiple regression analysis. However, good judgment and knowledge of the logical relationships involved must always be used as a guide to deciding which variables to include in the construction of a regression equation.

Standard Error of Estimate

As in simple two-variable regression analysis, a measure of dispersion or scatter around the regression plane or hyperplane can be used as an indicator of the error of estimation. Again, probability assumptions similar in principle to those of the simple regression model must be introduced. The following are the usual assumptions made in a linear multiple regression analysis, illustrated for the case of two independent variables:

(1) The conditional distributions of Y for given X_1 and X_2 are assumed to be normal.

(2) These conditional distributions are assumed to have equal standard deviations.

(3) The $Y - Y_c$ deviations are assumed to be independent of one another.

The variance around a regression hyperplane involving k variables, one of which is dependent and $k - 1$ are independent is

(10.61)
$$\overline{S^2_{Y.12\ldots(k-1)}} = \frac{\Sigma(Y - Y_c)^2}{n - k}$$

where n is the number of observations and k is the number of constants in the regression equation. The divisor $n - k$ represents the number of degrees of freedom, and its use provides an unbiased estimator of the population variance. The bar over the S^2 is conventionally used to denote an estimator which has been "corrected for sample bias" or "adjusted or corrected for degrees of freedom." The subscript notation to $\bar{S}^2$ lists the dependent variable to the left of the period and the $k - 1$ independent variables to the right. The subscripts $1, 2, \ldots, k - 1$ denote the variables $X_1, X_2, \ldots, X_{k-1}$, respectively. Hence, in our example involving the three variables Y_1, X_1, and X_2, the variance around the regression plane $Y_c = 3.92 + 2.50X_1 - 0.48X_2$ is given by

$$\text{(10.62)} \qquad \bar{S}^2_{Y.12} = \frac{\Sigma(Y - Y_c)^2}{n - 3}$$

The standard error of estimate, which is the square root of this variance, is

$$\text{(10.63)} \qquad \bar{S}_{Y.12} = \sqrt{\frac{\Sigma(Y - Y_c)^2}{n - 3}}$$

A short-cut formula for calculating this standard error of estimate in terms of deviations from means is the following:

$$\text{(10.64)} \qquad \bar{S}_{Y.12} = \sqrt{\frac{\Sigma y^2 - b_1\Sigma x_1 y - b_2\Sigma x_2 y}{n - 3}}$$

Substituting the values previously obtained in our illustrative problem into 10.64, we find

$$\bar{S}_{Y.12} = \sqrt{\frac{272 - (2.50)(102) - (-0.48)(57)}{7}}$$
$$= 2.52 \text{ (hundreds of dollars)}$$

The standard error measures the closeness of estimates derived from the regression equation to the actual observed values of Y. Hence, assuming for the moment that we are dealing with a large sample, and assuming normal distributions of Y values around the regression plane, we can assert that about 95% of the points are within $2\bar{S}_{Y.12}$ or within $2(2.52) = 5.04$ hundred dollars of the Y_c values computed from the regression equation. Also, prediction intervals for individual estimates of Y_c would be obtained in a manner similar to that of two-variable analysis indicated in (10.25), i.e., by calculating the prediction interval

$$\text{(10.65)} \qquad Y_c \pm z\bar{S}_{Y.12}$$

Hence, for a large sample, for the family with an income of \$6000 and consisting of four persons, for whom we earlier estimated consumer durables expenditures of 17.00 hundreds of dollars from the multiple regression equation (10.59), a 95% prediction interval (actually 95.5%), would be given by

$$\begin{aligned} Y_c \pm 2\bar{S}_{Y.12} &= 17.00 \pm 2(2.52) \\ &= 11.96 \text{ to } 22.04 \text{ (hundreds of dollars)} \end{aligned}$$

Therefore, with the assumptions indicated, we would assert that the family's expenditures on consumer durables would be included in the interval \$1196 to \$2204. In the long run, about 95% of such statements would be expected to be correct.

The preceding calculations were given to illustrate large sample procedures. Since the sample size in our illustrative example was only ten, the following corresponding interval using t values rather than z values should be computed:

(10.66)
$$Y_c \pm t\bar{S}_{Y.12}$$

The number of degrees of freedom for the t value is $n - 3$; the number used in the estimate, $\bar{S}_{Y.12}$, in this case is $10 - 3 = 7$. In Appendix A, Table A-6 for a 95% prediction interval (column headed .05), we find this t value for 7 degrees of freedom to be 2.365. Hence, the correct small sample prediction interval for the problem illustrated earlier is

$$\begin{aligned} Y_c \pm 2.365\bar{S}_{Y.12} &= 17.00 \pm 2.365(2.52) \\ &= 11.04 \text{ to } 22.96 \text{ (hundreds of dollars)} \end{aligned}$$

Multiple Coefficient of Determination

In two-variable correlation analysis, a measure of the degree of association between the two variables is given by the coefficient of determination, r^2, which was defined in (10.31) as

$$r^2 = 1 - \frac{s^2_{Y.X}}{s^2_Y}$$

As indicated earlier, r^2 measures the proportion of variance in the dependent variable, which is explained by the regression equation relating Y to X. Another way of stating it is that r^2 measures the proportion of variance in the Y variable, which is accounted for or is associated with the independent variable X.

An analogous measure, the *coefficient of multiple determination*, denoted R^2, with appropriate subscripts, quantifies the degree of correlation that exists when more than two variables are present. For the case of one dependent and two independent variables, the coefficient of determination, uncorrected for degrees of freedom, is defined as

(10.67)
$$R^2_{Y.12} = 1 - \frac{S^2_{Y.12}}{s^2_Y}$$

where

$$S^2_{Y.12} = \frac{\Sigma(Y - Y_c)^2}{n}$$

and

$$s^2_Y = \frac{\Sigma(Y - \bar{Y})^2}{n}$$

In keeping with the usual convention, the subscript notation in $R^2_{Y.12}$ lists the dependent variable to the left of the period and the independent variables to the right. As can be seen from these definitions, $S^2_{Y.12}$ is the variance of Y values around the regression plane and s^2_Y, previously referred to in Equation (10.33) as "total variance in Y," is the variance of Y values around their mean, $\bar{Y}$. Hence, in a manner completely similar to the interpretation of r^2, we may interpret $R^2_{Y.12}$ as measuring the proportion of variance in the dependent variable, which is explained by the regression equation relating Y to X_1 and X_2. Alternatively, it measures the proportion of variance in the Y variable, which is accounted for by all of the independent variables combined.

We illustrate the calculation of $R^2_{Y.12}$ for the example with which we have been working. $S^2_{Y.12}$ can be calculated from the short-cut formula

(10.68) $$S^2_{Y.12} = \frac{\Sigma y^2 - b_1\Sigma x_1 y - b_2\Sigma x_2 y}{n}$$

Substituting, we find

$$S^2_{Y.12} = \frac{(272) - (2.5)(102) - (-0.48)(57)}{10}$$
$$= 4.436$$

The total variance, or variance around $\bar{Y}$, s^2_Y can be computed from the short-cut formula

(10.69) $$s^2_Y = \frac{\Sigma Y^2}{n} - \bar{Y}^2$$

Substituting, we obtain

$$s^2_Y = \frac{3162}{10} - 17^2$$
$$= 27.20$$

Therefore, the coefficient of multiple determination, $R^2_{Y.12}$, is equal to

$$R^2_{Y.12} = 1 - \frac{4.436}{27.20} = .837$$

Thus, we have found that 83.7% of the variance in expenditures on consumer durables has been explained by the linear regression equation, which related that variable to family income and family size. Comparing this figure to the corresponding two-variable r^2 value of .832, previously obtained in Section 10.7 for the correlation between expenditures on consumer durables and family income, we find that the value of $R^2_{Y.12}$ is only .005 or about one-half of a percentage point higher than the figure for r^2. This means that the addition of the second independent variable, family size, X_2, has explained very little of the variance in consumer durables expenditures, Y, beyond that which was already accounted for by family income, X_1, alone. As we have noted earlier, the reason for this is the high correlation between the independent variables. Once family income has been taken into account, since family size moves together with that variable, family size can only poorly explain residual variation in consumer durables expenditures.

The usual practice is to adjust the coefficient of multiple determination for degrees of freedom. The adjusted coefficient of multiple determination in the three-variable case is given by

(10.70)
$$\overline{R^2_{Y.12}} = 1 - \left(\frac{S^2_{Y.12}}{s^2_Y}\right)\left(\frac{n-1}{n-k}\right)$$

In this equation, the effect of multiplying the ratio $S^2_{Y.12}/s^2_Y$ by the factor $(n-1)/n-k)$ is to replace the n's in the denominators of $S^2_{Y.12}$ and s^2_Y as defined in (10.67) by the appropriate numbers of degrees of freedom, $n-k$ and $n-1$, respectively.

Substituting numerical values for the illustrative problem into (10.70), we find

$$\overline{R^2_{Y.12}} = 1 - \left(\frac{4.436}{27.20}\right)\left(\frac{9}{7}\right)$$
$$= .790$$

An alternative calculation for $\overline{R^2_{Y.12}}$ is given by

(10.71)
$$\overline{R^2_{Y.12}} = 1 - (1 - R^2_{Y.12})\left(\frac{n-1}{n-k}\right)$$

The calculation for the illustrative problem is

$$\overline{R^2_{Y.12}} = 1 - (1 - .837)\left(\frac{9}{7}\right)$$
$$= .790$$

In modern computer programs for multiple regression and correlation analysis, it is standard practice to present coefficients of multiple determination which have been adjusted for degrees of freedom.

Two-Variable Correlation Coefficients

From the preceding discussion of the difficulties encountered in multiple correlation analysis when independent variables are intercorrelated, it is evident that it is good practice to compute coefficients of correlation or determination between each pair of independent variables that the analyst plans to enter into the regression equation. It is standard procedure in most multiple regression and correlation analysis computer programs to present a table of correlation coefficients for every pair of variables, including the dependent as well as all independent variables.

A convenient formula for calculating the correlation coefficient in terms of deviations from means, illustrated for the variables X_1 and X_2, is

(10.72)
$$r_{12} = \sqrt{\frac{(\Sigma x_1 x_2)^2}{(\Sigma x_1^2)(\Sigma x_2^2)}}$$

Substituting figures from Table 10-5 for the illustr ative problem, we find

$$r_{12} = \sqrt{\frac{27^2}{(46)(22)}} = .85$$

Similar calculations for the other two pairs of variables give the following results

$$r_{Y1} = \sqrt{\frac{102^2}{(46)(272)}} = .91$$

$$r_{Y2} = \sqrt{\frac{57^2}{(22)(272)}} = .74$$

In the print-out of computer programs, the correlation coefficients are usually presented in the form of a triangular table, such as is shown in Table 10-6. The 1.00's along the diagonal of this table indicate that the correlation coefficient of each variable with itself is 1.00, that is, each variable is perfectly and directly correlated with itself.

Table 10-6 Correlation Coefficients for Each Pair of the Three Variables: Expenditures on Consumer Durables, Y, Family Income, X_1, and Family Size, X_2.

	Y	X_1	X_2
Y	1.00		
X_1	.91	1.00	
X_2	.74	.85	1.00

Inferences About Population Net Regression Coefficients

In the preceding discussion of correlation and regression analysis, the various equations and measures were all stated in terms of sample values, rather than in terms of the corresponding population equations and characteristics. If the assumptions given at the beginning of the discussion of the standard error of estimate are met, then appropriate inferences and probability statements can be made concerning population parameters. In multiple regression analysis, a great deal of interest is centered on the reliability of the observed net regression coefficients. Just as in the two-variable case referred to in Section 10.8, where statistical inference about the population regression coefficient β was discussed, analogous hypothesis testing and estimation techniques are available for net regression coefficients, where three or more variables are involved.

In the two-variable problem, the regression coefficient b in the equation $\bar{Y}_X = a + bX$ is an estimate of the population parameter β in the population relationship $\mu_{Y.X} = \alpha + \beta X$. Correspondingly, the net regression coefficients in a three-variable problem, b_1 and b_2 in the equation $Y_c = a + b_1X_1 + b_2X_2$, are estimates of the parameters β_1 and β_2 in a population relationship denoted $\mu_{Y.12} = \alpha + \beta_1X_1 + \beta_2X_2$. The standard errors of the net regression coefficients which represent the estimated standard deviations of the sampling distributions of b_1 and b_2 values are given by

(10.73)
$$\hat{\sigma}_{b_1} = \frac{\bar{S}_{Y.12}}{\sqrt{\Sigma x_1^2(1 - r_{12}^2)}}$$

and

(10.74)
$$\hat{\sigma}_{b_2} = \frac{\bar{S}_{Y.12}}{\sqrt{\Sigma x_2^2(1 - r_{12}^2)}}$$

where all terms in (10.73) and (10.74) have the definitions stated above.

Substituting the required numerical values, we find

$$\hat{\sigma}_{b_1} = \frac{2.52}{\sqrt{46(1 - .72)}} = .70$$

and

$$\hat{\sigma}_{b_2} = \frac{2.52}{\sqrt{22(1 - .72)}} = 1.02$$

We can test hypotheses concerning β_1 and β_2 by computing t-statistics in the usual way

(10.75)
$$t_1 = \frac{b_1 - \beta_1}{\hat{\sigma}_{b_1}}$$

and

$$t_2 = \frac{b_2 - \beta_2}{\hat{\sigma}_{b_2}}$$

These t-statistics approach normality as the sample size and number of degrees of freedom become large.

Hence, to test the hypotheses that the net regression coefficients are equal to zero, that is, that family income and family size have no effect on consumer durables expenditures, or

$$H_0: \beta_1 = 0$$
$$H_1: \beta_1 \neq 0$$

and

$$H_0: \beta_2 = 0$$
$$H_1: \beta_2 \neq 0$$

we calculate

(10.76)
$$t_1 = \frac{b_1 - 0}{\hat{\sigma}_{b_1}} = \frac{b_1}{\hat{\sigma}_{b_1}}$$

and

$$t_2 = \frac{b_2 - 0}{\hat{\sigma}_{b_2}} = \frac{b_2}{\hat{\sigma}_{b_2}}$$

In the illustrative problem, we find

$$t_1 = \frac{2.50}{.70} = 3.57$$

and

$$t_2 = \frac{-.48}{1.02} = -.47$$

The number of degrees of freedom used to look up the critical t values for this test is $n - k$, which in this case is equal to $10 - 3 = 7$. This is the number of degrees of freedom used to estimate $\bar{S}_{Y.12}$ in the calculation of $\hat{\sigma}_{b_1}$ and $\hat{\sigma}_{b_2}$. The two-tailed critical t values at the 5 and 1% level of significance are ± 2.365 and ± 3.499, respectively (Appendix A, Table A-6). Thus, since for b_1, the computed t_1 value of 3.57 exceeds the positive critical values, we conclude that b_1 differs significantly from zero at both the 5 and 1% levels of significance. Therefore, we reject the null hypothesis that $\beta_1 = 0$. The computed t_2 value of $-.47$ for b_2 means that the b_2 value lies only .47 estimated standard errors below zero. Comparing this figure of $-.47$ with the critical values of -2.365 and -3.499, we conclude that b_2 does not differ significantly from zero at either the .05 or .01 level of significance. Hence, we accept the null hypothesis that $\beta_2 = 0$.

In summary, we conclude that the effect of family income, X_1, on consumer durables expenditures, Y, has been reliably measured, but that the influence of family size, X_2, has not. This result is consistent with the previous discussion of the difficulty of measuring the separate effects of two intercorrelated independent variables.

An important point concerning the interpretation of the results of a multiple regression analysis follows from the above discussion. If the basic purpose of computing a regression equation is to make predictions of values of the dependent variable, then the reliability of the individual net regression coefficients is of no consequence. On the other hand, if the purpose of the analysis is to accurately measure the separate effects of each of the independent variables on the dependent variable, then the reliability of the individual net regression coefficients is clearly of importance.

Other Measures in Multiple Regression Analysis

A number of other measures are sometimes calculated in a multiple regression and correlation analysis. Only brief reference will be made to them here.

The coefficient of multiple determination R^2 has been described as a measure of the effect of all of the independent variables combined on the dependent variable. More specifically, this coefficient measures the percentage of variance in the dependent variable that has been accounted for by all of the independent variables combined. The square root of the coefficient of multiple determination, $R = \sqrt{R^2}$, is referred to as the *coefficient of multiple correlation*. It is always assigned a plus sign. Since some of the individual independent variables may be positively correlated with the dependent variable and others negatively correlated, there would be no meaning in distinguishing between a positive and negative value for R. Analogous to the case of r^2 and r in two-variable analysis, R^2 is easier to interpret, since R^2 is a percentage figure whereas R is not.

It is possible to calculate measures which indicate the separate effect of each of the independent variables on the dependent variable, if the influence of all the other independent variables has been accounted for. For this purpose, it is conventional to compute so-called *coefficients of partial correlation*. For example, the partial correlation coefficient for family income in our illustrative problem, designated $r_{Y1.2}$ would show the partial correlation between Y and X_1 after the effect of X_2 on Y had been removed. The square of this coefficient, $r^2_{Y1.2}$, measures the reduction in variance brought about by introducing variable X_1 after X_2 has already been accounted for.

Sometimes it is difficult to compare the differences in net regression coefficients because the independent variables are stated in different units. For example, in the illustrative example, b_1 indicates the average difference in consumer durables expenditures, Y, with a *one unit difference in family income* X_1, whereas b_2 indicates the average difference in consumer durables expenditures, Y, with a *one unit difference in family size*. (In both cases, the other independent variable is held constant.) However, one unit differences in X_1 and X_2 are in different units, namely, \$1000 and one person, respectively. For improved comparability, the regression equation can be stated in a different form, where each of the variables is given in units of its own standard deviation. The transformed net regression coefficients are termed "beta coefficients." For example, in terms of beta coefficients, the linear regression equation for three variables would be

$$\textbf{(10.79)} \qquad \frac{Y}{s_Y} = a' + \beta_1 \frac{X_1}{s_{X_1}} + \beta_2 \frac{X_2}{s_{X_2}}$$

Thus, the beta coefficients are equal to

$$\textbf{(10.80)} \qquad \beta_1 = b_1 \frac{s_{X_1}}{s_Y}$$

and

$$\beta_2 = b_2 \frac{s_{X_2}}{s_Y}$$

As an illustration of the meaning of the beta coefficients, β_1 measures the number of standard deviations that Y_c changes with each change of one standard deviation in X_1.[9]

Selected General Considerations

A great deal of care must be exercised in the use of multiple regression and correlation techniques. In the development of the model, theoretical analysis, knowledge of the field of application, and logical judgment should aid in the selection of variables to be used in the study. Frequently, in business and economic applications, some of the relevant variables may not be easily quantifiable. Sometimes variables are not readily available, but must be constructed from different sets of data.

---

[9]The reader is warned that the battle of notation must be continually fought. We have encountered the use of beta (β) in this chapter for a population regression coefficient and in the present discussion for a beta coefficient. In Chapter 7, β denoted the probability of a Type II error. The specific meaning of this symbol must be determined from the context of the particular discussion.

In the discussion of multiple regression and correlation analysis, we have confined ourselves to the case of a linear model. The underlying assumptions should be checked for their validity. Simple graphic checks involve the examination of graphs of the dependent variable against each of the independent variables at the outset of the analysis and of plots of the $Y - Y_c$ deviations against each of the independent variables after fitting the regression equation. Sometimes transformations such as taking logarithms, reciprocals, or square roots of original observations may provide better adherence to original assumptions and better fits of regression equations to the data. Of course, frequently when a linear regression equation is used, it simply represents a convenient approximation to the unknown "true" relationship. Where linear relationships are inadequate, curvilinear regression equations may be required.

The quest to provide a good fit of the regression equation to the data leads to adding more and more independent variables. However, cost considerations, difficulties of providing data in the implementation and monitoring of the model, and the search for a reasonable simple model ("parsimony") point toward the use of as few independent variables as possible. Since no mechanistic statistical procedure exists to resolve this dilemma and many other problems of multiple regression and correlation analysis, subjective judgment inevitably plays a large role.

The dangers of extrapolation must be carefully guarded against. There are subtle difficulties in multiple as compared to two-variable analysis. Even within the range of the data, certain combinations of values of the independent variables may not have been observed. This means that statistically valid estimates of the dependent variable cannot be made for these combinations of values.

The advent of modern electronic computers has opened up a greater choice in selection of variables, the inclusion of larger numbers of variables, more options in performing transformations of variables, and more testing and experimentation with different types of statistical relationships. However, because of these increased possibilities, it becomes even more important that care and good judgment be exercised to avoid misuse of methods and misinterpretation of findings. A case study of the use of a series of multiple regression models developed on a computer for a marketing managerial decision problem in the area of forecasting is given in the next section.

10.12 A Case Study

The setting of this case study involves a large pharmaceutical company which has developed a new ethical drug product and has made the critical decision to launch it on the market. An ethical drug is one which can only be purchased upon prescription from a physician. A heavy investment has already been made in the development of the product, through the various stages of preliminary screening, more detailed evaluation, and final preparation for commercialization. Once the decision to market the product has been made, the marketing executive must decide on such matters as the amount of promotional spending, the allocation of the budget over a period of time, the amount of sales effort to be invested, and the duration of any special support for the project. Many other important decisions are initiated in other departments of the company, such as production, purchasing, and finance, as operations are geared to adjust to the new requirements. In this type of situation it is crucial that good short-term demand forecasts be provided in the initial months of marketing, say, for the first year of the product's life. Furthermore, it is important to know the expected demand levels at various points during this first year in order to plan the distribution of the promotion budget, the allocation of sales effort, and the timing of various types of special support for the product. This case study deals with the development of a sales forecasting model system based on the methodology of multiple regression analysis. The approach of this discussion is to indicate the general nature of project development. Thus, greater insight is given to some of the broader problems of the use of quantitative models in managerial applications than is possible when attention is directed solely to the specific mathematical or statistical techniques employed.

Let us consider first of all why any formal forecasting system was desired in the above situation. Typically, in the past, during the first few months that a new product was on the market, marketing management had found itself riddled with uncertainties and confronted with a flood of puzzling and often conflicting information. There was a flow of data, primarily from pharmaceutical market research firms, concerning the product's sales during these early months, the reactions of prescribing physicians, the success of competitive products, and the comparative promotional efforts exerted by the company and its competitors. However, some of this information was of dubious utility, and some gave conflicting evidence. How should the company sift this information, extract the important items, and make the correct decisions concerning the potential future success of the product?

A considerable amount of effort went into the problem definition phase and the determination of management objectives. It was decided that relatively short-term forecasts were desired for short-run decision-making purposes as opposed to long-term forecasts for long-run, broader planning purposes. Marketing management had to decide whether it was interested in predictions in dollars or in number of physical units, whether it required forecasts for periods of time, as for example, the last quarter of the first year after introduction, or at "time points," as for example, the monthly rate of sales during the twelfth month, or both. These basic definitional problems were resolved primarily from the viewpoint of the informational needs of the marketing vice president. In the very early months that the new product was on the market, this executive needed good short-term forecasts of its demand, say for the last quarter of the first year of the product's life and for the entire first year. Furthermore, it was important for him to know the expected levels of demand at various points during this first year in order to make decisions on the distribution of the promotional budget, the allocation of sales effort, etc.

It was recognized that the crucial point in these early forecasts was that an unsuccessful product be recognized quickly, and that a successful high-volume product be similarly discerned. The critical matter for decision making purposes is to place the product in its proper size class rather than to estimate dollar sales with extreme precision. Then, as time progresses during the year following introduction and as more current information becomes available, the marketing executive needs revised estimates of the new product's market position.

The first stage of the project involved a detailed investigation to determine the *feasibility* of constructing a sales forecasting model to operate during the early months after product introduction—a system which would provide early projections of future demand. A number of alternative methodological approaches were reviewed and rejected. Due to the nature of the case, traditional time series methods were not feasible because there was no past sales history of the new product which might be analyzed and projected. The technique finally selected was a series of least squares multiple regression equations, with the model using source data available during the early months of the new product's life.

In Chapter 5, we referred to the idea that in any statistical inquiry it is useful for the investigators to consider at the design stage what the ideal experiment might be. In the project under discussion, it was fruitful to consider the ideal regression equation for predicting new product sales. On the left-hand side of the equation would be the dependent variable, in this case, new product sales, say, for the fourth quarter of the first year after introduction. On the right-hand side would be variables which represent the determinants of this demand such as the underlying medical and economic factors, market variables which measure the relative preference for this product versus its competitors, etc. Probably, if a group of knowledgeable medical and marketing persons were assembled, they would not be able to agree on what these determinants were. However, even if agreement could be reached, the relevant information would simply not be available in a form suitable for inclusion in the type of model under discussion. Hence, as is invariably the case in such investigations, compromises with the ideal situation had to be made because of practical considerations.

It was decided that the methodology, similar to other forecasting tools, would use historical information. In this case the information was the data for other new pharmaceutical products, for the variables included, over a reasonable prior period of time. By measuring various factors related to sales of all new products, the existing relationships could be measured statistically. The specific variables for which data were obtainable could not all be considered actual determinants of sales. However, some of the variables, such as sales of the product in the first few months of the product's life reflected the operation of the underlying determinants of demand, and thus might represent good predictive factors.

The basic rationale of the regression equation approach was that if for all new products introduced over a substantial period of time there was a strong relationship between demand levels and the specific variables considered during the early months of a product's life, and if this relationship persisted, then the probable market activity for the new product currently under consideration could be estimated from the corresponding early-month variables.

Another comment on practical expedients may be noted. Why use a data base which consisted of all new products, that is, all types of pharmaceutical products for all companies? It would seem more reasonable to use only information about other new products which were similar to this one in some sense as the basic data from which to construct the model. For example, in the pharmaceutical industry, products are classified into more or less homogeneous groupings known as "therapeutic classes." Hence it might appear reasonable to use as a data base only figures relating to those new products which were in the same therapeutic class as this one and which were placed on the market during the past few years. This procedure proved not to be feasible, because there simply were not enough new products in the same therapeutic class to provide an adequate data base for the multiple regression analysis. The possibility of using only data concerning all new products produced by this particular company had to be rejected on the same grounds of insufficient data. Therefore, it was decided that the historical data base would consist of information on all new pharmaceutical products that had been placed on the market during the past few years. Intensive testing was carried out to evaluate the feasibility of constructing a workable forecasting system from this type of data base.

It was important in the initial phase of the study to define and evaluate the specific variables to be included in the model. The first step in this connection was an itemization of all candidate variables, dependent and independent, based on the subjective judgments of marketing research and other personnel. Knowledge of the marketing environment was essential for this task. The unit of association for the proposed regression analyses was the new product. A review of the sources of the necessary historical information revealed that considerable pertinent data were available for new products for the prior few years.

Approximately 30 factors were defined for possible inclusion in the model. Five different dependent variables were defined such as sales of the product during the fourth quarter of the first year of the product's life, referred to as "Fourth Quarter Sales"; another was "First Year Sales," etc. The 25 candidate independent variables included current market activity for new products, changes in market activity during the early months, attitudinal factors, and promotional expenditures. For example, among the factors considered were purchases of new products by drug stores, prescription activity, market-share measures, stocking and repurchase patterns of drugstores, measures of physician awareness and use, attitudes of physicians toward future use, expenditures on different types of advertising, and a variable which isolated differences in characteristics of certain product groups. In this as in other studies, the figures for a number of variables were not directly available, but had to be constructed from available data. For example, as indicated above, it was felt that promising independent variables for predictive purposes would include market-share measures. Such measures were not directly available, but had to be constructed. The market-share of a new product was defined basically as the proportion of total sales in its therapeutic class accounted for by that product. In some cases, it was necessary to redefine the therapeutic class to include other products which were competitive with the new product but were not encompassed in the standard definition of the class. It was also hypothesized that month-to-month change in market-share during the early months of the new product's life might be a good predictive factor, so this variable had to be calculated.

In this study, the data search and collection phase was time consuming and costly. Hence, numerous practical decisions were required. If two similar variables were available, but the data for one were costly to obtain relative to the second, figures were compiled only for the first. Careful checks had to be made of the consistency in definition and reporting of each variable over time and between different sources. In many cases, computational adjustments were required to obtain the necessary consistency. Data had to be checked for completeness. For example, certain variables had to be dropped from consideration because available data were incomplete. Clerical forms had to be devised for the recording and checking of the data. Subsequently the data were transferred onto computer tapes.

The preliminary experimentation and evaluation phase consisted of a series of computer runs in which a large number of multiple regression equations were calculated. The purposes of these test runs were to determine (a) the overall predictability of future product demand and the relative accuracy of estimates and (b) the relative importance of the individual variables considered. The tests consisted of experiments with various combinations of dependent and independent variables for different time periods. Specifically, a series of multiple regression equations were calculated from the data; that is, a set of little models was constructed on a test basis.

A computer program for stepwise multiple regression analysis which we will now describe was used to derive the prediction equations. In order to discuss the nature of this procedure, let us assume we have one dependent variable and ten possible independent variables, denoted Y and $X_1, X_2, \ldots, X_{10}$, respectively. The computer first calculates all of the two-variable correlation coefficients between every pair of variables. Then, it selects that independent variable which explains the greatest proportion of variance in the dependent variable. Assume that the independent variable is X_4. The least squares regression equation between Y and X_4, of the form $Y_c = a + bX_4$ is calculated along with the related standard errors and other measures. The computer then seeks among the remaining nine independent variables for the one which accounts for the greatest percentage of the unexplained variance in the dependent variable, Y. Assume that variable is X_6. The least squares regression equation between Y, X_4, and X_6 of the form $Y_c = a' + b_4X_4 + b_6X_6$ is then computed along with the supplementary measures. The program thus proceeds in a stepwise fashion, always including as the next independent variable the one which accounts for the greatest percentage of remaining unexplained variance. A cut-off point for inclusion of variables can be predetermined in terms of certain standard significance tests. A variety of stepwise and other types of computer programs for multiple regression analysis is available, and improvements in these procedures are constantly being made.

Scores of computer runs were made during the test and evaluation stage. This is a type of model construction procedure that simply would not have been feasible prior to the advent of modern electronic data processing equipment. The evaluation of the test results was very encouraging. The degree of prediction accuracy appeared to be very high, as reflected by the standard errors of estimate and correlation measures. There was indication of decided month-to-month improvement in forecasting ability during the early months. Better predictions were made for aggregate sales over a time period as compared to levels of sales at points in time. The reason for this was that the former type of variable is more stable and therefore easier to predict; seasonal and random factors tend to be offset over a longer time span. It was possible to draw certain conclusions about the individual variables under consideration. For example, it became apparent that the factors reflecting actual market activity in the early months were better predictors than those reflecting attitudes and demand generation effort. On the basis of the results obtained in the initial feasibility study, the decision was made to proceed with the next phase, the implementation of the model.

The implementation stage represented an extension of the initial effort culminating in the development of a final working model. A basic goal was a more detailed analysis of the variables under consideration in order to select the combinations which furnished the highest degree of predictability, and provided relatively simple and manageable model equations. The latter criteria were established so that the final model would be relatively easy to maintain and would be less vulnerable to potential future data inadequacies. As a result of considerable experimentation and analysis, an "initial forecasting system" was developed comprising five individual forecasts of aggregate demand for time periods and levels of demand at points in time. This system was then tested with prior new products, to see how it would have performed had it been available. On the basis of these past relationships, there appeared to be good predictability of future sales of new pharmaceutical products from factors available early in the lives of these products.

The next phase of the implementation study consisted of the construction of a system to produce revised forecasts as the time since product introduction increased. Essentially this system consisted of a series of regression equations based upon more current information than was included in the initial forecast. This forecasting system was deemed necessary because of the volatile nature of markets for new products. Important changes in market conditions frequently occur between the time at which the forecast is made and the time or period for which it is made.

A maintenance and control system should be established as part of the overall construction of any model system. In this case, this entailed periodic updating of the model by inclusion of new data and recalculation of coefficients in the forecasting system. A continuous review of the basic data was required. Source data should be monitored for the presence of any possible changes in definition, inadequacies, or atypical situations. If new factors become predominant, new source data become available, or other similar conditions evolve, it is appropriate to test for the usefulness of inclusion of these new variables in the system.

The forecasting model system proved its worth by providing accurate and useful predictions, and it made a number of valuable contributions to marketing management. The use of a formal model for forecasting purposes removed the elements of emotion and vested-interest influences from the forecast and based it on the most appropriate available current and past information. Furthermore, since the model provided a quantitative statement of the relationship of future sales to a variety of factors, it was possible to estimate the probable effects on sales of changes in these factors and the interactions among these factors by manipulating the model. Despite the possibilities of error in the model, the marketing executive is doubtless in a better position when he can combine judgment with the insights generated by this type of model than when he must rely on unaided judgment alone.

PROBLEMS

1. Scatter diagram and regression line showing relationship between earnings in 1968 and price per share at end of 1968 for selected common stocks.

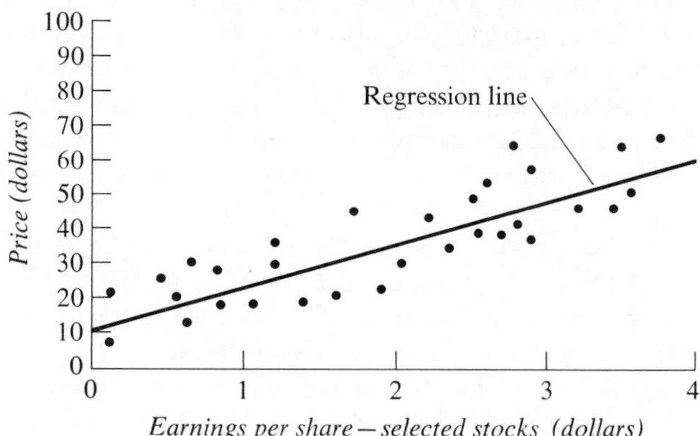

On the basis of the above chart, estimate the values of a and b in the equation of the line of regression $\bar{Y}_X = a + bX$. Use specific numbers.

2. A personnel director devised a test of manual dexterity which he wished to use to screen job applicants in order to predict performance in assembly work. To determine the effectiveness of the test he selected a sample of 50 workers in the assembly department, determined their outputs per hour and their scores on the test. The results of a regression analysis of his data were as follows:

   $\bar{Y}_X = 19.7 + 0.6X$
   X, the test scores, ranged from 25 to 74
   Y ranged from 31 to 59 units of output per hour
   $s_Y = 10.00$
   $s_{Y.X} = 6.0$

   (a) Would you be willing to conclude that a relationship exists between test scores and units of output per hour? Justify your answer statistically, using a .05 risk of a Type I error.
   (b) Assuming a relationship exists between the two variables, estimate with 95% confidence the output per hour of a job applicant who scored 50 on the dexterity test.

3. As personnel director of a large manufacturing firm, you are given the following information collected from a simple random sample of employees:

   $\bar{Y}_X = 50 + .4X$

X is score on aptitude test
Y is the quality rating by company officials at the end of two years of service
Unit of association is employee
X varies between 40 to 110
$n = 100$
$s_Y^2 = 50$
$s_{Y.X}^2 = 20$

(a) Do you believe there is any correlation (for all employees of the firm) between their score on the aptitude test and the quality rating by company officials at the end of two years of service? Justify statistically.
(b) In terms of this problem, explain precisely the meaning of the regression coefficient, $b = +.4$.
(c) Do you think an employee who scored 125 on the aptitude test will receive a quality rating (after two years of service) in excess of 100? Why or why not?
(d) Calculate r^2 and interpret your answer.

4. Explain in words the meaning of each of the following statements:
(a) $s_{Y.X}$ is usually smaller than s_Y.
(b) A high r^2 does not imply that an estimating equation will make excellent predictions.

5. A competent research worker concluded that two variables were correlated in the universe. Another competent research worker, using the same sample information, was unwilling to conclude that the two variables were correlated in the universe. Does this seem possible to you? Why or why not?

6. A least squares linear regression and correlation analysis was conducted on the data obtained for a simple random sample of 102 salesmen of the International Conglomerate Company with the following results:

$\bar{Y}_X = -10 + X$
X is age in years
Y is annual commissions in thousands of dollars
X ranges between 30 and 50 years
$n = 102$
$s_Y = 5$ (thousands of dollars)
$s_{Y.X} = 3$ (thousands of dollars)

(a) The sales manager objected to the results of the equation, because the value $a = -10$ did not seem reasonable to him. Explain briefly how you would reply to the sales manager.
(b) Interpret the regression coefficient, $b = 1$, *specifically in terms of this problem.*
(c) Would you conclude that there is any correlation *for all 30-to 50-year-old salesmen* of this company between age and annual commissions? Carry out the necessary computations to justify statistically.
(d) Would a 40-year-old salesman who received \$18,000 in annual sales

commissions be considered a "poor performer" or just an average salesman? Why?

7. A sample survey of 102 families gave a linear regression equation of

$$\bar{Y}_X = 8.0 - 0.2X$$

where Y was the percentage of income spent for medical care and X was the family income *in thousands of dollars*. Individual family incomes in the sample ranged from \$2500 to \$15,000. The following values were calculated:

$$s_Y = 0.6\%$$
$$s_{Y.X} = 0.4\%$$

(a) Compute the coefficient of determination, r^2, and state its meaning in terms of this problem.
(b) To draw inferences about the relationship between X and Y for all families in the universe, one could test the significance of r. What is the null hypothesis for this test? Distinguish clearly the conclusions made when the null hypotheses is (1) accepted and (2) rejected.
(c) A family with an income of \$10,000 spent \$1000 on medical care. Should this be considered an unusual expenditure? Justify statistically.

8. Based upon a sample of 32 of its salesmen, a large corporation finds the following relationship:

$$\bar{Y}_X = -4 + 20X$$
$$s_Y = 90$$
$$s_{Y.X} = 30$$

where X = number of calls on prospects by salesmen
Y = sales made, in hundreds of dollars
n = 32

(a) Explain precisely the meaning of the -4 and the 20 in the above equation.
(b) Calculate and interpret r^2 precisely in terms of this problem.
(c) Test the significance of r. Indicate clearly the hypothesis that is tested and give the meaning of your conclusion in terms of the problem.
(d) Distinguish clearly between the measures, s_Y, $s_{Y.X}$, and the standard error of r with particular reference to the distributions involved.

9. The following data represent observations on ten families randomly selected in a given poverty area:

Family	*Clothing Expenditures (Y) (hundreds of dollars)*	*Income (X) (hundreds of dollars)*	*Also*
A			$\Sigma X^2 = 10{,}000$
.	.	.	
.	.	.	$\Sigma Y^2 = 750$
.	.	.	
J			$\Sigma XY = 15{,}000$
	42	200	

(a) Estimate the constants of a linear equation from which family clothing expenditures can be estimated, given family income.
(b) What is the meaning of the term, "method of least squares"?

10. Assume that a random sample of 40 families in a certain city produced the following calculations and results on X, family income, in thousands of dollars, and Y, beef consumption, in pounds per week:

$$\Sigma X = 170.0 \qquad \Sigma Y = 169.1$$

$$\Sigma X^2 - \frac{(\Sigma X)^2}{n} = 285.0 \qquad \Sigma XY - \frac{(\Sigma X)(\Sigma Y)}{n} = 118.0$$

$$\Sigma Y^2 - \frac{(\Sigma Y)^2}{n} = 55.0 \qquad s_{Y.X} = .39$$

(a) Compute the constants of the least squares regression line, and interpret carefully the regression coefficient, b.
(b) Construct a prediction interval, using a .95 confidence coefficient, for the number of pounds of beef consumed by a given family with an income of $5000.
(c) Test the hypothesis that $\beta = 0$. What conclusion is reached? What does this imply about ρ^2?
(d) Briefly explain why the variance about the regression line cannot be greater than the variance about the mean of Y. (Ignore the $n - 1$ and $n - 2$ adjustments.)

11. Assume that from a random sample of 102 new products brought to the market, the following least squares regression equation was determined:

$$\bar{Y}_X = 1.0 + 4.0X$$

Y is "demand" measured by first year sales in millions of dollars
X is "awareness" measured by the proportion of consumers who had heard of the product by the third month after introduction of the product; $0 \leq X \leq 1$

$$s_Y^2 = 1$$
$$s_{Y.X}^2 = 0.2$$

(a) Would you be willing to conclude that a statistical relationship exists between demand and awareness? Justify your conclusion statistically using a .05 risk of Type I error.
(b) Explain in your own words the meaning of a Type I error specifically in terms of this problem.
(c) Assuming a relationship between demand and awareness, estimate with 98% confidence the demand for a new product for which 30% of consumers had heard of the product by the third month after introduction. You may assume normal conditional probability distributions for demand.
(d) Criticize or explain the following statements, which pertain to regression analysis in general:
(1) $s_{Y.X}$ is usually smaller than s_Y
(2) The regression coefficient, b, measures the amount of change in Y per unit change in X

12. The following results were obtained by correlating dollar volume of sales, Y, with number of employees, X, for a random sample of 66 marketing establishments of a given type. These firms had between 10 and 100 employees.

$$\bar{Y}_X \text{ (thousands of dollars)} = 40 + 12X$$
$$s_{Y.X} = 4 \text{ (thousands of dollars)}$$
$$r = +0.9$$

   (a) Interpret specifically in terms of the problem the value of the regression coefficient, $b = 12$.
   (b) Would you be willing to estimate the dollar volume of sales for an establishment with 200 employees? Discuss briefly.
   (c) Is it reasonable to conclude that the degree of correlation between volume of sales and number of employees is attributable to chance variation? Demonstrate statistically.

13. An insurance company wished to examine the relationship between income and amount of life insurance held by heads of families. The company drew a simple random sample of ten family heads and obtained the following results:

Family	*Amount of Life Insurance* (*$000 omitted*)	*Income* (*$000 omitted*)
A	9	4
B	20	8
C	22	9
D	15	8
E	17	8
F	30	12
G	18	6
H	25	10
I	10	6
J	20	9

   (a) Determine the linear regression equation using the method of least squares with income as the independent variable.
   (b) What is the meaning of the regression coefficient b in this case?
   (c) Test the hypothesis that the population regression coefficient is equal to zero. State your conclusion.
   (d) Compute the estimated standard error of estimate $\hat{\sigma}_{Y.X}$.
   (e) What is your estimate of the amount of life insurance carried by a family head from the same population whose income is $10,000? Give a 95.5% prediction interval around this estimate. For simplicity, assume the sample is large, although n is only equal to ten. Use the standard error calculated in (d).

14. The insurance company in Problem 13 decided to add a second independent variable, size of family, to the analysis. The following results were obtained:

Family	Y Amount of Life Insurance (\$000 omitted)	X_1 Income (\$000 omitted)	X_2 Family Size
A	9	4	3
B	20	8	4
C	22	9	5
D	15	8	3
E	17	8	3
F	30	12	7
G	18	6	2
H	25	10	6
I	10	6	3
J	20	9	4

(a) Determine the linear multiple regression equation using the method of least squares with income and family size as the independent variables.
(b) What are the meanings of the net regression coefficients, b_1 and b_2, in this case?
(c) Test the hypotheses that the net regression coefficients are equal to zero.
(d) Compute the standard error of estimate and the coefficient of multiple determination, both unadjusted and adjusted for degrees of freedom.

15. The sales department of a large manufacturing firm hired ten college graduates. Before training they each took two different aptitude tests. They were then given a year of training after which they were rated by a committee. The following data were the results of the tests and rating:

Trainee	Test I Score X_1	Test II Score X_2	Final Rating Y
A	74	40	91
B	59	41	72
C	83	45	95
D	76	43	90
E	69	40	82
F	88	47	98
G	71	37	80
H	69	36	75
I	61	34	74
J	70	37	79

(a) Compute the multiple linear regression equation by the method of least squares to estimate final rating from Test I score and Test II score.
(b) What is the meaning of the net regression coefficient b_1 in this particular case? How would this value differ in meaning from the regression coeffi-

cient in simple correlation between Test I score and final rating alone?

(c) Compute the adjusted standard error of estimate and interpret its meaning.

(d) Compute the adjusted coefficient of multiple determination and interpret its meaning.

(e) Estimate the final rating for an individual who scored 70 on Test I and 37 on Test II.

CHAPTER ELEVEN

Time Series

11.1 Introduction

The decisions of business management determine whether a firm will thrive and expand or whether its position will deteriorate, and in the worst instance, pass out of existence. These vital decisions are based on perceptions of future outcomes which will affect the payoffs of possible alternative courses of action. Since these outcomes occur in the future, they must be forecast. Thus, a financial manager in choosing among alternative investments for his funds must in some way assess the relative future profitabilities of these investments. A marketing manager, in choosing marketing strategies for his products, must make some assessment of future amounts and nature of demand for these products. Not only must businessmen forecast, but they must plan and think through the nature of the activities which will permit them to accomplish their objectives. However, it is clear that business planning and decision making are inseparable from forecasting. Prediction of sales, earnings, costs, production, inventories, purchases, and capital appropriations is the basic foundation of corporate planning and control.

There is considerable variation in company practice regarding methods of forecasting. For example, in customer demand forecasting, the range of prevailing methods includes informal "seat of the pants" estimating, executive panels and composite opinions, consensus of sales force opinions, combined user responses, statistical techniques, and various combinations of these methods.

Because of such factors as the stepped-up complexity of business operations, the need for greater accuracy and timeliness, the dependence of outcomes on so many different variables, and the demonstrated utility of the techniques, management is increasingly turning to formal models, such as those provided by statistical methods, for assistance in the difficult task of peering into the future. A very widely applied and extremely useful set of procedures is *time series analysis*. A time series is a set of statistical observations arranged in chronological order. There are two different kinds of time series. In the first, the items in the series represent a set of *independent observations* from the same probability distribution, taken over time. In such a series the value of an observation at any particular point in time does not depend upon the value of preceding observations. As an example of this type of series, let us consider an experiment in which we roll a pair of dice many times. If we note the order of the rolls, i.e., 1, 2, 3, . . . , and the corresponding totals obtained on these rolls, say, 8, 4, 12, . . . , we have recorded such a time series.

In the second type of time series, the one with which we are concerned in this chapter, the values of successive observations *are not independent* of the values of preceding observations. Examples include a weekly series of end-of-week stock prices, a monthly series of steel production, and an annual series of national income. In each of these series, the level at any point in time depends upon levels achieved in preceding periods. Such time series are essentially historical series, whose values at any point in time are the resultants of the interplay of large numbers of diverse economic, political, social, and other factors. New factors may enter, old ones may recede, and generally, there is a continual shifting in importance of the effect of the various factors. This means that in the case of time series dealing with economic, social, and political phenomena, the search for regularities is difficult because of modification in underlying institutions, structures, and processes.

A first step in the prediction of any series involves an examination of past observations. Time series analysis deals with the methods for analyzing these past data and for projecting them to obtain estimates of future values. The traditional or "classical" methods of time series analysis, which we will primarily discuss, are descriptive in nature and do not provide for probability statements concerning future events. On the other hand, econometric methods, which are beyond the scope of this text and are referred to only very briefly, are based on probabilistic models. Furthermore, these models attempt to spell out the explanatory variables and mechanisms which produce a given time series. However, at the present state of the art, it appears fair to state that the superiority of econometric models over less sophisticated forecasting techniques has yet to be demonstrated. Also, no completely suitable probabilistic methods of analyzing time series have been as yet constructed and perhaps never will. The time series models we shall discuss, although admittedly only approximate and not highly refined, have proven their worth when cautiously and sensibly applied. It is very important to realize that these methods cannot simply be used mechanistically but must at all times be supplemented by sound subjective judgment.

Although the preceding discussion has referred to the use of time series analysis for the purpose of forecasting and its usefulness for planning and control, these procedures also are often used for the simple purpose of historical description. Hence, for example, they may be usefully employed in an analysis in which interest centers upon the comparative differences in the nature of variations in different time series. The general nature of the classical time series model is described in the next section.

11.2 The Classical Time Series Model

If we wished to construct a mathematical model of an economic time series that was ideally satisfying from a philosophical point of view, we might seek to define and measure the many determinants of the variations in the time series and then proceed to state the mathematical relationships between these determinants and the particular series in question. However, the determinants of change in an economic time series are multitudinous, including such factors as changes in population, consumer tastes, technology, investment or capital-goods formation, weather, customs, and numerous other variables both of an economic and non-economic nature. The enormity and impracticability of the task of measuring all of the aforementioned factors and then relating them mathematically to an economic time series whose variations we wish to account for and predict militates against the use of this direct approach to time series analysis. Hence, it is not surprising that a more indirect and practical approach, such as classical methodology, has come into use. Classical time series analysis is essentially a descriptive method which attempts to break down an economic time series into distinct components which represent the effects of the operation of groups of explanatory factors such as those listed earlier. These component variations are

(1) Secular trend
(2) Cyclical fluctuations
(3) Seasonal variations
(4) Irregular movements

Secular trend refers to the smooth upward or downward movement which characterizes a time series over a long period of time. This type of movement is particularly reflective of the underlying continuity of fundamental demographic and economic phenomena. The word secular is derived from the Latin word *saeculum*, meaning a generation or age. Hence, secular trend movements are thought of as long-term movements, usually requiring a minimum of 15 or 20 years to describe. Secular trend, or simply trend as it is often referred to, is conceived of as a stable movement and requires a different type of model for its description than do the other components of time series.

Secular trend movements are attributable to factors such as population change, technological progress, and large-scale shifts in consumer tastes. For example, if we could examine a time series on the number of pairs of shoes produced in the United States extending annually, say, from the 1700's until the present, we would find an underlying trend of growth throughout the entire period, despite fluctuations around this general upward movement. If we compare recent figures with those near the beginning of the series, we find the recent numbers are much larger because of the increase in population, because of the technical advances in shoe producing equipment enabling vastly increased levels of production, and because of shifts in consumer tastes and levels of affluence which have meant a larger per capita requirement of shoes than in earlier times.

Cyclical fiuctuations, or business cycle movements, are recurrent up and down movements around secular trend levels which have a duration of anywhere from about 2 to 15 years. The duration of these cycles can be measured in terms of their turning points, or in other words, from trough to trough or peak to peak. These cycles are recurrent rather than strictly periodic. The amplitude and duration of cyclical fluctuations in industrial series differ from those of agricultural series, and there are differences within these categories and within individual series. Hence, cycles in durable goods activity generally display greater relative fluctuations than consumer goods activity and a particular time series of, say, consumer goods activity may possess business cycles which have considerable variations in both duration and amplitude.

Economists have produced a large number of explanations of business cycle fluctuations including external theories which seek the causes outside the economic system and internal theories which are in terms of factors within the economic system which lead to self-generating cycles. In theories of the latter type, periods of contracted business activity are viewed as containing within themselves the seeds or determinants of the following period of expansion, which similarly breeds the following period of contraction. In this connection, the terminology generally used is that of the late Wesley Mitchell of the National Bureau of Economic Research who distinguished various phases of the cycle. The period of *expansion* ends at the *peak*, or upper turning-point, moves into *contraction* which terminates at the lower turning-point, the *trough* or revival. Then these phases repeat themselves, albeit with different duration and amplitude. Most economists take an eclectic stand, admitting that there are both internal and external components in any comprehensive explanation of business cycle activity.

Since it is clear from the foregoing discussion that there is no single simple explanation of business cycle activity and that there are different types of cycles of varying length and size, it is not surprising that no highly accurate method of forecasting this type of activity has been devised. Indeed, no generally satisfactory mathematical model has been constructed for either the description or forecasting of these cycles, and perhaps never will be. Therefore, it is not surprising to find that classical time series analysis adopts a relatively rough approach to the statistical measurement of business cycle. The approach is a residual one; that is, after trend and seasonal variations have been eliminated from a time series, by definition, the remainder or residual is treated as being attributable to cyclical and irregular factors. Since the irregular movements are by their very nature erratic and not particularly tractable to statistical analysis, no explicit attempt is usually made to separate them from the cyclical movements, or vice versa. However, the cyclical fluctuations are generally large relative to these irregular movements and ordinarily no particular difficulty in description or analysis arises from this source.

Seasonal variations are periodic patterns of movement in a time series. Such variations are considered to be a type of cycle which completes itself within the period of a calendar year and then continues in a repetition of this basic pattern. The major factors in producing these annually repetitive patterns of seasonal variations are weather and customs, where the latter term is broadly interpreted to include observance of various holidays such as Easter and Christmas. Series of monthly data or quarters of the year are ordinarily used to examine these seasonal variations. Hence, regardless of trend or cyclical levels, one can observe in the United States that each year more ice cream is sold during the summer months than during the winter, whereas more fuel oil for home heating purposes is consumed in the winter than during the summer months. Both of these cases illustrate the effect of weather or climatic factors in determining seasonal patterns. Also, department store sales generally reveal a minor peak during the months in which Easter occurs and a larger peak in December, when Christmas occurs, reflecting the shopping customs of consumers associated with these dates.

Of course, changes can occur in seasonal patterns because of changing institutional and other factors. Hence, a change in the date of the automobile show can cause a change in the seasonal pattern of automobile sales. Similarly, the advent of refrigeration techniques with the corresponding widespread use of home refrigerators has brought about a change in the seasonal pattern of ice cream sales. The techniques of measurement of seasonal variations which we will discuss are particularly well suited to the measurement of relatively stable patterns of seasonal variations, but can be adapted to cases of changing seasonal movements as well.

Irregular movements are fluctuations in time series which are erratic in nature, and follow no regularly recurrent or other discernible pattern. These movements are sometimes referred to as *residual variations*, since, by definition, they represent what is left over in an economic time series after trend, cyclical, and seasonal elements have been accounted for. These irregular fluctuations result from sporadic, unsystematic occurrences such as wars, earthquakes, accidents, strikes, and the like. Whereas in the classical time series model, the elements of trend, cyclical, and seasonal variations are viewed as resulting from systematic influences leading to either gradual growth, decline, or recurrent movements, irregular movements are considered to be so erratic that it would be fruitless to attempt to describe them in terms of a formal model. Irregular movements can result from a large number of causes of widely differing impact.

11.3 Description of Secular Trend

As pointed out in the preceding section, the classical model involves the separate statistical treatment of the component elements of a time series. We shall begin our discussion by indicating how the description of the underlying secular trend is accomplished.

Before the trend of a particular time series can be determined, it is generally necessary to subject the data to some preliminary treatment. The amount of such adjustment depends somewhat on the time period for which the data are stated. For example, if the time series is in monthly form, certain reconciliations for calendar differences may be required. Hence, rather detailed procedures have been worked out to adjust for the shifting date of Easter between March and April. For certain economic series, for example, department store sales, measures of seasonal variations could be substantially influenced by the calendar month in which Easter falls. Also, it is often necessary to revise the monthly data to take account of the differing number of days per month. This may be accomplished by stating the data for each month on a per day basis by dividing the monthly figures by the number of days in the respective months, or by the number of working days per month.

Even when the original data are in annual form, which is often the case where primary interest is centered on the long term trend of the series, the data may require a considerable amount of preliminary treatment before a meaningful analysis can be carried out. Sometimes when the figures are in dollars, but the concern of the analysis is upon changes in physical volume, it is necessary to "deflate" the dollar value series by dividing it by an appropriate index number series. The rationale and procedures for this adjustment are discussed in the next chapter. Adjustments for changes in population size are often made by dividing the original series by population figures to state the series in per capita form. Frequently, comparisons of trends in these per capita figures are far more meaningful than corresponding comparisons in the unadjusted figures.

It is particularly important to scrutinize a time series and adjust it for differences in definitions of statistical units, the consistency and coverage of the reported data, and similar items. Some adjustments are relatively easy to make. Others are difficult and time-consuming. For example, if a time series is given in units of short tons up to a certain date and in long tons thereafter, the adjustment is obvious and simple to perform. On the other hand, if one is working with a time series of sales of a particular type of store, say "general stores," and the agency which produced these data changed the definition of a general store several times and had varying coverages, sometimes excluding firms below a certain size while including them at other times, the adjustments may be exceedingly difficult to make. It is important to realize that one cannot simply proceed in a mechanical fashion to analyze a time series. Careful and critical preliminary treatment of such data is required to insure the meaningfulness of the results.

Purposes for Fitting Trend Lines

The secular trend in a time series can be measured by the free-hand drawing of a line or curve which seems to fit the data, by fitting appropriate mathematical functions, or by the use of moving average methods. Moving averages are discussed later in this chapter in connection with seasonal variations.

A free-hand curve may be fitted to a time series by visual inspection. When this type of characterization of a trend line is employed, the investigator is usually interested in a quick description of the underlying growth or decline in a series, without any careful further analysis. In many instances, this rapid graphic method may suffice. However, it clearly has certain disadvantages. Different investigators would surely obtain different results for the same time series. Indeed, even the same analyst would probably not sketch in exactly the same trend line in two different attempts on the same series. This excessive amount of subjectivity in choice of a trend line is especially disadvantageous if further quantitative analysis is planned. For example, if one's purpose is to study cyclical fluctuations, a trend line may be fitted to a time series only for the purpose of eliminating trend. In this case, it is useful to have a mathematical method of describing the trend. Of course, this is not meant to imply that subjectivity is not present in the use of mathematically determined trend lines. In fact, subjective judgment plays an important part in that case too. However, at least if different investigators agree on the fitting of a specified type of mathematical trend line to the same set of data, they would obtain the same results. Other disadvantages of "free-hand" subjective curves are that analysts can impose their individual biases concerning trend configurations, and there is no known method of measuring or accounting for these biases. Also, in the case of free-hand trend lines, there is no explicit criterion of a "best fitting" line, as there is in the case of trend lines fitted by the method of least squares (see Section 11.4). Although other advantages might be cited for mathematically fitted trend lines, if a simple approach such as the free-hand method appears appropriate it should be used. That is, if the purpose of the analysis requires only a very rough estimate of the form of the trend, a free-hand curve may suffice. On the other hand, if more exact methods seem to be required, the fitting of mathematical functions to describe trend may be utilized. In the ensuing discussion, we will concentrate on mathematically fitted trend lines.

Even in the case of the mathematical measurement of secular trend, the purpose of the analysis is of considerable importance in the selection of the appropriate trend line. Several different types of purposes can be specified.

1. Trend lines may be fitted for the purpose of historical description. In this case interest is focused upon such matters as the nature of the underlying trend during the period studied, the rates of growth or decline and acceleration or deceleration in this growth or decline. Comparisons may be made of the historic growth or decline in a particular series with that of others.

If the purpose is historical description, any good fitting line will in general suffice. The line need not have logical implications for forecasting purposes, nor should it be evaluated primarily by characteristics which might be desirable for other purposes.

2. A second purpose is that of prediction or projection into the future. In this case, particularly if long-term projection is desired, the selected line should have logical implications when it is extended into the future. Hence, for example, a straight line of the form $Y = a + bX$, where X is time and Y is the variable to be projected, does not ordinarily possess such logical char-

acteristics. The equation of a straight line embodies within it the assumption of a constant growth or decline of b units per time period with no upper or lower limit. That is, if the sign of b is positive, the line, if projected indefinitely into the future, goes to plus infinity; if b is negative, the line goes to negative infinity. Unlimited increase or decrease is obviously not a logical assumption for economic or demographic phenomena. Indeed, in certain cases, even a value below zero is not meaningful. For example, series such as production or population cannot take on negative values. Hence, it would not be logical to project a negatively inclined straight line, which might describe a decreasing trend of production or sales, to the point where it drops below the X axis, yielding negative values.

This discussion is not meant to imply that a straight line or any other type of mathematical function can never be applied usefully for either short or long-term prediction purposes. On the contrary, there have been many useful applications of these types of functions for forecasting purposes. However, the analyst, when engaging in prediction, must always carefully weigh the implications of the models he projects into the future as regards their reasonableness for the phenomena being described and predicted.

3. A third purpose for which trend lines are fitted to economic data is to describe and eliminate secular movements from the series in order that the non-trend elements may be studied. Thus, if the analyst's primary interest is to study cyclical fluctuations, freeing the original data of trend enables him to examine cyclical movements undisturbed by the presence of the trend factor. For this purpose, any type of trend line which does a reasonably good job of bisecting the individual business cycles in the data would be appropriate. Although cycles can ordinarily be observed in economic data even though the trend element has not been removed, the elimination of trend brings the cyclical fluctuations into sharper focus. Particularly where further analysis of these cyclical oscillations is desired, as for example, in studying duration and amplitude of cycles, it is advisable to separate these fluctuations from the underlying trend movement.

Types of Secular Movements

There have been considerable variations in the secular movements of different economic and business series. Over long periods of time, some companies and industries have experienced periods of growth and then have gone into steep declines when more modern competitive processes and products have emerged in other companies and industries. Real gross national product in the United States has exhibited a relatively constant rate of growth of about 3% per year since the early part of this century. Since real gross national product is a measure of overall economic activity, it represents a type of average of all series for production of goods and services. Although many series have shown more or less similar trends to that of real gross national product, sharp divergences have also occurred. One example of these differential movements is that in recent years, the service sector of the economy has been growing relative to the agricultural sector. In fact, in the post World War II period while employment in the service industries generally was increasing, the numbers of persons employed in agriculture was decreasing.

Numerous studies have shown trends for a large number of American industries, which may be characterized as increasing at a decreasing percentage rate. Indeed, some investigators adapted growth curves originally used to describe biological growth to depict the past change of many industrial series. These growth curves, of which the *Gompertz* and *logistic* are the best known, are S-shaped for increasing series when plotted on graph paper with an arithmetic vertical scale, and are concave downward on a semilogarithmic chart. A semilogarithmic chart has a logarithmic vertical scale and an arithmetic horizontal scale. It is conventional in graphing time series on both arithmetic and semilogarithmic paper to plot the variable of interest on the vertical axis and the time on the horizontal axis.[1] The general shapes of such a growth curve when plotted on both arithmetic and semilogarithmic graph paper is depicted in Figure 11-1. The so-called "law of growth" has been used to describe this type of change over time in an industry. As can be seen on the arithmetic chart, in the early stages of the industry, the growth is slow at first, then proceeds through a period of rapid growth, where the

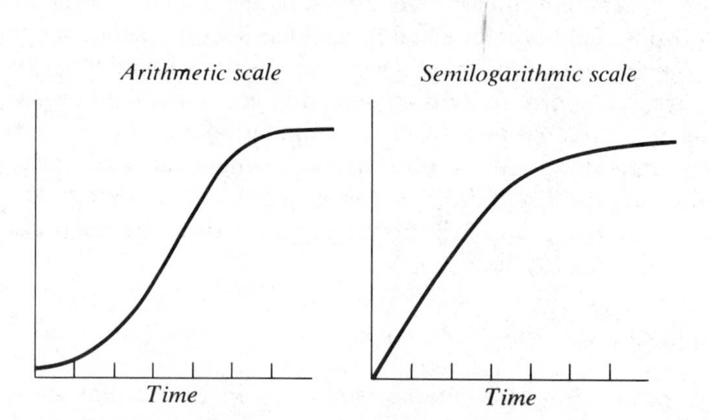

Figure 11-1 A growth curve plotted on arithmetic and semilogarithmic paper.

---

[1]The familiar arithmetic graph has an arithmetically-ruled vertical scale. Hence, equal vertical distances represent equal *amounts* of change. Semilogarithmic graphs have a logarithmically-ruled vertical scale. Here, equal vertical distances represent equal *percentage rates* of change. For example, on an arithmetic graph, a straight line inclined upwards depicts a series which is increasing with constant amounts of change. On a semilogarithmic graph, a straight line inclined upwards depicts a series which is increasing at a constant percentage rate.

series increases by increasing amounts. Then the industry moves through a point beyond which it increases by decreasing amounts and finally there is a tapering off into a period of "maturity." Throughout all stages, as seen on the semilogarithmic chart, although the industry is growing, the increases are at a decreasing percentage rate. Various reasons for this type of industrial growth were propounded by the investigators who discerned analogous changes in biological and industrial time series data. Since the equations of these growth curves are exponential in character, it is extremely difficult to fit such curves by the method of least squares described later in this chapter.[2]

As was indicated earlier, there are a great variety of types of secular movements in economic time series, and many of these trends cannot adequately be described nor projected by means of growth curves. The growth curves have a number of desirable characteristics. For example, they have finite lower and upper limits which are determined by the data to which the curves are fitted. However, no one family of curves is apt to be generally satisfactory for trend fitting purposes. In fact, the growth curves have been found to be quite inadequate for industrial growth prediction. The most commonly used polynomial type trend lines as fitted by the method of least squares are discussed in this chapter. The purpose of fitting, the goodness of fit obtained, knowledge of the growth and decay processes involved, and trial and error experimentation are all essential ingredients in the selection of the appropriate trend line.

11.4 The Fitting of Trend Lines by the Method of Least Squares

For situations in which it is desirable to have a mathematical equation to describe the secular trend of a time series, the most widely used method is the fitting of some type of polynomial function to the data. In this section, we illustrate the general method by means of very simple examples, fitting in turn a straight line and a second degree parabola by the method of least squares to time series data.

The Method of Least Squares

The method of least squares, when used to fit trend lines to time series data, is employed mainly because it is a simple, practical method which provides best fits according to a reasonable criterion. However, it should be recognized that the method of least squares does not have the same type of

---

[2]A special technique known as "the method of selected points" is ordinarily used for this purpose.

theoretical underpinning when applied to fitting trend lines as when used in regression and correlation analysis, as described in Chapter 10. The major difficulty is that the usual probabilistic assumptions that are present in regression and correlation analysis are simply not met in the case of time series data. For example, in the illustrative problem in Chapter 10 involving the relationship between consumer durables expenditures and income for a sample of families, there were two possible theoretical models. The first was one in which both consumer durables expenditures and income were random variables, the second, one in which income was a controlled variable, that is, families of prespecified incomes were selected and consumer durables expenditures was a random variable. Hence, in each model, the dependent variable was a random variable. Therefore, the model assumed conditional probability distributions of this random variable around the computed values of the dependent variable which fell along the regression line. These computed Y values were the means of the conditional probability distributions. A number of assumptions are implicit in this type of model. Deviations from the regression line are considered to be random errors describable by a probability distribution. The successive observations of the dependent variable are assumed to be independent. For example, Family C's expenditures were assumed to be independent of Family B's.

Clearly, in the fitting of secular trend lines to time series data, the probabilistic assumptions of the method of least squares are not met. If a trend line is fitted, for example, to an annual time series of department store sales, time is treated as the independent variable X and department store sales is the dependent variable Y. It is not reasonable to think of the deviation of actual sales in a given year from the computed trend value as a random error. Indeed, if the original data are annual, then deviations from trend would be considered to represent the operation of cyclical and irregular factors. Seasonal factors would not be present in annual data because by definition they complete themselves within a year. Finally, the assumption of independence is not met in the case of time series data. A department store's sales in a given year surely are not independent of what they were in the preceding year. In summary, returning to the point made at the outset of this discussion, the method of least squares when used to fit trend lines is employed primarily because of its practicality, simplicity, and good fit characteristics rather than because of its justification from a theoretical viewpoint.

Fitting an Arithmetic Straight-Line Trend

As an example, we will fit a straight line by the method of least squares to an annual series on value added by manufacture in the United States from 1949 to 1963. Although we wrote the equation of a straight line in the discussion of regression analysis in Chapter 10 as $\bar{Y}_X = a + bX$, in time series analysis we will use the equation

(11.1) $$Y_t = a + bx$$

The computed trend value is denoted Y_t, with the subscript t standing for trend. That is, Y_t is the computed trend figure for the time period x. In time series analysis, the computations can be simplified by transforming the X variable, which is the independent variable time, to a simpler variable with fewer digits. This is accomplished by stating the time variable in terms of deviations from the arithmetic mean time period, which is simply the middle time period.

The transformed time variable is denoted by lower case x. Hence in the illustrative example in Table 11-1, $x = 0$ in 1956, the middle year in the time series which runs from 1949 through 1963. The x values (or $X - \bar{X}$ figures) for years before and after 1956 are respectively $-1, -2, -3, \ldots$, and $1, 2, 3, \ldots$. The constants in the trend equation are interpreted in a similar way to those in the straight line discussed in regression analysis; a is the computed trend figure for the period when $x = 0$, in this case, 1956; b is the slope of the trend line, or the amount of change in Y_t per unit change in x, or per year in the present example. Because of the fact that $\Sigma x = 0$, the computation of the constants for the trend line is simpler than in the corresponding case of the straight-line regression equation. In Chapter 10, the normal equations for fitting a straight line were given as

(11.2) $$\Sigma Y = na + b\Sigma X$$

(11.3) $$\Sigma XY = a\Sigma X + b\Sigma X^2$$

In the least squares fitting of a straight-line trend equation, x is substituted for X. Since $\Sigma x = 0$, the normal equations become

(11.4) $$\Sigma Y = na$$

(11.5) $$\Sigma xY = b\Sigma x^2$$

Solving these equations for a and b gives

(11.6) $$a = \frac{\Sigma Y}{n} = \bar{Y}$$

(11.7) $$b = \frac{\Sigma xY}{\Sigma x^2}$$

Hence, the constant a is simply equal to the mean of the Y values and b is calculated by a division of two numbers easily determined from the original data. The calculations for fitting a straight-line trend to the time series on value added by manufacture are given in Table 11-1. Columns (2) through (5)

Table 11-1 Straight-Line Trend Fitted by the Method of Least Squares to Values Added by Manufacture in the United States, 1949–1963.

Year (1)	x (2)	*Value added by Manufacture*[a] *(billions of dollars)* Y (3)	xY (4)	x^2 (5)	Y_t (6)	*Percent of Trend* $\frac{Y}{Y_t}\cdot 100$ (7)
1949	−7	75.4	−527.8	49	84.7	89.0
1950	−6	89.7	−538.2	36	92.1	97.4
1951	−5	102.1	−510.5	25	99.5	102.6
1952	−4	109.2	−436.8	16	106.8	102.2
1953	−3	121.6	−364.8	9	114.2	106.5
1954	−2	117.0	−234.0	4	121.6	96.2
1955	−1	135.0	−135.0	1	129.0	104.7
1956	0	144.9	0	0	136.3	106.3
1957	1	147.8	147.8	1	143.7	102.9
1958	2	141.5	283.0	4	151.1	93.6
1959	3	161.5	484.5	9	158.5	101.9
1960	4	163.9	655.6	16	165.8	98.9
1961	5	164.3	821.5	25	173.2	94.9
1962	6	179.1	1074.6	36	180.6	99.2
1963	7	192.1	1344.7	49	188.0	102.2
Totals	0	2045.1	2064.6	280		

$$a = \frac{2045.1}{15} = 136.34$$

$$b = \frac{2064.6}{280} = 7.3736$$

$$Y_t = 136.34 + 7.3736X$$

$x = 0$ in 1956

x is in one year intervals; Y is in billions of dollars

[a]Value added is obtained by subtracting the cost of materials, supplies, containers, fuel, purchased electrical energy, and contract work from the value of shipments for products manufactured plus receipts for services rendered.

In general, the "value added" by a business firm is the sales of that firm minus its costs of materials and costs of products purchased from other firms. Hence, this difference represents the "value added" to the national product by this particular firm.

SOURCE: 1963 Census of Manufactures, U.S. Department of Commerce.

contain the basic computations for determining the values of a and b. As indicated in the calculation of these constants at the bottom of the table, $a = 136.34$ and $b = 7.3736$. The trend equation is $Y_t = 136.34 + 7.3736x$. An identification statement such as the one given below the trend equation that $x = 0$ in 1956 and Y is in billions of dollars should always accompany the equation, since it is not possible to fully interpret the meaning of the trend line without it. The trend figures are determined by substituting the appropriate values of x into the trend equation. Hence, for example, the trend figure for 1949 is

$$Y_{t,1949} = 136.34 + 7.3736\,(-7) = \$84.7 \text{ billions}$$

Since the b value measures the change in Y_t per year it can be added to each trend value to obtain the following year's figure. The trend figures are given in Column (6) of Table 11-1.

The trend line is graphed in Figure 11-2. Any two points can be plotted to determine the line. Interpreting the values of $a = 136.34$ and $b = 7.37$ (rounded), we have a computed trend figure for value added by manufacture in 1956 of \$136.34 billions and an increase in trend of \$7.37 billions per year. As can be seen from the graph, the trend line fits the data rather closely. Since the line was fitted by the method of least squares, the sum of the squared deviations of the actual data from the trend line is less than from any other straight line that could have been fitted to the data, and the total of the deviations above the line is equal to the total below the line.

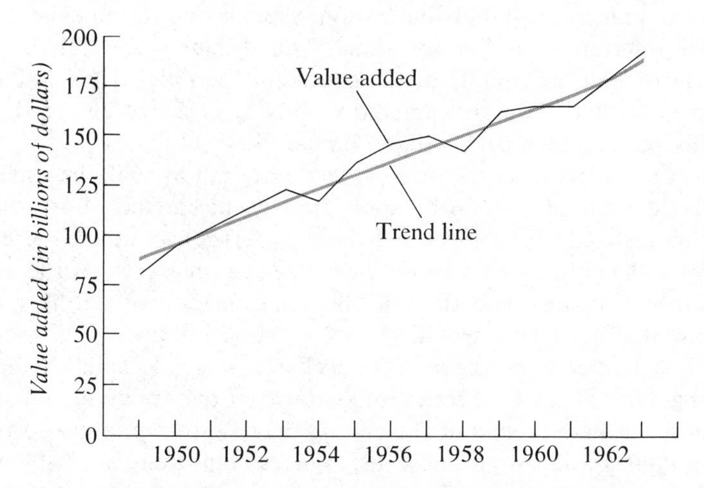

Figure 11-2 Straight-line trend fitted to value added by manufacture in the United States, 1949–1963.

A couple of technical points concerning the fitting procedure may be noted. Since the present illustration contained an odd number of years, the time period at which $x = 0$, or the x origin, coincided with one of the years of data, and the x values were stated as 1, 2, 3, . . . for years after the x origin and $-1, -2, -3, \ldots$ for years before the origin. On the other hand, if there had been an even number of years, the mean time period at which $x = 0$ would fall midway between the two central years. For example, suppose, there had been one less year of data and the value added figures were available only for 1949–1962. Then there would be 14 annual figures and $x = 0$ at 1955 1/2. The two central years 1955 and 1956 deviate from this origin by $-1/2$ and $+1/2$, respectively. To avoid the use of fractions, it is usual to state the deviations in terms of one-half year intervals rather than a year. Hence, the x values for 1956, 1957, 1958, . . . , would be 1, 3, 5, . . . , and for 1955, 1954, 1953, . . . , they would be $-1, -3, -5, \ldots$. The computation of the constants a and b would proceed in the usual way. However, now a would be interpreted as the computed trend figure for a time point midway between the two central years and the b value would be the amount of change in trend per one-half year.

A comment may be made on the meaning of a time period such as 1955 1/2. If the original data had been, say, end of year inventories, then the 1955 and 1956 figures would pertain to the end of December, 1955 and 1956, respectively. Hence, 1955 1/2 would refer to a time point halfway between these end of December figures or July 1, 1956. Figures such as these on inventories are often referred to as "point of time" data, since they pertain to specific time points. However, if the original annual data pertain to sales for an entire year, or as in our illustration, value added for an entire year, the figures are referred to as "period data." Such figures are viewed as being centered in time at the middle of the given time periods. Thus, value added figures for 1955 and 1956 are centered at July 1, 1955 and July 1, 1956, and 1955 1/2 is interpreted as pertaining to January 1, 1956.

If the time intervals of the original data were not annual, the transformed time variable x would have to be appropriately interpreted. For example, if the data were stated in the form of five-year averages and there were an odd number of such figures, then x would be in five-year intervals. If there were an even number of figures, and the non-fractional method of stating x referred to above were used, then x would be in 2-1/2-year intervals.

Another matter to consider in the fitting of a trend line is the beginning and ending dates of the time series in question. If the beginning date is at the trough of a business cycle and the ending date is at a cycle peak in a series which has an increasing trend, obviously the resultant trend line will tend to be too steeply inclined, overstating the average growth which has occurred. Analogously, an understatement of growth would tend to occur in a series with an increasing trend if the first figure in the series were at a cyclical peak and the last figure occurred at a trough. For this reason, it is good practice to attempt to have terminal dates which are at about the same stage of the business cycle. If the time series covers a very long period of time, this matter of terminal dates is not so important as for shorter periods, since for periods of substantial duration the effect of any two actual figures on the trend line will tend to be small.

Projection of the Trend Line

Projections of the computed trend line can be obtained by substituting the appropriate values of x into the trend equation. For example, if a projected trend figure for 1968 were desired for value added by manufacture, it would be computed by substituting $x = 12$ in the previously determined trend equation. Hence,

$$Y_{t,1968} = 136.34 + 7.3736(12) = \$224.8 \text{ billions}$$

A rougher estimate of this trend figure would be obtained by extending the straight line graphically in Figure 11-2 to the year 1968. It must be remembered that these projections are estimates of only the trend level in 1968 and not of the actual figure for value added by manufacture in that year. If a prediction of the latter figure were desired, estimates of the non-trend factors would have to be combined with the trend estimate. This means that a prediction of cyclical fluctuations would have to be made and incorporated with the trend figure. Accurate forecasts of this type are difficult to make over extended time periods. However, insofar as managerial applications of secular trend analysis are concerned, for long-range planning purposes, often all that is desired is a projection of the trend level of the economic variable of interest. For example, a good estimate of the trend of demand would be adequate for a business firm planning a plant expansion to anticipate demand many years into the future. Accompanying predictions of business cycle standings many years into the future would not be required, nor for that matter, would they be realistically feasible.

Cyclical Fluctuations

As was previously indicated, when a time series consists of annual data, it contains trend, cyclical, and irregular elements. The seasonal variations are absent, since they occur within a year. Hence, deviations of the actual annual data from a computed trend line are attributable to cyclical and irregular factors. Since the cyclical element is the dominant factor, a study of these deviations from trend essentially represents an examination of business cycle fluctuations. The deviations from trend are most easily observed by dividing the original data by the corresponding trend figures for the same time period. By convention, the result of this division of an original figure by a trend value is multiplied by 100 to express the figure as a percent of trend. Hence, if the original figure is exactly equal to the trend figure, the percent of trend is 100; if the original figure exceeds the trend value, the percent of trend is above 100; if the original figure is less than the trend value, the percent of trend is below 100.

The formula for percent of trend figures is

(11.8)
$$\text{Percent of trend} = \frac{Y}{Y_t} \cdot 100$$

where Y = annual time series data
Y_t = trend values

In summary, the original annual data contain trend, cyclical, and irregular factors. When converted to percent of trend, these numbers contain only cyclical and irregular movements, since the division by trend eliminates that factor. The rationale of this procedure is easily seen by using a so-called multiplicative model for the analysis. That is, the original annual figures are viewed as representing the combined effect of trend, cyclical, and irregular factors. In symbols, let T, C, and I represent trend, cyclical, and irregular factors, respectively, and Y and Y_t mean the same as in (11.8). Then dividing the original time series by the corresponding trend values yields

(11.9)
$$\frac{Y}{Y_t} = \frac{T \times C \times I}{T} = C \times I$$

The percents of trend for the series on value added by manufacture are given in Column (7) of Table 11-1 and are plotted in Figure 11-3. As may be seen from the chart, the underlying upward trend movement is no longer present. Instead, the percent of trend series fluctuates about the line labeled 100, which is the trend level. These percent of trends are sometimes referred to as cyclical relatives; that is, the original data are stated relative to the trend figure. Of course, strictly speaking, Y/Y_t is the cyclical relative, and the multiplication by 100 converts the relative to a percentage figure. Another way of depicting cyclical fluctuations is in terms of relative cyclical residuals, which are percentage deviations from trend, and are computed by the formula

(11.10)
$$\text{Relative cyclical residual} = \frac{Y - Y_t}{Y_t} \cdot 100$$

Hence, for example, if we refer to the value added data in Table 11-1 for 1963, the actual figure is 192.1, the computed trend value is 188.0, and the percent of trend is 102.2. The relative cyclical residual in this case is +2.2%, indicating that the actual value added figure is 2.2% above the trend figure because of cyclical and irregular factors. These residuals are positive or negative depending on whether the actual time series figures fall above or below the computed trend values. The graph of relative cyclical residuals is visually identical to that of the percent of trend values except that relative cyclical residuals are shown as fluctuations around a zero base line rather than around a base line of 100%.

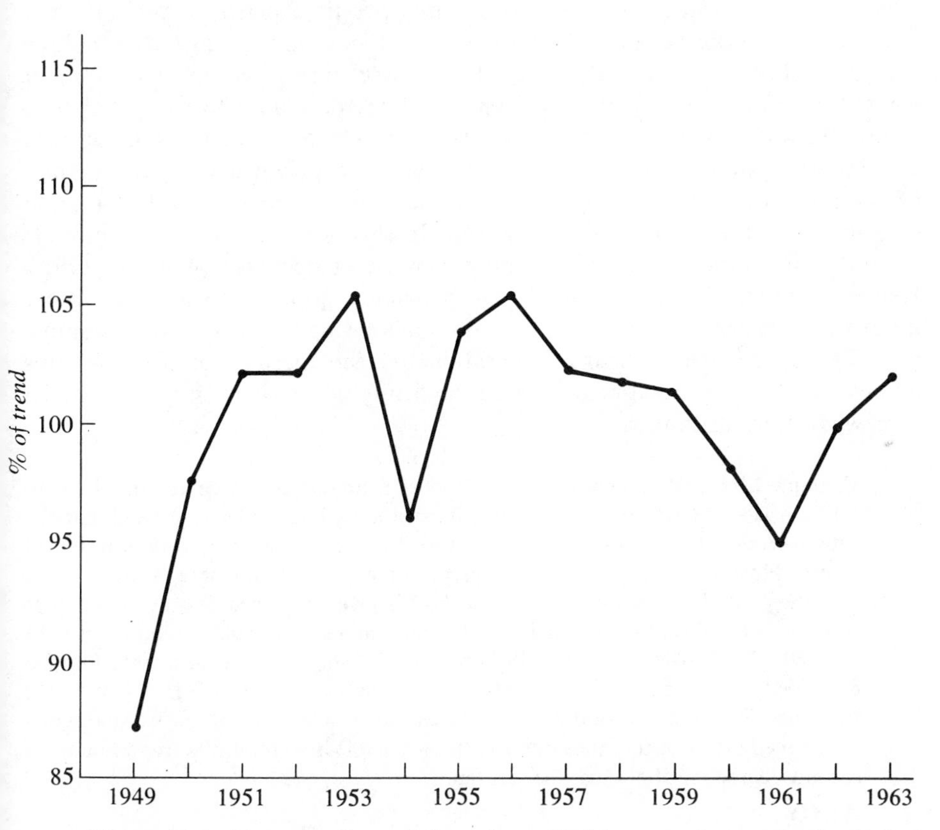

Figure 11-3 Percents of trend for value added by manufacture in the United States, 1949–1963.

The familiar charts of business cycle fluctuations that often appear in publications such as the financial pages of newspapers and business periodicals are usually graphs of either percents of trend or relative cyclical residuals. These charts may be studied for timing of peaks and troughs of cyclical activity, for amplitude of fluctuations, for duration of periods of expansion and contraction, and for other relevant items of interest to the business cycle analyst.

Fitting a Second Degree Trend Line

The preceding discussion on the fitting of a straight line pertains to the case in which the secular trend of the time series can be characterized as increasing or decreasing by constant amounts per time period. Actually very few economic time series exhibit this type of constant change over a long period of time, say, over a period of several business cycles. Therefore, it generally is necessary to fit other types of lines or curves to the given time series. It was indicated in the discussion of regression analysis in Section 10.10 that polynomial functions are particularly convenient to fit by the method of least squares. Frequently a second degree parabola provides a good description of the trend of a time series. In this type of curve, the amounts of change in the trend figures, Y_t, may increase or decrease per time period. Hence, a second degree parabola may provide a good fit to a series whose trend is increasing by increasing amounts, increasing by decreasing amounts, etc. The procedure of fitting a parabola by the method of least squares involves the same general principles as the fitting of a straight line, but entails somewhat more arithmetic.

Example 11-1 We illustrate the method of fitting a second degree parabola to a time series in terms of a very simple illustration. The reader is warned that the time period in this example is entirely too short to permit a valid description of trend. However, the illustration is given for expository purposes only to indicate the procedure involved. In Table 11-2 is given a time series on the number of persons employed in anthracite coal mining in a certain coal region from 1954 to 1960. This series is graphed in Figure 11-3. The trend of these data may be described as decreasing by decreasing amounts. As indicated in Chapter 10, the general form of a second degree parabola is $\bar{Y}_X = a + bX + cX^2$. Analogous to the method of stating the equation for a straight-line trend, the trend line for a second degree parabola may be written

(11.11) $$Y_t = a + bx + cx^2$$

where Y_t = the trend values
a, b, c = constants to be determined
x = deviations from the middle time period

The normal equations for fitting a second degree parabola were given in Chapter 10 (10.52). If the transformed variable x, representing deviations from the mean time period is substituted for X in Equations (10.52), since $\Sigma x = 0$, the normal equations become

(11.12) $$\Sigma Y = na + c\Sigma x^2$$

(11.13) $$\Sigma x^2 Y = a\Sigma x^2 + c\Sigma x^4$$

(11.14) $$b = \frac{\Sigma xY}{\Sigma x^2}$$

Table 11-2 Second Degree Parabola Fitted by the Method of Least Squares to the Number of Persons Employed in Anthracite Coal Mining in a Certain Coal Region, 1954–1960.

Year (1)	x (2)	*Number Employed (in thousands)* Y (3)	xY (4)	x^2Y (5)	x^2 (6)	x^4 (7)	Y_t (8)
1954	−3	83	−249	747	9	81	84
1955	−2	60	−120	240	4	16	62
1956	−1	54	−54	54	1	1	44
1957	0	21	0	0	0	0	30
1958	1	22	22	22	1	1	20
1959	2	13	26	52	4	16	14
1960	3	13	39	117	9	81	12
Totals	0	266	−336	1232	28	196	

Hence, the constant b is determined by the same equation as in fitting the straight line. The constants a and c are found by solving simultaneously the Equations (11.12) and (11.13).

In the present problem, since there are an odd number of years, $x = 0$ in the middle year, 1957. Solving for b by substituting the appropriate totals from Table 11-2, we have

$$b = \frac{-336}{28} = -12$$

Substituting into Equations (11.12) and (11.13) gives

$$266 = 7a + 28c$$
$$1232 = 28a + 196c$$

Dividing the second equation by 4 to equate the coefficients of a, we obtain

$$266 = 7a + 28c$$
$$308 = 7a + 49c$$

Subtracting the first equation from the second,

$$42 = 21c$$

and

$$c = \frac{42}{21} = 2$$

Substituting this value for c into the first equation

$$266 = 7a + 28(2)$$
$$a = 30$$

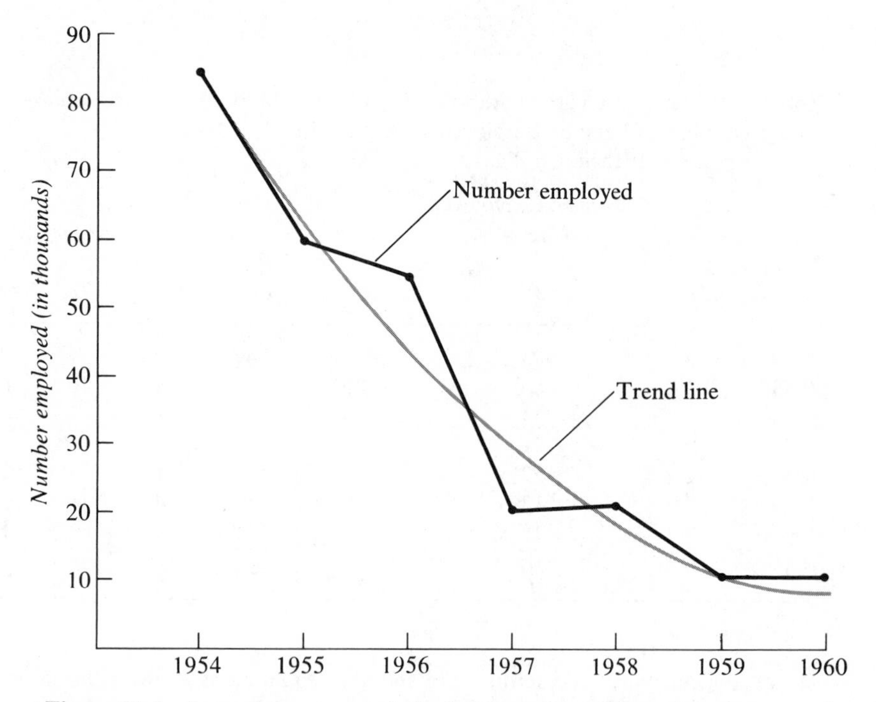

Figure 11-4 Second degree parabola fitted to the number of persons employed in anthracite coal mining in a certain coal region, 1954–1960.

Therefore, the equation of the second degree parabola fitted to the employment time series is

(11.15) $$Y_t = 30 - 12x + 2x^2$$

where $x = 0$ in 1957
x is in one-year intervals
Y is in thousands of persons

The trend figures, Y_t, shown in Column (8) of Table 11-2 are obtained by substituting the appropriate values of x into Equation (11.15). The constants a, b, and c may be interpreted as follows: a is the computed trend figure at the time origin, that is when $x = 0$; b is the slope of the parabola at the time origin; and c indicates the amount of acceleration or deceleration in the curve, or the amount by which the slope changes per time period.[3]

[3]In calculus terms, the derivative of the second degree parabola trend equation is

$$\frac{dY_t}{dx} = b + 2cx$$

Hence, the slope of the curve differs at each time period x. When $x = 0$, $\frac{dY_t}{dx} = b$. Therefore, the slope at the time origin is b. The second derivative is $\frac{d^2Y_t}{dx^2} = 2c$. Thus, the acceleration or rate of change in the slope is $2c$ per time period.

Although the second degree parabola appears from Figure 11-4 to provide a reasonably good fit to the data in this example, the dangers of a mechanistic projection of a trend line are clearly illustrated. The parabola would begin to turn upward after 1960, and the projected trend figure for each year would be higher than the preceding year's figure. Therefore, only if an analysis of all of the underlying factors determining the trend of this series revealed reasons for a reversal of the observed decline should one be willing to entertain the notion of extending the trend line into the future for forecasts, even for relatively short periods.

Fitting Logarithmic Trend Lines

As discussed earlier, the equations of trend lines embody assumptions concerning the type of change that takes place over time. Hence, the arithmetic straight line assumes a trend that increases or decreases by constant amounts, whereas the second degree parabola assumes that the change in these amounts of change is constant per unit time. It is often useful to describe the secular trend of an economic time series in terms of the percentage rates of change that are taking place. Logarithmic trend lines are useful for this purpose.

If a time series increases or decreases at exactly a constant percentage rate, a straight line fitted to the logarithms of the data constitutes a perfect fit. Some economic time series in the United States, as for example, gross national product, although not changing exactly at a constant rate, have exhibited trends of approximately constant percentage increases over substantial periods of time. The equation of the logarithmic straight line which would describe the trend of such series is

(11.16) $$\log Y_t = a + bx$$

The method of fitting this line is the same as for the arithmetic straight line, $Y_t = a + bx$, except that wherever Y appeared before $\log Y$ now appears. Hence, the values of the constants a and b are computed as follows

(11.17) $$a = \frac{\Sigma \log Y}{n}$$

(11.18) $$a = \frac{\Sigma x \log Y}{\Sigma x^2}$$

After a and b have been calculated, trend figures are determined by substituting values of x into the trend equation, computing $\log Y_t$, and taking the antilogarithm to obtain Y_t. Although we will not present another example to indicate the fitting process, since there really are no new principles involved, we will illustrate the calculation of a trend figure for the type of trend line under discussion.

Suppose the logarithmic trend line for a particular series had been determined to be

$$\log Y_t = 2.3657 + 0.0170x$$

Then the logarithm of the trend figure for the year in which $x = 2$ would be given by substituting this value of x into the trend equation to obtain

$$\log Y_t = 2.3657 + 0.0170(2)$$
$$\log Y_t = 2.3997$$

Taking the antilog of this value yields the trend figure

$$Y_t = \text{antilog } 2.3997 = 251.0$$

The rate of change implied by this trend line can be obtained by calculating antilog $b - 1$. For example, in the above illustration the antilog of the slope coefficient b is

$$\text{antilog } b = 1.040$$

This figure is the ratio of each trend figure to the preceding one. Subtracting 1.00 from this figure yields $1.04 - 1.00 = 0.04$. Hence, the trend figures increase by 4% per time period. If the series had been a declining one and the result of the above calculation, for example, was -0.04, this would mean the trend figures decrease by 4% per time period. Even though a time series may not exhibit a trend with constant rates of change throughout its entire extent, sometimes it can be broken down into segments during which the rate of change has been approximately constant. It is often useful in such cases to make comparisons of the rates of change similarly determined from different economic time series of interest.

Logarithmic second degree parabolas can also be fitted to time series in which the trend is increasing at an increasing percentage rate, increasing at a decreasing percentage rate, etc. However, ordinarily polynomials of third or higher degree are not fitted to time series in either arithmetic or logarithmic form. The reason is that such curves permit too many changes in direction and tend to follow the cyclical fluctuations in the data as well as the trend. Therefore, these curves often do not have the required characteristic of a trend line of depicting the smooth, continuous movement which underlies the cyclical swings in a time series.

In the attempt to find an appropriate trend line, the analyst should always plot the time series on both arithmetic and semi-logarithmic graph paper. These two types of graphs may aid him in determining whether an arithmetic or logarithmic line would provide a better description of the trend.

11.5 Measurement of Seasonal Variations

For long-range planning and decision making, in terms of time series components, executives of a business or governmental enterprise concentrate primarily on forecasts of secular trend movements. For intermediate planning periods, say from about two to five years, business cycle fluctuations are of critical importance, too. For shorter range planning, operational decision and control purposes, seasonal variations must also be taken into account.

Seasonal movements, as indicated in Section 11.2, are periodic patterns of variation in a time series. Strictly speaking, the terms "seasonal movements" or "seasonal variations" can be applied to any regularly repetitive movements which occur in a time series where the interval of time for completion of a cycle is one year or less. Hence, under this classification are subsumed movements such as daily cycles in utilization of electrical energy and the weekly cycles in the use of public transportation vehicles. However, seasonal movements generally refer to the annual repetitive patterns of economic activity which are associated with climatic and custom factors. As noted earlier, these movements are generally examined by using series of monthly or quarterly data.

Purpose of Analyzing Seasonal Variations

Just as was true in the case of the study of trend movements, seasonal variations may be studied because *interest is primarily centered upon these movements*, or they may be measured merely *in order that they may be eliminated*, so that business cycle fluctuations can be more clearly revealed. For example, as an illustration of the first purpose, a company might be interested in analyzing the seasonal variations in sales of a product it produces in order to iron out variations in production, scheduling, and in personnel requirements. Another reason a company's interest may be primarily focused on seasonal variations is to budget a predicted annual sales figure by monthly or quarterly periods based on observed seasonal patterns in the past.

On the other hand, as an illustration of the second purpose, an economist may wish to eliminate the usual month-to-month variations in series such as personal income, unemployment rates, and housing starts in order to study the underlying business cycle fluctuations present in these data.

Rationale of the Ratio-to-Moving Average Method

There are a number of techniques by which seasonal variations can be measured, but only the most widely used one, the so-called "ratio-to-moving average method" will be discussed here. It is most frequently applied to monthly data, but we will illustrate its use for a series of quarterly figures, thus reducing substantially the required number of computations.

It is helpful in acquiring an understanding of the rationale of the measurement of seasonal fluctuations to begin with the final product, the seasonal indices. The object of the calculations when the raw data are for quarterly periods and a stable or regular seasonal pattern is present is to obtain four seasonal indices, each one indicating the seasonal importance of a quarter of the year. The arithmetic mean of these four indices in 100.0. Hence, if the seasonal index for, say, the first quarter is 105, this means that the first quarter averages 5% higher than the average for the year as a whole. If the original data had been monthly, there would be twelve seasonal indices which average 100.0, and each index would indicate the seasonal importance of a particular month. These indices are descriptive of the recurrent seasonal pattern in the original series.

As an example of how these seasonal indices might be used, we can refer to the aforementioned purpose of budgeting a predicted annual sales figure, say, by quarterly periods. Suppose that $40,000,000 of sales of particular products was budgeted for the next year, or an average of $10,000,000 per quarter. If the quarterly seasonal indices based on an observed stable seasonal pattern in the past were 97.0, 110.0, 85.0, and 108.0, respectively, for the four quarters of the year, then the amounts of sales budgeted for each quarter would be

First quarter .97 × $10 million = $ 9.7 million
Second quarter 1.10 × 10 million = 11.0 million
Third quarter .85 × 10 million = 8.5 million
Fourth quarter 1.08 × 10 million = 10.8 million

The essential problem in the measurement of seasonal variations is eliminating from the original data the non-seasonal elements in order to isolate the stable seasonal component. In trend analysis, when annual data were used and it was desired to arrive at cyclical fluctuations, a similar problem existed. It was solved by obtaining measures of trend and using these as base line or reference figures. Deviations from trend were then measures of cyclical (and irregular) movements. Analogously, when we have monthly or quarterly original data, which consist of all of the components of trend, cycle, seasonal, and irregular movements, ideally we would like to obtain a series of base line figures which contain all of the non-seasonal elements. Then deviations from the base line would represent the pattern of seasonal variations. Unsurprisingly, this ideal method of measurement is not feasible. However, the practical method used is to obtain a series of moving averages which roughly include the trend and cycle components. Dividing the original data by these moving average figures eliminates the trend and cyclical elements and yields a series of figures which contain seasonal and irregular movements. These data are then averaged by months or by quarters so as to eliminate the irregular disturbances in order to isolate the seasonal factor. This method of describing a pattern of stable seasonal movements is explained below.

Ratio-to-Moving Average Method

In order to derive a set of seasonal indices from a series characterized by a stable seasonal pattern, about five to eight years of monthly or quarterly data are required. A stable seasonal pattern means that the peaks and troughs generally occur in the same months or quarters year after year.

The ratio-to-moving average method of computing seasonal indices for quarterly data may be summarized as consisting of the following steps:

1. Derive a four-quarter moving average which contains the trend and cyclical components present in the original quarterly series. A four-quarter moving average is simply an annual average of the original quarterly data successively advanced one quarter at a time. For example, the first moving average figure contains the first four quarters. Then the first quarter is dropped, and the second through fifth quarterly figures are averaged. The computation proceeds this way until the last moving average is calculated, containing the last four quarters of the original series. In the actual calculation an adjustment is made in order to center the moving average figures so their timing corresponds to that of the original data.

The reason these moving averages include the trend and cyclical components may perhaps be most easily understood by considering what these averages do not contain. Since they are annual averages, they do not contain seasonal movements, since such fluctuations, by definition, average out over a one-year period. Also, the irregular movements which tend to raise the figures for certain months or quarters and to lower them in others tend to cancel out when averaged over the year. Thus, only the trend and cyclical elements tend to be present in the moving averages.

2. Divide the original data for each quarter by the corresponding moving average figure. These "ratio-to-moving average" numbers contain only the seasonal and irregular movements, since the trend and cyclical components were eliminated in the division by the moving average.

3. Arrange the ratio-to-moving average figures by quarters, that is, all the first quarters in one group, all the second quarters in another, and so forth. Average these ratio-to-moving average figures for each quarter in an attempt to eliminate the irregular movements, and thus to isolate the stable seasonal component. The type of average used for this procedure is referred to as a "modified mean." This is an arithmetic mean of the ratio-to-moving average figures after dropping the highest and lowest extreme values.

4. Make an adjustment to force the four modified means to total 400 and thus average out to 100.0. The resultant four figures, one for each quarter of the year constitute the seasonal indices for the series in question.

In symbols, this procedure may be summarized as follows. Let Y be the original quarterly observations; MA the moving average figures; and T, C, S, I, the trend, cyclical, seasonal, and irregular components, respectively. Then, dividing the original data by the moving average values gives

$$\frac{Y}{MA} = \frac{T \times C \times S \times I}{T \times C} = S \times I \tag{11.19}$$

Averaging these ratio-to-moving average figures (Y/MA) accomplishes an elimination of the irregular movements which tend to make the Y/MA values too high in certain years and too low in others. Hence, if the eliminations of the non-seasonal elements were perfect, the final seasonal indices would reflect only the effect of seasonal variations. Of course, since the entire method is a rather rough and approximate procedure, the non-seasonal elements are generally not completely eliminated. The moving average usually contains the trend and *most* of the cyclical fluctuations. Therefore the cyclical component is usually not completely absent in the Y/MA values. Also, the modified means do not ordinarily remove all of the erratic disturbances attributed to the irregular component. Nevertheless, in the case of series with a stable seasonal pattern, the computed seasonal indices generally isolate the underlying seasonal pattern quite well.

In Table 11-3 is given a quarterly series of feed grain price index numbers of average prices received by farmers from 1959 to 1966. The base period of the index number series is 1957–1959. As is indicated in the section on index numbers later in this chapter, this means that the average level of prices during this period is designated as 100. Index figures above and below 100 represent price levels which are higher and lower, respectively, than during the base period. Examination of this series reveals that feed grain prices tend to be highest during the second and third quarters, that is, during the spring and summer months, and lowest during the first and fourth quarters, or during the fall and winter. The calculation of quarterly seasonal indices will be illustrated in terms of this series.

The feed grain price indices have been listed in Column (2) of Table 11-3 from the first quarter of 1959 through the second quarter of 1966. The inclusion of the first two quarters of 1966 permits the computation of the moving averages for all four quarters of 1965. Our first task is the calculation of the four quarter moving average. This moving average would simply be calculated as indicated above by averaging four quarters at a time, continually moving the average up by a quarter. However, because of a problem of centering of dates, a slightly different type of average, a so-called "two-of-a-our quarter moving average" is calculated. The problem is as follows. An verage of four quarterly figures would be centered halfway between the lating of the second and third figures and would thus not correspond to the late of either of those figures. For example, the average of the four quarters f 1959, the first figures shown in Column (2) of Table 11-3, would be centered nidway between the second and third quarter dates, or at the center of the ear, July 1, 1959. The original quarterly figures are centered at the middles f their respective time periods, or, for simplicity, say, February 15, May 15, ugust 15, and November 15. Hence, the dates of a simple four quarter noving average would not correspond to those of the original data. This roblem is easily solved by averaging the moving averages two at a time. For example, as we have seen, the first moving average obtainable from Table 1-3 is centered at July 1, 1959. The second moving average, which contains he last three quarters of 1959 and the first quarter of 1960, is centered at)ctober 1, 1959. Averaging these two figures yields a figure centered at ugust 15, the same as the dating of the third quarter.

The easiest way to calculate this properly centered moving average is iven in Columns (3) through (5) of Table 11-3. In Column (3) is given a four uarter moving total. The first figure, 391, is the total of the first four quarterly gures, 96, 103, 100, and 92. This figure is listed opposite the third quarter, 959, although actually it is centered at July 1. The next four quarter moving otal is obtained by dropping the figure for the first quarter, 1959 and including the first quarter, 1960 figure. Hence, 388 is the total of 103, 100, 92, nd 93. The total of 391 and 388 or 779 is the first entry in Column (4). This epresents the total for the eight months which would be present in the averging of the first two simple four quarter moving averages. Dividing this otal by 8 yields the first two-of-a-four quarter moving average figure of 97.38, roperly centered at the middle of the third quarter, 1959.[4]

The moving averages given in Column (5) of Table 11-3 are shown in igure 11-5 along with the original data. It is very useful to examine graphs n the calculation of seasonal indices because we can observe visually what is ccomplished in each major step of the procedure. We have noted earlier that he original data, if stated in monthly or quarterly form, contain all of the omponents of trend, cycle, seasonal, and irregular movements. Although the ime period is too short for trend to be revealed, we can observe in the series f feed grain price index numbers some effects of cyclical fluctuations as the ata move into a trough at the end of 1960 and continue in an expansion wing thereafter. The repetitive annual rhythm of the seasonal movements is learly discernible. Irregular movements are also present. The moving

[4]If the computations are carried out on a calculating machine, it is most efficient to place e reciprocal of 8, or 1/8 in the keyboard and then multiply it by the totals in Column (4) to eld the desired moving average.

Table 11-3 Feed Grain Index Numbers of Average Prices Received by Farmers by Quarters, 1959–1966: Computations for Seasonal Indices and Deseasonalizing of Original Data.

Quarter (1)	*Feed Grain Price Index Numbers* (1957–1959 = 100) (2)	*Four Quarter Moving Total* (3)	*Two-of-a-Four Quarter Moving Total* (4)	*Moving Average* Col(5) = Col(4) × 1/8 (5)	*Original Data as Percent of Moving Average* [Col(2) ÷ Col(5)] × 100 (6)	*Seasonal Index* (7)	*Deseasonalized Feed Grain Price Index Numbers* [Col(2) ÷ Col(7)] × 100 (8)
1959							
I	96					98.8	97.2
II	103					102.9	100.1
III	100	391	779	97.38	102.69	102.6	97.5
IV	92	388	771	96.38	95.46	95.7	96.1
1960							
I	93	383	762	95.25	97.64	98.8	94.1
II	98	379	752	94.00	104.26	102.9	95.2
III	96	373	744	93.00	103.23	102.6	93.6
IV	86	371	737	92.13	93.35	95.7	89.9
1961							
I	91	366	733	91.63	99.31	98.8	92.1
II	93	367	741	92.63	100.40	102.9	90.4
III	97	374	750	93.75	103.47	102.6	94.5
IV	93	376	756	94.50	98.41	95.7	97.2

1962							
I	93	380	759	94.88	98.02	98.8	94.1
II	97	379	758	94.75	102.37	102.9	94.3
III	96	379	762	95.25	100.79	102.6	93.6
IV	93	383	771	96.38	96.49	95.7	97.2
1963							
I	97	388	786	98.25	98.73	98.8	98.2
II	102	398	801	100.13	101.87	102.9	99.1
III	106	403	809	101.13	104.82	102.6	103.3
IV	98	406	814	101.75	96.31	95.7	102.4
1964							
I	100	408	813	101.63	98.40	98.8	101.2
II	104	405	813	101.63	102.33	102.9	101.1
III	103	408	823	102.88	100.12	102.6	100.4
IV	101	415	838	104.75	96.42	95.7	105.5
1965							
I	107	423	851	106.38	100.58	98.8	108.3
II	112	428	853	106.63	105.04	102.9	108.8
III	108	425	848	106.00	101.89	102.6	105.3
IV	98	423	842	105.25	93.11	95.7	102.4
1966							
I	105	419				98.8	106.3
II	108					102.9	105.0

SOURCE: Agricultural Handbook No. 325, U.S. Department of Agriculture, 1966.

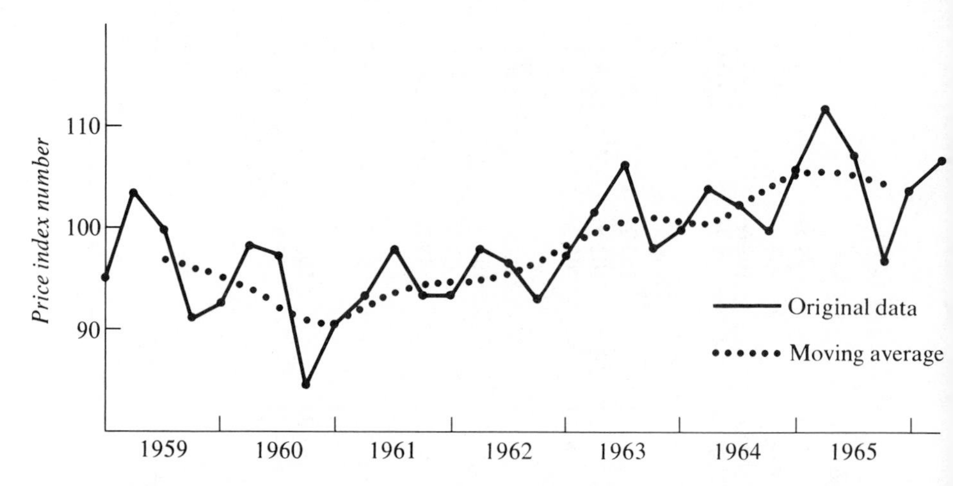

Figure 11-5 Feed grain index numbers of average prices received by farmers by quarters, 1959–1966.

average which runs smoothly through the original data can be observed to follow the cyclical fluctuations rather closely and if the series were long enough, we would be able to see how the moving average describes trend movements as well. Another way to view this point is to note that the seasonal variations and to a large degree the irregular movements are absent from the smooth line which traces the path of the moving average. It should be noted that there are no moving average figures corresponding to the first two and the last two quarters of original data. Correspondingly, if the original data were in monthly form and a twelve month moving average were computed, there would be no moving averages to correspond to the first six months of data nor to the last six months of data.

The "ratio-to-moving average" figures, or original data, Column (2), divided by the moving average, Column (5), are given in Column (6) of Table 11-3. As is customary, these figures have been multiplied by 100 to express them in percentage form. They are often referred to as "percent of moving average" values, and may be represented symbolically as $(Y/MA) \times 100$. These values are graphed in Figure 11-6. As can be seen in the graph, the trend and cyclical movements are no longer present in these figures. The 100-base line represents the level of the moving average or the trend-cycle base. The fluctuations above and below this base line clearly reveal the repetitive seasonal movement of feed grain prices. As noted earlier, the irregular component is also present in these figures.

The next step in the procedure involves the attempt to remove the effect of irregular movements from the $(Y/MA) \times 100$ values. This is accomplished

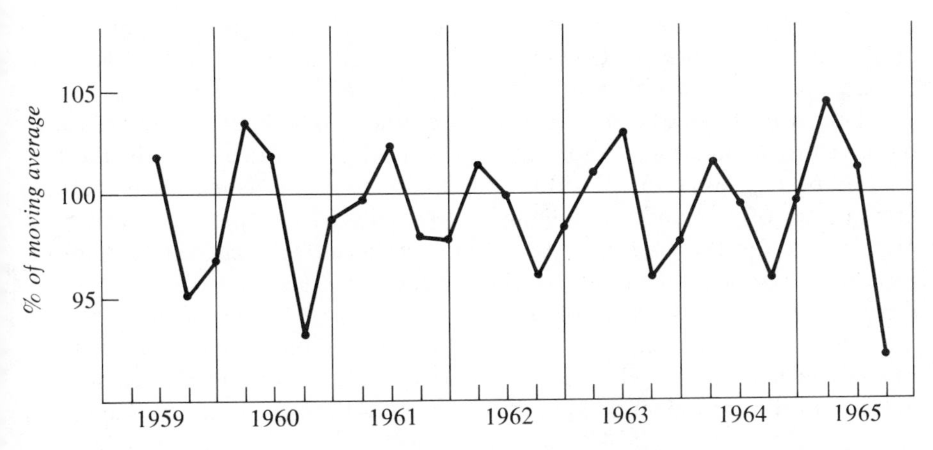

Figure 11-6 Percent of moving averages for feed grain price index numbers, 1959–1966.

by averaging the percents of moving average figures for the same quarter. That is, the first quarter $(Y/MA) \times 100$ values are averaged, the second quarter values are averaged, and so forth. The average customarily used in this procedure is a modified mean, which is simply the arithmetic mean of the percents of moving average figures for each quarter over the different years, after eliminating the lowest and highest figures. It is desirable to make these deletions particularly when the highest and lowest figures tend to be atypical because of erratic or irregular factors such as strikes, work stoppages, or other unusual occurrences.

The percent of moving average figures for each quarter are listed in Table 11-4. The highest and lowest figures have been designated as deleted by a line drawn through them, and the modified means of the remaining values are shown for each quarter. These means are 98.6, 102.7, 102.4, and 95.6, respectively, for the first through fourth quarters. The total of these modified means is 399.3. Since it is desirable that the four indices total 400, in order that they average 100%, each of them is multiplied by the adjustment factor of 400/399.3. This adjustment has the effect of forcing a total of 400 by raising each of the unadjusted figures by the same percentage. The final quarterly seasonal indices are shown on the bottom line of Table 11-4.

As indicated earlier, if interest centers on the pattern of seasonal variations itself, the four quarterly indices represent the final product of the analysis. On the other hand, sometimes the purpose of measuring seasonal variations is to eliminate them from the original data in order to examine, for example, the cyclical movements. The method of "deseasonalizing" the original data or adjusting these figures for seasonal movements is simply to divide them by the appropriate seasonal indices. This adjustment is shown in Table 11-3 for the feed grain price data by the division of the original figures in Column (2) by the seasonal indices in Column (7). The result is multiplied by 100, since the seasonal index is stated as a percentage rather than as a relative.

Let us illustrate the meaning of a deseasonalized figure by reference to the first line of figures in Table 11-3. The feed grain price index in the first quarter of 1959 was 96. Dividing this figure by the seasonal index for the first quarter of 98.8 and multiplying by 100 yields 97.2. This is the feed grain price index for the first quarter of 1959 adjusted for seasonal variations. *That is, it represents the level that food prices would have attained if there had not been the depressing effect of seasonality in the first quarter of the year.* All time series components, other than seasonal variations, are present in these deseasonalized figures. This idea can be expressed symbolically as follows in terms of the aforementioned multiplicative model of the time series analysis:

$$\frac{Y}{SI} = \frac{T \times C \times S \times I}{S} = T \times C \times I \tag{11.20}$$

The figures for the feed grain price index numbers adjusted for seasonal movements are graphed in Figure 11-7. It can be seen that the underlying cyclical movement is present in these data, irregular movements are indicated, and if

Table 11-4 Feed Grain Price Index Numbers: Calculation of Quarterly Seasonal Indices from Percent of Moving Average Figures.

	Percent of Moving Averages			
	Quarter			
	I	*II*	*III*	*IV*
1959			102.69	95.46
1960	~~97.64~~	104.26	103.23	93.35
1961	99.31	~~100.40~~	103.47	~~98.41~~
1962	98.02	102.37	100.79	96.49
1963	98.73	101.87	~~104.82~~	96.31
1964	98.40	102.33	~~100.12~~	96.42
1965	~~100.58~~	~~105.04~~	101.89	~~93.11~~
Modified Means	98.6	102.7	102.4	95.6

Total of Modified Means = 399.3

Adjustment Factor = 400/399.3 = 1.0018

Seasonal Indices

I	*II*	*III*	*IV*
98.8	102.9	102.6	95.7

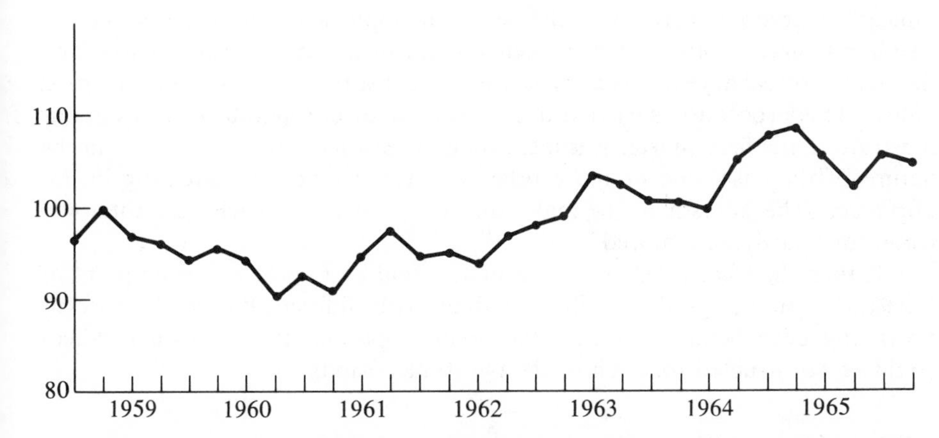

Figure 11-7 Deseasonalized figures for feed grain index numbers of average prices received by farmers by quarters, 1959–1966.

a sufficiently long period had been used, say, at least a couple of business cycles, the trend would also be apparent. It may be noted that as compared to the plot of the original data in Figure 11-5, most of the repetitive seasonal movements are no longer present in the deseasonalized figures. However, ordinarily, as in this case too, the adjustment for seasonality is not perfect. To the extent that seasonal indices do not completely portray the effect of seasonality, division of original data by seasonal indices will not entirely remove these influences.

Seasonal indices are often used for the adjustment purpose just discussed. Economic time series adjusted for seasonal variations are often charted in the *Federal Reserve Bulletin*, the *Survey of Current Business*, and other publications. Also quarterly gross national product figures are often given as "seasonally adjusted at annual rates." These are simply deseasonalized quarterly figures multiplied by four to state the result in annual terms.

Changing Seasonal Patterns

Although the ratio-to-moving average method explained above pertains to series which exhibit stable seasonal patterns, it is easily adaptable to the case of changing seasonal patterns as well. The method of adaptation will be briefly explained here, but no numerical examples will be given. The percents of moving average figures should be plotted for each quarter separately in order to detect evidence of changing seasonal patterns. Hence, for example, suppose a graph of first quarter percent of moving average figures exhibited a gradual upward drift, when plotted chronologically by years. This would indicate a gradual increase in the seasonal importance of that quarter. A free-hand curve is often fitted to such percent of moving average figures, and the values of each year given by the curve constitute the changing seasonal index. This procedure is carried out for each quarter separately. If, as in our illustration, the first quarter was increasing in seasonal importance, then in the nature of the case, one or more other quarters would be decreasing in importance. The adjustment to make the four indices for each year total 400 would ordinarily be required.

If there had been a sharp break in seasonal pattern at a specific point of time rather than a gradual shift over time, with different but stable patterns prevailing both before and after the break, separate stable seasonal indices could be constructed for each of the two time periods.

Other Measures of Seasonality

Although the ratio-to-moving average method of measuring seasonal variations is the most widely employed technique, there are several other techniques that are sometimes used. The so-called "graphic method" is essentially the same as the ratio-to-moving average technique except that it substitutes drawing a free-hand curve for the computation of a moving average to represent the trend and cyclical components of the original series. All other steps in the procedure are the same as in the ratio-to-moving average method.

Another technique is the link-relative method. In this procedure, the average relationship between successive time periods is used to describe the pattern of seasonality. Thus, for example, the standing of the second quarter relative to the first quarter is computed for each year. If a second quarter figure is higher than the corresponding first quarter figure, the link relative exceeds 100%. If the reverse is true, the link relative is less than 100. An average link relative is then obtained for each quarter-to-quarter comparison. These four numbers, when adjusted to total 400, constitute the seasonal indices.

A third method involves computing percentages of annual averages. That is, each quarter is expressed as a percentage of the average for the year. The averages of these figures for each quarter over all years are then the seasonal indices. Again, these indices are adjusted to total 400 or average to 100. Of course, all of the aforementioned methods could be applied to monthly or other suitable time series data as well as to quarterly figures.

All of these methods have advantages and disadvantages. As compared to the ratio-to-moving average technique, they all involve less computation, and are somewhat simpler procedures. On the other hand, they do not attempt to remove the non-seasonal elements as explicitly and as completely as does the ratio-to-moving average procedure. Nevertheless, these "quick and dirty" techniques may be quite effective in certain instances, particularly when used for short-term forecasts, say, of about three years or less.

Numerous electronic computer programs have been developed recently, particularly by the U.S. Bureau of the Census and the U.S. Bureau of Labor Statistics, for the computation of seasonal indices by the ratio-to-moving average method. They assume monthly data and compute changing seasonal indices, offering many refinements, useful options, and problem solving capabilities. Doubtless, in the future there will be increasing utilization of computers in connection with the general problem of the measurement of seasonality and the adjustment of economic time series for seasonal variations.

11.6 Forecasting Methods

We have seen how classical methods are used in analyzing the separate components of an economic time series. These methods involve an implicit assumption that the various components act independently of one another. For example, there were no specific procedures established for taking into account cyclical influences on seasonal variations, or secular changes in the structure of business cycles. Special procedures can be established to gauge some of these interactions, but basically the model used in classical time series analysis assumes that there are independent sources of variation in economic time series and measures these sources separately. This decomposition or separation process, although often very useful for descriptive or analytical purposes, is nevertheless artificial. Therefore, it is not surprising that for a complex problem such as economic forecasting, it virtually never suffices simply to make mechanistic extrapolations based on classical time series analysis alone. However, time series analysis frequently is a very helpful starting point and an extremely useful supplement to other analytical and judgmental methods of forecasting.

In short-term forecasting, often a combined trend-seasonal projection provides a convenient first step. For example, as a first approximation, say, in a company's forecast of next year's sales by months, a projection of a trend figure for annual sales might be obtained. Then this figure might be allocated among months based on an appropriate set of seasonal indices. Of course, the basic underlying assumption in this procedure is the persistence of the historical pattern of trend and seasonal variations of the sales of this company into the next year. A more complete forecast might involve superimposing a cyclical prediction as well. Thus, for example, again the first step may involve a projection of trend to obtain an annual sales figure. Then, an adjustment of this estimate may be made based on judgment with respect to recent cyclical growth rates. Suppose that the past few years represented the expansion phase of a business cycle and the cyclical growth rate for the economy during the next year was predicted at about 4% by a group of economists. Assume further that the company in question had found these forecasts in the past were quite accurate and applicable to the company's own cyclical growth rate—over and above its own forecast of trend levels. Then the company might increase its trend forecast by this 4% figure to obtain a trend-cycle prediction. Again, if predictions by months were required, a monthly average could be obtained from the trend-cycle forecast and seasonal indices could be applied to yield the monthly allocation. Ordinarily, no attempt would be made to predict the irregular movements.

Cyclical Forecasting and Business Indicators

Cyclical movements are more difficult to forecast than trend and seasonal elements. These cyclical fluctuations in a specific time series are strongly influenced by the general business cycle movements characteristic of large sectors of the overall economy. However, since there is considerable variability in the timing and amplitude with which many individual economic series trace out their cyclical swings, there is no simple mechanical method of projecting these movements.

Relatively "naive methods" such as the extension of the same percentage rate of increase or decrease in, say, sales as occurred last year or during the past few years are often made. These may be quite accurate, particularly if the period for which the forecast is made occurs during the same phase of the business cycle as the time periods from which the projections are made. However, the most difficult and most important items to forecast are the cyclical turning points at which reversals in direction occur. Obviously, managerial planning and implementation which has presupposed a continuation of a cyclical expansion phase can give rise to serious problems if an unpredicted cyclical downturn occurs during the planning period.

Many statistical series produced by governmental and private sources have been extensively used as business indicators. Some of these series represent activity in specific areas of the economy such as employment in nonagricultural establishments or average hours worked per week in manufacturing. Others are very broad measures of aggregate activity pertaining to the economy as a whole, as for example, gross national product and personal income. We have noted earlier that economic series, while exhibiting a certain amount of commonality in business cycle fluctuations, nevertheless display differences in timing and amplitude. The National Bureau of Economic Research has studied these differences carefully and has specified a number of time series as statistical indicators of cyclical revivals and recessions.

These time series have been classified into three groups. The first group consists of the so-called "*leading series*." These are series which have usually reached their cyclical turning points prior to the analogous turns in general economic activity. The group includes series such as the layoff rate in manufacturing; value of new orders, durable goods industries; and the common stock price index, industrials, rails, and utilities. The second group are series whose cyclical turns have roughly *coincided* with those of the general business cycle. Included are such series as the unemployment rate, the industrial production index, gross national product, and dollar sales of retail stores. Finally, the third group consists of the "*lagging series*," those whose arrivals at cyclical peaks and troughs usually lag behind those of the general business cycle. This group includes series such as plant and equipment expenditures, consumer installment debt, and bank interest on short-term business loans. It is worth noting that rational explanations stemming from economic theory can be given for the logic of the placement of the various series into the respective groups, in addition to the empirical observations themselves. These statistical indicators are adjusted for seasonal movements. They are published monthly in *Business Conditions Digest*, by the Bureau of the Census. Another publication which carries the National Bureau statistical indicators, as well as other time series with accompanying analyses, is *Economic Indicators*, published by the Council of Economic Advisors.

Probably the most widespread application of these cycle indicators is as an aid in the prediction of the timing of *cyclical turning points*. If, for example, most of the leading indicators move in an opposite direction from the prevailing phase of the cyclical activity, this is taken to be a possible harbinger of a cyclical turning point. A subsequent similar movement by a majority of the roughly coincident indices would be considered a confirmation of the fact that a cyclical turn was in progress. These cyclical indicators, like all other statistical tools, have their limitations and must be used carefully. They are not completely consistent in their timing, and leading indicators sometimes give incorrect signals of forthcoming turning points because of erratic fluctuations in individual series. Furthermore, it is not possible to predict, with any high degree of assurance, the length of time between a signal given by the leading series group of an impending cyclical turning point and the turning point itself. There has been considerable variation in this lead time during past cycles of business activity.

Diffusion Indices

Another cyclical forecasting aid which makes use of numbers of individual economic time series is the diffusion index. This type of measure utilizes the fact that various economic series attain their peak and trough levels at different points in time. A diffusion index attempts to measure the extent to which cyclical movements are diffused throughout the economy. The index, in its simplest form, is the percentage of seasonally adjusted series which are expanding at a given point in time and is ordinarily computed on a monthly basis. Hence, for example, suppose during a cyclical expansion phase, 80 out of 100 time series increased over a preceding month. The diffusion index would be 80%. As the expansion period lengthens in time and a peak in aggregate economic activity is approached, some of the previously increasing series will begin to decline. When the percentage of rising series drops below 50, a signal is indicated that a peak has been reached and that a contraction in aggregate activity is beginning. Usually, this type of diffusion index tends to lead somewhat the turning points of aggregate economic activity. However, false signals are often given as the index crosses the 50% point without the corresponding follow-up movement in general business conditions. Frequently, a few months must elapse before a turning point can be identified with a considerable degree of assurance.

This type of index clearly makes some simplifying assumptions about the phenomena it attempts to describe. It assumes that most economic series display fairly common cyclical fluctuations, but that there is also a considerable amount of spread in timing of turning points among individual series. The diffusion index also attributes equal importance to the individual series which are utilized in the determination of the percentage which are expanding. Despite these and other assumptions and limitations, diffusion indices are generally regarded by business men and economists as effective early warning signals for turning points in overall economic activity.

Other Forecasting Methods

Most individuals and companies engaged in forecasting do not depend upon any single method, but rather utilize a variety of different approaches. It stands to reason that if there is substantial agreement among a number of forecasts arrived at by relatively independent methods, greater reliance would be placed on this consensus than would have been on the results of any single technique.

Other methods of prediction range from very informal judgmental techniques to highly sophisticated mathematical models. At the informal end of this scale, for example, sales forecasts are sometimes derived from the combined outlooks of the sales force of a company, from panels of executive opinion, or from a composite of both of these. At the other end of the scale are the more formal mathematical models such as regression equations or complex econometric models. An example of the use of multiple regression equations in new product demand forecasting was given in the case study discussed in Chapter 10. There is widespread usage of various types of regression equations by which firms attempt to predict the movements of their own company's or industry's activity on the basis of relationships to other economic and demographic factors. Often, for example, a company's sales are predicted on the basis of relationships with other series whose movements precede those of the sales series to be forecasted.

Among the most formal and mathematically sophisticated methods of forecasting in current use are econometric models. An *econometric model* is a set of two or more simultaneous mathematical equations which describe the interrelationships among the variables in the system. Some of the more complex models in current use for prediction of movements in overall economic activity include dozens of individual equations. Special methods of solution for the parameters of these equation systems have been developed, since in many instances ordinary least squares techniques are not appropriate. These econometric models have been primarily used for prediction at the level of the economy as a whole, and for industries, but are coming into increasing use for prediction at the company level as well.

Management uses forecasts as an important ingredient of its planning, operational, and control functions. Invariably, no single method is relied upon but judgment is applied to the results of various forecasting methods. Often, formal prediction techniques make their greatest contribution by narrowing considerably the area within which intuitive judgment is applied.

PROBLEMS

1. Given the following data from the *ABC* Shirt Company:

$$Y = \text{Actual Sales, June 1968} = 46{,}500$$
$$Y_t = \text{Trend Value, June 1968} = 50{,}000$$
$$SI = \text{June Seasonal Index} = 90$$

   (a) Express seasonally adjusted sales as a percent of trend. What general factors account for the difference between your calculated value and 100%?
   (b) What is the meaning of the trend value?

2. A certain department store experiences marked seasonal variations in sales. The July seasonal index is 70, and the trend value for sales in July 1969 was $28,000. Do you think sales in July 1969 were closer to $28,000 or $19,600 ($28,000 × .7)? Discuss.

3. The following series shows the total national income in billions of dollars from 1933 to 1937:

Year	*National Income (billions of dollars)*
1933	35.0
1934	40.2
1935	44.0
1936	49.9
1937	55.0

(a) For each year, compute the Y_t values for the equation

$$Y_t = a + bx$$

(b) Are these Y_t values a good description of the secular trend of national income? Why or why not?

(c) Compute the relative cyclical residual for 1935 and explain what it means.

4. (a) In the percent-of-moving average method (ratio-to-moving average method) of determining seasonal indices, how is each of the non-seasonal elements removed? Explain.

   (b) Given the following data on milk production in a section of the United States during June, July, and August of 1969 and relevant seasonal indices on milk production:

Month	*Milk Production (thousands of lbs)*	*Seasonal Index*
June 1969	13,178	126.54
July 1969	12,663	117.90
August 1969	11,625	105.99

   (1) Would you attribute the decline in milk production during this period merely to seasonal variations? Specify the calculations you would make to answer this question, but do not perform these calculations.

   (2) Explain specifically how you would determine the effect of business cycle fluctuations on milk production. Specify any information that would be required in addition to that given above.

5. The trend equation for sales of the Expo Corporation is as follows:

$Y_t = 190 + .24x$
$x = 0$ in July 1963
x is in monthly intervals
Y is monthly sales in millions of dollars

Actual figures for certain months in 1967 are given below:

Month	*Actual Sales (millions of dollars)*	*Seasonal Index*
September	167	88
October	198	104
November	210	114
December	391	200

(a) What does the seasonal index of 88 for September mean?

(b) Isolate for December the effect of each component of a time series (trend, cycle and random, seasonal).

6. The following data pertain to the number of automobiles sold by the Valley Rudentino Corporation:

Year	*Quarter*	*No. of Autos Sold*
1960	1st	152
	2nd	277
	3rd	203
	4th	174
1961	1st	205
	2nd	363
	3rd	255
	4th	182
1962	1st	171
	2nd	325
	3rd	233
	4th	180
1963	1st	202
	2nd	396
	3rd	274
	4th	238
1964	1st	212
	2nd	350
	3rd	246
	4th	208
1965	1st	241
	2nd	453
	3rd	362
	4th	355

(a) Using the ratio-to-moving average method, determine constant seasonal indices for each of the four quarters.
(b) Do you think constant seasonal indices should be employed in this problem? Why or why not?
(c) Assuming that constant seasonal indices are appropriate, adjust the quarterly sales figures between 1960 and 1965 for seasonal variations.
(d) Assume the trend in sales for the Valley Rudentino Corporation can be described by the following equation:

$$Y_t = 200 + 25x$$
$x = 0$ in 2nd quarter of 1960
x is in one-year intervals
Y is the number of autos sold

Do you see any evidence of cycles in the data between 1960 and 1965? What is the basis of your answer?

7. The following table presents the consumption of electric power in the United States:

Year	*Consumption (billion kilowatt hours)*
1910	20
1920	57
1930	116
1940	182
1950	396
1960	832

   (a) What was the average amount of increase per decade in the above series between 1910 and 1960? (Do not use trend line.)
   (b) What was the average percentage increase per decade between 1910 and 1960? (Do not use trend line.)
   (c) Do you think the figures you calculated in (a) and (b) are typical for the series? Explain.
   (d) Fit a linear trend line to (1) the natural numbers and (2) the logarithms of the above data by the method of least squares.
   (e) Interpret the meaning of the constants of the trend equation to the natural numbers specifically in terms of this problem.
   (f) Are the answers to (a) and (b) consistent with the slopes obtained in (d)? Why or why not?
   (g) In 1932, consumption of electric power was 100 billion kilowatt-hours. Compute and interpret both the absolute cyclical residual and the relative cyclical residual for that year.

8. Given the following values for a particular series: actual value, December 1967 = 320 units; seasonal index, December = 200. $Y_t = 150 + 5.0x - 0.2x^2$, with $x = 0$ at June 15, 1960; x is in one year units.
   (a) What is the meaning of the seasonal index, 200?
   (b) What is the relative cyclical residual for December 1967 adjusted for seasonal variation?
   (c) What does the relative cyclical residual mean?

9. Observe the following information for the production of the M & N Company.

Month and Year	*Production (in 1000's)*	*S I*
January 1969	40	110
February 1969	38	100
March 1969	36	90

Trend Equation: $\text{Log } Y_t = 1.5500 + 0.0135x - .0007x^2$

$x = 0$ in December 1968
x is in one-month intervals
Y is production in 1000 units

(a) Summarize in words the way in which the trend of the above series is changing (based upon the equation given).
(b) Would it be proper to conclude that the M & N Company declined cyclically between January and March 1969? Show all relevant calculations.
(c) Explain the meaning of the January seasonal index (110).
(d) Would you be willing to use the above equation for forecasting? Discuss.

10. The following trend equation resulted from the fitting of a least squares parabola to the natural numbers of the size of the labor force in a Southern county.

$Y_t = 49.17 + 4.23x - 0.19x^2$
$x = 0$ in 1940
x is in 2-1/2-year intervals
Y is the size of the labor force in thousands

(a) Assume the above trend line is "a good fit." What generalizations can you make concerning the way in which the labor force of this county has grown in absolute amounts? Concerning the percentage rate at which it has grown?
(b) In part (a) above, you were instructed to assume the trend line was "a good fit." However, the actual size of the labor force in 1965 was 92,073. This is rather striking evidence that the equation is not "a good fit." Do you agree? Discuss.

11. The following sentences refer to the ratio-to-moving average method of measuring seasonal variation when applied to United States monthly gasoline sales from 1950 to 1969. Insert in the blank space of each sentence the letter corresponding to the phrase that will complete the sentence most appropriately.
(a) A twelve month moving total was computed because______.
 (1) Trend is thus eliminated.
 (2) This will give column totals equal to 1200.
 (3) Seasonal variation cancels out over a period of twelve months.
(b) A two-item total is then taken of the twelve month totals in order to______.
 (1) Obtain moving average figures for the first and last six months.
 (2) Center the moving average properly.
 (3) Eliminate the rest of the random movements.
(c) This two-item total of a twelve month moving total is divided by______ to get the centered twelve month moving average.
 (1) 14
 (2) 2
 (3) 12
 (4) 24
(d) This moving average contains______.
 (1) All of the trend, most of the cycle, all of the seasonal variation, and some irregular (random) variation.

(2) All of the trend, most of the cycle, and possibly some irregular (random) variation.
(3) Most of the trend, most of the cycle, and all of the irregular (random) variation.

(e) The original data are then divided by the moving average figures. These specific seasonal relatives, Y/MA values, contain______.
(1) All of the seasonal, possibly some of the cycle, and practically all of the irregular.
(2) Seasonal only.
(3) All of the trend, most of the cycle, and none of the irregular.

(f) Modified means are taken of the specific seasonal relatives, Y/MA in order to______.
(1) Eliminate from the specific seasonal relatives, Y/MA values, the nonseasonal elements.
(2) Get rid of seasonal elements.
(3) Eliminate the trend in the specific seasonal relatives, Y/MA.
(4) Compensate for a changing seasonal pattern.

(g) To adjust the original data for seasonal variation, one computes______.
(1) Original data times seasonal index.
(2) Seasonal index divided by original data.
(3) Original data divided by seasonal index.

C H A P T E R   T W E L V E

Index Numbers

12.1 The Nature and Use of Index Numbers

The preceding sections have dealt with methods for describing and analyzing variations in economic activity over time. Another important method for summarizing change in economic variables over time is the use of index numbers. In its simplest form, an index number is nothing more than a relative number, or a "relative" which expresses the relationship between two figures, where one of the figures is used as a base. For example, in a time series of prices of a particular commodity, the prices may be expressed as price relatives by dividing every figure by the price in a base period. In the calculation of economic indices, it is conventional to state the relative numbers as percentages, where the base period figure is 100 (%). If only a single time series is involved, as in our illustration of prices of a single commodity, the result is referred to as a simple index, or merely as a series of price relatives. On the other hand, if in each time period the price relatives for several different commodities are all combined into a single summary figure, these summary figures constitute a composite index number series. The problems of construction and interpretation of simple indices are minor compared to those of composite indices. Consequently, our discussion will pertain solely to composite indices. In keeping with general practice, we will ordinarily employ the term "index number" to mean "composite index number." In this context, then, the term "index number" refers to a summary measure which states a relative comparison between groups of related items.

Although our discussion refers to index numbers for time series, such measures may be constructed for spatial comparisons as well. Hence, for example, variations in levels of food prices among different cities for the same time period might be measured in terms of index numbers, where the price level for a particular city represents the base figure of 100.

Series of index numbers are extremely useful in the study and analysis of economic activity. Every economy, regardless of the political and social structure of the environment within which it operates is engaged in the production, distribution, and consumption of goods and services. Convenient methods of aggregation, averaging, and approximation are required to summarize the myriad of individual activities and transactions which take place. Index numbers have proved to be very useful tools in this connection. Thus, we find indices of industrial production, agricultural production, stock market prices, wholesale prices, consumer prices, prices of exports and imports, incomes of various types, and so forth in common use. A convenient classification for economic indices is in terms of indices of price, quantity, or value. The present discussion concentrates primarily on price indices because most of the problems of construction, interpretation, and use of indices may be illustrated in terms of such measures. After presenting some of the general problems of index number construction, we deal with the computational methods of construction of index numbers, using the illustrative data of a simple example.

12.2 General Problems of Index Number Construction

In a brief treatment, it is not feasible to discuss all of the relevant problems of index number construction. However, many of the important matters are subsumed under the following categories: (1) selection of items to be included, (2) choice of a base period, (3) mathematical method of construction, and (4) weights to be used. We will discuss these problems in the order given.

Selection of Items to be Included

In the construction of price indices as in other problems involving statistical methods, the definition of the problem and the statistical universe to be investigated are of paramount importance. Most of the widely used price index number series are produced by governmental agencies or sizable private organizations and are used in a large variety of ways. Hence, it is not feasible to state a simple purpose for each price index from which a clear definition of the problem and statistical universe might follow. However, every index attempts to answer meaningful questions, and it is these general purposes of an index that determine the specific items to be included. As an example, let us consider the Consumer Price Index produced by the Bureau of Labor Statistics (BLS) of the U.S. Department of Labor. This index is used in a great many ways and provides the basis for a great many economic decisions. For example, fluctuations in the wages of about 3 million workers are partially based on changes that occur in the index figures. The series is watched by monetary authorities as an indicator of inflationary or deflationary movements of prices. Even such matters as the size of alimony payments, rentals on commercial buildings, and the proportion of certain estates which must be invested in stocks have been tied to movements in this index number series. Inevitably, such series have many limitations when applied for different purposes. However, an index attempts to answer a question concerning the average movement of certain prices over time. The specific nature of this question about price movements determines the items to be included in the index. Similarly, many of the limitations of the use of the index for the aforementioned widely different purposes stem from what the index does and does not attempt to measure.

Let us pursue the illustration of the Consumer Price Index. Essentially, what this index attempts to measure is how much it would cost at retail to purchase a particular combination of goods and services compared to what it would have cost in a base period. More specifically, the combination of goods and services consists of items selected to represent a typical "market basket" of purchases by city wage earners and city clerical workers and their families. These families are considered to have "moderate incomes." The relevant universe comprises about 40% of the U.S. population. Hence, the index does not attempt to describe changes in prices of purchases by low income families, high income families, farm families, or the families of business men or professional people. By means of periodic consumer surveys, the Bureau determines the goods and services purchased by the specified families and how these families spread their spending among these items. In summary, the general question the index purports to answer determines the items to be included. Obviously, if the indices have other purposes, as for example, indices of export prices or agricultural prices, very different lists of items would be included.

However, even when the general purpose of an index is clearly defined, many problems remain concerning the choice of items to be included. In the case of the Consumer Price Index, the BLS has determined that there are about 2000 items that moderate income city wage earner families purchase. However, the BLS includes only about 400 of these goods and services, having found that these few hundred accurately reflect the average change in the cost of the entire market basket. The choice of the commodities to be included in a price index is ordinarily not determined by usual sampling procedures. Each good and service cannot be considered as a random sampling unit equally as representative as any other unit. Rather, an attempt is made to include practically all of the most important items, and by pricing these, to obtain a representative portrayal of the movement of the entire population of prices. If subgroup indices are required, as for example, indices of food, housing, medical care, etc., as well as an overall consumers' price index, more items must be included than if only the overall index were desired. After the decisions have been made concerning the commodities to be included, sophisticated sampling procedures are often utilized to determine the specific prices that will be included.

Choice of a Base Period

A second problem in the construction of a price index is the choice of a base period, that is, a period whose level of prices represents the base from which changes in prices are measured. As indicated earlier, the level of prices in the base period is taken as 100%. Price levels in non-base periods are stated as percentages of the base period level. The base period may be a conventional calendar time interval such as a month or a year, or even a period of years. It is usually considered advisable to use a time period which is "normal" as regards levels of prices. Of course, it is virtually impossible to devise a meaningful definition of what constitutes "normality" in almost any area of economic experience. However, the criterion of normality of prices in the choice of a base period implies operationally that the time period selected should not be one which is at or near the peaks or troughs of price fluctuations. Actually, there is nothing mathematically incorrect about using as a base a period when price levels were unusually low or high. The point is that the use of such time intervals as bases tends to produce distorted concepts, since comparisons are made with atypical periods.

The use of a period of years as a base provides an averaging effect on year-to-year variations. Any particular year may have relatively unique influences present, but if, say, a three-to-five-year base period is used, these will tend to be evened out. Most of the United States governmental indices have used such time intervals as base periods, as for example, 1935–1939, 1947–1949, and 1957–1959.

Another point in the choice of a base time interval is suggested by the aforementioned three time periods. That is, it is desirable that the base period be not too far away in time from the present. The further away we move from the base period the dimmer are our recollections of economic conditions prevailing at that time. Consequently, comparisons with these remote periods tend to lose significance and to become rather tenuous in meaning. Therefore, producers of index number series, such as United States governmental agencies, shift their base periods every decade or so, in order that comparisons may be made with a base time interval in the recent past. Furthermore, it is desirable to shift the base from time to time because a period which previously may have been thought of as normal or average may no longer be so considered after a long lapse of time.

Other considerations may also be involved in choosing a base period from an index. If a number of important existing indices have a certain base period, it is desirable for purposes of ease of comparability for newly constructed indices to use the same time periods. Also, as new commodities are developed and indices are revised to include them, it becomes desirable to shift the base period to a time interval which reflects the newer economic environment.

Mathematical Method of Construction

There are two basic types of price indices from the viewpoint of mathematical method of construction, the *aggregative* type and the *average of relatives*. These terms are quite descriptive of the methods used because in the aggregative type of index, prices are aggregated or summed in a non-base period and are compared with a similar aggregate of prices in the base period; or the prices in a non-base period are weighted by quantities and these products are aggregated and compared with a similar aggregate for the base period. On the other hand, in an average of relatives index, the first step involves the computation of a price relative for each commodity by dividing its price in a non-base period by the price in a base period. Then an average of these price relatives is calculated. Analogously to the situation for aggregative indices, weights may be applied to the individual price relatives in computing the index.

Weights

Since the individual items included in an index cannot ordinarily be considered to have equal importance, weighting schemes are usually employed to attempt to give each item its proper influence in the index number calculation. In the case of a price index, the relative importance of the individual commodities changes over time, primarily because the underlying conditions of supply and demand are changing. This poses a basic problem of index number construction to which there is no really satisfactory solution.

A good bit of the difficulty concerning the use of weights in a price index arises from different conceptions as to what such an index is supposed to measure. Keynes and other economists were interested in the utility and levels of satisfaction derived from goods and services. Hence, they referred to a type of cost of living index which would measure how much money would have to be expended in one period to yield the same amount of satisfaction or utility as in another period. Assumptions were required concerning the similarity of the groups of people at the two time points as well as their standards of taste. The practical difficulties in constructing such an index are obvious, and it is not surprising that no such measures are currently in existence. The types of measures which are most feasible are those referred to in the preceding sub-section relating to mathematical methods of construction; namely indices which pertain to the changing cost of a fixed combination of goods and services or to the variation in average levels of prices.

In the case of the *aggregative* type of price index, the weights employed are *quantities*. These may be quantities of goods and services consumed, produced, exported, imported, or may represent other pertinent quantities such as numbers of shares of stock traded or outstanding. On the other hand, in *averages of relatives* indices, the weights used are *values*. These may similarly be values of goods and services consumed, produced, etc. Although it is somewhat of an oversimplification to identify quantity weights with aggregative indices and value weights with averages of relatives, it is a good operating generalization for our purposes. In the illustrative examples which follow, we will see why these weighting systems are plausible for the respective types of price indices.

Even shifting from indices which attempt to deal with levels of satisfaction to price indices of either the aggregative or averages of relatives variety does not eliminate or solve the problem of weighting. Questions remain, such as whether base period weights, current period weights, or some average of the two should be used. Such problems are dealt with in the sections which follow.

12.3 Aggregative Price Indices

In this section, we consider the construction and interpretation of unweighted and weighted aggregative price indices. The unweighted index is discussed first.

Unweighted Relative of Aggregates Index

In order to illustrate the various types of price indices, we will consider the artificial problem of constructing a price index for a list of only four food commodities. The base period will be 1960, and we will be interested in the change which took place in these prices from 1960 to 1967 for a typical family of four that purchased these products at retail prices in a certain city. As was indicated earlier, the universe and other basic elements of the problem should be very carefully defined. However, we will purposely leave these matters very indefinite, and will concentrate on the methods of construction and interpretation of the various indices. Hence, these indices will deliver different answers to our vaguely worded problem. In Table 12-1 are shown the basic data of the problem, and the calculation of the unweighted relative of aggregates index, also known as the simple aggregative index. As indicated at the bottom of Table 12-1, the prices per unit are summed (or aggregated) for each year. Then, one year is selected as a base, in our case, 1960. The price index for any given year is obtained by dividing the sum of prices for that year by the similar sum for the base period. The resulting figure is multiplied by 100 to express the index in percentage form. Hence, the index takes the value 100 in the base period. If the symbol P_0 is used to denote the price in a base period and P_n the price in a non-base period, the general formula for the unweighted relative of aggregates index may be expressed as follows:

Unweighted Relative of Aggregates Price Index

(12.1)
$$\frac{\Sigma P_n}{\Sigma P_0} \cdot 100$$

Let us interpret the index figure of 129.3 for 1967. It would have cost \$2.05 in 1960 to have purchased one pound of coffee, one loaf of bread, one dozen eggs, and one pound of hamburger. The corresponding cost in 1967 was \$2.65. Expressing \$2.65 as a percentage of \$2.05, we find that in 1967 it would have cost 129.3% of the cost in 1960 to have purchased one unit each of the specified commodities. Stated in terms of percentage change, it would have cost 29.3% *more* in 1967 than in 1960 to have purchased the stated bill of goods.

The interpretation of the unweighted relative of aggregates index is very straightforward. However, this type of index suffers from the serious limitation that it is unduly influenced by high priced commodities. The total of prices in 1960 and 1967, respectively, were \$2.05 and \$2.65, an increase of \$0.60. If we added to the list of commodities one which declined from \$4.00 to \$3.00 per unit from 1960 to 1967, the totals for 1960 and 1967 would then

Table 12-1 Calculation of the Unweighted Relative of Aggregates Index for Food Prices, 1960 and 1967.

Food Commodity	*Unit Price* 1960 P_{60}	1967 P_{67}
Coffee (pound)	\$0.70	\$0.85
Bread (loaf)	0.25	0.35
Eggs (dozen)	0.50	0.65
Hamburger (pound)	0.60	0.80
	\$2.05	\$2.65

Unweighted Relative of Aggregates Index for 1967, on 1960 base

$$\frac{\Sigma P_{67}}{\Sigma P_{60}} \cdot 100 = \frac{\$2.65}{\$2.05} \times 100 = 129.3$$

For 1960, on 1960 base

$$\frac{\Sigma P_{60}}{\Sigma P_{60}} \cdot 100 = \frac{\$2.05}{\$2.05} \times 100 = 100.0$$

become \$6.05 and \$5.65. Hence the price index figure for 1967 would be 93.4, indicating a decline in prices of 6.6%. Although the prices of four commodities increased and only one decreased, the overall index shows a decline, because of the dominance of the one high priced commodity. Furthermore, this high priced commodity may be one which is relatively unimportant in the consumption pattern of the group to which the index pertains. Clearly, this type of so-called "unweighted index" is one which has an inherent haphazard weighting scheme, as indicated above.

Another deficiency of this type of index is the arbitrary nature of its calculation because of the quoted units for which the prices are stated. For example, if the price of eggs were stated per half-dozen rather than per dozen or if any of the other prices were stated on a different basis, the calculated price index figure would change. However, even if all of the prices were stated for the same quoted unit of each commodity, say, per pound, the problems concerning the inherent haphazard weighting scheme would still remain. In this case, the index would be dominated by the commodities which happened to have high prices on a per pound basis. These may be the very commodities which are purchased least, because of their expensive nature. Because of the difficulties of converting a simple aggregative index into an economically meaningful measure, the need for applying explicit weights is apparent. We now turn to weighted aggregative price indices.

Weighted Relative of Aggregates Indices

In order to attribute the appropriate importance to each of the items included in an aggregative index, some reasonable weighting plan must be used. The weights to be used depend on the purposes of the index calculation, that is, on the economic question which the index attempts to answer. In the case of a consumer food price index such as the one we have been discussing, reasonable weights would be given by the amounts of the individual food commodities purchased by the consumer units to whom the indices pertain. These would constitute so-called "quantity weights," since they represent quantities of commodities purchased. The specific types of quantities to be used in an aggregative index would depend, of course, on the economic nature of the index computed. Hence an aggregative index of export prices would use quantities of commodities and services exported, an index of import prices would use quantities imported, and so forth.

Table 12-2 shows the prices of the same food commodities given in Table 12-1, but also quantities consumed during the base period, 1960. Specifically, these figures given in the column Q_{60} (the symbol Q denotes quantity) represent average quantities consumed per week in 1960 by the consumer units to which the index pertains. Hence, they indicate an average consumption of one pound of coffee, three loaves of bread, etc. The figures given under the column labeled $P_{60}Q_{60}$ indicate the dollar expenditures for the quantities purchased in 1960. Correspondingly, the numbers under the column headed $P_{67}Q_{60}$ specify what it would have cost to purchase these amounts of food in 1967. Hence the sums, $\Sigma P_{60}Q_{60} = \$3.05$ and $\Sigma P_{67}Q_{60} = \$4.00$, indicate what it could have cost to purchase the specified quantities of food commodities in 1960 and 1967, respectively. The index number for 1967 on a 1960 base is given by expressing the figure for $\Sigma P_{67}Q_{60}$ as a percentage of the $\Sigma P_{60}Q_{60}$ figure, yielding in this case a figure of 131.1, as shown at the bottom of Table 12-2. Of course, the index number for the base period 1960 would be 100.0.

What this type of index measures is the change in the total cost of a fixed bill of goods. For example, in this case, the 131.1 figure indicates that in 1967 it would have cost 131.1% of what it cost in 1960 to purchase the weekly market basket of commodities representing an average consumption pattern in 1960. Roughly speaking, this indicates an average price rise of 31.1% for this food market basket from 1960 to 1967. Referring back to the corresponding index figure for the simple or unweighted index of 129.3%, we see that it is quite close to the 131.1 figure for the weighted index. The

Table 12-2 Calculation of the Weighted Relative of Aggregates Index for Food Prices, Using Base Period Quantities Consumed as Weights (Laspeyres Method).

	Unit Price		*Quantity*		
	1960	1967	1960		
Food Commodity	P_{60}	P_{67}	Q_{60}	$P_{60}Q_{60}$	$P_{67}Q_{60}$
Coffee (pound)	\$0.70	\$0.85	1	\$0.70	\$0.85
Bread (loaf)	0.25	0.35	3	0.75	1.05
Eggs (dozen)	0.50	0.65	2	1.00	1.30
Hamburger (pound)	0.60	0.80	1	0.60	0.80
				\$3.05	\$4.00

Weighted Relative of Aggregates Index, with Base Period Weights:

For 1967, on 1960 base

$$\frac{\Sigma P_{67}Q_{60}}{\Sigma P_{60}Q_{60}} \cdot 100 = \frac{\$4.00}{\$3.05} \times 100 = 131.1$$

reason for the closeness of the two figures is that in our example we have assumed that the prices of all four commodities have moved in the same direction with the percentage changes all falling between about 20 to 40%. On the other hand, if there had been more dispersion in price movements, for example, if some prices increased while some decreased, as is often actually the case, the weighted index would have tended to differ more from the unweighted one.

The weighted aggregative index using base period weights is also known as the Laspeyres index. The general formula for this type of index may be expressed as follows:

Weighted Relative of Aggregates Price Index,
Base Period Weights (Laspeyres Method)

(12.2)
$$\frac{\Sigma P_n Q_0}{\Sigma P_0 Q_0} \cdot 100$$

where P_0 = price in a base period
P_n = price in a non-base period
Q_0 = quantity in a base period

The basic dilemma posed by the use of any weighting system is clearly illustrated by a consideration of the Laspeyres index. Since an aggregative price index attempts to measure price changes and contains data on both prices and quantities, it appears logical to hold the quantity factor constant in order to isolate change attributable to price movements. If both prices and quantities were permitted to vary, their changes would be entangled and it would not be possible to ascertain that part of the movement due to price changes. However, by keeping quantities fixed as of the base period in a consumer price index, the Laspeyres index assumes a frozen consumption pattern. As time goes on, this becomes a more and more unrealistic and untenable assumption. The consumption pattern of the current period would seem to represent a more realistic set of weights from the economic viewpoint.

However, let us consider the implications of the use of an aggregative index using current period (non-base period) weights. This type of index is known as the Paasche method. The general formula for a Paasche index is

Weighted Relative of Aggregates Price Index, Current Period Weights (Paasche Method)

(12.3)
$$\frac{\Sigma P_n Q_n}{\Sigma P_0 Q_n} \cdot 100$$

Hence, if such an index is prepared on an annual basis, the weights would have to change each year, since they would consist of current year quantity figures. The 1961 Paasche index would be computed by the formula $\Sigma P_{61} Q_{61}/\Sigma P_{60} Q_{61}$, the 1962 index would be $\Sigma P_{62} Q_{62}/\Sigma P_{60} Q_{62}$, and so forth. The interpretation of any one of the resulting figures in terms of price change from the base period, assuming the consumption pattern of the current period, is clear. However, the use of changing current period weights destroys the possibility of obtaining unequivocal measures of year-to-year price change. For example, if the Paasche formulas for the 1961 and 1962 indices given above are compared, it will be noted that both prices and quantities have changed. Therefore, no clear statement can be made about price movements from 1961 to 1962. Thus, the use of current year weights makes year-to-year comparisons of price changes impossible.

Another practical disadvantage of using current period weights is the necessity of obtaining a new set of weights in each period. Let us consider the U.S. Bureau of Labor Statistics Consumer Price Index as an example of the difficulty of obtaining such weights. In order to obtain an appropriate set of weights for this index, the BLS conducts a massive sample survey of the expenditure patterns of families in a large number of cities. Such surveys have been carried out at about ten-year intervals. They are large-scale, expensive undertakings. From a practical standpoint, it would be simply infeasible for such surveys to be conducted at, say one-year or more frequent time intervals. Because of these disadvantages, the current period weighted aggregative method is not used in any well known price index number series.

In summary, because of the above-mentioned considerations, and other factors as well, probably the most generally satisfactory type of price index is the weighted relative of aggregates index, using a fixed set of weights. The term "fixed set of weights" rather than "base period weights" is used here, because the weights may pertain to a period which is somewhat different from the time period which represents the base for measuring price changes. For example, the base period for the Consumer Price Index is 1957–1959, whereas the weights were derived from a 1960–1961 survey of consumer expenditures. The BLS revises its weighting system about every ten years, and also changes the reference base period for the measurement of price changes with about the same frequency. This procedure constitutes a workable solution to the dilemma of needing to retain constant weights in order to isolate price change, and requiring up-to-date weights in order to have a recent realistic description of consumption patterns.

The weighted relative of aggregates index using a fixed set of weights described in the preceding paragraph is referred to as the *fixed-weight aggregative index* and is defined by the formula

Weighted Relative of Aggregates Price Index with Fixed Weights

$$\textbf{(12.4)} \qquad \frac{\Sigma P_n Q_f}{\Sigma P_0 Q_f}$$

where Q_f denotes a fixed set of quantity weights. The Laspeyres method may be viewed as a special case of this index in which the period to which the weights refer is the same as the base period for prices. In order to clarify discussion of the two different time periods, the term "weight base" is used for the period to which the quantity weights pertain, whereas the term "reference base" is used to designate the time period from which price changes are measured. Of course, a distinct advantage of a fixed-weight aggregative index is that the reference base period for measuring price changes may be changed without a corresponding change in the weight base. In certain instances, this is a useful and practical procedure, particularly in the case of some U.S. government indices which utilize data from censuses or large-scale sample surveys for changes in weights.

12.4 Average of Relatives Indices

Just as in the case of aggregative indices, averages of relatives indices may be either unweighted or weighted. We consider first the unweighted indices, using the same data on prices as were used in the preceding section.

Unweighted Arithmetic Mean of Relatives Index

The price data previously shown in Tables 12-1 and 12-2 are given in Table 12-3. The first step in the calculation of any average of relatives price index is the calculation of price relatives in which the price of each commodity is expressed as a percentage of the price in the base period. These price relatives for 1967 on a 1960 base, denoted $(P_{67}/P_{60}) \times 100$ are shown in Column (4) of Table 12-3. Theoretically, once the price relatives are obtained, any average, including the arithmetic mean, the geometric mean, median, mode, etc., could conceivably be used as a measure of their central tendency. In fact, only the arithmetic mean and geometric mean are usually considered for both unweighted and weighted indices. At the bottom of Table 12-3 are shown the calculations of both the unweighted arithmetic mean of relatives and the unweighted geometric mean of relatives for 1967 on a 1960 base. The formulas are Equations (3.2) and (3.20), with the price relatives representing the items to be averaged. Stated in general form, the formulas for these unweighted averages of relatives are:

Unweighted Arithmetic Mean of Relatives Index

$$\frac{\Sigma\left(\frac{P_n}{P_0}\cdot 100\right)}{n} \tag{12.5}$$

and

Unweighted Geometric Mean of Relatives Index

$$\text{antilog}\,\frac{\Sigma \log\left(\frac{P_n}{P_0}\cdot 100\right)}{n} \tag{12.6}$$

where

$$\frac{P_n}{P_0}\cdot 100 = \text{the price relative for a commodity or service}$$

$$n = \text{the number of commodities and services}$$

The numerical values of these two indices are very close. Again this may by explained in terms of the moderate dispersion in the individual price relatives, and, in particular, in terms of the fact that there were no extreme values. Although these are both unweighted indices, just as in the case of the unweighted aggregative index, there is, in fact, an inherent weighting pattern present. It is useful to consider the implications of this inherent weighting system. In the unweighted arithmetic mean of relatives, percentage increases are balanced off against equal percentage decreases. For example, if we consider two commodities, one whose price increased by 10% and one whose price declined by 10% from 1960 to 1967, the respective price relatives for 1967 on a 1960 base would be 110 and 90. The unweighted arithmetic mean of these two figures is 100, indicating that, on the average, prices have remained unchanged.

On the other hand, an unweighted geometric mean balances off reciprocal

Table 12-3 Calculation of the Unweighted Arithmetic Mean and Unweighted Geometric Mean of Relatives Index of Food Prices for 1967 on a 1960 Base.

Food Commodity (1)	Unit Price 1960 P_{60} (2)	Unit Price 1967 P_{67} (3)	Price Relative $\frac{P_{67}}{P_{60}} \cdot 100$ (4)	Logarithm of Price Relative $\log\left(\frac{P_{67}}{P_{60}} \cdot 100\right)$ (5)
Coffee (pound)	\$0.70	\$0.85	121.4	2.0842
Bread (loaf)	0.25	0.35	140.0	2.1461
Eggs (dozen)	0.50	0.65	130.0	2.1139
Hamburger (pound)	0.60	0.80	133.3	2.1249
			524.7	8.4691

Unweighted Arithmetic Mean of Relatives Index for 1967, on a 1960 Base

$$\frac{\Sigma\left(\frac{P_{67}}{P_{60}} \cdot 100\right)}{4} = \frac{524.7}{4} = 131.2$$

Unweighted Geometric Mean of Relatives Index for 1967, on a 1960 Base

$$\text{antilog}\,\frac{\Sigma \log\left(\frac{P_{67}}{P_{60}} \cdot 100\right)}{n} = \text{antilog}\,\frac{8.4691}{4} = \text{antilog}\ 2.1173 = 131.0$$

price changes. For example, if we consider two commodities, one whose price doubled from 1960 to 1967, and one whose price halved over the same period, the respective price relatives are 50 and 200. The geometric mean of these two relatives is 100, again indicating that, on the average, prices have remained unchanged.

In summary, the unweighted arithmetic mean attaches the same weight to equal percentage changes in opposite directions, whereas the unweighted geometric mean attaches equal weight to reciprocal price changes. However, neither of the two methods provides for an explicit weighting in terms of the importance of the commodities whose prices have changed. Since it is widely recognized that explicit weighting is required to permit the individual items in an index to exert their proper influence, virtually none of the important governmental or private organization price indices are of the "unweighted" variety. We now turn to a consideration of weighted average of relatives indices.

Weighted Arithmetic Mean of Relatives Indices

Although both the arithmetic and geometric means can be used for calculating weighted averages of relatives, in fact, only the weighted arithmetic mean is ordinarily employed. The reasons include the fact that the arithmetic mean is a much more generally understood average, and also tends to yield a more intuitively reasonable picture of the change in cost of a market basket of goods than does the geometric mean. The latter point follows from the aforementioned characteristic of the geometric mean of balancing reciprocal price changes. Only weighted means of relatives will be discussed here.

The general formula for a weighted arithmetic mean of price relatives is

Weighted Arithmetic Mean of Relatives, General Form

(12.7)
$$\frac{\Sigma\left(\frac{P_n}{P_0} \cdot 100\right) w}{\Sigma w}$$

where w = the weight applied to the price relatives

Customarily, the weights used in this type of index are values, such as values consumed, produced, purchased, sold, etc. For example, in the type of food price index we have used as our illustrative problem, the weights would be values consumed, that is, the dollar expenditures on the individual food commodities by the typical family to whom the index pertains. It seems reasonable that the importance attached to the price change for each commodity be indicated by the amounts spent on these commodities. In the field of index number construction, value = price $\times$ quantity. For example, if a commodity has a price of \$.10 and the quantity consumed is three units, then the *value* of the commodity consumed is \$.10 $\times$ 3 = \$.30. Since prices and quantities can pertain to either a base period or a current period, the following systems of weights are all possibilities: P_0Q_0, P_0Q_n, P_nQ_0, and P_nQ_n. The weights P_0Q_0 and P_nQ_n are, respectively, base period values and current period values, the other two are mixtures of base and current period prices and quantities. Interestingly, the weighting systems P_0Q_0 and P_0Q_n, when used in the weighted arithmetic mean of relatives, result in indices which are algebraically identical to the Laspeyres and Paasche aggregative indices, respectively. This point is illustrated in (12.8), where base period weights P_0Q_0 are used in the weighted arithmetic mean of relatives.

Weighted Arithmetic Mean of Relatives, with Base Period Value Weights

(12.8)
$$\frac{\Sigma\left(\frac{P_n}{P_0}\right) P_0Q_0}{\Sigma P_0Q_0} \cdot 100 = \frac{\Sigma P_nQ_0}{\Sigma P_0Q_0} \cdot 100$$

As is clear from (12.8), the P_0's in the numerator cancel, yielding the Laspeyres index. The calculation of the weighted arithmetic mean of relatives, using 1960 base period value weights is given in Table 12-4, for the data of our illustrative problem. The numerical value of the index is, of course, exactly the same as that obtained previously for the weighted aggregative index with base period quantity weights (Laspeyres Method) in Table 12-2.

Table 12-4 Calculation of the Weighted Arithmetic Mean of Relatives Index of Food Prices for 1967 on a 1960 Base, Using Base Period Value Weights.

Food Commodity (1)	*Prices* 1960 P_{60} (2)	1967 P_{67} (3)	*Price Relatives* $\frac{P_{67}}{P_{60}}\cdot 100$ (4)	*Quantity* 1960 Q_{60} (5)	$P_{60}Q_{60}$ (6)	*Weighted Price Relatives* *Col* (4) × *Col* (6) $\frac{P_{67}}{P_{60}}\cdot 100\,P_{60}Q_{60}$
Coffee (pound)	$0.70	$0.85	121.4	1	$0.70	$ 84.980
Bread (loaf)	0.25	0.35	140.0	3	0.75	105.000
Eggs (dozen)	0.50	0.65	130.0	2	1.00	130.000
Hamburger (pound)	0.60	0.80	133.3	1	0.60	79.980
					$3.05	$399.960

Weighted Arithmetic Mean of Relatives, Base Period Value Weights for 1967, on 1960 Base

$$\frac{\Sigma\left(\frac{P_{67}}{P_{60}}\cdot 100\right)(P_{60}Q_{60})}{\Sigma P_{60}Q_{60}} = \frac{\$399.960}{\$3.05} = 131.1$$

Since the two indices in (12.8) are algebraically identical, it would seem immaterial which is used, but there are instances when it is more feasible to compute one rather than the other. For example, it is more convenient to use the weighted average of relatives than the Laspeyres index when value weights are easier to obtain than quantity weights; when the basic price data are more easily obtainable in the form of relatives than absolute values, and when an overall index is broken down into a number of component indices and there is a desire for comparison of the individual components in the form of relatives. As an illustration of the first of these situations, it is usually easier for manufacturing firms to furnish value of production weights in the form of "value added by manufacturing" (sales minus cost of raw materials) than to provide detailed data on quantities produced.

As indicated earlier, the Paasche index and the weighted arithmetic mean of relatives with a P_0Q_n weighting system are algebraically identical. The reasons given for the wider use of the Laspeyres than the Paasche index analogously apply to a similarly wider usage of weighted means of relatives with a P_0Q_0 weighting system than with a P_0Q_n scheme. The other two possible value weighting systems P_nQ_0 and P_nQ_n create interpretational difficulties, and therefore are not utilized in any of the important indices.

12.5 Quantity Indices

The discussion in the preceding sections has referred to price indices. Another important group of summary measures of economic change is represented by *quantity indices.* Such indices measure changes in physical *quantities* such as the volume of industrial production, physical volume of imports and exports, quantities of goods and services consumed, volume of stock transactions, etc. In virtually all currently used *quantity indices*, what is actually measured is the change in the *value* of a set of goods from the base period to the current period attributed to changes in *quantities* only, prices being held constant. This corresponds to the interpretation of what is measured in weighted price indices as being the change in the *value* of a set of goods from the base period to the current period and is attributed to changes in *prices* only, quantities being held constant. The same types of procedures used for the calculation of price indices are also employed to obtain quantity indices. Except for the case of the unweighted aggregate index, where its calculation for a quantity index would not be meaningful, corresponding quantity indices may be obtained by interchanging P's and Q's in the formulas given earlier in this chapter.

Unweighted averages of relatives quantity indices can be determined by establishing quantity relatives $Q_n/Q_0 \cdot 100$ and calculating the arithmetic or geometric mean of these figures. As indicated in the preceding paragraph, an unweighted aggregative quantity index would not be meaningful. The reason is that it does not make sense to add up quantities which are stated in different units.

As was true for price indices, weighted quantity indices are preferable to unweighted ones. A weighted aggregative index of the Laspeyres type is given by the following formula:

Weighted Relative of Aggregates Quantity Index, Base Period Weights (*Laspeyres Method*

(12.9)
$$\frac{\Sigma Q_nP_0}{\Sigma Q_0P_0} \cdot 100$$

Just as the corresponding Laspeyres price index measures the change in price levels from a base period assuming a fixed set of quantities produced or consumed in the base period, etc., this quantity index measures change in quantities produced or consumed, etc., assuming a fixed set of prices which existed in the base period. Paralleling the corresponding situation for price indices, the weighted average of relatives quantity index, using base period value weights given in (12.10) is algebraically identical to this Laspeyres index.

Weighted Arithmetic Mean of Relatives Quantity Index, Base Period Value Weights

$$\frac{\Sigma\left(\frac{Q_n}{Q_0}\cdot 100\right)Q_0P_0}{\Sigma Q_0P_0} \tag{12.10}$$

Let us interpret the meaning of these two equivalent weighted indices by considering the Laspeyres version, given in (12.9). Also, we continue with the assumption that the raw data refer to quantities of food items consumed (during a week) and prices paid by a typical family in an urban area. The numerator of the index shows the value of the specified food items consumed in year n at base year prices. The denominator refers to the value of the food items consumed in the base year. Suppose a figure of 125 resulted from such an index. Since prices were kept constant, the increase would be solely attributable to an average increase of 25% in the quantity of these food items consumed.

FRB Index of Industrial Production

Probably the most widely used and best known quantity index in the United States is the Federal Reserve Board (FRB) Index of Industrial Production. This index measures changes in the physical volume of output of manufacturing, mining, and utilities. In addition to the overall index of industrial production, component indices are published by industry groupings such as Manufactures and Minerals, and by sub-components such as Durable Manufactures and Nondurable Manufactures. Separate indices are reported for the output of consumer goods, output of equipment for business and government use, and of materials. Following the groupings used by the last Standard Industrial Classification of the U.S. Bureau of the Budget, indices are also prepared for 25 major industrial groups and 175 subgroups. The indices are issued monthly, utilizing 1957–1959 as a reference base and 1957 as a weight base. The Index of Industrial Production is closely watched by businessmen, economists, financial analysts, and others as a major indicator of the physical output of the economy.

The method of construction is the weighted arithmetic mean of relatives, using the aforementioned base periods. Numerous problems have had to be resolved concerning both the quantity relatives and value weights. Many industries cannot easily provide physical output data for the quantity relatives. Therefore, related data are sometimes used instead which tend to move more or less parallel to output, such as shipments and man-hours worked. The weights used are value-added data, which at the individual company level represent the sales of the firm minus all purchases of materials and services from other business firms. The reason that value-added rather than value of final production weights are used is to avoid the problem of double counting. For example, if the value of final product were used for a steel company which sells its steel to an automobile company, and the value of the final product of the automobile company were also used, there would be double-counting of the steel which went into the making of the automobile. Hence, the weights used follow the value-added approach in which the value of so-called "intermediate products" that are produced at all stages prior to the final product are excluded. From the viewpoint of the economist, the value-added of a firm is conceptually equivalent to the total of its factor of production payments—wages, interest, rent, and profits. (See footnote to Table 11-1.)

12.6 Deflation of Value Series by Price Indices

One of the most useful applications of price indices is to adjust series of dollar figures for changes in levels of prices. The result of this adjustment procedure, known as "deflation," is to restate the original dollar figures in terms of so-called "constant dollars." The rationale of the procedure is illustrated in terms of the simple example given in Table 12-5. In Column (2) are shown average (arithmetic mean) weekly wage figures for factory workers in a large city in 1959 and 1967. Such unadjusted dollar figures are usually referred to as stated in "current dollars." In Column (3) is shown a consumer price index for the given city, on a reference base period of 1959. The notation (1959 = 100) is a conventional method of specifying the base period. For simplicity of interpretation, let us assume the consumer price index was computed by the Laspeyres method. As we note from Column (2), average

Table 12-5 Calculation of Average Weekly Wages in 1959 Constant Dollars for Factor Workers in a Large City, 1959 and 1967.

Year (*1*)	*Average Weekly Wages* (2)	*Consumer Price Index* (*1959* = 100) (*3*)	*"Real" Average Weekly Wages* (*1959 constant dollars*) (*4*)
1959	$ 90.00	100	$ 90.00
1967	120.00	115	104.35

weekly wages of the given workers has increased from \$90.00 in 1959 to \$120.00 in 1967, a gain of 33-1/3%. However, can these workers purchase 33-1/3% more goods and services with this increased income? If prices of all of these goods and services had remained unchanged between 1959 and 1967, all other things being equal, the answer would be "yes." But, as can be seen in Column (3), prices rose 15% over the period. To determine what average weekly wages are in terms of 1959 constant dollars (dollars with 1959 purchasing power) we carry out the division \$120.00/1.15 = \$105.35. That is, we divide the 1967 weekly wage figure in current dollars by the 1967 consumer price index stated as a decimal figure (around a base of 1.00 rather than 100) to obtain the \$104.35 figure for average weekly wages in 1959 constant dollars. As indicated in the heading of Column (4), the result of this adjustment for price change is referred to as "real wages," in this case "real average weekly wages." The implication of the term "real" is that a portion of the increase in wages in dollars is absorbed by the increase in prices. The adjustment attempts to isolate the "real change," in terms of the volume of goods and services which the weekly wages can purchase at base year prices. In summary, the dollar value figures in Column (2) are divided by the price index figures in Column (3) (stated on a base of 1.00) to obtain real value figures in Column (4). The same procedure would have been followed if there has been a series of figures in Columns (2) and (3), say annually, rather than just the current and base period figures of the example. This process of dividing a dollar value figure by a price index is referred to as a deflation of the current dollar value series, whether a decrease or an increase occurs in going from figures in current dollars to constant dollars.

The rationale of the deflation procedure stems from the basic relationship of value = price × quantity. The weekly wages in current dollars are value figures. They may be viewed as value aggregates composed of a sum of prices of labor times quantities of such labor. By dividing such a figure by a price index, an attempt is made to isolate the change attributable to the concept of quantity or physical volume. Hence, we may think of the real average weekly wage figures as reflecting the changes in quantities of goods over which the wage figures have command.

This deflation procedure is very widely used in business and economics. One interesting application is in connection with attempts to measure economic well-being and economic growth. For example, in comparing growth rates among countries, frequently one of the most important indicators used is per capita growth in real gross national product. The division of gross national product by population to yield per capita figures may be viewed as an adjustment for differences in population size. The division of the gross national product figures by a relevant price index to obtain real gross national product is an adjustment for change in price levels. The resultant figures for per capita real gross national product are extremely useful measures of physical volume of production.

Of course, there are numerous limitations to the use of the deflation procedure. For example, in the weekly wages illustration, the "market basket" of commodities and services implicit in the consumer price index may not refer specifically to the factory workers to whom the weekly wages pertain. Secondly, even if the index had been constructed for this specific group of factory workers, the index is, after all, only an average. Hence, it is subject to all the interpretational problems of any such measure of central tendency. Furthermore, inferences from such data must be used with care. For example, even if there has been an increase in real average weekly wages from one period to another, we obviously cannot immediately infer an increase in economic welfare for the factory worker group in question. In the later period there may be a less equitable distribution of this income, taxes may be higher leading to lower disposable income, and so forth. Nevertheless, despite such limitations and caveats, the deflation procedure is a very useful, practical, and widely utilized tool of business and economic analysis.

In the above illustration, we have seen how a price index may be used to remove from a value aggregate the change which is attributable to price movements. Another way to view the deflation procedure is as a method of adjustment of a value figure for changes in the purchasing power of money. In this connection, it is important to note that a *purchasing power index* is conceptually the *reciprocal of a price index.* For example, assume you have $20 to purchase shoes in a certain year when a pair of shoes costs $10. The $20 enables you to purchase two pairs of shoes. Suppose, in a later year the price of shoes has risen to $20. You now can only purchase one pair of shoes. Let us imagine a price index composed solely of the price of this pair of shoes. If the earlier year is the base period, the base period price index is 100 and the later period figure is 200. On the other hand, if the price of these shoes has doubled, the purchasing power of the dollar relative to shoes has halved. Hence, a purchasing power index which was at a level of 100 in the earlier base year should stand at a level of 50 in the later period. If the indices are stated around 1.00 rather than around 100 in the base period, the reciprocal relationship can be expressed as $2 \times 1/2 = 1$. That is, the doubling in the price index and the halving in the purchasing power index are reciprocals. It is this relationship between price and purchasing power indices which is implied in comparative popular statements of the nature that a dollar today is worth only (say) fifty cents in terms of the dollar in some earlier period.

Other Types of Deflation

Deflation procedures are sometimes used for purposes other than the adjustment of value series for price changes. One interesting example is the deflation of agricultural crop yield figures by a weather index in order to arrive at adjusted yields which reflect changes due to technology, that is, changes which are the result of improved production and efficiency techniques. The effect of weather is removed by the deflation process. The United States Department of Agriculture as a result of years of research on the effect of weather on corn crops has constructed a yearly weather index since 1929. This index has a value of 100 for a year when weather factors are "neutral," that is, have neither a favorable nor unfavorable effect on corn yields. In Table 12-6 is given an illustration of the use of this weather index for deflation of corn yields from 1960 to 1962. In a manner analogous to the procedure

Table 12-6 Deflation of Yields in the Corn Belt by a Weather Index, 1960–1962.

Years	*Actual Yields (in bushels per acre)*	*Weather Index*	*Adjusted Yields (in bushels per acre)*
1960	64.6	105	61.5
1961	74.1	108	68.6
1962	77.2	106	72.8

for deflation of value series, actual yields are divided by the weather index (stated on a base of 1.00 rather than 100) to give adjusted yields.

12.7 Selected Considerations in the Use of Index Numbers

Numerous problems arise in connection with the use of index numbers for analysis and decision purposes. A few of these are discussed below.

Shifting the Base

For a variety of reasons, it frequently becomes necessary to change the reference base of an index number series from one time period to another without returning to the original raw data and recomputing the entire series. This change of reference base period is usually referred to as "shifting the base." For example, it may be desired to compare several index number series which have been computed on different base periods. Particularly if the several series are to be shown on the same graph, it may be desirable for them to have the same base period. In other situations, the shifting of a base period may simply reflect the desire to state the series in terms of a more recent time period. The procedure for accomplishing the shift is simple and is illustrated in the following example in which a housing price index with a reference base of 1960 is shifted to a new base period of 1966.

In the following table, the original price index is shown in the first column stated on a base period of 1960. The shift to a 1966 base period is accom-

	Housing Price Index (1960 = 100)	*Housing Price Index* (1966 = 100)
1959	98.2	89.3
1960	100.0	90.9
1961	101.7	92.5
1962	103.9	94.5
1963	104.6	95.1
1964	106.8	97.1
1965	108.2	98.4
1966	110.0	100.0
1967	112.1	101.9

plished by dividing each figure in the original series by the index number for the desired new base period stated in decimal form. Hence, in this illustration, each index number on the old base of 1960 is divided by 1.10, the figure for 1966 stated as a decimal. Thus, the index number for 1959 shifted to the new base of 1966 is $98.2/1.10 = 89.3$; for 1960, the new figure is $100.0/1.10 = 90.9$, etc.

It may be noted that the relationships among the new index figures after the base is shifted are the same as in the old series. For example, the index number for 1960 exceeds that of 1959 by the same percentage in both series. That is, $100.0/98.2 = 90.9/89.3 = 1.0183$, and so forth throughout the two series. However, a subtlety arises concerning the weighting scheme. Let us suppose that the old series was computed using a Laspeyres type index. Hence, both the reference base and weight base periods are 1960. The base shifting procedure of dividing the series by the 1966 index number changes the reference period, but the weights still pertain to 1960. That is, the raw data that were originally collected for quantity weights pertained to 1960. The mere procedure of dividing the index numbers in the old series by one of its members does nothing to change these weights. Indeed, to obtain new weights for 1966 would involve a new data collection process. In summary, the new series has been shifted to a reference base period of 1966 for measuring price changes, but the weights are fixed at 1960. This point may be demonstrated algebraically as follows. Consider the Laspeyres price index for 1961 computed on the reference base of 1960. The formula for the computation of this figure may be written as $\Sigma P_{61} Q_{60}/\Sigma P_{60} Q_{60}$. Correspondingly, the formula for the 1966 price index on the 1960 base is $\Sigma P_{66} Q_{60}/\Sigma P_{60} Q_{60}$. The multiplication by 100 has been dropped in these formulas to simplify the discussion. To obtain the new price index figure for 1961 on a 1966 base, we divide the old 1961 figure by the old 1966 figure.

$$\frac{\Sigma P_{61} Q_{60}}{\Sigma P_{60} Q_{60}} \div \frac{\Sigma P_{66} Q_{60}}{\Sigma P_{60} Q_{60}} = \frac{\Sigma P_{61} Q_{60}}{\Sigma P_{66} Q_{60}}$$

Since the $\Sigma P_{60} Q_{60}$'s cancel, we note that the resultant index figure for 1961 is stated on a reference base of 1966, but the weights still pertain to 1960.

Despite the fact that weights are not changed by the simple base shifting procedure discussed here, this method of shifting reference bases is widely employed. It often represents the only practical way of base shifting, and analysts ordinarily do not view as a matter of serious concern the fact that the weighting system remains unchanged.

Splicing

Sometimes an index number series is available for a period of time, and then undergoes substantial revision including a shift in the reference base period. In these cases, if it is desired to obtain a continuous series going back through the time period of the older series prior to the shift in base, the old and revised series must be "spliced" together. Splicing involves a similar arithmetic procedure to that described earlier for shifting a base. For example, assume that a price index number series was revised by inclusion of certain new products, exclusion of some old products and change in definition of some other products. In the following table is shown such an old series on a reference base of 1960 and the revised series on a base of 1962. There must be an overlapping period for the old and revised series to provide for the splicing

	Old Price Index (1960 = 100)	*Revised Price Index* (1962 = 100)	*Spliced Price Index* (1962 = 100)
1959	95.8		88.7
1960	100.0		92.6
1961	104.3		96.6
1962	108.0	100.0	100.0
1963		103.5	103.5
1964		106.2	106.2
1965		107.8	107.8

or linking of the two series. The period of overlap in the example is 1962. The splicing of the two series to obtain a continuous series stated on the new base of 1962 is accomplished by dividing each figure in the old series by the old index figure for 1962, stated in decimal form. Hence, each figure in the old series is divided by 1.08. This states the old series on the new base of 1962.

The resulting spliced series is indicated in the last column of the above table. Had it been desired to state the continuous series on the basis of the old series and the old reference base of 1960, each figure in the revised index would be multiplied by 1.08.

Although the arithmetic procedure involved in splicing is very simple, the interpretation of the resulting continuous series may be extremely difficult, particularly if long time periods are involved. For example, it is difficult to specify precisely what is measured if, say, a price index in a later period contains prices of frozen foods, clothing made from synthetic fibers, television and similar recently developed products, whereas the spliced indices for an earlier period before these products were on the market did not contain these products. However, despite such conceptual difficulties, splicing frequently represents the only practical method of providing for comparability in similar phenomena measured by indices over different time periods.

Quality Changes

In the construction of an index such as the Consumer Price Index, the basic data on prices are collected by well trained investigators who price goods for which detailed specifications have been made. Rigidity is essential with respect to pricing the same items in the same stores. However, over time, as a result of technological and other improvements, there often is a corresponding improvement in the quality of many commodities. It is very difficult and in many cases impossible to make suitable adjustments in a price index for quality changes. The artificial but practical procedure adopted by the Bureau of Labor Statistics is to consider a product's quality improved only if changes have occurred that increase the cost of producing the product. Hence, for example, an automobile tire is not considered to have been improved if it delivers increased mileage at the same cost of production. Because of such actual improvements in product quality, many analysts feel that over reasonably long periods indices such as the Consumer Price Index which have shown steady rises in price levels overstate actual price increases in terms of a fixed market basket of goods.

Uses of Indices

Index number series are widely used in connection with business and governmental decision making and analysis. One of the best known applications of a price index is the use of the Consumer Price Index as an escalator in collective bargaining contracts. In this connection, over two million workers were covered by contracts which specify periodic changes in wage rates depending upon the amount by which the Consumer Price Index moves up or down. The Bureau of Labor Statistics Wholesale Price Index is similarly used for escalation clauses in contracts between business firms.

A great deal of use of index numbers is made at the company, industry, and overall economy level. In certain industries it is standard practice to key changes in selling prices to changes in indices of prices of raw materials and wage earnings. Assessments of past trends and current status and projection of future economic activity are made on the basis of appropriate indices. Economists follow many of the various indices for purposes of appraisal of performance of the economy and for analyzing its structure and behavior.

PROBLEMS

1. Data on mean weekly earnings of workers in a certain industry and values of the Consumer Price Index, CPI, for 1967 and 1968 are presented:

Year	*Mean Weekly Earnings*	*Consumer Price Index* (1957–59 = 100)
1967	$91.20	120
1968	95.00	125

Did average weekly earnings, when deflated by the CPI, increase from 1967 to 1968? Show your calculations and briefly explain your results.

2. United States Gross National Product, GNP, has been estimated at $447.3 billion in 1958 and at $628.7 billion in 1964. During this period an appropriate deflator (price index) rose from 100 to 108.9.
   (a) Calculate the percentage increase in current-dollar GNP during the period.
   (b) Calculate the percentage increase in constant-dollar GNP (1958 dollars).

3. (a) Assume that Gross National Product of a country has been increasing at an average rate of 3% per year. Would you agree that this may also be stated as: "GNP is increasing at an average rate of 30% per decade?"
   (b) Given the following information:

March 1968	
Manufacturers' Sales	$34.6 billion
Price Index (1960 = 100)	115
Seasonal Index	104

Assume that the price index and seasonal index are entirely appropriate to use. Express manufacturers' sales for March 1968 in seasonally adjusted, constant-dollar (1960 dollars) terms.

4. Assume the figures given below represent the Disposable Income and Consumer Price Index of a certain country.

	Disposable Income (billions of dollars)	*CPI* (1955 = 100)
1950	$100	90
1955	160	100
1960	180	120
1965	220	125

   (a) Adjust the above series on disposable income so that it reflects changes in disposable income in 1960 constant-dollars.
   (b) What was the average rate of change per year in disposable income in current-dollars during the period 1950–1965?

5. A price index of two commodities is to be constructed from the following data:

	Unit Price		*Quantities Consumed*	
Commodity	1968	1969	1968	1969
A	$1.00	$.50	3	2
B	.30	.60	7	8

A simple unweighted arithmetic mean of the two price relatives for 1969 on a 1968 base indicates that prices in 1969 were, on the average, 25% higher than in 1968. A simple unweighted arithmetic mean of the two price relatives for 1968 on a 1969 base indicates that prices in 1968 were, on the average, 25% higher than in 1969.
   (a) How do you explain these paradoxical results?
   (b) Compute what you consider to be the most generally satisfactory price index for 1969 using 1968 as a base year. You may use any of the above data you deem appropriate.
   (c) Explain precisely the meaning of the answer obtained from your calculation in part (b).

6. The Jones Metal Company uses three raw materials in its business. Given below are the average prices and the quantities consumed of these three products in 1963 and 1969.

	1963		1969	
Product	*Price*	*Quantity*	*Price*	*Quantity*
A	$20	20	$25	30
B	1	100	2	120
C	5	50	6	70

   (a) Compute an appropriate weighted relative of aggregates price index for 1969 on a 1963 base.

(b) A competitor reported that his 1969 index for the same products (1963 base year) was 120. Would you conclude that Jones paid more per unit in 1969 than his competitor? Why?

7. Assume the following is an index number series for department store prices in a certain city:

City Department Store Index

1957	100
1966	270
1967	230
1968	290
1969	310

As the analyst for a large department store you are asked to construct your own index for this store using a survey taken in 1961 which showed that the average store customer spent $250 on clothing, $50 on furniture, and $100 on all other items in that year. You also know the following:

Average Prices (in dollars)

	1957	1961	1968
Clothing	20	25	40
Furniture	40	50	60
All Others	8	10	11

Using the weighted relative of aggregate price index with base period weights
(a) Calculate the index for 1957, 1961, and 1968 if 1961 is the base year.
(b) Compare the increase in price levels for your store from 1957 to 1968 with the corresponding increase as determined from the city department store index.

8. A small electrical company produces three models of household exhaust fans. Average unit selling prices and quantities sold in 1965 and 1968 were as follows.

	1965		1968	
Model	*Price*	*Quantity (in 1000's)*	*Price*	*Quantity (in 1000's)*
Economy	$20	12	$22	15
Model *B*	30	4	36	4
Model *A*	35	8	40	9

(a) Calculate the index of fan prices for 1965 on a base year of 1968. Use the arithmetic mean of relatives method with base period weights.
(b) Explain, in words understandable to a layman, precisely what the value of your index means.

9. (a) Compute an index of apple prices for the data below by the weighted arithmetic mean of relatives method with 1966 = 100, using base year weights.
   (b) Compute a price index by the weighted aggregate method, using base year weights and the same base year.

	Price (dollars per bushel)		*Production (millions of bushels)*	
	Winesap	*Macintosh*	*Winesap*	*Macintosh*
1966	$2.45	$2.15	1.28	1.40
1967	2.52	2.23	1.32	1.47
1968	2.60	2.41	1.31	1.52

10. (a) Using the data of Problem 9, compute an index of apple *production* by the weighted arithmetic mean of relatives method, with 1966 = 100 using base year weights.
    (b) Compute an index by the weighted aggregate method, on the same base.
    (c) Compare your results in (a) and (b).

11. A company's gross sales in 1962 were $10,000,000 and in 1965 were $15,000,000. It uses the following price index as a price deflator.

	Price Index
1959	100
1962	125
1965	150

By what percentage did "real gross sales" increase from 1962 to 1965?

CHAPTER THIRTEEN

Decision Making Using Prior Information

13.1 Introduction

As has been indicated in preceding chapters of this text, in recent years there has been a reorientation of classical or traditional statistical inference, with the modern emphasis being placed on the problem of decision making under conditions of uncertainty. This modern formulation has come to be known as *statistical decision theory* or *Bayesian decision theory*. The latter term is often used as a means of emphasizing the role of Bayes' theorem in this type of decision analysis. The two ways of referring to modern decision analysis have come to be used interchangeably and will be so used henceforth in this book. Statistical decision theory has developed into an important model for the making of rational selections among alternative courses of action when information is incomplete and uncertain. It is a prescriptive theory rather than a descriptive one. That is, it presents the principles and methods for making the best decisions under specified conditions, but it does not purport to present a description of how actual decisions are made in the real world.

13.2 Structure of the Decision-Making Problem

Managerial decision making has become increasingly complex as the economy of the United States and the business units within it have grown larger and more intricate. However, Bayesian decision theory commences with the assumption that regardless of the type of decision—whether it involves long-range or short-range consequences; whether it is in finance, production, or marketing, or some other area; whether it is at a relatively high or low level of managerial responsibility—certain common characteristics of the decision problem can be discerned. These characteristics constitute the formal description of the problem and provide the structure for a solution. The decision problem under study may be represented by a model in terms of the following elements:

(1) The decision maker—he is charged with the responsibility for making the decision. The decision maker is viewed as an entity, and may be a single individual, a corporation, a government agency, etc.

(2) Alternative courses of action—the decision involves a selection among two or more alternative courses of action, referred to simply as "acts." The problem is to choose the best of these alternative acts. Sometimes the decision maker's problem is to choose the best of alternative "strategies," where each strategy is a decision rule indicating which act should be taken upon observation of a specific type of experimental or sample information.

(3) Events—occurrences which affect the achievement of the objectives. These are viewed as lying outside the control of the decision maker who does not know for certain which event will occur. The events constitute a mutually exclusive and complete set of outcomes. Hence, one and only one of the specified events can occur. Events are also synonymously referred to as "states of nature," "states of the world," or simply "outcomes."

(4) Payoff—a measure of net benefit received by the decision maker. These payoffs are summarized in a so-called "payoff table" or "payoff matrix," which displays the consequences of the choice of action selected and the event which occurs.

(5) Uncertainty—the indefiniteness concerning which events or states of nature will occur. This uncertainty is indicated in terms of probabilities

assigned to events. One of the distinguishing characteristics of Bayesian decision theory is the use of personalistic or subjective probabilities as well as other types of probabilities for these assignments.

The payoff table, expressed symbolically in general terms, is given in Table 13-1. It is assumed that there are n alternative acts, denoted $A_1, A_2, \ldots, A_n$. These different possible courses of action are listed as column headings in the table. There are m possible events or states of nature denoted $\theta_1, \theta_2, \ldots, \theta_m$. The payoffs resulting from each act and event combination are designated by the symbol u with appropriate subscripts. The letter u has been used because it is the first letter of the word "utility." The net benefit or payoff of selecting an act and having a state of nature occur can be treated most generally in terms of the utility of this consequence to the decision maker. How these utilities are arrived at is a technical matter that is discussed subsequently in this chapter. In summary, the utility of selecting act A_1 and having event θ_1 occur is denoted u_{11}; the utility of selecting act A_2 and having event θ_1 occur is u_{12}, and so forth. It may be noted that the first subscript in these utilities indicates the event that prevails and the second subscript denotes the act chosen. A convenient general notation is the symbol u_{ij}, which denotes the utility of selecting act A_j if subsequently event θ_i occurs. It is a common convention to denote the rows of a table (or matrix) by i, where i can take on values $1, 2, \ldots, m$ and the columns by j, where j can take on values $1, 2, \ldots, n$.

Table 13-1 The Payoff Table.

	Acts			
Events	A_1	A_2	$\cdots$	A_n
θ_1	u_{11}	u_{12}	$\cdots$	u_{1n}
θ_2	u_{21}	u_{22}	$\cdots$	u_{2n}
.	.	.	$\cdots$	.
.	.	.	$\cdots$	.
.	.	.	$\cdots$	.
θ_m	u_{m1}	u_{m2}	$\cdots$	u_{mn}

If the event which will occur were known with certainty beforehand by the decision maker, for example, θ_3, then he would merely have to look along row θ_3 in the payoff table and select that act which yields the greatest payoff. However, in the real world, the context of decision problems is such that since the states of nature lie beyond the control of the decision maker, he ordinarily does not know with certainty which specific event will occur. The choice of the best course of action in the face of this uncertainty is the crux of the decision maker's problem.

13.3 An Illustrative Example

In order to illustrate the ideas discussed in the preceding section, we will take a simplified business decision problem. This problem will also be continued in later sections to exemplify other principles of decision analysis. A man has invented a new device and has patented it. A bank is willing to lend him the money to manufacture the device himself. After some preliminary investigation, it is decided that the next five years is a suitable planning period for the comparison of payoffs from this invention. According to the inventor's analysis, if sales are strong, he anticipates profits of $800,000 over the next five years; if sales are average he expects to make $200,000; and if weak to lose $50,000. A company, Nationwide Enterprises, Inc., has offered to purchase the patent rights from him. Based on the royalty arrangement offered to him, the inventor estimates that if he sells the patent rights and if sales are strong he can anticipate a net profit of $400,000; if sales are average, $70,000; and if weak, $10,000. The payoff table for the inventor's problem is given in Table 13-2.

Table 13-2 Payoff Table for the Inventor's Problem. (In units of $10,000 profit.)

Events	A_1 *Manufacture Device Himself*	A_2 *Sell Patent Rights*
θ_1: Strong Sales	$80	$40
θ_2: Average Sales	20	7
θ_3: Weak Sales	−5	1

In this problem, the alternative acts, denoted A_1 and A_2, respectively, are for the inventor to manufacture the device himself or to sell the patent rights. The events or states of nature denoted θ_1, θ_2, and θ_3, respectively, are strong sales, average sales, and weak sales for the five-year planning period. The payoffs are in terms of the net profits which would accrue to the inventor under each act–event combination. In order to keep the numbers rather simple in this problem, the payoffs have been stated in units of $10,000; hence a net profit of $800,000 has been recorded as $80, a net loss of $50,000 has been entered as −$5, etc.

Some comment on the nature of the payoff figures, events, and alternative acts in decision problems and in the inventor's problem in particular is required. First of all, the appropriate payoff figures to be compared for different alternatives must be in the same units. For example, if the payoffs under one act are stated in units of time, whereas for another they are in terms of dollars, the difficulties in selecting the better act are obvious. Secondly, since the decision as to which is the best act depends on *future differences* in *net benefits* between alternatives, the present and future flows of benefits and costs associated with each alternative are the pertinent figures to be used in calculating the payoffs. Past benefits and costs are irrelevant, and benefits and costs which continue regardless of whether a particular act is chosen should not be included in the payoff calculations for that act. If the payoff figures are in terms of dollars, as in the inventor's problem, all outflows of cash should be subtracted from inflows of cash to arrive at the net benefit in terms of cash flows derived from each act.[1] In other problems where the payoff figures may be expressed in dollar cost terms, all inflows of cash should be subtracted from outflows of cash to yield the pertinent cost figures for each alternative.

The types of events or states of nature used in the inventor's problem are, of course, very simplified. As a general proposition, there is an unlimited number of possible events that could occur in the future relating to such matters as the customers, technological change, competitors, etc., which lie beyond the decision maker's control. These may all be considered to be states of nature which affect the potential payoffs of the alternative decisions to be made. However, in order to cut our way through the maze of complexities involved, and to construct a manageable framework of analysis for the problem, we can think of the variable "demand" as the resultant of all of these other underlying factors.

In the inventor's problem, three different levels of demand have been distinguished; namely, strong, average, and weak. It is helpful in this regard to think of "demand" as a random variable. In the present problem, demand is a discrete random variable which can take on three possible values. Demand could have been conceived of as a discrete random variable taking on any finite or infinite number of values. For example, it could have been

---

[1]It is good practice in the comparison of economic alternatives to compare the present values of discounted cash flows or, what amounts to the same thing, equivalent annual rates of return. Both of these methods take into account the time value of money; that is, the fact that a dollar received today is worth more than a dollar received in some future period. These are conceptually the types of monetary payoff values which should appear in the payoff table. This point is amply discussed in standard texts dealing with economy studies or investment analysis. To avoid a lengthy tangential discussion, we will not elaborate on the point here.

stated in numbers of units demanded or, say, in hundreds of thousands of units demanded. No new methodological problems arise; however, the amount of arithmetic required would obviously depend on the detail in which the events are stated. Demand can also be treated as a continuous rather than discrete random variable. The conceptual framework of the solution to the decision problem remains the same, but the required mathematics differs somewhat from the case where the events are stated in the form of a discrete random variable. Often the ability to state all of the relevant states of nature requires the exercise of imagination, experience, and systematic investigation.

The listing of alternative courses of action frequently also requires some imagination, experience, and systematic investigation. In business decision problems, the alternative acts are really the possible alternative uses of the decision maker's resources. In the inventor's problem, the alternatives were in terms of manufacturing the device himself or selling the patent rights. In other problems, the alternatives might be in terms of different possible investments, different types and sizes of plants to construct, the decision to merge or not to merge with another company, etc. Often, imagination is required to discover or ferret out possible alternative courses of action which might have otherwise been overlooked. Some acts or entire classes of acts can often be eliminated from consideration in a decision problem, simply because they are in violation of laws or existing social mores. In other situations, acts may not have been included in a list of alternatives to be considered, not because of their conscious rejection by the decision maker, but because of lack of awareness on the part of the decision maker. This lack of awareness may stem from a failure on the part of the decision maker to think in terms of change from conventional methods of operation and organization. It is important in any decision problem for creative thought to be exercised in the discovery and recognition of possible courses of action, because any optimizing technique can only choose the best course of action among those considered. If there is a better act which has not been included in the list of alternatives to be analyzed, clearly it cannot be chosen.

13.4 Criteria of Choice

Assuming the inventor in our illustrative problem has carried out the thinking, experiments, data collection, etc., required to construct the payoff matrix (Table 13-2), how should he now compare the alternative acts? Neither act is preferable to the other under all states of nature. For example, if event θ_1 occurs, that is, if sales are strong, the inventor would be better off to manufacture the device himself (act A_1), realizing a profit of \$800,000, as compared to selling the patent rights (act A_2), which would yield a profit of only \$400,000.

On the other hand, if event θ_3 occurs, and sales are weak, the preferable course of action would be to sell the patent rights, thereby earning a profit of $10,000, as compared to a loss of $50,000. If the inventor knew with *certainty* which event was going to occur, his decision procedure would be very simple. He would merely have to look along the row represented by that event and select the act which yielded the highest payoff. However, it is the uncertainty with regard to which state of nature will prevail that makes the decision problem an interesting one.

Maximin Criterion

A number of different criteria for selecting the best act have been suggested. One of the earliest suggestions, made by the mathematical statistician Abraham Wald,[2] is known as the *maximin* criterion. Under this method, the decision maker is supposed to assume that once he has chosen a course of action, nature might be malevolent and hence might select the state of nature which minimizes the decision maker's payoff. According to Wald, the decision maker should choose the act which maximizes his payoff under this pessimistic assumption concerning nature's activity. In other words, Wald suggested that a selection of the "best of the worst" is a reasonable form of protection. Applying this criterion to our illustration, if the decision maker chose act A_1, nature would cause event θ_3 to occur and the payoff would be a loss of $50,000. If the decision maker chose A_2, nature would again cause θ_3 to occur, since that would yield the worst payoff, in this case, a profit of $10,000. Comparing these worst or minimum payoffs, we have

	Action	
	A_1	A_2
Minimum Payoffs (in units of $10,000)	-5	1

The decision maker is now supposed to do the best he can in the face of this sort of perverse nature, and select that act which yields the greatest minimum payoff, act A_2. That is, he should sell the patent rights, for which the minimum payoff is $10,000. Thus, the proposed decision procedure is to choose the act which yields the maximum of the minimum payoffs—hence the use of the term "maximin."

Obviously, the maximin is a very pessimistic type of criterion. It is not reasonable to suppose that the usual businessman would or should make his decisions in this way. By following this decision procedure, he would always

[2]Abraham Wald, *Statistical Decision Functions* (New York: John Wiley and Sons, 1950).

be concentrating on the worst things that could happen to him. In most situations, the maximin criterion would freeze the businessman into complete inaction, and would imply that he should go out of business entirely. For example, let us consider an inventory stocking problem, where the events are possible levels of demand, the acts are possible stocking levels, that is, the numbers of items to be stocked, and the payoffs are in terms of profits. If zero items are stocked, the payoffs will be zero for every level of demand. For each of the other numbers of items stocked, we can assume that for some levels of demand, losses will occur. Since the worst that can happen if no items are stocked is that no profit will be made, whereas under all other courses of action the possibility of a loss exists, the maximin criterion would require the businessman to carry no stock or, in effect, go out of business. Such a procedure is not necessarily irrational, and it might be consistent with the attitudes toward risk of certain people. However, for the businessman who is willing to take some risks in the pursuit of his objectives, such an arbitrary decision rule would be completely unacceptable. A number of other decision criteria have been suggested by various writers, but, to avoid a lengthy digression, they will not be discussed here.[3]

It seems reasonable to argue that a decision maker should take into account the probabilities of occurrence of the different possible states of nature. As an extreme example, if the state of nature which results in the minimum payoff for a given act has (say) only one chance in a million of occurring, it would seem unwise to concentrate on this possible occurrence. The decision procedures we will focus upon include the probabilities of states of nature as an important part of the problem.

Expected Profit Under Uncertainty

In a realistic decision problem, it would be reasonable to suppose that a decision maker would have some idea of the likelihood of occurrence of the various states of nature and that this knowledge would help him choose a course of action. For example, in our illustrative problem, if the decision maker felt very confident that sales would be strong this would tend to move him toward manufacturing the device himself, since the payoff under that act would exceed that of selling the patent rights. By the same reasoning, if he were very confident that sales would be weak, he would be influenced to sell the patent rights. If there are many possible events and many possible courses of action, the problem becomes complex, and the decision maker clearly needs some orderly method of processing all the relevant information.

---

[3]See, for example, Chapter 5 of D. W. Miller and M. K. Starr, *Executive Decisions and Operations Research* (Englewood Cliffs, N.J.: Prentice-Hall, 1960).

Such a systematic procedure is provided by the computation of the *expected* monetary value of each course of action, and the selection of that act which yields the highest of these expected values. As we shall see, this procedure yields reasonable results in a wide class of decision problems. Furthermore, we will see how this method can be adjusted for the computation of expected utilities rather than expected monetary values in cases where the maximization of expected monetary values is not an appropriate criterion of choice.

We now return to the inventor's problem to illustrate the calculations for decision making by maximization of the expected monetary value criterion. In this case, the maximization takes the form of selecting that act which yields the largest expected profit. Let us assume that the inventor carries out the following probability assignment procedure. On the basis of extensive investigation of the experience with similar devices in the past, and on the basis of interviews with experts, the inventor comes to the conclusion that the odds are 50-50 that sales will be average; that is, the event we previously designated as θ_2 will occur. Furthermore, he concludes that it is somewhat less likely that sales will be strong (event θ_1) than that they will be weak (event θ_3). On this basis, the inventor assigns the following subjective probability distribution to the events in question:

Events	*Probability*
θ_1: Strong sales	0.2
θ_2: Average sales	0.5
θ_3: Weak sales	0.3
	1.0

In order to determine the basis for choice between the inventor's manufacturing the device himself (act A_1) and selling the patent rights (act A_2), we compute the expected profit for each of these courses of action. These calculations are shown in Table 13-3. As indicated in that table, profit is treated as a random variable, which takes on different values depending upon which event occurs. We compute its expected value in the usual way, according to Equation (4.1). As noted earlier in Chapter 4, the "expected value of an act" is the weighted average of the payoffs under that act, where the weights are the probabilities of the various events that can occur.

We see from Table 13-3 that the inventor's expected profit if he manufactures the device himself is \$245,000, whereas if he sells the patent rights, his expected profit is only \$118,000. If he acts on the basis of maximizing his expected profit, the inventor would select act A_1; that is, he would manufacture the device himself.

It is useful to have a brief term to refer to the expected benefit of choosing the optimal act under conditions of uncertainty. We shall refer to the expected value of the monetary payoff of the best act as the *expected profit under uncertainty*. Hence, in the foregoing problem, the expected profit under uncertainty is equal to $245,000.

We can summarize the method of calculating the *expected profit under uncertainty* as follows:

(1) Calculate the expected profit for each act as the weighted average of the profits under that act, where the weights are the probabilities of the various events that can occur.

(2) The expected profit under uncertainty is the maximum of the expected profits calculated under Step (1).

We now turn to an analysis of the same problem from another important point of view, that of "opportunity loss." The relationship between the results obtained by the two alternative methods of solution, that is, comparison of expected payoffs and comparison of expected opportunity losses is important in statistical decision theory.

Table 13-3 Inventor's Expected Profits (in units of $10,000 profit).

Act A_1: Manufacture Device Himself

Events	*Probability*	*Profit*	*Weighted Profit*
θ_1: Strong sales	0.2	$80	$16.0
θ_2: Average sales	0.5	20	10.0
θ_3: Weak sales	0.3	−5	−1.5
	1.0		$24.5

Expected Profit = $24.5 (ten thousands of dollars)
= $245,000

Act A_2: Sell Patent Rights

Events	*Probability*	*Profit*	*Weighted Profit*
θ_1: Strong sales	0.2	$40	$ 8.0
θ_2: Average sales	0.5	7	3.5
θ_3: Weak sales	0.3	1	0.3
	1.0		$11.8

Expected Profit = $11.8 (ten thousands of dollars)
= $118,000

Expected Opportunity Loss

A useful concept in the analysis of decisions under uncertainty is that of "opportunity loss." An opportunity loss is the loss incurred because of failure to take the best action available. Opportunity losses are calculated separately from each event which might occur. Given the occurrence of a specific event, we can determine the best act available. For a given event, the opportunity loss of an act is the difference between the payoff of that act and the payoff for the best act that could have been selected. Thus, for example, in the inventor's problem, if event θ_1 (strong sales) occurs, the best act is A_1 for which the payoff is \$80 (in units of ten thousand dollars). The opportunity loss of that act is \$80 − \$80 = \$0. The opportunity loss of Act A_2 is \$80 − \$40 = \$40.

It is conventional to state opportunity losses as positive numbers. Hence, when the original payoff table is in terms of profit figures, for each event, the payoffs of the various acts are subtracted from the payoffs of the best act available to obtain opportunity loss figures. On the other hand, if the original payoff table is in terms of cost figures, the cost of the best act is subtracted from the costs of the various acts to give opportunity loss figures. It is convenient to asterisk the payoff of the best act for each event in the original payoff table so as to denote that opportunity losses are measured from these figures. Both the original payoff table and the opportunity loss table are given in Table 13-4 for the inventor's problem.

We can now proceed with the calculation of expected opportunity loss in a completely analogous manner to the calculation of expected profits. Again, we use the probabilities of events as weights and determine the weighted average opportunity loss for each act. Our goal is to select that act which yields the *minimum* expected opportunity loss. The calculation of the expected opportunity losses for the two acts in the inventor's problem is given in Table 13-5. In this and subsequent chapters, the symbols OL will be used to denote opportunity loss, and EOL to represent expected opportunity loss. Hence, EOL(A_1) and EOL(A_2) denote the expected opportunity losses of acts A_1 and A_2, respectively. The inventor's EOL if he manufactures the device himself is \$18,000, and if he sells the patent rights, his EOL is \$163,000. If he selects the act which minimizes his EOL, he will choose A_1, that is, to manufacture the device himself. This is the same act that he selected under the criterion of maximizing expected profit. It can be proved that the best act according to the criterion of maximizing expected profit is also best if the decision maker follows the criterion of minimizing expected opportunity loss. The relationship between the maximum expected profit and the minimum expected opportunity loss will be examined later. It should be noted that

Table 13-4 Payoff Table and Opportunity Loss Table for the Inventor's Problem (in units of $10,000)

	Payoff Table		*Opportunity Loss Table*	
	Acts		*Acts*	
Events	A_1	A_2	A_1	A_2
θ_1: Strong sales	$80*	$40	$0	$40
θ_2: Average sales	20*	7	0	13
θ_3: Weak sales	−5	1*	6	0

Table 13-5 Expected Opportunity Losses in the Inventor's Problem (in units of $10,000).

Act A_1: Manufacture Device Himself

Events	*Probability*	*Opportunity Loss*	*Weighted Opportunity Loss*
θ_1: Strong sales	0.2	0	0
θ_2: Average sales	0.5	0	0
θ_3: Weak sales	0.3	6	1.8
	1.0		1.8

$$\text{EOL}(A_1) = 1.8 \text{ (ten thousands of dollars)} = \$18{,}000$$

Act A_2: Sell Patent Rights

Events	*Probability*	*Opportunity Loss*	*Weighted Opportunity Loss*
θ_1: Strong sales	0.2	40	8.0
θ_2: Average sales	0.5	13	6.5
θ_3: Weak sales	0.3	6	1.8
	1.0		16.3

$$\text{EOL}(A_2) = 16.3 \text{ (ten thousands of dollars)} = \$163{,}000$$

opportunity losses are not losses in the accountant's sense of profit and loss, because as we have seen, they even occur where only profits of different actions are compared for a given state. They represent foregone opportunities rather than incurred monetary losses.

Minimax Opportunity Loss

It was noted earlier that various criteria of choice have been suggested for the decision problem. One that has been advanced in terms of opportunity losses is that of "minimax opportunity loss." Under this method, the decision maker selects that act which minimizes the worst possible opportunity loss he can incur among the various acts. As with the maximin criterion which was applied to payoffs, the minimax criterion for opportunity loss takes a pessimistic view toward which states of nature will occur. Once the opportunity loss table has been prepared as in Table 13-4, the decision maker determines for each act the largest opportunity loss he can incur. For example, in the inventor's problem, under act A_2, it is \$400,000. The decision maker then proceeds to choose that act for which these worst possible losses are the least. That is, he minimizes among the maximum losses; hence, the use of the term "minimax." In the inventor's problem, the decision maker would choose act A_1, since the maximum possible opportunity loss \$60,000 under this course of action is less than the corresponding worst loss under act A_2, \$400,000. This criterion is also sometimes referred to as "minimax regret" where the opportunity losses are viewed as measures of regret for the taking of less than the best courses of action.[4]

In our illustrative problem, it happens that the minimax opportunity loss course of action, act A_1, is the same decision as would be made under the criterion of maximizing expected profit. However, this would not necessarily be so. The minimax loss criterion like the maximin payoff rule singles out for each course of action the worst consequence that can befall the decision maker, and then attempts to minimize this damage. As was true for the maximin payoff criterion, the minimax loss viewpoint yields results in many instances which imply that a businessman who faces risky ventures should simply go out of business.

Throughout the remainder of this text, we will use the Baysian decision theory criterion of maximizing expected profit or its equivalent, minimizing expected opportunity loss.

13.5 Expected Value of Perfect Information

In our discussion thus far, we have considered situations in which the decision maker chooses among alternative courses of action on the basis of *prior information* without attempting to gather further information before he

[4]See L. J. Savage, "The Theory of Statistical Decision," *Journal of the American Statistical Association*, **46,** 55–57, 1951.

makes his decision. That is, the probabilities used in computing the expected value of each act, as shown in Table 13-3, are termed "prior probabilities" to indicate that they represent probabilities established prior to obtaining additional information through sampling. The procedure of calculating expected values of each act based on these prior probabilities and selecting the optimal act is referred to in Bayesian decision theory as *prior analysis.* In Chapter 14 we consider how courses of action may be compared after these prior probabilities are revised on the basis of sample information, experimental data, or information resulting from tests of any sort. However, the analysis carried out to this point provides a yardstick for measuring the value of perfect information concerning which events will occur. This yardstick will be referred to as the "expected value of perfect information." In order to determine this value, we must calculate the *expected profit with perfect information.* Then, if we subtract the *expected profit under uncertainty,* whose calculation we previously examined, we will have the *expected value of perfect information.* These concepts will be explained in terms of the inventor's problem. We begin with the idea of expected profit with perfect information.

The calculation of the expected profit of acting with perfect information is based on the expected payoff if the decision maker has access to a perfect predictor. It is assumed that if this perfect predictor forecasts that a particular event will occur, then indeed that event will occur. The expected payoff under these conditions for the inventor's problem is given in Table 13-6. In order to understand the meaning of this calculation, it is necessary to adopt a long-run relative frequency point of view. If the forecaster says the event "strong sales" will prevail, the decision maker can look along that row in the payoff table and select the act that yields the highest profit. In the case of strong sales, the best act is A_1, which yields a profit of $800,000. Hence the figure $80 is entered under the profit column in Table 13-6. The same procedure is used to obtain the payoffs for each of the other possible events. The probabilities shown in the next column are the original probability assignments to the three states of nature. From a relative frequency viewpoint, these probabilities are now interpreted as the proportion of times the perfect predictor would forecast that each of the given states of nature would occur if the present situation were faced repetitively. Each time the predictor makes his forecast the decision maker selects the optimal payoff. The expected profit with perfect information is then calculated as shown in Table 13-6 by weighting these best payoffs by the probabilities and totaling the products. The expected profit with perfect information in the inventor's problem is $263,000. This figure can be interpreted as the average profit the inventor could realize from this type of device if he were faced with this decision problem repeatedly under identical conditions, and if he always took the best action after receiving the forecast of the perfect indicator. Expected profit with perfect information has sometimes been called the "expected profit under certainty," but this term

Table 13-6 Calculation of Expected Profit with Perfect Information for the Inventor's Problem (in units of $10,000 profit).

Predicted Event	*Profit*	*Probability*	*Weighted Profit*
θ_1: Strong sales	$80	0.2	$16.0
θ_2: Average sales	20	0.5	10.0
θ_3: Weak sales	1	0.3	.3
			$26.3

Expected Profit with Perfect Information = $26.3 (ten thousands of dollars)
= $263,000

is clearly somewhat misleading. The inventor is not *certain* to earn any one profit figure. The expected profit with perfect information is to be interpreted as indicated in this discussion.

The expected value of perfect information, abbreviated as EVPI, is defined as *the expected profit with perfect information* minus *the expected profit under uncertainty*. Its calculation is shown in Table 13-7.

The interpretation of the EVPI is clear from its calculation. In the inventor's problem, his expected payoff if he selects the optimal act under conditions of uncertainty is $245,000. (See Table 13-3.) On the other hand, if the perfect predictor were available and the inventor acted according to his predictions, the expected payoff would be $263,000. (See Table 13-6.) The difference of $18,000 represents the increase in profit attributable to the use of the perfect forecaster, and thus, the expected value of perfect information.

The expected opportunity loss of selecting the optimal act under conditions of uncertainty in the inventor's problem was also shown earlier to be $18,000. That is, this figure represented the minimum value among the expected opportunity losses associated with each act. As shown in Table 13-7, this figure is equal to the expected value of perfect information. It can be

Table 13-7 Calculation of Expected Value of Perfect Information for the Inventor's Problem.

Expected Profit with Perfect Information	$263,000
Less: Expected Profit under Uncertainty	245,000
Expected Value of Perfect Information (EVPI)	$ 18,000

EVPI = EOL of the Optimal Act under Uncertainty = $18,000

mathematically proved that this equality is true in general. Another term used for the expected opportunity loss of the optimal act under uncertainty is the *cost of uncertainty*. This term highlights the "cost" attached to the making of a decision under conditions of uncertainty. Expected profit would be larger if a perfect predictor were available and this uncertainty were removed. Hence this cost of uncertainty is also equal to the expected value of perfect information. In summary, the following three quantities are equivalent:

Expected value of perfect information
Expected opportunity loss of the optimal act under uncertainty
Cost of uncertainty

13.6 Decision Diagram Representation

It is useful to represent the structure of a decision problem under uncertainty by a "decision tree diagram," "decision diagram," or briefly "tree." The problem can be depicted in terms of a series of choices made in alternating order by the decision maker and "Chance." Forks at which the decision maker is in control of choice are referred to as *decision forks;* those at which Chance is in control as *chance forks*. Decision forks will be represented by a little square, whereas no special designation will be used for chance forks. Forks may also be referred to as branching points or junctures.

A simplified decision diagram is given for the inventor's problem in Figure 13-1. After explaining this skeletonized version, we will insert some additional information to obtain a completed diagram. As we can see from Figure 13-1, the first choice is the decision maker's at branching point 1. He can follow either branch A_1 or branch A_2; that is, he can choose either act A_1 or A_2. Assuming he follows path A_1, he comes to another juncture, which is a chance fork. Chance now determines whether the event which will occur is θ_1, θ_2, or θ_3. If Chance takes him down the θ_1 path, the terminal payoff is \$800,000; the corresponding payoffs are indicated for the other paths. An analogous interpretation holds if he chooses to follow branch A_2. Thus the decision diagram depicts the basic structure of the decision problem in schematic form. In Figure 13-2, additional information is superimposed on the diagram to represent the analysis and solution to the problem.

The decision analysis process represented by Figure 13-2 (and in other decision diagrams to be considered at later points) is known as *backward induction*. We imagine ourselves as located at the right-hand side of the tree diagram, where the monetary payoffs are. Let us consider first the upper three paths denoted θ_1, θ_2, and θ_3. To the right of these symbols we enter the respective probability assignments 0.2, 0.5, and 0.3 as given in Table 13-3.

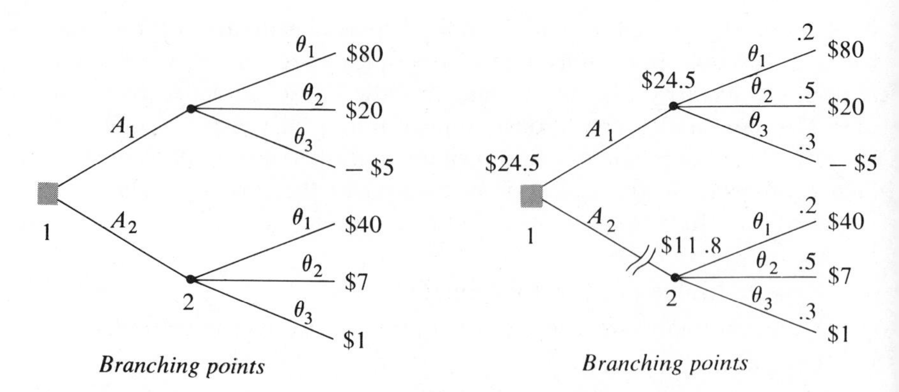

Figure 13-1 Simplified decision diagram for inventor's problem (payoffs are in units of $10,000 profit).

Figure 13-2 Decision diagram for inventor's problem (payoffs are in units of $10,000 profit).

These represent the probabilities assigned by Chance to following these three paths, after the decision maker has selected act A_1. Moving back to the chance fork from which these three paths emanate, we can calculate the expected monetary value of being located at that fork. This expected monetary value is $24.5 (in units of $10,000 as are the other obvious corresponding numbers) and is calculated in the usual way, that is,

$$\$24.5 = (.2)(\$80) + (.5)(\$20) + (.3)(-\$5).$$

This figure is entered at the chance fork under discussion. It represents the value of standing at that fork after choosing act A_1, as Chance is about to select one of the three paths. The analogous figure entered at the lower chance fork is $11.8. Therefore, imagining ourselves as being transferred back to branching point 1, where the little square represents a fork at which the decision maker can make a choice, we have the alternatives of selecting act A_1 or act A_2. Each of these acts leads us down a path at the end of which is a risky option whose expected profit has been indicated. Since following path A_1 yields a higher expected payoff than path A_2, we block off A_2 as a non-optimal course of action. This is indicated on the diagram by the two vertical lines. Hence, A_1 is the optimal course of action, and it has the indicated expected payoff of $24.5.

Thus, the decision tree diagram reproduces in compact schematic form the analysis given in Table 13-3. An analogous diagram could be constructed in terms of opportunity losses to reproduce the analysis of Table 13-5. However, it is much more customary to use tree diagrams to portray analyses in terms of payoffs rather than opportunity losses, and we will follow the usual practice in this text.

13.7 Decision Making Based on Expected Utility

In the decision analysis discussed up to this point, the criterion of choice was the maximization of expected monetary value. This criterion can be interpreted as a test of preferredness which selects as the optimal act the one which yields the greatest long-run average profit. That is, in a decision problem such as our illustrative example involving the inventor's choices, the optimal act is the one which would result in the largest long-run average profit if the same decision had to be made repeatedly under identical environmental conditions. In general, in such decision making situations, as the number of repetitions becomes large, the observed average payoff approaches the theoretical expected payoff. Gamblers, baseball managers, insurance companies, and others who engage in what is colloquially called "playing the percentages," may often be characterized as using the aforementioned criterion. However, many of the most important personal and business decisions are made under unique sets of conditions and in some of these occasions it may not be realistic to think in terms of many repetitions of the same decision situation. Indeed, in the business world, many of management's most important decisions are unique, high-risk, high-stake choice situations, whereas the less important, routine, repetitive decisions are ones that are customarily delegated to subordinates. Therefore, it is useful to have an apparatus for dealing with one-time decision making. Utility theory, which we discuss in this section, provides such an apparatus, as well as providing a logical method for repetitive decision making too.

Whether an individual, a corporation, or other entity would be willing to make decisions on the basis of the expected monetary value criterion depends upon the decision maker's attitude toward risk situations. Several simple choice situations are presented in Table 13-8 to illustrate that in choosing between two alternative acts we might select the one which has the lower expected value. The reason we might make such a choice is our feeling that the incremental expected gain of the act with greater expected monetary values does not sufficiently reward us for the additional risk involved.

In Table 13-8 are given three choice situations for alternative acts grouped in pairs. For each pair a decision must be made between the two alternatives.

The illustrative choices are to be made once and only once. That is, there is to be no repetition of the decision experiment. We shall assume for simplicity that all monetary payoffs are tax free. Suppose you choose acts C_2,

Table 13-8 Alternative Courses of Action with Different Expected Monetary Payoffs.

C_1: Receive \$0 for certain. That is, you are certain to incur neither a gain nor a loss.	or	C_2: Receive \$.60 with probability 1/2 and lose \$.40 with probability 1/2.
C_3: Receive \$0 for certain. That is, you are certain to incur neither a gain nor a loss.	or	C_4: Receive \$60,000 with probability 1/2 and lose \$40,000 with probability 1/2.
C_5: Receive a \$1,000,000 gift for certain.	or	C_6: Receive \$2,100,000 with probability 1/2 and receive \$0 with probability 1/2.

C_3, and C_5. In the case of the choice between C_1 and C_2, you might argue as follows. "The expected value of act C_1 is \$0; the expected value of C_2 is $E(C_2) = (1/2)(\$.60) + (1/2)(-\$.40) = \$.10$. C_2 has the higher expected value, and since if I incur the loss of \$.40, I can sustain such a loss with equanimity, I am willing to accept the risk involved in the selection of this course of action."

A useful way of viewing the choice between C_1 and C_2 is to think of C_2 as an option in which a fair coin is tossed. If it lands "heads" you receive a payment of \$.60. If it lands "tails" you must pay \$.40. On the other hand, the choice of act C_1 means that you are unwilling to play the game involved in flipping the coin; hence, you neither lose nor gain anything.

On the other hand, you might very well choose act C_3 rather than C_4, even though the respective expected values are

$$E(C_3) = \$0$$

$$E(C_4) = (1/2)(\$60{,}000) + (1/2)(-\$40{,}000) = \$10{,}000$$

In this case, you might reason that even though act C_4 has the higher expected monetary value, a calamity of no mean proportions would occur if the coin landed tails, and you incurred a loss (say, a debt) of \$40,000. Your present level of assets might cause you to view such a loss as intolerable. Hence, you would refuse to play the game. If you look at the difference between the two choices just discussed, C_1 versus C_2 and C_3 versus C_4, you will note that the only difference is that in C_2 we had the payoffs \$.60 and $-\$.40$. In C_4 the decimal point has been moved five places to the right for each of these numbers, making the monetary gains and losses much larger than in C_2. In all other respects, the wording of the choice between C_1 and C_2 and between C_3 and

C_4 is the same. Nevertheless, as we shall note after the ensuing discussion of the choice between acts C_5 and C_6, it is not necessarily irrational to select act C_2 over C_1 where C_2 has the greater expected monetary value and C_3 over C_4 where C_3 has the smaller expected monetary value.

In the choice between acts C_5 and C_6, you would probably select act C_5, which has the lower expected monetary value. That is, most people would doubtless prefer a gift of \$1,000,000 for certain to a 50-50 chance at \$2,100,000 and \$0, for which the expected payoff is

$$E(C_6) = (1/2)(\$2{,}100{,}000) + (1/2)(\$0) = \$1{,}050{,}000$$

In this case, you might argue that you would much prefer to have the \$1,000,000 for certain, and go home to contemplate your good fortune in peace than to play a game where on the flip of a coin you might receive nothing at all. You might further feel that there are relatively few things that you could do with \$2,100,000 that you could not accomplish with \$1,000,000. Hence, the incremental satisfaction to be derived even from winning on the toss of the coin in C_6 might not convince you to take the risk involved as compared to the "sure thing" of \$1,000,000 in the selection of act C_5.

From the above discussion, we may conclude that it is logical to depart sometimes from the criterion of maximizing expected monetary values in making choices in risk situations. We cannot specify how a person *should* choose among alternative courses of action involving monetary payoffs, given only the type of information contained in Table 13-8. His decisions will clearly depend upon his *attitude toward risk*, which in turn will depend on a combination of factors such as his level of assets, his liking or distaste for gambling, his psycho-emotional constitution, etc. Singling out the factor of level of assets, for example, it is evident that a large corporation with a substantial level of assets may choose to undertake certain risky ventures that a smaller corporation with smaller assets would avoid. In the case of the larger corporation an outcome of a loss of a certain number of dollars might represent an unfortunate occurrence but as a practical matter would not materially change the nature of operation of the business, whereas in the case of the smaller corporation a loss of the same magnitude might constitute a catastrophe, and might require the liquidation of the business. Hence, large and small corporations do and should have different attitudes toward risk. It may be noted that reverse attitudes toward risky ventures to those just indicated might be present in comparing a dynamic management of a small company with a highly conservative management of a large company.

To recapitulate, we can summarize the problem concerning decision making in problems involving payoffs which depend upon risky outcomes. Monetary payoffs are sometimes inappropriate as a calculation device, and it appears appropriate to substitute some other set of values or "numeraire" which reflects the decision maker's attitude toward risk. A clever approach to this problem has been furnished by Von Neumann and Morganstern, who developed the so-called Von Neumann and Morganstern utility measure. In the next section, we consider how these utilities may be derived, and the procedures for using them in decision analysis.

Construction of Utility Functions

We have seen that in certain risk situations we might prefer one course of action to another even though the first act has a lower expected monetary value. In the language of decision theory, the reason for preferring the first act is that it possesses greater expected utility than does the second act.[5] The procedure used to establish the utility function of a decision maker requires him to respond to a series of choices in each of which he receives with certainty an amount of money denoted B (for benefit) as opposed to a gamble in which he would receive an amount B_1 with probability p and an amount B_2 with probability $1 - p$. The question the decision maker must answer is what probability p he would require for consequence B_1 in order for him to be indifferent between receiving B for certain and partaking in the gamble involving the receipt of B_1 with probability p and B_2 with probability $1 - p$. This probability assessment provides the assignment of a utility index to the monetary value B. The data obtained from the series of questions posed to the decision maker result in a set of utility-money pairs which can be plotted on a graph and constitutes the decision maker's utility function for money. We will illustrate the procedure for constructing an individual's utility function by returning to one of the examples given in Table 13-8.

Assume we ask the decision maker which of the following he prefers:

C_3: Receive \$0 for certain. or C_4: Receive \$60,000 with probability 1/2
lose \$40,000 with probability 1/2.

Suppose he responds that he prefers to receive \$0 for certain. Our task then is to find out what probability he would require for the receipt of the \$60,000 to make him just indifferent between the gamble and the certain receipt of \$0. This will enable us to determine the utility he assigns to \$0. The first step is the arbitrary assignment of "utilities" to the monetary consequences in the gamble as, for example,

$$U(\$60{,}000) = 1$$

$$U(-\$40{,}000) = 0$$

---

[5]The term "utility" as used by Von Neumann and Morganstern and as used in this text differs from the economist's use of the same word. In traditional economics, utility referred to the inherent satisfaction delivered by a commodity and was measured in terms of psychic gains and losses. On the other hand, Von Neumann and Morganstern conceived of utility as a measure of value used in the assessment of situations involving risk, which provides a basis for choice making. The two concepts can give rise to widely differing numerical measures of utility.

where the symbol U denotes "utility" and $U(\$60{,}000) = 1$ is read "the utility of \$60,000 is equal to one." It should be emphasized that the assignment of the numbers 0 and 1 as the utilities of the lowest and highest outcomes of the gamble is entirely arbitrary. Any other numbers could have been assigned, just so the utility assigned to the higher monetary outcome is greater than that assigned to the lower outcome. Thus, the utility scale has an arbitrary zero point.

The expected utility of the indicated gamble is

$$E[U(C_4)] = (1/2)[U(\$60{,}000)] + (1/2)[U(-\$40{,}000)]$$
$$= (1/2)(1) + (1/2)(0) = 1/2$$

Therefore, since the decision maker has indicated that he prefers \$0 for certain to this gamble, it follows that the utility he assigns to \$0 is greater than 1/2 or $U(\$0) > 1/2$. In order to aid the decision maker in deciding how much greater than 1/2 the utility is which he assigns to the monetary outcome of \$0, we introduce the concept of a hypothetical urn for use in calibrating his utility assessment.

Let us assume we have an urn with 100 balls in it, 50 of which are black and 50 white. The balls are identical in all other respects. Further, we assume that if a ball is drawn at random from the urn and its color is black, the decision maker receives a payoff of \$60,000. On the other hand, if the ball is white, his payoff is −\$40,000. We now have constructed a physical counterpart of the gamble denoted C_4. The question now is, "Retaining the total number of balls in the calibrating urn at 100, but varying the composition in terms of the number of black and white balls, how many black balls would be required for a decision maker to say that he is indifferent between receiving \$0 for certain and participating in the gamble?" With 50 black balls in the urn, the decision maker prefers \$0 for certain. With 100 black balls (and zero white balls), the decision maker would obviously prefer the gamble, since it would result in a payoff of \$60,000 with certainty. For some number of black balls between 50 and 100 the decision maker should be indifferent, that is, he would be at the threshold beyond which he would prefer the gamble to the certainty of the \$0 payoff. Suppose we begin replacing white balls by black balls, and for some time the decision maker is still unwilling to participate in the gamble. Finally, when there are 70 black balls and 30 white balls, he announces that the point of indifference has been reached. We now can calculate the utility he has assigned to \$0 as follows:

$$U(\$0) = 0.70[U(\$60{,}000)] + 0.30[U(-\$40{,}000)] = 0.70(1) + 0.30(0) = 0.70$$

This utility calculation is a particular case of the general relationship

$$U(B) = pU(B_1) + (1 - p)U(B_2) \qquad \textbf{(13.1)}$$

where B is an amount of money received for certain and B_1 and B_2 are component prizes received in a gamble with probabilities p and $1 - p$, respectively.

We now have determined three money-utility pairs: (−$40,000, 0), ($60,000, 1), and ($0, 0.70), where the first figure in the ordered pair represents a monetary payoff in dollars and the second figure the utility index assigned to this amount. The utility figures for other monetary payoffs between −$40,000 and $60,000 can be assessed in exactly the same way as for $0, assuming the patience of our long-suffering decision maker holds out. Of course, as a practical matter, a relatively small number of points could be determined and the rest of the function interpolated. Suppose the utility function shown in Figure 13-3 results from the indifference probabilities assigned by the decision maker in the set of gambles proposed to him. The one point whose determination was illustrated ($0, 0.70) is depicted on the graph. This utility function can now be used to evaluate risk alternatives which might be presented to the decision maker. He can calculate the expected utility of an alternative by reading off the utility figure corresponding to each monetary outcome and then weighting these utilities by the probabilities that pertain to the outcomes. In other words, the utility figures can now be used by the decision maker in place of the original monetary values, for calculation of expected utilities, whereas for a person with his type of utility function, calculation of expected monetary values is clearly an inadequate guide for decision making.

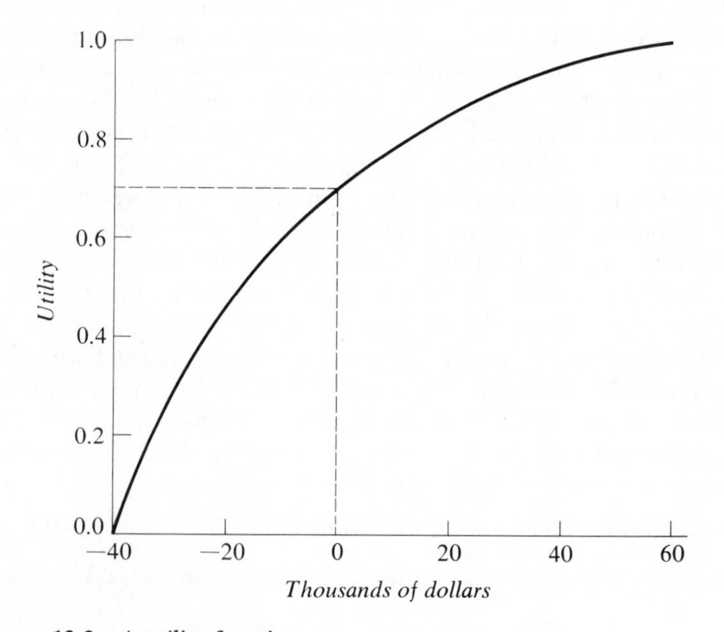

Figure 13-3 A utility function.

Characteristics and Types of Utility Functions

The utility function depicted in Figure 13-3 rises consistently from the lower left to the upper right side of the chart. That is, the utility curve has a positive slope throughout its extent. This is a general characteristic of utility functions; it simply implies that people ordinarily attach greater utility to a larger sum of money than to a smaller sum.[6] Economists have noted this psychological trait in traditional demand theory and have referred to it as a "positive marginal utility for money." The concave downward shape shown in Figure 13-3 illustrates the utility curve of an individual who has a diminishing marginal utility for money, although the marginal utility is always positive. This type of utility curve is characteristic of a "risk avoider," and is so indicated in Figure 13-4a. A person characterized by such a utility curve would prefer a small but certain monetary gain to a gamble whose expected monetary value is greater but may involve a large but unlikely gain, or a large and not unlikely loss. The linear function in Figure 13-4b depicts the behavior of a person who is "neutral" or "indifferent" to risk. For such a person every increment of, say, a thousand dollars has an associated constant increment in utility. This type of individual would use the criterion of *maximizing expected monetary value* in decision making because by so doing he would also *maximize expected utility.* That this is so is easy to demonstrate. Suppose an individual's utility function is linear with respect to money and is,

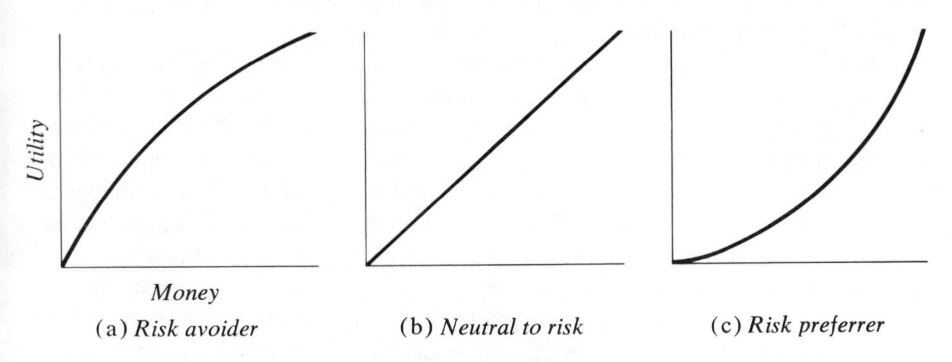

(a) *Risk avoider* (b) *Neutral to risk* (c) *Risk preferrer*

Figure 13-4 Various types of utility functions.

[6]The almost infinite variety of types of human behavior is attested to by the fact that conduct which runs counter to a generalization of this sort is even occasionally observed. Recently, newspapers carried an account of an heir to a fortune of $30 million who committed suicide at the age of 23, indicating in a final letter that his great wealth prevented him from living a normal life.

therefore, characterized by the following functional relationship in some relevant range of money payoffs:

$$U(x) = a + bx \tag{13.2}$$

where x = an amount of money
$U(x)$ = the utility of the specified amount of money

Then, using Rule 3 of Appendix C to take expected values in Equation (13.2), we get

$$E[U(x)] = a + bE(x) \tag{13.3}$$

Hence, if we view the monetary payoff as a random variable, the expected utility of x dollars is a function only of the expected monetary value, $E(x)$, since a and b are constants. It follows that if we maximize the right-hand side of Equation (13.3), or in other words maximize the expected monetary payoff, we also will maximize the expected utility of this monetary amount. This is an important principle which we will use throughout the remainder of the text. Conversely, it also is true that if we use the maximizing of expected monetary value as a decision criterion for an individual, we are assuming that his utility function for money is linear over the range of money under consideration.

In Figure 13-4c is shown the utility curve for a "risk preferrer" or "risk lover." This type of person willingly accepts gambles which have a smaller expected monetary value than an alternative payoff received with certainty. In the case of such an individual, the attractiveness of a possibly large payoff in the gamble tends to outweigh the fact that the probability of such a payoff may indeed be very small.

Empirical research suggests that most individuals have utility functions in which for small changes in money amounts the slope does not change very much. Hence, over these ranges of money outcomes, the utility function may approximately be considered as linear and as having a constant slope. However, in considering courses of action in which one of the consequences is very adverse or in which one of the payoffs is very large, individuals can be expected to depart from the maximization of expected monetary values as a guide to decision making. In later chapters, for simplicity of presentation, we will ordinarily use the criterion of maximizing expected monetary value, although as we have seen the appropriate criterion is in terms of maximizing expected utilities. For many business decisions, where the monetary consequences may represent only a small fraction of the total assets of the business unit, the use of maximization of expected monetary payoff may constitute a reasonable approximation to the decision-making criterion of maximization of expected utility. In other words, in such cases, the utility function may often be treated as approximately linear over the range of monetary payoffs considered.

Illustrative Examples

A couple of examples follow which illustrate the use of utility functions for decision purposes. In each case, the decision maker is assumed to have a "risk avoider" type of utility curve. These problems review a number of principles, and illustrate the use of utility functions which have been stated in the form of algebraic functions.

Example 13-1 Assume that Mr. Gordon has a risk-avoidance type of utility function for amounts of money Z between $-\$100$ and $+\$100$. By curve fitting, it has been determined that his utility function over the indicated monetary range can be expressed in the form of a second degree parabola as follows:

$$U(Z) = Z - .002Z^2 \qquad -\$100 \leq Z \leq \$100$$

The graph of this function is given in Figure 13-5, and the calculation of these points is given in the accompanying table.

Mr. Gordon is presented a lottery whose money prizes depend on the outcome of a random variable Z which has the following probability distribution:

z	$f(z)$
−100	.1
0	.2
50	.4
100	.3
	1.0

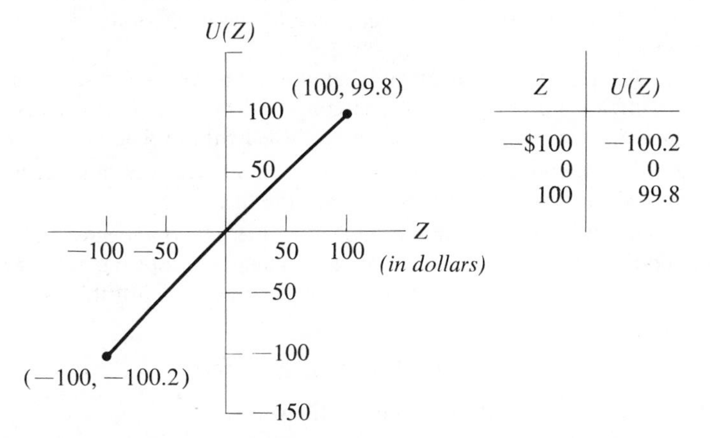

Z	U(Z)
−$100	−100.2
0	0
100	99.8

Figure 13-5 Graph of Mr. Gordon's utility function.

Given that the utility of a lottery is equal to the expected utility of its component prizes, should Mr. Gordon accept the lottery?

Solution: Let us first obtain a general expression for the utility of the lottery, which we denote $U(L)$.

(13.4) $$U(L) = E[U(Z)] = E[Z - .002Z^2]$$

By Rule 3 of Appendix C, the expected value in Equation (13.4) equals

(13.5) $$U(L) = E[Z - .002Z^2] = E(Z) - .002E(Z^2)$$

By Equations (4.1) and (4.2),

$$E(Z) = \sum_z zf(z) \quad \text{and} \quad E(Z^2) = \sum_z z^2f(z)$$

Hence, we can calculate these expected values from the given probability distribution as follows:

z	z^2	$f(z)$	$zf(z)$	$z^2f(z)$
−100	10,000	.1	−10	1000
0	0	.2	0	0
50	2500	.4	20	1000
100	10,000	.3	30	3000
		1.0	40	5000

$$E(Z) = 40 \qquad E(Z^2) = 5000$$

Substituting these values for $E(Z)$ and $E(Z^2)$ into Equation (13.5) gives $U(L) = E(Z) - .002E(Z^2) = 40 - .002(5000) = 40 - 10 = 30$ utility units. Since the utility of not accepting the lottery equals $U(\text{not accepting lottery}) = E[U(\$0)] = 0 - .002(0)^2 = 0$ utility units, and the utility of accepting the lottery equals $U(L) = 30$ utility units, Mr. Gordon should accept the lottery.

Comments:

(1) A lottery is the same type of situation that we have referred to previously as a gamble. Formally, it may be conceived of as a risky proposal in which component prizes $B_1, B_2, \ldots, B_n$ are awarded with probabilities $p_1, p_2, \ldots, p_n$.

(2) The utilities calculated in this problem are referred to in terms of "utility units." Often the term "utiles" is used for this purpose.

(3) It can be shown that if an individual has a second degree parabolic utility function, the utility to him of a lottery depends on both the *expected value* and *variance* of possible monetary outcomes. For example, in this problem, Equation (13.5) is

$$U(L) = E(Z) - .002E(Z^2)$$

Let us evaluate $E(Z^2)$. By Equation (4.12) we have

$$\sigma^2(Z) = E(Z^2) - [E(Z)]^2$$

Adding $[E(Z)]^2$ to both sides of this equation and transferring expressions to the other side of the equal sign gives

$$E(Z^2) = \sigma^2(Z) + [E(Z)]^2$$

Substituting this expression for $E(Z^2)$ into Equation (13.5), we have

$$U(L) = E(Z) - .002[\sigma^2(Z) + [E(Z)]^2]$$

Therefore, the utility of the lottery is a function of both the expected value and the variance of monetary outcomes. This finding gives us some insight into choice making behavior. *If an individual has a linear utility function, he acts consistently if he takes into account only the expected value of monetary outcomes.* On the other hand, we have observed that in certain situations, maximization of expected monetary values may be an inadequate criterion. This means that the decision maker finds that variance and perhaps other moments of the distribution of monetary payoffs are relevant. *In the case of a person with a second-degree parabolic utility function, we have found that the decision maker acts consistently if he takes into account both the expected value and variance of monetary consequences.*

(4) It may be noted in this problem that the utility scale did not extend simply from zero to one. As indicated previously, the zero point on the utility scale is arbitrary, as are the units in which we choose to measure utility.

Example 13-2 An individual has the following utility function for an amount Z dollars:

$$U(Z) = Z - 0.1Z^2 \qquad -\$100 \leq Z \leq \$5$$

Carry out the necessary calculations to indicate this individual's rankings of the following three proposals:

(a) A lottery whose prize Z depends on X as follows:

$$Z = X - 10$$

X is a random variable denoting the number of successes in a Bernoulli process with probability of success $p = 1/5$, and number of trials,

$$n = 50$$

(b) A lottery whose prize Z depends upon X as follows:

$$Z = X - 2$$

where X is a random variable denoting the number of successes in a Poisson process with parameter $\mu = 3$.

(c) A gift of one dollar for certain.

Solution: (a) The utility of the lottery is

$$U(L) = E[U(Z)] = E(Z) - .1E(Z^2)$$

It is necessary to determine the values for $E(Z)$ and $E(Z^2)$.

$$E(Z) = E(X - 10) = E(X) - 10$$

Since X is binomially distributed and pertains to *number* of successes, $E(X) = np$ and $\sigma^2(X) = npq$. Substituting the value of $E(X)$ gives

$$E(Z) = np - 10 = (50)(1/5) - 10 = 0$$

The value for $E(Z^2)$ can be obtained from

$$E(Z^2) = \sigma^2(Z) + [E(Z)]^2$$

Evaluating $\sigma^2(Z)$ yields

$$\sigma^2(Z) = \sigma^2(X - 10) = \sigma^2(X) = npq = 50(1/5)(4/5) = 8$$

Hence,

$$E(Z^2) = 8 + (0)^2 = 8$$

Therefore

$$U(L) = E(Z) - .1E(Z^2) = 0 - (.1)(8) = -0.8$$

(b) $U(L) = E(Z) - .1E(Z^2)$
Again, the values for $E(Z)$ and $E(Z^2)$ must be obtained.

$$E(Z) = E(X - 2) = E(X) - 2 = 3 - 2 = 1$$

Evaluating $\sigma^2(Z)$ yields

$$\sigma^2(Z) = \sigma^2(X - 2) = \sigma^2(X)$$

Since in the Poisson distribution, the variance is equal to the expected value, we have

$$\sigma^2(Z) = \sigma^2(X) = E(X) = 3$$

Hence,

$$E(Z^2) = \sigma^2(Z) + [E(Z)]^2 = 3 + (1)^2 = 4$$

and

$$U(L) = E(Z) - .1E(Z^2) = 1 - (.1)(4) = 0.6$$

(c) $U(L) = E[U(Z)] = E(Z - .1Z^2)$

$$U(L) = E[1 - (.1)(1)^2] = E(0.9) = 0.9$$

Therefore, in summary, the proposal in (c) is preferable to that in (b), which is preferable to that in (a), since $0.9 > 0.6 > -0.8$.

Assumptions Underlying Utility Theory

The utility measure we have discussed was derived by evoking from the decision maker his preferences between sums of money obtainable with certainty and lotteries or gambles involving a set of basic alternative monetary outcomes. This procedure entails a number of assumptions.

It is assumed that an individual when faced with the types of choices discussed can determine whether an act, say A_1 is preferable to another act, A_2; whether these acts are indifferently regarded or whether A_2 is preferred to A_1. If A_1 is preferred to A_2, then the utility assigned to A_1 should exceed the utility assigned to A_2.

Another behavioral assumption is that if the individual prefers A_1 to A_2 and he also prefers A_2 to A_3, then he should prefer A_1 to A_3. This is referred to as the principle of *transitivity*. The assumption extends also to indifference relationships. Hence, if the decision maker is indifferent between A_1 and A_2 and between A_2 and A_3, he should be indifferent between A_1 and A_3.

Furthermore, it is assumed that if a payoff or consequence of an act is replaced by another, and the individual is indifferent between the former and new consequences, he should also be indifferent between the old and new acts. This is often referred to as the principle of *substitution*.

Finally, it is assumed that the utility function is bounded. This means that utility cannot increase or decrease without limit. As a practical matter, this simply means that the range of possible monetary values is limited. For example, at the lower end the range may be limited by a bankruptcy condition.

It may be argued that human beings do not always exhibit the type of consistency in their choice behavior that is implied by these assumptions. However, the point is that if in the construction of an individual's utility function it is observed that he is behaving inconsistently and these incongruities are indicated to him, and if he is "reasonable" or "rational," he should adjust his choices accordingly. If he insists on being irrational and refuses to adjust his choices which violate the underlying assumptions of utility theory, then a utility function cannot be constructed for him and he cannot use maximization of expected utility as a criterion of rationality in his choice making. It is important to keep in mind that the type of theory discussed here does not purport to describe the way people actually *do behave* in the real world, but rather specifies how they *should behave* if their decisions are to be consistent with their own expressed judgments as to preferences among consequences. Indeed, it may be argued that since human beings are fallible and do make mistakes, it is useful to have normative procedures which police their behavior and provide ways in which it can be improved.

A Brief Note on Scales

Von Neumann-Morganstern utility scales are examples of *interval scales*. Such scales have a constant unit of measurement, but an arbitrary zero point. Differences between scale values can be expressed as multiples of one another, but individual values cannot.

Example 13-3 The familiar scales for temperature are examples of interval scales. We cannot say that 100°C is twice as hot as 50°C. The corresponding Fahrenheit measures would not exhibit a ratio of 2 to 1. On the other hand, we can say that the intervals or differences between 100°C and 50°C and 75°C and 50°C are in a two-to-one ratio. Thus, using the relationship $F = (9/5)C + 32°$:

$$C = 100°; F = (9/5)(100°) + 32° = 212°$$

$$C = 75°; F = (9/5)(75°) + 32° = 167°$$

$$C = 50°; F = (9/5)(50°) + 32° = 122°$$

The difference between 100° and 50° = 50°; between 75° and 50° = 25°. The ratio of 50° to 25° is two-to-one.

The difference between 212° and 122° = 90°; between 167° and 122° = 45°. The ratio of 90° to 45° is two-to-one.

In decision making using utility measures, if a different zero point and a different scale are selected, the same choices will be made. A constant can be added to each utility value, and each utility value can be multiplied by a constant, without changing the properties of the utility function. Thus, if a is any constant and b is a positive constant, and x is an amount of money,

$$U_2(x) = a + bU_1(x)$$

and $U_2(x)$ is as legitimate a measure of utility as $U_1(x)$.

This property is usually summarized by the statement that interval scales are "unique up to a positive linear transformation." $(Y = a + bx; b > 0)$

PROBLEMS

1. If possible states of nature are: competitor will set his price
   (a) higher,
   (b) the same, or
   (c) lower,
   what is wrong with assessing prior probabilities as .6, .3, and .2, respectively?
2. A new appliance store finds that in its first week of business it sold five major appliances, ten home appliances, and 30 small appliances. Based solely on this past knowledge, what prior probability distribution would you formulate for the type of appliance to be sold?
3. Explain the meaning of expected value of perfect information.
4. Explain the difference between expected opportunity loss and expected value of perfect information
5. Given an opportunity loss table, can you compute the corresponding payoff table? Explain why or why not.

6. Given the following payoff matrix measured in utility units:

Projected Sales Levels	Buy Specified Interest in Plywood, Inc. 100%	50%	25%
$30 Million	50	25	17
15 Million	35	15	5
10 Million	−6	−2	0
5 Million	−8	−4	−2

Construct the corresponding opportunity loss table.

7. The following is a payoff matrix in units of $1000:

Competitor's Price	Price the item at A_1 $0.90	A_2 $0.95	A_3 $1.00	A_4 $1.05
S_1: $1.00	10	6	3	1
S_2: 0.95	5	8	4	6
S_3: 0.90	12	9	8	5
S_4: 0.85	8	10	12	14

The prior probabilities are

State of Nature	Probability
S_1	.3
S_2	.2
S_3	.4
S_4	.1

Compute the EVPI by two different methods.

8. The following is a payoff table in units of $1000:

Demand is	Action A_1	A_2	A_3	A_4
S_1: Above average	18	15	16	11
S_2: Average	8	12	12	10
S_3: Below average	2	5	3	8

where A_1 = Keep store open weekdays, evenings, and Saturday
A_2 = Keep store open weekdays plus Wednesday evening
A_3 = Keep store open weekdays and Saturday
A_4 = Keep store open only weekdays

The prior probability distribution of demand is

S_i	$P(S_i)$
S_1	.5
S_2	.3
S_3	.2

(a) Find the expected profit under certainty.
(b) Find the expected profit under uncertainty.
(c) How much would you pay for information which yields the true state of nature?

9. Trivia Press, Inc. has been offered an opportunity to publish a new novel. If the novel is a success, the firm can expect to earn $8 million over the next five years; if a failure, to lose $4 million over the next five years. After reading the novel, the publisher assesses the probability of success as 1/3. Should he publish the book? What is the expected value of perfect information?

10. An advertising firm submits for acceptance a campaign costing $55,000. The company's marketing manager estimates that if the campaign is received well by the public, profits will increase by $175,000; if it is received moderately well, profits will increase by $55,000; and if it is received poorly, profits will remain unchanged. Compute the appropriate opportunity loss table.

11. Assume that there are ten urns, seven of type *A* and three of type *B*. Type *A* urns contain five white balls and five black balls. Type *B* urns contain eight white balls and two black balls. One of the ten urns is to be selected at random. You are required to guess whether the urn selected is of type *A* or *B*. Assume you are willing to act on the basis of expected monetary value. You will receive payoffs and penalties according to the payoff table given below.

True State of Nature	*Your Guess* Type *A*	Type *B*
Type *A*	+$500	−$ 40
Type *B*	− 300	+ 800

Find and interpret the expected value of perfect information.

12. As personnel manager of Lemon Motors you must decide whether to hire a new salesman. Depending on his performance, the payoff to the firm is

Sales	*Payoff*
High	$10,000
Average	3000
Low	−13,000

(a) If judging from his application, you feel the probabilities attached to his possible performances are high, .3; average, .4; and low, .3; should you hire him?

(b) How much would you be willing to pay a perfect predictor to tell you what the salesman's performance would be?

13. An operations research team is trying to decide whether to put the predictions of the ten leading investment advice newsletters into an information system it is building. The cost of including the predictions is $4850 a year. It is estimated that in 20 decisions to be made in a year, the added information would result in a new decision only once. However, the change in decision would result on the average in a saving of $75,000. Should the team include the newsletter in its information system?

14. A company has $100,000 available to invest. The company can either build a new plant or put the money in the bank at 4% interest. If business conditions remain good the company expects to make 10% on its investment in a new plant, but if there is a recession, the investment is expected to return only 2%. What probability must management assign to the occurrence of a recession to make the two investments equally attractive? Assume that the only two possible states for business conditions are "good" and "recession."

15. As marketing manager of a firm, you are trying to decide whether to open a new region for a product. Success of the product depends on demand in the new region. If demand is high, you expect to gain $100,000, if average $10,000, and if low to lose $80,000. From your knowledge of the region and your product you feel the chances are four out of ten that sales will be average, and equally likely that they will be high or low. Should you open the new region? How much would you be willing to pay to know the true state of nature?

16. In Problem 10, if the president of the company, after examining the proposed campaign, feels that the probabilities that it would be received "well" and "moderately well" are .4 and .2, respectively, what is the expected opportunity loss for each action? What is the optimal decision?

17. As manager of a plant you must decide to invest $25,000 in either a cost reduction program or a new advertising campaign. Assume that you know the cost reduction program will increase the profit-to-sales ratio, from the present 10% to 11%. The sales campaign, if successful, is expected to increase the present $2 million of sales by 12%. The probability that the campaign will be successful is .8. What is your best course of action?

18. A farmer is trying to decide whether to irrigate his 25 acres of cropland this year or next. If he irrigates this year, he will have to borrow $5000 for two years and pay a total of $600 in interest charges. If he waits until next year, he expects to have to borrow only $2500 for one year at 6% simple interest. An irrigation system is of value to the farmer only if there is a drought. If there is a mild drought, he expects irrigation to result in an increased yield of one ton per acre compared to output without the system, and if there is a severe drought, an increase of three tons per acre. The farmer's crop sells for $18 a ton. If he feels the probability of a mild drought each year is .40

and a severe drought .25, should he irrigate this year or wait until next year? (Assume the irrigation system has an infinite lifetime.)

19. A brewer presently packages beer in old style cans. He is debating whether to change the packaging of his beer for next year. He can adopt A_1, an easy-open aluminum can; A_2, a lift-top can; A_3, a new wide-mouth screw-top bottle; or retain A_4, the same old style cans. Profits resulting from each move depend on what the brewer's competitor does for the next year. The payoff matrix and prior probabilities, measured in \$10,000 units, are as follows:

Prior Probability	*Competitor Uses*	*Action* A_1	A_2	A_3	A_4
.5	Old Style Bottles	15	14	13	16
.2	Easy-Open Cans	12	11	10	8
.1	Lift-Top Cans	6	9	8	6
.2	Screw-Top Bottles	5	6	8	5

(a) Find the expected opportunity loss for each act.
(b) Determine EVPI.
(c) Determine the optimal decision.

20. R.B.A., Inc. is given the opportunity to submit a closed bid to the government to build certain electronic equipment. An exâmination of similar proposals made in the past revealed that the average profit per successful bid was \$175,000, and that R.B.A., Inc. received the contract (i.e., had the lowest bid) on 10% of its submitted bids. The cost of preparing a bid is, on the average, \$10,000. Should R.B.A., Inc. prepare a bid?

21. In Problem 20, suppose R.B.A., Inc. chose to prepare a bid. For this particular proposal, assume the company finds it can submit only the following four bids: \$1,600,000, \$1,700,000, \$1,800,000, or \$1,900,000. At \$1,600,000, expected profit is \$160,000. Each successive bid yields an increase in profit equal to the increase in the bid. From an examination of past accounting records, R.B.A., Inc. assesses the probabilities that the bids will be the lowest ones to be .4, .3, .2, .1, respectively. Which bid should be submitted?

22. (a) The expected monetary return of the decision to buy life insurance is negative. Thus it is irrational to buy life insurance. Do you agree or disagree? Explain.
    (b) If the A.T. & T. Corporation does not carry automobile insurance, why do you think this is so?

23. If the following prospects have the given utilities:

Prospect	*Utiles*
A	10
B	8
C	5
D	3
E	2

would you prefer C for certain to
(a) a chance of getting A with .4 probability and E with .6 probability?
(b) a chance of getting B with .5 probability and E with .5 probability?
(c) a chance of getting A with .3 probability and D with .7 probability?
(d) a chance of getting B with .4 probability and E with .6 probability?

24. You have a choice of placing your money in the bank and receiving interest equal to ten utiles or investing in Rerox stock. With a probability of .4, Rerox will yield gains equal to 45 utiles and with a probability of .6, it will cause a loss of 15 utiles.
(a) What is the expected utility of the prospect, "buy the stock"?
(b) Should you buy Rerox or put the money in the bank?

25. Assume X is a random variable which represents the number of successes in a binomially distributed process with parameters $p = .5$ and $n = 100$. A lottery is proposed whose payoff Y depends on X as follows:

$$Y = X + 5$$

The lottery costs 30 utiles. If the individual has the utility function

$$U(Y) = Y - .005Y^2 \qquad -40 < Y < 100$$

should he accept the lottery? Why or why not?

26. An investor is considering buying a franchised furniture business. He estimates that the business will yield either a loss of \$50,000 or a profit of \$100,000, \$200,000, or \$500,000 every five years, with probabilities .5, .2, .1, .2, respectively. The investor's utility function is found to be

Dollars	*Utiles*	*Dollars*	*Utiles*	*Dollars*	*Utiles*
−50,000	−40	25,000	1.2	150,000	7.5
−37,500	−10	50,000	2.5	200,000	10.0
−25,000	−4.2	75,000	5.8	250,000	12.5
−12,500	−2	100,000	6.0	375,000	26.0
0	0	125,000	6.2	500,000	40.0

(a) Graph his utility function and interpret the shape of the curve.
(b) Based on expected utility value, should he buy or not?
(c) Suppose he is informed that he can buy either the whole franchise, 3/4, 1/2, or 1/4 (i.e., if he purchases 1/4, he receives 1/4 of all profits and pays 1/4 of all losses). What is his best investment decision?

27. Drillwell Oil Company is debating what it should do with an option on a parcel of land. If it takes the option, the firm can drill with 100% interest or with 50% interest (i.e., all costs and profits are split with another firm). It costs \$50,000 to drill a well and \$20,000 to operate a producing well until it is dry. The oil is worth \$1 a barrel. Assume the well is either dry or produces 200,000 or 500,000 barrels of oil. The firm assesses the probability of each outcome as .8, .1, and .1, respectively.
(a) Based on expected monetary return, what is the best action?

(b) Suppose Drillwell's management has the following utility function:

Dollars	*Utiles*
−50,000	−30
−25,000	−10
65,000	25
130,000	60
215,000	120
430,000	200

What is the best decision based on expected utility?

28. I.R.S. has audited your last year's income tax and has sent you a bill for \$225 for back taxes. You now have the choice of paying the bill or disputing the audit. If you dispute it, it will cost you \$20 for an accountant's fee to prepare your case. After preliminary talks with your accountant, you feel the chances of your winning the dispute are five in one hundred.
    (a) Should you dispute the case based on monetary expectations?
    (b) Assume large losses of money are disastrous to you as a struggling student. This is reflected in your utility function, which indicates that $U(-\$20)$ is −4 utiles, $U(-\$225)$ is −425 utiles, and $U(-\$245)$ is −440 utiles. Based on expected utility, what is your best course of action?

29. A drug manufacturer has developed a new drug named Thalidimous. Tests have shown it to be extremely effective with almost no side effects. However, it has only been tested for three years and long range side effects are really unknown. The research department feels the probability the drug will have any serious long-range effects is 1 in 100. The Food and Drug Administration (FDA) must first clear the drug for sale. Assume the FDA evaluates the loss to society because of serious long-range side effects as −900,000 utiles, the gain to society because of the use of the drug as 8,000 utiles and the gain attributable to the economic advantages of production of a new drug as 1000 utiles. If the FDA accepts the firm's appraisal of the probability of long-range side effects should it "accept" the drug?

30. A firm is offered a contract to develop a special turbine engine. The contract stipulates that if the engine is not developed within three years from the date of the contract, the contract is void. If the engine is developed in time, the expected profit is \$450,000; if not, the expected loss is \$2,250,000. The research and development department is 90% sure it will be able to develop the engine.
    (a) If the firm is neutral to risk (i.e., makes decisions solely on an expected monetary value criterion), what is the expected opportunity loss of each action? What is the optimal action?
    (b) If the firm is not neutral to risk and has the following utility function, measured in units of \$10,000,

$$U(X) = X - .01X^2$$

what is the optimal action?

(c) Would you expect a contract like this to be taken by a large firm such as General Electric or a small research and development firm? Explain.

31. Show that if a utility function has the form

$$U(Y) = Y - bY^2 \qquad b \geq 0 \text{ and } Y < 1/2b$$

where Y is a random variable, the expected utility of Y can be written

$$E[U(Y)] = e - b(v + e^2)$$

where

e is the expected value of Y
v is the variance of Y

32. An individual has the following utility function for an amount Z dollars

$$U(Z) = Z - .05Z^2 \qquad \text{for } -10 < Z < +10$$

Carry out the necessary calculations and indicate this individual's ranking of the following three proposals:
(a) A lottery whose prize Z depends upon X as follows:

$$Z = X - 10$$

where X is binomially distributed with $p = 1/5$ and $n = 50$.
(b) A lottery whose prize Z depends upon X as follows:

$$Z = X - 2$$

where X has the Poisson distribution with $\mu = 3$.
(c) A gift of $.50.

33. A retailer must decide how much inventory he should carry, which is dependent upon demand. Since the stock is perishable, a loss occurs when he is overstocked. Because of space limitations, the retailer can stock at most five items. The cost per item is $1 and the selling price $5. The profit table and probabilities are

	Number	*Number of Units Stocked*					
Probability	*Demanded*	0	1	2	3	4	5
2/20	0	$0	$−1	$−2	$−3	$−4	$−5
3/20	1	0	4	3	2	1	0
5/20	2	0	4	8	7	6	5
5/20	3	0	4	8	12	11	10
4/20	4	0	4	8	12	16	15
1/20	5	0	4	8	12	16	20

What is the optimal stocking level and what is the retailer's expected profit?

CHAPTER FOURTEEN

Decision Making Using Both Prior and Sample Information

14.1 Introduction

The discussion in Chapter 13 may be referred to as *prior analysis*, that is, decision making in which expected payoffs of acts are computed on the basis of prior probabilities. In this chapter we discuss *posterior analysis*, in which expected payoffs are calculated with the use of *posterior probabilities*, which are revisions of prior probabilities on the basis of additional sample or experimental evidence. Bayes' theorem is utilized to accomplish the revision of the prior probabilities. The terms "prior" and "posterior" in this context are relative ones. For example, subjective prior probabilities may be revised to incorporate the additional evidence of a particular sample. The revised probabilities then constitute posterior probabilities. If these probabilities are in turn revised on the basis of another sample, they represent prior probabilities relative to the new sample information, and the revised probabilities are "posteriors."

The basic purpose of attempting to incorporate more evidence through sampling is to reduce the expected cost of uncertainty. If the expected cost of uncertainty (or the expected opportunity loss of the optimal act) is high, then it will ordinarily be wise to engage in sampling. Sampling in this context is understood to include statistical sampling, experimentation, testing, and any other methods used to acquire additional information.

14.2 Posterior Analysis

The general method of incorporating sample evidence into the decision making process will be illustrated in terms of the inventor problem of the preceding chapter. The example is somewhat artificial, because no particular sample size is assumed. However, the problem illustrates the general method of posterior analysis in a very straightforward way. Then, a more realistic problem will be considered in which the additional evidence incorporated with the prior probabilities is based on a sample whose size is specified.

Suppose in the problem discussed in Chapter 13, the inventor decided not to rely solely on prior probabilities concerning the demand for his new device, but to have a market research organization conduct a sample survey of potential consumers to gather additional evidence for the probable level of sales for his product. Let us assume that the survey can result in three types of sample results denoted x_1, x_2, and x_3, corresponding to the three states of nature, sales levels, θ_1, θ_2, and θ_3. Specifically, the possible results may be:

x_1: sample indicates strong sales

x_2: sample indicates average sales

x_3: sample indicates weak sales

The survey is conducted and the sample gives an indication of an average level of sales, that is, x_2 is observed. Assume that on the basis of previous surveys of this type, the market research organization can assess the reliability of the sample evidence in the following terms. In the past, when the actual level of sales after a new device was placed on the market was average, sample surveys properly indicated an average level of demand 80% of the time. However, when the actual level was strong sales, about 10% of the sample surveys incorrectly indicated demand as average and when the actual level was weak sales, about 20% of the sample surveys gave an indication of average sales. These relative frequencies represent conditional probabilities of the sample evidence "average sales," given the three possible underlying events concerning sales level, and can be symbolized as follows:

$$P(x_2 \mid \theta_1) = 0.1$$

$$P(x_2 \mid \theta_2) = 0.8$$

$$P(x_2 \mid \theta_3) = 0.2$$

The revision by means of Bayes' theorem of the prior probabilities assigned to the three sales levels on the basis of the observed sample evidence x_2 (average sales), is given in Table 14-1. In terms of Equation (1.10) for Bayes' theorem, x_2 plays the role of B, the sample observation; and θ_i replaces A_i, the possible events, or states of nature. In the usual way, after the joint probabilities are calculated, they are divided by their total, in this case, 0.48, to yield posterior or revised probabilities for the possible events. The effect of the weighting given to the sample evidence by Bayes' theorem in the revision

Table 14-1 Computation of Posterior Probabilities in the Inventor's Problem for the Sample Indication of an Average Level of Sales.

Events θ_i	*Prior Probability* $P(\theta_i)$	*Conditional Probability* $P(x_2 \vert \theta_i)$	*Joint Probability* $P(\theta_i)\, P(x_2 \vert \theta_i)$	*Posterior Probability* $P(\theta_i \vert x_2)$
θ_1: Strong Sales	0.2	0.1	0.02	0.042
θ_2: Average Sales	0.5	0.8	0.40	0.833
θ_3: Weak Sales	0.3	0.2	0.06	0.125
	1.0		0.48	1.000

of the prior probabilities may be noted by comparing the posterior probabilities with the corresponding "priors" in Table 14-1. With a sample indication of average sales, the prior probability of the event "average sales," 0.5 was revised upward to 0.833. Correspondingly, the probabilities of events "strong sales" and "weak sales" declined from 0.2 to 0.042 and from 0.3 to 0.125, respectively.

Decision Making After the Observation of Sample Evidence

The revised probabilities calculated in Table 14-1 can now be used to compute the "posterior expected profits" of the inventor's alternative courses of action. In Table 13-3, expected payoffs were computed based on the subjective prior probabilities assigned to the possible events. These can now be referred to as "prior expected profits." The calculation of the posterior expected profits (using the revised or posterior probabilities as weights) is displayed in Table 14-2. It is customary to denote prior probabilities as $P_0(\theta_i)$ and posterior probabilities as $P_1(\theta_i)$. That is, the subscript zero is used to denote prior probabilities and the subscript one to signify posterior probabilities.

Since the posterior expected profit of act A_1 exceeds that of A_2, the better of the two courses of action remains that of the inventor manufacturing the device himself. However, after the sample indication of "average sales," the expected profit of act A_1 has decreased from $245,000 based on the prior probabilities to $193,950 based on the revised probabilities. Also, the difference in the expected profits of the two acts has narrowed somewhat. The $245,000 and $193,950 figures are respectively the *prior expected profit under* uncertainty and the *posterior expected profit under uncertainty*. It is entirely possible for the optimal course of action under a posterior analysis to change from that of the prior analysis. In the present example, if the sample indication had been "weak sales," with appropriate conditional probabilities, it

Table 14-2 Calculation of Posterior Expected Profits in the Inventor's Problem Using Revised Probabilities of Events (in units of $10,000).

Act A_1: Manufacture Device Himself

Events	*Probability* $P_1(\theta_i)$	*Profit*	*Weighted Profit*
θ_1: Strong Sales	0.042	$80	3.360
θ_2: Average Sales	0.833	20	16.660
θ_3: Weak Sales	0.125	− 5	− .625
	1.000		19.395

Posterior Expected Profit A_1 = $19.395 (ten thousands of dollars) = $193,950

Act A_2: Sell Patent Rights

Events	*Probability* $P_1(\theta_i)$	*Profit*	*Weighted Profit*
θ_1: Strong Sales	0.042	$40	1.680
θ_2: Average Sales	0.833	7	5.831
θ_3: Weak Sales	0.125	1	0.125
	1.000		7.636

Posterior Expected Profit A_2 = 7.636 (ten thousands of dollars) = $76,360 would have been possible for the posterior expected profit of A_2 to have exceeded that of A_1. (Assume some figures and demonstrate this point.)

Insight can be gained into the cost of uncertainty and the value of obtaining additional information by calculating the "posterior expected value of perfect information," which is simply the expected payoff, using posterior probabilities of decision making in conjunction with a perfect predictor. We can now refer to the EVPI calculated in Chapter 13 (Tables 13-5 and 13-7) as the prior EVPI. This prior EVPI was $18,000. Therefore, it would have been worthwhile for the decision maker to pay up to $18,000 for perfect information to eliminate his uncertainty concerning states of nature. Since no sample could be expected to yield perfect information, the decision maker does not have as yet a clear guide as to the worth of obtaining additional information through sampling. Expected value of sample information is discussed in Chapter 15. However, the *prior* EVPI of $18,000 sets an upper limit for the worth of obtaining perfect information and thus eliminating uncertainty concerning events. After the decision maker obtains additional information through sampling, he can calculate the *posterior* EVPI. The change that occurs is useful in evaluating the decision to be made and the worth of attempting to obtain further information.

The posterior EVPI is computed to be $7500 as shown in Table 14-3. Analogously to prior analysis, the posterior EVPI is calculated by subtracting the posterior expected profit under uncertainty from the posterior expected profit under certainty. The alternative determination of the posterior EVPI as the expected opportunity loss of the optimal act using posterior probabilities is given in Table 14-4. The only difference between this calculation and the similar calculation in Table 13-5 for the prior expected opportunity loss for act A_1 is the substitution of posterior for prior probabilities.

In summary, the EVPI has been reduced from $18,000 to $7500 by the information obtained from the sample. Whereas the decision maker should have been willing to pay up to $18,000 for a perfect predictor prior to having the sample information, the expected value of perfect information is only $7500 after the sample indication of "average sales." In other words, the decision maker has reduced his cost of uncertainty, and in view of the information already gained from the survey the availability of a perfect forecaster is not as valuable as it was prior to the sample survey. Instead of a decrease occurring from the prior EVPI to the posterior EVPI as in this problem, there might very well have been an increase. This could occur if there were a marked difference between the posterior and prior probability distributions, and a reversal took place in the optimal act after the incorporation of sample information. Such an increase in the EVPI after inclusion of knowledge gained from sampling can be interpreted to mean that the doubt concerning the decision has been increased because of the additional evidence.

Table 14-3 Calculation of the Posterior Expected Value of Perfect Information for the Inventor's Problem (in units of $10,000).

Events	*Profit*	*Posterior Probability*	*Weighted Profit*
θ_1: Strong Sales	$80	0.042	3.360
θ_2: Average Sales	20	0.833	16.660
θ_3: Weak Sales	1	0.125	0.125
		1.000	20.145

Posterior Expected Profit with Perfect Information = 20.145 (ten thousands of dollars)
= $201,450

Posterior Expected Profit with Perfect Information =	$201,450
Less: Posterior Expected Profit under Uncertainty =	193,950
Posterior EVPI	$ 7500

Table 14-4 Posterior Expected Opportunity Loss of the Optimal Act in the Investor's Problem (in units of $10,000).

Act A_1: Manufacture Device Himself

Events	*Probability*	*Opportunity Loss*	*Weighted Opportunity Loss*
θ_1: Strong Sales	0.042	0	0
θ_2: Average Sales	0.833	0	0
θ_3: Weak Sales	0.125	6	0.75
			0.75

Posterior EOL of the Optimal Act = Posterior EVPI = 0.75 (ten thousands of dollars)
= $7500

Of course, the determination of the prior and posterior EVPI's by alternative methods is not necessary in practice, but the computations have been shown here to indicate the relationships involved.

An Acceptance Sampling Illustration

As another illustration of posterior analysis, we will consider a problem in acceptance sampling of a manufactured product. Let us assume the Renny

Corporation inspects incoming lots of articles produced by a supplier in order to determine whether to accept or reject these lots. In the past, incoming lots from this supplier have contained either 10%, 20% or 30% defective articles. On a relative frequency basis, lots with these percentages of defectives have occurred: 50%, 30%, and 20% of the time, respectively. The Renny Corporation feels justified in using these past percentages as prior probabilities for another lot which has just been delivered by the supplier. A simple random sample of ten units is drawn with replacement from the incoming lot and two defectives are found.

The Renny Corporation has found from past experience that it should accept lots that have 10% defectives and should reject those that have 20 or 30% defectives. That is, because of the costs of rework, it was not economical to accept lots with more then 10% defectives. On the basis of a careful analysis of past costs, the Renny Corporation constructed the payoff matrix in terms of opportunity losses shown in Table 14-5. The two possible courses

Table 14-5 Payoff Matrix Showing Opportunity Losses for Accepting and Rejecting Lots with Specified Proportions of Defectives.

Event	*Act*	
(p = Lot Proportion Defective)	A_1 *Reject*	A_2 *Accept*
.10	\$200	0
.20	0	\$100
.30	0	\$200

of action are act A_1, to reject the incoming lot, and act A_2, to accept the incoming lot. The three possible states of nature are the lot proportion defectives .10, .20, and .30. It will be more convenient to consider the lot defectiveness in terms of decimals than percentages. The lot proportion defective is denoted p, and will be treated as a discrete random variable which can take on the three given values. Of course, it is rather unrealistic to assume that an incoming lot can only be characterized by a .10, .20, or .30 fraction defectiveness. However, for convenience, that assumption will be made here. The same general principles would hold if a more realistic assumption were made, as, for example, that the proportion defective could take on values at one-percentage-point intervals, namely, 0.00, 0.01, . . . , 1.00. We now proceed to apply to this problem some of the principles of decision analysis we have learned.

Suppose the Renny Corporation had to take action concerning acceptance of the present lot before the drawing of the sample of ten units from this lot. Assuming the firm is willing to make its decision on the basis of prior information, what action should it take? It seems reasonable that in the absence of additional information, the company should use the past relative frequencies of lot proportion defectives as prior probability assignments. The prior analysis of the company's two courses of action based on expected opportunity losses is given in Table 14-6.[1] As shown in the table, using the past relative frequencies as prior probabilities, the expected opportunity loss of rejecting the

Table 14-6 Prior Expected Opportunity Losses for the Renny Corporation Problem.

Act A_1: Reject the Lot

Event p	*Prior Probability* $P_0(p)$	*Opportunity Loss*	*Weighted Opportunity Loss*
.10	.50	$200	$100
.20	.30	0	0
.30	.20	0	0
	1.00		$100

$\text{EOL}(A_1) = \$100$

Act A_2: Accept the Lot

Event p	*Prior Probability* $P_0(p)$	*Opportunity Loss*	*Weighted Opportunity Loss*
.10	.50	$ 0	$ 0
.20	.30	100	30
.30	.20	200	40
	1.00		$70

$\text{EOL}(A_2) = \$70$
Prior EVPI $= \$70$

[1]The notation $P_0(p)$ is used in Table 14-6 for the prior probability distribution of the random variable p and $P_1(p)$ is used in Table 14-7 for the corresponding posterior probability distribution. Also, the random variable is referred to by the lower case p in this and subsequent illustrations. This is a departure from the convention of using capital letters to denote random variables and lower case letters to represent the values taken on by the random variable. The reason for this exception in notation is to avoid confusion because of the common use of the capital P to mean "probability."

lot is \$100, of accepting the lot, \$70. Hence, the optimal act is A_2, to accept the lot. As noted earlier, the \$70 figure is the prior EVPI, since it represents the expected opportunity loss of the optimal act, using the prior probability distribution.

We turn now to the posterior analysis. Assume the Renny Corporation proceeds to draw the simple random sample of ten units with replacement from the incoming lot, and observes two defectives. Taking into account this sample evidence, what is the optimal course of action?

Following the same general procedure as in the inventor's problem, we can use the sample evidence to revise the prior probabilities assigned to the possible lot proportions of defectives. The application of Bayes' theorem for this purpose is shown in Table 14-7. The conditional probabilities shown in the third column of Table 14-7 are often referred to as "likelihoods." That is, they represent the likelihoods of obtaining two defectives in ten units in a simple random sample drawn with replacement from the assumed incoming lots. When the basic random variable is the parameter p of a Bernoulli process, as in this problem, the likelihoods of the observed "number of successes" in the sample are computed by the binomial distribution. The likelihood figures in Table 14-7 were obtained from Appendix A, Table A-1. The notation $P(X = 2 \mid n = 10, p)$ means "the probability that the random variable "number of successes" is equal to two in ten trials of a Bernoulli process whose parameter is p." We will use this type of symbolism in this and other problems for likelihood calculations. With the evidence of two defectives in a sample of ten units, or 20% defectives in the sample, the prior probability that the lot contains 20% defective is revised upward from .30 to .3869, as indicated in Table 14-7. Correspondingly, the probabilities that the lots contain 10 or 20% defectives are revised downwards.

Expected payoffs of the two acts can now be recomputed using the revised or posterior probabilities. These computations are shown in Table 14-8. The optimal act is still A_2, to accept the lot. However, the posterior

Table 14-7 Computation of Posterior Probabilities in the Renny Corporation Problem Incorporating Evidence Based on a Sample of Size 10.

Events p	*Prior Probability* $P_0(p)$	*Conditional Probability* $P(X = 2 \mid n = 10, p)$	*Joint Probability* $P_0(p)P(X = 2 \mid n = 10, p)$	*Posterior Probability* $P_1(p)$
.10	.50	.1937	.09685	.4136
.20	.30	.3020	.09060	.3869
.30	.20	.2335	.04670	.1995
	1.00		.23415	1.0000

expected opportunity losses of the two acts are much closer together than were the prior ones. Furthermore, the posterior expected opportunity loss of the optimal act A_2 is \$78.59, which represents an increase from the prior expected opportunity loss of the optimal act, \$70. In other words, the posterior EVPI now exceeds the prior EVPI, which indicates that the value of having a perfect predictor available has increased.

Effect of Sample Size

We can use this acceptance sampling problem to illustrate an important point in Bayesian decision theory, namely, the effect of sample size on the posterior probability distribution. Suppose that instead of a simple random sample of size 10 being drawn from the incoming lot, a similar sample of 100 units was drawn. Furthermore, let us assume that 20 defectives were found in the sample. In other words, the fraction of defectives in this larger sample is .20, just as it was in the smaller sample of ten units. The computation of

Table 14-8 Posterior Expected Opportunity Losses in the Renny Corporation Problem.

Act A_1: Reject the Lot

Event p	*Posterior Probability* $P_1(p)$	*Opportunity Loss*	*Weighted Opportunity Loss*
.10	.4136	\$200	\$82.72
.20	.3869	0	0
.30	.1995	0	0
	1.0000		\$82.72

$\text{EOL}(A_1) = \$82.72$

Act A_2: Accept the Lot

Event p	*Posterior Probability* $P_1(p)$	*Opportunity Loss*	*Weighted Opportunity Loss*
.10	.4136	\$ 0	\$ 0
.20	.3869	100	38.69
.30	.1995	200	39.90
	1.0000		\$78.59

$\text{EOL}(A_2) = \$78.59$
Posterior EVPI = \$78.59

posterior probabilities by the use of Bayes' theorem and the information from the sample of size 100 is given in Table 14-9. The conditional probabilities in Table 14-9 were obtained from a table of binomial probabilities which includes $n = 100$.

While .30 was the prior probability assigned to the state of nature that the incoming lot proportion defective was 0.20, as indicated in the last column of Table 14-9, the revised probability is .9336. Hence, because of the implicit weight given to the sample evidence by Bayes' theorem, it is much more probable that the defective lot fraction is .20 according to the posterior distribution than according to the prior distribution. Furthermore, comparing the posterior probability distributions in Tables 14-7 and 14-9, a much higher probability (.9336) is assigned to the event $p = .20$ after 20% defectives have been observed in a sample of size 100 than when that percentage of defectives is found in a sample of size 10 (.3869). A rough generalization of this result is that as sample size increases, the posterior distribution of the random variable "proportion defective" is more and more influenced by the sample evidence and less and less by the prior distribution.

Prior and Posterior Means

Let us consider a somewhat more formal explanation of the relationship between the prior distribution, sample evidence, and the posterior distribution. This can be given in terms of the change that takes place between the mean of the prior distribution and the mean of the corresponding posterior distribution. For brevity, we will refer to the mean of a prior distribution of a random variable representing states of nature as a "prior mean" or "prior expected value" and the corresponding mean of a posterior distribution as a "posterior mean" or "posterior expected value." The prior mean is obtained by the usual method for computing the mean of any probability distribution. The calculation of the prior mean for the acceptance sampling problem is given in Table

Table 14-9 Computation of Posterior Probabilities in the Renny Corporation Problem Incorporating Evidence Based on a Sample of Size 100.

Events p	*Prior Probability* $P_0(p)$	*Conditional Probability* $P(X = 20 \mid n = 100, p)$	*Joint Probability* $P_0(p)P(X = 20 \mid n = 100, p)$	*Posterior Probability* $P_1(p)$
.10	.50	.0012	.00060	.0188
.20	.30	.0993	.02979	.9336
.30	.20	.0076	.00152	.0476
	1.00		.03191	1.0000

Table 14-10 Calculation of the Prior Mean for the Defective Proportion in the Renny Corporation Problem.

Events p	Prior Probability $P_0(p)$	$pP_0(p)$
.10	.50	.05
.20	.30	.06
.30	.20	.06
	1.00	.17

Prior Mean $= E_0(p) = .17$ defectives

14-10. Thus, the prior mean is equal to .17 defective articles. The notation $E_0(p)$ is used for the prior mean, the letter E denoting "expected value" and the subscript zero denoting "prior distribution." Analogously, the mean of the posterior distribution is denoted $E_1(p)$. The corresponding computation of the posterior means for the cases in which the posterior distributions reflect sample evidence of two defectives in a sample of ten units and 20 defectives in a sample of 100 units are given in Table 14-11.

Rounding off the results obtained in Table 14-11, we observe posterior means of .179 defectives utilizing the sample evidence of two defectives in a sample of ten units, and .203 using sample evidence of 20 defectives in a sample

Table 14-11 Calculation of Posterior Means for the Defective Proportion in the Renny Corporation Problem.

Posterior Distribution Incorporating Sample Evidence $X = 2$, $n = 10$			Posterior Distribution Incorporating Sample Evidence $X = 20$, $n = 100$		
Events p	Posterior Probability $P_1(p)$	$pP_1(p)$	p	Posterior Probability $P_1(p)$	$pP_1(p)$
.10	.4136	.04136	.10	.0188	.00188
.20	.3869	.07738	.20	.9336	.18672
.30	.1995	.05985	.30	.0476	.01428
	1.000	.17859		1.0000	.20288

Posterior Mean $= E_1(p) = .17859$ Posterior Mean $= E_1(p) = .20288$

of 100. Hence, in the case of the smaller sample size, the posterior mean lies closer to the prior mean of .17 defectives than to the sample evidence of .20 defectives. On the other hand, when the larger sample is employed, the posterior mean falls closer to the sample evidence of .20 defectives than to the mean of the prior distribution.[2] This empirical finding is in keeping with the previously given statement that as the sample size increases, the posterior dis-

Table 14-12 Posterior Expected Opportunity Losses in the Renny Corporation Problem: The Posterior Probabilities Incorporate Sample Evidence $X = 20$, $n = 100$.

Act A_1: Reject the Lot

Event p	*Posterior Probability* $P_i(p)$	*Opportunity Loss*	*Weighted Opportunity Loss*
.10	.0188	$200	$3.76
.20	.9336	0	0
.30	.0476	0	0
	1.0000		$3.76

EOL(A_1) = $3.76

Act A_2: Accept the Lot

Event p	*Posterior Probability* $P_1(p)$	*Opportunity Loss*	*Weighted Opportunity Loss*
.10	.0188	$ 0	$ 0
.20	.9336	100	93.36
.30	.0476	200	9.52
	1.0000		$102.88

EOL(A_2) = $102.88
Posterior EVPI = $3.76

[2]The posterior mean of .203 exceeds both the values of the prior mean of .17 and the sample evidence of .20 defectives. An examination of the results for $p = .10$ and $p = .30$ in Table 14-9 gives the reason. The likelihood of 20 defectives in a sample of 100 for $p = .30$ is more than six times the similar likelihood for $p = .10$. Despite a lower prior probability for $p = .30$ than for $p = .10$, the joint probability and therefore the posterior probability in the case of .30 far exceed those for $p = .10$. The net effect is to pull $E_1(p)$ somewhat closer to .30 than to .10.

tribution is progressively more influenced by the sample evidence and less by the prior distribution.

It is instructive to determine the optimal act using posterior probabilities which incorporate the evidence of 20 defectives in a sample of size 100. Table 14-12 gives the posterior expected opportunity losses based on the aforementioned probabilities. The posterior expected opportunity losses of act A_1 and A_2 are further apart than in any of the preceding cases. The optimal act is A_1, to reject the lot, with a posterior expected opportunity loss of only \$3.76 as compared to the corresponding figure of \$102.88 for A_2. Comparing this decision in favor of act A_1 to the analogous choice of act A_2 based on prior expected opportunity losses, we see here an illustration of a situation where a reversal of decision takes place because of evidence observed in a sample. The very low figure of \$3.76, which can be interpreted as the posterior EVPI, indicates that after the observation of 20 defectives in a sample of 100 units, it would not be wise to spend much money accumulating additional evidence before making the decision concerning acceptance or rejection of the lot.

14.3 Two-Action Problems with Linear Payoff Functions

In all of the Bayesian decision theory problems discussed thus far, choices between alternative courses of action were based on comparisons of expected payoffs either in monetary or utility units. In both prior and posterior analysis, events can be represented by the values of a random variable, and probabilities can be assigned to these values. Then for each act, these probabilities are applied as weights to payoffs for each value of the random variable and summed to yield expected payoffs. In this type of analysis, it does not matter how many acts are involved nor how the payoffs vary with the value of the random variable. In fact, heretofore in our discussion, the figures in the payoff table have been taken as given at the outset of the problem without any examination of how they were derived. Parenthetically, the reader is reminded that payoffs may either be stated in the form of profits or costs, depending upon the nature of the problem.

In this section, we discuss a somewhat special type of problem in which only two actions are involved and in which payoff is a linear function of the basic random variable which represents events. Simple but effective methods are available for handling this type of problem. By considering alternative methods of solution, we can get some interesting insights into the particular information which is of the greatest importance for decision making in these problems.

As a first example of a two-action situation with linear payoff functions, we consider the following problem. An importer of specialty foods is trying to decide whether or not to market a rather unique type of cocktail onion which is imported from Japan. Since he sells exclusively by mail order, his "market" consists of a list of 100,000 names and addresses. If he decides to market the onions, he will buy them in bulk and package them himself. His packaging and handling costs are

Fixed cost for set up:	\$1000
Labor, material, and mailing costs:	\$.80 per jar

The importer will sell the onions at \$1.00 per jar.

Suppose we designate the two courses of action in this problem as A_1: market the product, and A_2: do not market the product, and we let p be the proportion of the 100,000 potential customers on the list who will buy the onions if they are made available. Then there is some "break-even value" of p above which the optimal act is A_1, to market the product, and below which the optimal act is A_2, not to market it. Let us denote this break-even value of p as p^*. At p^*, the packager is indifferent between acts A_1 and A_2. In a problem of this type, we can determine the value of p^* by obtaining a profit function for each act in terms of the basic random variable p, setting these profit functions equal to one another, and solving for p. This procedure yields the break-even value p^*, because it gives a p value such that the payoffs of the two acts, A_1 and A_2, are equal.

Comparison of Expected Profits

The profit function for act A_1 may be expressed as

(14.1) $$\pi_1 = -\$1000 + 100{,}000p(\$1.00 - \$.80)$$

where

π_1 = the profit derived from act A_1
p = proportion of the 100,000 potential customers who will buy the product if it is made available

The subtraction of \$1000 represents the fixed cost for set-up. The factor $100{,}000p$ is the number of jars sold if the proportion of the 100,000 potential customers who purchase it is p. The \$0.20 represented by the factor (\$1.00 − \$.80) is the profit per jar exclusive of fixed cost. Multiplying out the second term on the right-hand side of Equation (14.1) yields the simple expression:

(14.2) $$\pi_1 = -\$1000 + \$20{,}000p$$

The profit function for act A_2 is very simple, because regardless of the value of p, if the product is not marketed, profit will be equal to zero dollars. Hence, we express the profit function for act A_2 as

(14.3) $$\pi_2 = \$0$$

where π_2 = the profit obtained from act A_2

To derive p^*, the break-even value of p, we equate π_1 and π_2, since this operation yields the value of p for which the profits from acts A_1 and A_2 are equal. Equating Equations (14.2) and (14.3), we have

$$0 = -\$1000 + \$20{,}000p$$

Thus,

$$p^* = \frac{\$1000}{\$20{,}000} = 0.05$$

In summary, if exactly 5% of the customers on the mailing list would purchase the product, the importer would be indifferent between act A_1, to market the product, and act A_2, not to market it.

A break-even chart for the profits of acts A_1 and A_2 is given in Figure 14-1. As shown in this graph, the profit functions for act A_1 (Equation 14.2) and A_2 (Equation 14.3) intersect at $p^* = .05$, the break-even value of p.[3]

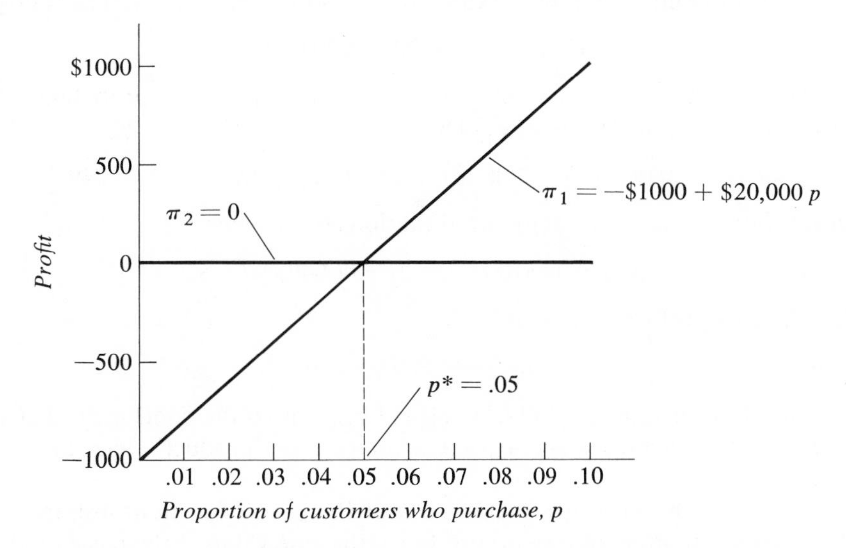

Figure 14-1 Profit functions in the importer's problem showing the break-even value of p.

---

[3]Actually, p^* is most accurately referred to as an "indifference point" rather than as a "break-even point" or "break-even value," because the decision maker is indifferent between acts A_1 and A_2 at this point. However, because the break-even terminology is commonly used in statistical decision theory, it will be retained here.

For p values above .05, act A_1 yields a profit and is therefore preferable to A_2, whose payoff is \$0. On the other hand, as shown in the chart, for p values less than .05, act A_1 results in a loss, and therefore A_2 is the better course of action.

Suppose that from his experience with similar products in the past, the importer had arrived at the following prior probability distribution of p:

p	$P_0(p)$
.02	.10
.04	.40
.06	.45
.08	.05

If the importer had to make his decision without the opportunity to sample the mailing list to test the "marketability" of the onions, what would be the optimal act and the expected profit from this course of action? The most direct way to obtain an answer to this question is simply to compute the expected profit of each act and to compare these values. The expected profit of act A_1 is obtained by taking expected values in Equation (14.2) as follows:

(14.4) $$E(\pi_1) = -\$1000 + \$20{,}000E(p)$$

The expected value of p, $E(p)$, is derived from the prior probability distribution of p in the usual way. Thus,

$$E(p) = (.02)(.10) + (.04)(.40) + (.06)(.45) + (.08)(.05) = .049$$

Substituting this value for $E(p)$ into Equation (14.4) yields

(14.5) $$E(\pi_1) = -\$1000 + \$20{,}000(.049) = -\$20$$

The expected profit of act A_2 is

(14.6) $$E(\pi_2) = E(\$0) = \$0$$

A comparison of Equations (14.5) and (14.6) leads to the conclusion that the better course of action is act A_2, not to market the product, since $E(\pi_2)$ exceeds $E(\pi_1)$.

This problem illustrates the fact that if the payoff functions are linear in a two-action situation and we do not know the true value of the basic random variable, we can establish a decision procedure solely in terms of the *expected value* of the random variable. For example, in this problem, the decision procedure would be

$$\text{Choose act } A_2 \text{ if } E(p) < .05$$

$$\text{Choose act } A_1 \text{ if } E(p) > .05$$

We would be indifferent to the two courses of action at $E(p) = p^* = .05$. Of course, this analysis assumes the decision maker has a linear utility function for money. Therefore, he maximizes expected utility by maximizing expected profit.

If p were known rather than considered as a random variable, the same sort of decision procedure would be used, except that p rather than $E(p)$ would be compared to p^*. If the payoff function were non-linear, the decision rule could not be expressed solely in terms of the expected value of the basic random variable. For example, if the profit functions had been of the form $\pi = a + bp + cp^2$, then $E(\pi) = a + bE(p) + cE(p^2)$. Hence the decision rule would involve $E(p^2)$ as well as $E(p)$.

Comparison of Expected Opportunity Losses

Although the concept of the break-even value of p was introduced in the preceding paragraphs, this p^* value was not directly involved in the calculation of the preferred act by the comparison of *expected profits* method. On the other hand, if acts A_1 and A_2 are compared in terms of their *expected opportunity losses*, the break-even value of the random variable pertaining to states of nature is an integral part of the calculations. Again, as in the preceding chapter, we point out that, in practice, the decision maker need not perform calculations by several different methods in order to choose among two possible actions. However, increased insight into Bayesian decision analysis for a two-action problem is given by considering the comparison of the expected opportunity loss method. Therefore, we will also analyze the importer's problem from the standpoint of these losses.

As in previous problems involving opportunity losses, attention is focused on the possibility that the decision maker may make an incorrect decision. In doing so, he foregoes profits that he may have made or incurs costs he would not have had had he made the correct decision. Table 14-13 gives the opportunity loss matrix for the importer's problem. The possible states of nature or events are indicated in terms of values of p which are below or above the break-even value of $p^* = .05$. The equal sign has been included in the state-

Table 14-13 Opportunity Loss Table for the Importer's Problem.

	Opportunity Losses	
Events	*Market the Product* A_1	*Do not Market the Product* A_2
$p \leq .05$	\$20,000 $(.05 - p)$	\$0
$p > .05$	\$0	\$20,000 $(p - .05)$

ment of the event $p \leq .05$, although it is immaterial whether it is included there or in the other event, or excluded from both, since when $p = .05$, opportunity loss equals zero under acts A_1 and A_2, and the decision maker is indifferent as to a choice between these two acts. However, for a later comparison between traditional hypothesis testing methods and Bayesian decision analysis, it is convenient to state the events as indicated. Turning to the entries in Table 14-13, we see that the opportunity loss is zero if act A_2 is taken and p is less than .05 and is zero if act A_1 is chosen and p exceeds .05. The other two entries are "opportunity loss functions" for choosing incorrect courses of action. In Bayesian decision analysis, these opportunity loss functions are often conventionally referred to simply as "loss functions." As an illustration, we will discuss the nature of the loss function in the lower right-hand corner of Table 14-13, that is, the loss function when act A_2 is chosen and the event that occurs is $p > .05$.

Let $\text{OL}(A_2 \mid p)$ denote the conditional opportunity loss of act A_2, given that p exceeds .05. Then the following algebraic expression gives the value of $\text{OL}(A_2 \mid p)$ for $p > .05$:

(14.7) $$\text{OL}(A_2 \mid p) = \$20{,}000(p - .05) \qquad \text{for } p > .05$$

When $p = .05$, the opportunity loss of selecting act A_2 is zero. For $p > .05$, Equation (14.7) gives the opportunity loss of choosing A_2 as a positive number in dollars. Thus, for example, if $p = .06$, Equation (14.7) gives

$$\text{OL}(A_2 \mid p = .06) = \$20{,}000(.06 - .05) = \$200$$

The \$200 figure is the "regret" or opportunity loss of selecting act A_2; that is, not marketing the cocktail onions, when $p = .06$. Hence, the packager by selecting the wrong act, A_2, foregoes the opportunity to realize a profit of \$200 he would have made with $p = .06$ had he selected the other act, A_1. For each one percentage point increase of p, there is an additional \$200 opportunity loss. The corresponding loss function for act A_1 is

(14.8) $$\text{OL}(A_1 \mid p) = \$20{,}000(.05 - p) \qquad \text{for } p \leq .05$$

Each loss function has two pieces, one below the break-even value, $p^* = .05$, and one above that figure. In Figure 14-2 are given the graphs of the opportunity loss functions for acts A_1 and A_2 in the importer's problem.[4] In Figure 14-3 the two loss functions are shown on the same graph. For simplicity, the portions of the functions where opportunity losses are equal to zero have not been included. A comparison of Figure 14-3, which is in terms of opportunity losses, with Figure 14-1, which is in terms of profits of the two acts, indicates that essentially the same information is depicted. Whereas the first chart is

---

[4]For brevity of notation, $\text{OL}(A_1 \mid p)$ for $p \leq .05$ and $\text{OL}(A_2 \mid p)$ for $p > .05$ are shown as $\text{OL}(A_1 \mid p \leq .05)$ and $\text{OL}(A_2 \mid p > .05)$ in Figures 14-2 and 14-3.

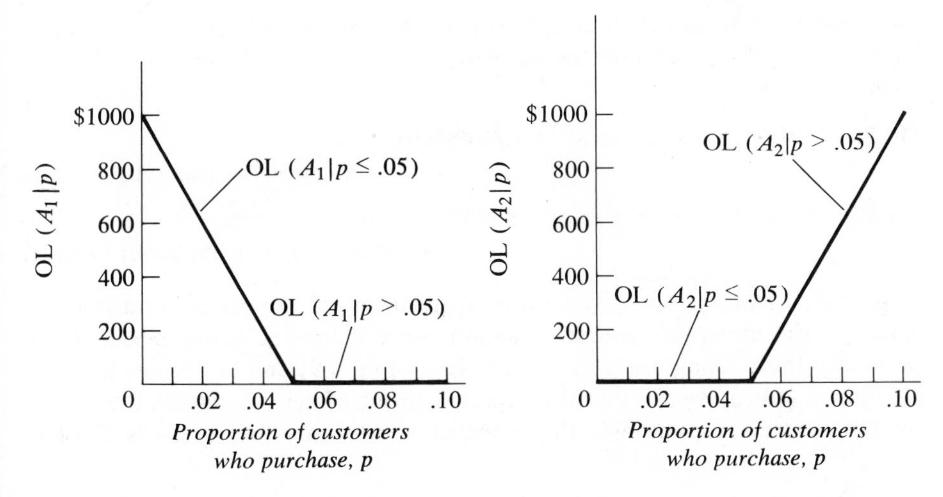

Figure 14-2 Opportunity loss functions for acts A_1 and A_2 in the importer's problem.

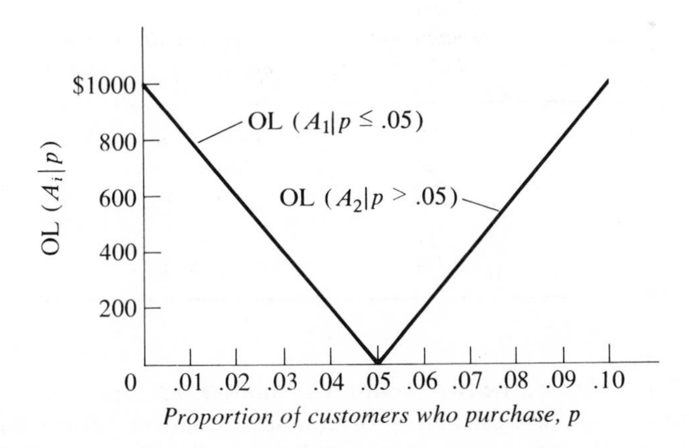

Figure 14-3 Opportunity loss functions for Acts A_1 and A_2 shown on the same graph.

in units of dollars of profit resulting from the two acts, the latter is in terms of differences between these profits at each value of the basic random variable p.

We can now carry out a comparison of the two acts in terms of expected opportunity losses in an analogous way to the comparison in terms of expected profits. Hence, we assume the same prior probability distribution of p utilized in the profit comparison and use Equations (14.7) and (14.8) to calculate opportunity losses at each value of p. Table 14-14 shows these opportunity losses for acts A_1 and A_2 and the prior probability distribution of p. Calculating expected opportunity losses in the usual way, we obtain the following results:

$$\text{(14.9)} \quad \text{EOL}(A_1) = (.10)(\$600) + (.40)(\$200) + (.45)(\$0) + (.05)(\$0) = \$140$$

$$\text{(14.10)} \quad \text{EOL}(A_2) = (.10)(\$0) + (.40)(\$0) + (.45)(\$200) + (.05)(\$600) = \$120$$

Thus, assuming the given prior distribution of the random variable p, act A_2 is the preferable course of action, since it has the lower expected opportunity loss. Comparing the results given in (14.9) and (14.10) to those in (14.5) and (14.6), we see that the expected profit of act A_2 exceeds that of A_1 by $20, and correspondingly the expected opportunity loss of A_2 is $20 less than that of A_1. Hence, the two methods of analysis are simply two equivalent ways of comparing the desirability of courses of action. Since businessmen are more familiar with comparisons of alternatives in terms of profits, the expected profits method is doubtless the more appropriate method of presentation to management. On the other hand, it has been traditional in statistical decision theory to emphasize the concept of "regrets" or "opportunity losses," and in theoretical presentations, the criterion of minimization of expected opportunity losses is more frequently encountered.

Table 14-14 Prior Probability Distribution and Opportunity Losses in the Importer's Problem.

Event p	*Prior Probability* $P_0(p)$	*Opportunity Losses* *Act* A_1	*Act* A_2
.02	.10	$600	$ 0
.04	.40	200	0
.06	.45	0	200
.08	.05	0	600
	1.00		

Decision Tree for the Importer's Problem

The decision analysis in the importer's problem is summarized in a simple tree diagram in Figure 14-4. The data are given in terms of opportunity

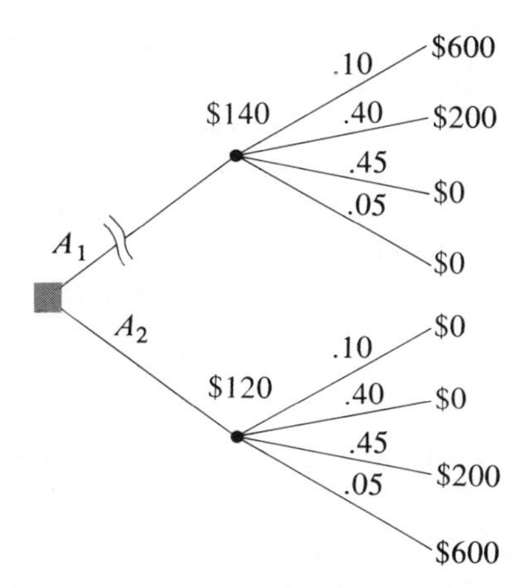

Figure 14-4 Decision diagram for the importer's problem in terms of opportunity losses.

losses (and probabilities) shown in Table 14-14, but, of course, an analogous diagram could be drawn in terms of profits. As shown in Figure 14-4, act A_1 has been blocked off, since under the given probability distribution A_2 is the preferable decision.

Posterior Analysis for the Importer's Problem

The solution to the importer's problem given in the preceding sections is an example of prior analysis, since a prior probability distribution was used in obtaining expected profits and expected opportunity losses. However, the term "prior" was ordinarily not used in our discussion of expected profits and opportunity losses in order not to clutter up the terminology. A similar posterior analysis can be carried out using a probability distribution revised on the basis of sample information. For example, suppose the importer decided to test the "marketability" of the onions. He sent advertisements to 100 persons selected at random from the mailing list and six of these people purchased the onions. Given this information, what is now the best decision?

With the above sample data, we can compute a revised probability distribution of p and apply the same decision procedure as previously. In Columns (1) through (5) of Table 14-15 are shown the calculations for revising the prior probability distribution. The results of the multiplication of the values of p in Column (1) by the posterior probabilities in Column (5) are shown in Column (6). The sum of Column (6) is the posterior expected value of p, denoted $E_1(p)$. Since $E_1(p) = .0538$, and this value exceeds the break-even value of $p^* = .05$, the best act now is A_1, to market the onions. A comparison of the two acts in terms of expected profits or expected opportunity losses can be carried out exactly as in the prior analysis, except that posterior probabilities would be used wherever prior probabilities appeared previously. We will not perform those calculations here. It may be noted that this problem is an illustration of a situation in which the most desirable course of action was changed on the basis of the sample information. Whereas A_2 was the preferable action without sampling, A_1 became the better act in the light of the sample data.

Table 14-15 Calculation of the Posterior Probability Distribution and Posterior Expected Value of the Random Variable p in the Importer's Problem.

(1) Event p	(2) Prior Probability $P_0(p)$	(3) Conditional Probability $P(X = 6 \mid p)$	(4) Joint Probability $P_0(p)P(X = 6 \mid p)$	(5) Posterior Probability $P_1(p)$	(6) Column (1) × Column (5) $pP_1(p)$
.02	.10	.0114	.00114	.01	.0002
.04	.40	.1052	.04208	.34	.0136
.06	.45	.1657	.07457	.60	.0360
.08	.05	.1233	.00617	.05	.0040
			.12396	1.00	.0538

$$E_1(p) = .0538$$

Comparison of Expected Costs

A two-action problem in which the payoffs are in terms of linear cost functions can be solved in a similar way to the preceding problem in which payoffs were expressed as linear profit functions. For example, suppose that a company has analyzed its costs for producing a run of 100 units of an assembly component by two different production methods as follows:

$$c_1 = \$300 + \$2000p$$

and

$$c_2 = \$100 + \$7000p$$

where c_1 = the cost of Method 1
c_2 = the cost of Method 2
p = the proportion of defective assemblies produced

As can be seen from these equations, the increase in cost as the proportion of defectives increases is greater for Method 2 than Method 1. We will assume in this problem that past experience suggests the same prior probability distribution of defectiveness applies to the two processes. However, the differential in cost because of production of defectives is primarily due to the difference in type and amount of rework required. It may be assumed that the final product produced by the two methods is essentially the same. Hence, the decision maker's objective is to select the method which yields the lower expected cost.

The break-even value of p can be determined by equating c_1 and c_2:

$$\$300 + \$2000p = \$100 + \$7000p$$

Solving this equation yields the break-even value

$$p^* = 0.04$$

A break-even chart for the costs of Methods 1 and 2 is given in Figure 14-5.

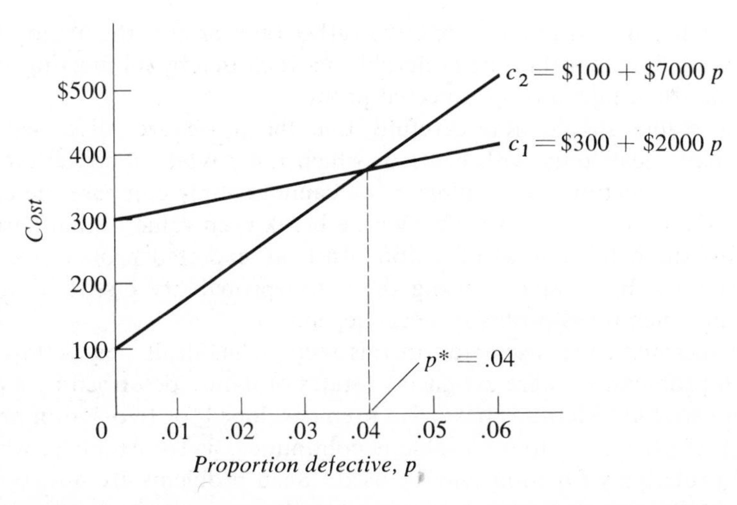

Figure 14-5 Cost functions for two production methods showing the break-even value of p.

As can be seen in Figure 14-5, the cost functions intersect at $p^* = .04$, as determined analytically from the two equations. For p values lower than .04, Method 2 yields the lower cost, whereas for p values above .04, Method 1 is preferable. We leave as an exercise the comparison of expected costs for the two methods and the comparison of expected opportunity losses in a prior analysis. Assume a simple prior probability distribution to carry out the problem.

The general decision procedure for two-action problems under uncertainty with linear payoffs is summarized below in terms of profit functions.

(1) Obtain the profit functions for each act in terms of the basic random variable for the underlying states of nature.

(2) Determine the expected profits of each act in terms of the expected value of the basic random variable by computing expected values for each profit function. This yields linear functions of the following general forms, for example, if p is the basic random variable:

$$E(\pi_1) = a_1 + b_1 E(p)$$

$$E(\pi_2) = a_2 + b_2 E(p)$$

(3) Determine the expected value of the random variable from the appropriate prior or posterior probability distribution and substitute into the equations obtained in Step (2).

(4) Select the act which has the higher expected profit.

If the problem is in terms of costs rather than profits, the analogous procedure is used, except that the preferable act is chosen by minimizing expected cost rather than maximizing expected profit.

The reader should note carefully that the procedures discussed in this section have dealt only with cases in which the payoff functions are linear. If the payoff functions are nonlinear, we cannot simply compare the expected value of the basic random variable with a break-even value of that variable to determine the better course of action. Instead, expected profits or costs for each act must be computed using the entire probability distribution of the events on which these profits or costs depend.

Furthermore, the discussion in this section has dealt only with the case in which probabilities were assigned to states of nature described by a discrete random variable. Methods have also been developed for two-action problems in which the basic random variable is continuous, as for example, where the normal probability function may be used. Such problems are not considered here.

PROBLEMS

1. Given:

State of Nature	$P(\theta)$	$P(X \mid \theta)$	$P(\theta)P(X \mid \theta)$	$P(\theta \mid X)$
θ_1: Housing starts will increase next year	.5	.6		
θ_2: Housing starts will remain at the same level or will decline		.4		

Fill in the blanks and interpret the data, if X is the result of a survey of 100 construction companies.

2. A prior probability function is

S	$P_0(S)$
S_1	.1
S_2	.2
S_3	.3
S_4	.4

and a sample observation X occurs which has the following properties: $P(X|S_1) = .8$, $P(X|S_2) = .6$, $P(X|S_3) = .5$, and $P(X|S_4) = .2$. What are the revised probabilities?

3. A firm is trying to decide whether to embark on a new advertising campaign. Management assigns the following prior probability distribution:

State of Nature	*Probability*
S_1: Successful	.5
S_2: Unsuccessful	.5

A sample result is observed which has the following probability of occurring:

.3 if S_1 is true,
.6 if S_2 is true.

Revise the prior probability distribution in light of this new information.

4. In a certain situation, before sampling, the best act is A_1 and the EVPI is \$100. After a sample was drawn the best act was still A_1 and the revised EVPI was \$50. The cost of sampling was \$20. Can you conclude that the actual value of the sample information to the decision maker was \$30? Explain your answer.

5. Let p be the true percentage of customers who will purchase a new product. Assume p can take on the values given below with the respective prior probabilities.

p	$P_0(p)$
.10	.8
.20	.2

A simple random sample of ten customers is drawn. The customers are asked whether they would purchase the product. What are the revised prior probabilities if

(a) one customer would purchase?
(b) two customers would purchase?
(c) three customers would purchase?

6. Let μ equal the average number of pedestrians hurt per month in a process for which the Poisson distribution is a suitable model. The state of nature, μ, can assume the values given below with the respective prior probabilities.

μ	$P_0(\mu)$
5	.6
6	.4

If during a particular month the number of pedestrians hurt was six, what are the revised probabilities?

7. There are two actions to take, A_1 and A_2, and there are two states of nature S_1 and S_2. A_1 is preferred if S_1 is true, and A_2 is preferred if S_2 is true. If the prior probabilities are $P(S_1) = .7$ and $P(S_2) = .3$ and you observe a sample observation S such that the $P(S|S_1) = .9$ and $P(S|S_2) = .2$, can you conclude that A_1 is the best act? Explain your answer.

8. A small retail company is considering putting in a credit system. Let p = proportion of new accounts that will be uncollectable if a credit system is installed. The states of nature and their respective prior probabilities are

p	$P_0(p)$
.01	.1
.05	.4
.10	.4
.15	.1

Assume a credit system is installed. What are the revised probabilities if credit is extended to 100 customers and

(a) four are uncollectable accounts?
(b) twelve are uncollectable accounts?

Use a normal curve approximation.

9. Given

State of Nature	$P(X \mid \theta)$	$P(\theta \mid X)$
θ_1	.6	.5
θ_2	.4	.5

where θ_1 is "product is as good as or superior to competitor's" and θ_2 is "product is worse than competitor's." An experiment was run with the results given above. What was the prior probability distribution before sampling?

10. Let p be the proportion of defective cigarette lighters in a lot offered to you by a jobber. Given

p	$P_0(p)$	*Opportunity Loss of Accepting Lot*
.10	.5	0
.20	.3	100
.30	.2	200

A sample of ten is taken and two are found defective. What is the expected opportunity loss of the action "accept,"
(a) if action is taken before sampling?
(b) if action is taken after sampling?

11. Management's prior probability assessment of demand for a newly developed product is: high, .6; low, .2; and average, .2. A survey, taken to help determine the true demand for the product, indicates demand is average. The reliability of the survey is such that it will indicate "average" demand 70% of the time when it is really high, 95% of the time when it is really average, and 10% of the time when it is actually low. In light of this information, what would be the reassessed probabilities of the three states of nature?

12. In Problem 11, Chapter 13, suppose two balls are drawn without replacement from the urn which was previously selected. Both balls are white. Compute a revised expected value of perfect information.

13. In Problem 9, Chapter 13, the publisher sends a copy of the book to ten critics for their opinions. The results are slightly unfavorable; six out of ten dislike the book. If the book should be a failure, the probability that a critic would dislike it is .8, and if the book should be a success, there is a 50-50 chance that a critic would like it. In view of this added information, would you recommend publication of the book?

14. It costs a baker \$.75 to produce a pie which sells for \$1.00. If a pie is produced but not sold it will spoil and will have to be thrown away. Write a mathematical expression for profit as a function of x, the number of pies produced, and p, the proportion of produced items which are sold. Find the break-even value of p.

15. It costs \$.75 to produce an item which sells for \$1.00 if perfect, but for only \$.50 if defective. Total production is 10,000 items per month. Let p be the proportion of the production per month which is defective. Write a mathematical expression for profit and find the break-even value for p.

16. In Problem 15, assume the present proportion of defectives being produced is .25. Management is considering installing a new machine which will increase the cost of production to \$.80 per item, but this new machine is supposed to be more reliable. Management assesses the probability distribution of the reliability of the machine $(1 - p)$ as

$(1-p)$ Reliability	Probability
.75	.3
.85	.4
.95	.3

Should the new machine be installed based on

(a) the above information alone?

(b) the above information and the fact that a test run of the new machine produced 20 defective items out of 200? (Use a normal curve approximation.)

17. A manufacturing firm is considering the purchase of 100,000 special relays to use in production of its machinery. The loss due to a defective relay is $2.50. A new firm has just started producing a type of relay which is reported to be more reliable than the previously supplied relays, but is similar in other respects. The reliability (i.e., the probability that the item is good, $1-p$) of the previously supplied relays was .99; management assigns the following probabilities to the reliability of the new firm's relays:

Reliability	Probability
.995	1/3
.990	1/3
.980	1/3

What action should the manufacturing firm take

(a) on this information alone?

(b) if a sample of 500 relays from the new firm contains three defectives? (Use a normal curve approximation.)

18. A large firm is contemplating the purchase of 25,000 ball-point pens. Supplier *A* offers the pens at $.48 each, guarantees each pen, and will replace all defective pens free. Supplier *B* offers the pens at $.45 each with no guarantee, but offers to replace defective pens with good pens for $.40. Management estimates the loss in labor time and inconvenience due to a defective pen as $.10. Moreover it feels the probability distribution of the rate of defectives of supplier *B* is

p	$P_0(p)$
.03	.1
.05	.3
.07	.6

A random sample of 200 pens is obtained free from *B* and nine pens are found defective. (The pens must be returned.)

(a) Write out the cost function of each act (buy from *A*, and buy from *B*). Find the break-even value of p.

(b) Given the sample information, what is the best act? Use a normal curve approximation.

19. An employment manager is deciding whether to hire a new salesman. He would have to pay the man \$10,080 a year salary, plus 10% commission on gross sales. The average salesman calls on 720 customers a year and the average sale grosses \$1000. The profit to the firm considering everything but the cost of paying the salesman is 30% of gross sales. If p is the proportion of customers seen by the salesman that result in sales,
   (a) Write out the profit function and find the break-even value of p.
   (b) Write out the opportunity loss function for each of the two acts "hire" and "not hire" the new salesman.

20. Assume that in Problem 19 you are given

State of Nature p	$P_0(p)$	P(*test indicates "average salesman"* $\mid p$)
.04	.2	.1
.06	.2	.1
.08	.2	.3
.10	.2	.8
.12	.2	.9

   (a) What is the best decision and its expected profit based on prior beliefs about p?
   (b) You subject the salesman to a series of tests which cost \$75 to try to determine his true sales ability. The test results indicate he is an "average salesman." Using this information, should the salesman be hired, and what is the revised expected profit?

21. A large chain of supermarkets requires 24,000 fluorescent light bulbs for its stores, and the manager is looking for a source of supply. Supplier A offers bulbs at a price of \$2.00 per bulb and will replace the defective bulbs with guaranteed good ones at \$2.00 each. Supplier B offers the bulbs at \$2.50 per bulb and guarantees to replace all defectives free. Assume the proportion of defective bulbs produced by both suppliers is the same but unknown.
   (a) What defective rate would make the management indifferent as to its source of supply?
   (b) From past experience with fluorescent bulbs, the prior distribution of the proportion of defective bulbs is given as

Proportion Defective	*Probability*
.15	.1
.20	.2
.25	.3
.30	.3
.35	.1

   From which supplier should the manager purchase?

(c) The management is given a random sample of 20 bulbs without cost. The bulbs are tested on a special machine, and three are found to be defective. What is the best act now, and what is the EVPI?

22. You are a dealer in gems, which you cut, polish, and sell for industrial use and for use in making jewelry. You are offered a lot of 2000 large gems at a price of $300 each ($600,000 in all). If they are of good quality, each gem may be cut and polished at a cost of $200 and sold as jewelry for a price of $1300. However, if they are only of industrial quality, they may be cut and polished at a cost of $20 and sold for a price of $120. (Be very careful with zeroes in this problem.) You have bought uncut stones from this merchant previously and know his source of supply. From your experience, your prior beliefs about the quality of the lot offered you are described by the following probability distribution:

Proportion of Good Quality Gems in Lot p	*Prior Probability* $P_0(p)$
.10	.3
.20	.4
.30	.2
.40	.1

(a) Let p be the proportion of good quality gems in the lot. Write an expression for your profit as a function of p, assuming that you buy the lot.
(b) There are two actions in this problem; a_1, buy the lot; and, a_2, do not buy the lot. Using the result you obtained in (a), write an opportunity loss function for each act.
(c) If you must make up your mind without examining the stones, what should you do and why?
(d) If you take a sample of 100 gems and find 32 to be of good quality, what is the revised expected value of the proportion of good quality stones and what then is the best decision? Use a normal curve approximation.

23. The government is trying to decide whether to build a reservoir in Dodd or Todd County. Dodd County has a population of 800,000 and Todd 1,200,000. The cost of building the reservoir is the same in each location. The immediate expected benefit of the reservoir is computed as $10 per person in the county immediately *after* the reservoir is built. Building the reservoir will displace people. The loss due to displacement is estimated at $50 per person. The reservoir in Dodd will displace 15,000 people, and in Todd it will displace 92,000 people. Of those displaced, the percentage of people who will move out of Todd County is between 10 and 20% and out of Dodd County is between 5 and 15%. From economic studies and comparisons with previous reservoir sites, the prior probabilities of the proportion that will move out are estimated for each county as follows:

θ	Dodd $Po(\theta)$	Todd $Po(\theta)$
.05	.3	.0
.10	.4	.3
.15	.3	.4
.20	.0	.3

(a) Write a mathematical expression for the benefit derived from each reservoir, letting x represent the proportion of displaced people who move out of the county.
(b) Based on prior beliefs, where should the reservoir be built?
(c) In each county a random sample of 20 persons was drawn. These people were asked whether they would move out of the county if displaced. Three people in Dodd and five in Todd said that they would move. Based on this new information, where should the reservoir be built?

24. There are two states of nature and two alternative actions open to a firm. The payoff matrix in units of thousands of dollars, and initial probabilities are

Action	State of Nature S_1	S_2	
A_1	100	10	$P(S_1) = .3$
A_2	20	30	$P(S_2) = .7$

The firm is planning to construct an information system, i.e., an organized system of data collection, storage, and analysis. It can construct two possible information structures A and B. The systems yield information of either type M_1 or M_2. The reliability (i.e., $P(M_i|S_j)$) of the information from each system is as follows:

	System A M_1	M_2
S_1	.9	.1
S_2	.2	.8

	System B M_1	M_2
S_1	.6	.4
S_2	.4	.6

(a) What is the maximum amount the firm should pay for an information system?
(b) Given that you receive information of type M_1 from system A, what are the revised prior probabilities, the best action, and the revised EVPI?
(c) Given that you receive information of type M_1 from system B, what are the revised prior probabilities, the best action, and the revised EVPI?

CHAPTER FIFTEEN

Devising Optimal Strategies Prior to Sampling

15.1 Introduction

In Chapter 13, we considered *prior analysis*, a method for decision making *prior* to the incorporation of additional information obtained through sampling or experimentation. In Chapter 14, we studied *posterior analysis*, a corresponding method for decision making *after* additional information has been obtained through sampling or experimentation. In posterior analysis, a sample is drawn and the best act is chosen taking into account both the sample data and prior probabilities of the events which affect payoffs. In this chapter, we are concerned with *preposterior analysis*, which answers the question whether it is worthwhile to collect sample or experimental data at all. If such collection of data is worthwhile, the method further specifies the best courses of action for each possible type of sample or experimental outcome, and how large a sample to take. This type of analysis leads to a more complex but more interesting view of decision making than those we have considered so far. In both prior and posterior analysis, a final decision is made among the alternative courses of action utilizing the information at hand. Such a decision, which makes a final disposition of the choice of a best act, is referred to in Bayesian analysis as a terminal decision. The act itself is referred to as a terminal act. In many business decision-making situations, the wise course of action is not to choose a terminal act, but rather to delay making a terminal decision in order to obtain further information. Preposterior analysis, in addition to answering the questions of whether additional information should be obtained and if so, how much, also delineates the optimal decision rule to employ based on the possible types of evidence that can be produced by the additional information if it is gathered. An obvious difficulty in this type of decision procedure is that the specific outcome of a sample (or additional information) is unknown prior to the taking of the sample. Yet, in the nature of the case, the decision as to whether sample information will be worthwhile must be made prior to the actual drawing of the sample. Since there is a cost associated with acquiring additional knowledge through a sampling process, it only pays to take a sample if the anticipated worth of this sample information exceeds its cost. The methods discussed in this chapter provide a procedure for determining the *expected value of this sample information*.

The possibility of delaying a decision in order to acquire more information as opposed to choosing a terminal act opens up the further possibility of a multistage or *sequential decision process*. After taking the first sample, the question again arises whether a terminal decision should be made now or whether another sample should be drawn. As long as the anticipated incremental worth of another sample exceeds its incremental cost, in keeping with general economic principles, it pays to continue sampling. Conceptually we can envision the possibilities of a sample of one, two, or three items, and so forth. It seems reasonable that there must be some optimal sample size, depending upon the environmental circumstances beyond which further increases are uneconomical. However, we will begin by considering a simpler problem involving only a single stage sample of fixed size before turning to the more complex problem of sequential decision making. In Section 15.2, we will discuss two examples of preposterior analysis involving a choice between a terminal decision without sampling and a decision to examine a fixed size single sample and then select a terminal act.

15.2 Preposterior Analysis

A preposterior analysis, as the name implies, is an investigation that must be carried out *before* sample information is obtained and therefore prior to the availability of posterior probabilities based on a particular sample outcome. However, this type of analysis takes into account all possible sample results and computes the expected worth (or expected opportunity loss) of a strategy which assumes the best acts are selected dependent upon the type of sample information observed. We will illustrate preposterior analysis by an oversimplified example in order that the basic principles of the procedure may be conveyed without getting bogged down in computational detail.

A Marketing Example

The A. B. Westerhoff Company, a consumer products manufacturing firm, considered the marketing of a new product it had developed. However, the company wanted to appraise the advisability of engaging a market research firm to help determine whether sufficient consumer demand existed to warrant placing the product on the market. The market research firm offered to conduct a nationwide survey of consumers to obtain an appropriate indication of the market for the product. The fee for the survey was \$15,000. The A. B. Westerhoff Company also wished to carry out an analysis which would specify whether it was better to act now on the basis of prior betting odds as to the success or failure of the product and estimated payoffs, or to engage the market research firm and then act on the basis of the survey indication.

The company analyzed the situation as follows. There were two available actions

a_1: Market the product

a_2: Do not market the product

The states of nature or possible events were defined as

θ_1: Successful product

θ_2: Unsuccessful product

Although only two states of nature for success of the product are used in this example, many states could be employed to indicate degree of success. Similarly, more than two courses of action might have been considered.

The company decided to view the problem in terms of opportunity losses of incorrect action. Based on appropriate estimates, the opportunity loss matrix (in thousands of dollars) shown in Table 15-1 was constructed.

The company, on the basis of its past experience with products of this type, assessed the odds that the product would be a failure as three-to-one. That is, the company assigned prior probabilities to the success and failure of the product as follows:

$$P(\theta_1) = P(\text{Successful product}) = 1/4 = .25$$

$$P(\theta_2) = P(\text{Unsuccessful product}) = 3/4 = .75$$

Table 15-1 Payoff Table Showing Opportunity Losses for the A. B. Westerhoff Company Problem (in thousands of dollars).

	Market a_1	*Do Not Market* a_2
θ_1: Product Is a Success	$ 0	$200
θ_2: Product Is a Failure	160	0

Table 15-2 Calculation of Prior Expected Opportunity Losses for the A. B. Westerhoff Company Problem (in thousands of dollars).

a_1: *Market the Product*

Events	*Probability* $P(\theta_i)$	*Opportunity Loss*	*Weighted Opportunity Loss*
θ_1: Successful Product	.25	$ 0	$ 0
θ_2: Unsuccessful Product	.75	160	120
	1.00		$120

$\text{EOL}(a_1) = \$120$ (thousand)

a_2: *Do Not Market the Product*

Events	*Probability* $P(\theta_i)$	*Opportunity Loss*	*Weighted Opportunity Loss*
θ_1: Successful Product	.25	$200	$ 50
θ_2: Unsuccessful Product	.75	0	0
	1.00		$ 50

$\text{EOL}(a_2) = \$50$ (thousand)

EOL of the Optimal Act $= \text{EOL}(a_2) = \$50$ (thousand)

Hence, the prior analysis given in Table 15-2 was conducted to determine the optimal course of action if no additional information was obtained. As shown in Table 15-2, the optimal course of action was a_2 (do not market). The expected opportunity loss for this course of action was $50,000. A decision diagram of this prior analysis is given in Figure 15-1. We read the tree

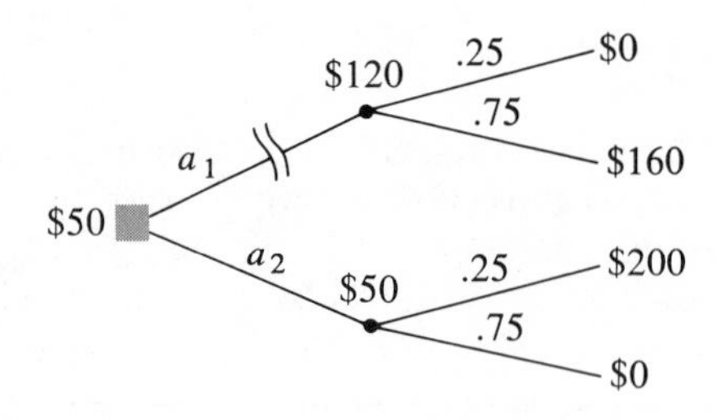

Figure 15-1 Decision diagram for action by the A. B. Westerhoff Company if no survey is conducted. (All losses are in thousands of dollars)

in the usual way, using the method of backward induction, discussed in Section 13.6, beginning at the right-hand side of the tree and proceeding inward. The tree summarizes the information shown in Table 15-2. At the right-hand tips are payoffs for the occurrence of the two states of nature θ_1 and θ_2 if actions a_1 and a_2 are taken. On the branches leading to the payoff figures are the prior probabilities of the two states of nature, .25 and .75. The expected opportunity losses of actions a_1 and a_2 ($120 thousand and $50 thousand) obtained by weighting the payoffs by these probabilities are displayed at the two appropriate nodes. Branch a_1 is blocked off, since it represents a non-optimal course of action. The expected payoff of action a_2, $50 thousand, is entered at the initial node, designating the expected value of selecting the optimal course of action. As usual, a small square has been used to indicate a node at which the decision maker makes a choice.

The A. B. Westerhoff Company then turned its attention to the problem of analyzing its expected opportunity losses if its action was based on the survey results obtained by the market research firm. It requested the market research firm to indicate the nature and reliability of the results that would be supplied by the consumer survey. The market research firm replied that the survey would yield one of the following three types of indications:

X_1: Favorable

X_2: Intermediate

X_3: Unfavorable

That is, an X_1 indication meant an observed level of consumer demand in the survey which was favorable to the success of the product; an X_3 result was unfavorable, and an X_2 indication meant a situation falling between levels X_1 and X_3 and was classified as "intermediate." Although we are using the verbal classifications of "favorable," "intermediate," and "unfavorable" here for simplicity of reference, each of these indications will be understood to represent a specific numerical range of demand. How this type of sample is used in decision making is explained subsequently.

As a description of the anticipated *reliability* of the survey indications, the market research firm supplied the array of conditional probabilities shown in Table 15-3. The entries in this table are values of $P(X_j \mid \theta_i)$ based on past relative frequencies in similar types of surveys. That is, they represent the conditional probabilities of each type of sample evidence given that the product was successful or unsuccessful. For example, the $P(X_1 \mid \theta_1) = .72$ entry in the upper left-hand corner of the table means the probability is 72% that the survey will yield a favorable indication, given a successful new product. The subscript j is used for the sample indication, denoted X_j, analogous to the use of the subscript i for the states of nature, denoted θ_i. It may not always be feasible to use past relative frequencies as the conditional probability assignments as we are doing here. The use of past relative frequencies as probabilities for future events always involves the assumption of a continuation of the same conditions as existed when the relative frequencies were established. As we have seen in other examples, probability functions such as the binomial distribution are often appropriate for computing these "likelihoods" or conditional probabilities of sample results.

The A. B. Westerhoff Company analysts decided to carry out a preposterior analysis in order to determine whether it was worthwhile engaging the market research firm to conduct the survey. For this purpose, it was necessary to compare the expected opportunity loss of purchasing the survey and then selecting a terminal act to the expected opportunity loss of terminal action without the survey. The latter expected loss figure was previously obtained from the prior analysis. As an intermediate step, the analysts computed the joint probability distribution of the sample evidence X_j and the events θ_i, as shown in Table 15-4. These joint probabilities were obtained by multiplying the prior (marginal) probabilities by the appropriate conditional probabilities. For example, .18, the upper left-hand entry in the joint probability distribution in Table 15-4 was obtained by multiplying .25 by .72. In symbols, $P(X_1 \cap \theta_1) = P(\theta_1)P(X_1 \mid \theta_1)$, etc.

Table 15-3 Conditional Probabilities of Three Types of Survey Evidence for the A. B. Westerhoff Company Problem.

	Conditional Probabilities $P(X_j \mid \theta_i)$			
Events	X_1	X_2	X_3	*Total*
θ_1: Successful Product	.72	.16	.12	1.00
θ_2: Unsuccessful Product	.08	.12	.80	1.00

Table 15-4 Calculation of the Joint Probability Distribution of Survey Evidence and Events for the A. B. Westerhoff Company Problem.

Events θ_i	*Prior Probabilities* $P(\theta_i)$	*Conditional Probabilities* $P(X_j \mid \theta_i)$			*Joint Probabilities* $P(X_j \cap \theta_i)$			*Total*
		X_1	X_2	X_3	X_1	X_2	X_3	
θ_1	.25	.72	.16	.12	.18	.04	.03	.25
θ_2	.75	.08	.12	.80	.06	.09	.60	.75
Total	1.00				.24	.13	.63	1.00

It is useful to consider the following interesting points about the joint probability distribution in order to understand the subsequent analysis. As in any joint bivariate probability distribution, the totals in the margins of the table are marginal probabilities. The row totals are the prior probabilities $P(\theta_1) = .25$ and $P(\theta_2) = .75$. The column totals are the marginal probabilities of the survey evidence; that is, $P(X_1) = .24$, $P(X_2) = .13$, and $P(X_3) = .63$. Conditional probabilities of events, given the survey evidence, that is, probabilities of the form $P(\theta_i \mid X_j)$, can be computed by dividing the joint probabilities by the appropriate column totals, $P(X_j)$. For example, the probability of a successful product, given a favorable survey indication, is given by

$$P(\theta_1 | X_1) = \frac{P(\theta_1 \cap X_1)}{P(X_1)} = \frac{.18}{.24}$$

and so forth. These probabilities can be viewed as revised or posterior probability assignments to the events θ_1 and θ_2, given the survey evidence X_1, X_2, and X_3, respectively. The calculation of these posterior probabilities, represents an application of Bayes' Theorem, Equation (1.10), as is shown in the following method of symbolizing the probability $P(\theta_1 \mid X_1)$ just calculated:

$$P(\theta_1 | X_1) = \frac{P(\theta_1 \cap X_1)}{P(X_1)} = \frac{P(\theta_1)P(X_1 | \theta_1)}{\sum_{i=1}^{2} P(\theta_i)P(X_1 | \theta_i)}$$

That is, the joint probability $P(\theta_1 \cap X_1) = .18$ was calculated in Table 15-4 by multiplying the marginal probability $P(\theta_1) = .25$ by the conditional probability $P(X_1 \mid \theta_1) = .72$; the marginal probability $P(X_1) = .24$ was obtained by adding the joint probabilities $P(\theta_1)P(X_1 \mid \theta_1) = .18$ and $P(\theta_2)P(X_1 \mid \theta_2) = .06$, or somewhat more formally, $P(X_1) = \sum_{i=1}^{2} P(\theta_i)P(X_1 \mid \theta_i)$.

The A. B. Westerhoff Company proceeded with its analysis and constructed the decision diagram shown in Figure 15-2. All monetary figures are in thousands of dollars. Returning to the beginning of the problem, the first choice is to purchase the survey or not to purchase it. Therefore, starting at node (a) at the left and following the "no survey" branch of the tree, we move along to node (b). From node (b) to the right is reproduced the decision tree depicted in Figure 15-1 for the prior analysis with no survey. As indicated earlier, the $50 entry at node (b) is the expected opportunity loss of choosing the optimal terminal act without conducting the survey.

Expected Value of Sample Information

On the other hand, suppose at the outset (at node (a)), the decision is to conduct the survey. Making this decision, we move down to branch point (c). The results of the survey then determine which branch to follow. The three branches representing X_1, X_2, and X_3 types of survey information emanating from node (c) are marked with their respective marginal probabilities

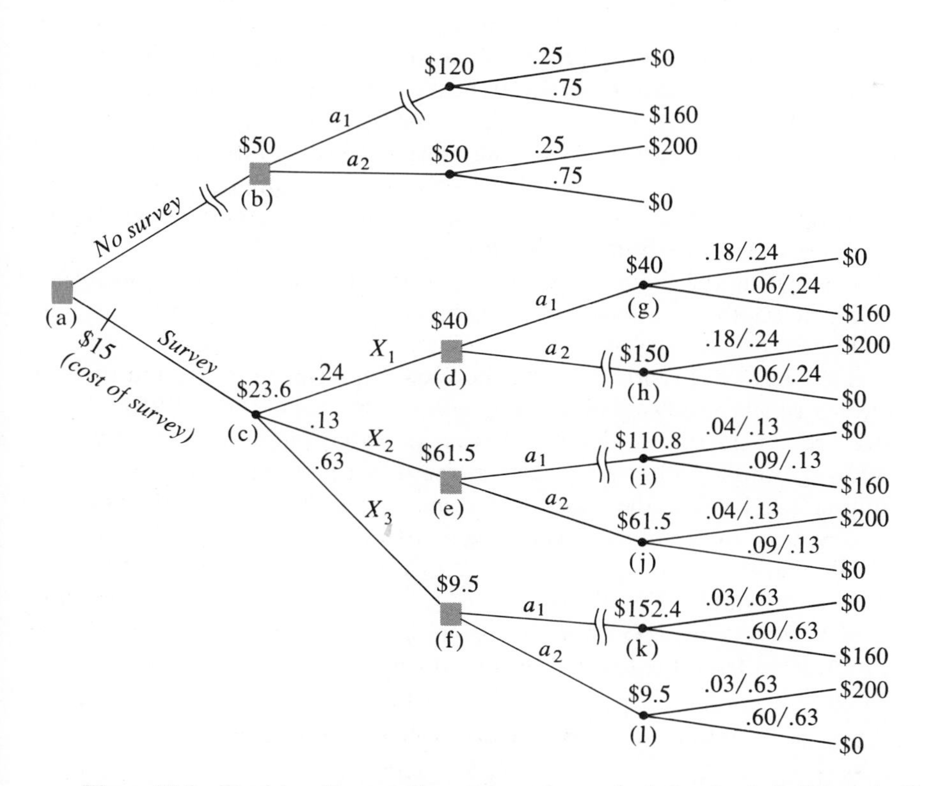

Figure 15-2 Decision diagram for preposterior analysis for the A. B. Westerhoff Company problem (in thousands of dollars).

.24, .13, and .63. Suppose type X_1 information were observed, that is, a "favorable" indication. Moving ahead to branch point (d) we can either take action a_1 or a_2, that is, to market or not to market the product. If act a_1 is selected, we move to node (g); if act a_2 is selected, to node (h). At these points, the probability questions that must be answered are of the a posteriori or revised probability variety. For example, the probabilities shown on the two branches stemming from node (g), if act a_1 is chosen, .18/.24 and .06/.24, are the conditional (posterior) probabilities, given type X_1 information, that the product is successful or unsuccessful, or in symbols, $P(\theta_1 \mid X_1)$ and $P(\theta_2 \mid X_1)$, respectively. Thus, they represent revised probabilities of these two events, after having observed a particular type of sample evidence. These conditional probabilities are calculated from Table 15-4, as indicated earlier by dividing joint probabilities by the appropriate marginal probabilities. Now looking forward from node (g) and using the posterior probabilities, .18/.24 and .06/.24, as weights applied to the losses attached to the two events "successful product" and "unsuccessful product," we obtain an expected payoff of an opportunity loss of \$40 for act a_1. This figure is entered at node (g). Comparing it with the corresponding figure of \$150 for a_2, we block off act a_2 as being nonoptimal. Therefore \$40 is carried down to node (d), representing the payoff for the optimal act upon observing type X_1 information. Similar calculations yield \$61.5 and \$9.5 at nodes (e) and (f) for X_2 and X_3 types of information. Weighting these three payoffs by the marginal probabilities of type X_1, X_2, and X_3 indications, .24, .13, and .63, respectively, we obtain a loss of \$23.6 as the expected payoff of conducting the survey and taking optimal action after the observation of the sample evidence.

Comparing the \$23.6 figure with the \$50 obtained under the "no survey" option, we see that it would be worthwhile to pay up to \$50 − \$23.6 = \$26.4 (thousands) for the survey.

This difference represented by the \$26.4 (thousands) is referred to as the *expected value of sample information*, and we will denote it as EVSI. Hence, if we are considering a choice between immediate terminal action without obtaining sample information and a decision to sample and then select a terminal act, EVSI is the *expected amount by which the terminal opportunity loss is reduced by the information to be derived from the sample*. This EVSI is a gross figure, since it has not taken into account the cost of obtaining the survey information. To calculate the *expected net gain of sample information*, which we denote as ENGS, the cost of obtaining the sample information must be subtracted from the expected value of this sample information. Hence in general,

(15.1) $$\text{ENGS} = \text{EVSI} - \text{Cost of Sample Information}$$

and in this problem,

$$\text{ENGS} = \$26.4 - \$15 = \$11.4$$

In conclusion, since the expected value of sample information was $26,400 and the cost of the survey was $15,000, the ENGS was $11,400. Therefore, the A. B. Westerhoff Company decided that it was worthwhile to engage the market research firm to conduct the survey.

It is important to recognize that the EVSI and ENGS computations have been made with respect to the particular prior probability distribution used in the analysis. If different prior probabilities were used, the survey would have had correspondingly different EVSI and ENGS values. Sensitivity analysis, discussed in Section 15.5, can be used to test how sensitive the alternative actions are to the size of the prior probabilities.

Some Considerations in Preposterior Analysis

A few points concerning preposterior analysis which arise from a consideration of the foregoing example are worthy of mention. First of all, we note that in this problem, the optimal action if no survey were conducted was a_2, not to market the product. On the other hand, as seen from Figure 15-2, if the survey were carried out and a favorable indication of demand, X_1, were obtained, the best course of action would be a_1, to market the product. The fact that for at least one of the survey outcomes it was possible for the decision maker to change his selection of an act is what gives the survey some value. Clearly, in general, if a decision maker's course of action cannot be modified regardless of the experimental outcome, then the experiment is without value.[1]

Second, the calculations in the preceding problem were carried out in terms of opportunity losses. If the payoffs had been expressed in terms of profits, the obvious equivalent analysis would have been required. For example, instead of the EVSI being the expected amount by which the terminal opportunity loss is reduced by the information to be derived from the sample, it would be the *expected amount by which the terminal profit is increased* by this information. Hence, expected profit of terminal action without sampling and expected profit of choosing a terminal action after sampling would be calculated. The first of these quantities would be subtracted from the second to yield the same EVSI figure obtained in the opportunity loss analysis.

Third, in this problem, the terms "sample" and "survey" were used interchangeably. Actually, even if the survey had represented a complete enumeration, the same analysis could have been carried out. Thus, the term "sample" in this context is used in a very general sense to include any sample from size one up to a complete census. Indeed, the sample outcomes X_1, X_2, and X_3 need not have arisen from a statistical sampling procedure but more generally

[1]If a decision maker is of the "My mind is made up, don't bother me with the facts" variety, and his mind can really not be altered by "the facts," then it is worthless to obtain the information in question.

could represent any set of experimental outcomes. Therefore, a preposterior analysis may be viewed as yielding the "expected value of experimental information" rather than the "expected value of sample information." However, since the latter term is conventionally used in Bayesian decision analysis, we have resisted the temptation of adopting a new term.

Finally, another level of generalization suggests itself. Only one survey of a fixed size was considered in this problem. Many surveys of different types and different sizes could possibly have been considered. The preposterior expected opportunity loss would then be calculated for each of these different surveys. The minimum of these figures would be subtracted from the prior expected opportunity loss to yield the EVSI. In order to obtain the ENGS, the total expected opportunity loss for each survey is calculated by adding the cost of the survey to the corresponding preposterior expected opportunity loss. Then these figures are subtracted from the prior expected loss. The maximum of these differences is the ENGS, since it represents the expected net gain of the survey with the lowest total expected opportunity loss. Although the theoretical number of possible surveys or experiments that might be considered is infinite, obviously there would ordinarily be a delimitation at the outset based on factors of practicability or feasibility. On the other hand, the use of computers increases considerably the number of possible alternatives that can practically be compared.

15.3 Extensive and Normal Form Analyses

The type of preposterior investigation carried out in the A. B. Westerhoff Company problem is known as *extensive form analysis*. It is perhaps easiest to characterize this type of analysis in terms of a decision tree diagram, such as is shown in Figure 15-2. In that diagram, as indicated previously, a prior analysis is given in the upper part of the tree with the resultant prior expected opportunity loss figure entered at node (b). Similarly, an extensive form analysis is given in the lower part of the tree with the resultant preposterior expected opportunity loss entered at node (c).

For purposes of comparison, we summarize the procedures for prior, posterior, and extensive form preposterior analysis. This comparison assumes experimental (sample) information is collected by a single-stage procedure as in the A. B. Westerhoff Company problem. Each type of analysis may be thought of as starting at the right-hand side of a decision tree diagram, and then proceeding inward by the process referred to earlier as backward induction.

1. *Prior analysis*
   (a) Sketch a tree and depict states of nature at the right-hand tips.

   (b) Assign payoffs to each of the states for each possible action.
   (c) Assign prior probabilities to each state of nature.
   (d) Calculate expected terminal payoffs for each act.
   (e) Select the act with the highest expected terminal payoff.
2. *Posterior analysis*
   (a) Do steps (a) and (b) as in prior analysis.
   (b) Assign posterior probabilities to each state of nature based on a specific outcome of the experimental information.
   (c) Do steps (d) and (e) as in prior analysis.
3. *Extensive form—preposterior analysis*
   (a) Do steps (a) and (b) as in prior analysis.
   (b) Assign marginal probabilities to each possible experimental (sample) outcome.
   (c) Assign posterior probabilities to each state of nature given specific experimental outcomes.
   (d) For each experimental outcome, carry out the posterior analysis as in 2 (c).
   (e) Weight the expected terminal payoffs of the best act for each type of experimental information by the marginal probability of occurrence of that type of information, and add these products to yield an overall preposterior expected payoff.
   (f) The difference between the preposterior expected payoff and a prior expected payoff calculated as in 1 above is the EVSI.

Concept of a Strategy

If there were many possible experiments, rather than one as assumed in the foregoing listing, the extensive form analysis also specifies which of the possible experiments should be carried out. Secondly, the extensive form analysis supplies a decision rule which selects an optimal act for each possible outcome of the chosen experiment. For example, in Figure 15-2, this decision rule was as follows:

$$X_1 \rightarrow a_1$$

$$X_2 \rightarrow a_2$$

$$X_3 \rightarrow a_2$$

where $X_1 \rightarrow a_1$ means that if experimental outcome X_1 is observed, select act a_1, etc. Such a decision rule is referred to as a *strategy*. Mathematically, a strategy can be defined as a function in which the elements of the domain are possible experimental outcomes and the elements of the range are possible acts. We now turn to an alternative procedure to extensive form analysis, known as *normal form analysis*, in which the problem commences with a listing of all of the possible strategies that might be employed. Normal form analysis then proceeds to make an explicit comparison of the worth of all of these strategies to arrive at the same optimal strategy as was derived in extensive form analysis.

Normal Form Analysis

In the preceding section, extensive form analysis was applied to the A. B. Westerhoff Company problem. In order to indicate the nature of the normal form procedure and the relationship between extensive and normal forms of analysis, the same problem will now be solved by the latter method. Then some factors relating to the choice between the two alternative procedures will be given.

All possible strategies or decision rules for the A. B. Westerhoff Company problem are listed in Table 15-5. The strategies, denoted $s_1, s_2, \ldots, s_8$, indicate the acts that are taken in response to each sample outcome.

Table 15-5 A Listing of All Possible Strategies in the A. B. Westerhoff Company Problem.

Sample Outcome	s_1	s_2	s_3	s_4	s_5	s_6	s_7	s_8
X_1	a_1	a_1	a_1	a_1	a_2	a_2	a_2	a_2
X_2	a_1	a_1	a_2	a_2	a_1	a_1	a_2	a_2
X_3	a_1	a_2	a_1	a_2	a_1	a_2	a_1	a_2

Hence, for example, strategy s_3 is

$$X_1 \rightarrow a_1$$

$$X_2 \rightarrow a_2$$

$$X_3 \rightarrow a_1$$

and s_4 is the one previously described. In this problem there are eight possible strategies. There are three sample outcomes, X_1, X_2, and X_3, and two possible acts, a_1 and a_2. In general, the number of possible strategies is given by n^r, where n denotes the number of acts and r is the number of sample outcomes. Hence, in this case, there are $2^3 = 8$ possible strategies.

It is clear from an examination of Table 15-5 that some of the strategies are not very sensible. For example, strategies s_1 and s_8 select the same act regardless of the experimental outcome (survey indication). Strategy s_5

selects act a_2, not to market the product, if the "favorable" survey indication X_1 is obtained, but perversely it would market the product if the intermediate or "unfavorable" indications X_2 or X_3 are obtained. On the other hand, strategies s_2 and s_4 seem to be quite logical and would probably be the only ones a reasonable person would seriously consider on an intuitive basis if he contemplated using the experimental information at all.

Continuing with the normal form analysis, we now compute the expected payoff of each possible strategy. The method consists in calculating for each strategy the expected opportunity loss conditional on the occurrence of each state of nature. These conditional expected losses are referred to in Bayesian decision theory as *risk*. We will use that term here or its equivalent "*conditional expected losses*." The weighted average or expected value of these risks, using prior probabilities of states as weights, yields the *expected opportunity loss* or *expected risk* of the strategy. In Table 15-6 is shown the calcula-

Table 15-6 Calculation of Risks or Conditional Expected Opportunity Loss for Strategies s_2 and s_4 (in thousands of dollars).

Strategy s_2

States of Nature	*Opportunity Losses*		*Probabilities of Action $s_2(a_1, a_1, a_2)$*			*Risk (Conditional Expected Loss)*
	a_1	a_2	a_1	a_1	a_2	$R(s_2; \theta_i)$
θ_1	\$ 0	\$200	.72	.16	.12	\$24.00
θ_2	160	0	.08	.12	.80	32.00

$$R(s_2; \theta_1) = .72(\$0) + .16(\$0) + .12(\$200) = \$24.00$$
$$R(s_2; \theta_2) = .08(\$160) + .12(\$160) + .80(\$0) = \$32.00$$

Strategy s_4

States of Nature	*Opportunity Losses*		*Probabilities of Action $s_1(a_1, a_2, a_2)$*			*Risk (Conditional Expected Loss)*
	a_1	a_2	a_1	a_2	a_2	$R(s_4; \theta_i)$
θ_1	\$ 0	\$200	.72	.16	.12	\$56.00
θ_2	160	0	.08	.12	.80	12.80

$$R(s_4; \theta_1) = .72(\$0) + .16(\$200) + .12(\$200) = \$56.00$$
$$R(s_4; \theta_2) = .08(\$160) + .12(\$0) + .80(\$0) = \$12.80$$

tion of the risks associated with strategies s_2 and s_4. For example, let us consider the calculations for strategy s_2. In the left-hand portion of the table is given the payoff table in terms of opportunity losses. In the next section are shown "probabilities of action," which are probabilities of taking actions specified by the given strategy based on the observations of the possible types of sample evidence. A convenient notation for designating a strategy is given in the caption of this section of the table. Hence, the notation $s_2(a_1, a_1, a_2)$ means strategy s_2 consists in taking act a_1 if the sample indication X_1 is observed; again selecting a_1 if X_2 is observed; and choosing a_2 if X_3 is observed. That is, the first element within the parentheses denotes the action to be taken upon observing the first sample indication, the second element specifies the action to be taken upon observing the second sample indication, etc.

The risk or expected opportunity loss, given the occurrence of a specific state of nature, is calculated by multiplying these probabilities of action by the respective losses incurred if these actions are taken, and summing the products. Thus, the risk or expected opportunity loss associated with the use of strategy s_2, given that state of nature θ_1 occurs, denoted $R(s_2; \theta_1)$, is calculated to be \$24.00, as shown below the appropriate portion of Table 15-6. Correspondingly, the expected opportunity loss of strategy s_2, conditional on the occurrence of θ_2, denoted $R(s_2; \theta_2)$, is calculated to be \$32.00. The analogous calculation of risks for strategy s_4 is given in the lower part of Table 15-6.

A revealing alternative way of thinking about these risks or conditional expected losses is that they are simply *calculations for each state of nature of the loss in taking the wrong act times the probability that the wrong act will be taken*. For example, let us consider $R(s_2; \theta_1)$, the conditional expected loss of strategy s_2, given that θ_1 occurs. Now, if θ_1 occurs, the correct (best) course of action is a_1; the incorrect act is a_2. The \$24.00 figure for $R(s_2; \theta_1)$ is merely \$200, the loss of taking act a_2 times .12, the total probability of selecting a_2 under strategy s_2, given the occurrence of θ_1. Similarly, the \$32.00 figure for $R(s_2; \theta_2)$ is equal to \$160, the loss of taking act a_1 times .20, the total probability of selecting a_1 under strategy s_2, conditional on the occurrence of θ_2. The risk calculations of Table 15-6 are shown in Table 15-7, utilizing this alternative conception.

A summary of the risks or conditional expected opportunity losses for all eight strategies is given in Table 15-8. Let us review the interpretation of these figures, taking, as an example, strategy s_4. In a relative frequency sense, that is, in a large number of identical situations of successful new products, θ_1, if survey evidence X_1, X_2, and X_3 occur with the specified probabilities and if strategy s_4 is employed, the average opportunity loss per product would be \$56.00. A similar interpretation holds for unsuccessful products, θ_2, with an average loss of \$12.80. Now, we can weight these average losses by the prior probabilities of successful and unsuccessful products to obtain an overall expected opportunity loss for strategy s_4. The prior probabilities were given earlier as $P(\theta_1) = .25$ and $P(\theta_2) = .75$. Since θ_1 occurs with probability .25 and the conditional expected loss if θ_1 occurs is \$56.00, and since θ_2 occurs with probability .75 and the conditional expected loss if θ_2 occurs is \$12.80, the expected opportunity loss of using strategy s_4 is

$$\text{EOL}(s_4) = (.25)(\$56.00) + (.75)(\$12.80) = \$23.60$$

In symbols, we have

$$\text{EOL}(s_4) = P(\theta_1)R(s_4;\, \theta_1) + P(\theta_2)R(s_4;\, \theta_2)$$

Stating this relationship in general form, the expected opportunity loss of the kth strategy, s_k, is

$$\textbf{(15.2)}\quad \text{EOL}(s_k) = P(\theta_1)R(s_k;\, \theta_1) + P(\theta_2)R(s_k;\, \theta_2) + \cdots + P(\theta_m)R(s_k;\, \theta_m) = \sum_{i=1}^{m} P(\theta_i)R(s_k;\, \theta_i)$$

The decision rule for which this expected opportunity loss is a minimum is known as the "Bayes' strategy."

The expected opportunity losses of the eight strategies in the A. B. Westerhoff Company problem as calculated from Equation (15.2) are given in Table 15-9. The optimal strategy, or the one for which the expected

Table 15-7 Alternative Calculation of Risks or Conditional Expected Opportunity Losses for Strategies s_2 and s_4 (in thousands of dollars).

Strategy $s_2(a_1, a_1, a_2)$

States of Nature	*Opportunity Loss of Wrong Act*	*Probability of Wrong Act Given θ_i*	*Conditional Expected Opportunity Loss $R(s_2; \theta_i)$*
θ_1	\$200	.12	\$24.00
θ_2	160	.20	32.00

Strategy $s_4(a_1, a_2, a_2)$

States of Nature	*Opportunity Loss of Wrong Act*	*Probability of Wrong Act Given θ_i*	*Conditional Expected Opportunity Loss $R(s_4; \theta_i)$*
θ_1	\$200	.28	\$56.00
θ_2	160	.08	12.80

Table 15-8 Risks or Conditional Expected Opportunity Losses for the Eight Strategies in the A. B. Westerhoff Company Problem (in thousands of dollars).

States of Nature	*Strategies*							
	s_1	s_2	s_3	s_4	s_5	s_6	s_7	s_8
θ_1	$ 0	$24.00	$32.00	$56.00	$144.00	$168.00	$176.00	$200.00
θ_2	160.00	32.00	140.80	12.80	147.20	19.20	128.00	0

Table 15-9 Expected Opportunity Losses of the Eight Strategies in the A. B. Westerhoff Company Problem.

Strategy	*Expected Opportunity Loss*
s_1	$120.00
s_2	30.00
s_3	113.60
s_4	23.60*
s_5	146.40
s_6	56.40
s_7	140.00
s_8	50.00

opportunity loss is least, is s_4. An asterisk has been placed beside the $23.60 expected opportunity loss associated with decision rule s_4 to indicate that it is the minimum loss figure.

In summary, using normal form analysis, the optimal strategy in this problem is $s_4(a_1, a_2, a_2)$; this strategy has an expected opportunity loss of $23.60 (thousands). If we refer to Figure 15-2, we see that this is exactly the same solution arrived at by extensive form analysis. In that figure the $23.6 is shown at node (c) and the optimal strategy (a_1, a_2, a_2) is arrived at by noting for the survey outcomes X_1, X_2, and X_3 the forks which have not been blocked off.

We can now summarize the steps involved in a normal form analysis, in which sample evidence is obtained from a single sample (experiment) as in the foregoing example.

Normal form analysis

(a) List all possible strategies in terms of actions to be taken upon observation of sample outcomes.

(b) Calculate the conditional expected opportunity loss (risk) for each state of nature. The probabilities to be used in this calculation are conditional probabilities of sample outcomes, given states of nature.

(c) Compute the (unconditional) expected opportunity loss of each strategy by weighting the conditional expected opportunity losses by the prior probabilities of states of nature.

(d) Select the strategy which has the minimum expected opportunity loss.

This summary of normal form analysis has been given in terms of a single experiment. If more than one experiment were conducted, the decision maker should carry out steps (a) through (d) above and should then select that experiment which yields the lowest expected opportunity loss. Also, the summary has been expressed in terms of opportunity losses. If payoffs of utility or profits were used, the same procedures would be followed except that the decision maker would maximize expected utility or profit rather than minimize expected opportunity loss.

15.4 Comparison of Extensive and Normal Form Analyses

As we have seen, extensive and normal forms of analysis are equivalent approaches. In both procedures, an expected opportunity loss (or expected profit) is calculated before experimental results are observed. This expected payoff anticipates the selection of optimal acts after the observation of experimental outcomes. However, there are differences between the two types of analysis in the way in which the various components of the procedures are performed. These differences give rise to advantages and disadvantages in the two forms of analysis.

As was clear from the A. B. Westerhoff Company example, the extensive form solution can be calculated more rapidly. This follows from the fact that *in the extensive form approach it is not necessary to carry out expected loss calculations for every possible decision rule.* Because of the blocking off of non-optimal courses of action posterior to the observation of sample evidence, only the expected loss for the optimal strategy need be carried out. In some practical problems the number of decision rules or strategies that must be evaluated in a normal form analysis may be very large. For example, in the problem discussed in this chapter, the number of possible strategies was $2^3 = 8$, because there were two acts and three experimental outcomes. If there had been three acts, $n = 3$, and four experimental outcomes, $r = 4$, there would have been $n^r = 3^4 = 81$ strategies for which normal form calculations would be required.

On the other hand, the normal form of analysis may appeal more to those who feel uneasy about making the subjective probability assessments involved in preposterior analysis. In many problems, the conditional probabilities of sample outcomes, given states of nature may be based on relative frequencies, as in the A. B. Westerhoff Company case, or on an appropriate probability distribution, as in another problem discussed later in this chapter. However, the prior probabilities, $P(\theta_i)$, will most likely represent subjective or judgmental assignments in most real problems. In normal form analysis, these prior probabilities are applied as the last step in the calculations. Therefore, it is possible to proceed all the way to this point without introducing subjective probability assessments. It is then possible to judge how sensitive the choice of the optimal strategy is to the prior probability assignments. That is, we can determine by how much the magnitudes of the prior probabilities may be permitted to change without a shift occurring in the optimal decision rule to be employed. This procedure is discussed below.

15.5 Sensitivity Analysis

In the new product decision problem discussed in this chapter, the point was made that there were only two decision rules which appeared reasonable on an intuitive basis, namely, s_2 and s_4. As can be seen in Table 15-9, these strategies yielded the two lowest expected opportunity losses, with figures of \$30.00 and \$23.60, respectively, for s_2 and s_4. However, we can observe in Table 15-6 that although the conditional expected loss for s_2 is lower than s_4 if θ_1 occurs (\$24.00 versus \$56.00, respectively), the opposite is true if θ_2 occurs (\$32.00 versus \$12.80, respectively). Since these conditional expected losses were weighted by prior probabilities of states of nature, $P(\theta_1)$ and $P(\theta_2)$, it is clear that to a considerable extent, the choice between strategies s_2 and s_4 is dependent on the magnitudes of these prior probabilities. We can test how *sensitive* the choice between decision rules s_2 and s_4 is to the size of these prior probabilities. This test is an example of *sensitivity analysis*, that is, a study which tests how sensitive the solution to a decision problem is to changes in the data for the variables of the problem.

In the present problem, a sensitivity test can be accomplished by solving for a *break-even value for the prior probability of one of the two states of nature* such that the expected opportunity losses of the two strategies will be equal.

If the prior probability rises above this break-even value, s_2 becomes the optimal act rather than s_4. We illustrate this procedure in the new product decision problem. As indicated earlier, the (unconditional) expected opportunity losses of the two strategies in question were computed as follows:

$$\text{EOL}(s_2) = (.25)(\$24.00) + (.75)(\$32.00) = \$30.00$$

$$\text{EOL}(s_4) = (.25)(\$56.00) + (.75)(\$12.80) = \$23.60$$

If we now substitute p for the .25 value of $P(\theta_1)$, and $1 - p$ for the .75 figure of $P(\theta_2)$, we can solve for the value of p for which EOL (s_2) is equal to EOL(s_4). Making this substitution and equating the resultant expressions for expected opportunity loss, we have

$$p(\$24.00) + (1 - p)(\$32.00) = p(\$56.00) + (1 - p)(\$12.80)$$

Carrying out the multiplications, we obtain

$$\$24.00p + \$32.00 - \$32.00p = \$56.00p + \$12.80 - \$12.80p$$

Collecting terms, we find the break-even value of p

$$\$51.20\,p = \$\ 19.20$$

$$p = \frac{19.20}{51.20} = 0.375$$

In summary, we conclude that if $p = P(\theta_1) = 0.375$, then the expected opportunity losses of strategies s_2 and s_4 would be equal, and we would have no preference between them. In the A. B. Westerhoff Company new product decision problem, $p = P(\theta_1) = .25$. Hence, we observe that this subjective prior probability assignment could have varied up to a value of .375 and s_4 would still have been a better strategy than s_2. However, for p values in excess of .375, s_2 has a lower expected opportunity loss, and is therefore the better rule.

Of course, break-even values of p could be determined between s_4 and other strategies as well. In the case of a strategy such as s_5, whose conditional expected values given θ_1 and θ_2 are each higher than the corresponding figures for s_4 (\$144.00 versus \$56.00 and \$147.20 versus \$12.80, respectively), it is impossible for the weighted average or expected opportunity loss to be less than that of s_4, regardless of the weights p and $1 - p$. In a situation such as this, the strategy s_4 is said to *dominate* s_5. As a general definition, a strategy, say s_1, is said to be a *dominating strategy* with respect to s_2 if

$$R(\theta; s_1) \leq R(\theta; s_2) \qquad \text{for all values of } \theta$$

and

$$R(\theta; s_1) < R(\theta; s_2) \qquad \text{for at least one value of } \theta$$

In words, s_1 is said to dominate s_2 if the conditional expected losses (risks) for s_1 are *equal to or less than* the corresponding figures for s_2 for every state of nature, *and* the conditional expected loss for s_1 is *less than* the corresponding figure for s_2 for at least one value of θ.

The preceding discussion illustrated the use of sensitivity analysis to determine the effects of variations in prior probabilities on the selection of a best act. We could also determine the effects of changes in the entries in the payoff matrix on our choice of a best act. In general, in any decision analysis problem, it is very useful to test how sensitive the solution is to changes in all of the important variables of the problem. In problems involving numerous states of nature, experimental outcomes, and courses of action, many calculations may be required to carry out such an analysis. The use of computers in such situations is usually a practical necessity.

15.6 An Acceptance Sampling Example

As another illustration of preposterior analysis, we will return to the problem of acceptance sampling of manufactured product discussed in Section 14.2. Since the problem was posed there as one in posterior analysis, we will make the necessary changes to convert it to a problem in preposterior analysis. We assume as previously a situation in which the Renny Corporation inspects incoming lots of articles produced by a supplier in order to determine whether to accept or reject these lots. We also retain the assumptions that in the past, incoming lots from this supplier have contained either 10%, 20%, or 30% defectives and that on a relative frequency basis, lots with these percentages of defectives have occurred 50%, 30%, and 20% of the time, respectively. The same payoff matrix as shown in Table 14-5 is assumed. Table 14-6 showed the calculation of prior expected opportunity losses for the two acts:

a_1: Reject the incoming lot[2]

a_2: Accept the incoming lot

if action were taken without sampling. For convenience, this information on events, prior probabilities, payoffs, and prior expected opportunity losses is summarized in Table 15-10. The preferable action without sampling is act a_2, "accept the incoming lot," which has a prior expected opportunity loss of \$70 as opposed to act a_1, "reject the incoming lot," which has a prior expected opportunity loss of \$100.

---

[2]The alternative actions have been denoted here by lower case rather than capital letters to maintain consistency with the notation we have used in specifying strategies.

Let us now assume that we would like to know whether it is better to make the decision concerning acceptance or rejection of the incoming lot without sampling or to draw and inspect a random sample of items from the incoming shipment and then make the decision. We recognize this problem as one of preposterior analysis, specifically one which requires an evaluation of the *expected value of sample information* (EVSI) and the *expected net gain of sampling* (ENGS). We will assume that a sample of two articles drawn without replacement is contemplated and the cost of sampling and inspecting these two articles is \$5. Hence, our problem is one of choice between a terminal decision without sampling and a decision to examine a fixed size sample of two articles and then select a terminal act.

Table 15-10 Basic Data for Acceptance Sampling Problem for the Renny Corporation.

Event (*p*-*Lot Proportion Defective*)	*Prior Probability* $P_0(p)$	*Opportunity Losses* a_1 *Reject*	a_2 *Accept*
.10	.50	\$200	\$ 0
.20	.30	0	100
.30	.20	0	200
	1.00		

Prior EOL(a_1) = (.50)(\$200) + (.30)(\$0) + (.20)(\$0) = \$100
Prior EOL(a_2) = (.50)(\$0) + (.30)(\$100) + (.20)(\$200) = \$70

Of course, a sample of only two articles may be entirely too small to be realistic in many cases. On the other hand, in the case of certain manufactured complex assemblies of large unit cost, where the testing procedure is destructive, as for example, in missile testing or in space probing where the test vehicle is not retrievable, it may only be feasible to test a very small number of items. However, the principles illustrated in this problem are perfectly general. The assumption of a larger sample size would merely increase the computational burden. Furthermore, in Section 15.3 we will consider appropriate decision theory procedures for determining an optimal sample size.

Again as in the preceding example in this section, we will solve the problem in two different ways, first through the use of extensive form analysis, and then by the normal form.

Extensive Form Analysis

The decision diagram for the preposterior analysis of the Renny Corporation problem is given in Figure 15-3. As indicated in the legend, the alternative actions in this problem have been denoted:

$$a_1\text{: Reject the lot}$$

$$a_2\text{: Accept the lot}$$

The possible sample outcomes based on a random sample of two articles drawn from the incoming lot are denoted:

$$X = 0 \text{ Defectives}$$

$$X = 1 \text{ Defective}$$

$$X = 2 \text{ Defectives}$$

Starting at node (a), the two basic choices are shown, not to engage in sampling inspection, or to sample and inspect two articles from the incoming lot. The prior analysis for the "no sampling" choice is shown in the usual way in the upper portion of the tree. As previously observed in Table 15-10, the prior expected opportunity loss of rejection, a_1, is \$100 and of acceptance, a_2, it is \$70. Hence, action a_1 has been blocked off on the branch emanating from node (b) as a non-optimal act, and the lower figure of \$70 for a_2 has been entered at node (b). This \$70 figure is the expected opportunity loss of choosing the optimal act without sampling.

On the other hand, the alternative choice is to sample and inspect two items prior to making the terminal decision of acceptance or rejection of the lot. To carry out the calculations for this part of the analysis, we begin at the right-hand tips of the tree and proceed inwards by backward induction in the usual way. The expected values of \$116.60 and \$55.80 shown at nodes (g) and (h), for example, are the expected opportunity losses of taking acts a_1 and a_2, respectively, after having observed zero defectives, $X = 0$, in the sample of two articles. The probabilities used to calculate these expected values are the conditional (posterior) probabilities, given $X = 0$, that the lot proportion defective, p, is equal to .10, .20, and .30, respectively. In symbols

$$P(p = .10 \mid X = 0) = .583$$

$$P(p = .20 \mid X = 0) = .276$$

$$P(p = .30 \mid X = 0) = .141$$

The calculation of these posterior probabilities is shown in Table 15-12 and will be explained shortly. Since acceptance of the lot, act a_2, has a lower

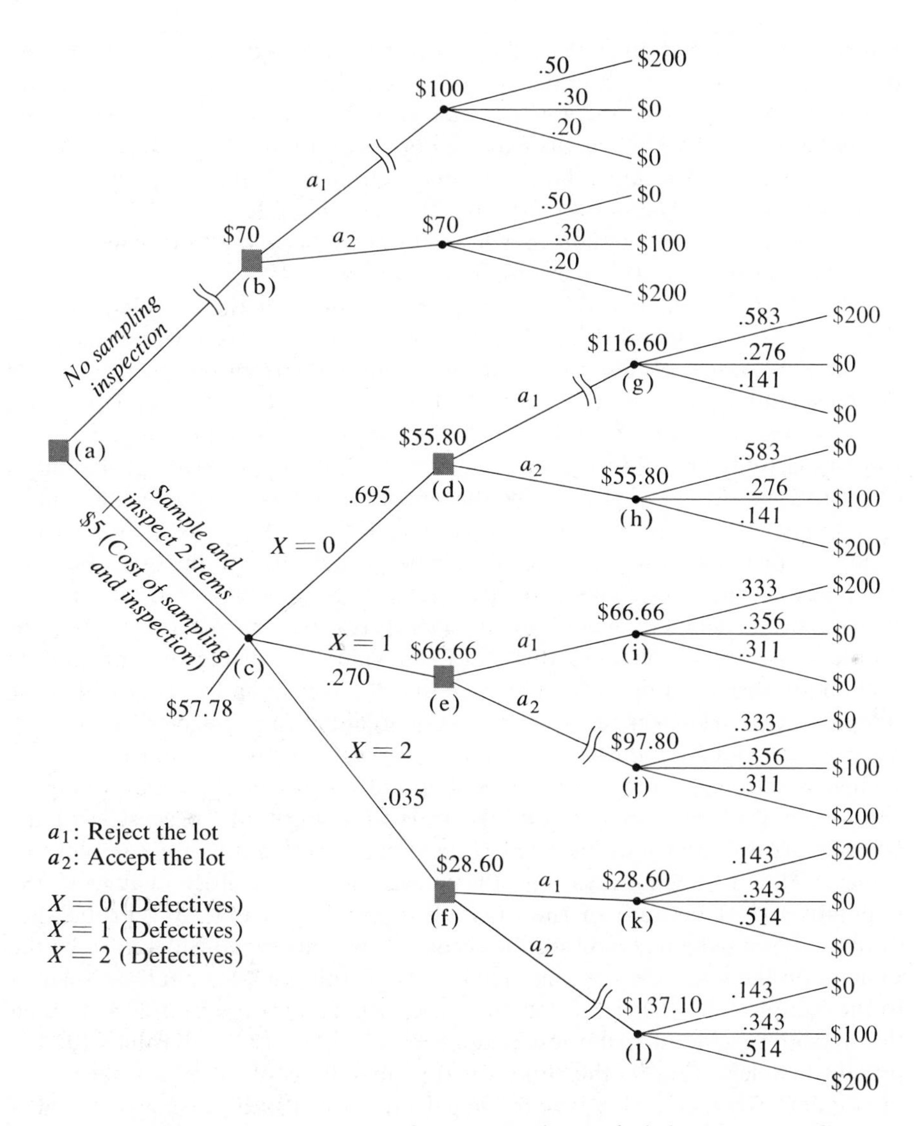

Figure 15-3 Decision diagram for preposterior analysis in the Renny Corporation problem.

expected loss than does rejection, act a_1, given that zero defectives have been observed in the sample, the lower figure, \$55.80, is entered at node (d). Also, act a_1 is blocked off as being non-optimal. Corresponding figures of \$66.66 and \$28.60 for expected losses of optimal acts after observing one and two defectives are shown at nodes (e) and (f), respectively. Rejection, act a_1, is shown to be optimal after observing either one or two defectives. Weighting the three payoffs shown at nodes (d), (e), and (f) by the marginal probabilities of observing zero, one, and two defectives, .695, .270, and .035, respectively, we find a figure of $57.78 as the expected opportunity loss of sampling and inspecting two articles and taking terminal action after observing the sample evidence. Before proceeding with the calculation of EVSI and ENGS, we pause to consider the method of calculating the marginal and posterior probabilities referred to in this paragraph.

In order to calculate the marginal probabilities of the sample evidence, $P(X = 0)$, $P(X = 1)$, and $P(X = 2)$, and the posterior probabilities of the form $P(p \mid X)$, we must obtain the joint probability distribution of the sample evidence and events, that is, $P(X \cap p)$, for $X = 0$, 1, and 2, and $p = .10$, $p = .20$, and $p = .30$. As in the preceding A. B. Westerhoff Company problem, these joint probabilities are obtained by multiplying prior probabilities of events, in this case, $P_0(p)$, by the appropriate conditional probabilities of sample evidence given events, in this case, $P(X \mid p)$. In the A. B. Westerhoff Company problem, it was assumed that these probabilities of the type P(sample outcomes | events) were based on past relative frequencies of occurrence in surveys previously conducted by a market research firm. In the present problem, we must calculate these $P(X \mid p)$ values by using an appropriate probability distribution. Let us assume that the incoming lot from which the sample of two articles is drawn is very large relative to the size of the sample. Then, as previously indicated in Section 2.4, we can assume that even if the sample is selected without replacement, the drawings of the articles may be considered for practical purposes as trials of a Bernoulli process, and the *binomial distribution* may be used to calculate probabilities of sample outcomes. That is, we may assume that since there is so little change in the population (lot) because of the drawing of the first article, the probability of obtaining a defective item on the second draw may be considered to be the same as on the first. On the other hand, if the incoming lot is not large relative to the sample size, that is, if it is not at least ten times the sample size, then the hypergeometric distribution is appropriate. The $P(X \mid p)$ values for the present problem, that is, the conditional probabilities of observing zero, one, or two defectives, given lot proportion defectives of .10, .20, and .30 calculated by the use of the binomial distribution are shown in Table 15-11. As indicated in the table, the probabilities of zero, one, and two defectives in a sample of size two from a lot which contains .10 defectives are given by the respective terms of the binomial whose parameters are $n = 2$, $p = .10$, and corresponding calculations provide the probabilities for the cases in which $p = .20$ and .30.

Table 15-12 shows the computation of the joint probability distribution $P(X \cap p)$. This joint distribution is obtained by multiplying the prior probabilities, $P_0(p)$ by the conditional probabilities $P(X \mid p)$ derived in Table 15-11.

Table 15-11 Conditional Probabilities of Specified Numbers of Defectives in a Sample of Two Articles from an Incoming Lot in the Renny Corporation Problem.

$$P(X = 0 \mid p = .10) = (.90)^2 = .81$$
$$P(X = 1 \mid p = .10) = 2(.90)(.10) = .18$$
$$P(X = 2 \mid p = .10) = (.10)^2 = .01$$
$$1.00$$

$$P(X = 0 \mid p = .20) = (.80)^2 = .64$$
$$P(X = 1 \mid p = .20) = 2(.80)(.20) = .32$$
$$P(X = 2 \mid p = .20) = (.20)^2 = .04$$
$$1.00$$

$$P(X = 0 \mid p = .30) = (.70)^2 = .49$$
$$P(X = 1 \mid p = .30) = 2(.70)(.30) = .42$$
$$P(X = 2 \mid p = .30) = (.30)^2 = .09$$
$$1.00$$

For example, .405, the upper left-hand entry in the joint probability distribution in Table 15-12, was obtained by multiplying .50 by .81. In symbols, $P(X = 0 \cap p = .10) = P_0(p = .10)\, P(X = 0 \mid p = .10)$. We can now obtain the marginal probabilities of the sample outcomes of zero, one, and two defectives. The column totals in the joint probability distribution are $P(X = 0) = .695$, $P(X = 1) = .270$, and $P(X = 2) = .035$, respectively. These are the marginal probabilities entered in the decision diagram in Figure 15-3, on the branches emanating from node (c).

The calculations of posterior or revised probabilities of lot proportion defectives given the sample outcomes $X = 0$, 1, and 2 are shown in Table 15-13. These posterior probabilities were obtained by dividing the joint probabilities by the appropriate column totals, $P(X)$. For example, the probability that the incoming lot contains .10 defectives, given that zero defectives were observed in the sample of two articles, is

$$P(p = .10 \mid X = 0) = \frac{P(X = 0 \cap p = .10)}{P(X = 0)} = \frac{.405}{.695} = .583$$

As previously indicated in connection with Table 15-4, the calculation of these posterior probabilities represents an application of Bayes' Theorem. The posterior probabilities are shown in Figure 15-3 on the branches stemming from nodes (g), (h), (i), (j), (k), and (1).

We can now complete the extensive form analysis. As we noted earlier, the \$70 at node (b) in Figure 15-3 represents the expected loss of terminal action without sampling. Correspondingly, the \$57.78 at node (c) is the expected loss of sampling and inspecting two articles and then taking terminal action. Hence, the *expected value of sample information* is

$$\text{EVSI} = \$70 - \$57.78 = \$12.22$$

Table 15-12 Calculation of the Joint Probability Distribution of Sample Outcomes and Events in the Renny Corporation Problem.

Lot Proportion Defective (p)	*Prior Probabilities* $P_0(p)$	*Conditional Probabilities* $P(X \mid p)$ $X = 0$	$X = 1$	$X = 2$	*Joint Probabilities* $P(X \cap p)$ $X = 0$	$X = 1$	$X = 2$	*Total*
.10	.50	.81	.18	.01	.405	.090	.005	.50
.20	.30	.64	.32	.04	.192	.096	.012	.30
.30	.20	.49	.42	.09	.098	.084	.018	.20
					.695	.270	.035	1.00

Table 15-13 Calculation of the Posterior Probabilities of Lot Proportion Defectives in the Renny Corporation Problem.

$$P(p = .10 \mid X = 0) = .405/.695 = .583$$
$$P(p = .20 \mid X = 0) = .192/.695 = .276$$
$$P(p = .30 \mid X = 0) = .098/.695 = .141$$
$$1.000$$

$$P(p = .10 \mid X = 1) = .090/.270 = .333$$
$$P(p = .20 \mid X = 1) = .096/.270 = .356$$
$$P(p = .30 \mid X = 1) = .084/.270 = .311$$
$$1.000$$

$$P(p = .10 \mid X = 2) = .005/.035 = .143$$
$$P(p = .20 \mid X = 2) = .012/.035 = .343$$
$$P(p = .30 \mid X = 2) = .018/.035 = .514$$
$$1.000$$

Thus, the expected amount by which the terminal opportunity loss of action without sampling is reduced by the procedure of sampling two articles and taking action after inspection of the sample outcomes is \$12.22. Since the sampling and inspection of two articles costs \$5.00, the *expected net gain of sample information* is

$$\text{ENGS} = \$12.22 - \$5.00 = \$7.22$$

Therefore, since the expected net gain of sample information is \$7.22 for terminal action after sampling and inspecting two articles as compared to terminal action without sampling, it pays the Renny Company to follow the former course of action.

Normal Form Analysis

We now turn to the normal form analysis of the Renny Corporation problem which we have just solved by extensive form procedures. As in the A. B. Westerhoff problem, we commence the normal form analysis by listing all possible strategies or decision rules. These are shown in Table 15-14. We can see on an intuitive basis that some of the strategies are illogical and would therefore be apt to have relatively large expected opportunity losses associated with them. For example, strategy s_2 (a_1, a_1, a_2), rejects the incoming lot if the sample of two articles yields zero or one defectives, but accepts the lot if two defectives are observed. Strategy s_4 (a_1, a_2, a_2), also employs a perverse type of logic, since it rejects the lot on an observation of zero defectives but accepts it if one or two defectives are observed. Strategy s_1 (a_1, a_1, a_1), rejects the lot and $s_8(a_2, a_2, a_2)$ accepts the lot regardless of the

Table 15-14 A Listing of All Possible Strategies in the Renny Corporation Problem.

Sample Outcome (Number of Defectives)	Strategy							
	s_1	s_2	s_3	s_4	s_5	s_6	s_7	s_8
$X = 0$	a_1	a_1	a_1	a_1	a_2	a_2	a_2	a_2
$X = 1$	a_1	a_1	a_2	a_2	a_1	a_1	a_2	a_2
$X = 2$	a_1	a_2	a_1	a_2	a_1	a_2	a_1	a_2

a_1: Reject the lot
a_2: Accept the lot

sample observation. The two most logical decision rules appear to be strategy $s_5(a_2, a_1, a_1)$, which accepts the lot if zero defectives are observed in the sample and rejects the lot if one or two defectives are observed, and strategy $s_7(a_2, a_2, a_1)$, which accepts the lot if zero or one defectives are observed and rejects the lot if two defectives are found.

The next step is to calculate for each strategy the risks or conditional expected losses associated with the occurrence of each event (lot proportion defective). We will illustrate these calculations for strategies s_5 and s_7 by computing for each proportion defective the *loss* of taking the wrong act times the *probability* of taking the wrong act. These risks, denoted $R(s_5; p)$ and $R(s_7; p)$, are shown in Table 15-15. For example, let us consider strategy $s_5(a_2, a_1, a_1)$. The risk or conditional expected loss if the lot proportion defective is .10 is equal to $R(s_5; p = .10) = \$200 \times .19$ by the following reasoning. Referring back to the original payoff matrix in Table 15-10, we recall that the correct course of action if $p = .10$ is to accept, and if $p = .20$ or .30, to reject the lot. Strategy $s_5(a_2, a_1, a_1)$, accepts the lot on the observation of zero defectives and rejects it otherwise. From Table 15-12 we find a conditional probability of .81 of observing zero defectives, given that the lot

Table 15-15 Calculation of Risks or Conditional Expected Opportunity Losses for Strategies s_5 and s_7.

Strategy $s_5(a_2, a_1, a_1)$

Events (Lot Proportion Defective) p	*Opportunity Loss of Wrong Act*	*Probability of Wrong Act Given p*	*Risk (Conditional Expected Loss) $R(s_5; p)$*
.10	\$200	.19	\$38
.20	100	.64	64
.30	200	.49	98

Strategy $s_7(a_2, a_2, a_1)$

Events (Lot Proportion Defective) p	*Opportunity Loss of Wrong Act*	*Probability of Wrong Act Given p*	*Risk (Conditional Expected Loss) $R(s_7; p)$*
.10	\$200	.01	\$ 2
.20	100	.96	96
.30	200	.91	182

proportion defective is $p = .10$. Hence, using strategy s_5, the probability of making the wrong decision, given a lot proportion defective of $p = .10$ is $1 - .81 = .19$. The loss associated with rejection if $p = .10$ is \$200 (Table 15-10). Therefore, as shown in Table 15-15, the risk of strategy s_5, given $p = .10$, is $R(s_5;\ p = .10) = \$200 \times .19 = \$38$. Analogous calculations produce the other risks shown in Table 15-15. A summary of the risks for all eight strategies is presented in Table 15-16.

We can now calculate the expected loss of each strategy by Equation (15.2). Adapting the notation of that equation to that used in the present problem, we obtain as the expected opportunity loss of the kth strategy

$$\text{EOL}(s_k) = \sum_p P_0(p)R(s_k; p) \tag{15.3}$$

That is, for each strategy, we weight the conditional expected losses associated with each lot proportion defective (Table 15-16), by the prior probabilities of such incoming lots (Table 15-12).

Hence, the expected opportunity losses (expected risks) of strategies s_5 and s_7 are:

$$\text{EOL}(s_5) = (.50)(\$38) + (.30)(\$64) + (.20)(\$98) = \$57.80$$
$$\text{EOL}(s_7) = (.50)(\$\ 2) + (.30)(\$96) + (.20)(\$182) = \$66.20$$

Table 15-16 Risks or Conditional Expected Opportunity Losses for the Eight Strategies in the Renny Corporation Problem.

Events (Lot Proportion Defective) p	*Strategies* s_1	s_2	s_3	s_4	s_5	s_6	s_7	s_8
.10	\$200	\$198	\$164	\$162	\$38	\$36	\$ 2	\$ 0
.20	0	4	32	36	64	68	96	100
.30	0	18	84	102	98	116	182	200

The expected opportunity losses of all eight strategies are given in Table 15-17. The optimal strategy is seen to be $s_5(a_2, a_1, a_1)$, since it has the minimum expected opportunity loss of \$57.80. This strategy, which accepts the lot if zero defectives are observed in the sample and rejects the lot if one or two defectives are found, is seen to be the same as the one found in the preceding extensive form analysis depicted in the decision diagram in Figure 15-3. The minor difference between the expected opportunity loss of the optimal strategy, \$57.78 in the extensive form analysis and \$57.80 in the normal form analysis, is attributable to rounding of decimal places.

Table 15-17 Expected Opportunity Losses of the Eight Strategies in the Renny Corporation Problem.

Strategy	*Expected Opportunity Loss*
s_1	$100.00
s_2	103.80
s_3	108.40
s_4	112.20
s_5	57.80*
s_6	61.60
s_7	66.20
s_8	70.00

15.7 Optimal Sample Size

In the above acceptance sampling problem, we indicated how the *expected* value of *sample information* can be derived prior to the actual drawing of a sample. This EVSI figure was obtained by subtracting the expected opportunity loss of the best terminal act without sampling from the expected loss of the optimal strategy with sampling. More precisely, the latter expected value is the expected opportunity loss of a decision to sample and then take optimal terminal action after observation of the sample outcome. In that problem, we assumed a fixed or predetermined sample size of two articles. However, as was previously indicated, pp. 705, the method of analysis presented is a general one, and the only practical effect of an assumption of a larger sample size would have been an increase in the computational burden. Nevertheless, the question remains, "Can an *optimal sample size* be derived in a problem such as the one presented?" The answer is "yes," and the general method for obtaining such an optimal value follows.

As might be suspected intuitively, an increase in sample size brings about an increase in the EVSI. However, an increase in sample size also results in an increase in the cost of sampling. (We are using the term "cost of sampling" here to mean the total cost of sampling and inspection.) Therefore, what we would like to do is to find the sample size for which the difference between the EVSI and the cost of sampling is the largest. An equivalent and more convenient approach is to minimize *total loss*, where the total loss associated with any sample size n is defined as

(15.4) Total Loss = Cost of Sampling + Expected Opportunity Loss of the Optimal Strategy

Let us consider how the quantities in Equation (15.4) might be calculated. The cost of sampling would ordinarily not be difficult to calculate. In many instances, this cost may be entirely variable, that is, proportional to the number of articles sampled. In that case, the cost of sampling would be equal to

(15.5) $$C = vn$$

where C = cost of sampling
v = cost of sampling each unit
n = number of units in the sample

Hence, in the Renny Corporation example, where the cost of sampling was \$5.00 for two articles, if costs were entirely variable, the cost of sampling each unit would be \$2.50. Hence, $v = \$2.50$ and $n = 2$. The cost of sampling for a sample of size ten would be $C = \$2.50 \times 10 = \$25.00$, etc.

On the other hand, in certain situations, a portion of the total cost of sampling might be fixed and the remaining part variable. In this case, the cost of sampling would be

(15.6) $$C = f + vn$$

where C = cost of sampling
f = fixed cost
v = cost of sampling each unit
n = number of units in the sample

Thus, for example, in a case in which the fixed cost of sampling is \$10, and the variable cost per unit is \$2, the cost of sampling 20 units would be

$$C = \$10 + (\$2)(20) = \$50$$

Other formulas can be derived for more complex situations.

Turning now to the other term in Equation (15.4) for *total loss*, namely, the *expected opportunity loss of the optimal strategy*, we pause first to introduce more convenient notation. Let us assume we are dealing with an acceptance sampling problem in which the sample size is ten. Although in the normal form analysis of the Renny Corporation problem, we considered every possible strategy for the acceptance of the incoming lot, it can be shown that the only strategies worth considering as potential optimal decision rules are those which accept the incoming lot if a certain number or less of defectives are observed, and which reject the lot, otherwise. As previously indicated in Example 2-8 of Chapter 2, it is conventional to refer to this critical number as "the acceptance number," denoted c. Hence, in the present problem in which the sample size is ten, the possible values for the acceptance number c are 0, 1, 2, . . . , 10. For example, if $c = 0$, the lot is accepted if zero defectives are observed in a sample of size $n = 10$, and rejected otherwise. If $c = 1$, the lot is accepted if *one or less* defectives are observed, and rejected otherwise. If $c = 2$, the lot is accepted if *two or less* defectives are observed, and rejected otherwise, etc. Therefore, we can characterize the optimal strategy for each sample size by two figures: c, the acceptance number, and n the sample size. For example, the c, n pair, denoted (c, n), for an optimal strategy with an acceptance number $c = 2$ and sample size $n = 10$ would be (2, 10). In the Renny Corporation problem, the c, n pair was (0, 2) for the optimal strategy s_5.

Theoretically, we could calculate for every possible sample size the expected opportunity loss of the optimal strategy (c, n). Adding this figure to the cost of sampling, we would obtain the total loss associated with each sample size. We would then select as the optimal sample size the one which yielded the *minimum total loss*.

At first blush, this might seem to be an impossible procedure because the sample size could conceivably take on any positive integral value up to infinity. However, n must be a finite value because of the cost of sampling involved. The expected value of sample information (EVSI) cannot exceed the expected value of perfect information (EVPI). As we saw in Section 13.3, EVPI is the expected opportunity loss of the optimal act prior to sampling. That is, it is the expected loss of the best terminal action without sampling. Since the value of sampling cannot exceed that EVPI figure, it would never be worthwhile to take a sample so large that the cost of sampling exceeded this EVPI figure. Hence, n for the optimal sample size will be a finite number.

The acceptance sampling type of problem we have been discussing is one which can be described as involving "binomial sampling." That is, the conditional probabilities of sample outcomes of the form $P(X \mid p)$ were calculated in the Renny Corporation problem by the binomial probability distribution. In this type of problem, the optimal strategies for the various sample sizes must be calculated by the methods we have indicated, which may be characterized as trial and error procedures. That is, there is no simple general formula which enables us to derive the optimal (c, n) pairs for every problem. Therefore, there may be a considerable amount of calculation involved to determine the optimal sample size. For sufficiently large and important problems, the use of computers may represent the only practical method to carry out the computations.

15.8 General Comments

In this chapter we have discussed decision-making procedures for the selection of optimal strategies prior to the obtaining of experimental data or sample information. The method used, that is, preposterior analysis, anticipates the adoption of best actions after the observation of the experimental or sample results. The general principles of this type of analysis have been discussed using two examples. The first illustration involved a new product introduction. In that case, experimental evidence in the form of results of a sample survey was obtainable for revising prior probabilities of occurrence of the states of nature "successful product" and "unsuccessful product." The second example involved a decision concerning acceptance or rejection of an incoming lot. Here, information in the form of results of a random sample drawn from the lot was obtainable for revising prior probabilities of the states of nature "lot proportion defectives." In the first example, an empirical joint frequency distribution on the success or lack of it of past new products and survey results provided information for the calculation of probabilities required for the problem solution. In the second example, the decision depended on incoming lot proportion defectiveness, denoted p. The observable sample data were represented by the random variable "number of defectives," denoted X, in a sample drawn from the incoming lot. This random variable was binomially distributed. In both of these problems, the states of nature and the experimental or sample evidence were discrete random variables. Another class of problems is comprised by situations in which the states of nature and sample evidence are represented by continuous random variables. For example, the decision may depend on a parameter μ, the mean of a population, and the observed sample evidence may be a sample mean, $\bar{X}$. The mathematics required for the solution of this class of problems is outside the scope of this book. However, the same basic principles discussed in this chapter for extensive form and normal form analysis are applicable, regardless of whether the random variables representing states of nature and experimental outcomes are discrete or continuous.

PROBLEMS

1. Define EVSI and expected net gain from sample information and compare the two concepts.
2. Define and contrast the extensive and normal form approaches.
3. Let θ_1 be the state "it will rain tomorrow" and θ_2 "it will not rain tomorrow." The weatherman will predict either X, "it will rain" or Y, "it will not rain." Fill in the missing entries of the following table.

θ	$P(\theta)$	$P(X \mid \theta)$	$P(Y \mid \theta)$	$P(\theta)P(X \mid \theta)$	$P(\theta)P(Y \mid \theta)$	$P(\theta \mid X)$	$P(\theta \mid Y)$
θ_1	.3	.8					
θ_2	.7	.3					

4. Let the two possible states of nature be θ_1: the average level of stock market prices will advance, θ_2: the average level of stock market prices will stay the same or decrease. Your prior belief is that there is a 60% chance that market prices will advance. In a week the forecast of a well-known econometric

model will be published. Either an advance or no advance will be indicated. The model is correct 80% of the time. Find the posterior probabilities for θ for each of these indications.

5. Given the following illustration of extensive analysis with payoffs in terms of opportunity losses

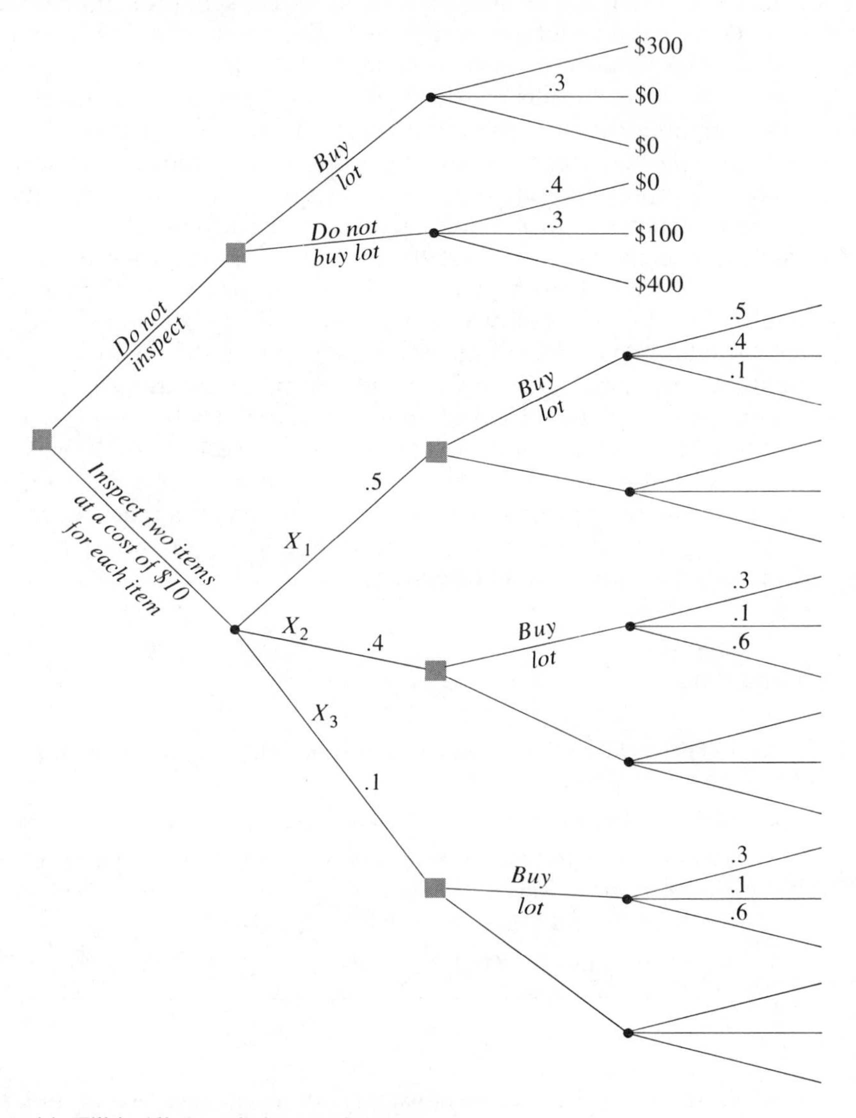

(a) Fill in all the missing entries.
(b) Interpret each part of the tree.
(c) Find the EVPI before and after sampling.

(d) Find the EVSI.
(e) Find the expected net gain from sampling.

6. A motel located near the site of a soon-to-be-opened world's fair is contemplating the construction of some temporary extra rooms. The rooms are of no value after the fair is closed since the present size of the motel is more than adequate for normal demand. Let θ_1 be "demand for the temporary rooms is at least enough to cover the cost of building them," and θ_2 be, "demand for the temporary rooms is not sufficient to warrant construction." A survey can be taken which would yield two possible indications X_1 or X_2. If the following is known:

State of Nature	$P_0(\theta)$	$P(X_1 \mid \theta)$	*Opportunity Loss Table (in "Utiles")*	*Build*	*Do Not Build*
θ_1	.5	.55	θ_1	0	6
θ_2	.5	.50	θ_2	5	0

(a) Find:
   (1) the optimal act before sampling.
   (2) the optimal act if X_1 occurs.
   (3) the optimal act if X_2 occurs.
   (4) $P(X_1)$ and $P(X_2)$.
   (5) the EVSI
(b) Should the survey be taken?

7. A mutual fund is contemplating the sale of a certain common stock. A study can be made which will yield two possible results, X_1 or X_2. Given the following information

State of Nature	$P_0(\theta)$	$P(X_1 \mid \theta)$	$P(X_2 \mid \theta)$
θ_1: Stock Price Up	.4	.6	.4
θ_2: Stock Price Down or Same	.6	.9	.1

Opportunity Loss Table (Units of \$10,000)

	θ_1	θ_2
Sell	7	0
Do Not Sell	0	7

(a) Find:
   (1) the optimal act before sampling
   (2) the optimal act if X_1 results.
   (3) the optimal act if X_2 results.
   (4) $P(X_1)$ and $P(X_2)$.

(5) the EVSI.

(b) If the cost of the study is $1500, what is the expected net gain from the study?

8. Assume there are three possible states, S_1, business will be a success, S_2, business will have limited success, and S_3, business will not be a success; with prior probabilities of occurrence .58, .25, and .17, respectively. The following is the opportunity loss table, in thousands of dollars.

Opportunity Loss Table

	S_1	S_2	S_3
Invest in Business	0	0	500
Do Not Invest	400	100	0

An investigation costing $10,000 is run for which there are three outcomes X, Y, and Z. The probabilities of these outcomes are .6, .3, and .1, respectively. The posterior probabilities of S_i, given the respective outcomes are

	X	Y	Z
S_1	.8	.3	.1
S_2	.1	.5	.4
S_3	.1	.2	.5
	1.0	1.0	1.0

Draw a decision tree diagram and make all the necessary entries on the branches dealing with outcome X.

9. A certain manufacturing process produces lots of 500 units each. Either 10% or 30% of the 500 items are defective in each lot. The quality control department inspects each lot before shipment. If accepted, the lot produces a profit of $500, but if rejected, the lot is sold for scrap at cost. If a 30% defective lot is sent out, confidence in the company is lost and the firm estimates its loss in good will and hence future orders, at $1500. In the past 80% of the lots produced contained 50 defectives. It costs $3 to test an item, but the test is not destructive (i.e., if the item was good, it still is good after the test and can be sold). What is the expected net gain from running a test on three items drawn at random and what is the best decision rule? Use the binomial probability distribution to compute conditional probabilities.

10. A pharmaceutical firm classifies its drugs as low volume, medium volume, and high volume products. The firm introduces many new drugs each year. When a new drug is released, an initial marketing plan (i.e., stocking of the drug, advertising, etc.) is put into effect. It is important that the firm initiate the proper market plan. For example, if a drug has a high volume potential but a low volume market plan is initiated, the drug will not realize its potential. The opportunity loss table for different marketing plans is

Marketing Plan Used	State of Nature Low Volume	Medium Volume	High Volume
Low	$ 0	$200,000	$700,000
Medium	200,000	0	500,000
High	400,000	200,000	0

A new drug, Novatol, is to be released. The marketing vice president feels the probability that it will be high volume is .2 and low volume .1. A forecasting model has been developed which has the following reliability (that is, P(indication/state of nature))

Model Indicates	State of Nature Low Volume	Medium Volume	High Volume
Low	.85	.10	.05
Medium	.10	.80	.10
High	.05	.10	.85

What is the expected value of information from the forecasting model for the drug Novatol?

11. In Problem 24 of Chapter 13, you are offered Stock, Inc. investment advice, which is correct 70% of the time. What is the maximum amount you should pay for Stock, Inc. advice on the predicted price movement of Rerox common stock?

12. In Problem 13, Chapter 14, assume the cost to the publisher was $200 to solicit the opinion of a professional critic.
    (a) If the publisher solicits the opinion of just one critic, how much is the expected net gain from that critic's advice?
    (b) What is the expected net gain of querying two critics?

13. A large manufacturing firm is planning to build a new plant in either Georgia or Pennsylvania. The cost of the required land would be $750,000 less in Georgia due to the inexpensiveness of the land and special tax incentives. However, in Pennsylvania a skilled labor force would be available, whereas that might not be the case in Georgia. It costs the company $5000 to train and pay an employee until he is productive. It would be cheaper to move a skilled labor force into the area, but it is against company policy to relocate labor. The firm needs 500 skilled workers. Assume that either 300, 350, or 400 skilled employees can be hired in Georgia and the rest would have to be trained. Management feels that the probabilities of hiring these numbers of employees are .3, .4, and .3, respectively. A survey of the labor situation can be undertaken. The reliability of the survey is

Survey Indicates	*State of Nature* 300	350	400
300	.8	.3	.1
350	.1	.6	.2
400	.1	.1	.7

What is the maximum amount that the firm should pay for the survey?

14. In Problem 7 of this chapter, suppose the mutual fund felt that the prior probability the common stock price would increase was .1 instead of .4. What would be the expected net gain from the study? Compare it to the expected net gain if the prior probability is .4.

15. In Problem 13 of Chapter 14, at what value of the prior probability of success would the decision concerning publication change?

16. A mail order house with a fixed "market" of 100,000 people is deciding whether to sell a new line of goods. If more than 30% of its customers will purchase, the company should market the line and if less than 30% purchase, it should not. For simplicity, assume that either 20% or 40% will purchase the new line. The manager of the firm believes the probability that only 20% of the market will buy is .6. The payoff table (in units of $10,000) is

	20% Will Purchase	*40% Will Purchase*
Market New Line	−3	5
Do Not Market New Line	0	0

(a) One person is selected at random from the 100,000 and asked whether he will purchase. What is the expected value of this information?
(b) Suppose that two persons are selected at random from the 100,000 and asked whether they will purchase. What is the expected value of this information?

17. For Problem 5 of this chapter, construct a table showing all possible strategies and select the optimal strategy if the two items are inspected.

18. In Problem 8 of this chapter, the following strategy table has been constructed. Fill in those strategies that are missing and select the optimal strategy.

Sample Outcome	s_1	s_2	s_3	s_4	s_5	s_6	s_7	s_8
X	a_1	a_1	a_1	a_1	a_2	a_2	a_2	a_2
Y		a_2	a_1	a_1		a_2	a_1	a_1
Z		a_1	a_2	a_1		a_1	a_2	a_1

a_1 denotes "invest in business"
a_2 denotes "do not invest"

19. In Problem 10 of this chapter there are 27 strategies. Assume the optimal strategy is one of the following three:

Model Indication	s_5	s_{15}	s_{23}
Low	L	M	L
Medium	M	M	M
High	M	H	H

L, M, and H mean low, medium, and high market plans used, respectively. Compute conditional expected losses and expected opportunity losses to determine which of these three strategies is best.

20. The following table pertains to Problem 13 of this chapter. Fill in the missing strategies.

Sample Outcomes	s_1	s_2	s_3	s_4	s_5	s_6	s_7	s_8
300	a_1	a_1	a_1	a_1	a_2	a_2	a_2	a_2
350	a_1	a_1	a_2		a_1	a_1		a_2
400		a_1	a_2			a_1		a_1

a_1 denotes "build in Pennsylvania"
a_2 denotes "build in Georgia"

Compute the conditional expected losses as well as the expected opportunity losses for strategies s_1, s_3, and s_4. Which strategy is optimal?

21. Suppose there are three states of nature θ_1, θ_2, and θ_3; two actions, a_1 and a_2; and three outcomes from a forecast, X, Y, and Z. The outcome X indicates that θ_1 will occur, Y indicates θ_2 will occur, and Z indicates θ_3 will occur. The opportunity loss table is

	a_1	a_2
θ_1	0	1300
θ_2	300	0
θ_3	800	0

A forecasting model has been developed which has the following reliability:

	Model Indicates		
State of Nature	X	Y	Z
θ_1	.75	.15	.10
θ_2	.09	.50	.41
θ_3	.07	.09	.84

Of the eight strategies only two are plausible: $s_3(a_1, a_2, a_2)$ and $s_5(a_1, a_1, a_2)$. Use normal form analysis to determine the optimal strategy. In computing the conditional expected losses for the two strategies, use the *loss* of taking the wrong act times the probability of taking the wrong act. The prior probabilities for θ_1, θ_2, and θ_3 are .35, .30, and .35, respectively.

CHAPTER SIXTEEN

Sequential Decision-Making Procedures

16.1 Introduction

We have considered decision-making procedures based solely on prior knowledge, on an incorporation of experimental or sample information with this prior knowledge, and on the devising of optimal strategies prior to obtaining experimental or sample data. Whenever information from sampling was involved, we dealt with a *single sample of fixed size*. In this chapter, we discuss decision-making procedures in which a choice is made after some information has been obtained between taking terminal action or obtaining further information. If the better course of action is to obtain further information, and this is done, a decision has to be made again between taking terminal action or gathering even more information. These decisions proceed sequentially until finally, terminal action is taken. Such a multistage series of choices is referred to as a *sequential decision-making procedure*.

The additional information obtained at each stage may consist of a single observation or a group of observations. For example, an illustration of the single observation case occurs in quality control work, where a lot may be inspected by taking successive observations of one item each. After each item, a choice is made between (1) stopping sampling and making a terminal decision to reject or accept the lot, and (2) drawing an additional item before again deciding whether to take a terminal act or to continue sampling.

A corresponding illustration involving groups of observations also occurs in quality control work. Instead of successive observations of one item each, samples of n items are drawn at every stage. We will illustrate the procedure by assuming $n = 10$. Upon the drawing of the first sample of ten items, if a very small number of defectives is present, say zero defectives, the lot is accepted. If a large number of defectives is present, say six or more, the lot is rejected. If one through five defectives are observed, another sample of ten items is drawn. Now, if a small cumulative number of defectives is present in the two samples of ten items each, say a total of one defective, the lot may be accepted. On the other hand, a large cumulative number of defectives would result in rejection of the lot, and an intermediate number would require the drawing of another sample. Finally, a stage would be reached at which only two decisions are possible, rejection or acceptance of the lot.

If the lot is of very good quality, that is, contains a small percentage of defectives, it will tend to be accepted very quickly, perhaps on the first or second sample. Similarly, if the lot is of very poor quality, that is, contains a large percentage of defectives, it will tend to be rejected very quickly. Greater numbers of samples would be entailed for lots of intermediate quality. Although theoretically an extremely large number of samples might be needed for a lot of intermediate quality, in actual practice, the procedure is truncated and a decision for acceptance or rejection is *forced* after a specified number of samples.

A major advantage of these sequential sampling plans is that, on the average, they require fewer observations than single sampling plans (fixed sample size) to achieve the same levels of Type I and Type II errors. In classical statistics, sequential sampling plans are designed to meet prescribed risks of making Type I and Type II errors. As was true of other procedures in classical statistical inference, discussed in Chapters 6, 7, 8, and 9, costs or profits are not explicitly included in the analysis. On the other hand, Bayesian sequential decision procedures, which are the type discussed in this chapter, do explicitly incorporate costs and profits. Extensive form analysis, as discussed in Chapter 15, is used in the Bayesian sequential decision procedure with calculations of new posterior probabilities as further information is acquired at each successive stage.

We will illustrate Bayesian sequential decision procedures by the following new product development problem.

16.2 A New Product Development Problem

A company was considering the development of a new product. Sufficient research had been carried out so that a possible course of action was to proceed to direct commercialization of the product. However, the company had had several unsuccessful products in the past. Therefore, deciding to analyze the development of the product quantitatively, the firm selected Bayesian sequential decision procedures for that purpose.

The company faced the full array of problems typical of the development of any new product. Assuming that the first stage of development was carried out, the firm could decide at that point, based on the information derived, to proceed directly to commercialization and to introduce the product, or it could terminate the project completely and decide not to introduce the product. On the other hand, it could choose to proceed to the second development stage, generating additional information concerning the product's chances for success. Again at the end of the second development stage, in the light of the new information, it would face the same set of decisions as at the end of the first stage. This sequential decision procedure would continue for as many development stages as the firm had under consideration. Often in fields such as marketing and production, the earliest development stages represent preliminary screening procedures, whereas later ones entail more costly detailed screening and intensive investigative efforts.

Assumptions

In order to keep the exposition simple, we will make a number of simplifying assumptions in our illustrative problem. First of all, we will disregard the time value of money in computing expected payoffs. That is, we will ignore the fact that a dollar received one year from today is worth less than a dollar today. In a more realistic analysis, this fact should be taken into account, and appropriate discount factors should be applied to both costs and payoffs pertaining to future time periods. Second, we will assume that only two stages of development are under consideration and that the company has decided to move ahead into the first stage. Third, as in some of our earlier problems, we will assume only two states of nature, namely, θ_1: the product is successful, and θ_2: the product is unsuccessful, and two alternative actions, A_1: introduce the product, and A_2: do not introduce the product. Fourth, we will assume that delays in marketing the product occasioned by the length of the development periods do not alter the payoffs. This will forestall possible confusion arising out of changing payoff figures. However, in a real problem, the effects of competition and other factors which might bring about changes in payoff figures because of delays in introducing the product should be taken into account. Finally, as in previous problems, we are presupposing that a linear utility function for money over the range of payoffs considered is a suitable approximation. Therefore, it is meaningful to use the maximization of expected monetary payoffs as a criterion of preferredness among alternative courses of action.

Expected Value of Perfect Information

We will refer to the person who has responsibility for the decision process as the "decision maker." We begin the analysis by assuming the decision maker has assessed payoffs in terms of net profit in dollars and has prior betting odds of 50 : 50 on the success of the product. These data are summarized in Table 16-1. The prior expected profits of the two courses of action have been calculated and are shown at the bottom of the table. As indicated in the table, the preferable course of action based on prior betting odds is to introduce the product (act A_1), yielding an expected payoff of $500,000 as opposed to $0, if the product is not introduced (act A_2). The *expected profit under uncertainty* is equal to $500,000, since that is the expected payoff of selecting the optimal act under conditions of uncertainty. Using methods discussed in Chapter 13, we now calculate the *expected value of perfect information* (EVPI), in order to evaluate the upper limit of the worth of obtaining additional information. Table 16-2 shows this computation by the two

Table 16-1 Calculation of Prior Expected Profits in the New Product Development Problem.

Events θ_i	*Prior Probability* $P(\theta_i)$	*Introduce Product* A_1	*Do Not Introduce Product* A_2
θ_1: Successful Product	.50	$2,000,000	$0
θ_2: Unsuccessful Product	.50	− 1,000,000	0
	1.00		

Prior Expected Profit (A_1) = (.50)($2,000,000) + (.50)(−$1,000,000) = $500,000
Prior Expected Profit (A_2) = (.50)($0) + (.50)($0) = $0

Expected Profit Under Uncertainty = $500,000

methods previously discussed. Since the expected value of perfect information is high, that is, \$500,000, it appears reasonable to investigate whether the information gathering process of sequential development is worthwhile.

Decision Tree Diagram

As indicated earlier, it was assumed in this problem that the company had decided to proceed with first stage development. At the end of the problem, we will return to the question of whether the company should undertake the development process at all or move directly to commercialization. However, assuming that the company enters first-stage development, the analysis must indicate whether it is better to take an optimal terminal action after gathering the information at this stage or to move on to second-stage development. The information gathering costs are \$20,000 for first-stage development and \$100,000 for second stage development. The first-stage information is 60% reliable. This means that if the product were successful, the probability is .60 that the first-stage information would indicate a successful product and .40 that it would indicate an unsuccessful product. Second-stage information is 70% reliable in the same sense.

Figures 16-1 and 16-2 show the tree diagram for the sequential decision procedure. The tree has been shown in two sections to clarify the presenta-

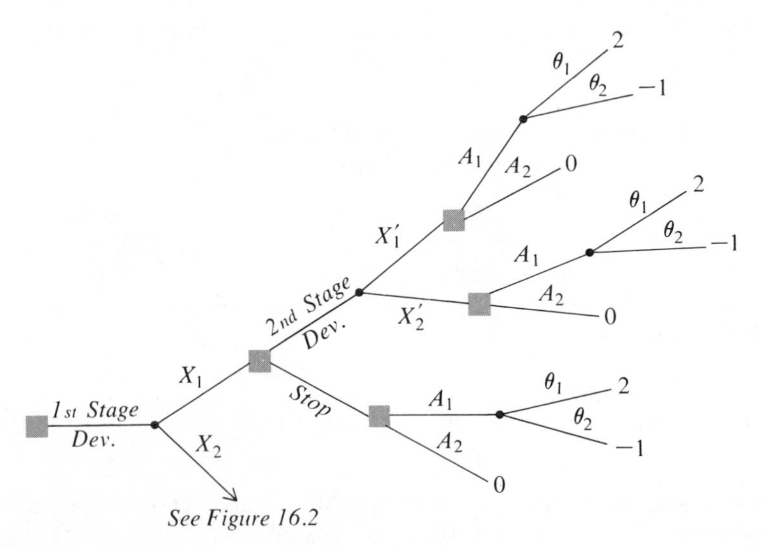

Figure 16-1 Decision tree diagram, assuming a type X_1 indication is obtained in first-stage development (all payoffs are in millions of dollars).

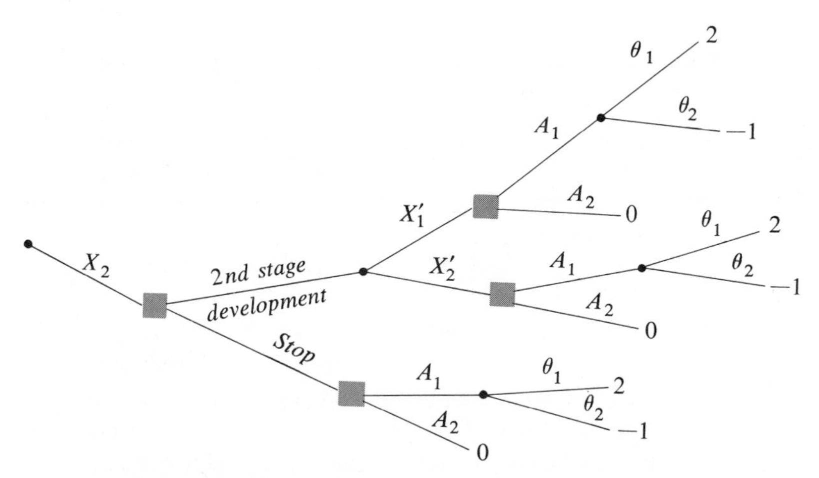

Figure 16-2 Decision tree diagram, assuming a type X_2 indication is obtained in first-stage development (all payoffs are in millions of dollars).

tion. We will trace through Figure 16-1 to indicate the decision process. Two types of information can be obtained from first-stage development:

X_1: An indication that the product will be successful

X_2: An indication that the product will be unsuccessful

From Figure 16-1, we see that if type X_1 information is the outcome, the decision maker has two choices. He can stop the development process and make a terminal decision or he can proceed with second-stage development. If he stops development, he can choose terminal action A_1 or A_2. If he selects A_2, his payoff is \$0. If he selects A_1, state of nature θ_1 or θ_2 will eventuate, that is, the product will either be successful or unsuccessful with the payoffs indicated in Table 16-1.

On the other hand, if the company enters second-stage development, again two types of indications may be given, X_1', and X_2' (X_1 prime and X_2 prime). These symbols have the same meaning as X_1 and X_2, respectively, except that the prime signifies information generated at the second stage. At this point, we have assumed that the decision maker must take a terminal action. If a third stage of development were contemplated, the diagram would continue as at the earlier stages.

A similar description of the tree diagram applies to Figure 16-2, which pertains to the case where X_2 information is observed as a result of first-stage development.

Table 16-2 Calculation of the Expected Value of Perfect Information.

Events	*Prior Probability* $P(\theta_i)$	*Profit*	*Weighted Profit*
θ_1: Successful Product	.50	\$2,000,000	\$1,000,000
θ_2: Unsuccessful Product	.50	0	0
			\$1,000,000

	Expected Profit with Perfect Information	= \$1,000,000
Less:	Expected Profit under Uncertainty	= 500,000
	Expected Value of Perfect Information	= \$ 500,000

Events	*Prior Probability*	*Opportunity Losses* Introduce Product A_1	*Opportunity Losses* Do Not Introduce Product A_2
θ_1: Successful Product	.50	\$ 0	\$2,000,000
θ_2: Unsuccessful Product	.50	1,000,000	0

$$\text{EOL}(A_1) = (.50)(\$0) + (.50)(\$1{,}000{,}000) = \$500{,}000$$
$$\text{EOL}(A_2) = (.50)(\$2{,}000{,}000) + (.50)(\$0) = \$1{,}000{,}000$$

$$\text{Expected Value of Perfect Information} = \text{Minimum EOL}(A_j) = \$500{,}000$$

Sequential Decision Analysis

We can now proceed with the sequential decision analysis, entering the appropriate numerical quantities in the decision tree. Since there is a possibility of information being generated at two development stages before a terminal choice is made, we must calculate posterior probabilities, which incorporate the information outcomes at each of the successive stages. The posterior probabilities of the events θ_1 and θ_2 after the information indication X_1 or X_2 has been observed become the prior probabilities of these events at the second stage. These prior probabilities are in turn revised to incorporate the information indications X_1' and X_2' generated from the second development stage. The revised figures represent the posterior probabilities after the second development stage. As usual, Bayes' Theorem is used to accomplish the revision of probabilities. These calculations are given in Tables 16-3 and 16-4. The conditional probabilities in these tables are based on the previously stated assumption that first-stage information is 60% reliable and second-stage information is 70% reliable.

At this point, we will interrupt the explanation of the sequential decision problem in order to illustrate an important and very interesting aspect of Bayes' Theorem. In Tables 16-3 and 16-4 we used Bayes' Theorem to revise prior probabilities in a sequential manner. For example, there were subjective probabilities $P(\theta_i)$ assigned to the states of nature θ_1 and θ_2. If X_1 was observed at the first stage, posterior probabilities were calculated incorporating this information. These posterior probabilities became the prior probabilities for the second stage. If X_1' was observed at the second stage, posterior probabilities were again calculated to reflect this information. Now, an interesting question arises. Suppose we had revised prior probabilities not one step at a time, but had revised our original subjective probabilities on the basis of all accumulated experimental information in one step. For example, suppose we were to revise the original subjective probabilities in one step in the light of the sequence of information X_1, X_1'. That is, we would compute $P(\theta_i \mid X_1 \cap X_1')$, which for simplicity, we denote $P(\theta_i \mid X_1X_1')$. How would these probabilities compare with the corresponding posterior probabilities

Table 16-3 Computation of Posterior Probabilities Which Incorporate First-Stage Development Information.

X_1 Observed at First Stage

Events θ_i	*Prior Probabilities* $P(\theta_i)$	*Conditional Probabilities* $P(X_1 \mid \theta_i)$	*Joint Probabilities* $P(\theta_i)P(X_1 \mid \theta_i)$	*Posterior Probabilities* $P(\theta_i \mid X_1)$
θ_1	.50	.60	.30	.60
θ_2	.50	.40	.20	.40
			$P(X_1) = .50$	1.00

X_2 Observed at Second Stage

θ_i	$P(\theta_i)$	$P(X_2 \mid \theta_i)$	$P(\theta_i)P(X_2 \mid \theta_i)$	$P(\theta_i \mid X_2)$
θ_1	.50	.40	.20	.40
θ_2	.50	.60	.30	.60
			$P(X_2) = .50$	1.00

Table 16-4 Computation of Posterior Probabilities Which Incorporate First- and Second-Stage Development Information.

X_1 Observed at First Stage; X_1' at Second Stage

Events θ_i	*Prior Probabilities* $P(\theta_i)$	*Conditional Probabilities* $P(X_1' \mid \theta_i)$	*Joint Probabilities* $P(\theta_i)P(X_1' \mid \theta_i)$	*Posterior Probabilities* $P(\theta_i \mid X_1')$
θ_1	.60	.70	.42	.78
θ_2	.40	.30	.12	.22
			$P(X_1') = .54$	1.00

X_1 Observed at First Stage; X_2' at Second Stage

θ_i	$P(\theta_i)$	$P(X_2' \mid \theta_i)$	$P(\theta_i)P(X_2' \mid \theta_i)$	$P(\theta_i \mid X_2')$
θ_1	.60	.30	.18	.39
θ_2	.40	.70	.28	.61
			$P(X_2') = .46$	1.00

X_2 Observed at First Stage; X_1' at Second Stage

θ_i	$P(\theta_i)$	$P(X_1' \mid \theta_i)$	$P(\theta_i)P(X_1' \mid \theta_i)$	$P(\theta_i \mid X_1')$
θ_1	.40	.70	.28	.61
θ_2	.60	.30	.18	.39
			$P(X_1') = .46$	1.00

X_2 Observed at First Stage; X_2' at Second Stage

θ_i	$P(\theta_i)$	$P(X_2' \mid \theta_i)$	$P(\theta_i)P(X_2' \mid \theta_i)$	$P(\theta_i \mid X_2')$
θ_1	.40	.30	.12	.22
θ_2	.60	.70	.42	.78
			$P(X_2') = .54$	1.00

resulting from the one step at a time revision? The answer is that the posterior probability distributions would be identical. The alternative one-step calculation of posterior probabilities on the basis of all cumulative experimental evidence is given in Table 16-5. These probabilities are seen to be identical with the corresponding posterior probabilities in the last column of Table 16-4.

Returning now to the sequential decision problem, we enter the probabilities derived in Tables 16-3 and 16-4 at the appropriate places in the decision tree diagrams of Figures 16-1 and 16-2. Then, we use the method of *backward induction*, which, as we have seen in Section 13.6, involves starting at the right-hand side of the decision tree diagram and proceeding inward. We compute expected values and select optimal acts at each stage, moving inward until we return to the beginning of the tree. Thus, we evaluate the

Table 16-5 Alternative One-step Calculation of Posterior Probabilities.

X_1 Observed at First Stage; X_1' at Second Stage

Events θ_i	Prior Probabilities $P(\theta_i)$	Conditional Probabilities $P(X_1X_1' \mid \theta_i)$	Joint Probabilities $P(\theta_i)P(X_1X_1' \mid \theta_i)$	Posterior Probabilities $P(\theta_i \mid X_1X_1')$
θ_1	.50	(.60)(.70) = .42	.21	.78
θ_2	.50	(.40)(.30) = .12	.06	.22
			.27	1.00

X_1 Observed at First Stage; X_2' at Second Stage

θ_i	$P(\theta_i)$	$P(X_1X_2' \mid \theta_i)$	$P(\theta_i)P(X_1X_2' \mid \theta_i)$	$P(\theta_i \mid X_1X_2')$
θ_1	.50	(.60)(.30) = .18	.09	.39
θ_2	.50	(.40)(.70) = .28	.14	.61
			.23	1.00

X_2 Observed at First Stage; X_1' at Second Stage

θ_i	$P(\theta_i)$	$P(X_2X_1' \mid \theta_i)$	$P(\theta_i)P(X_2X_1' \mid \theta_i)$	$P(\theta_i \mid X_2X_1')$
θ_1	.50	(.40)(.70) = .28	.14	.61
θ_2	.50	(.60)(.30) = .18	.09	.39
			.23	1.00

X_2 Observed at First Stage; X_2' at Second Stage

θ_i	$P(\theta_i)$	$P(X_2X_2' \mid \theta_i)$	$P(\theta_i)P(X_2X_2' \mid \theta_i)$	$P(\theta_i \mid X_2X_2')$
θ_1	.50	(.40)(.30) = .12	.06	.22
θ_2	.50	(.60)(.70) = .42	.21	.78
			.27	1.00

best courses of action at the later stages in order to determine the best moves at earlier stages. Costs of obtaining information must be subtracted at each stage to ascertain whether to stop or continue development.

The evaluations of the portions of the decision tree diagram shown in Figures 16-1 and 16-2 are given in Figures 16-3 and 16-4, respectively. All payoff figures are in millions of dollars. The posterior probabilities assigned to the events θ_1 and θ_2 are the appropriate figures from Tables 16-3 and 16-4, depending on whether post-first-stage or post-second-stage probabilities are relevant. The marginal probabilities of the information indications also come from Tables 16-3 and 16-4. For example, the probabilities assigned to the observing of X_1' and X_2' information at the second stage after X_1 has been observed in first-stage development are $P(X_1') = .54$ and $P(X_2') = .46$, respectively. As can be seen from Table 16-4, these are marginal probabilities of the two types of information, computed after allowance has been made for the revision of prior probabilities in the light of information obtained at the first stage.

Expected values are computed in the usual way, using the appropriate marginal and posterior probabilities to weight payoffs. As can be seen from Figures 16-3 and 16-4, the decision maker would proceed as follows. If favorable information is obtained from first-stage development, that is, X_1 is observed, he should stop further development and take action A_1, that is,

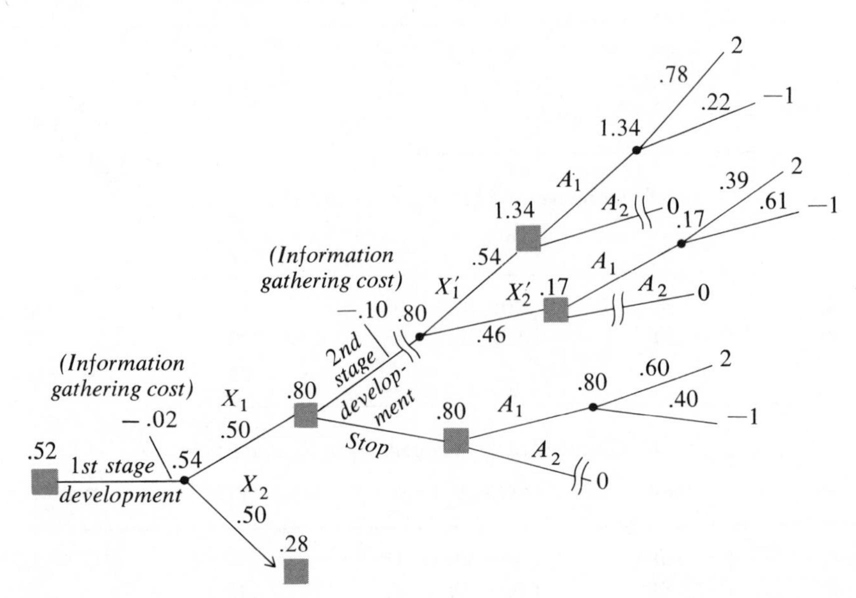

Figure 16-3 Evaluation of tree diagram depicted in Figure 16.1 (all payoffs are in millions of dollars).

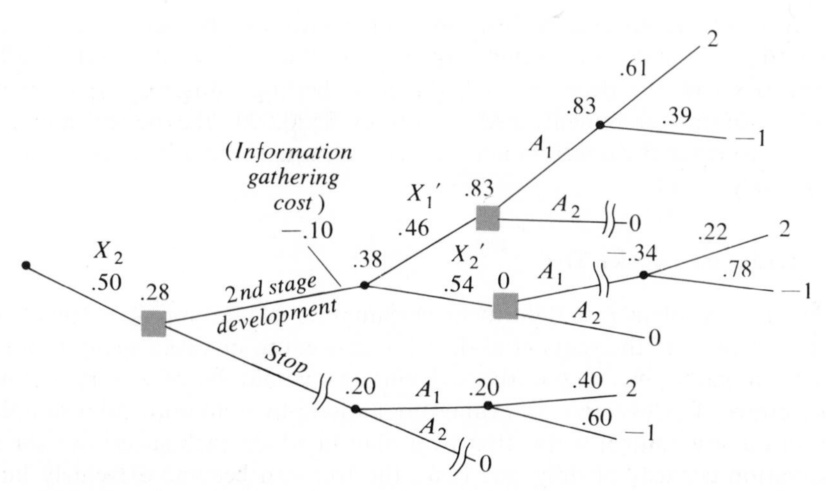

Figure 16-4 Evaluation of tree diagram depicted in Fig. 16.2 (all payoffs are in millions of dollars).

introduce the product (See Figure 16-3). The expected payoff of this course of action is \$.80 million. On the other hand, if X_2 is observed on the first stage, he should proceed to second-stage development. If a favorable indication results from second-stage development, that is, X_1' is observed, he should take action A_1 and introduce the product. However, if an unfavorable indication results, that is, X_2', then the product should not be introduced. The expected payoff of proceeding into second-stage development, if X_2 is observed at the first stage, is \$280,000. This figure is obtained by subtracting \$100,000, the cost of gathering information by engaging in second-stage development from \$380,000, the gross expected payoff of proceeding into that development stage. (See Figure 16-4.)

Weighting the expected payoffs of taking the best action after observing type X_1 and X_2 information (\$.80 million and \$280,000, respectively) by the marginal probabilities of observing these types of information (.50 and .50) yields \$540,000 as the gross expected payoff of proceeding into first-stage development. Subtracting \$20,000, the cost of gathering information by pursuing first-stage development yields an expected payoff of \$520,000 for entering first-stage development and selecting optimal courses of action thereafter. Since we assumed at the outset that the decision was made to enter first-stage development, the preceding analysis indicates the expected payoff associated with that course of action and the optimal moves in the sequential decision process thereafter.

However, suppose we wished to compare the advisability of entering the development process with the alternative of proceeding directly to commercialization of the product without any information gathering. Returning to Table 16-1, we find an expected payoff of $500,000 if action is taken solely on the basis of the decision maker's prior betting odds. Comparing this payoff with the aforementioned payoff of $520,000, the better course of action is to enter the development process, although the difference in payoffs is relatively small.

Truncating the Decision Tree

In the problem just described, the number of development stages was limited to two. If the sequential decision procedure involved many more information gathering stages, the calculations would become very detailed and tedious. For example, in evaluating an item-by-item sequential sampling plan in quality control work, that is, a plan in which each successive sample observation consists of only one item, the tree can become extremely large. The general principle for determining when to end the sampling procedure is to stop wherever the cost of an additional observation exceeds the opportunity loss of optimal terminal action. Sometimes the use of computers may be required to carry out the computations. Whether or not computers are used, it is doubtlessly desirable to truncate or cut the tree. The probability of reaching some of the higher level stages becomes very small, particularly if the number of stages is large. Most of the tree will have already been terminated at lower levels. Guesses can be made at the payoffs pertaining to the remaining high-level positions, and the tree may thus be truncated at arbitrary levels. Sensitivity analysis can also be used in which widely different payoffs (or opportunity losses) may be assumed for some of the remaining high-level positions. It will probably be observed that the effect of these widely different assumptions on calculated expected terminal payoffs is negligible. Hence, a substantial reduction in calculations may be effected at virtually no loss in accuracy.

General Remarks

The sequential decision procedures discussed in this chapter did not really introduce any new principles. However, they represent a powerful framework of analysis for determining what decisions should be made and when they should be made. This point is perhaps most clearly noted in the case of multi-stage business decision problems where a delay in decision is a definite alternative course of action. For example, if "delay one year" is one of the acts to be considered, it is evaluated in the light of all relevant future decisions. That is, the expected payoff assigned to "delay one year" takes into account the relationship between this decision and the future decisions which stem from it. Curiously, as we have seen, actions are planned in an optimal fashion from any point forward by a backward induction technique of problem solution.

Although in this chapter, a new product development problem has been used as an example, Bayesian sequential decision procedures are clearly applicable to a wide variety of situations. Wherever sequential sampling or information gathering is involved, as in quality control inspection of product, market research surveys, pilot investigations of various types, etc., these techniques may be effectively employed.

PROBLEMS

1. Explain the difference between a single-stage decision problem and a sequential decision problem.
2. Construct and solve a realistic example of a two-stage decision problem.
3. Let two possible states of nature be θ_1; stock market prices will advance and θ_2, stock market prices will stay at the same level or will decrease. Your prior belief is that there is a 60% chance of an advance. In a week the results of Econometric Model *A* will be published. Either an advance or no advance will be indicated. The model is correct 80% of the time. Likewise, Econometric Model *B* will be published one week after Model *A* and it is also correct 80% of the time. Suppose both models indicate that stock market prices will advance. Compute posterior probabilities which incorporate information from both econometric models.
4. The management committee of the H & B Company is contemplating the addition of a plant extension. The total costs will be $700,000 but if demand for the company's products is higher than average, incremental profit will be $1,800,000. If demand is not higher than average, revenues will remain unchanged. There is a 50% chance that demand will be higher than average so the committee has determined that at least one sample test of the market should also be taken at a cost of $90,000. Information from the first test sample is 60% reliable. This means that if the demand is high, the probability is .60 that information from the first test sample will indicate a high demand. Information from the second test sample is 80% reliable in the same sense. Draw a decision tree for this problem, and make all of the necessary entries on the tree. There are only two states of nature; demand will be higher than average or not higher than average.
5. The "Stop and Eat" Corporation is a chain of drive-in restaurants. A vice-president of the chain is considering whether to locate a new restaurant on Route 22 just outside of Altoona, Pa. He feels that if this particular restaurant is successful, the company can make a profit over the planning period of $2,000,000, but if it is not successful, it will lose $700,000. The probabilities

of these occurrences are .4 and .6, respectively. It is known that a sample of the demand in the area must be taken but whether one sample or two sequential samples should be taken is still undecided. The reliability of the first sample is .65 and of the second sample .80. The cost of the first sample is $70,000 and the second is $100,000. Draw a tree diagram for this problem and make all the necessary entries on the tree. Interpret the results of the diagram.

6. The ABC Investment Corporation wishes to undertake a direct investment on the island of Atlantis. In order to appraise the situation, the company feels that some research must be carried out. The research will yield one of two indications: X_1, favorable for investment and X_2, unfavorable for investment. Prior probabilities had been determined to be .4 that the country would not sink into the ocean and .6 that the country will sink. If it invests, the ABC Corporation will either earn $10 million or lose $5 million depending on whether the island sinks. The reliability of the research is .60 and will cost $500,000. The president of the corporation feels that more extensive research should be conducted after the results of the earlier research are determined. This extra extensive research will cost an additional $700,000, but its reliability is .90. Draw a decision tree for this problem and determine when it is advisable for the company to undertake the extra extensive research.

7. A mutual fund is contemplating the sale of a certain stock. A study can be made which would yield two possible results, X_1 or X_2, where X_1 indicates that the stock price will increase over a planning period and X_2 indicates the price will go down or remain the same.

State of Nature	*Prior Probabilities* $P(\theta_i)$	*Conditional Probabilities* $P(X_1 \mid \theta_i)$	$P(X_2 \mid \theta_i)$
θ_1: Stock price up	.4	.6	.4
θ_2: Stock price down or same	.6	.1	.9

Opportunity Loss Table
(*Units of* $10,000)

	θ_1	θ_2
Sell	7	0
Don't Sell	0	7

A second study can also be made which has a reliability of .7. The mutual fund has decided to make the first survey which costs $4000 but, as yet, is undecided about the second study which costs $12,000. Using a tree diagram, determine whether it is advisable to carry out the second study. Also, determine whether carrying out any kind of study is better than not carrying out any at all.

8. In Problem 6 of Chapter 15 suppose that a second survey is also possible. The reliability of the second survey is .65 and will cost .80 utiles. If the cost of the first survey is .40 utiles, draw a decision tree diagram showing all possible occurrences with respective probabilities and expected utility gains or losses.

9. Assume that in Problem 8 of Chapter 15 a second investigation costing \$50,000 can also be conducted. There are three outcomes: X_1, Y_1, and Z_1. The probabilities of such outcomes given the respective states of nature are

	X_1	Y_1	Z_1
S_1	.90	.05	.05
S_2	.10	.80	.10
S_3	.05	.10	.85

Draw a decision tree diagram and make all the necessary entries.

10. A home applicance manufacturing firm is considering the marketing of a new product. Before marketing a new product, the firm sends out mail questionnaires to its panel members consisting largely of housewives selected at random. The survey results are then classified into three categories: favorable, uncertain, and unfavorable. After the first survey is completed, a second survey is undertaken a couple of months later using non-panel members selected at random. Based upon either the first or both survey results, the firm decides whether or not to market the new product in question.

In detail, the situation is the following. There are two possible states of nature. These are

Q_1: New product is a success

Q_2: New product is a failure

There are two available actions. They are

D_1: Market a new product

D_2: Do not market the new product

The firm's opportunity losses (in units of \$10,000) due to incorrect action are assumed as follows:

	D_1 (*Market*)	D_2 (*Do Not Market*)
Q_1 (Success)	\$ 0	\$100
Q_2 (Failure)	120	0

The prior probabilities of success and failure are .25 and .75, respectively, and the conditional probabilities of observing (X_1) favorable; (X_2) uncertain; and

(X_3) unfavorable survey results on the first sample when Q is the true state of nature are shown in the following table:

	X_1	X_2	X_3
Q_1	.72	.16	.12
Q_2	.08	.12	.80

A similar table for the second sample is

	X_1	X_2	X_3
Q_1	.82	.10	.08
Q_2	.04	.06	.90

Each sample costs $75,000.

Draw a decision tree for the problem. Make all necessary entries and interpret the results.

CHAPTER SEVENTEEN

Comparison of Classical Statistical Inference and Bayesian Decision Theory

17.1 Introduction

Topics in classical statistical inference and Bayesian decision theory have been discussed in preceding chapters of this text. The two main classes of problems treated in statistical inference are hypothesis testing and confidence interval estimation. Bayesian decision theory is concerned with methods for choosing among alternative courses of action under conditions of uncertainty. Although the terminology of statistical inference and Bayesian decision theory differ, there are many similarities in the structure of the problems to which they address themselves and in their methods of analysis. However, there are important differences particularly in their methods of analysis, which are a matter of continuing discussion and debate. It is the purpose of this chapter to compare some aspects of these two important fields of statistical analysis. In Sections 17.2 and 17.3, we consider illustrative problems which present a comparison of classical hypothesis testing methods and Bayesian decision theory. In Section 17.4, we turn to a comparison of classical and Bayesian estimation procedures. In conclusion, in Section 17.5, we present some general comments on the areas of common ground and differences between these two schools of thought.

To introduce the comparison, let us consider a standard hypothesis testing problem. Suppose we wish to test the null hypothesis, $H_0: p \leq p_0$, where p_0 is a known or hypothesized population proportion, against the alternative hypothesis, $H_1: p > p_0$. For example, we might test the hypothesis that the proportion of defectives in a shipment of manufactured product, p, is equal to or less than .03 against the alternative hypothesis that p is greater than .03. Using classical hypothesis testing methods, we could design a decision rule, which would tell us whether to accept or to reject the null hypothesis on the basis of a random sample drawn from the shipment. We would fix α, the desired maximum probability of making a Type I error, and through the use of the power curve we could determine the risks of making Type II errors for values of p within the alternative hypothesis H_1. Table 17-1 summarizes the relationship between actions concerning these hypotheses and the truth or falsity of the hypotheses. For convenience, the table is given in terms of the null hypothesis, H_0. However, it is understood that when H_0 is true, H_1 is false and when H_0 is false, H_1 is true. As earlier, we refer to the truth or falsity of H_0 as the prevailing "state of nature." Also, as indicated in the column headings of the table, the symbols a_1 and a_2 denote the actions "accept H_0" and "reject H_0," respectively.

Table 17-1 Relationships Between Actions Concerning a Null Hypothesis and the Truth or Falsity of the Hypothesis.

State of Nature	*Action Concerning the Null Hypothesis* a_1: *Accept* H_0	a_2: *Reject* H_0
H_0 is True	No Error	Type I Error
H_0 is False	Type II Error	No Error

We see that the structure of this hypothesis testing problem includes (1) states of nature representing the truth or falsity of the null hypothesis, (2) actions a_1 and a_2 which accept or reject the null hypothesis, and (3) sample or experimental data which, when examined in the light of a decision rule, lead to one of the actions indicated under (2).

Let us rephrase the example in terms of Bayesian decision theory. We are dealing with a two-action problem involving acts a_1 and a_2, where the states of nature are the possible proportion of defectives, p. Although p varies along a continuum, and hence may be considered as a continuous random variable, we assume, for comparative purposes, that only two states of nature are distinguished, $\theta_1 : p \leq .03$ and $\theta_2 : p > .03$. Hence, these two states of nature, θ_1 and θ_2, correspond to those of the classical hypothesis testing problem, H_0 is true and H_0 is false, respectively. Finally, a random sample can be drawn from the shipment, and the observed sample or experimental data can be used to help decide between a_1 and a_2 as the better action. Therefore, the same three components of the structures of the decision theory problem are present as were discussed in the hypothesis testing problem, (1) states of nature, (2) alternative actions, and (3) experimental data which aid in the choice of actions. Furthermore, Table 17-2 is a payoff table for this problem in terms of opportunity losses and is similar to Table 17-1. The symbols $L(a_2 \mid \theta_1)$ and $L(a_1 \mid \theta_2)$ denote the opportunity loss of action a_2 given that θ_1 is the true state of nature and a_1 given that θ_2 is the true state of nature. The zeroes in the other two cells of the table indicate that there is no opportunity loss when the correct action is taken for the specified states of nature. Actually, payoffs would ordinarily be treated as a function of p and would vary with p. However, as indicated earlier, we have assumed in this discussion that only two states of nature are distinguished.

Table 17-2 Payoff Table in Terms of Opportunity Losses for the Two-Action Problem.

	Actions	
State of Nature	a_1	a_2
θ_1	0	$L(a_2 \mid \theta_1)$
θ_2	$L(a_1 \mid \theta_2)$	0

Now, let us consider the differences in the two approaches. In hypothesis testing, the choice of the significance level α establishes the decision rule and is thus the overriding feature in the choice between alternative actions. In symbols, $\alpha = P(a_2 \mid H_0$ is true). That is, α is the conditional probability of rejecting the null hypothesis given that it is true. Hence, a major criterion of choice among actions in hypothesis testing is the relative frequency of occurrence of this type of error. But how is α chosen? In many applications, conventional significance levels such as .05 and .01 are used uncritically with little or no thought being given to underlying considerations. However, it would be unfair to criticize a methodological approach simply because there are misuses of it. In classical statistics, the investigator is supposed to consider the relative seriousness of both Type I and Type II errors in establishing alternative hypotheses and significance levels at which these hypotheses are to be tested. Also, the investigator is aided by prior knowledge concerning the likelihood that H_0 and H_1 are true. For example, in the problem just discussed, why did the investigator not set up the null hypothesis as, say, $H_0: p \leq .001$ or $H_0: p \leq .60$? In this particular acceptance sampling problem he may know that two hypotheses such as these would be utterly ridiculous because of the extremely low and extremely high proportion of defectives implied. Hence, he may feel that it is virtually certain that the first of these hypotheses is false and that the second is true. Consequently, it would not be useful to set up the hypotheses in these forms.

Prior knowledge concerning the likelihood of truth of the competing hypotheses also helps the investigator in establishing the significance level. Hence, if he feels that it is very likely that the null hypothesis is true, he will tend to set α at a very low figure, in order to maintain a low probability of erroneously rejecting that hypothesis.

However, advocates of Bayesian decision theory criticize these classical hypothesis testing procedures for informality and for excessive reliance on unaided intuition and judgment. The Bayesians argue that their decision theory structure represents a logical extension of classical hypothesis testing, since it explicitly provides for the assignment of prior probability distributions to states of nature and incorporates losses into the formal structure of the problem. These decision theorists contend that losses are supposed to be considered in classical hypothesis testing in evaluating the relative seriousness of Type I and Type II errors, but ask, "How can they be considered if no explicit loss function is formulated?"

We turn now to an illustrative problem in an acceptance sampling setting which contains a comparison of the Bayesian approach with classical hypothesis testing. The problem demonstrates that if tests of hypotheses are conducted in the usual manner of establishing decision rules of rejecting or failing to reject hypotheses at pre-set levels of significance, non-optimal decisions may be made.

17.2 A Comparative Problem

Let us assume a situation in which a company inspects incoming lots of articles produced by a particular supplier. Acceptance sampling inspection is carried out to decide whether to accept or reject these incoming lots by selecting a single random sample of n articles from each lot. As in previous problems of this type, we will make the simplifying assumption that lots of only a few levels of proportions defective are produced, in this case, .02, .05, and .08 defective. On the basis of an analysis of past costs, the company constructed the payoff table in terms of opportunity losses depicted in Table 17-3. From long experience, the firm has determined that lots which contain .02 defectives are "good" and should be accepted. Hence, as indicated in Table 17-3, "accept" is the best act in the case of a .02 defective lot, and the opportunity loss in that case is $0. On the other hand, "reject" is the optimal act for lots which contain .05 and .08 defectives, and correspondingly the opportunity loss is $0 for such correct action.

On the basis of past performance, it has been determined that 50% of the supplier's lots are 2% defective, 25% are 5% defective, and 25% are 8% defective. In the absence of any further information, these past relative frequencies are adopted as the prior probabilities that such lots will be submitted by the supplier for acceptance or rejection.

Table 17-3 Payoff Matrix Showing Opportunity Losses for Actions of Acceptance and Rejection.

States of Nature (p = *lot proportion defective*)	*Prior Probability*	*Act* a_1 *Reject*	a_2 *Accept*
.02	.50	$200	$ 0
.05	.25	0	300
.08	.25	0	500
	1.00		

In order to compare the Bayesian decision theory and traditional hypothesis testing approaches, we will first carry out a study of possible single sampling plans (See p. 725) by extensive and normal form preposterior analyses. The result of these analyses will be a determination of the optimal sampling plan or strategy. Then, a hypothesis testing solution will be given, and a comparison will be made of the two approaches.

A decision tree diagram is given in Figure 17-1, beginning with the decision to sample and inspect n items. We move to branch point (b), where the results of the sampling inspection then determine which branch to follow. The possible results of sampling have been classified into three categories. The number of defectives, denoted X, may have been equal to or less than some number c_1, where $c_1 < n$. It may have been greater than c_1, but equal to or less than c_2, where $c_1 < c_2 < n$. Finally, the number of defectives may have been greater than c_2. These three types of results, for purposes of brevity, are

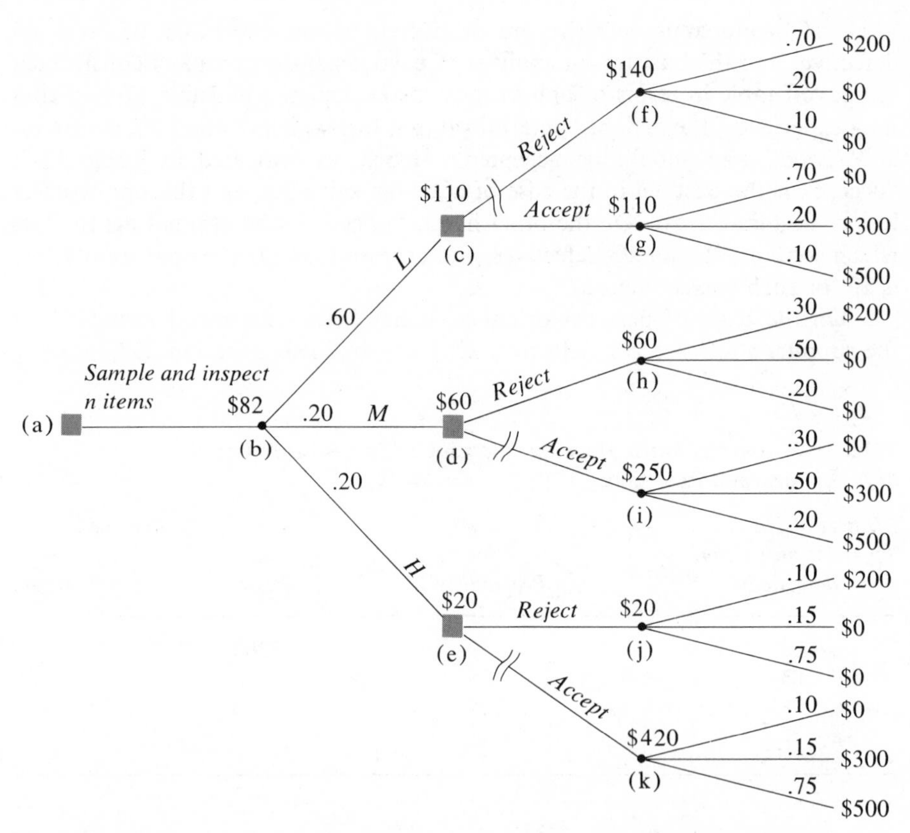

Figure 17-1 Decision tree diagram for the acceptance sampling problem.

referred to as Type L (low), Type M (middle), and Type H (high) information, respectively. In Table 17-4, a joint frequency distribution is given for sample results and states of nature. We will assume that these frequencies were derived from a large number of past observations and therefore may be taken to represent probability in the relative frequency sense.[1] For example, in the past, in .42 of the lots inspected, the number of defectives observed was c_1 or less, and the lots contained .02 defectives. In terms of marginal frequencies, Type L information ($X \leq c_1$) was observed in .60 of the lots, Type M ($c_1 < X \leq c_2$) was observed in .20 of the lots, and Type H ($X > c_2$) was observed in .20 of the lots.

[1]Alternatively, the conditional probabilities of sample results (likelihoods), derivable from this table may be thought of as having been calculated from an appropriate probability distribution such as the binomial or hypergeometric. However, the basic methodological discussion remains unchanged.

Returning to the decision tree, we observe the three branches representing L, M, and H types of information emanating from node (b) marked with their respective probabilities, .60, .20, and .20. We will give a brief explanation of the usual extensive form analysis, using Type L information as an example. If Type L information is observed, we move to branch point (c), where we can either accept or reject the lot. If we reject, we move to node (f); if we accept, to node (g). The probabilities shown on the three branches stemming from (f), .70, .20, and .10 are the posterior probabilities, given Type L information, that the lots contain .02, .05, and .08 defectives, respectively. The calculation of these posterior probabilities by Bayes' theorem is given in Table 17-5. These probabilities can also be derived from Table 17-4 by dividing joint probabilities by the appropriate marginal probabilities, for example, .70 = .42/.60.

We now use the standard backward induction technique (See Section 13.6) to obtain the expected opportunity loss of the optimal strategy. Looking forward from node (f) and using the posterior probabilities .70, .20, and .10 as weights attached to the three states of nature, .02, .05, and .08 defective lots,

Table 17-4 Joint Frequency Distribution of Sample Results and States of Nature.

	Sample Results			
States of Nature (p = lot proportion defective)	*Type L* $X \leq c_1$	*Type M* $c_1 < X \leq c_2$	*Type H* $X > c_2$	*Total*
.02	.42	.06	.02	.50
.05	.12	.10	.03	.25
.08	.06	.04	.15	.25
	.60	.20	.20	1.00

Table 17-5 Calculation of Posterior Probabilities of States of Nature Given.

Type L Information ($X \leq c_1$)

States of Nature (p = lot proportion defective)	*Prior Probability* $P_0(p)$	*Conditional Probability* $P(L \mid p)$	*Joint Probability* $P_0(p)P(L \mid p)$	*Posterior Probability* $P_1(p)$
.02	.50	.84	.42	.70
.05	.25	.48	.12	.20
.08	.25	.24	.06	.10
			.60	1.00

we obtain an expected opportunity loss of \$140 for the act "reject." Comparing this figure with the corresponding one of \$110 for "accept," we block off the action "reject" as being non-optimal. Therefore \$110 is carried down to node (c), representing the payoff for the optimal act upon observing Type L ($X \leq c_1$) information. Similar calculations yield \$60 and \$20 at (d) and (e) for Types M and H information. Weighting these three payoffs by the marginal probabilities of obtaining Types L, M, and H information, we obtain a loss of \$82 as the expected payoff of sampling and inspecting n items. The cost of sampling and inspection would then have to be added if, for example, we wished to make a comparison with the expected loss of terminal action without sampling. However, for our comparison of the Bayesian decision theory approach with hypothesis testing, we will focus attention on the \$82 figure, which has been entered at node (b). We note, in summary, that \$82 is the expected loss of the optimal strategy which accepts the lot if Type L information is observed and rejects it, otherwise.

We turn now to normal form analysis in which all possible decision rules or strategies will be considered as a means of commenting on traditional hypothesis testing procedures. There are eight possible strategies implicit in the decision tree diagram shown in Figure 17-1. These are enumerated in Table 17-6. An R denotes reject; an A denotes accept. Therefore, for example, strategy s_3 means accept the lot if Type L or Type M information is observed, that is, if c_2 or fewer defectives are found. Strategy s_4 signifies acceptance if c_1 or fewer defectives are observed. Therefore, a choice between strategies s_3 and s_4 means, in acceptance sampling terms, a selection between a single sampling plan with an acceptance number of c_2 and one with an acceptance number of c_1. The conclusion of the extensive form analysis was that s_4 is the optimal strategy, that is, s_4 has the minimum expected opportunity loss.

As in previous problems, certain of the decision rules do not make much sense. For example, strategy s_1 would reject the lot and strategy s_2 would accept the lot, regardless of the type of information revealed by the sample. Strategy s_6 would reject a lot if a small number of defectives ($X \leq c_1$) were

Table 17-6 Possible Decision Rules Based on Information Derived from Single Samples of Size n.

Sample Information	s_1	s_2	s_3	s_4	s_5	s_6	s_7	s_8
Type L ($X \leq c_1$)	R	A	A	A	A	R	R	R
Type M ($c_1 < X \leq c_2$)	R	A	A	R	R	A	R	A
Type H ($X > c_2$)	R	A	R	R	A	A	A	R

observed in the sample, but would accept the lot for larger numbers of defectives ($X > c_1$). The only strategies that appear to be at all logical are s_3 and s_4.

Now, let us suppose this problem had been approached from the standpoint of a null hypothesis testing procedure. The alternative hypotheses would be

$$H_0: p = .02$$

$$H_1: p = .05 \text{ or } .08$$

Acceptance or rejection of the null hypothesis, H_0, would mean acceptance or rejection of the lot, respectively. As indicated earlier, the company conducting the acceptance sampling wishes to accept lots which are .02 defective and to reject, otherwise. Hence, the rejection of a good lot, that is, one that contains .02 defectives, constitutes a Type I error. Let us assume the company decides to test the null hypothesis at a preselected .05 level of significance. That is, the company specifies that it wants to reject lots which contain .02 defectives no more than five times in 100. We will examine what this selection of $\alpha = .05$ implies concerning the choice of a decision rule.

Power curves may be plotted for each of the strategies or decision rules given in Table 17-6. However, the only ones we will show are for s_3 and s_4. As implied earlier, none of the other strategies are worthy of further consideration. The power curves are depicted in Figure 17-2.

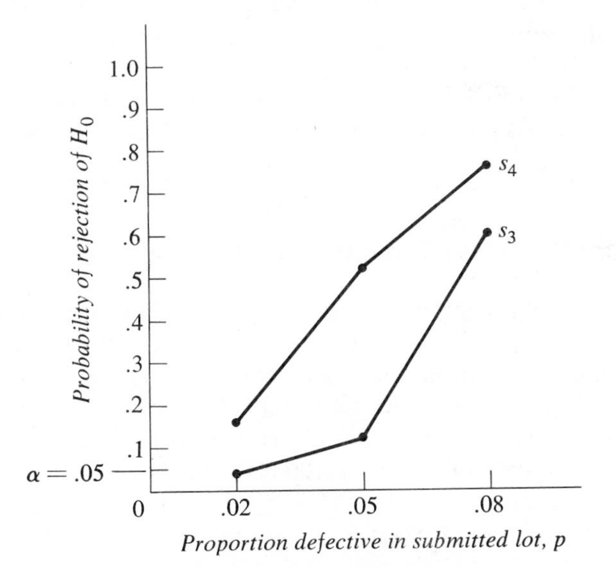

Figure 17-2 Power curves for strategies s_3 and s_4.

Let us consider how the power curves are plotted by taking the points for $p = .02$ as an example. Strategy s_3 accepts H_0 (accepts the lot), if Type L or Type M information is observed. From Table 17-4, we find that the conditional probability of observing Type L or Type M information, given $p = .02$, is $.42/.50 + .06/.50 = .96$. Therefore, the probability that H_0 will be accepted, given $p = .02$, is .96. Hence, the probability that H_0 will be rejected, given $p = .02$, is $1 - .96 = .04$. Symbolically, for strategy s_3, $P(\text{Rejection of } H_0 \mid p = .02) = .04$. Analogously, we find that for strategy s_4, $P(\text{Rejection of } H_0 \mid p = .02) = 1 - .42/.50 = .16$.

Now, if we impose the condition of a .05 significance level, that is, lots containing .02 defectives should be rejected no more than 5% of the time, we find that strategy s_3 meets this criterion but strategy s_4 does not! Therefore, traditional hypothesis testing procedures would require the use of strategy s_3, which has been shown to be non-optimal. Looking at Figure 17-1, we can see why this is so. Under strategy s_3, if Type M information is observed, the lot must be accepted, incurring an expected loss of \$250, whereas under strategy s_4, if Type M information is observed, the lot is rejected, with a loss of only \$60. In summary, the expected opportunity losses of the two strategies are

$$\text{EOL}(s_3) = (.60)(\$110) + (.20)(\$250) + (.20)(\$20) = \$120$$

$$\text{EOL}(s_4) = (.60)(\$110) + (.20)(\$60) + (.20)(\$20) = \$82$$

The major criticism of traditional hypothesis testing procedures implied by this example is that too much burden is placed on significance levels as a means of deciding between alternative acts. *Specifically, the inclusion of economic costs, or more generally, opportunity losses is not a standard procedure in the decision-making process.*

Another illustrative problem is given in the next section, which more thoroughly contrasts the two sets of procedures. This is followed by a general comparative discussion.

17.3 A More Complete Comparative Problem

As another illustration of the comparison between the classical and Bayesian approaches to decision making, we return to the two-action problem discussed earlier in Section 14.3. For convenience, we repeat some of the information in the problem. An importer wants to decide whether to market jars of a unique type of cocktail onion. He sells exclusively by mail order and his "market" consists of a list of 100,000 names and addresses. His two courses of action are

A_1: Market the product

A_2: Do not market the product

The symbol p denotes the proportion of the 100,000 potential customers on the list who will buy the onions if they are made available. The break-even value of p, denoted p^* has been determined to be .05. Hence, if less than .05 of the potential customers purchase the onions, the importer will lose money, at .05 he breaks even, if more than .05 purchase, he earns a profit. Table 14-13 showing his opportunity losses as a function of p is reproduced in Table 17-7.

We will again assume that the importer decides to test the marketability of his product by sending advertisements to 100 persons selected at random from his mailing list. Let us consider first of all the type of decision rule that would be applied if we tested the null hypothesis $H_0: p \leq .05$ with $\alpha = .05$, that is, at the conventional 5% level of significance against the alternative, $H_1: p > .05$. The standard error of the sample proportion is

$$\sigma_{\bar{p}} = \sqrt{\frac{pq}{n}} = \sqrt{\frac{(.05)(.95)}{100}} = .0218$$

Table 17-7 Opportunity Loss Table for the Importer's Problem.

	Opportunity Losses	
Events	*Market the Product* A_1	*Do Not Market the Product* A_2
$p \leq .05$	\$20,000 $(.05 - p)$	\$0
$p > .05$	\$0	\$20,000 $(p - .05)$

Using the normal distribution for a one-tailed test, we find a critical z value of 1.65 in Table A-5, Appendix A. That is, .05 of the total area in a normal curve lies to the right of a value 1.65 standard deviations above the mean. Hence, the critical $\bar{p}$ value, denoted $\bar{p}_c$, above which H_0 is rejected is

$$\bar{p}_c = .05 + 1.65(.0218) = .086$$

The normal sampling distribution and critical $\bar{p}$ value are depicted in Figure 17-3.

Thus, the importer's decision rule is as follows. Send the advertisements to the sample of 100 potential customers. If 8.6% or fewer persons purchase the product, accept H_0 and do not market the product. If more than 8.6% purchase, reject H_0 and market the product. It is convenient to work in terms of numbers of persons rather than proportion of persons. In numbers of persons, the critical value would be 8.6 persons. The same result would have been obtained if we had computed the standard error, $\sigma_{n\bar{p}} = \sqrt{npq}$, and had worked the problem entirely in terms of numbers of persons, rather than

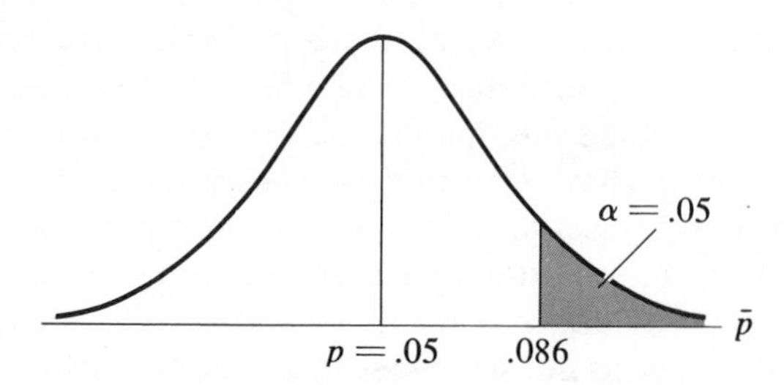

Figure 17-3 Sampling distribution for testing H_0: $p \leq .05$ at $\alpha = .05$.

proportions. Since the number of persons in the sample of 100 who purchase the product, denoted X, can only be a whole number, we will round off the critical value of 8.6 to 9. Of course, this has the effect of somewhat lowering the level of significance at which the test is performed. We can see this by observing in Figure 17-3 that if the critical value is moved to the right, a smaller area remains in the right-hand tail. We can now summarize the decision rule as follows:

(1) If $X \leq 9$, do not market the product
(2) If $X > 9$, market the product

We will summarize this decision rule even more succinctly by referring to it entirely by the notation, $X \leq 9$. That is, this one symbol will be understood to mean the decision rule given in (1) and (2). For purposes of comparison, we will consider the implications of this decision rule and two others, $X \leq 2$ and $X \leq 7$. Of course, any number of decision rules might conceivably be compared, but we are arbitrarily selecting the indicated three. Hence, in terms of the problem, we will compare the following strategies or decision rules. Advertisements for the product will be sent to 100 persons selected at random from the mailing list. If X, the number of buyers, is equal to or less than 2, 7, or 9, the decision will be market the product. Otherwise, do not market.

Classical Analysis

In classical hypothesis testing, we evaluate and choose among decision rules on the basis of probabilities of error. As we have seen, the power curve summarizes these probabilities for every value the population parameter, in this case p, can take on. The power curves for the three decision rules are shown in the three upper panels of Figure 17-4. The probabilities of rejecting the null hypothesis of this problem using the three decision rules are given in

Table 17-8 for a few selected values of p. These are the p values over which the prior probability distribution was defined in the original problem discussed in Chapter 14.

It would be useful in the ensuing discussion for the reader to fix firmly in his mind that the null hypothesis is true for $p \leq .05$ and false for $p > .05$. As can be seen from the power curves in Figure 17-4 and the figures in Table 17-8, the probabilities of *rejecting* the null hypothesis when $p < .05$ are greater for decision rule $X \leq 2$ than for $X \leq 7$ and those for $X \leq 7$ are greater than for $X \leq 9$ (for the same p values). Hence, the probabilities of committing Type I errors, that is, of rejecting the null hypothesis when it is true, are greatest for the rule $X \leq 2$, smaller for $X \leq 7$, and smallest for $X \leq 9$. On the other hand, the probabilities of *accepting* the null hypothesis when $p > .05$ are least for decision rule $X \leq 2$, larger for $X \leq 7$, and largest for $X \leq 9$. Hence, the probabilities of committing Type II errors, that is, of accepting

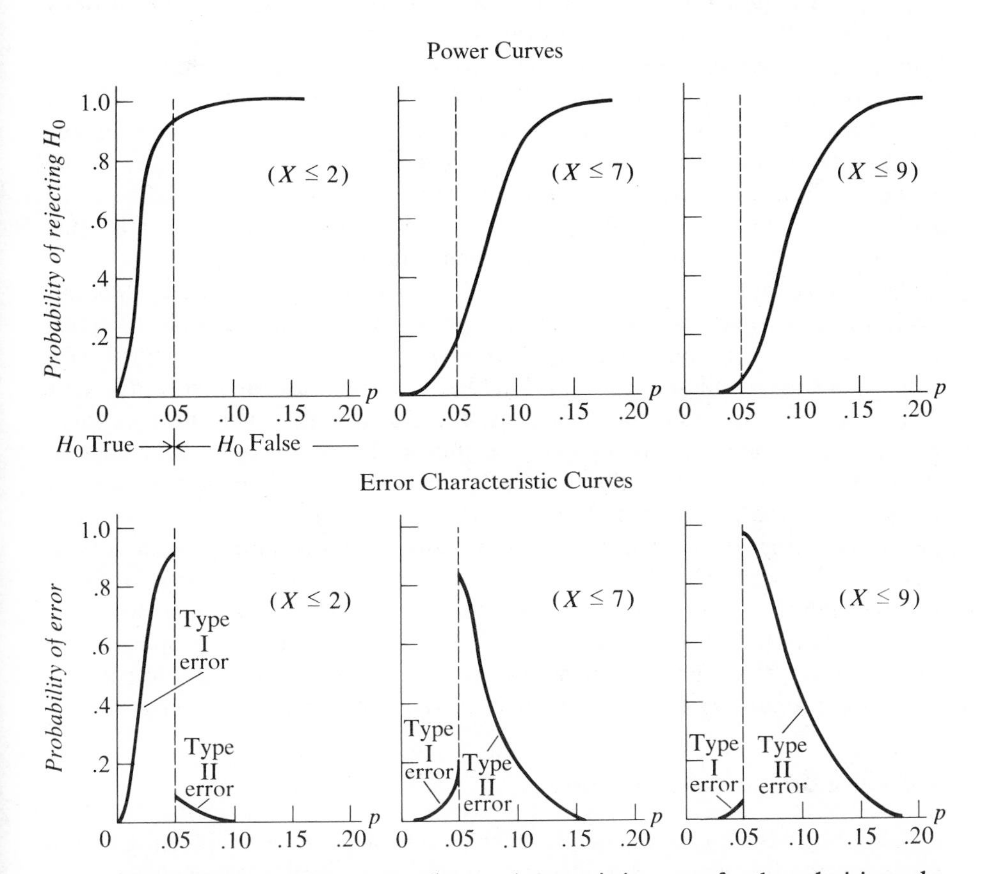

Figure 17-4 Power curves and error characteristic curves for three decision rules.

Table 17-8 Probability of Rejecting the Null Hypothesis, $H_0: p \leq .05$ in the Importer's Problem for Three Decision Rules.

p	$X \leq 2$ P(Reject H_0)	$X \leq 7$ P(Reject H_0)	$X \leq 9$ P(Reject H_0)
.02	.5000	.0002	.0000
.04	.8461	.0571	.0054
.06	.9545	.3372	.1020
.08	.9864	.6443	.3557

the null hypothesis when it is false, are least for the rule $X \leq 2$, larger for $X \leq 7$, and largest for $X \leq 9$. It is easiest to see these relationships visually on the so-called "error characteristic curve," which is a plot of the Type I and Type II errors. The error characteristic curves for the three decision rules are shown in the lower panels of Figure 17-4. It may be noted that the error characteristic curve has the same ordinates as the power curve in the region where the null hypothesis is true, but shows the complements of the power curve in the region where the null hypothesis is false. Thus, the error characteristic curve shows the probabilities of rejecting H_0 when it is true, and the probabilities of accepting H_0 when it is false.

We can now see how difficult it is in practice to choose among different decision rules by formally considering only conditional probabilities of error, as in classical hypothesis testing. For example, rule $X \leq 2$ is worst with respect to the commission of Type I errors, but best with the respect to Type II errors. The decision maker is supposed to take the seriousness of these errors into account but as we know in this problem, the opportunity losses increase as p departs from the break-even value $p^* = .05$. Bayesian decision theorists ask the pointed question, "How can the decision maker process the necessary information on a judgmental basis if he does not have a formal structure for including these losses?" We turn now to a Bayesian decision theory analysis of this problem, which takes into account not only probabilities of error but the consequences of these errors in terms of opportunity losses. Also, as we have seen previously, the Bayesian approach introduces the prior probability distribution of occurrence of states of nature.

Bayesian Decision Theory Approach

We begin the Bayesian analysis of this decision problem by plotting on the same charts the error characteristic curves previously obtained and the opportunity loss functions. These charts are depicted in the upper three panels of Figure 17-5. The opportunity loss function shows the loss of incorrect action as a function of p, the population proportion that will purchase the product if it is made available. The opportunity loss function is shown in tabular form in Table 17-7 and was previously graphed in Figure 14-3. The scale for opportunity losses is indicated on the right-hand side of the three upper panels of Figure 17-5. On each of these panels, the opportunity loss at $p^* = .05$ is zero. The line to the left of $p^* = .05$ depicts the opportunity losses of making Type I errors, or in terms of the problem, the opportunity losses of marketing the product when it should not be marketed. The line to the right of $p^* = .05$ shows the opportunity losses of making Type II errors, or in terms of the problem, the opportunity losses of not marketing the product when it should be.

As can be seen in the upper panels of Figure 17-5, for p values *close* to the break-even figure of .05, the probability of error is high, but the opportunity loss is low. Therefore, the "risk" or conditional expected (opportunity) loss tends to be *low*. As observed in Section 15.3, the risk or conditional expected loss is obtained by multiplying the opportunity loss of taking the

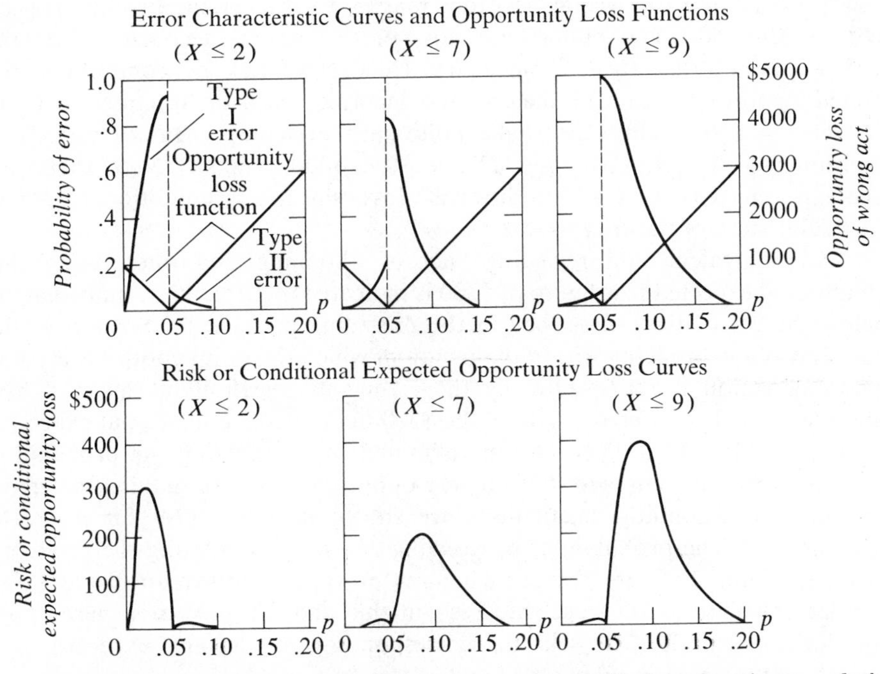

Figure 17-5 Error characteristic curves, opportunity loss functions, and risk curves for three decision rules.

wrong act by the probability of taking the wrong act. For p values *far away* from the break-even figure, the probability of error is low, but the opportunity loss is high. Again, the multiplication tends to result in a relatively *low* conditional expected loss. On the other hand, for the *intervening* values of p, neither close to nor far removed from the break-even level, the probability of error is relatively high, and the opportunity loss of incorrect action is high. Hence, the conditional expected opportunity losses for these intermediate values of p tend to be relatively *high*. The curves for these conditional expected opportunity losses (or "risk curves") are plotted in the lower three panels of Figure 17-5 for the three decision rules. As can be seen, these curves exhibit shapes implied by the preceding discussion.

It is instructive to consider the calculation of a few of the conditional expected loss figures. The computations are shown in Table 17-9 for the decision rule $X \leq 2$, for p values of .02, .04, .06, and .08. Let us consider the case in which $p = .02$. The probability that two or fewer persons ($X \leq 2$) in the sample of 100 who received advertisements will purchase the cocktail onions is .50. Hence the probability of making the correct decision, not to market the product (accept H_0), is .50. Since the opportunity loss of making the correct decision is \$0, the risk or conditional expected opportunity loss of making the correct decision is (.50)(\$0) = \$0. Correspondingly, the probability of making the wrong decision, that is, to market the product (reject H_0), is also .50. The opportunity loss of this wrong decision is \$20,000 (.05 − .02) = \$600. (See Table 17-7) Hence, the risk or conditional expected opportunity loss of making the incorrect decision is equal to (.50) × (\$600) = \$300. Thus, the total conditional expected opportunity loss of the decision rule $X \leq 2$, given $p = .02$, is \$0 + \$300 = \$300. Note that this figure is referred to as a "conditional" loss, since it is conditional on the particular state of nature $p = .02$.

An equivalent and revealing way of obtaining and interpreting the conditional expected loss figure of \$300 is indicated in Table 17-9, immediately below the calculations described in the preceding paragraph. Since $p = .02$ and thus is a figure lying within the region in which the null hypothesis is true, incorrect action is represented by the erroneous rejection of this null hypothesis, or a Type I error. Hence the \$300 figure is the conditional expected opportunity loss of a Type I error, obtained by multiplying the probability of committing such an error (.50) by the opportunity loss of making the error (\$600). Corresponding calculations are shown in Table 17-9 for $p = .04$, .06, and .08. The probabilities of rejecting H_0 (or of marketing the product) given in Table 17-9 are the same figures previously shown in Table 17-8. Similar calculations to those discussed in this and the preceding paragraph for decision rule $X \leq 2$ yield the corresponding conditional expected loss figures for the decision rules $X \leq 7$ and $X \leq 9$.

Risk or conditional expected opportunity loss curves for the three de-

Table 17-9 Risks or Conditional Expected Opportunity Losses for Decision Rule $X \leq 2$.

Given p = .02

Decision	*Probability of Decision*	*Opportunity Loss*	*Risk or Conditional Expected Opportunity Loss*
Do Not Market (Accept Ho)	.50	$ 0	$ 0
Market (Reject Ho)	.50	600	300
			$300

Conditional EOL (Type I Error) = *P*(Type I Error)(Loss of Type I Error)
= (.50)($600) = $300

Given p = .04

Decision	*Probability of Decision*	*Opportunity Loss*	*Risk or Conditional Expected Opportunity Loss*
Do Not Market (Accept Ho)	.1539	$ 0	$
Market (Reject Ho)	.8461	200	169.22
			$169.22

Conditional EOL (Type I Error) = *P*(Type I Error)(Loss of Type I Error)
= (.8461)($200) = $169.22

Given p = .06

Decision	*Probability of Decision*	*Opportunity Loss*	*Risk or Conditional Expected Opportunity Loss*
Do Not Market (Accept Ho)	.0455	$200	$9.10
Market (Reject Ho)	.9545	0	0
			$9.10

Conditional EOL (Type II Error) = *P*(Type II Error)(Loss of Type II Error)
= (.0455)($200) = $9.10

Given p = .08

Decision	*Probability of Decision*	*Opportunity Loss*	*Risk or Conditional Expected Opportunity Loss*
Do Not Market (Accept Ho)	.0136	$600	$8.16
Market (Reject Ho)	.9864	0	0
			$8.16

Conditional EOL (Type II Error) = *P*(Type II Error)(Loss of Type II Error)
= (.0136)($600) = $8.16

cision rules are graphed in the lower three panels of Figure 17-5. As can be seen from the chart, the conditional expected losses for p values less than .05 (for Type I errors) are generally greater for decision rule $X \leq 2$ than for the other two strategies, whereas the conditional expected losses for p values greater than .05 (for Type II errors) are generally greater for decision rule $X \leq 9$ than for the other two strategies. We recall that decision rule $X \leq 9$ is the one that corresponds to the conventional hypothesis test at the .05 level of significance. As can be seen from the risk curve for that decision rule, the conditional expected opportunity losses of Type II errors for certain values of p are enormously larger than losses for Type I errors. It is this sort of combined effect of probabilities of error and opportunity losses that is difficult to take into account in setting significance levels in classical hypothesis testing.

To recapitulate, we now have three decision rules for which we have calculated conditional expected opportunity losses for each state of nature, p. In order to evaluate these decision rules, we must now take into account the probability of occurrence of the states of nature. Using the prior probabilities of the p values as weights by which to multiply the conditional expected opportunity losses, we can obtain unconditional expected opportunity losses. The best decision rule is the one with the lowest total unconditional expected opportunity loss. The calculation of these unconditional expected opportunity losses for the three decision rules is given in Table 17-10.

As can be observed from Table 17-10, the decision rule, among the three considered, which yields the smallest (unconditional) expected opportunity loss is $X \leq 7$. In fact, the expected loss associated with the rule $X \leq 7$ ($74.90) is about 25% less than the corresponding figure for the rule associated with $X \leq 9$, or a .05 significance level ($100.58). Hence, in terms of the problem, the importer is better advised to market the jars of cocktail onions if he observes more than seven purchasers in his sample of 100 than if he observes more than nine such purchasers. Again, we observe, as in the problem in the

Table 17-10 Unconditional Expected Opportunity Losses for Three Decision Rules.

State of Nature	*Prior Probabilities*	*Conditional Expected Opportunity Losses*			*Unconditional Expected Opportunity Losses*		
p	$P_0(p)$	$X \leq 2$	$X \leq 7$	$X \leq 9$	$X \leq 2$	$X \leq 7$	$X \leq 9$
.02	.10	$300.00	$ .12	$ 0.00	$ 30.00	$ .01	$ 0.00
.04	.40	169.22	11.42	1.08	67.69	4.57	.43
.05	.00	0	0	0	0	0	0
.06	.45	9.10	132.56	179.60	4.10	59.65	80.82
.08	.05	8.16	213.42	386.58	.41	10.67	19.33
	1.00				$102.20	$74.90	$100.58

preceding section, that non-optimal decisions can be made using classical hypothesis testing methods which do not take into account loss functions and prior probability distributions of states of nature.

A few points are worthy of note in the preceding problem. First, the prior probability distribution involved only a few states of nature (p values). There could have been a much larger number. The basic principles are the same. Only the tediousness of arithmetic would have been greater with more p values. Second, it may be observed also that p was treated as a *discrete* random variable. Methods also exist for handling the case in which the basic random variable is continuous. Third, as in many previous problems, we used minimization of expected opportunity loss expressed in *monetary values* as the criterion for the selection of the best course of action. As always, the corresponding maximization of expected *utility* is the conceptually desirable procedure. The use of expected monetary values implies a linear utility function for money over the range of values considered. Fourth, only three decision rules were considered. Conceptually, all possible decision rules for the predetermined sample size of 100 could have been considered. In practice, most of the rules for large values of X could be dismissed from consideration very rapidly. A few calculations would demonstrate that such rules would have large (unconditional) expected losses. For problems involving large numbers of possible decision rules, computer computations may be required. Fifth, in this problem the sample size was predetermined. As we have seen in earlier chapters (See Section 15.7), Bayesian decision theory has a systematic approach to the question, "Should we sample at all, and if so, how much?" On the other hand, classical procedures do not really have any generally satisfactory method of answering that question. In practice, for a sample of a given size, the same classical hypothesis testing procedures which we considered in Chapter 7 are used to set up decision rules after a sample has been drawn as in establishing a rule prior to the selection of a sample. Classical procedures also have no real practical method for deciding the optimal sample size. Finally, in this problem, a linear opportunity loss function was assumed. If a nonlinear function were assumed, the same general double-humped curve (referred to by some Bayesian decision theorists as the "butterfly curve") for conditional expected opportunity loss would have been observed, provided only that opportunity losses increase as p departs farther and farther from the break-even figure p^*.

17.4 Classical and Bayesian Estimation

In the preceding two sections, comparisons were made between hypothesis testing procedures in classical statistical inference and the corresponding approaches in Bayesian decision theory. In this section, a comparison is made between the estimation techniques in the two approaches.

In Chapter 8, a brief description was given of classical point estimation techniques, that is, methods in which a population parameter value is estimated by a single statistic computed from the observations in a sample. For example, the mean of a sample may be used as the best single estimate of a population mean. In most practical problems, it is not sufficient to have merely a point estimate. If we were given two different point estimates of a population parameter, and no further information, we could not distinguish the degree of reliability to be placed upon these estimates. Yet one estimate might be based on a sample of size 10,000 and the other on a sample of size ten. Clearly, these estimates differ greatly in reliability. As we have seen, traditional statistics handles the problem of indicating reliability by the use of the confidence interval procedure. In this section we will compare this classical technique to the corresponding approach of Bayesian statistics. However, before making this comparison, we will pause for a comment on point estimation techniques in the two approaches.

Point Estimation

In Sections 8.2 and 8.3, criteria of goodness of estimation and the classical method of maximum likelihood estimation were discussed. We have become familiar with point estimators which derive from these methods, such as the observed sample proportion of successes $\bar{p}$ in a Bernoulli process, which is used as the estimator of the population parameter p, and the observed sample mean $\bar{x}$ in a process described by the normal distribution, which is used as the estimator of the population mean μ_x.

Bayesian decision theory takes a different approach to the problem of point estimation. It views estimation as a straightforward problem of decision making. The estimator is the decision rule, the estimate is the action, and the possible values that the population parameter can assume are the states of nature. For example, the sample mean $\bar{x}$ might be the estimator (decision rule), 10.6 might be the estimate (action), and the possible values that the population mean μ_x can assume are the parameter values (states of nature). In this formulation, the unknown population parameter is treated as a random variable.

To clarify the method, we will introduce some notation. Let θ be the true value of the parameter we want to estimate and $\hat{\theta}$ the estimate or action. Then, a loss is involved if the value of $\hat{\theta}$ differs from θ and the amount of the loss is some function of the difference between $\hat{\theta}$ and θ. Hence, two possible loss functions might be

(17.1) $$L(\hat{\theta}; \theta) = |\hat{\theta} - \theta|$$

and

(17.2) $$L(\hat{\theta}; \theta)^2 = (\hat{\theta} - \theta)^2$$

where $L(\hat{\theta}\,;\theta)$ is the loss involved in estimating (taking action) $\hat{\theta}$ when the parameter value (state of nature) is θ.

Somewhat more generally, the loss functions (17.1) and (17.2) may be written as

(17.3) $$L(\hat{\theta};\theta) = k(\theta)|\hat{\theta} - \theta|$$

and

(17.4) $$L(\hat{\theta};\theta) = k(\theta)(\hat{\theta} - \theta)^2$$

respectively, where $k(\theta)$ is a constant for a particular value of θ. This constant may be in money units, utility units, and so forth. For simplicity, in the ensuing discussion, we will assume $k(\theta) = 1$ utile. Therefore, we are dealing with functions of the form of (17.1) and (17.2), and the losses are given in units of utility.

Expression (17.1) is referred to as a linear loss function; (17.2) as a quadratic loss function (also squared error loss function). The nature of these functions can be illustrated by simple examples. Assume the true value of the parameter θ is 10. Suppose we consider the losses involved if we estimate this parameter incorrectly as $\hat{\theta} = 11$ and $\hat{\theta} = 12$. For these two estimates, the linear loss function, (17.1), is, respectively,

$$L(11;10) = |11 - 10| = 1$$

and

$$L(12;10) = |12 - 10| = 2$$

On the other hand, the quadratic loss function, (17.2), is equal to

$$L(11;10) = (11 - 10)^2 = 1$$

and

$$L(12;10) = (12 - 10)^2 = 4$$

In other words, in the linear case the loss in overestimating by two units is *twice* as much as in overestimating by one unit. In the quadratic case, the loss in overestimating by two units is *four* times as much as in overestimating by one unit. It may be noted that in both functions, an underestimate of a given size, say two units, is equally as serious as an overestimate of the same size. Such loss functions are referred to as symmetrical.

The ideas of the two aforementioned loss functions were referred to earlier in Section 3.18 in somewhat different forms. There, we were concerned with guessing the value of an observation selected at random from a frequency distribution. The penalty of an incorrect guess or estimate was referred to as the "cost of error." That "cost" corresponds to "loss" in the present discussion. It was pointed out that if the cost of error varies directly with the size of error regardless of sign (the linear loss function, (17.1)), the median is the "best guess," since it minimizes average absolute deviations. On the other hand, if the cost of error varies according to the square of the error (the quadratic loss function, (17.2)), the mean should be the estimated value, since the average of the squared deviations about it is less than around any other figure. It is of interest to note that least-squares methods of estimation in classical statistics assume a quadratic loss function, since they obtain estimates for which the average squared error is minimized. Whether or not this is an appropriate loss function for the particular problem involved is rarely investigated.

The Bayesian method of point estimation begins with the setting up of whatever loss function appears to be appropriate. Then these losses are used in the standard decision procedure. Risks (conditional expected losses) are computed for each decision rule, or estimator. Prior probabilities are assigned to states of nature, or parameter values. Expected risks are computed for each decision rule. Then the estimator for which the expected risk is the least is the one chosen.

No Bayesian point estimators will be derived here, but one result is of particular interest. If the parameter p of a Bernoulli process is estimated, using a squared error loss function, and a uniform or rectangular (continuous) prior probability distribution for p is assumed, that is, all values between zero and one are assumed to be equally likely, then the Bayesian estimator of p, denoted $\hat{p}$, is

(17.5)
$$\hat{p} = \frac{X + 1}{n + 2}$$

where X = the number of successes and
n = the number of trials.[2]

It turns out that this value of $\hat{p}$ is also the mean or expected value of the posterior distribution of p if the prior distribution of p is assumed to be rectangular (and continuous) and the sample evidence is an observation of X successes in n trials. We recall that the maximum likelihood estimate of p is simply the observed proportion of successes X/n. If the sample size n is large, these two estimates are approximately equal. Furthermore, there are other prior probability distributions besides the rectangular (uniform) distribution for which the mean of the posterior distribution will have a difference of this order of magnitude when compared with the maximum likelihood estimator. This brings out a very interesting point. From the Bayesian point of view, the standard use of the maximum likelihood estimate in such situations carries with it assumptions concerning the nature of the prior distribu-

[2]See A. M. Mood and F. A. Graybill, *Introduction to the Theory of Statistics* (New York: McGraw-Hill, 1963, 2nd ed.) pp. 189–190.

tion of p. The Bayesian decision analyst would argue that some of these prior distributions are very unreasonable in the context of particular problems. For example, a rectangular prior distribution implies that all values of the parameter (in its admissible range) are equally likely. Such an implication may be quite unrealistic based upon the prior knowledge of the individual carrying out the estimation.

Interval Estimation

We turn now to a consideration of interval estimation in classical and Bayesian statistics. We have seen that in estimating a population parameter in classical confidence interval estimation an interval is set up on the basis of a sample of n observations and a so-called "confidence coefficient" is associated with this interval. Suppose, for example, in the importer's problem, we wanted to make a confidence interval estimate of p, the proportion of all potential customers on the mailing list who would purchase the jars of cocktail onions if advertisements were sent to them, and let us assume we want to make this estimate on the basis of the proportion, $\bar{p}$, who purchased in a simple random sample of 100 drawn from the list. We could establish (say) a 95% confidence interval around $\bar{p}$ for the estimation of p in the usual way. Let us review the interpretation of this confidence interval. According to the classical school, it is definitely *incorrect* to say that the probability is 95% that the parameter is included in the interval. It is argued that the population parameter is a particular value, and therefore cannot be considered to be a random variable. Indeed, in all of classical statistics, it is forbidden to make conditional probability statements about a population parameter given the value of a sample statistic, e.g., $P(p/\bar{p})$. The permissible types of statements concern conditional probabilities of sample statistics given the value of a population parameter. For example, in a problem involving a Bernoulli process, we could compute probabilities of the type, $P(\bar{p}/p)$.

Returning to the importer's problem, from the classical viewpoint, the confidence interval estimate of p cannot be interpreted as a probability statement about the proportion of the 100,000 potential customers on the mailing list who would purchase the product. Since the interval is considered to be the random variable, the confidence coefficient refers to the concept that 95% of the intervals so constructed would bracket or include the true value of the population parameter. Thus, on a relative frequency basis, 95% of the statements made on the basis of such intervals would be correct. Furthermore, in keeping with the classical viewpoint, only the evidence of this particular sample can be used in establishing the confidence interval. Prior knowledge of any sort cannot be made a part of the estimation procedure. Finally, just as in hypothesis testing, the way in which the sample observations are to be used must be decided upon prior to the examination of these observations.

The Bayesian's approach to this general problem stands in sharp contrast to the classical procedure. He argues that if the value of the population parameter is unknown, then it can and should be treated as a random variable. In a setting such as the importer's problem, we would view the population parameter p as a basic random variable affecting a decision that must be made. Hence, we would be willing to compute conditional probabilities of the type $P(p \mid \bar{p})$. Furthermore, we would state that these conditional probabilities are the ones which are relevant to the decision maker rather than those of the form $P(\bar{p} \mid p)$. For example, in problems similar to that of the importer's, we might be interested in the probability that at least a certain proportion of the population would purchase the product based on the sample evidence. We would not be interested in the reverse conditional probability concerning a proportion in the sample given some postulated value for the population. The Bayesian decision analyst would argue that the confidence interval information is not particularly relevant. The decision maker is not interested in the proportion of correct statements that would be made in the long run, but rather in making a correct decision in this particular case.

The Bayesian approach also maintains that it is not wise to restrict oneself to the evidence of the particular sample which has been drawn but that the sample evidence should be incorporated with prior information through the use of Bayes' theorem to produce a posterior probability distribution. This leads to the Bayesian approach to the problem which classical inference solves by confidence interval estimation. The Bayesian procedure begins with the assignment of a prior probability distribution to the parameter being estimated. For example, in the importer's problem, a prior probability distribution is assigned to the random variable p (p. 668). Then a sample is drawn and the sample evidence is used to revise the prior probability distribution. This revision generates a posterior probability distribution. Thus, in the importer's problem in Chapter 14, it was assumed that six persons in the sample of 100 purchased the product. The revision of the prior probability distribution on the basis of this sample evidence yielded the posterior distribu-

Table 17-11 Posterior Probability Distribution of the Random Variable p in the Importer's Problem.

Event p	*Posterior Probability* $p_1(p)$
.02	.01
.04	.34
.06	.60
.08	.05

tion shown in Table 14-15, which is reproduced in Table 17-11 for convenience. Hence, the Bayesian analyst would make probability statements about the random variable p based on this distribution. As we have pointed out previously, the particular distribution in the importer's problem may be unrealistic because there are only four values of p. However, statements such as the following can be made. The probability is .94 that p lies between .04 and .06 (it is either .04 or .06, in this case). The probability is .95 that the value of p is .06 or less, etc. In a more realistic problem we may have a large number of possible values for p if that random variable is discrete, or we may have a probability density function over p if the random variable is continuous. The principle remains the same. If the prior distribution was a subjective probability distribution, then the posterior probabilities similarly represent degrees of belief or betting odds.

An interesting result occurs which is analogous to a relationship indicated earlier between classical and Bayesian point estimation when a rectangular prior distribution was assumed. If a rectangular distribution is assumed for the random variable p, and if the sample size is large, then there is a close coincidence between the statements made under the two schools of thought. Specifically, for example, the posterior probability that p lies in a .95 confidence interval is approximately .95. A prior distribution which is roughly rectangular or uniform is often referred to by Bayesian decision theorists as a "diffuse" or a "gentle" prior distribution. Such a distribution implies roughly equal likelihood of occurrence of all values of the random variable in its admissible range. This type of distribution is thought of as an appropriate subjective prior distribution when the decision maker has virtually complete ignorance of the value of the parameter being estimated. Doubtless, such states of almost complete lack of knowledge about parameter values are rare. Hence, Bayesian decision theorists argue that the uncritical use of confidence interval estimates may imply unreasonable assumptions about the investigator's prior knowledge concerning the parameter being estimated.

17.5 Some Remarks on Classical and Bayesian Statistics

As might be surmised from the material in this chapter, a considerable amount of controversy has arisen between those adhering to the classical or orthodox school of statistics and those advocating the Bayesian viewpoint. This type of controversy is characteristic of lively, dynamic, intellectually challenging fields, and that description is certainly accurate of the field of statistical analysis. In this section, we will comment on some of the areas of common ground and some of the points of difference between the two schools of thought.

Despite differences in terminology, both schools conceptualize a problem of decision making in which there are states of nature and actions which must be taken in the light of sample or experimental evidence about these states of nature. Both schools use conditional probabilities of sample outcomes, given states of nature (population parameters) for the decision process. These conditional probabilities provide the error characteristic curve on the basis of which the classicist chooses his decision rule. Informally, he is supposed to take into account the relative seriousness of Type I and Type II errors by considering the entire error characteristic curve, but since losses virtually always vary with population parameter values, it is not clear how he can actually do this.

The Bayesian approach supplements or completes the classical analysis by formally providing a loss function which specifies the seriousness of errors in selecting acts, and by assigning prior probabilities to states of nature either on an objective or subjective basis. However, serious measurement problems are clearly present both in the establishment of loss functions and prior probability distributions.

It might be noted that some classicists have affirmed that hypothesis testing is not a decision problem, but rather one of drawing conclusions or inferences. However, other classical adherents have specifically formulated hypothesis testing as an action problem. In any event, it is not always clear whether a problem is one of inference or decision making.

An important area of disagreement between non-Bayesian and Bayesian analysts is the matter of subjective prior probability distributions. The non-Bayesians argue that the only legitimate types of probabilities are "objective" or relative frequency of occurrence probabilities. They find it difficult to accept the idea that subjective or personalistic probabilities should be processed together with relative frequencies, as in the Bayesian's use of Bayes' theorem, to arrive at posterior probabilities. The Bayesian argues that in actual decision making we do exactly that type of analysis. We have prior betting-odds on events that influence the payoffs of our actions. On the observation of sample or experimental information, we revise these prior betting-odds. This is an argument centered upon "descriptive" behavior, that is, a purported description of how people actually behave. However, the Bayesian goes further, and says that his procedures are "normative" or "prescriptive," that is, they specify how a reasonable person *should* choose among alternatives to be consistent with his own evaluations of payoffs and degrees of belief attached to uncertain events. He also argues that if we rigidly maintain that only objective probabilities have meaning, we prevent ourselves from handling some of the most important uncertainties involved in problems of decision making. This latter point is surely a cogent one, particularly in areas such as business and economic decision making.

The problem of how to assign prior probabilities is troublesome, even to convinced Bayesians, and is a subject of on-going research. There are unresolved problems involved in determining whether all events should be considered equally likely under ignorance, how to pose questions to a decision maker to derive his betting-odds distribution, or more generally, how best to quantify judgments about uncertainty.

The Bayesian turns the tables on the orthodox school, which accuses him of excessive subjectivity, and directs a similar charge against classical statistics. The choices of hypotheses to test, probability distributions to use, significance and confidence levels, and what data to collect in order to obtain a relative frequency distribution are all inextricably interwoven with subjective judgments.

The preceding indication of some of the points of disagreement between the classical and Bayesian schools tends to emphasize a polarization of points of view. However, the fact is that even within each of these schools there are philosophical and methodological disagreements, as well. The years ahead promise to be exciting and productive ones for the entire field of statistical analysis for decision making.

PROBLEMS

1. Discuss the similarities and differences between Bayesian decision theory and classical statistical inference. On what grounds do advocates of Bayesian decision theory criticize classical hypothesis testing procedures?

2. Suppose the following table is a joint frequency distribution for a particular problem:

	Sample Results			
States of Nature	*Type L*	*Type M*	*Type H*	*Total*
θ_1	.36	.07	.01	.44
θ_2	.10	.20	.02	.32
θ_3	.04	.03	.17	.24
Total	.50	.30	.20	1.00

The only strategies that make sense are s_2 (A, R, R) and s_7 (A, A, R) where A means accept H_0 and R means reject H_0. As an example, s_2 (A, R, R) means accept H_0 if type L information is observed, but reject H_0 if type M or H information is observed. Construct power curves for both of these strategies. If the probability of a Type I error has been set at .05, which strategy satisfies the requirement?

The alternative hypotheses are

$$H_0 : \theta = \theta_1$$

$$H_1 : \theta = \theta_2 \text{ or } \theta_3$$

3. Complete the following joint frequency distribution table. Determine which of the following strategies satisfy the requirement that Type I error be no larger than .10: s_3 (R, R, A), s_5 (R, A, A). Draw power curves for both strategies. The hypotheses and the symbols have the same meaning as in Problem 2.

States of Nature	*Sample Results* Type L	Type M	Type H	Total
θ_1	.10	—	.06	.23
θ_2	.04	—	—	.35
θ_3	—	.09	.30	—
Total	.17	—	.45	1.00

4. Describe an error characteristic curve and state how it differs from a power curve.

5. Suppose that in the problem in Chapter 17 dealing with the importer, the following table has been constructed showing the probability of rejecting the null hypothesis, $H_0: p \leq .05$, for three decision rules:

p	$X \leq c_1$ P(rejecting H_0)	$X \leq c_2$ P(rejecting H_0)	$X \leq c_3$ P(rejecting H_0)
.02	.45	.04	.00
.04	.67	.10	.03
.06	.88	.43	.10
.08	.94	.72	.21

Construct power curves and error characteristic curves for the three decision rules.

6. Describe a risk or conditional expected opportunity loss curve. What does it show?

7. Suppose the following table shows the opportunity losses for Problem 5 of this chapter.

Events	Market the Product A_1	Do Not Market the Product A_2
$p \leq .05$	\$30,000 $(.05 - p)$	\$ 0
$p > .05$	0	30,000 $(p - .05)$

Calculate the risk or conditional expected opportunity losses for decision rule $X \leq c_2$. Draw the risk or conditional expected opportunity loss curve pertaining to the decision.

8. Calculate the unconditional expected opportunity losses for the three decision rules in Problems 5 and 7 of this chapter. Which of these three rules yields the smallest expected opportunity loss? The prior probabilities for the p values are

p	$P_0(p)$
.02	.15
.04	.35
.06	.40
.08	.10
	1.00

Bibliography

1. Probability

Feller, W., *An Introduction to Probability Theory and Its Applications*, Vols. 1 and 2. New York, N.Y.: John Wiley & Sons, Inc., 1957, 1966.

Goldberg, S., *Probability: An Introduction*. Englewood Cliffs, N.J.: Prentice-Hall, Inc., 1960.

Hodges, J., and E. Lehman, *Basic Concepts of Probability and Statistics*. San Francisco, Cal.: Holden-Day, Inc., 1964.

Kemeny, J. G., H. Mirkil, J. L. Snell, and G. L. Thompson, *Finite Mathematical Structures*. Englewood Cliffs, N.J.: Prentice-Hall, Inc., 1958.

Mosteller, F., R. Rourke, and G. Thomas, Jr., *Probability and Statistics*. Reading, Mass.: Addison-Wesley Publishing Co., Inc., 1961.

Parzen, E., *Modern Probability Theory and Its Applications*. New York, N.Y.: John Wiley & Sons, Inc., 1960.

2. Classical Statistics

Anderson, R., and T. Bancroft, *Statistical Theory in Research*. New York, N.Y.: McGraw-Hill Book Company, 1952.

Clelland, R., J. deCani, F. Brown, J. Bursk, and D. Murray, *Basic Statistics with Business Applications*. New York, N.Y.: John Wiley & Sons, Inc., 1966.

Dixon, W., and F. Massey, Jr., *Introduction to Statistical Analysis*. 3rd ed., New York, N.Y.: McGraw-Hill Book Co., 1969.

Ehrenfeld, S., and S. Littauer, *Introduction to Statistical Method*. New York, N.Y.: McGraw-Hill Book Co., 1964.

Ezekiel, M., and K. Fox, *Methods of Correlation and Regression Analysis*. 3rd ed., New York, N.Y.: John Wiley & Sons, Inc., 1959.

Fox, K., *Intermediate Economic Statistics*. New York, N.Y.: John Wiley & Sons, Inc., 1968.

Freund, J., and F. Williams, *Modern Business Statistics*. 2nd ed., Englewood Cliffs, N.J.: Prentice-Hall, Inc., 1969.

Mood, A., and F. Graybill, *Introduction to the Theory of Statistics*. New York, N.Y.: McGraw-Hill Book Co., 1963.

Neter, J., and W. Wasserman, *Fundamental Statistics for Business and Economics*. 3rd ed., Boston Mass.: Allyn and Bacon, Inc., 1966.

Peters, William S., *Readings in Applied Statistics*. Englewood Cliffs, N.J.: Prentice-Hall, Inc., 1969.

Richmond, S., *Statistical Analysis*. 2nd ed., New York, N.Y.: The Ronald Press Co., 1964.

Summers, G., and W. Peters, *Statistical Analysis for Decision Making*. Englewood Cliffs, New Jersey, Prentice-Hall, Inc., 1968.

Yule, G., and M. Kendall, *An Introduction to the Theory of Statistics*. 14th ed., New York, N.Y.: Hafner Publishing Co., 1950.

3. Decision Theory

Chernoff, H., and L. Moses, *Elementary Decision Theory*. New York, N.Y.: John Wiley & Sons, Inc., 1959.

Forester, J., *Statistical Selection of Business Strategies*. Homewood, Ill.: Richard D. Irwin, Inc., 1968.

Hadley, G., *Introduction to Probability and Statistical Decision Theory*. San Francisco, Cal.: Holden-Day, Inc., 1967.

Lindley, D. V., *Introduction to Probability and Statistics from a Bayesian Viewpoint*, Part 2, *Inference*. New York, N.Y.: Cambridge University Press, 1965.

Pratt, J., H. Raiffa, and R. Schlaifer, *Introduction to Statistical Decision Theory*. New York, N.Y.: McGraw-Hill Book Company, Inc., 1965.

Raiffa, H., *Decision Analysis, Introductory Lectures on Choices Under Uncertainty*. Reading, Mass.: Addison-Wesley Publishing Co., Inc., 1968.

Raiffa, H., and R. Schlaifer, *Applied Statistical Decision Theory*. Cambridge, Mass.: Division of Research, Graduate School of Business Administration, Harvard University, 1961.

Schlaifer, R., *Probability and Statistics for Business Decisions*. New York, N.Y.: McGraw-Hill Book Co., Inc., 1959.

Wald, A., *Statistical Decision Functions*. New York, N.Y.: John Wiley & Sons, Inc., 1950.

4. Statistical Tables

Hald, A., *Statistical Tables and Formulas*. New York, N.Y.: John Wiley & Sons, Inc., 1952.

Military Standard 105D. Washington, D.C.: U.S. Government Printing Office, 1963.

National Bureau of Standards, *Tables of the Binomial Distribution*. Washington, D.C.: U.S. Government Printing Office, 1950.

Owen, D., *Handbook of Statistical Tables*. Reading, Mass.: Addison-Wesley, 1962.

Pearson, E.S., and Hartley, H. O., *Biometrika Tables for Statisticians*. Cambridge: Cambridge University Press, 1954.

RAND Corporation, *A Million Random Digits with 100,000 Normal Deviates*. New York, N.Y.: Free Press of Glencoe, 1955.

5. Dictionary of Statistical Terms

Freund, J., and F. Williams, *Dictionary/Outline of Basic Statistics*. New York, N.Y.: McGraw-Hill Book Co., 1966.

APPENDIX A

Table A.1 Selected Values of the Binomial Cumulative Distribution Function

$$F(c) = P(X \leq c) = \sum_{x=0}^{c} \binom{n}{x} (1-p)^{n-x} p^x$$

Example If $p = 0.20$, $n = 7$, $c = 2$, then $F(2) = P(X \leq 2) = 0.8520$.

		p									
n	c	0.05	0.10	0.15	0.20	0.25	0.30	0.35	0.40	0.45	0.50
2	0	0.9025	0.8100	0.7225	0.6400	0.5625	0.4900	0.4225	0.3600	0.3025	0.2500
	1	0.9975	0.9900	0.9775	0.9600	0.9375	0.9100	0.8775	0.8400	0.7975	0.7500
3	0	0.8574	0.7290	0.6141	0.5120	0.4219	0.3430	0.2746	0.2160	0.1664	0.1250
	1	0.9928	0.9720	0.9392	0.8960	0.8438	0.7840	0.7182	0.6480	0.5748	0.5000
	2	0.9999	0.9990	0.9966	0.9920	0.9844	0.9730	0.9571	0.9360	0.9089	0.8750
4	0	0.8145	0.6561	0.5220	0.4096	0.3164	0.2401	0.1785	0.1296	0.0915	0.0625
	1	0.9860	0.9477	0.8905	0.8192	0.7383	0.6517	0.5630	0.4752	0.3910	0.3125
	2	0.9995	0.9963	0.9880	0.9728	0.9492	0.9163	0.8735	0.8208	0.7585	0.6875
	3	1.0000	0.9999	0.9995	0.9984	0.9961	0.9919	0.9850	0.9744	0.9590	0.9375
5	0	0.7738	0.5905	0.4437	0.3277	0.2373	0.1681	0.1160	0.0778	0.0503	0.0312
	1	0.9774	0.9185	0.8352	0.7373	0.6323	0.5282	0.4284	0.3370	0.2562	0.1875
	2	0.9988	0.9914	0.9734	0.9421	0.8965	0.8369	0.7648	0.6826	0.5931	0.5000
	3	1.0000	0.9995	0.9978	0.9933	0.9844	0.9692	0.9460	0.9130	0.8688	0.8125
	4	1.0000	1.0000	0.9999	0.9997	0.9990	0.9976	0.9947	0.9898	0.9815	0.9688
6	0	0.7351	0.5314	0.3771	0.2621	0.1780	0.1176	0.0754	0.0467	0.0277	0.0156
	1	0.9672	0.8857	0.7765	0.6554	0.5339	0.4202	0.3191	0.2333	0.1636	0.1094
	2	0.9978	0.9842	0.9527	0.9011	0.8306	0.7443	0.6471	0.5443	0.4415	0.3438
	3	0.9999	0.9987	0.9941	0.9830	0.9624	0.9295	0.8826	0.8208	0.7447	0.6562
	4	1.0000	0.9999	0.9996	0.9984	0.9954	0.9891	0.9777	0.9590	0.9308	0.8906
	5	1.0000	1.0000	1.0000	0.9999	0.9998	0.9993	0.9982	0.9959	0.9917	0.9844
7	0	0.6983	0.4783	0.3206	0.2097	0.1335	0.0824	0.0490	0.0280	0.0152	0.0078
	1	0.9556	0.8503	0.7166	0.5767	0.4449	0.3294	0.2338	0.1586	0.1024	0.0625
	2	0.9962	0.9743	0.9262	0.8520	0.7564	0.6471	0.5323	0.4199	0.3164	0.2266
	3	0.9998	0.9973	0.9879	0.9667	0.9294	0.8740	0.8002	0.7102	0.6083	0.5000
	4	1.0000	0.9998	0.9988	0.9953	0.9871	0.9712	0.9444	0.9037	0.8471	0.7734
	5	1.0000	1.0000	0.9999	0.9996	0.9987	0.9962	0.9910	0.9812	0.9643	0.9375
	6	1.0000	1.0000	1.0000	1.0000	0.9999	0.9998	0.9994	0.9984	0.9963	0.9922
8	0	0.6634	0.4305	0.2725	0.1678	0.1001	0.0576	0.0319	0.0168	0.0084	0.0039
	1	0.9428	0.8131	0.6572	0.5033	0.3671	0.2553	0.1691	0.1064	0.0632	0.0352
	2	0.9942	0.9619	0.8948	0.7969	0.6785	0.5518	0.4278	0.3154	0.2201	0.1445
	3	0.9996	0.9950	0.9786	0.9437	0.8862	0.8059	0.7064	0.5941	0.4770	0.3633
	4	1.0000	0.9996	0.9971	0.9896	0.9727	0.9420	0.8939	0.8263	0.7396	0.6367
	5	1.0000	1.0000	0.9998	0.9988	0.9958	0.9887	0.9747	0.9502	0.9115	0.8555
	6	1.0000	1.0000	1.0000	0.9999	0.9996	0.9987	0.9964	0.9915	0.9819	0.9648
	7	1.0000	1.0000	1.0000	1.0000	1.0000	0.9999	0.9998	0.9993	0.9983	0.9961
9	0	0.6302	0.3874	0.2316	0.1342	0.0751	0.0404	0.0207	0.0101	0.0046	0.0020
	1	0.9288	0.7748	0.5995	0.4362	0.3003	0.1960	0.1211	0.0705	0.0385	0.0195
	2	0.9916	0.9470	0.8591	0.7382	0.6007	0.4628	0.3373	0.2318	0.1495	0.0898
	3	0.9994	0.9917	0.9661	0.9144	0.8343	0.7297	0.6089	0.4826	0.3614	0.2539
	4	1.0000	0.9991	0.9944	0.9804	0.9511	0.9012	0.8283	0.7334	0.6214	0.5000
	5	1.0000	0.9999	0.9994	0.9969	0.9900	0.9747	0.9464	0.9006	0.8342	0.7461
	6	1.0000	1.0000	1.0000	0.9997	0.9987	0.9957	0.9888	0.9750	0.9502	0.9102
	7	1.0000	1.0000	1.0000	1.0000	0.9999	0.9996	0.9986	0.9962	0.9909	0.9805
	8	1.0000	1.0000	1.0000	1.0000	1.0000	1.0000	0.9999	0.9997	0.9992	0.9980

Source: From Irwin Miller and John E. Freund, *Probability and Statistics for Engineers*, © 1965 by Prentice-Hall, Inc.

Table A.1 (continued)

n	c	0.05	0.10	0.15	0.20	0.25	0.30	0.35	0.40	0.45	0.50
						p					
10	0	0.5987	0.3487	0.1969	0.1074	0.0563	0.0282	0.0135	0.0060	0.0025	0.0010
	1	0.9139	0.7361	0.5443	0.3758	0.2440	0.1493	0.0860	0.0464	0.0232	0.0107
	2	0.9885	0.9298	0.8202	0.6778	0.5256	0.3828	0.2616	0.1673	0.0996	0.0547
	3	0.9990	0.9872	0.9500	0.8791	0.7759	0.6496	0.5138	0.3823	0.2660	0.1719
	4	0.9999	0.9984	0.9901	0.9672	0.9219	0.8497	0.7515	0.6331	0.5044	0.3770
	5	1.0000	0.9999	0.9986	0.9936	0.9803	0.9527	0.9051	0.8338	0.7384	0.6230
	6	1.0000	1.0000	0.9999	0.9991	0.9965	0.9894	0.9740	0.9452.	0.8980	0.8281
	7	1.0000	1.0000	1.0000	0.9999	0.9996	0.9984	0.9952	0.9877	0.9726	0.9453
	8	1.0000	1.0000	1.0000	1.0000	1.0000	0.9999	0.9995	0.9983	0.9955	0.9893
	9	1.0000	1.0000	1.0000	1.0000	1.0000	1.0000	1.0000	0.9999	0.9997	0.9990
11	0	0.5688	0.3138	0.1673	0.0859	0.0422	0.0198	0.0088	0.0036	0.0014	0.0005
	1	0.8981	0.6974	0.4922	0.3221	0.1971	0.1130	0.0606	0.0302	0.0139	0.0059
	2	0.9848	0.9104	0.7788	0.6174	0.4552	0.3127	0.2001	0.1189	0.0652	0.0327
	3	0.9984	0.9815	0.9306	0.8389	0.7133	0.5696	0.4256	0.2963	0.1911	0.1133
	4	0.9999	0.9972	0.9841	0.9496	0.8854	0.7897	0.6683	0.5328	0.3971	0.2744
	5	1.0000	0.9997	0.9973	0.9883	0.9657	0.9218	0.8513	0.7535	0.6331	0.5000
	6	1.0000	1.0000	0.9997	0.9980	0.9924	0.9784	0.9499	0.9006	0.8262	0.7256
	7	1.0000	1.0000	1.0000	0.9998	0.9988	0.9957	0.9878	0.9707	0.9390	0.8867
	8	1.0000	1.0000	1.0000	1.0000	0.9999	0.9994	0.9980	0.9941	0.9852	0.9673
	9	1.0000	1.0000	1.0000	1.0000	1.0000	1.0000	0.9998	0.9993	0.9978	0.9941
	10	1.0000	1.0000	1.0000	1.0000	1.0000	1.0000	1.0000	1.0000	0.9998	0.9995
12	0	0.5404	0.2824	0.1422	0.0687	0.0317	0.0138	0.0057	0.0022	0.0008	0.0002
	1	0.8816	0.6590	0.4435	0.2749	0.1584	0.0850	0.0424	0.0196	0.0083	0.0032
	2	0.9804	0.8891	0.7358	0.5583	0,3907	0.2528	0.1513	0.0834	0.0421	0.0193
	3	0.9978	0.9744	0.9078	0.7946	0.6488	0.4925	0.3467	0.2253	0.1345	0.0730
	4	0.9998	0.9957	0.9761	0.9274	0.8424	0.7237	0.5833	0.4382	0.3044	0.1938
	5	1.0000	0.9995	0.9954	0.9806	0.9456	0.8822	0.7873	0.6652	0.5269	0.3872
	6	1.0000	0.9999	0.9993	0.9961	0.9857	0.9614	0.9154	0.8418	0.7393	0.6128
	7	1.0000	1.0000	0.9999	0.9994	0.9972	0.9905	0.9745	0.9427	0.8883	0.8062
	8	1.0000	1.0000	1.0000	0.9999	0.9996	0.9983	0.9944	0.9847	0.9644	0.9270
	9	1.0000	1.0000	1.0000	1.0000	1.0000	0.9998	0.9992	0.9972	0.9921	0.9807
	10	1.0000	1.0000	1.0000	1.0000	1.0000	1.0000	0.9999	0.9997	0.9989	0.9968
	11	1.0000	1.0000	1.0000	1.0000	1.0000	1.0000	1.0000	1.0000	0.9999	0.9998
13	0	0.5133	0.2542	0.1209	0.0550	0.0238	0.0097	0.0037	0.0013	0.0004	0.0001
	1	0.8646	0.6213	0.3983	0.2336	0.1267	0.0637	0.0296	0.0126	0.0049	0.0017
	2	0.9755	0.8661	0.6920	0.5017	0.3326	0.2025	0.1132	0.0579	0.0269	0.0112
	3	0.9969	0.9658	0.8820	0.7473	0.5843	0.4206	0.2783	0.1686	0.0929	0.0461
	4	0.9997	0.9935	0.9658	0.9009	0.7940	0.6543	0.5005	0.3530	0.2279	0.1334
	5	1.0000	0.9991	0.9925	0.9700	0.9198	0.8346	0.7159	0.5744	0.4268	0.2905
	6	1.0000	0.9999	0.9987	0.9930	0.9757	0.9376	0.8705	0.7712	0.6437	0.5000
	7	1.0000	1.0000	0.9998	0.9988	0.9944	0.9818	0.9538	0.9023	0.8212	0.7095
	8	1.0000	1.0000	1.0000	0.9998	0.9990	0.9960	0.9874	0.9679	0.9302	0.8666
	9	1.0000	1.0000	1.0000	1.0000	0.9999	0.9993	0.9975	0.9922	0.9797	0.9539
	10	1.0000	1.0000	1.0000	1.0000	1.0000	0.9999	0.9997	0.9987	0.9959	0.9888
	11	1.0000	1.0000	1.0000	1.0000	1.0000	1.0000	1.0000	0.9999	0.9995	0.9983
	12	1.0000	1.0000	1.0000	1.0000	1.0000	1.0000	1.0000	1.0000	1.0000	0.9999
14	0	0.4877	0.2288	0.1028	0.0440	0.0178	0.0068	0.0024	0.0008	0.0002	0.0001
	1	0.8470	0.5846	0.3567	0.1979	0.1010	0.0475	0.0205	0.0081	0.0029	0.0009

Table A.1 (continued)

n	c	p = 0.05	0.10	0.15	0.20	0.25	0.30	0.35	0.40	0.45	0.50
14	2	0.9699	0.8416	0.6479	0.4481	0.2811	0.1608	0.0839	0.0398	0.0170	0.0065
	3	0.9958	0.9559	0.8535	0.6982	0.5213	0.3552	0.2205	0.1243	0.0632	0.0287
	4	0.9996	0.9908	0.9533	0.8702	0.7415	0.5842	0.4227	0.2793	0.1672	0.0898
	5	1.0000	0.9985	0.9885	0.9561	0.8883	0.7805	0.6405	0.4859	0.3373	0.2120
	6	1.0000	0.9998	0.9978	0.9884	0.9617	0.9067	0.8164	0.6925	0.5461	0.3953
	7	1.0000	1.0000	0.9997	0.9976	0.9897	0.9685	0.9247	0.8499	0.7414	0.6047
	8	1.0000	1.0000	1.0000	0.9996	0.9978	0.9917	0.9757	0.9417	0.8811	0.7880
	9	1.0000	1.0000	1.0000	1.0000	0.9997	0.9983	0.9940	0.9825	0.9574	0.9102
	10	1.0000	1.0000	1.0000	1.0000	1.0000	0.9998	0.9989	0.9961	0.9886	0.9713
	11	1.0000	1.0000	1.0000	1.0000	1.0000	1.0000	0.9999	0.9994	0.9978	0.9935
	12	1.0000	1.0000	1.0000	1.0000	1.0000	1.0000	1.0000	0.9999	0.9997	0.9991
	13	1.0000	1.0000	1.0000	1.0000	1.0000	1.0000	1.0000	1.0000	1.0000	0.9999
15	0	0.4633	0.2059	0.0874	0.0352	0.0134	0.0047	0.0016	0.0005	0.0001	0.0000
	1	0.8290	0.5490	0.3186	0.1671	0.0802	0.0353	0.0142	0.0052	0.0017	0.0005
	2	0.9638	0.8159	0.6042	0.3980	0.2361	0.1268	0.0617	0.0271	0.0107	0.0037
	3	0.9945	0.9444	0.8227	0.6482	0.4613	0.2969	0.1727	0.0905	0.0424	0.0176
	4	0.9994	0.9873	0.9383	0.8358	0.6865	0.5155	0.3519	0.2173	0.1204	0.0592
	5	0.9999	0.9978	0.9832	0.9389	0.8516	0.7216	0.5643	0.4032	0.2608	0.1509
	6	1.0000	0.9997	0.9964	0.9819	0.9434	0.8689	0.7548	0.6098	0.4522	0.3036
	7	1.0000	1.0000	0.9996	0.9958	0.9827	0.9500	0.8868	0.7869	0.6535	0.5000
	8	1.0000	1.0000	0.9999	0.9992	0.9958	0.9848	0.9578	0.9050	0.8182	0.6964
	9	1.0000	1.0000	1.0000	0.9999	0.9992	0.9963	0.9876	0.9662	0.9231	0.8491
	10	1.0000	1.0000	1.0000	1.0000	0.9999	0.9993	0.9972	0.9907	0.9745	0.9408
	11	1.0000	1.0000	1.0000	1.0000	1.0000	0.9999	0.9995	0.9981	0.9937	0.9824
	12	1.0000	1.0000	1.0000	1.0000	1.0000	1.0000	0.9999	0.9997	0.9989	0.9963
	13	1.0000	1.0000	1.0000	1.0000	1.0000	1.0000	1.0000	1.0000	0.9999	0.9995
	14	1.0000	1.0000	1.0000	1.0000	1.0000	1.0000	1.0000	1.0000	1.0000	1.0000
16	0	0.4401	0.1853	0.0743	0.0281	0.0100	0.0033	0.0010	0.0003	0.0001	0.0000
	1	0.8108	0.5147	0.2839	0.1407	0.0635	0.0261	0.0098	0.0033	0.0010	0.0003
	2	0.9571	0.7892	0.5614	0.3518	0.1971	0.0994	0.0451	0.0183	0.0066	0.0021
	3	0.9930	0.9316	0.7899	0.5981	0.4050	0.2459	0.1339	0.0651	0.0281	0.0106
	4	0.9991	0.9830	0.9209	0.7982	0.6302	0.4499	0.2892	0.1666	0.0853	0.0384
	5	0.9999	0.9967	0.9765	0.9183	0.8103	0.6598	0.4900	0.3288	0.1976	0.1051
	6	1.0000	0.9995	0.9944	0.9733	0.9204	0.8247	0.6881	0.5272	0.3660	0.2272
	7	1.0000	0.9999	0.9989	0.9930	0.9729	0.9256	0.8406	0.7161	0.5629	0.4018
	8	1.0000	1.0000	0.9998	0.9985	0.9925	0.9743	0.9329	0.8577	0.7441	0.5982
	9	1.0000	1.0000	1.0000	0.9998	0.9984	0.9929	0.9771	0.9417	0.8759	0.7728
	10	1.0000	1.0000	1.0000	1.0000	0.9997	0.9984	0.9938	0.9809	0.9514	0.8949
	11	1.0000	1.0000	1.0000	1.0000	1.0000	0.9997	0.9987	0.9951	0.9851	0.9616
	12	1.0000	1.0000	1.0000	1.0000	1.0000	1.0000	0.9998	0.9991	0.9965	0.9894
	13	1.0000	1.0000	1.0000	1.0000	1.0000	1.0000	1.0000	0.9999	0.9994	0.9979
	14	1.0000	1.0000	1.0000	1.0000	1.0000	1.0000	1.0000	1.0000	1.0000	0.9997
	15	1.0000	1.0000	1.0000	1.0000	1.0000	1.0000	1.0000	1.0000	1.0000	1.0000
17	0	0.4181	0.1668	0.0631	0.0225	0.0075	0.0023	0.0007	0.0002	0.0000	0.0000
	1	0.7922	0.4818	0.2525	0.1182	0.0501	0.0193	0.0067	0.0021	0.0006	0.0001
	2	0.9497	0.7618	0.5198	0.3096	0.1637	0.0774	0.0327	0.0123	0.0041	0.0012
	3	0.9912	0.9174	0.7556	0.5489	0.3530	0.2019	0.1028	0.0464	0.0184	0.0064
	4	0.9988	0.9779	0.9013	0.7582	0.5739	0.3887	0.2348	0.1260	0.0596	0.0245

Table A.1 (continued)

		p									
n	c	0.05	0.10	0.15	0.20	0.25	0.30	0.35	0.40	0.45	0.50
17	5	0.9999	0.9953	0.9681	0.8943	0.7653	0.5968	0.4197	0.2639	0.1471	0.0717
	6	1.0000	0.9992	0.9917	0.9623	0.8929	0.7752	0.6188	0.4478	0.2902	0.1662
	7	1.0000	0.9999	0.9983	0.9891	0.9598	0.8954	0.7872	0.6405	0.4743	0.3145
	8	1.0000	1.0000	0.9997	0.9974	0.9876	0.9597	0.9006	0.8011	0.6626	0.5000
	9	1.0000	1.0000	1.0000	0.9995	0.9969	0.9873	0.9617	0.9081	0.8166	0.6855
	10	1.0000	1.0000	1.0000	0.9999	0.9994	0.9968	0.9880	0.9652	0.9174	0.8338
	11	1.0000	1.0000	1.0000	1.0000	0.9999	0.9993	0.9970	0.9894	0.9699	0.9283
	12	1.0000	1.0000	1.0000	1.0000	1.0000	0.9999	0.9994	0.9975	0.9914	0.9755
	13	1.0000	1.0000	1.0000	1.0000	1.0000	1.0000	0.9999	0.9995	0.9981	0.9936
	14	1.0000	1.0000	1.0000	1.0000	1.0000	1.0000	1.0000	0.9999	0.9997	0.9988
	15	1.0000	1.0000	1.0000	1.0000	1.0000	1.0000	1.0000	1.0000	1.0000	0.9999
	16	1.0000	1.0000	1.0000	1.0000	1.0000	1.0000	1.0000	1.0000	1.0000	1.0000
18	0	0.3972	0.1501	0.0536	0.0180	0.0056	0.0016	0.0004	0.0001	0.0000	0.0000
	1	0.7735	0.4503	0.2241	0.0991	0.0395	0.0142	0.0046	0.0013	0.0003	0.0001
	2	0.9419	0.7338	0.4797	0.2713	0.1353	0.0600	0.0236	0.0082	0.0025	0.0007
	3	0.9891	0.9018	0.7202	0.5010	0.3057	0.1646	0.0783	0.0328	0.0120	0.0038
	4	0.9985	0.9718	0.8794	0.7164	0.5187	0.3327	0.1886	0.0942	0.0411	0.0154
	5	0.9998	0.9936	0.9581	0.8671	0.7175	0.5344	0.3550	0.2088	0.1077	0.0481
	6	1.0000	0.9988	0.9882	0.9487	0.8610	0.7217	0.5491	0.3743	0.2258	0.1189
	7	1.0000	0.9998	0.9973	0.9837	0.9431	0.8593	0.7283	0.5634	0.3915	0.2403
	8	1.0000	1.0000	0.9995	0.9957	0.9807	0.9404	0.8609	0.7368	0.5778	0.4073
	9	1.0000	1.0000-	0.9999	0.9991	0.9946	0.9790	0.9403	0.8653	0.7473	0.5927
	10	1.0000	1.0000	1.0000	0.9998	0.9988	0.9939	0.9788	0.9424	0.8720	0.7597
	11	1.0000	1.0000	1.0000	1.0000	0.9998	0.9986	0.9938	0.9797	0.9463	0.8811
	12	1.0000	1.0000	1.0000	1.0000	1.0000	0.9997	0.9986	0.9942	0.9817	0.9519
	13	1.0000	1.0000	1.0000	1.0000	1.0000	1.0000	0.9997	0.9987	0.9951	0.9846
	14	1.0000	1.0000	1.0000	1.0000	1.0000	1.0000	1.0000	0.9998	0.9990	0.9962
	15	1.0000	1.0000	1.0000	1.0000	1.0000	1.0000	1.0000	1.0000	0.9999	0.9993
	16	1.0000	1.0000	1.0000	1.0000	1.0000	1.0000	1.0000	1.0000	1.0000	0.9999
19	0	0.3774	0.1351	0.0456	0.0144	0.0042	0.0011	0.0003	0.0001	0.0000	0.0000
	1	0.7547	0.4203	0.1985	0.0829	0.0310	0.0104	0.0031	0.0008	0.0002	0.0000
	2	0.9335	0.7054	0.4413	0.2369	0.1113	0.0462	0.0170	0.0055	0.0015	0.0004
	3	0.9868	0.8850	0.6841	0.4551	0.2630	0.1332	0.0591	0.0230	0.0077	0.0022
	4	0.9980	0.9648	0.8556	0.6733	0.4654	0.2822	0.1500	0.0696	0.0280	0.0096
	5	0.9998	0.9914	0.9463	0.8369	0.6678	0.4739	0.2968	0.1629	0.0777	0.0318
	6	1.0000	0.9983	0.9837	0.9324	0.8251	0.6655	0.4812	0.3081	0.1727	0.0835
	7	1.0000	0.9997	0.9959	0.9767	0.9225	0.8180	0.6656	0.4878	0.3169	0.1796
	8	1.0000	1.0000	0.9992	0.9933	0.9713	0.9161	0.8145	0.6675	0.4940	0.3238
	9	1.0000	1.0000	0.9999	0.9984	0.9911	0.9674	0.9125	0.8139	0.6710	0.5000
	10	1.0000	1.0000	1.0000	0.9997	0.9977	0.9895	0.9653	0.9115	0.8159	0.6762
	11	1.0000	1.0000	1.0000	1.0000	0.9995	0.9972	0.9886	0.9648	0.9129	0.8204
	12	1.0000	1.0000	1.0000	1.0000	0.9999	0.9994	0.9969	0.9884	0.9658	0.9165
	13	1.0000	1.0000	1.0000	1.0000	1.0000	0.9999	0.9993	0.9969	0.9891	0.9682
	14	1.0000	1.0000	1.0000	1.0000	1.0000	1.0000	0.9999	0.9994	0.9972	0.9904
	15	1.0000	1.0000	1.0000	1.0000	1.0000	1.0000	1.0000	0.9999	0.9995	0.9978
	16	1.0000	1.0000	1.0000	1.0000	1.0000	1.0000	1.0000	1.0000	0.9999	0.9996
	17	1.0000	1.0000	1.0000	1.0000	1.0000	1.0000	1.0000	1.0000	1.0000	1.0000

Table A.1 (continued)

n	c	0.05	0.10	0.15	0.20	0.25	0.30	0.35	0.40	0.45	0.50
20	0	0.3585	0.1216	0.0388	0.0115	0.0032	0.0008	0.0002	0.0000	0.0000	0.0000
	1	0.7358	0.3917	0.1756	0.0692	0.0243	0.0076	0.0021	0.0005	0.0001	0.0000
	2	0.9245	0.6769	0.4049	0.2061	0.0913	0.0355	0.0121	0.0036	0.0009	0.0002
	3	0.9841	0.8670	0.6477	0.4114	0.2252	0.1071	0.0444	0.0160	0.0049	0.0013
	4	0.9974	0.9568	0.8298	0.6296	0.4148	0.2375	0.1182	0.0510	0.0189	0.0059
	5	0.9997	0.9887	0.9327	0.8042	0.6172	0.4164	0.2454	0.1256	0.0553	0.0207
	6	1.0000	0.9976	0.9781	0.9133	0.7858	0.6080	0.4166	0.2500	0.1299	0.0577
	7	1.0000	0.9996	0.9941	0.9679	0.8982	0.7723	0.6010	0.4159	0.2520	0.1316
	8	1.0000	0.9999	0.9987	0.9900	0.9591	0.8867	0.7624	0.5956	0.4143	0.2517
	9	1.0000	1.0000	0.9998	0.9974	0.9861	0.9520	0.8782	0.7553	0.5914	0.4119
	10	1.0000	1.0000	1.0000	0.9994	0.9961	0.9829	0.9468	0.8725	0.7507	0.5881
	11	1.0000	1.0000	1.0000	0.9999	0.9991	0.9949	0.9804	0.9435	0.8692	0.7483
	12	1.0000	1.0000	1.0000	1.0000	0.9998	0.9987	0.9940	0.9790	0.9420	0.8684
	13	1.0000	1.0000	1.0000	1.0000	1.0000	0.9997	0.9985	0.9935	0.9786	0.9423
	14	1.0000	1.0000	1.0000	1.0000	1.0000	1.0000	0.9997	0.9984	0.9936	0.9793
	15	1.0000	1.0000	1.0000	1.0000	1.0000	1.0000	1.0000	0.9997	0.9985	0.9941
	16	1.0000	1.0000	1.0000	1.0000	1.0000	1.0000	1.0000	1.0000	0.9997	0.9987
	17	1.0000	1.0000	1.0000	1.0000	1.0000	1.0000	1.0000	1.0000	1.0000	0.9998
	18	1.0000	1.0000	1.0000	1.0000	1.0000	1.0000	1.0000	1.0000	1.0000	1.0000

Table A.2 Coefficients of the Binomial Distribution

Example If $n = 8$ and $x = 6$, $\binom{8}{6} = 28$.

This table gives the value of $\binom{n}{x}$ in $\binom{n}{x}q^{n-x}p^x$, the general term of $(q + p)^n$.

n	$\binom{n}{0}$	$\binom{n}{1}$	$\binom{n}{2}$	$\binom{n}{3}$	$\binom{n}{4}$	$\binom{n}{5}$	$\binom{n}{6}$	$\binom{n}{7}$	$\binom{n}{8}$	$\binom{n}{9}$	$\binom{n}{10}$
0	1										
1	1	1									
2	1	2	1								
3	1	3	3	1							
4	1	4	6	4	1						
5	1	5	10	10	5	1					
6	1	6	15	20	15	6	1				
7	1	7	21	35	35	21	7	1			
8	1	8	28	56	70	56	28	8	1		
9	1	9	36	84	126	126	84	36	9	1	
10	1	10	45	120	210	252	210	120	45	10	1
11	1	11	55	165	330	462	462	330	165	55	11
12	1	12	66	220	495	792	924	792	495	220	66
13	1	13	78	286	715	1287	1716	1716	1287	715	286
14	1	14	91	364	1001	2002	3003	3432	3003	2002	1001
15	1	15	105	455	1365	3003	5005	6435	6435	5005	3003
16	1	16	120	560	1820	4368	8008	11440	12870	11440	8008
17	1	17	136	680	2380	6188	12376	19448	24310	24310	19448
18	1	18	153	816	3060	8568	18564	31824	43758	48620	43758
19	1	19	171	969	3876	11628	27132	50388	75582	92378	92378
20	1	20	190	1140	4845	15504	38760	77520	125970	167960	184756

Table A.3 Selected Values of the Poisson Cumulative Distribution

$$F(c) = P(X \leq c) = \sum_{x=0}^{c} \frac{\mu^x e^{-\mu}}{x!}$$

Example If $\mu = 1.00$, then $F(2) = P(X \leq 2) = 0.920$.

$\mu \backslash c$	0	1	2	3	4	5	6	7	8	9
0.02	0.980	1.000								
0.04	0.961	0.999	1.000							
0.06	0.942	0.998	1.000							
0.08	0.923	0.997	1.000							
0.10	0.905	0.995	1.000							
0.15	0.861	0.990	0.999	1.000						
0.20	0.819	0.982	0.999	1.000						
0.25	0.779	0.974	0.998	1.000						
0.30	0.741	0.963	0.996	1.000						
0.35	0.705	0.951	0.994	1.000						
0.40	0.670	0.938	0.992	0.999	1.000					
0.45	0.638	0.925	0.989	0.999	1.000					
0.50	0.607	0.910	0.986	0.998	1.000					
0.55	0.577	0.894	0.982	0.998	1.000					
0.60	0.549	0.878	0.977	0.997	1.000					
0.65	0.522	0.861	0.972	0.996	0.999	1.000				
0.70	0.497	0.844	0.966	0.994	0.999	1.000				
0.75	0.472	0.827	0.959	0.993	0.999	1.000				
0.80	0.449	0.809	0.953	0.991	0.999	1.000				
0.85	0.427	0.791	0.945	0.989	0.998	1.000				
0.90	0.407	0.772	0.937	0.987	0.998	1.000				
0.95	0.387	0.754	0.929	0.984	0.997	1.000				
1.00	0.368	0.736	0.920	0.981	0.996	0.999	1.000			
1.1	0.333	0.699	0.900	0.974	0.995	0.999	1.000			
1.2	0.301	0.663	0.879	0.966	0.992	0.998	1.000			
1.3	0.273	0.627	0.857	0.957	0.989	0.998	1.000			
1.4	0.247	0.592	0.833	0.946	0.986	0.997	0.999	1.000		
1.5	0.223	0.558	0.809	0.934	0.981	0.996	0.999	1.000		
1.6	0.202	0.525	0.783	0.921	0.976	0.994	0.999	1.000		
1.7	0.183	0.493	0.757	0.907	0.970	0.992	0.998	1.000		
1.8	0.165	0.463	0.731	0.891	0.964	0.990	0.997	0.999	1.000	
1.9	0.150	0.434	0.704	0.875	0.956	0.987	0.997	0.999	1.000	
2.0	0.135	0.406	0.677	0.857	0.947	0.983	0.995	0.999	1.000	
2.2	0.111	0.355	0.623	0.819	0.928	0.975	0.993	0.998	1.000	
2.4	0.091	0.308	0.570	0.779	0.904	0.964	0.988	0.997	0.999	1.000
2.6	0.074	0.267	0.518	0.736	0.877	0.951	0.983	0.995	0.999	1.000
2.8	0.061	0.231	0.469	0.692	0.848	0.935	0.976	0.992	0.998	0.999
3.0	0.050	0.199	0.423	0.647	0.815	0.916	0.966	0.988	0.996	0.999

Source: From Statistical Quality Control, by Eugenc L. Grant. Copyright 1964 by McGraw-Hill Book Company. Used with permission of McGraw-Hill Book Company.

Table A.3 (continued)

$\mu \backslash c$	0	1	2	3	4	5	6	7	8	9
3.2	0.041	0.171	0.380	0.603	0.781	0.895	0.955	0.983	0.994	0.998
3.4	0.033	0.147	0.340	0.558	0.744	0.871	0.942	0.977	0.992	0.997
3.6	0.027	0.126	0.303	0.515	0.706	0.844	0.927	0.969	0.988	0.996
3.8	0.022	0.107	0.269	0.473	0.668	0.816	0.909	0.960	0.984	0.994
4.0	0.018	0.092	0.238	0.433	0.629	0.785	0.889	0.949	0.979	0.992
4.2	0.015	0.078	0.210	0.395	0.590	0.753	0.867	0.936	0.972	0.989
4.4	0.012	0.066	0.185	0.359	0.551	0.720	0.844	0.921	0.964	0.985
4.6	0.010	0.056	0.163	0.326	0.513	0.686	0.818	0.905	0.955	0.980
4.8	0.008	0.048	0.143	0.294	0.476	0.651	0.791	0.887	0.944	0.975
5.0	0.007	0.040	0.125	0.265	0.440	0.616	0.762	0.867	0.932	0.968
5.2	0.006	0.034	0.109	0.238	0.406	0.581	0.732	0.845	0.918	0.960
5.4	0.005	0.029	0.095	0.213	0.373	0.546	0.702	0.822	0.903	0.951
5.6	0.004	0.024	0.082	0.191	0.342	0.512	0.670	0.797	0.886	0.941
5.8	0.003	0.021	0.072	0.170	0.313	0.478	0.638	0.771	0.867	0.929
6.0	0.002	0.017	0.062	0.151	0.285	0.446	0.606	0.744	0.847	0.916

	10	11	12	13	14	15	16
2.8	1.000						
3.0	1.000						
3.2	1.000						
3.4	0.999	1.000					
3.6	0.999	1.000					
3.8	0.998	0.999	1.000				
4.0	0.997	0.999	1.000				
4.2	0.996	0.999	1.000				
4.4	0.994	0.998	0.999	1.000			
4.6	0.992	0.997	0.999	1.000			
4.8	0.990	0.996	0.999	1.000			
5.0	0.986	0.995	0.998	0.999	1.000		
5.2	0.982	0.993	0.997	0.999	1.000		
5.4	0.977	0.990	0.996	0.999	1.000		
5.6	0.972	0.988	0.995	0.998	0.999	1.000	
5.8	0.965	0.984	0.993	0.997	0.999	1.000	
6.0	0.957	0.980	0.991	0.996	0.999	0.999	1.000

$\mu \backslash c$	0	1	2	3	4	5	6	7	8	9
6.2	0.002	0.015	0.054	0.134	0.259	0.414	0.574	0.716	0.826	0.902
6.4	0.002	0.012	0.046	0.119	0.235	0.384	0.542	0.687	0.803	0.886
6.6	0.001	0.010	0.040	0.105	0.213	0.355	0.511	0.658	0.780	0.869
6.8	0.001	0.009	0.034	0.093	0.192	0.327	0.480	0.628	0.755	0.850
7.0	0.001	0.007	0.030	0.082	0.173	0.301	0.450	0.599	0.729	0.830
7.2	0.001	0.006	0.025	0.072	0.156	0.276	0.420	0.569	0.703	0.810
7.4	0.001	0.005	0.022	0.063	0.140	0.253	0.392	0.539	0.676	0.788
7.6	0.001	0.004	0.019	0.055	0.125	0.231	0.365	0.510	0.648	0.765
7.8	0.000	0.004	0.016	0.048	0.112	0.210	0.338	0.481	0.620	0.741

Table A.3 (continued)

$\mu\backslash c$	0	1	2	3	4	5	6	7	8	9
8.0	0.000	0.003	0.014	0.042	0.100	0.191	0.313	0.453	0.593	0.717
8.5	0.000	0.002	0.009	0.030	0.074	0.150	0.256	0.386	0.523	0.653
9.0	0.000	0.001	0.006	0.021	0.055	0.116	0.207	0.324	0.456	0.587
9.5	0.000	0.001	0.004	0.015	0.040	0.089	0.165	0.269	0.392	0.522
10.0	0.000	0.000	0.003	0.010	0.029	0.067	0.130	0.220	0.333	0.458

	10	11	12	13	14	15	16	17	18	19
6.2	0.949	0.975	0.989	0.995	0.998	0.999	1.000			
6.4	0.939	0.969	0.986	0.994	0.997	0.999	1.000			
6.6	0.927	0.963	0.982	0.992	0.997	0.999	0.999	1.000		
6.8	0.915	0.955	0.978	0.990	0.996	0.998	0.999	1.000		
7.0	0.901	0.947	0.973	0.987	0.994	0.998	0.999	1.000		
7.2	0.887	0.937	0.967	0.984	0.993	0.997	0.999	0.999	1.000	
7.4	0.871	0.926	0.961	0.980	0.991	0.996	0.998	0.999	1.000	
7.6	0.854	0.915	0.954	0.976	0.989	0.995	0.998	0.999	1.000	
7.8	0.835	0.902	0.945	0.971	0.986	0.993	0.997	0.999	1.000	
8.0	0.816	0.888	0.936	0.966	0.983	0.992	0.996	0.998	0.999	1.000
8.5	0.763	0.849	0.909	0.949	0.973	0.986	0.993	0.997	0.999	0.999
9.0	0.706	0.803	0.876	0.926	0.959	0.978	0.989	0.995	0.998	0.999
9.5	0.645	0.752	0.836	0.898	0.940	0.967	0.982	0.991	0.996	0.998
10.0	0.583	0.697	0.792	0.864	0.917	0.951	0.973	0.986	0.993	0.997

	20	21	22
8.5	1.000		
9.0	1.000		
9.5	0.999	1.000	
10.0	0.998	0.999	1.000

$\mu\backslash c$	0	1	2	3	4	5	6	7	8	9
10.5	0.000	0.000	0.002	0.007	0.021	0.050	0.102	0.179	0.279	0.397
11.0	0.000	0.000	0.001	0.005	0.015	0.038	0.079	0.143	0.232	0.341
11.5	0.000	0.000	0.001	0.003	0.011	0.028	0.060	0.114	0.191	0.289
12.0	0.000	0.000	0.001	0.002	0.008	0.020	0.046	0.090	0.155	0.242
12.5	0.000	0.000	0.000	0.002	0.005	0.015	0.035	0.070	0.125	0.201
13.0	0.000	0.000	0.000	0.001	0.004	0.011	0.026	0.054	0.100	0.166
13.5	0.000	0.000	0.000	0.001	0.003	0.008	0.019	0.041	0.079	0.135
14.0	0.000	0.000	0.000	0.000	0.002	0.006	0.014	0.032	0.062	0.109
14.5	0.000	0.000	0.000	0.000	0.001	0.004	0.010	0.024	0.048	0.088
15.0	0.000	0.000	0.000	0.000	0.001	0.003	0.008	0.018	0.037	0.070

	10	11	12	13	14	15	16	17	18	19
10.5	0.521	0.639	0.742	0.825	0.888	0.932	0.960	0.978	0.988	0.994
11.0	0.460	0.579	0.689	0.781	0.854	0.907	0.944	0.968	0.982	0.991
11.5	0.402	0.520	0.633	0.733	0.815	0.878	0.924	0.954	0.974	0.986
12.0	0.347	0.462	0.576	0.682	0.772	0.844	0.899	0.937	0.963	0.979
12.5	0.297	0.406	0.519	0.628	0.725	0.806	0.869	0.916	0.948	0.969

Table A.3 (continued)

$\mu \backslash c$	10	11	12	13	14	15	16	17	18	19
13.0	0.252	0.353	0.463	0.573	0.675	0.764	0.835	0.890	0.930	0.957
13.5	0.211	0.304	0.409	0.518	0.623	0.718	0.798	0.861	0.908	0.942
14.0	0.176	0.260	0.358	0.464	0.570	0.669	0.756	0.827	0.883	0.923
14.5	0.145	0.220	0.311	0.413	0.518	0.619	0.711	0.790	0.853	0.901
15.0	0.118	0.185	0.268	0.363	0.466	0.568	0.664	0.749	0.819	0.875

	20	21	22	23	24	25	26	27	28	29
10.5	0.997	0.999	0.999	1.000						
11.0	0.995	0.998	0.999	1.000						
11.5	0.992	0.996	0.998	0.999	1.000					
12.0	0.988	0.994	0.997	0.999	0.999	1.000				
12.5	0.983	0.991	0.995	0.998	0.999	0.999	1.000			
13.0	0.975	0.986	0.992	0.996	0.998	0.999	1.000			
13.5	0.965	0.980	0.989	0.994	0.997	0.998	0.999	1.000		
14.0	0.952	0.971	0.983	0.991	0.995	0.997	0.999	0.999	1.000	
14.5	0.936	0.960	0.976	0.986	0.992	0.996	0.998	0.999	0.999	1.000
15.0	0.917	0.947	0.967	0.981	0.989	0.994	0.997	0.998	0.999	1.000

$\mu \backslash c$	4	5	6	7	8	9	10	11	12	13
16	0.000	0.001	0.004	0.010	0.022	0.043	0.077	0.127	0.193	0.275
17	0.000	0.001	0.002	0.005	0.013	0.026	0.049	0.085	0.135	0.201
18	0.000	0.000	0.001	0.003	0.007	0.015	0.030	0.055	0.092	0.143
19	0.000	0.000	0.001	0.002	0.004	0.009	0.018	0.035	0.061	0.098
20	0.000	0.000	0.000	0.001	0.002	0.005	0.011	0.021	0.039	0.066
21	0.000	0.000	0.000	0.000	0.001	0.003	0.006	0.013	0.025	0.043
22	0.000	0.000	0.000	0.000	0.001	0.002	0.004	0.008	0.015	0.028
23	0.000	0.000	0.000	0.000	0.000	0.001	0.002	0.004	0.009	0.017
24	0.000	0.000	0.000	0.000	0.000	0.000	0.001	0.003	0.005	0.011
25	0.000	0.000	0.000	0.000	0.000	0.000	0.001	0.001	0.003	0.006

	14	15	16	17	18	19	20	21	22	23
16	0.368	0.467	0.566	0.659	0.742	0.812	0.868	0.911	0.942	0.963
17	0.281	0.371	0.468	0.564	0.655	0.736	0.805	0.861	0.905	0.937
18	0.208	0.287	0.375	0.469	0.562	0.651	0.731	0.799	0.855	0.899
19	0.150	0.215	0.292	0.378	0.469	0.561	0.647	0.725	0.793	0.849
20	0.105	0.157	0.221	0.297	0.381	0.470	0.559	0.644	0.721	0.787
21	0.072	0.111	0.163	0.227	0.302	0.384	0.471	0.558	0.640	0.716
22	0.048	0.077	0.117	0.169	0.232	0.306	0.387	0.472	0.556	0.637
23	0.031	0.052	0.082	0.123	0.175	0.238	0.310	0.389	0.472	0.555
24	0.020	0.034	0.056	0.087	0.128	0.180	0.243	0.314	0.392	0.473
25	0.012	0.022	0.038	0.060	0.092	0.134	0.185	0.247	0.318	0.394

Table A.3 (continued)

	24	25	26	27	28	29	30	31	32	33
16	0.978	0.987	0.993	0.996	0.998	0.999	0.999	1.000		
17	0.959	0.975	0.985	0.991	0.995	0.997	0.999	0.999	1.000	
18	0.932	0.955	0.972	0.983	0.990	0.994	0.997	0.998	0.999	1.000
19	0.893	0.927	0.951	0.969	0.980	0.988	0.993	0.996	0.998	0.999
20	0.843	0.888	0.922	0.948	0.966	0.978	0.987	0.992	0.995	0.997
21	0.782	0.838	0.883	0.917	0.944	0.963	0.976	0.985	0.991	0.994
22	0.712	0.777	0.832	0.877	0.913	0.940	0.959	0.973	0.983	0.989
23	0.635	0.708	0.772	0.827	0.873	0.908	0.936	0.956	0.971	0.981
24	0.554	0.632	0.704	0.768	0.823	0.868	0.904	0.932	0.953	0.969
25	0.473	0.553	0.629	0.700	0.763	0.818	0.863	0.900	0.929	0.950

	34	35	36	37	38	39	40	41	42	43
19	0.999	1.000								
20	0.999	0.999	1.000							
21	0.997	0.998	0.999	0.999	1.000					
22	0.994	0.996	0.998	0.999	0.999	1.000				
23	0.988	0.993	0.996	0.997	0.999	0.999	1.000			
24	0.979	0.987	0.992	0.995	0.997	0.998	0.999	0.999	1.000	
25	0.966	0.978	0.985	0.991	0.994	0.997	0.998	0.999	0.999	1.000

Table A.4 Four-Place Common Logarithms

N	0	1	2	3	4	5	6	7	8	9	Proportional Parts 1	2	3	4	5	6	7	8	9
10	0000	0043	0086	0128	0170	0212	0253	0294	0334	0374	4	8	12	17	21	25	29	33	37
11	0414	0453	0492	0531	0569	0607	0645	0682	0719	0755	4	8	11	15	19	23	26	30	34
12	0792	0828	0864	0899	0934	0969	1004	1038	1072	1106	3	7	10	14	17	21	24	28	31
13	1139	1173	1206	1239	1271	1303	1335	1367	1399	1430	3	6	10	13	16	19	23	26	29
14	1461	1492	1523	1553	1584	1614	1644	1673	1703	1732	3	6	9	12	15	18	21	24	27
15	1761	1790	1818	1847	1875	1903	1931	1959	1987	2014	3	6	8	11	14	17	20	22	25
16	2041	2068	2095	2122	2148	2175	2201	2227	2253	2279	3	5	8	11	13	16	18	21	24
17	2304	2330	2355	2380	2405	2430	2455	2480	2504	2529	2	5	7	10	12	15	17	20	22
18	2553	2577	2601	2625	2648	2672	2695	2718	2742	2765	2	5	7	9	12	14	16	19	21
19	2788	2810	2833	2856	2878	2900	2923	2945	2967	2989	2	4	7	9	11	13	16	18	20
20	3010	3032	3054	3075	3096	3118	3139	3160	3181	3201	2	4	6	8	11	13	15	17	19
21	3222	3243	3263	3284	3304	3324	3345	3365	3385	3404	2	4	6	8	10	12	14	16	18
22	3424	3444	3464	3483	3502	3522	3541	3560	3579	3598	2	4	6	8	10	12	14	15	17
23	3617	3636	3655	3674	3692	3711	3729	3747	3766	3784	2	4	6	7	9	11	13	15	17
24	3802	3820	3838	3856	3874	3892	3909	3927	3945	3962	2	4	5	7	9	11	12	14	16
25	3979	3997	4014	4031	4048	4065	4082	4099	4116	4133	2	3	5	7	9	10	12	14	15
26	4150	4166	4183	4200	4216	4232	4249	4265	4281	4298	2	3	5	7	8	10	11	13	15
27	4314	4330	4346	4362	4378	4393	4409	4425	4440	4456	2	3	5	6	8	9	11	13	14
28	4472	4487	4502	4518	4533	4548	4564	4579	4594	4609	2	3	5	6	8	9	11	12	14
29	4624	4639	4654	4669	4683	4698	4713	4728	4742	4757	1	3	4	6	7	9	10	12	13
30	4771	4786	4800	4814	4829	4843	4857	4871	4886	4900	1	3	4	6	7	9	10	11	13
31	4914	4928	4942	4955	4969	4983	4997	5011	5024	5038	1	3	4	6	7	8	10	11	12
32	5051	5065	5079	5092	5105	5119	5132	5145	5159	5172	1	3	4	5	7	8	9	11	12
33	5185	5198	5211	5224	5237	5250	5263	5276	5289	5302	1	3	4	5	6	8	9	10	12
34	5315	5328	5340	5353	5366	5378	5391	5403	5416	5428	1	3	4	5	6	8	9	10	11
35	5441	5453	5465	5478	5490	5502	5514	5527	5539	5551	1	2	4	5	6	7	9	10	11
36	5563	5575	5587	5599	5611	5623	5635	5647	5658	5670	1	2	4	5	6	7	8	10	11
37	5682	5694	5705	5717	5729	5740	5752	5763	5775	5786	1	2	3	5	6	7	8	9	10
38	5798	5809	5821	5832	5843	5855	5866	5877	5888	5899	1	2	3	5	6	7	8	9	10
39	5911	5922	5933	5944	5955	5966	5977	5988	5999	6010	1	2	3	4	5	7	8	9	10
40	6021	6031	6042	6053	6064	6075	6085	6096	6107	6117	1	2	3	4	5	6	8	9	10
41	6128	6138	6149	6160	6170	6180	6191	6201	6212	6222	1	2	3	4	5	6	7	8	9
42	6232	6243	6253	6263	6274	6284	6294	6304	6314	6325	1	2	3	4	5	6	7	8	9
43	6335	6345	6355	6365	6375	6385	6395	6405	6415	6425	1	2	3	4	5	6	7	8	9
44	6435	6444	6454	6464	6474	6484	6493	6503	6513	6522	1	2	3	4	5	6	7	8	9
45	6532	6542	6551	6561	6571	6580	6590	6599	6609	6618	1	2	3	4	5	6	7	8	9
46	6628	6637	6646	6656	6665	6675	6684	6693	6702	6712	1	2	3	4	5	6	7	7	8
47	6721	6730	6739	6749	6758	6767	6776	6785	6794	6803	1	2	3	4	5	5	6	7	8
48	6812	6821	6830	6839	6848	6857	6866	6875	6884	6893	1	2	3	4	4	5	6	7	8
49	6902	6911	6920	6928	6937	6946	6955	6964	6972	6981	1	2	3	4	4	5	6	7	8
50	6990	6998	7007	7016	7024	7033	7042	7050	7059	7067	1	2	3	3	4	5	6	7	8
51	7076	7084	7093	7101	7110	7118	7126	7135	7143	7152	1	2	3	3	4	5	6	7	8
52	7160	7168	7177	7185	7193	7202	7210	7218	7226	7235	1	2	2	3	4	5	6	7	7
53	7243	7251	7259	7267	7275	7284	7292	7300	7308	7316	1	2	2	3	4	5	6	6	7
54	7324	7332	7340	7348	7356	7364	7372	7380	7388	7396	1	2	2	3	4	5	6	6	7
N	0	1	2	3	4	5	6	7	8	9	1	2	3	4	5	6	7	8	9

Table A.4 (continued)

N	0	1	2	3	4	5	6	7	8	9	Proportional Parts								
											1	2	3	4	5	6	7	8	9
55	7404	7412	7419	7427	7435	7443	7451	7459	7466	7474	1	2	2	3	4	5	5	6	7
56	7482	7490	7497	7505	7513	7520	7528	7536	7543	7551	1	2	2	3	4	5	5	6	7
57	7559	7566	7574	7582	7589	7597	7604	7612	7619	7627	1	2	2	3	4	5	5	6	7
58	7634	7642	7649	7657	7664	7672	7679	7686	7694	7701	1	1	2	3	4	4	5	6	7
59	7709	7716	7723	7731	7738	7745	7752	7760	7767	7774	1	1	2	3	4	4	5	6	7
60	7782	7789	7796	7803	7810	7818	7825	7832	7839	7846	1	1	2	3	4	4	5	6	6
61	7853	7860	7868	7875	7882	7889	7896	7903	7910	7917	1	1	2	3	4	4	5	6	6
62	7924	7931	7938	7945	7952	7959	7966	7973	7980	7987	1	1	2	3	3	4	5	6	6
63	7993	8000	8007	8014	8021	8028	8035	8041	8048	8055	1	1	2	3	3	4	5	5	6
64	8062	8069	8075	8082	8089	8096	8102	8109	8116	8122	1	1	2	3	3	4	5	5	6
65	8129	8136	8142	8149	8156	8162	8169	8176	8182	8189	1	1	2	3	3	4	5	5	6
66	8195	8202	8209	8215	8222	8228	8235	8241	8248	8254	1	1	2	3	3	4	5	5	6
67	8261	8267	8274	8280	8287	8293	8299	8306	8312	8319	1	1	2	3	3	4	5	5	6
68	8325	8331	8338	8344	8351	8357	8363	8370	8376	8382	1	1	2	3	3	4	4	5	6
69	8388	8395	8401	8407	8414	8420	8426	8432	8439	8445	1	1	2	2	3	4	4	5	6
70	8451	8457	8463	8470	8476	8482	8488	8494	8500	8506	1	1	2	2	3	4	4	5	6
71	8513	8519	8525	8531	8537	8543	8549	8555	8561	8567	1	1	2	2	3	4	4	5	5
72	8573	8579	8585	8591	8597	8603	8609	8615	8621	8627	1	1	2	2	3	4	4	5	5
73	8633	8639	8645	8651	8657	8663	8669	8675	8681	8686	1	1	2	2	3	4	4	5	5
74	8692	8698	8704	8710	8716	8722	8727	8733	8739	8745	1	1	2	2	3	4	4	5	5
75	8751	8756	8762	8768	8774	8779	8785	8791	8797	8802	1	1	2	2	3	3	4	5	5
76	8808	8814	8820	8825	8831	8837	8842	8848	8854	8859	1	1	2	2	3	3	4	5	5
77	8865	8871	8876	8882	8887	8893	8899	8904	8910	8915	1	1	2	2	3	3	4	4	5
78	8921	8927	8932	8938	8943	8949	8954	8960	8965	8971	1	1	2	2	3	3	4	4	5
79	8976	8982	8987	8993	8998	9004	9009	9015	9020	9025	1	1	2	2	3	3	4	4	5
80	9031	9036	9042	9047	9053	9058	9063	9069	9074	9079	1	1	2	2	3	3	4	4	5
81	9085	9090	9096	9101	9106	9112	9117	9122	9128	9133	1	1	2	2	3	3	4	4	5
82	9138	9143	9149	9154	9159	9165	9170	9175	9180	9186	1	1	2	2	3	3	4	4	5
83	9191	9196	9201	9206	9212	9217	9222	9227	9232	9238	1	1	2	2	3	3	4	4	5
84	9243	9248	9253	9258	9263	9269	9274	9279	9284	9289	1	1	2	2	3	3	4	4	5
85	9294	9299	9304	9309	9315	9320	9325	9330	9335	9340	1	1	2	2	3	3	4	4	5
86	9345	9350	9355	9360	9365	9370	9375	9380	9385	9390	1	1	2	2	3	3	4	4	5
87	9395	9400	9405	9410	9415	9420	9425	9430	9435	9440	0	1	1	2	2	3	3	4	4
88	9445	9450	9455	9460	9465	9469	9474	9479	9484	9489	0	1	1	2	2	3	3	4	4
89	9494	9499	9504	9509	9513	9518	9523	9528	9533	9538	0	1	1	2	2	3	3	4	4
90	9542	9547	9552	9557	9562	9566	9571	9576	9581	9586	0	1	1	2	2	3	3	4	4
91	9590	9595	9600	9605	9609	9614	9619	9624	9628	9633	0	1	1	2	2	3	3	4	4
92	9638	9643	9647	9652	9657	9661	9666	9671	9675	9680	0	1	1	2	2	3	3	4	4
93	9685	9689	9694	9699	9703	9708	9713	9717	9722	9727	0	1	1	2	2	3	3	4	4
94	9731	9736	9741	9745	9750	9754	9759	9763	9768	9773	0	1	1	2	2	3	3	4	4
95	9777	9782	9786	9791	9795	9800	9805	9809	9814	9818	0	1	1	2	2	3	3	4	4
96	9823	9827	9832	9836	9841	9845	9850	9854	9859	9863	0	1	1	2	2	3	3	4	4
97	9868	9872	9877	9881	9886	9890	9894	9899	9903	9908	0	1	1	2	2	3	3	4	4
98	9912	9917	9921	9926	9930	9934	9939	9943	9948	9952	0	1	1	2	2	3	3	4	4
99	9956	9961	9965	9969	9974	9978	9983	9987	9991	9996	0	1	1	2	2	3	3	3	4
N	0	1	2	3	4	5	6	7	8	9	1	2	3	4	5	6	7	8	9

Table A.5 Areas Under the Standard Normal Probability Distribution Between the Mean and Successive Values of z.

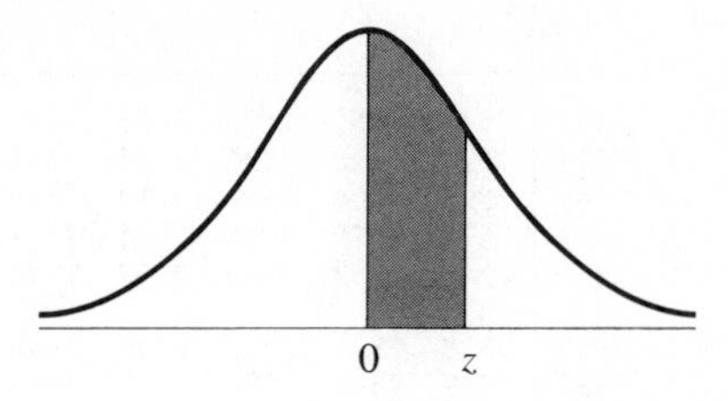

Example If $z = 1.00$, then the area between the mean and this value of z is 0.3413.

z	.00	.01	.02	.03	.04	.05	.06	.07	.08	.09
0.0	.0000	.0040	.0080	.0120	.0160	.0199	.0239	.0279	.0319	.0359
0.1	.0398	.0438	.0478	.0517	.0557	.0596	.0636	.0675	.0714	.0753
0.2	.0793	.0832	.0871	.0910	.0948	.0987	.1026	.1064	.1103	.1141
0.3	.1179	.1217	.1255	.1293	.1331	.1368	.1406	.1443	.1480	.1517
0.4	.1554	.1591	.1628	.1664	.1700	.1736	.1772	.1808	.1844	.1879
0.5	.1915	.1950	.1985	.2019	.2054	.2088	.2123	.2157	.2190	.2224
0.6	.2257	.2291	.2324	.2357	.2389	.2422	.2454	.2486	.2518	.2549
0.7	.2580	.2612	.2642	.2673	.2704	.2734	.2764	.2794	.2823	.2852
0.8	.2881	.2910	.2939	.2967	.2995	.3023	.3051	.3078	.3106	.3133
0.9	.3159	.3186	.3212	.3238	.3264	.3289	.3315	.3340	.3365	.3389
1.0	.3413	.3438	.3461	.3485	.3508	.3531	.3554	.3577	.3599	.3621
1.1	.3643	.3665	.3686	.3708	.3729	.3749	.3770	.3790	.3810	.3830
1.2	.3849	.3869	.3888	.3907	.3925	.3944	.3962	.3980	.3997	.4015
1.3	.4032	.4049	.4066	.4082	.4099	.4115	.4131	.4147	.4162	.4177
1.4	.4192	.4207	.4222	.4236	.4251	.4265	.4279	.4292	.4306	.4319
1.5	.4332	.4345	.4357	.4370	.4382	.4394	.4406	.4418	.4429	.4441
1.6	.4452	.4463	.4474	.4484	.4495	.4505	.4515	.4525	.4535	.4545
1.7	.4554	.4564	.4573	.4582	.4591	.4599	.4608	.4616	.4625	.4633
1.8	.4641	.4649	.4656	.4664	.4671	.4678	.4686	.4693	.4699	.4706
1.9	.4713	.4719	.4726	.4732	.4738	.4744	.4750	.4756	.4761	.4767
2.0	.4772	.4778	.4783	.4788	.4793	.4798	.4803	.4808	.4812	.4817
2.1	.4821	.4826	.4830	.4834	.4838	.4842	.4846	.4850	.4854	.4857
2.2	.4861	.4864	.4868	.4871	.4875	.4878	.4881	.4884	.4887	.4890
2.3	.4893	.4896	.4898	.4901	.4904	.4906	.4909	.4911	.4913	.4916
2.4	.4918	.4920	.4922	.4925	.4927	.4929	.4931	.4932	.4934	.4936
2.5	.4938	.4940	.4941	.4943	.4945	.4946	.4948	.4949	.4951	.4952
2.6	.4953	.4955	.4956	.4957	.4959	.4960	.4961	.4962	.4963	.4964
2.7	.4965	.4966	.4967	.4968	.4969	.4970	.4971	.4972	.4973	.4974
2.8	.4974	.4975	.4976	.4977	.4977	.4978	.4979	.4979	.4980	.4981
2.9	.4981	.4982	.4982	.4983	.4984	.4984	.4985	.4985	.4986	.4986
3.0	.49865	.4987	.4987	.4988	.4988	.4989	.4989	.4989	.4990	.4990
4.0	.49997									

Table A.6 Student's t-Distribution

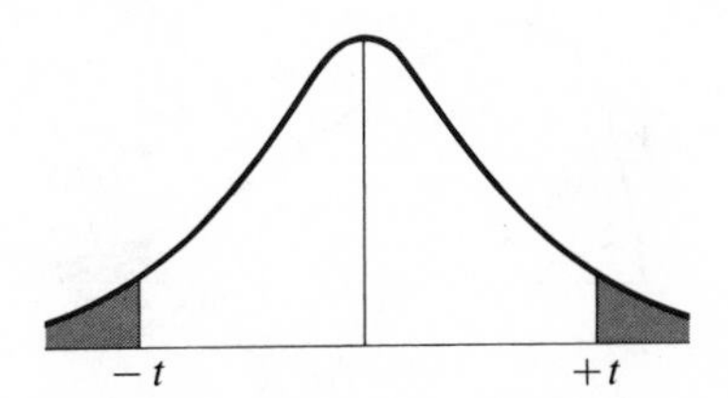

Example For 15 degrees of freedom, the t-value which corresponds to an area of .05 in both tails combined is 2.131.

Degrees of Freedom	*Area in Both Tails Combined*			
	.10	.05	.02	.01
1	6.314	12.706	31.821	63.657
2	2.920	4.303	6.965	9.925
3	2.353	3.182	4.541	5.841
4	2.132	2.776	3.747	4.604
5	2.015	2.571	3.365	4.032
6	1.943	2.447	3.143	3.707
7	1.895	2.365	2.998	3.499
8	1.860	2.306	2.896	3.355
9	1.833	2.262	2.821	3.250
10	1.812	2.228	2.764	3.169
11	1.796	2.201	2.718	3.106
12	1.782	2.179	2.681	3.055
13	1.771	2.160	2.650	3.012
14	1.761	2.145	2.624	2.977
15	1.753	2.131	2.602	2.947
16	1.746	2.120	2.583	2.921
17	1.740	2.110	2.567	2.898
18	1.734	2.101	2.552	2.878
19	1.729	2.093	2.539	2.861
20	1.725	2.086	2.528	2.845
21	1.721	2.080	2.518	2.831
22	1.717	2.074	2.508	2.819
23	1.714	2.069	2.500	2.807
24	1.711	2.064	2.492	2.797
25	1.708	2.060	2.485	2.787
26	1.706	2.056	2.479	2.779
27	1.703	2.052	2.473	2.771
28	1.701	2.048	2.467	2.763
29	1.699	2.045	2.462	2.756
30	1.697	2.042	2.457	2.750
40	1.684	2.021	2.423	2.704
60	1.671	2.000	2.390	2.660
120	1.658	1.980	2.358	2.617
Normal Distribution	1.645	1.960	2.326	2.576

Source: Table A.6 is taken from Table III of Fisher and Yates: *Statistical Tables for Biological, Agricultural and Medical Research*, published by Oliver and Boyd Ltd., Edinburgh, and by permission of the authors and publishers.

Table A.7 Chi Square (χ^2) Distribution

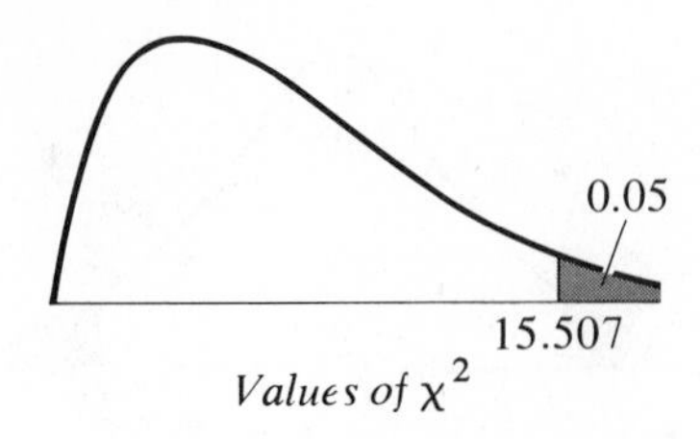

Example In a chi square distribution with $\nu = 8$ degrees of freedom, the area to the right of a chi square value of 15.507 is 0.05.

Degrees of Freedom	*Area in Right Tail*				
v	.20	.10	.05	.02	.01
1	1.642	2.706	3.841	5.412	6.635
2	3.219	4.605	5.991	7.824	9.210
3	4.642	6.251	7.815	9.837	11.345
4	5.989	7.779	9.488	11.668	13.277
5	7.289	9.236	11.070	13.388	15.086
6	8.558	10.645	12.592	15.033	16.812
7	9.803	12.017	14.067	16.622	18.475
8	11.030	13.362	15.507	18.168	20.090
9	12.242	14.684	16.919	19.679	21.666
10	13.442	15.987	18.307	21.161	23.209
11	14.631	17.275	19.675	22.618	24.725
12	15.812	18.549	21.026	24.054	26.217
13	16.985	19.812	22.362	25.472	27.688
14	18.151	21.064	23.685	26.873	29.141
15	19.311	22.307	24.996	28.259	30.578
16	20.465	23.542	26.296	29.633	32.000
17	21.615	24.769	27.587	30.995	33.409
18	22.760	25.989	28.869	32.346	34.805
19	23.900	27.204	30.144	33.687	36.191
20	25.038	28.412	31.410	35.020	37.566
21	26.171	29.615	32.671	36.343	38.932
22	27.301	30.813	33.924	37.659	40.289
23	28.429	32.007	35.172	38.968	41.638
24	29.553	33.196	36.415	40.270	42.980
25	30.675	34.382	37.652	41.566	44.314
26	31.795	35.563	38.885	42.856	45.642
27	32.912	36.741	40.113	44.140	46.963
28	34.027	37.916	41.337	45.419	48.278
29	35.139	39.087	42.557	46.693	49.588
30	36.250	40.256	43.773	47.962	50.892

Source: Table A.7 is taken from Table IV of Fisher and Yates: *Statistical Tables for Biological, Agricultural and Medical Research,* published by Oliver and Boyd Ltd., Edinburgh, and by permission of the authors and publishers.

Table A.8 F-Distribution

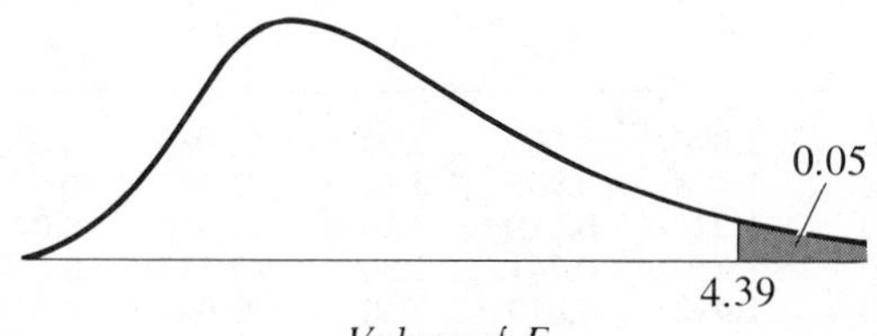

Example In an F-distribution with $\nu_1 = 5$ and $\nu_2 = 6$ degrees of freedom, the area to the right of an F value of 4.39 is 0.05. The value on the F-scale to the right of which lies .05 of the area is in lightface type. The value on the F-scale to the right of which lies .01 of the area is in boldface type. ν_1 = number of degrees of freedom for numerator; ν_2 = number of degrees of freedom for denominator.

ν_2 \ ν_1	1	2	3	4	5	6	7	8	9	10	20	30	40	50	100	200	∞	ν_1 / ν_2
1	161	200	216	225	230	234	237	239	241	242	248	250	251	252	253	254	254	1
	4,052	**4,999**	**5,403**	**5,625**	**5,764**	**5,859**	**5,928**	**5,981**	**6,022**	**6,056**	**6,208**	**6,261**	**6,286**	**6,302**	**6,334**	**6,352**	**6,366**	
2	18.51	19.00	19.16	19.25	19.30	19.33	19.36	19.37	19.38	19.39	19.44	19.46	19.47	19.47	19.49	19.49	19.50	2
	98.49	**99.00**	**99.17**	**99.25**	**99.30**	**99.33**	**99.36**	**99.37**	**99.39**	**99.40**	**99.45**	**99.47**	**99.48**	**99.48**	**99.49**	**99.49**	**99.50**	
3	10.13	9.55	9.28	9.12	9.01	8.94	8.88	8.84	8.81	8.78	8.66	8.62	8.60	8.58	8.56	8.54	8.53	3
	34.12	**30.82**	**29.46**	**28.71**	**28.24**	**27.91**	**27.67**	**27.49**	**27.34**	**27.23**	**26.69**	**26.50**	**26.41**	**26.35**	**26.23**	**26.18**	**26.12**	
4	7.71	6.94	6.59	6.39	6.26	6.16	6.09	6.04	6.00	5.96	5.80	5.74	5.71	5.70	5.66	5.65	5.63	4
	21.20	**18.00**	**16.69**	**15.98**	**15.52**	**15.21**	**14.98**	**14.80**	**14.66**	**14.54**	**14.02**	**13.83**	**13.74**	**13.69**	**13.57**	**13.52**	**13.46**	
5	6.61	5.79	5.41	5.19	5.05	4.95	4.88	4.82	4.78	4.74	4.56	4.50	4.46	4.44	4.40	4.38	4.36	5
	16.26	**13.27**	**12.06**	**11.39**	**10.97**	**10.67**	**10.45**	**10.29**	**10.15**	**10.05**	**9.55**	**9.38**	**9.29**	**9.24**	**9.13**	**9.07**	**9.02**	
6	5.99	5.14	4.76	4.53	4.39	4.28	4.21	4.15	4.10	4.06	3.87	3.81	3.77	3.75	3.71	3.69	3.67	6
	13.74	**10.92**	**9.78**	**9.15**	**8.75**	**8.47**	**8.26**	**8.10**	**7.98**	**7.87**	**7.39**	**7.23**	**7.14**	**7.09**	**6.99**	**6.94**	**6.88**	
7	5.59	4.74	4.35	4.12	3.97	3.87	3.79	3.73	3.68	3.63	3.44	3.38	3.34	3.32	3.28	3.25	3.23	7
	12.25	**9.55**	**8.45**	**7.85**	**7.46**	**7.19**	**7.00**	**6.84**	**6.71**	**6.62**	**6.15**	**5.98**	**5.90**	**5.85**	**5.75**	**5.70**	**5.65**	
8	5.32	4.46	4.07	3.84	3.69	3.58	3.50	3.44	3.39	3.34	3.15	3.08	3.05	3.03	2.98	2.96	2.93	8
	11.26	**8.65**	**7.59**	**7.01**	**6.63**	**6.37**	**6.19**	**6.03**	**5.91**	**5.82**	**5.36**	**5.20**	**5.11**	**5.06**	**4.96**	**4.91**	**4.86**	
9	5.12	4.26	3.86	3.63	3.48	3.37	3.29	3.23	3.18	3.13	2.93	2.86	2.82	2.80	2.76	2.73	2.71	9
	10.56	**8.02**	**6.99**	**6.42**	**6.06**	**5.80**	**5.62**	**5.47**	**5.35**	**5.26**	**4.80**	**4.64**	**4.56**	**4.51**	**4.41**	**4.36**	**4.31**	
10	4.96	4.10	3.71	3.48	3.33	3.22	3.14	3.07	3.02	2.97	2.77	2.70	2.67	2.64	2.59	2.56	2.54	10
	10.04	**7.56**	**6.55**	**5.99**	**5.64**	**5.39**	**5.21**	**5.06**	**4.95**	**4.85**	**4.41**	**4.25**	**4.17**	**4.12**	**4.01**	**3.96**	**3.91**	
20	4.35	3.49	3.10	2.87	2.71	2.60	2.52	2.45	2.40	2.35	2.12	2.04	1.99	1.96	1.90	1.87	1.84	20
	8.10	**5.85**	**4.94**	**4.43**	**4.10**	**3.87**	**3.71**	**3.56**	**3.45**	**3.37**	**2.94**	**2.77**	**2.69**	**2.63**	**2.53**	**2.47**	**2.42**	
30	4.17	3.32	2.92	2.69	2.53	2.42	2.34	2.27	2.21	2.16	1.93	1.84	1.79	1.76	1.69	1.66	1.62	30
	7.56	**5.39**	**4.51**	**4.02**	**3.70**	**3.47**	**3.30**	**3.17**	**3.06**	**2.98**	**2.55**	**2.38**	**2.29**	**2.24**	**2.13**	**2.07**	**2.01**	
40	4.08	3.23	2.84	2.61	2.45	2.34	2.25	2.18	2.12	2.07	1.84	1.74	1.69	1.66	1.59	1.55	1.51	40
	7.31	**5.18**	**4.31**	**3.83**	**3.51**	**3.29**	**3.12**	**2.99**	**2.88**	**2.80**	**2.37**	**2.20**	**2.11**	**2.05**	**1.94**	**1.88**	**1.81**	
50	4.03	3.18	2.79	2.56	2.40	2.29	2.20	2.13	2.07	2.02	1.78	1.69	1.63	1.60	1.52	1.48	1.44	50
	7.17	**5.06**	**4.20**	**3.72**	**3.41**	**3.18**	**3.02**	**2.88**	**2.78**	**2.70**	**2.26**	**2.10**	**2.00**	**1.94**	**1.82**	**1.76**	**1.68**	
100	3.94	3.09	2.70	2.46	2.30	2.19	2.10	2.03	1.97	1.92	1.68	1.57	1.51	1.48	1.39	1.34	1.28	100
	6.90	**4.82**	**3.98**	**3.51**	**3.20**	**2.99**	**2.82**	**2.69**	**2.59**	**2.51**	**2.06**	**1.89**	**1.79**	**1.73**	**1.59**	**1.51**	**1.43**	
200	3.89	3.04	2.65	2.41	2.26	2.14	2.05	1.98	1.92	1.87	1.62	1.52	1.45	1.42	1.32	1.26	1.19	200
	6.76	**4.71**	**3.88**	**3.41**	**3.11**	**2.90**	**2.73**	**2.60**	**2.50**	**2.41**	**1.97**	**1.79**	**1.69**	**1.62**	**1.48**	**1.39**	**1.28**	
∞	3.84	2.99	2.60	2.37	2.21	2.09	2.01	1.94	1.88	1.83	1.57	1.46	1.40	1.35	1.24	1.17	1.00	∞
	6.64	**4.60**	**3.78**	**3.32**	**3.02**	**2.80**	**2.64**	**2.51**	**2.41**	**2.32**	**1.87**	**1.69**	**1.59**	**1.52**	**1.36**	**1.25**	**1.00**	

Source: This table is abridged by permission from *Statistical Methods*, 5th ed., by George W. Snedecor, © 1956 by The Iowa State University Press.

Table A.9 Table of Exponential Functions

x	e^x	e^{-x}	x	e^x	e^{-x}
0.00	1.000	1.000	3.00	20.086	0.050
0.10	1.105	0.905	3.10	22.198	0.045
0.20	1.221	0.819	3.20	24.533	0.041
0.30	1.350	0.741	3.30	27.113	0.037
0.40	1.492	0.670	3.40	29.964	0.033
0.50	1.649	0.607	3.50	33.115	0.030
0.60	1.822	0.549	3.60	36.598	0.027
0.70	2.014	0.497	3.70	40.447	0.025
0.80	2.226	0.449	3.80	44.701	0.022
0.90	2.460	0.407	3.90	49.402	0.020
1.00	2.718	0.368	4.00	54.598	0.018
1.10	3.004	0.333	4.10	60.340	0.017
1.20	3.320	0.301	4.20	66.686	0.015
1.30	3.669	0.273	4.30	73.700	0.014
1.40	4.055	0.247	4.40	81.451	0.012
1.50	4.482	0.223	4.50	90.017	0.011
1.60	4.953	0.202	4.60	99.484	0.010
1.70	5.474	0.183	4.70	109.95	0.009
1.80	6.050	0.165	4.80	121.51	0.008
1.90	6.686	0.150	4.90	134.29	0.007
2.00	7.389	0.135	5.00	148.41	0.007
2.10	8.166	0.122	5.10	164.02	0.006
2.20	9.025	0.111	5.20	181.27	0.006
2.30	9.974	0.100	5.30	200.34	0.005
2.40	11.023	0.091	5.40	221.41	0.005
2.50	12.182	0.082	5.50	244.69	0.004
2.60	13.464	0.074	5.60	270.43	0.004
2.70	14.880	0.067	5.70	298.87	0.003
2.80	16.445	0.061	5.80	330.30	0.003
2.90	18.174	0.055	5.90	365.04	0.003
3.00	20.086	0.050	6.00	403.43	0.002

Table A.10 Squares, Square Roots, and Reciprocals

n	n^2	$\sqrt{n}$	$1/n$	n	n^2	$\sqrt{n}$	$1/n$
				50	2 500	7.071 068	.02000000
1	1	1.000 000	1.0000000	51	2 601	7.141 428	.01960784
2	4	1.414 214	.5000000	52	2 704	7.211 103	.01923077
3	9	1.732 051	.3333333	53	2 809	7.280 110	.01886792
4	16	2.000 000	.2500000	54	2 916	7.348 469	.01851852
5	25	2.236 068	.2000000	55	3 025	7.416 198	.01818182
6	36	2.449 490	.1666667	56	3 136	7.483 315	.01785714
7	49	2.645 751	.1428571	57	3 249	7.549 834	.01754386
8	64	2.828 427	.1250000	58	3 364	7.615 773	.01724138
9	81	3.000 000	.1111111	59	3 481	7.681 146	.01694915
10	100	3.162 278	.1000000	60	3 600	7.745 967	.01666667
11	121	3.316 625	.09090909	61	3 721	7.810 250	.01639344
12	144	3.464 102	.08333333	62	3 844	7.874 008	.01612903
13	169	3.605 551	.07692308	63	3 969	7.937 254	.01587302
14	196	3.741 657	.07142857	64	4 096	8.000 000	.01562500
15	225	3.872 983	.06666667	65	4 225	8.062 258	.01538462
16	256	4.000 000	.06250000	66	4 356	8.124 038	.01515152
17	289	4.123 106	.05882353	67	4 489	8.185 353	.01492537
18	324	4.242 641	.05555556	68	4 624	8.246 211	.01470588
19	361	4.358 899	.05263158	69	4 761	8.306 624	.01449275
20	400	4.472 136	.05000000	70	4 900	8.366 600	.01428571
21	441	4.582 576	.04761905	71	5 041	8.426 150	.01408451
22	484	4.690 416	.04545455	72	5 184	8.485 281	.01388889
23	529	4.795 832	.04347826	73	5 329	8.544 004	.01369863
24	576	4.898 979	.04166667	74	5 476	8.602 325	.01351351
25	625	5.000 000	.04000000	75	5 625	8.660 254	.01333333
26	676	5.099 020	.03846154	76	5 776	8.717 798	.01315789
27	729	5.196 152	.03703704	77	5 929	8.774 964	.01298701
28	784	5.291 503	.03571429	78	6 084	8.831 761	.01282051
29	841	5.385 165	.03448276	79	6 241	8.888 194	.01265823
30	900	5.477 226	.03333333	80	6 400	8.944 272	.01250000
31	961	5.567 764	.03225806	81	6 561	9.000 000	.01234568
32	1 024	5.656 854	.03125000	82	6 724	9.055 385	.01219512
33	1 089	5.744 563	.03030303	83	6 889	9.110 434	.01204819
34	1 156	5.830 952	.02941176	84	7 056	9.165 151	.01190476
35	1 225	5.916 080	.02857143	85	7 225	9.219 544	.01176471
36	1 296	6.000 000	.02777778	86	7 396	9.273 618	.01162791
37	1 369	6.082 763	.02702703	87	7 569	9.327 379	.01149425
38	1 444	6.164 414	.02631579	88	7 744	9.380 832	.01136364
39	1 521	6.244 998	.02564103	89	7 921	9.433 981	.01123596
40	1 600	6.324 555	.02500000	90	8 100	9.486 833	.01111111
41	1 681	6.403 124	.02439024	91	8 281	9.539 392	.01098901
42	1 764	6.480 741	.02380952	92	8 464	9.591 663	.01086957
43	1 849	6.557 439	.02325581	93	8 649	9.643 651	.01075269
44	1 936	6.633 250	.02272727	94	8 836	9.695 360	.01063830
45	2 025	6.708 204	.02222222	95	9 025	9.746 794	.01052632
46	2 116	6.782 330	.02173913	96	9 216	9.797 959	.01041667
47	2 209	6.855 655	.02127660	97	9 409	9.848 858	.01030928
48	2 304	6.928 203	.02083333	98	9 604	9.899 495	.01020408
49	2 401	7.000 000	.02040816	99	9 801	9.949 874	.01010101
50	2 500	7.071 068	.02000000	100	10 000	10.00000	.01000000

Source: Frederick E. Croxton and Dudley J. Cowden, Practical Business Statistics, 3rd edition,

Table A.10 (continued)

n	n^2	$\sqrt{n}$	$1/n$ .0	n	n^2	$\sqrt{n}$	$1/n$ .00
100	10 000	10.00000	10000000	150	22 500	12.24745	6666667
101	10 201	10.04988	09900990	151	22 801	12.28821	6622517
102	10 404	10.09950	09803922	152	23 104	12.32883	6578947
103	10 609	10.14889	09708738	153	23 409	12.36932	6535948
104	10 816	10.19804	09615385	154	23 716	12.40967	6493506
105	11 025	10.24695	09523810	155	24 025	12.44990	6451613
106	11 236	10.29563	09433962	156	24 336	12.49000	6410256
107	11 449	10.34408	09345794	157	24 649	12.52996	6369427
108	11 664	10.39230	09259259	158	24 964	12.56981	6329114
109	11 881	10.44031	09174312	159	25 281	12.60952	6289308
110	12 100	10.48809	09090909	160	25 600	12.64911	6250000
111	12 321	10.53565	09009009	161	25 921	12.68858	6211180
112	12 544	10.58301	08928571	162	26 244	12.72792	6172840
113	12 769	10.63015	08849558	163	26 569	12.76715	6134969
114	12 996	10.67708	08771930	164	26 896	12.80625	6097561
115	13 225	10.72381	08695652	165	27 225	12.84523	6060606
116	13 456	10.77033	08620690	166	27 556	12.88410	6024096
117	13 689	10.81665	08547009	167	27 889	12.92285	5988024
118	13 924	10.86278	08474576	168	28 224	12.96148	5952381
119	14 161	10.90871	08403361	169	28 561	13.00000	5917160
120	14 400	10.95445	08333333	170	28 900	13.03840	5882353
121	14 641	11.00000	08264463	171	29 241	13.07670	5847953
122	14 884	11.04536	08196721	172	29 584	13.11488	5813953
123	15 129	11.09054	08130081	1,3	29 929	13.15295	5780347
124	15 376	11.13553	08064516	174	30 276	13.19091	5747126
125	15 625	11.18034	08000000	175	30 625	13.22876	5714286
126	15 876	11.22497	07936508	176	30 976	13.26650	5681818
127	16 129	11.26943	07874016	177	31 329	13.30413	5649718
128	16 384	11.31371	07812500	178	31 684	13.34166	5617978
129	16 641	11.35782	07751938	179	32 041	13.37909	5586592
130	16 900	11.40175	07692308	180	32 400	13.41641	5555556
131	17 161	11.44552	07633588	181	32 761	13.45362	5524862
132	17 424	11.48913	07575758	182	33 124	13.49074	5494505
133	17 689	11.53256	07518797	183	33 489	13.52775	5464481
134	17 956	11.57584	07462687	184	33 856	13.56466	5434783
135	18 225	11.61895	07407407	185	34 225	13.60147	5405405
136	18 496	11.66190	07352941	186	34 596	13.63818	5376344
137	18 769	11.70470	07299270	187	34 969	13.67479	5347594
138	19 044	11.74734	07246377	188	35 344	13.71131	5319149
139	19 321	11.78983	07194245	189	35 721	13.74773	5291005
140	19 600	11.83216	07142857	190	36 100	13.78405	5263158
141	19 881	11.87434	07092199	191	36 481	13.82027	5235602
142	20 164	11.91638	07042254	192	36 864	13.85641	5208333
143	20 449	11.95826	06993007	193	37 249	13.89244	5181347
144	20 736	12.00000	06944444	194	37 636	13.92839	5154639
145	21 025	12.04159	06896552	195	38 025	13.96424	5128205
146	21 316	12.08305	06849315	196	38 416	14.00000	5102041
147	21 609	12.12436	06802721	197	38 809	14.03567	5076142
148	21 904	12.16553	06756757	198	39 204	14.07125	5050505
149	22 201	12.20656	06711409	199	39 601	14.10674	5025126
150	22 500	12.24745	06666667	200	40 000	14.14214	5000000

Table A.10 (continued)

n	n^2	$\sqrt{n}$	$1/n$ .00
200	40 000	14.14214	5000000
201	40 401	14.17745	4975124
202	40 804	14.21267	4950495
203	41 209	14.24781	4926108
204	41 616	14.28286	4901961
205	42 025	14.31782	4878049
206	42 436	14.35270	4854369
207	42 849	14.38749	4830918
208	43 264	14.42221	4807692
209	43 681	14.45683	4784689
210	44 100	14.49138	4761905
211	44 521	14.52584	4739336
212	44 944	14.56022	4716981
213	45 369	14.59452	4694836
214	45 796	14.62874	4672897
215	46 225	14.66288	4651163
216	46 656	14.69694	4629630
217	47 089	14.73092	4608295
218	47 524	14.76482	4587156
219	47 961	14.79865	4566210
220	48 400	14.83240	4545455
221	48 841	14.86607	4524887
222	49 284	14.89966	4504505
223	49 729	14.93318	4484305
224	50 176	14.96663	4464286
225	50 625	15.00000	4444444
226	51 076	15.03330	4424779
227	51 529	15.06652	4405286
228	51 984	15.09967	4385965
229	52 441	15.13275	4366812
230	52 900	15.16575	4347826
231	53 361	15.19868	4329004
232	53 824	15.23155	4310345
233	54 289	15.26434	4291845
234	54 756	15.29706	4273504
235	55 225	15.32971	4255319
236	55 696	15.36229	4237288
237	56 169	15.39480	4219409
238	56 644	15.42725	4201681
239	57 121	15.45962	4184100
240	57 600	15.49193	4166667
241	58 081	15.52417	4149378
242	58 564	15.55635	4132231
243	59 049	15.58846	4115226
244	59 536	15.62050	4098361
245	60 025	15.65248	4081633
246	60 516	15.68439	4065041
247	61 009	15.71623	4048583
248	61 504	15.74802	4032258
249	62 001	15.77973	4016064
250	62 500	15.81139	4000000

n	n^2	$\sqrt{n}$	$1/n$ .00
250	62 500	15.81139	4000000
251	63 001	15.84298	3984064
252	63 504	15.87451	3968254
253	64.009	15.90597	3952569
254	64 516	15.93738	3937008
255	65 025	15.96872	3921569
256	65 536	16.00000	3906250
257	66 049	16.03122	3891051
258	66 564	16.06238	3875969
259	67 081	16.09348	3861004
260	67 600	16.12452	3846154
261	68 121	16.15549	3831418
262	68 644	16.18641	3816794
263	69 169	16.21727	3802281
264	69 696	16.24808	3787879
265	70 225	16.27882	3773585
266	70 756	16.30951	3759398
267	71 289	16.34013	3745318
268	71 824	16.37071	3731343
269	72 361	16.40122	3717472
270	72 900	16.43168	3703704
271	73 441	16.46208	3690037
272	73 984	16.49242	3676471
273	74 529	16.52271	3663004
274	75 076	16.55295	3649635
275	75 625	16.58312	3636364
276	76 176	16.61325	3623188
277	76 729	16.64332	3610108
278	77 284	16.67333	3597122
279	77 841	16.70329	3584229
280	78 400	16.73320	3571429
281	78 961	16.76305	3558719
282	79 524	16.79286	3546099
283	80 089	16.82260	3533569
284	80 656	16.85230	3521127
285	81 225	16.88194	3508772
286	81 796	16.91153	3496503
287	82 369	16.94107	3484321
288	82 944	16.97056	3472222
289	83 521	17.00000	3460208
290	84 100	17.02939	3448276
291	84 681	17.05872	3436426
292	85 264	17.08801	3424658
293	85 849	17.11724	3412969
294	86 436	17.14643	3401361
295	87 025	17.17556	3389831
296	87 616	17.20465	3378378
297	88 209	17.23369	3367003
298	88 804	17.26268	3355705
299	89 401	17.29162	3344482
300	90 000	17.32051	3333333

Table A.10 (continued)

n	n^2	$\sqrt{n}$	$1/n$ .00	n	n^2	$\sqrt{n}$	$1/n$ .00
300	90 000	17.32051	3333333	350	122 500	18.70829	2857143
301	90 601	17.34935	3322259	351	123 201	18.73499	2849003
302	91 204	17.37815	3311258	352	123 904	18.76166	2840909
303	91 809	17.40690	3300330	353	124 609	18.78829	2832861
304	92 416	17.43560	3289474	354	125 316	18.81489	2824859
305	93 025	17.46425	3278689	355	126 025	18.84144	2816901
306	93 636	17.49286	3267974	356	126 736	18.86796	2808989
307	94 249	17.52142	3257329	357	127 449	18.89444	2801120
308	94 864	17.54993	3246753	358	128 164	18.92089	2793296
309	95 481	17.57840	3236246	359	128 881	18.94730	2785515
310	96 100	17.60682	3225806	360	129 600	18.97367	2777778
311	96 721	17.63519	3215434	361	130 321	19.00000	2770083
312	97 344	17.66352	3205128	362	131 044	19.02630	2762431
313	97 969	17.69181	3194888	363	131 769	19.05256	2754821
314	98 596	17.72005	3184713	364	132 496	19.07878	2747253
315	99 225	17.74824	3174603	365	133 225	19.10497	2739726
316	99 856	17.77639	3164557	366	133 956	19.13113	2732240
317	100 489	17.80449	3154574	367	134 689	19.15724	2724796
318	101 124	17.83255	3144654	368	135 424	19.18333	2717391
319	101 761	17.86057	3134796	369	136 161	19.20937	2710027
320	102 400	17.88854	3125000	370	136 900	19.23538	2702703
321	103 041	17.91647	3115265	371	137 641	19.26136	2695418
322	103 684	17.94436	3105590	372	138 384	19.28730	2688172
323	104 329	17.97220	3095975	373	139 129	19.31321	2680965
324	104 976	18.00000	3086420	374	139 876	19.33908	2673797
325	105 625	18.02776	3076923	375	140 625	19.36492	2666667
326	106 276	18.05547	3067485	376	141 376	19.39072	2659574
327	106 929	18.08314	3058104	377	142 129	19.41649	2652520
328	107 584	18.11077	3048780	378	142 884	19.44222	2645503
329	108 241	18.13836	3039514	379	143 641	19.46792	2638522
330	108 900	18.16590	3030303	380	144 400	19.49359	2631579
331	109 561	18.19341	3021148	381	145 161	19.51922	2624672
332	110 224	18 22087	3012048	382	145 924	19.54483	2617801
333	110 889	18.24829	3003003	383	146 689	19.57039	2610966
334	111 556	18.27567	2994012	384	147 456	19.59592	2604167
335	112 225	18.30301	2985075	385	148 225	19.62142	2597403
336	112 896	18.33030	2976190	386	148 996	19.64688	2590674
337	113 569	18.35756	2967359	387	149 769	19.67232	2583979
338	114 244	18.38478	2958580	388	150 544	19.69772	2577320
339	114 921	18.41195	2949853	389	151 321	19.72308	2570694
340	115 600	18.43909	2941176	390	152 100	19.74842	2564103
341	116 281	18.46619	2932551	391	152 881	19.77372	2557545
342	116 964	18.49324	2923977	392	153 664	19.79899	2551020
343	117 649	18.52026	2915452	393	154 449	19.82423	2544529
344	118 336	18.54724	2906977	394	155 236	19.84943	2538071
345	119 025	18.57418	2898551	395	156 025	19.87461	2531646
346	119 716	18.60108	2890173	396	156 816	19.89975	2525253
347	120 409	18.62794	2881844	397	157 609	19.92486	2518892
348	121 104	18.65476	2873563	398	158 404	19.94994	2512563
349	121 801	18.68154	2865330	399	159 201	19.97498	2506266
350	122 500	18.70829	2857143	400	160 000	20.00000	2500000

Table A.10 (continued)

n	n^2	$\sqrt{n}$	$1/n$ .00
400	160 000	20.00000	2500000
401	160 801	20.02498	2493766
402	161 604	20.04994	2487562
403	162 409	20.07486	2481390
404	163 216	20.09975	2475248
405	164 025	20.12461	2469136
406	164 836	20.14944	2463054
407	165 649	20.17424	2457002
408	166 464	20.19901	2450980
409	167 281	20.22375	2444988
410	168 100	20.24846	2439024
411	168 921	20.27313	2433090
412	169 744	20.29778	2427184
413	170 569	20.32240	2421308
414	171 396	20.34699	2415459
415	172 225	20.37155	2409639
416	173 056	20.39608	2403846
417	173 889	20.42058	2398082
418	174 724	20.44505	2392344
419	175 561	20.46949	2386635
420	176 400	20.49390	2380952
421	177 241	20.51828	2375297
422	178 084	20.54264	2369668
423	178 929	20.56696	2364066
424	179 776	20.59126	2358491
425	180 625	20.61553	2352941
426	181 476	20.63977	2347418
427	182 329	20.66398	2341920
428	183 184	20.68816	2336449
429	184 041	20.71232	2331002
430	184 900	20.73644	2325581
431	185 761	20.76054	2320186
432	186 624	20.78461	2314815
433	187 489	20.80865	2309469
434	188 356	20.83267	2304147
435	189 225	20.85665	2298851
436	190 096	20.88061	2293578
437	190 969	20.90454	2288330
438	191 844	20.92845	2283105
439	192 721	20.95233	2277904
440	193 600	20.97618	2272727
441	194 481	21.00000	2267574
442	195 364	21.02380	2262443
443	196 249	21.04757	2257336
444	197 136	21.07131	2252252
445	198 025	21.09502	2247191
446	198 916	21.11871	2242152
447	199 809	21.14237	2237136
448	200 704	21.16601	2232143
449	201 601	21.18962	2227171
450	202 500	21.21320	2222222

n	n^2	$\sqrt{n}$	$1/n$ .00
450	202 500	21.21320	2222222
451	203 401	21.23676	2217295
452	204 304	21.26029	2212389
453	205 209	21.28380	2207506
454	206 116	21.30728	2202643
455	207 025	21.33073	2197802
456	207 936	21.35416	2192982
457	208 849	21.37756	2188184
458	209 764	21.40093	2183406
459	210 681	21.42429	2178649
460	211 600	21.44761	2173913
461	212 521	21.47091	2169197
462	213 444	21.49419	2164502
463	214 369	21.51743	2159827
464	215 296	21.54066	2155172
465	216 225	21.56386	2150538
466	217 156	21.58703	2145923
467	218 089	21.61018	2141328
468	219 024	21.63331	2136752
469	219 961	21.65641	2132196
470	220 900	21.67948	2127660
471	221 841	21.70253	2123142
472	222 784	21.72556	2118644
473	223 729	21.74856	2114165
474	224 676	21.77154	2109705
475	225 625	21.79449	2105263
476	226 576	21.81742	2100840
477	227 529	21.84033	2096436
478	228 484	21.86321	2092050
479	229 441	21.88607	2087683
480	230 400	21.90890	2083333
481	231 361	21.93171	2079002
482	232 324	21.95450	2074689
483	233 289	21.97726	2070393
484	234 256	22.00000	2066116
485	235 225	22.02272	2061856
486	236 196	22.04541	2057613
487	237 169	22.06808	2053388
488	238 144	22.09072	2049180
489	239 121	22.11334	2044990
490	240 100	22.13594	2040816
491	241 081	22.15852	2036660
492	242 064	22.18107	2032520
493	243 049	22.20360	2028398
494	244 036	22.22611	2024291
495	245 025	22.24860	2020202
496	246 016	22.27106	2016129
497	247 009	22.29350	2012072
498	248 004	22.31591	2008032
499	249 001	22.33831	2004008
500	250 000	22.36068	2000000

Table A.10 (continued)

n	n^2	$\sqrt{n}$	$1/n$ .00	n	n^2	$\sqrt{n}$	$1/n$ .00
500	250 000	22.36068	2000000	550	302 500	23.45208	1818182
501	251 001	22.38303	1996008	551	303 601	23.47339	1814882
502	252 004	22.40536	1992032	552	304 704	23.49468	1811594
503	253 009	22.42766	1988072	553	305 809	23.51595	1808318
504	254 016	22.44994	1984127	554	306 916	23.53720	1805054
505	255 025	22.47221	1980198	555	308 025	23.55844	1801802
506	256 036	22.49444	1976285	556	309 136	23.57965	1798561
507	257 049	22.51666	1972387	557	310 249	23.60085	1795332
508	258 064	22.53886	1968504	558	311 364	23.62202	1792115
509	259 081	22.56103	1964637	559	312 481	23.64318	1788909
510	260 100	22.58318	1960784	560	313 600	23.66432	1785714
511	261 121	22.60531	1956947	561	314 721	23.68544	1782531
512	262 144	22.62742	1953125	562	315 844	23.70654	1779359
513	263 169	22.64950	1949318	563	316 969	23.72762	1776199
514	264 196	22.67157	1945525	564	318 096	23.74868	1773050
515	265 225	22.69361	1941748	565	319 225	23.76973	1769912
516	266 256	22.71563	1937984	566	320 356	23.79075	1766784
517	267 289	22.73763	1934236	567	321 489	23.81176	1763668
518	268 324	22.75961	1930502	568	322 624	23.83275	1760563
519	269 361	22.78157	1926782	569	323 761	23.85372	1757469
520	270 400	22.80351	1923077	570	324 900	23.87467	1754386
521	271 441	22.82542	1919386	571	326 041	23.89561	1751313
522	272 484	22.84732	1915709	572	327 184	23.91652	1748252
523	273 529	22.86919	1912046	573	328 329	23.93742	1745201
524	274 576	22.89105	1908397	574	329 476	23.95830	1742160
525	275 625	22.91288	1904762	575	330 625	23.97916	1739130
526	276 676	22.93469	1901141	576	331 776	24.00000	1736111
527	277 729	22.95648	1897533	577	332 929	24.02082	1733102
528	278 784	22.97825	1893939	578	334 084	24.04163	1730104
529	279 841	23.00000	1890359	579	335 241	24.06242	1727116
530	280 900	23.02173	1886792	580	336 400	24.08319	1724138
531	281 961	23.04344	1883239	581	337 561	24.10394	1721170
532	283 024	23.06513	1879699	582	338 724	24.12468	1718213
533	284 089	23.08679	1876173	583	339 889	24.14539	1715266
534	285 156	23.10844	1872659	584	341 056	24.16609	1712329
535	286 225	23.13007	1869159	585	342 225	24.18677	1709402
536	287 296	23.15167	1865672	586	343 396	24.20744	1706485
537	288 369	23.17326	1862197	587	344 569	24.22808	1703578
538	289 444	23.19483	1858736	588	345 744	24.24871	1700680
539	290 521	23.21637	1855288	589	346 921	24.26932	1697793
540	291 600	23.23790	1851852	590	348 100	24.28992	1694915
541	292 681	23.25941	1848429	591	349 281	24.31049	1692047
542	293 764	23.28089	1845018	592	350 464	24.33105	1689189
543	294 849	23.30236	1841621	593	351 649	24.35159	1686341
544	295 936	23.32381	1838235	594	352 836	24.37212	1683502
545	297 025	23.34524	1834862	595	354 025	24.39262	1680672
546	298 116	23.36664	1831502	596	355 216	24.41311	1677852
547	299 209	23.38803	1828154	597	356 409	24.43358	1675042
548	300 304	23.40940	1824818	598	357 604	24.45404	1672241
549	301 401	23.43075	1821494	599	358 801	24.47448	1669449
550	302 500	23.45208	1818182	600	360 000	24.49490	1666667

Table A.10 (continued)

n	n^2	$\sqrt{n}$	$1/n$ .00
600	360 000	24.49490	1666667
601	361 201	24.51530	1663894
602	362 404	24.53569	1661130
603	363 609	24.55606	1658375
604	364 816	24.57641	1655629
605	366 025	24.59675	1652893
606	367 236	24.61707	1650165
607	368 449	24.63737	1647446
608	369 664	24.65766	1644737
609	370 881	24.67793	1642036
610	372 100	24.69818	1639344
611	373 321	24.71841	1636661
612	374 544	24.73863	1633987
613	375 769	24.75884	1631321
614	376 996	24.77902	1628664
615	378 225	24.79919	1626016
616	379 456	24.81935	1623377
617	380 689	24.83948	1620746
618	381 924	24.85961	1618123
619	383 161	24.87971	1615509
620	384 400	24.89980	1612903
621	385 641	24.91987	1610306
622	386 884	24.93993	1607717
623	388 129	24.95997	1605136
624	389 376	24.97999	1602564
625	390 625	25.00000	1600000
626	391 876	25.01999	1597444
627	393 129	25.03997	1594896
628	394 384	25.05993	1592357
629	395 641	25.07987	1589825
630	396 900	25.09980	1587302
631	398 161	25.11971	1584786
632	399 424	25.13961	1582278
633	400 689	25.15949	1579779
634	401 956	25.17936	1577287
635	403 225	25.19921	1574803
636	404 496	25.21904	1572327
637	405 769	25.23886	1569859
638	407 044	25.25866	1567398
639	408 321	25.27845	1564945
640	409 600	25.29822	1562500
641	410 881	25.31798	1560062
642	412 164	25.33772	1557632
643	413 449	25.35744	1555210
644	414 736	25.37716	1552795
645	416 025	25.39685	1550388
646	417 316	25.41653	1547988
647	418 609	25.43619	1545595
648	419 904	25.45584	1543210
649	421 201	25.47548	1540832
650	422 500	25.49510	1538462

n	n^2	$\sqrt{n}$	$1/n$ .00
650	422 500	25.49510	1538462
651	423 801	25.51470	1536098
652	425 104	25.53429	1533742
653	426 409	25.55386	1531394
654	427 716	25.57342	1529052
655	429 025	25.59297	1526718
656	430 336	25.61250	1524390
657	431 649	25.63201	1522070
658	432 964	25.65151	1519757
659	434 281	25.67100	1517451
660	435 600	25.69047	1515152
661	436 921	25.70992	1512859
662	438 244	25.72936	1510574
663	439 569	25.74879	1508296
664	440 896	25.76820	1506024
665	442 225	25.78759	1503759
666	443 556	25.80698	1501502
667	444 889	25.82634	1499250
668	446 224	25.84570	1497006
669	447 561	25.86503	1494768
670	448 900	25.88436	1492537
671	450 241	25.90367	1490313
672	451 584	25.92296	1488095
673	452 929	25.94224	1485884
674	454 276	25.96151	1483680
675	455 625	25.98076	1481481
676	456 976	26.00000	1479290
677	458 329	26.01922	1477105
678	459 684	26.03843	1474926
679	461 041	26.05763	1472754
680	462 400	26.07681	1470588
681	463 761	26.09598	1468429
682	465 124	26.11513	1466276
683	466 489	26.13427	1464129
684	467 856	26.15339	1461988
685	469 225	26.17250	1459854
686	470 596	26.19160	1457726
687	471 969	26.21068	1455604
688	473 344	26.22975	1453488
689	474 721	26.24881	1451379
690	476 100	26.26785	1449275
691	477 481	26.28688	1447178
692	478 864	26.30589	1445087
693	480 249	26.32489	1443001
694	481 636	26.34388	1440922
695	483 025	26.36285	1438849
696	484 416	26.38181	1436782
697	485 809	26.40076	1434720
698	487 204	26.41969	1432665
699	488 601	26.43861	1430615
700	490 000	26.45751	1428571

Table A.10 (continued)

n	n^2	$\sqrt{n}$	$1/n$ .00	n	n^2	$\sqrt{n}$	$1/n$ .00
700	490 000	26.45751	1428571	750	562 500	27.38613	1333333
701	491 401	26.47640	1426534	751	564 001	27.40438	1331558
702	492 804	26.49528	1424501	752	565 504	27.42262	1329787
703	494 209	26.51415	1422475	753	567 009	27.44085	1328021
704	495 616	26.53300	1420455	754	568 516	27.45906	1326260
705	497 025	26.55184	1418440	755	570 025	27.47726	1324503
706	498 436	26.57066	1416431	756	571 536	27.49545	1322751
707	499 849	26.58947	1414427	757	573 049	27.51363	1321004
708	501 264	26.60827	1412429	758	574 564	27.53180	1319261
709	502 681	26.62705	1410437	759	576 081	27.54995	1317523
710	504 100	26.64583	1408451	760	577 600	27.56810	1315789
711	505 521	26.66458	1406470	761	579 121	27.58623	1314060
712	506 944	26.68333	1404494	762	580 644	27.60435	1312336
713	508 369	26.70206	1402525	763	582 169	27.62245	1310616
714	509 796	26.72078	1400560	764	583 696	27.64055	1308901
715	511 225	26.73948	1398601	765	585 225	27.65863	1307190
716	512 656	26.75818	1396648	766	586 756	27.67671	1305483
717	514 089	26.77686	1394700	767	588 289	27.69476	1303781
718	515 524	26.79552	1392758	768	589 824	27.71281	1302083
719	516 961	26.81418	1390821	769	591 361	27.73085	1300390
720	518 400	26.83282	1388889	770	592 900	27.74887	1298701
721	519 841	26.85144	1386963	771	594 441	27.76689	1297017
722	521 284	26.87006	1385042	772	595 984	27.78489	1295337
723	522 729	26.88866	1383126	773	597 529	27.80288	1293661
724	524 176	26.90725	1381215	774	599 076	27.82086	1291990
725	525 625	26.92582	1379310	775	600 625	27.83882	1290323
726	527 076	26.94439	1377410	776	602 176	27.85678	1288660
727	528 529	26.96294	1375516	777	603 729	27.87472	1287001
728	529 984	26.98148	1373626	778	605 284	27.89265	1285347
729	531 441	27.00000	1371742	779	606 841	27.91057	1283697
730	532 900	27.01851	1369863	780	608 400	27.92848	1282051
731	534 361	27.03701	1367989	781	609 961	27.94638	1280410
732	535 824	27.05550	1366120	782	611 524	27.96426	1278772
733	537 289	27.07397	1364256	783	613 089	27.98214	1277139
734	538 756	27.09243	1362398	784	614 656	28.00000	1275510
735	540 225	27.11088	1360544	785	616 225	28.01785	1273885
736	541 696	27.12932	1358696	786	617 796	28.03569	1272265
737	543 169	27.14774	1356852	787	619 369	28.05352	1270648
738	544 644	27.16616	1355014	788	620 944	28.07134	1269036
739	546 121	27.18455	1353180	789	622 521	28.08914	1267427
740	547 600	27.20294	1351351	790	624 100	28.10694	1265823
741	549 081	27.22132	1349528	791	625 681	28.12472	1264223
742	550 564	27.23968	1347709	792	627 264	28.14249	1262626
743	552 049	27.25803	1345895	793	628 849	28.16026	1261034
744	553 536	27.27636	1344086	794	630 436	28.17801	1259446
745	555 025	27.29469	1342282	795	632 025	28.19574	1257862
746	556 516	27.31300	1340483	796	633 616	28.21347	1256281
747	558 009	27.33130	1338688	797	635 209	28.23119	1254705
748	559 504	27.34959	1336898	798	636 804	28.24889	1253133
749	561 001	27.36786	1335113	799	638 401	28.26659	1251564
750	562 500	27.38613	1333333	800	640 000	28.28427	1250000

Table A.10 (continued)

n	n^2	$\sqrt{n}$	$1/n$ .00	n	n^2	$\sqrt{n}$	$1/n$ .00
800	640 000	28.28427	1250000	850	722 500	29.15476	1176471
801	641 601	28.30194	1248439	851	724 201	29.17190	1175088
802	643 204	28.31960	1246883	852	725 904	29.18904	1173709
803	644 809	28.33725	1245330	853	727 609	29.20616	1172333
804	646 416	28.35489	1243781	854	729 316	29.22328	1170960
805	648 025	28.37252	1242236	855	731 025	29.24038	1169591
806	649 636	28.39014	1240695	856	732 736	29.25748	1168224
807	651 249	28.40775	1239157	857	734 449	29.27456	1166861
808	652 864	28.42534	1237624	858	736 164	29.29164	1165501
809	654 481	28.44293	1236094	859	737 881	29.30870	1164144
810	656 100	28.46050	1234568	860	739 600	29.32576	1162791
811	657 721	28.47806	1233046	861	741 321	29.34280	1161440
812	659 344	28.49561	1231527	862	743 044	29.35984	1160093
813	660 969	28.51315	1230012	863	744 769	29.37686	1158749
814	662 596	28.53069	1228501	864	746 496	29.39388	1157407
815	664 225	28.54820	1226994	865	748 225	29.41088	1156069
816	665 856	28.56571	1225490	866	749 956	29.42788	1154734
817	667 489	28.58321	1223990	867	751 689	29.44486	1153403
818	669 124	28.60070	1222494	868	753 424	29.46184	1152074
819	670 761	28.61818	1221001	869	755 161	29.47881	1150748
820	672 400	28.63564	1219512	870	756.900	29.49576	1149425
821	674 041	28.65310	1218027	871	758 641	29.51271	1148106
822	675 684	28.67054	1216545	872	760 384	29.52965	1146789
823	677 329	28.68798	1215067	873	762 129	29.54657	1145475
824	678 976	28.70540	1213592	874	763 876	29.56349	1144165
825	680 625	28.72281	1212121	875	765 625	29.58040	1142857
826	682 276	28.74022	1210654	876	767 376	29.59730	1141553
827	683 929	28.75761	1209190	877	769 129	29.61419	1140251
828	685 584	28.77499	1207729	878	770 884	29.63106	1138952
829	687 241	28.79236	1206273	879	772 641	29.64793	1137656
830	688 900	28.80972	1204819	880	774 400	29.66479	1136364
831	690 561	28.82707	1203369	881	776 161	29.68164	1135074
832	692 224	28.84441	1201923	882	777 924	29.69848	1133787
833	693 889	28.86174	1200480	883	779 689	29.71532	1132503
834	695 556	28.87906	1199041	884	781 456	29.73214	1131222
835	697 225	28.89637	1197605	885	783 225	29.74895	1129944
836	698 896	28.91366	1196172	886	784 996	29.76575	1128668
837	700 569	28.93095	1194743	887	786 769	29.78255	1127396
838	702 244	28.94823	1193317	888	788 544	29.79933	1126126
839	703 921	28.96550	1191895	889	790 321	29.81610	1124859
840	705 600	28.98275	1190476	890	792 100	29.83287	1123596
841	707 281	29.00000	1189061	891	793 881	29.84962	1122334
842	708 964	29.01724	1187648	892	795 664	29.86637	1121076
843	710 649	29.03446	1186240	893	797 449	29.88311	1119821
844	712 336	29.05168	1184834	894	799 236	29.89983	1118568
845	714 025	29.06888	1183432	895	801 025	29.91655	1117318
846	715 716	29.08608	1182033	896	802 816	29.93326	1116071
847	717 409	29.10326	1180638	897	804 609	29.94996	1114827
848	719 104	29.12044	1179245	898	806 404	29.96665	1113586
849	720 801	29.13760	1177856	899	808 201	29.98333	1112347
850	722 500	29.15476	1176471	900	810 000	30.00000	1111111

Table A.10 (continued)

n	n^2	$\sqrt{n}$	$1/n$ .00
900	810 000	30.00000	1111111
901	811 801	30.01666	1109878
902	813 604	30.03331	1108647
903	815 409	30.04996	1107420
904	817 216	30.06659	1106195
905	819 025	30.08322	1104972
906	820 836	30.09983	1103753
907	822 649	30.11644	1102536
908	824 464	30.13304	1101322
909	826 281	30.14963	1100110
910	828 100	30.16621	1098901
911	829 921	30.18278	1097695
912	831 744	30.19934	1096491
913	833 569	30.21589	1095290
914	835 396	30.23243	1094092
915	837 225	30.24897	1092896
916	839 056	30.26549	1091703
917	840 889	30.28201	1090513
918	842 724	30.29851	1089325
919	844 561	30.31501	1088139
920	846 400	30.33150	1086957
921	848 241	30.34798	1085776
922	850 084	30.36445	1084599
923	851 929	30.38092	1083424
924	853 776	30.39737	1082251
925	855 625	30.41381	1081081
926	857 476	30.43025	1079914
927	859 329	30.44667	1078749
928	861 184	30.46309	1077586
929	863 041	30.47950	1076426
930	864 900	30.49590	1075269
931	866 761	30.51229	1074114
932	868 624	30.52868	1072961
933	870 489	30.54505	1071811
934	872 356	30.56141	1070664
935	874 225	30.57777	1069519
936	876 096	30.59412	1068376
937	877 969	30.61046	1067236
938	879 844	30.62679	1066098
939	881 721	30.64311	1064963
940	883 600	30.65942	1063830
941	885 481	30.67572	1062699
942	887 364	30.69202	1061571
943	889 249	30.70831	1060445
944	891 136	30.72458	1059322
945	893 025	30.74085	1058201
946	894 916	30.75711	1057082
947	896 809	30.77337	1055966
948	898 704	30.78961	1054852
949	900 601	30.80584	1053741
950	902 500	30.82207	1052632

n	n^2	$\sqrt{n}$	$1/n$ .00
950	902 500	30.82207	1052632
951	904 401	30.83829	1051525
952	906 304	30.85450	1050420
953	908 209	30.87070	1049318
954	910 116	30.88689	1048218
955	912 025	30.90307	1047120
956	913 936	30.91925	1046025
957	915 849	30.93542	1044932
958	917 764	30.95158	1043841
959	919 681	30.96773	1042753
960	921 600	30.98387	1041667
961	923 521	31.00000	1040583
962	925 444	31.01612	1039501
963	927 369	31.03224	1038422
964	929 296	31.04835	1037344
965	931 225	31.06445	1036269
966	933 156	31.08054	1035197
967	935 089	31.09662	1034126
968	937 024	31.11270	1033058
969	938 961	31.12876	1031992
970	940 900	31.14482	1030928
971	942 841	31.16087	1029866
972	944 784	31.17691	1028807
973	946 729	31.19295	1027749
974	948 676	31.20897	1026694
975	950 625	31.22499	1025641
976	952 576	31.24100	1024590
977	954 529	31.25700	1023541
978	956 484	31.27299	1022495
979	958 441	31.28898	1021450
980	960 400	31.30495	1020408
981	962 361	31.32092	1019368
982	964 324	31.33688	1018330
983	966 289	31.35283	1017294
984	968 256	31.36877	1016260
985	970 225	31.38471	1015228
986	972 196	31.40064	1014199
987	974 169	31.41656	1013171
988	976 144	31.43247	1012146
989	978 121	31.44837	1011122
990	980 100	31.46427	1010101
991	982 081	31.48015	1009082
992	984 064	31.49603	1008065
993	986 049	31.51190	1007049
994	988 036	31.52777	1006036
995	990 025	31.54362	1005025
996	992 016	31.55947	1004016
997	994 009	31.57531	1003009
998	996 004	31.59114	1002004
999	998 001	31.60696	1001001
1000	1 000 000	31.62278	1000000

APPENDIX B

Symbols, Subscripts, and Summations

In statistics, *symbols* such as X, Y, and Z are used to represent different sets of data. Hence, if we have data for five families, we might let

$$X = \text{family income}$$
$$Y = \text{family clothing expenditures}$$
$$Z = \text{family savings}$$

Subscripts are used to represent individual observations within these sets of data. Thus, we write X_i to represent the income of the ith family, where i takes on the values 1, 2, 3, 4, and 5. In this notation X_1, X_2, X_3, X_4, and X_5 stand for the incomes of the first family, the second family, etc. The data are arranged in some order, such as by size of income, the order in which the data were gathered, or any other way suitable to the purposes or convenience of the investigator. The subscript i is a variable used to index the individual data observations. Continuing with the example, X_i, Y_i, and Z_i represent the income, clothing expenditures, and savings of the ith family. For example, X_2 represents the income of the second family, Y_2 clothing expenditures of the second family (same family), and Z_5 the savings of the fifth family.

Now, let us suppose that we have data for two different samples, say the net worths of 100 corporations and the test scores of 20 students. To refer to individual observations in these samples, we can let X_i denote the net worth of the ith corporation, where i assumes values from 1 to 100. This latter idea is indicated by the notation $i = 1, 2, 3, \ldots, 100$. Also we can let Y_j denote the test score of the jth student, where $j = 1, 2, 3, \ldots, 20$. Thus, the different subscript letters make it clear that different samples are involved. Letters such as X, Y, and Z are generally used to represent the different variables or types of measurements involved, whereas subscripts such as i, j, k, and l are used to designate individual observations.

We now turn to the method of expressing summations of sets of data.

Suppose we want to add a set of four observations, denoted X_1, X_2, X_3, and X_4. A convenient way of designating this addition is

$$\sum_{i=1}^{4} X_i = X_1 + X_2 + X_3 + X_4$$

where the symbol $\sum$ (Greek capital "sigma") means "the sum of." Hence, the symbol

$$\sum_{i=1}^{4} X_i$$

is read "the sum of the X_i's, i going from 1 to 4." For example, if $X_i = 3$, $X_2 = 1, X_3 = 10$, and $X_4 = 5$,

$$\sum_{i=1}^{4} X_i = 3 + 1 + 10 + 5 = 19$$

In general, if there are n observations, we write

$$\sum_{i=1}^{n} X_i = X_1 + X_2 + \cdots + X_n$$

Example 1 Let $X_1 = -2, X_2 = 3, X_3 = 5$. Find

(a) $$\sum_{i=1}^{3} X_i$$

(b) $$\sum_{j=1}^{3} X_j^2$$

(c) $$\sum_{j=1}^{3} (2X_j + 3)$$

Solution:

(a) $$\sum_{i=1}^{3} X_i = X_1 + X_2 + X_3 = -2 + 3 + 5 = 6$$

(b) $$\sum_{j=1}^{3} X_j^2 = X_1^2 + X_2^2 + X_3^2 = (-2)^2 + (3)^2 + (5)^2 = 38$$

(c) $$\begin{aligned}\sum_{j=1}^{3} (2X_j + 3) &= (2X_1 + 3) + (2X_2 + 3) + (2X_3 + 3)\\ &= (-4 + 3) + (6 + 3) + (10 + 3)\\ &= -1 + 9 + 13 = 21\end{aligned}$$

Example 2 Prove

(a)
$$\sum_{i=1}^{n} aX_i = a\sum_{i=1}^{n} X_i$$

(b)
$$\sum_{i=1}^{n} a = na$$

(c)
$$\sum_{i=1}^{n} (X_i + Y_i) = \sum_{i=1}^{n} X_i + \sum_{i=1}^{n} Y_i$$

where a is a constant.

Solution:

(a)
$$\begin{aligned}\sum_{i=1}^{n} aX_i &= aX_1 + aX_2 + \cdots + aX_n \\ &= a(X_1 + X_2 + \cdots + X_n) \\ &= a\sum_{i=1}^{n} X_i\end{aligned}$$

(b)
$$\begin{aligned}\sum_{i=1}^{n} a &= a\sum_{i=1}^{n} 1 \\ &= a\underbrace{(1 + 1 + \cdots + 1)}_{n \text{ terms}} \\ &= na\end{aligned}$$

(c)
$$\begin{aligned}\sum_{i=1}^{n} (X_i + Y_i) &= X_1 + Y_1 + X_2 + Y_2 + \cdots + X_n + Y_n \\ &= (X_1 + X_2 + \cdots + X_n) + (Y_1 + Y_2 + \cdots + Y_n) \\ &= \sum_{i=1}^{n} X_i + \sum_{i=1}^{n} Y_i\end{aligned}$$

These three summation properties have been indicated as Rules 1, 2, and 3 in Appendix B.

Double summations are used to indicate summations of more than one variable, where different subscript indexes are involved. For example, the symbol

$$\sum_{j=1}^{3}\sum_{i=1}^{2} X_iY_j$$

means "the sum of the products of X_i and Y_j where $i = 1, 2$ and $j = 1, 2, 3$." Thus, we can write

$$\sum_{j=1}^{3}\sum_{i=1}^{2} X_iY_j = X_1Y_1 + X_2Y_1 + X_1Y_2 + X_2Y_2 + X_1Y_3 + X_2Y_3$$

Simplified Summation Notations

In this text, simplified summation notations are often used in which subscripts are eliminated. Thus, for example, ΣX, ΣX^2, and ΣY^2 are used instead of

$$\sum_{i=1}^{n} X_i \quad \sum_{i=1}^{n} X_i^2 \quad \text{and} \quad \sum_{i=1}^{n} Y_i^2 \text{ respectively.}$$

Also in this text, subscripts have ordinarily been dropped in the case of probability distributions. For example, consider the following discrete probability distribution:

x_i	$f(x_i)$
$x_1 = 0$	.2
$x_2 = 1$	.3
$x_3 = 2$	.5
	1.0

The statement that the sum of the probabilities is equal to one is given by

$$\sum_{i=1}^{3} f(x_i) = 1$$

However, we have used the following customary simplified notation:

x	$f(x)$
0	.2
1	.3
2	.5
	1.0

The corresponding summation statement is

$$\sum_{x} f(x) = 1$$

where $\sum_{x}$ means "sum over all values of x." The notation is also often further simplified by writing

$$\sum f(x) = 1$$

Summation Properties

Rule 1

$$\sum_{i=1}^{n} aX_i = a\sum_{i=1}^{n} X_i$$

Rule 2

$$\sum_{i=1}^{n} a = \underbrace{a + a + \cdots + a}_{n \text{ terms}} = na$$

Rule 3

$$\sum_{i=1}^{n} (X_i + Y_i) = \sum_{i=1}^{n} X_i + \sum_{i=1}^{n} Y_i$$

APPENDIX C

Properties of Expected Values and Variances

In keeping with notational conventions used in this text, a's and b's represent constants, whereas X's represent random variables.

Rule 1 $E(a) = a$

The expected value of a constant is equal to that constant.

Rule 2 $E(bX) = bE(X)$

The expected value of a constant times a random variable is equal to the constant times the expected value of the random variable.

Rule 3 $E(a + bX) = a + bE(X)$

Rule 3 combines Rules 1 and 2. A brief proof is given for Rule 3 to illustrate a general method of proofs for expected values.

Let X denote a discrete random variable which takes on values $x_1, x_2, \ldots, x_i, \ldots, x_n$ with probabilities $f(x_1), f(x_2), \ldots, f(x_i), \ldots, f(x_n)$.

Then, using the definition of an expected value given in Equation (4.2) in Chapter 4, we have

$$(1)\quad E(a + bX) = \sum_{i=1}^{n} (a + bx_i) f(x_i) = \sum_{i=1}^{n} af(x_i) + \sum_{i=1}^{n} b(x_i) f(x_i)$$

$$(2)\quad = a\sum_{i=1}^{n} f(x_i) + b\sum_{i=1}^{n} x_i f(x_i)$$

$$(3)\quad = a(1) + bE(X) = a + bE(X)$$

Rule 4 $E(X_1 + X_2 + \cdots + X_n) = E(X_1) + E(X_2) + \cdots + E(X_n)$ where $X_1, X_2, \ldots, X_n$ are random variables.

The expected value of a sum equals the sum of the expected values. The X_i's are not restricted in any way. That is, they may either be independent or dependent.

Expressing this rule in somewhat different symbols, we have

$$E\left[\sum_{i=1}^{n} (X_i)\right] = \sum_{i=1}^{n} [E(X_i)]$$

Treating the expected value and summation symbols as operators, that is, as defining specific operations on the X_i's, we have the result that the summation sign and expected value symbol are interchangeable operators.

Rule 5 $\sigma^2(a) = 0$
The variance of a constant is equal to zero.

Rule 6 $\sigma^2(bX) = b^2\sigma^2(X)$
The variance of a constant times a random variable is equal to the constant squared times the variance of the random variable.

Rule 7 $\sigma^2(a + bX) = b^2\sigma^2(X)$
Rule 7 combines Rules 5 and 6. As in the case of Rule 3 for the expected value, a simple application of the definition of a variance yields the desired result. The proof is left to the reader as an exercise.

Rule 8 $\sigma^2(X_1 + X_2 + \cdots + X_n) = \sigma^2(X_1) + \sigma^2(X_2) + \cdots + \sigma^2(X_n)$ where $X_1, X_2, \ldots, X_n$ are independent random variables, that is, every pair of X_i's is independent.

Thus, if the X_i's are independent, the variance of a sum is equal to the sum of the variances.

Expressing this rule in summation terminology, we obtain

$$\sigma^2\left[\sum_{i=1}^{n} (X_i)\right] = \sum_{i=1}^{n} [\sigma^2(X_i)]$$

Hence, the variance and summation symbols are interchangeable operators if the X_i's are independent.

Rule 9 $\sigma^2(a_1X_1 + a_2X_2) = a_1^2\sigma^2(X_1) + a_2^2\sigma^2(X_2)$ if X_1 and X_2 are independent.

Rule 9 is derived by applying Rules 7 and 8.

Special cases of Rule 9 are given as Rules 10 and 11.
In Rule 10, $a_1 = +1$ and $a_2 = +1$.
In Rule 11, $a_1 = +1$, but $a_2 = -1$.

Rule 10 $\sigma^2(X_1 + X_2) = \sigma^2(X_1) + \sigma^2(X_2)$ if X_1 and X_2 are independent.

Rule 11 $\sigma^2(X_1 - X_2) = \sigma^2(X_1) + \sigma^2(X_2)$ if X_1 and X_2 are independent.

Rule 12 $\sigma^2(\overline{X}) = \dfrac{\sigma_X^2}{n}$

where X is a random variable, μ_x and σ_x are its mean and standard deviation, respectively, and $\overline{X}$ is the arithmetic mean in a sample of n independent observations of X. If $X_1, X_2, \ldots, X_n$ denote the n observations, then

$$\overline{X} = \frac{1}{n}\sum_{i=1}^{n} X_i$$

This rule may be proven in a few steps.

$$(1) \qquad \sigma^2(\overline{X}) = \sigma^2\left(\frac{1}{n}\sum_{i=1}^{n} X_i\right) = \frac{1}{n^2}\sigma^2\left(\sum_{i=1}^{n} X_i\right) \qquad \text{(by Rule 6)}$$

$$(2) \qquad = \frac{1}{n^2}\sum_{i=1}^{n} [\sigma^2(X_i)] \qquad \text{(by Rule 8)}$$

But since every X_i has the same probability distribution as X, then $\sigma^2(X_i) = \sigma^2(X)$ for each i. Hence,

$$(3) \qquad \sum_{i=1}^{n} [\sigma^2(X_i)] = n\sigma_X^2$$

Substituting (3) into (2) gives

$$(4) \qquad \sigma^2(\overline{X}) = \left(\frac{1}{n^2}\right)(n\sigma_X^2) = \frac{\sigma_X^2}{n}$$

which completes the proof.

Let us express Rule 12 in the language and symbolism of sampling theory. If a simple random sample of size n is drawn from an infinite population (or a finite population with replacement), with standard deviation σ_X, the variance of the sample mean is given by

$$(5) \qquad \sigma_{\overline{X}}^2 = \frac{\sigma_X^2}{n}$$

Rule 13 $E(\bar{X}) = \mu_x$
where the same conditions prevail as in Rule 12, that is, X is a random variable, μ_X and σ_X are its mean and standard deviation, respectively, etc.

This rule is easily proven as follows:

(1) $$E(\bar{X}) = E\left(\frac{1}{n}\sum_{i=1}^{n} X_i\right)$$

$$= \frac{1}{n}\left[E\left(\sum_{i=1}^{n} X_i\right)\right] \qquad \text{(by Rule 2)}$$

$$= \frac{1}{n}\left[\sum_{i=1}^{n} E(X_i)\right] \qquad \text{(by Rule 4)}$$

But since every X_i has the same probability distribution as X, then $E(X_i) = E(X)$ for each i. Hence,

(3) $$\sum_{i=1}^{n} E(X_i) = nE(X) = n\mu_x$$

Substituting (3) into (2) gives

(4) $$E(\bar{X}) = \left(\frac{1}{n}\right)(n\mu_x) = \mu_x$$

As in Rule 12, let us express this result in terms of sampling theory. If a simple random sample of size n is drawn from an infinite population (or a finite population with replacement) with mean μ_X, the expected value (arithmetic mean) of the sample mean is given by

(5) $$E(\bar{X}) = \mu_{\bar{x}} = \mu_x$$

Rule 14 $$E\left[\frac{n}{n-1}s_X^2\right] = E\left[\frac{\Sigma(X_i - \bar{X})^2}{n-1}\right] = \sigma_X^2$$

where the same conditions prevail as in Rules 12 and 13 and

$$s_X^2 = \frac{1}{n}\sum_{i=1}^{n} (X_i - \bar{X})^2$$

This result states that the expected value of the indicated statistic is equal to the population variance, σ_X^2. The significance of Rule 14 is that in order to obtain an unbiased estimator of the population variance, the sample variance, s_X^2, defined as above must be multiplied by

$$\frac{n}{n-1}$$

Index

G

H

I

K

L

M

N

T

U

V

W

B 1
C 2
D 3
E 4
F 5
G 6
H 7
I 8
J 9

Probability and Statistical Symbols Used in This Textbook	Meaning
$P(A_1 \cap A_2)$	Joint probability of the occurrence of events A_1 and A_2 (1.3)
$P(A_2 \mid A_1)$	Conditional probability of event A_2 given the event A_1 (1.3)
$P_m(n; n_1, n_2, \ldots, n_k)$	Number of distinguishable arrangements that can be formed of n objects, taken n at a time, where n_1 are of type 1, n_2 of type 2, and n_k are of type k (1.5)
P_n	Price in a non-base period in an index number formula (12.3)
P_0	Price in a base period in an index number formula (12.3)
$P_0(p)$	Prior probability distribution of the random variable p (14.2)
$P_1(p)$	Posterior probability distribution of the random variable p (14.2)
$P_0(\theta_i)$	Prior probability of state of nature θ_i (14.2)
$P_1(\theta_i)$	Posterior probability of state of nature θ_i (14.2)
$P(\theta_i \mid X_j)$	Posterior probability of state θ_i given sample evidence X_j (15.2)
$P(X_j \mid \theta_i)$	Conditional probability of sample evidence X_j given state of nature θ_i (15.2)
${}_nP_x$	Number of permutations of n objects taken x at a time (1.5)
p	Probability of success on a given trial of a Bernoulli process (binomial distribution); also used as a population proportion of successes (2.4)
$\bar{p}$	Weighted mean of two sample percentages (7.3)
$\bar{p} = \dfrac{x}{n}$	Proportion of successes in a sample of size n (6.2)
p^*	Break-even (indifference) point of p; the value of p for which the payoffs of two alternative acts are equal (14.3)
π_1, π_2	Profits derived from acts A_1 and A_2, respectively (14.3)
Q_f	A fixed set of quantity weights in a price index (12.3)
Q_n	Quantity in a non-base period in an index number formula (12.3)
Q_0	Quantity in a base period in an index number formula (12.3)
$q = 1 - p$	Probability of failure on a given trial of a Bernoulli process (binomial distribution); also used as the population proportion of failures (2.4)
$R(s_1; \theta_1)$	Risk or expected opportunity loss associated with the use of strategy s_1 given that state of nature θ_1 occurs (15.3)
$R^2_{Y.12}$	Sample coefficient of multiple determination, not adjusted for degrees of freedom; this case involves two independent variables X_1 and X_2 and the dependent variable Y (10.11)
$\bar{R}^2_{Y.12}$	Sample coefficient of multiple determination, adjusted for degrees of freedom; this case involves two independent variables X_1 and X_2 and the dependent variable Y (10.11)
r_{Y1}, r_{Y2}, r_{12}	Sample correlation coefficients for the variables Y and X_1, Y and X_2, and X_1 and X_2, respectively (10.11)
r	Number of rows in a contingency table or in an arrangement of data to which an analysis of variance is applied (9.2)
r	Sample correlation coefficient (10.7)
r^2	Sample coefficient of determination (10.
$r_{Y1.2}$	Partial correlation coefficient between Y and X_1 after the effect of X_2 on Y has been removed
ρ	Population coefficient of correlation (10
ρ^2	Population coefficient of determination (10.7)
S	Set or sample space (1.2)
$S = \{a, b, c\}$	Set S consists of the elements a, b, and c (1.2)
S	Effect of the seasonal factor in time series analysis (11.5)
SI	Seasonal Index (11.5)
SS_b	Between-column sum of squares (9.4)
SS_t	Total sum of squares (9.4)
SS_w	Within-column sum of squares (9.4)
$S^2_{Y.12}$	Sample variance around a regression plane involving two independent variables X_1 and X and the dependent variable Y; this variance is not adjusted for degrees of freedom (10.11)
$S^2_{Y.12\ldots(k-1)}$	Sample variance around a regression hyperplane involving $k - 1$ independent variables $X_1, X_2, \ldots, X_{k-1}$ and the dependent variable Y; this variance is adjusted for degrees of freedom (10.11)
$s = s_x$	Standard deviation of a sample (3.16)
$s^2 = s_x^2$	Variance of a sample (3.16)
$s_1, s_2, \ldots$	Strategies s_1, s_2, etc. (15.3)
s_d	Standard deviation of the differences between pairs of observations made on the same individuals or objects (7.6) $s_d = \sqrt{(\Sigma(d - \bar{d})^2)/(n - 1)}$
$s_{\bar{d}}$	Standard error of $\bar{d}$, where $\bar{d}$ is the mean difference of pairs of observations made on the same individuals or objects (7.6)
s_i	Standard deviation of the sample from the ith stratum (8.6)
$s_{\bar{p}}$	Estimated standard error of a proportion (8.4)
$s_{\bar{p}_1 - \bar{p}_2}$	Estimated or approximate standard error of the difference between two sample proportions (7.3)